Lecture Notes in Electrical Engineering

Volume 405

About this Series

"Lecture Notes in Electrical Engineering (LNEE)" is a book series which reports the latest research and developments in Electrical Engineering, namely:

- Communication, Networks, and Information Theory
- Computer Engineering
- Signal, Image, Speech and Information Processing
- Circuits and Systems
- Bioengineering

LNEE publishes authored monographs and contributed volumes which present cutting edge research information as well as new perspectives on classical fields, while maintaining Springer's high standards of academic excellence. Also considered for publication are lecture materials, proceedings, and other related materials of exceptionally high quality and interest. The subject matter should be original and timely, reporting the latest research and developments in all areas of electrical engineering.

The audience for the books in LNEE consists of advanced level students, researchers, and industry professionals working at the forefront of their fields. Much like Springer's other Lecture Notes series, LNEE will be distributed through Springer's print and electronic publishing channels.

More information about this series at http://www.springer.com/series/7818

Yingmin Jia · Junping Du
Weicun Zhang · Hongbo Li
Editors

Proceedings of 2016 Chinese Intelligent Systems Conference

Volume II

Springer

Editors
Yingmin Jia
Beihang University
Beijing
China

Weicun Zhang
University of Science and Technology Beijing
Beijing
China

Junping Du
Beijing University of Posts and Telecommunications
Beijing
China

Hongbo Li
Tsinghua University
Beijing
China

ISSN 1876-1100 ISSN 1876-1119 (electronic)
Lecture Notes in Electrical Engineering
ISBN 978-981-10-2334-7 ISBN 978-981-10-2335-4 (eBook)
DOI 10.1007/978-981-10-2335-4

Library of Congress Control Number: 2016948594

Printed on acid-free paper

This Springer imprint is published by Springer Nature
The registered company is Springer Nature Singapore Pte Ltd.
The registered company address is: 152 Beach Road, #22-06/08 Gateway East, Singapore 189721, Singapore

Contents

Autonomous Control of Mobile Robots for Opening Doors Based on Multi-sensor Fusion

Xiaomei Ma and Chaoli Wang

Abstract This paper presents a method of opening doors by mobile robots autonomously. Considering that the existing fingerprint lock will collapse when finger injures, the use of robot shows its durability and convenience, and it avoids carrying even losing keys trouble, and incarnates people's awareness of intelligent robot era. In this robot opening door system, a mobile phone is applied to send open-door command, the robot receives the command via Wi-Fi, then plans path based on laser sensor and encoder sensors to move to the door position, and aims at the door handle based on visual servo, and finally opens the door. Experiments have proved the validity and feasibility of the presented method. Meanwhile, we are discussing other applications of this method.

Keywords Service robot · Path planning · Multi-sensor fusion · Inverse kinematics

1 Introduction

Nowadays, the application of mobile robots is expanding extremely, not only in industry, agriculture, national defense, service industry, but also in mine, hunting, rescue, radiation, and space field such as harmful and dangerous occasions, and all these fields receive very good application [1]. Being an active research area for a long time, autonomous robotics has attracted more and more attention [2]. Taking sweeping robot for example, it comes into millions of households and is accepted by more and more ordinary people.

X. Ma (✉) · C. Wang
Control Science and Engineering, University of Shanghai for Science and Technology, Shanghai 200093, China
e-mail: april_2016@163.com

C. Wang
e-mail: clclwang@126.com

Y. Jia et al. (eds.), *Proceedings of 2016 Chinese Intelligent Systems Conference*, Lecture Notes in Electrical Engineering 405,
DOI 10.1007/978-981-10-2335-4_1

Navigation technology is one of the core technologies of mobile robots. The main navigation methods include: magnetic navigation, inertial navigation, GPS navigation, road navigation, visual navigation, voice navigation, and so on. GPS navigation systems can be found in motor vehicles and other various land-based vehicles. Paper [3] concludes that the accuracy of the GPS position fixes has a significant impact on the relative contributions that each dead-reckoning navigation sensor error makes. The RGB-D (Kinect-style) camera provides high quality synchronized videos of both color and depth, and dramatically increases robotic object recognition, manipulation, navigation, and interaction capabilities [4, 5]. Paper [6] presents an incremental method for concurrent mapping and localization for mobile robots equipped with 2D laser range finders, and illustrates that accurate maps of large, cyclic environments can be generated even in the absence of any odometric data. This paper adopts laser positioning and navigation to resolve path planning problem in the environment with complex obstacles.

This paper focuses on the navigation based on the laser scanner and encoder sensors, the positioning of the door handle based on Kinect, and the inverse kinematics of the manipulator.

2 System Description

This system is designed for mobile robots to open doors autonomously as long as users send open command. Figure 1 is a flowchart showing an overall operation of the system.

The difficulties mainly lie in the path planning and navigation of mobile robot in the environment with complex obstacles, the alignment of the door handle with robot, and the inverse kinematics of the manipulator.

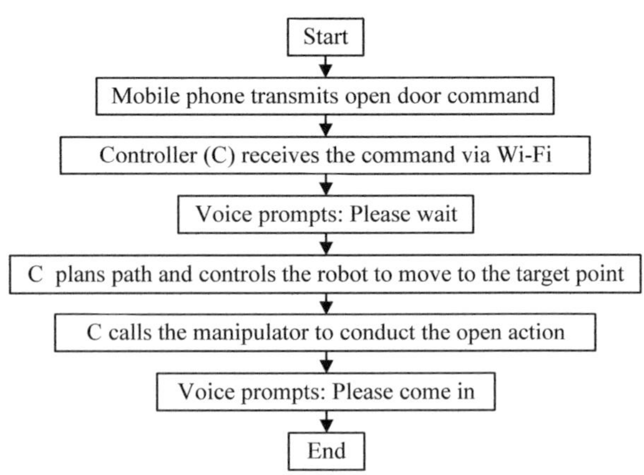

Fig. 1 The overall operation of the system

3 Laser Positioning and Navigation

The indoor map information is stored in the configuration file in advance, If in a new environment, the configuration file needs to be modified.

In order to achieve navigation and path planning of mobile robot in the environment with complex obstacles, this paper combines laser scanner with encoder sensors to achieve the best positioning effect. Laser scanner data and the robot current pose are used to update the map.

A-star algorithm is used to plan the path [7–9]. The planned path, which are a series of inflection points are recorded. The speed and the motion direction of the robot is calculated according to the robot's current pose and the target (door) position, the robot is controlled to move along the inflection points until it arrives at the target point. The navigation and path planning program interface is shown in Fig. 2.

Left is map display part, which is responsible for the real-time display of the map and path. Right top is the laser scanner device connection part. Right center is motion control part, which is designed to control the robot to move forward, backward, turn left, turn right, and stop. Right bottom is target point input part. Robot pose part is used to set the robot current pose.

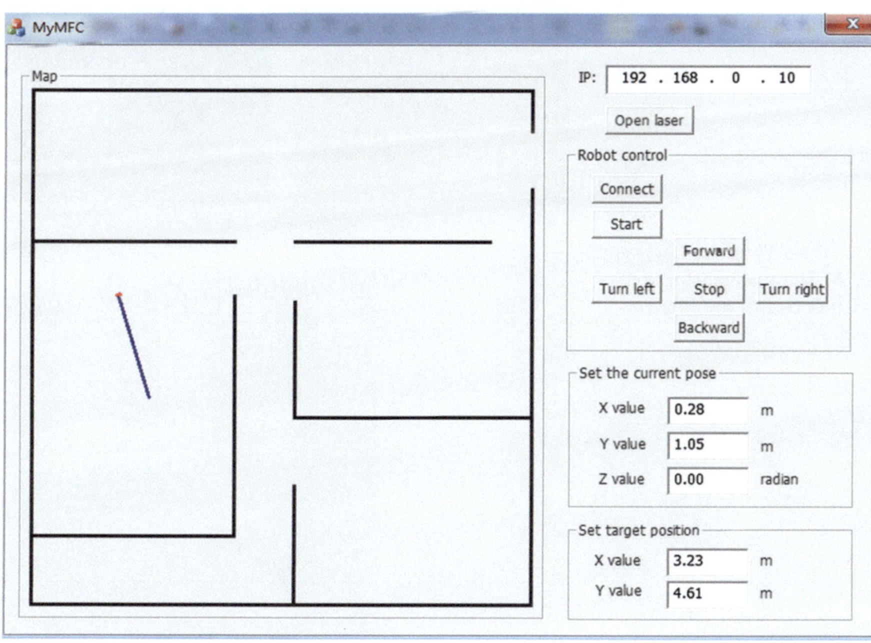

Fig. 2 The navigation and path planning program interface

4 Control Algorithm of the Manipulator

The open action is realized by manipulator rotating the door handle. Manipulator aligns door handle is the premise. However, manipulator usually misaligns the door handle when robot arrives at the door position.

This paper adopts a Kinect to obtain the target position. Since Kinect is RGB-D style, it provides both color and depth information. A red square mark is used to solve the alignment problem. The centroid of the red square is installed at the same vertical line with the door handle. The alignment is achieved by aligning the manipulator with the red marker. By handling the video captured by Kinect, robot is controlled to move right or left to align the red mark. Figure 3 shows the red mark and its extraction. Mathematical morphology is applied in the process of digital image processing [10–12].

The deployed manipulator is four DOF. D-H method is used to establish the coordinate frames [13]. The manipulator with D-H link coordinate frames is shown in Fig. 4.

Table 1 displays the D-H link parameter. Coordinate frame $i-1$ can be transformed into i through the following consecutive relative movement:

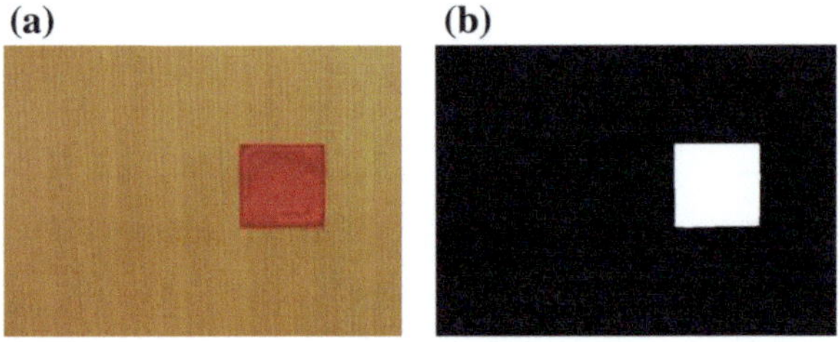

Fig. 3 The red mark and its extraction

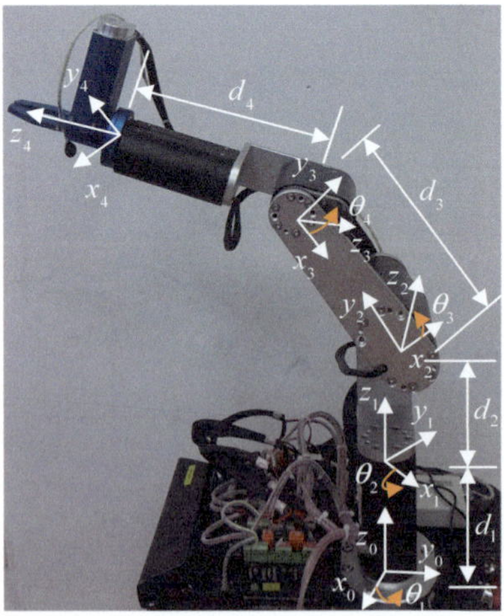

Fig. 4 The manipulator with D-H link coordinate frames

Table 1 D-H parameter

i	a_i	α_i	d_i	θ_i
1	0	0	d_1 (constant)	θ_1
2	0	$\pi/2$	d_2 (constant)	θ_2
3	0	0	d_3 (constant)	θ_3
4	0	$\pi/2$	d_4 (constant)	θ_4

Step 1 Translate d_i from x_{i-1} to x_i along z_{i-1}.
Step 2 Rotate θ_i angle from x_{i-1} to x_i about z_{i-1}, $\theta_i \in (-\pi, \pi]$.
Step 3 Translate a_i from z_{i-1} to z_i along x_i.
Step 4 Rotate angle α_i from z_{i-1} to z_i about x_i. $\alpha_i \in (-\pi, \pi]$.

The homogeneous transformation matrix of the continuous relative transformation is defined as

$$A_i^{i-1} = \text{Trans}_z(d_i)\text{Rot}_z(\theta_i)\text{Trans}_x(a_i)\text{Rot}_x(\alpha_i) = \begin{bmatrix} c_i & -c\alpha_i s_i & s\alpha_i s_i & a_i c_i \\ s_i & c\alpha_i c_i & -s\alpha_i c_i & a_i s_i \\ 0 & s\alpha_i & c\alpha_i & d_i \\ 0 & 0 & 0 & 1 \end{bmatrix} \tag{1}$$

where $s_i \triangleq \sin\theta_i, c_i \triangleq \cos\theta_i, s\alpha_i \triangleq \sin\alpha_i, c\alpha_i \triangleq \cos\alpha_i$.

Substituting D-H parameter in Table 1 into Eq. (1) leads to

$$A_4^0 = A_1^0 A_2^1 A_3^2 A_4^3 = \begin{bmatrix} c_{12}c_{34} & s_{12} & c_{12}s_{34} & d_3 s_{12} + d_4 s_{12} \\ s_{12}c_{34} & -c_{12} & s_{12}s_{34} & -d_3 c_{12} - d_4 c_{12} \\ s_{34} & 0 & -c_{34} & d_1 + d_2 \\ 0 & 0 & 0 & 1 \end{bmatrix} \tag{2}$$

where $c_{ij} \triangleq \cos(\theta_i + \theta_j), s_{ij} \triangleq \sin(\theta_i + \theta_j)$.

It needs to control the gripper of the manipulator to move to the door handle position. The pose of gripper can be obtained as long as the robot aligns the door handle with the known height of door handle and the distance between it and robot which is obtained from the Kinect. Namely, A_4^0 is known, each joint variable $q = [q_1, q_2, q_3, q_4]^T$ should be solved according to the joint conversion relationship, which belongs to classic inverse kinematics of robot [13].

The joints 2 and 3 of the manipulator are mainly controlled. The simplified model is shown in Fig. 5. The angle coordinates of the two joints (q_1, q_2) are solved from the coordinates of the end of the manipulator (z_1, y_1) in working space.

Doing geometric analysis according to Fig. 5 leads to Eq. (3)

$$\begin{cases} z_1 = l_1 \sin(q_1) + l_2 \sin(q_1 + q_2) \\ y_1 = l_1 \cos(q_1) + l_2 \cos(q_1 + q_2) \end{cases} \tag{3}$$

Solving Eq. (3) leads to Eqs. (4) and (5)

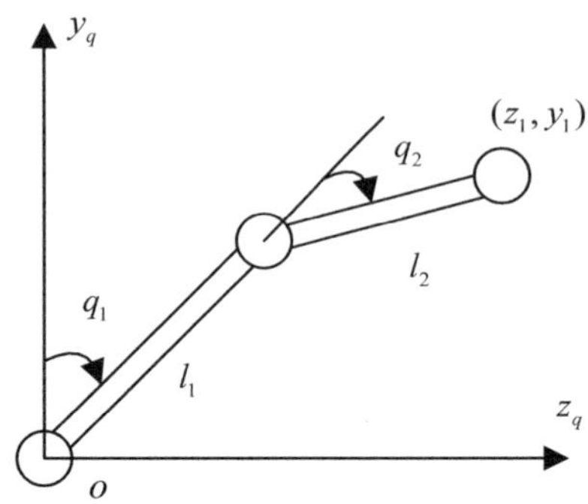

Fig. 5 The simplified two-DOF structure

$$q_1 = \arccos\left(\frac{z_1^2 + y_1^2 - l_1^2 - l_2^2}{2l_1 l_2}\right) \tag{4}$$

$$q_2 = \begin{cases} p_1 - p_2, q_1 > 0 \\ p_1 + p_2, q_1 \le 0 \end{cases} \tag{5}$$

where $p_1 = \arctan\frac{y_1}{z_1}, p_2 = \arccos\left(\frac{z_1^2 + y_1^2 - l_1^2 - l_2^2}{2l_1\sqrt{z_1^2 + y_1^2}}\right)$.

5 Experiment Results

In order to verify the effectiveness of the proposed method, a mobile robot equipped with two 70 W DC motors and optical encoders with a precision level of 500 pulses rotation were used.

As shown in Fig. 6, the mobile robot is equipped with a laser scanner, a Kinect, a four-DOF manipulator with a single-DOF gripper. The UST-10LX model laser scanner communicates with the interface via internet. The 2D plane measuring range of the laser scanner is 270° with angular resolution of 0.25°. The maximum measuring distance is 30 m. The main controller was designed with a PC. The robot can be controlled by two motors to move forward, backward, left, right, and stop. Figure 6 shows the movie snapshot of the process of opening the door.

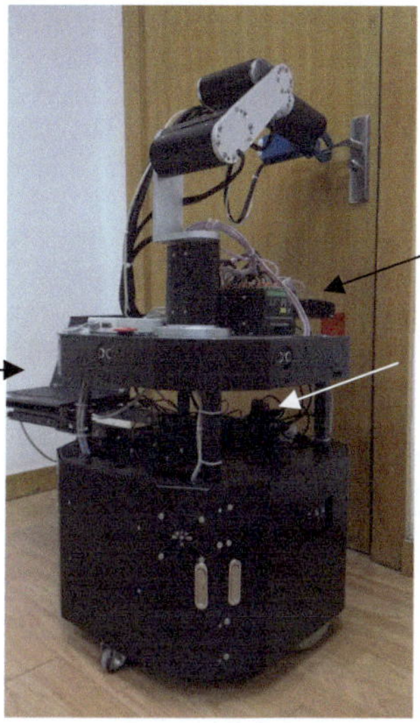

Fig. 6 The mobile robot for experiment

6 Conclusion

This paper presents a method of opening doors by mobile robots autonomously. In the robot opening door system, all users need to do is to apply a mobile phone to send the open command. The robot receives the command via Wi-Fi, then plans path to move to the door position based on laser scanner and encoder sensors, and aims at the door handle based on visual servo, finally opens the door autonomously. The use of robot shows its advantage over other methods. It is not only durable but also convenient. Besides, sometimes we come back home without keys, this method avoids carrying even losing keys trouble. In addition, with more and more autonomous robots coming into millions of households, the use of robot incarnates people's awareness of intelligent robot era and pursuit of science and fashion. Experiments have proved the validity and feasibility of the presented method. Meanwhile, we are discussing other applications of this method.

References

1. Xu BG, Yin YX, Zhou MJ (2007) Present situation and Prospect of intelligent mobile robot [J]. Robot Technol Appl 2:29–34
2. Sakagami Y, Watanabe R, Aoyama C, et al. (2002) The intelligent ASIMO: system overview and integration. IEEE/RSJ Int Conf Intell Robots Syst 3:2478–2483
3. Abbott E, Powell D (1999) Land-vehicle navigation using GPS. Proc IEEE 87(1):145–162
4. Lai K, Bo L, Ren X, et al. (2011) A large-scale hierarchical multi-view RGB-D object dataset. In: IEEE International conference on robotics and automation, pp 1817–1824

5. Sturm J, Engelhard N, Endres F, et al. (2012) A benchmark for the evaluation of RGB-D SLAM systems. In: 2012 IEEE/RSJ international conference on intelligent robots and systems, pp 573–580
6. Thrun S, Burgard W, Fox D (2000) A real-time algorithm for mobile robot mapping with applications to multi-robot and 3D mapping. In: IEEE international conference on robotics and automation, vol 1, pp 321–328
7. Cheng LP, Liu CX, Yan B (2014) Improved hierarchical A-star algorithm for optimal parking path planning of the large parking lot. In: 2014 IEEE international conference on information and automation, pp 695–698
8. Zhang Y, Tang GJ, Chen L (2012) Improved A-star algorithm for time dependent vehicle routing problem. Control Eng China 19(5):750–756
9. Yao J, Lin C, Xie X, et al. (2010) Path planning for virtual human motion using improved A* star algorithm. IN: 2010 seventh international conference on information technology: new generations, pp 1154–1158
10. Cui Y (2002) Image processing and analysis: the method and application of mathematical morphology [M]. Science Press, Beijing
11. Li JS, Li XH (2007) Digital image processing [M]. Tsinghua University Press, Beijing, pp 225–248
12. Zhu XC, Liu F, Hu D (2014) Digital image processing and image communication [M]. Beijing university of posts and telecommunications press, Beijing
13. Huo W (2005) Robots dynamics and control [M]. Higher Education Press, Beijing

Linear Active Disturbance Rejection Control Approach for Load Frequency Control Problem Using Diminishing Step Fruit Fly Algorithm

Congzhi Huang and Yan Li

Abstract The diminishing step fruit fly optimization algorithm (DS-FOA) is employed to optimize the performance of the load frequency control (LFC) problem by employing the linear active disturbance rejection control (LADRC) approach. First of all, the LFC problem taking into account the case of a single generator supplying power to a single service area is presented. Second, the general LADRC solution to the problem is given, where the diminishing step fruit fly optimization algorithm (DS-FOA) is employed to optimize the performance of the system with the approach. Third, the performances of the system with the following three control approaches are compared, including the traditional PID control approach, the normal LADRC approach, and the DS-FOA optimized LADRC approach. With the proposed LADRC approach, the system performance is much better than that of the traditional PID controller, and a much better performance is achieved with the proposed DS-FOA optimized LADRC approach. The performance superiority of the proposed approach is also validated by the frequency domain analysis results given.

Keywords Fruit fly optimization algorithm (FOA) · Load frequency control (LFC) · Linear active disturbance rejection control (LADRC) · Power system

1 Introduction

The problem to control the active power output of generating units responding to the disturbances in grid frequency and tie line power exchange within the designated limitation is denoted by the load frequency control issue [1]. Nowadays, it has

C. Huang (✉) · Y. Li
School of Control and Computer Engineering, North China Electric Power University, Beijing 102206, China
e-mail: hcz190@ncepu.edu.cn

Y. Li
e-mail: liyan_kj@ncepu.edu.cn

Y. Jia et al. (eds.), *Proceedings of 2016 Chinese Intelligent Systems Conference*, Lecture Notes in Electrical Engineering 405,
DOI 10.1007/978-981-10-2335-4_2

become a hot topic with the aggressive development of smart grid all over the world. The traditional PID control strategy is usually employed to solve the LFC problem. The particle swarm optimized PID control approach is proposed in [2], and the other approaches refer to the model predictive control [3], adaptive fuzzy logic control [4], robust $H\infty$ control [5], and so on.

However, the performance improvement of the system with the traditional PID control approach is usually limited, while most of the advanced control approaches are difficult to put into practice due to its complex structure. To solve the disturbance rejection control problem widespread in industry, the active disturbance rejection control approach by Han [6] was brought into being. Its core idea is that the total disturbance, including all the internal uncertainties and external disturbance can be estimated with the aid of an extended state observer constructed in real time, and then cancelled by the nonlinear PD control law [7]. However, there were too many parameters to be tuned in the original ADRC approach, and thus it is often difficult to conduct parameter tuning work and put it into practice. Fortunately, the linear version of the ADRC approach emerges at a historic moment in [8], which largely simplified the parameters tuning procedure and thus quickly found a number of applications in practical engineering, see the integrated guidance and control system [9], and the hot strip width and gauge regulation problem [10].

In this paper, a general LADRC approach is introduced, and the DS-FOA is employed to tune the LADRC approach. In addition, the LFC problem for the single area power system with a non-reheated turbine is given for an example. A fourth-order LADRC approach is used to illustrate the effectiveness of the proposed approach. In order to improve the system performance, the DS-FOA is employed to tune the parameters in the LADRC approach, the result is also compared with that of the traditional PID approach in [11] and the normal LADRC approach in [12].

2 LFC Problem Formulation

A linear model of a single area power system with a non-reheated turbine is shown in Fig. 1.

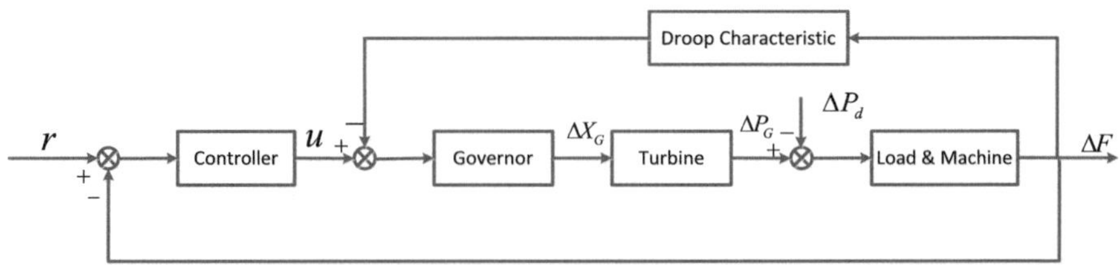

Fig. 1 Linear model of a single area power system

As can be seen from Fig. 1, r is the reference input, u is the controller output, and ΔF is the frequency deviation, which is also the controlled process output in the load frequency control problem. ΔP_d represents the external load disturbance, ΔP_G represents the incremental quantity in generator output, and ΔX_G the incremental quantity in governor valve position.

The plant discussed in this paper is the same with that in [11] consisting of the following three parts: governor, turbine, and load and machine. The dynamics of the governor can be described as $G_g(s) = 1/(T_g s + 1)$, and the turbine model is often represented by $G_t(s) = 1/(T_t s + 1)$. The transfer function of the load and machine is denoted by $G_m(s) = K_p/(T_p s + 1)$. In addition, the droop characteristic is a feedback controller with the gain of $1/R$ so as to improve the system performance.

Since the LFC problem for the power system under consideration is expressed only to relatively small changes in load, so the controller can be used to ensure the frequency deviation as small as possible under small load changes.

3 Linear Active Disturbance Rejection Control Solution

A. *Linear Active Disturbance Rejection Control*

The typical block diagram for the system with the LADRC approach is shown in Fig. 2.

As shown in Fig. 2, an ESO based on the process input and output data is constructed to estimate the process output its finite time derivatives, and the total disturbance. Consider the following general controlled process:

$$y^{(n)} = g(t, y, \dot{y}, \ldots, u, \dot{u}, \ldots, w) + bu, \tag{1}$$

where y is the controlled process output variable, u is the control input, and w represents the external disturbance. Taking the estimation value of b as b_0, it can be rewritten as:

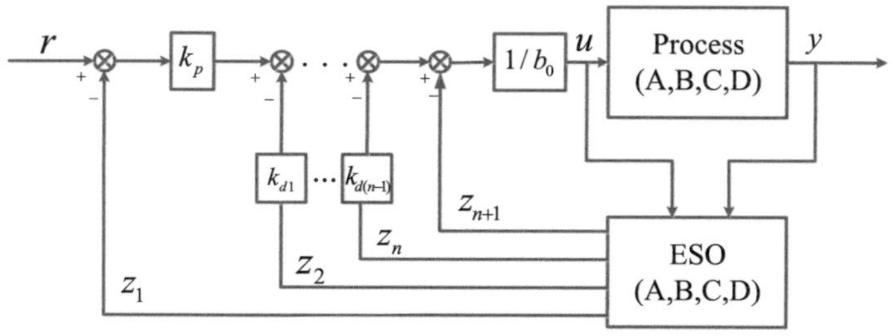

Fig. 2 Block diagram of control system with the LADRC approach

$$y^{(n)} = g(t, y, \dot{y}, \ldots, u, \dot{u}, \ldots, w) + (b - b_0)u + b_0 u = f + b_0 u, \tag{2}$$

where $f = g(t, y, \dot{y}, \ldots, u, \dot{u}, \ldots, w) + (b - b_0)u$ is defined as the total disturbance, including the unknown dynamics as well as all the external disturbances. Define an augmented state vector $x = [x_1, x_2, \ldots, x_n, x_{n+1}]^T$, the extended state equation of the original process is given as follows:

$$\begin{cases} \dot{x}_1 = x_2 \\ \ldots \\ \dot{x}_n = x_{n+1} + b_0 u, \\ \dot{x}_{n+1} = h \\ y = x_1 \end{cases} \tag{3}$$

where the state variables $x_1, x_2, \ldots, x_n$ are the process state, $x_{n+1} = f$ added as an augmented state, and $h = f'$ is the unknown variable representing the first order time derivative of the total disturbance.

Then, the following extended state observer is constructed:

$$\begin{cases} \dot{z}_1 = z_2 + \beta_1 (y(t) - z_1) \\ \ldots \\ \dot{z}_n = z_{n+1} + \beta_n (y(t) - z_1) + b_0 u' \\ z_{n+1} = \beta_{n+1} (y(t) - z_1) \end{cases} \tag{4}$$

where the observer gain is denoted by $L = [\beta_1, \beta_2, \ldots, \beta_n, \beta_{n+1}]^T$. With appropriate observer gains, the following state variables can be tracked accurately:

$$\begin{aligned} &z_1(t) \to y(t), z_2(t) \to \dot{y}(t), \ldots, z_n(t) \to y^{(n-1)}(t) \\ &z_{n+1}(t) \to f \end{aligned} \tag{5}$$

In order to simplify the parameter tuning procedure of the ESO, all the observer poles are placed at the same place $-\omega o$, where the observer bandwidth w_o is the only parameter to be tuned. The characteristic equation of the ESO is given as:

$$s^n + \beta_1 s^{n-1} + \cdots + \beta_{n-1} s + \beta_n = (s + \omega_o)^n. \tag{6}$$

Then, the ESO gain L can be obtained as follows: $\beta_1 = n\omega_o$, $\beta_2 = 0.5n(n-1)\omega_o^2, \ldots$, $\beta_o = n\omega_o^n$, $\beta_{n+1} = \omega_o^{n+1}$. With the well-tuned ESO, the control law is then designed:

$$u = \frac{-z_{n+1} + u_0}{b_0} \tag{7}$$

This control law reduces the original plant to a cascaded integrator, which can be easily controlled by a PD controller as follows:

$$u_0 = k_p(r - z_1) - k_{d1}z_2 - \cdots - k_{d(n-1)}z_n \tag{8}$$

After the gains are selected in order to place all the roots of the closed-loop characteristic equation at the same place $-\omega_c$, where w_c is the controller bandwidth.

$$s^n + k_{d(n-1)}s^{n-1} + \cdots + k_{d1}s + k_p = (s + \omega_c)^n, \tag{9}$$

where the controller parameters k_p, $k_{d1},\ldots,$ $k_{d(n-1)}$ are chosen as follows: $k_p = \omega_c^n$, $k_{d1} = \omega_c^{n-1},\ldots,$ $k_{d(n-1)} = n\omega_c$.

Now, there are only three parameters to be tuned in the LADRC approach: b_0, ω_c and ω_o. In the following section, the FOA will be employed to tune them so as to achieve better performance.

B. *Fruit Fly Optimization Algorithm*

The fruit fly optimization algorithm (FOA) proposed by Pan [13] is a kind of swarm intelligent algorithm. Compared with the traditional particle swarm optimization (PSO) algorithm, the FOA can achieve an optimal speed. But it also has its own disadvantages: it is easy to fall into a local optimal solution, and it is easy to become premature. Thus there are many scholars trying to improve the traditional FOA. The diminishing step fruit fly optimization algorithm (DS-FOA) is proposed in [14] to improve the global searching ability at the beginning and ensure the local optimization ability later. In the DS-FOA, the searching step length L changes from a constant variable into a decline variable:

$$L = L_0 - \frac{L_0(G-1)}{G_{\max}}, \tag{10}$$

where L_0 is the initial searching step length, G is the current optimization iteration number and $G_{\max}$ *is* the max optimization iteration number. The tuning procedure of the DS-FOA optimized LADRC approach is given as follows:

Step 1 Initialization. Set the max optimization iteration number $G_{\max}$, and the flies population size P.

Step 2 Generate the initial locations of the fruit fly swarm randomly, $\theta_0 = [(x_{01}, y_{01}), (x_{02}, y_{02}), (x_{03}, y_{03})]$.

Step 3 Use (10)–(12) to assign each fruit fly a direction and distance randomly, and change the searching step L.

$$X(i) = X_{0i} + L * \text{rand}(1,1) \tag{11}$$

$$Y(i) = Y_{0i} + L * \text{rand}(1,1) \tag{12}$$

Step 4 Take the values of the parameters b_0, ω_c and ω_o into the LADRC approach, and obtain the system error $e(t) = y(t) - r(t)$, where y is the system output, and r is the set point. The performance index ITAE (Integration of Time multiplied by Absolute Error) can be selected as the fitness function.

$$J = \int_0^{\infty} t|e(t)|\mathrm{d}t \tag{13}$$

Based on Eq. (13), the value of performance index J can be easily obtained.

Step 5 Compute the minimum value of $J_{\min}$ and the location of b_0, ω_c and ωo: $\theta_i = [(x_{i1}, y_{i1}), (x_{i2}, y_{i2}), (x_{i3}, y_{i3})]$

Step 6 Once $J_{\min,}$ is obtained, update the initial location as $\theta_0 = [(x_{ib0}, y_{ib0}), (x_{i\omega c}, y_{i\omega c}), (x_{i\omega o}, y_{i\omega o})]$.

Step 7 Repeat step 3–6. When it comes to achieve maximum iteration number then run out of circulation, and get the final values of b_0, ω_c and ω_o.

4 Illustrative Examples

The nominal parameters of the non-reheated turbine plant in Fig. 1 are chosen as follows: $K_p = 120$, $T_p = 20$, $T_t = 0.3$, $T_g = 0.08$, and $R = 2.4$. With the given nominal parameters, the process model with droop characteristic can be obtained as $G_p(s) = 250/(s^3 + 15.88s^2 + 42.46s + 106.2)$. To verify the robustness of the proposed approach, the parameters of the system are assumed to vary by ±50 %. $1/T_t \in [2.564, 4.762]$, $1/T_g \in [9.615, 17.857]$, $1/T_p \in [0.033, 0.1]$, $K_p/T_p \in [4, 12]$, $1/RT_g \in [3.081, 10.639]$.

The PID controller $K(s) = 0.4036 + 0.6356/s + 0.1832s$ was the one adopted in [11]. The LADRC tuning parameters in [12] are selected as follows: the estimation value $b = 250$, the observer bandwidth $\omega_o = 5.15$ and the controller bandwidth $\omega_c = 280$.

In the DS-FOA optimized LADRC approach, the population size is set as 30, the number of iterations is chosen as 20, and the initial searching step length is equal to 1, and the simulation results are presented in Fig. 3. Figure 4 shows the optimization processes of DS-FOA. Table 1 shows the optimization result.

A. *Disturbance Rejection Test*

After conducting a disturbance rejection test, the performance superiority of the optimized LADRC approach can be verified. With the load demand $\Delta P_d = 0.01$, the responses of the power system with non-reheated turbine and the DS-FOA optimized LADRC approach are shown and compared with the PID and LADRC performance from [11, 12] in Fig. 4.

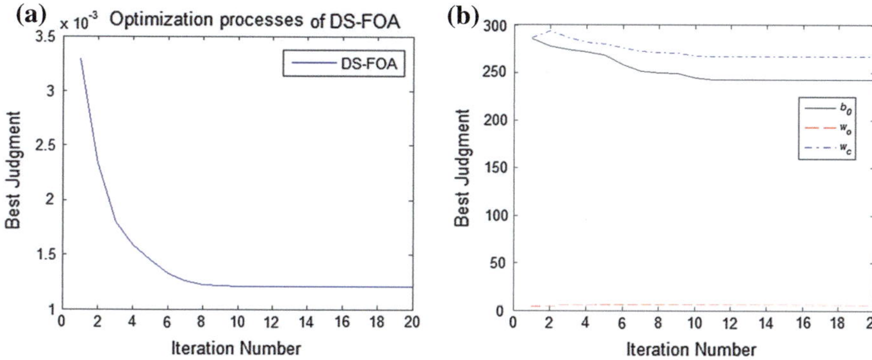

ITAE Optimization processes; LADRC parameters optimization processes

Fig. 3 Optimization processes of DS-FOA

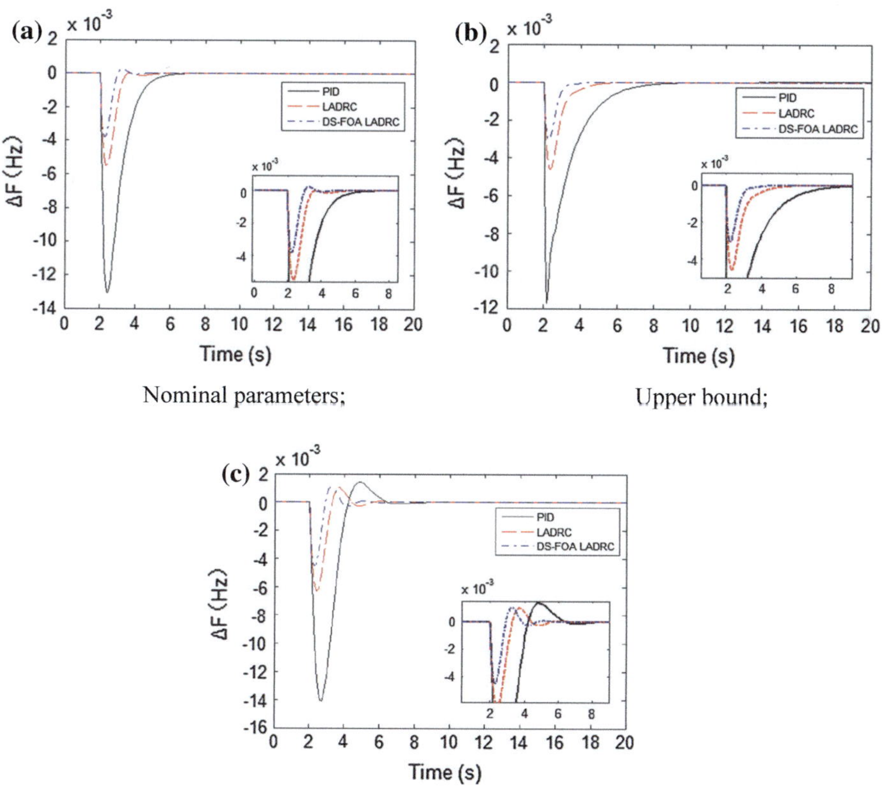

Nominal parameters; Upper bound; Lower bound

Fig. 4 The responses of power system with a non-reheated turbine and the following three control approaches: PID, LADRC and DS-FOA optimized LADRC approach

Table 1 Result of DS-FOA optimization LADRC parameters

b_0	ω_c	ω_o	ITAE
242.6164	267.1187	6.3466	0.001208

As shown in Fig. 4, the performance of the system with the fourth-order LADRC is much better than that of PID controller. In addition, by using DS-FOA optimize LADRC parameters, the system shows a much better performance.

B. *Performances Indices*

To further compare the performance of the system with the DS-FOA optimized LADRC approach with that of the PID controller and the normal LADRC approach, the performance indices ITAE are computed and listed in Table 2.

In this table, it is shown that with the DS-FOA optimized LADRC approach in the system, the performance indices are much smaller than the other two control approaches, which explicitly indicates that the performance superiority of the proposed approach.

C. *Open Loop Transfer Function Frequency Domain Analysis*

For the unit negative feedback control system, the actuator is assumed to be linear and its gain is equal to 1. Its open loop transfer function is $G_{cloop}(s) = G_p * G_c$, where G_p is plant transfer function and G_c is controller transfer function. It is easy to know the transfer function of plant G_p and the PID controller G_{cPID} from [11].

$$G_{cPID}(s) = 0.4036 + \frac{0.6356}{s} + 0.1832s \tag{14}$$

In [15], the transfer function of LADRC has been summarized. For this paper, the transfer function of LADRC and DS-FOA LADRC can be expressed as follows:

$$G_{\mathrm{CLADRC}}(s) = \frac{2.195\times10^7 s^4 + 4.522\times10^8 s^3 + 3.493\times10^9 s^2 + 1.199\times10^{10} s + 1.544\times10^{10}}{250s^3 + 2.195\times10^7 s^2 + 2.68\times10^7 s + 2.693\times10^7} \tag{15}$$

$$G_{\mathrm{CDLADRC}}(s) = \frac{1.849\times10^7 s^4 + 4.602\times10^8 s^3 + 4.296\times10^9 s^2 + 1.782\times10^{10} s + 2.773\times10^{10}}{240s^3 + 1.849\times10^7 s^2 + 2.371\times10^7 s + 2.389\times10^7} \tag{16}$$

Table 2 The Performance Indices ITAE

Controller	Optimized LADRC	LADRC	PID
ITAE	0.00121/0.00110/0.00197	0.0069/0.0077/0.0104	0.0302/0.0365/0.0329

Note ITAE performance indices are presented in order as the system with the normal/upper bound/lower bound parameters

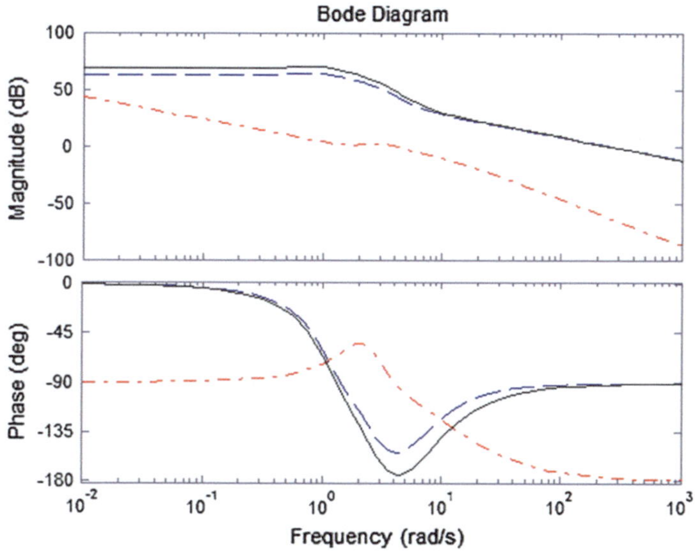

Fig. 5 Frequency domain analysis of PID controller, LADRC and DS-FOA optimized LADRC (*solid* LADRC; *dashed* DS-FOA optimized LADRC; *dash dotted* PID)

Table 3 Open loop transfer function frequency domain analysis

Controller	PM dB (Cut-off frequency rad/sec)
PID	92.3273 (3.8749)
LADRC	89.0296 (249.8372)
DS-FOA LADRC	88.0407 (249.9800)

The Bode plots for the open loop transfer functions of the system with the three controllers are shown in Fig. 5.

As can be seen from Fig. 5 and Table 3, all the three controllers make the system stable. The LADRC increase the cut-off frequency to achieve a better response speed than the traditional PID control approach. By using the DS-FOA optimized LADRC parameters, the system has a higher phase margin in middle frequency so that it has a better dynamic characteristics, which also proves the performance superiority of the proposed approach.

5 Conclusion

In this paper, the DS-FOA optimized LADRC approach is proposed, and its performance superiority is validated by the simulation example. By using DS-FOA tuning the three control parameters of LADRC, b_0, ω_o and ω_c, the LFC problem in this paper is solved. The performance of LADRC demonstrates the effectiveness and robustness of the proposed LADRC approach. In addition, by using the DS-FOA, the better parameters of LADRC can be found and the controller achieved better effect and stronger robustness. However, its deficiency is that the performance of the FOA can be improved further, and the optimized parameters of LADRC by DS-FOA for the reheated turbine plant and hydroturbine plant remains a difficult and challenging problem to be further investigated.

Acknowledgments This work is supported by National Natural Science Foundation of China under Grant 61304041, and the Fundamental Research Funds for the Central Universities 2016ZZD03.

References

1. Kundur P (1994) Power system stability and control. McGraw-Hill, New York
2. Kouba NEY, Menaa M, Hasni M, et al. (2014) Optimal load frequency control in interconnected power system using PID controller based on particle swarm optimization. In: International conference on electrical sciences and technologies in Maghreb, pp 1–8
3. Shi X, Hu J, Yu J, et al. (2015) A novel load frequency control strategy based on model predictive control. In: IEEE power and energy society general meeting, pp 1–5
4. Yousef H, Al-Kharusi K, Albadi MH et al (2014) Load frequency control of a multi-area power system: an adaptive fuzzy logic approach. IEEE Trans Power Syst 29(29):1822–1830
5. Ning C (2016) Robust *H* load-frequency control in interconnected power systems. IET Control Theory Appl 10(1):67–75
6. Han J (1998) Auto-disturbances-rejection controller and its applications. Control Decis 13 (1):19–23
7. Han J (2009) From PID to active disturbance rejection control. IEEE Trans Ind Electron 56 (3):900–906
8. Gao Z (2003) Scaling and parameterization based controller tuning. Proc Am Control Conf 6:4989–4996
9. Xue W, Huang C, Huang Y (2013) Design methods for the integrated guidance and control system. Control Theory Appl 30(12):1511–1520
10. Wang L, Tong C, Li Q, Yin Y, Gao Z, Zheng Q (2012) A practical decoupling control solution for hot strip width and gauge regulation based on active disturbance rejection. Control Theory Appl 29(11):1471–1478
11. Tan W (2010) Unified tuning of PID load frequency controller for power systems via IMC. IEEE Trans Power Syst 25(1):341–350
12. Hang C, Zheng Q (2014) Application of linear active disturbance rejection control to power system load frequency control. Int J Intell Control Syst 19(1):1–7
13. Pan W (2012) A new fruit fly optimization algorithm: taking the financial distress model as an example. Knowl Based Syst 26:69–74
14. Ning J, Wang B, Li H, Baohua X (2014) Research on and application of diminishing step fruit fly optimization algorithm. J Shenzhen Univ Sci Eng 31(4):367–373
15. Hang C, Gao Z (2013) On transfer function representation and frequency response of linear active disturbance rejection control. In: Proceedings of Chinese control conference, pp 72–77

Local Zernike Moment and Multiscale Patch-Based LPQ for Face Recognition

Xiaoyu Sun, Xiaoyan Fu, Zhuhong Shao, Yuanyuan Shang and Hui Ding

Abstract In this paper, a novel feature extraction method combining Zernike moment with multiscale patch-based local phase quantization is introduced, which can deal with the problem of uncontrolled image conditions in face recognition, such as expressions, blur, occlusion, and illumination changes (EBOI). First, the Zernike moments are computed around each pixel other than the whole image and then double moment images are, respectively, constructed from the real and imaginary parts. Subsequently, multiscale patch-based local phase quantization descriptor is utilized for the non-overlapping patches of moment images to obtain the texture information. Afterward, the support vector machine (SVM) is employed for classification. Experimental results performed on ORL, JAFFE, and AR databases clearly show that the LZM-MPLPQ method outperforms the state-of-the-art methods and achieves better robustness against severe conditions abovementioned.

Keywords Local Zernike moment · Local phase quantization · Face recognition · EBOI

1 Introduction

With the wide applications of surveillance, automation, and intelligent devices, face recognition (FR) has become one of the most active research areas of computer vision. Even though there are many methods proposed in FR [1–3] during the last

X. Sun · X. Fu (✉) · Z. Shao · Y. Shang · H. Ding
College of Information and Engineering, Capital Normal University, Beijing 100048, China
e-mail: fuxiaosg@163.com

X. Fu · Z. Shao · Y. Shang · H. Ding
Beijing Advanced Innovation Center for Imaging Technology, Capital Normal University, Beijing 100048, China

Y. Shang · H. Ding
Beijing Key Laboratory of Electronic System Reliability Technology, Capital Normal University, Beijing 100048, China

Y. Jia et al. (eds.), *Proceedings of 2016 Chinese Intelligent Systems Conference*, Lecture Notes in Electrical Engineering 405,
DOI 10.1007/978-981-10-2335-4_3

two decades, it is still a challenging problem to adapt different conditions such as expressions, blur, occlusion, and illumination changes (EBOI).

Generally, a FR system is constituted of feature extraction and pattern classification. It is of great significance to choose proper method to extract features. Global and local features are two classes of feature extraction methodologies. Since local features exemplified by LBP [4] and Gabor [5] have better affine invariance and illumination insensitivity compared with global features, they are usually applied to the FR system with changeable image variations. However, LBP gets poor antinoise ability while Gabor method suffers from high computational cost because of processing 40 components (5 scales and 8 orientations) for every facial image. In recent years, many other local feature methods have been proposed to achieve better performance. For example, Ojansivu and Heikkila [6] presented the local phase quantization method (LPQ) in discrete Fourier transform domain. To further improve the performance, LPQ family is enriched by patch-based LPQ [7] and adaptive LPQ [8]. Moreover, Chan [9] proposed an efficient method combining several multiscale feature descriptors, which achieved high classification rate and blur tolerance. On the other hand, the usages of Zernike moment (ZM) invariants [10, 11], which possess the properties of rotation invariance and less information redundancy, enabled to achieve successful recognition against expression variations [12] and occlusions [13]. However, ZM-based methods are usually used as global features, which negatively impact recognition rate in the images with illumination and blur.

In this work, a novel local face feature extraction method with the hybridization of local Zernike moment and multiscale patch-based LPQ is proposed, which is termed as LZM-MPLPQ and shows better robustness against the EBOI conditions. First, the Zernike moments are computed around each pixel to generate the moment images, which can better deal with expression changes and occlusion. Then, calculating by varying the filter size and combined histograms of all the LPQ regions obtained by dividing moment image into non-overlapping patches, we have the multiscale patch-based LPQ descriptor. MPLPQ is used to get stronger robustness to blur and illumination variations. In the proposed method, it is worth emphasizing that the real and imaginary components of ZM, which are extracted to avoid phase information redundancy considering the MPLPQ, are available to gain contour and shading information similar to the magnitude and phase components.

The proposed method is introduced in Sect. 2. Experimental results performed on several datasets are provided in Sect. 3, and Sect. 4 concludes the paper.

2 The Proposed Method

To develop a novel face extraction method that is insensitive to EBOI, we propose to use a combination with two descriptors: local Zernike moment and local phase quantization. And we investigate how they are useful for face recognition invariant to environment conditions. In the following, we introduce these methods in detail.

2.1 Local Zernike Moment

Let $f(x, y)$ be a two-dimensional grayscale image. The definition of orthogonal Zernike moment (ZM) with order n and repetition m is described as

$$Z_{nm} = \frac{n+1}{\pi}\int_{0}^{2\pi}\int_{0}^{1} f(r,\theta)R_{nm}(\rho)e^{-jm\theta}r\mathrm{d}r\mathrm{d}\theta \tag{1}$$

where $f(r, \theta)$ is the transformation of $f(x, y)$ in the polar coordinate, $R_{nm}(\rho)$ represents the Zernike radial polynomial, which is defined by

$$R_{nm}(\rho) = \sum_{s=0}^{(n-|m|)/2} \frac{(-1)^s \rho^{n-2s}(n-s)!}{s!((n+|m|)/2-s)!((n-|m|)/2-s)!} \tag{2}$$

where n is the order and m is the number of iterations satisfying $|m| < n$ and n -$|m|$ is even.

For digital images, the discrete form of ZM is given by

$$Z_{nm} = \frac{n+1}{\pi}\sum_{x=0}^{N-1}\sum_{y=0}^{N-1} f(x,y)R_{nm}(\rho)e^{-jm\theta} \tag{3}$$

where N is the number of pixels in each coordinates of the image, and x and y satisfy $x^2 + y^2 \leq 1$. The parameters ρ and θ are calculated by the mapping transformation as

$$\rho = \sqrt{x^2 + y^2}, \theta = \tan^{-1}(y/x)$$

Inspired by the method reported in [10], we use local Zernike moment (LZM) to the feature extraction. The local Zernike moment is defined as

$$Z_{nm}^{k} = \sum_{p,q=-(k-1)/2}^{(k-1)/2} f(i-p, j-q)R_{nm}^{k}(\rho)e^{-jm\theta} \tag{4}$$

where k is the kernel size. Different from [10], where two layers of LZM for both phase and magnitude components have been applied, our method computes the moment at every pixel of image, and then utilize the real and imaginary parts of the moment to generate two moment images only once. Experiment results in Table 1 clearly show that real and imaginary histograms can get better performance than phase and magnitude histograms using our method. Therefore, histograms are extracted from real and imaginary parts instead of the phase and magnitude ones in [10]. The real and imaginary parts of LZM representation are shown in Fig. 1a. It seems that imaginary parts are more robust to illumination.

Table 1 Recognition rates using local Zernike moment based on phase (LZM-P), magnitude (LZM-M), phase and magnitude (LZM-PM), real part (LZM-R), imaginary part (LZM-I), and real and imaginary (LZM-IR) parts on ORL database

	LZM-P	LZM-M	LZM-PM	LZM-R	LZM-I	LZM-IR
Rate (%)	85.08	98.75	97.49	99.26	98.01	99.75

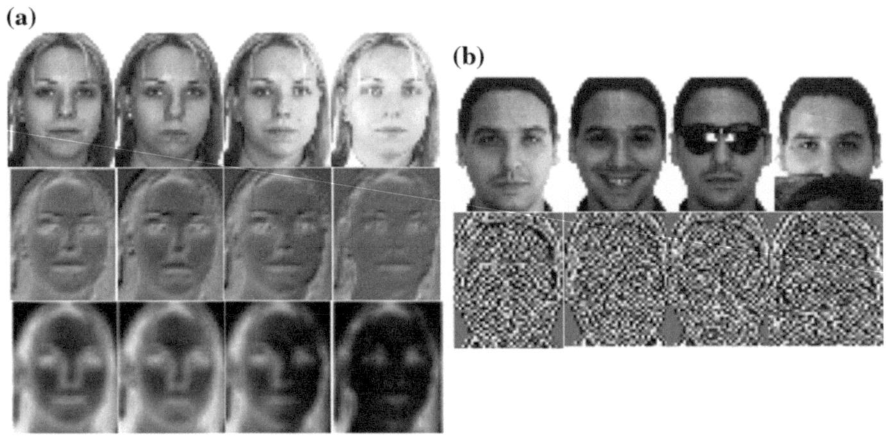

Fig. 1 **a** Local Zernike moment representation of face images with four illumination conditions. The images in each column stand for the input image, the imaginary part of Z^3_{64}, and the real part of Z^3_{64}. **b** LPQ representation of face images with illumination, expression (smile), wearing glasses, and occlusion

2.2 Local Phase Quantization

LPQ [6] operator has been broadly used in facial analysis based on blur invariant property. The phase information is examined in local neighborhood N_x at each pixel position x of the image $f(x)$, using a short-term Fourier transform (STFT), defined by

$$F(u,x) = \sum_{y \in N_x} f(y) W(y-x) e^{-j2\pi u^T y} \tag{5}$$

where $W(y\text{-}x)$ is a window function defining the neighborhood N_x centered by x, and u is the frequency. The local Fourier coefficients are computed at four frequency points $u_1 = [d, 0]^T$, $u_2 = [0, d]^T$, $u_3 = [d, d]^T$, $u_4 = [d, -d]^T$, where d is a sufficiently small frequency satisfying blur invariant condition. For each pixel position, the result of $F(x)$ is described as

$$F(x) = [F(u_1, x), F(u_2, x), F(u_3, x), F(u_4, x)] \tag{6}$$

Next, the phase information in the Fourier coefficients $G_x = [\mathrm{Re}\{F_x\}, \mathrm{Im}\{F_x\}]$ is computed, i.e., $g(x) \in \{g_1(x), g_2(x), \ldots, g_N(x)\}$, and vectors are quantized as

$$q_j = \begin{cases} 1 & \text{if } g_j(x) \geq x \\ 0 & \text{otherwise} \end{cases} \tag{7}$$

The resulting eight binary coefficients q_j are represented as integer values between 0–255 using binary coding

$$f_{\mathrm{LPQ}} = \sum_{j=1}^{8} q_j(x) 2^{j-1} \tag{8}$$

Finally, a 256-dimensional histogram of these integer values from all image positions is composed and we use this feature vector in classification. The LPQ descriptors of face image with different conditions are shown in Fig. 1b.

2.2.1 LZM-MPLPQ Method

The objective of this paper is to explore a novel local feature extraction methodology to get strong robustness to the image conditions of EBOI simultaneously. The proposed method LZM-MPLPQ is illustrated in Fig. 2.

First, LZM is calculated to obtain real and imaginary parts of the input image to generate moment images. The selection of parameters for all LZM descriptors, like the order of the polynomials n, the number of iterations m, and neighborhood size s, has significant influence on the final performance. In Sect. 3, we will discuss this problem in detail. Next, LPQ-based methods are applied to moment images

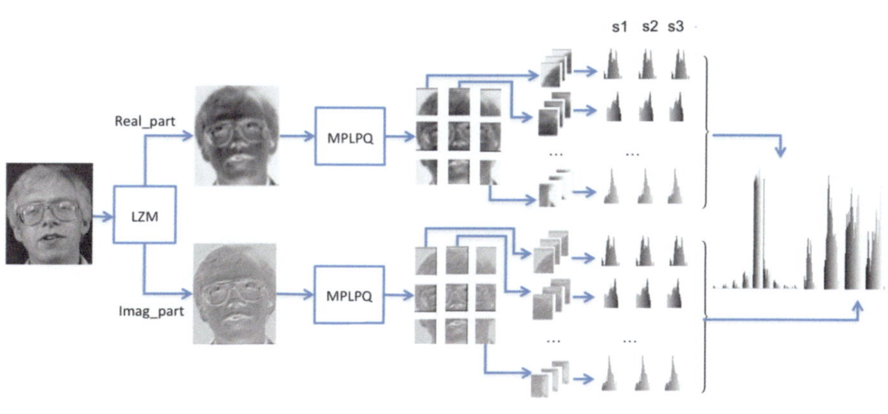

Fig. 2 Flowchart of the proposal

successively, that is, each image is divided into several non-overlapping patches (M_1, M_2,…, M_p) and three-scale LPQ are computed for each patch. Then the histogram of an image is computed as

$$h=[h_1,h_2,\ldots,h_p]|h_i=[h_i^1,h_i^2,\ldots,h_i^s]|h_i^j=\mathrm{LPQ}_w(M_i) \tag{9}$$

where $i \in [1, p]$, $j \in [1, s]$ and w is the size of the LPQ filter, $w = 2j + 1$. The final LZM-MPLPQ face descriptor is

$$H=[h_{\mathrm{real}},h_{\mathrm{imag}}] \tag{10}$$

where h_{real} and h_{imag}, respectively, denote real and imaginary part histograms, and H is the final histogram sequence.

Different from other existing ZM-based or LPQ-based methods, the proposed LZM-MPLPQ has three advantages. (1) Since LPQ has been proved to be robust to blur, we improve the insensitivity to both blur and illumination changes using different filter scales followed by dividing images into several non-overlapping sub-regions. (2) Real and imaginary components of LZM we extracted contain finer details of information than the phase or magnitude. (3) The integration of moment properties (affine variance, less redundancy, etc.) and blur variance positively enhance the robustness to EBOI.

3 Experimental Results

To evaluate the performance of the proposed method, a set of experiments are carried out on ORL [14], JAFFE [15], and AR [16] databases. SVM [17] is applied for face classification. Moreover, three state-of-the-art methods (Mono + PLPQ [7], MLPQ [9], and LZM [10]) are compared with our method in the experiment. Tenfold cross validation is used to obtain average classification rate to better indicate the performance.

3.1 Experimental Face Databases

ORL database contains 40 distinct persons and each person has 10 different images with varying lighting, facial expressions, and facial details. JAFFE database includes 213 grayscale images of Japanese female, which emphasizes on the 7 categories of expressions such as anger, disgust, happiness, neutral, fear, sadness, and surprise. In order to verify the robustness of all the varying conditions of EBOI, AR database (contains 4,000 color images corresponding to 126 subjects) is used. The images of three databases are cropped (using the eyes coordinates from the databases) and resized to the same size respectively (Fig. 3).

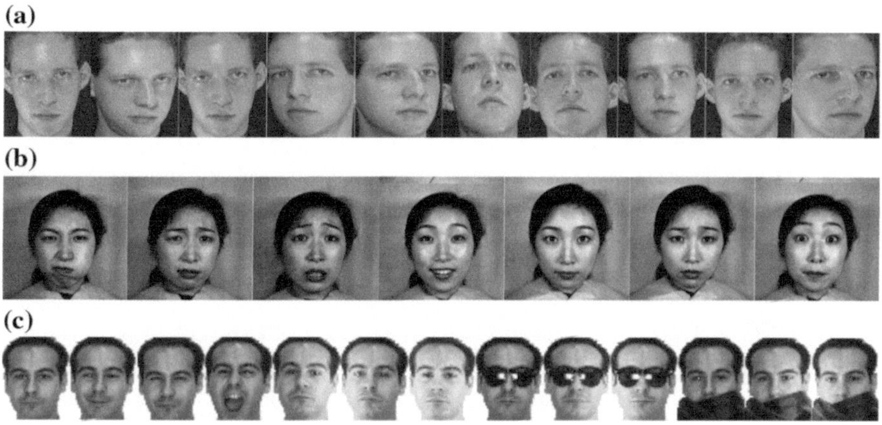

Fig. 3 Sample face images from **a** ORL, **b** JAFFE, **c** AR databases

3.2 Recognition Results

To explore the optimal parameters in LZM-MPLPQ method, such as the number of iterations m and the order of the polynomials n in local Zernike moment, 29 group experiments have been conducted. The results of $Z_{1,1} \sim Z_{9,9}$ on ORL database are shown in Table 2. It is very clear that the highest recognition rate of 99.75 % is achieved when $m = 4$ and $n = 6$. Moreover, Table 3 shows that both patches of multiscale LPQ and size of local neighborhood of LZM can also affect the final result when m and n are invariant. It can be observed that the best number of patches is 2 × 2 and scale size is 3 × 3 on ORL database. Totally, the results illustrate that local Zernike moment is insensitive to the different parameters.

To further evaluate the effect of different image conditions of EBOI in FR, we compare our approach with three existing methods on ORL, JAFFE, and AR databases. The results of different methods are presented in Table 4. It is obvious that our method can handle illumination change on ORL database and expression

Table 2 Recognition rates of different LZM performed on ORL database

	$m = 0$	$m = 1$	$m = 2$	$m = 3$	$\mathbf{m = 4}$	$m = 5$	$m = 6$	$m = 7$	$m = 8$	$m = 9$
$n = 1$	–	98.74	–	–	–	–	–	–	–	–
$n = 2$	98.50	–	92.79	–	–	–	–	–	–	–
$n = 3$	–	98.02	–	98.76	–	–	–	–	–	–
$n = 4$	98.00	–	94.53	–	99.74	–	–	–	–	–
$n = 5$	–	96.25	–	99.02	–	98.98	–	–	–	–
$\mathbf{n = 6}$	94.50	–	94.02	–	**99.75**	–	94.45	–	–	–
$n = 7$	–	96.53	–	99.01	–	98.29	–	99.03	–	–
$n = 8$	97.25	–	93.36	–	99.51	–	93.54	–	99.23	–
$n = 9$	–	99.23	–	99.00	–	98.49	–	98.26	–	97.79

The highest recognition rate and optimal parameters were shown in bold

Table 3 The accuracy of different local neighborhood scales of LZM (rows) and patches of MLPQ (columns) performed on ORL database

	No-patch	**2 × 2**	3 × 3	4 × 4	5 × 5	6 × 6	7 × 7	8 × 8	9 × 9
3 × 3	87.88	**99.75**	99.02	98.76	98.00	96.97	96.02	94.74	92.76
5 × 5	78.75	98.99	98.51	97.76	97.50	97.01	96.26	93.01	90.99
7 × 7	69.34	69.34	97.46	98.25	97.01	96.51	95.51	90.99	89.20
9 × 9	78.26	98.78	98.53	98.25	97.23	95.49	96.03	95.74	90.24

The highest recognition rate and optimal parameters were shown in bold

Table 4 Comparisons of recognition rates using different methods

Method	Mono + PLPQ	LZM	MLPQ	Our
ORL	94.69	93.18	95.50	99.75
JAFFE	99.26	94.44	88.67	100
AR	94.58	85.56	87.13	95.51

variations effectively on JAFFE database. Moreover, the results on AR database show that our method achieves better performance than other methods. It indicates that the LZM-MPLPQ is tolerated well to occlusions (sun glasses and scarf). Therefore, the LZM-MPLPQ method outperforms the state-of-the-art methods when the image conditions at EBOI.

4 Conclusion

In this paper, we proposed a novel local feature extraction method based on LZM and multiscale patch-based LPQ. With the LZM-MPLPQ method for feature extraction, a simple but powerful FR system has been also presented. The performance of our method was compared to the state-of-the-art methods in ORL, JAFFE, and AR databases. Experimental results showed that our method improved the recognition rate and significantly reduced the impact of the image variations of EBOI in FR simultaneously. For the future work, we plan to further investigate the usage of our method in face analysis-related studies, such as color face recognition, face verification, and facial expression recognition.

References

1. Belhumeur PN, Hespanha JP, Kriegman DJ (1997) Eigen-faces vs fisherfaces: recognition using class specific linear projection. IEEE Trans Pattern Anal Mach Intell 19(7):711–720
2. Zhang WC, Shan SG, Gao W et al (2005) Local Gabor binary pattern histogram sequence (LGBPHS): a novel non-statistical model for face representation and recognition. In: IEEE international conference on computer vision, pp 786–791

3. Chan CH, Kittler J, Messer K (2007) Multiscale local binary pattern histograms for face recognition. In: International conference on biometrics, pp 809–818
4. Ahonen T, Hadid A, Pietika M (2006) ¨inen. face description with local binary patterns: application to face recognition. IEEE Trans Pattern Anal Mach Intell 28(12):2037–2041
5. Zhang B, Shan S, Chen X et al (2007) Histogram of Gabor phase patterns (HGPP): a novel object representation approach for face recognition. IEEE Trans Image Process 16(1):57–68
6. Ojansivu V, Heikkila J (2008) Blur insensitive texture classification using local phase quantization. In: International conference on image and signal processing, pp 236–243
7. Nguyen HT, Caplier A (2014) Patch based local phase quantization of monogenic components for face recognition. In: IEEE international conference on image processing, pp 229–233
8. Heikkila J, Ojansivu V, Rahtu E (2010) Improved blur insensitivity for decorrelated local phase quantization. In: IEEE international conference on pattern recognition, pp 818–821
9. Chan CH, Tahir MA, Kittler J et al (2013) Multiscale local phase quantization for robust component-based face recognition using kernel fusion of multiple descriptors. IEEE Trans Pattern Anal Mach Intell 35(5):1164–1177
10. Sariyanidi E, Dag˘lı V, Tek SC et al (2012) Local Zernike moments: a new representation for face recognition. In: IEEE international conference on image processing, pp 585–588
11. Dai XB, Liu TL, Shu HZ et al (2014) Pseudo-Zernike moment invariants to blur degradation and their use in image recognition. In: Intelligent science and intelligent data engineering, pp 90–97
12. Lajevardi SM, Hussain ZM (2010) Higher order orthogonal moments for invariant facial expression recognition. IEEE Trans Digital Signal Process 20(6):1771–1779
13. Kanan HR, Faez K, Gao YH (2008) Face recognition using adaptively weighted patch PZM array from a single exemplar image per person. IEEE Trans Pattern Recogn 41(12):3799–3812
14. Samaria FS, Harter AC (1994) Parameterisation of a stochastic model for human face identification. In: IEEE workshop on applications of computer vision, pp 138–142
15. Lyons MJ, Akamatsu S, Kamachi M et al (1998) Coding facial expressions with Gabor wavelets. In: IEEE international conference on automatic face and gesture recognition, pp 200–205
16. Martinez A, Benavente R (1998) The AR Database. CVC Technical Report No. 24
17. Chang CC, Lin CJ (2011) LIBSVM: a library for support vector machines. ACM Trans Intell Syst Technol 2(3):389–396

Design and Control of the Upright Controllable Force Sub-system for the Suspended Gravity Compensation System

Jiao Jia, Yingmin Jia and Shihao Sun

Abstract This paper introduces the design and control of the suspended gravity compensation system's upright controllable force subsystem. The upright subsystem consists of the transmission compensation module, the buffer module, and the control part. The transmission compensation module includes the transmission components. The buffer module includes the compression spring and its fittings. The control part contains servo motor, tension sensor, and relative displacement sensor. The subsystem can follow the object upright motion actively and provide constant force to compensate the gravity of the object.

Keywords Lagrange method · Controllable force · Robust control · Interference suppression · Active follow

1 Introduction

One of the vital differences between the space environment and the ground laboratory environment is that the space is microgravity. In order to represent the spacecraft motion in space on the ground and improve the fidelity of ground verification of guidance and control system, the microgravity environment should be built. Up to now, the gravity compensation methods have been developed such as weight loss, neutral buoyancy, air-bearing table, and suspension system [1–5]. There are both advantages and disadvantages of these approaches. Free falling objects [1] and parabolic flight [2] are the common weight loss methods, but they suffer from short time, limited space, and costly. Neutral buoyancy [3] has the disadvantages of large damping, high maintenance cost, and only applying to low speed situation, and air-bearing table method [4] could only provide five degrees of

J. Jia (✉) · Y. Jia · S. Sun
The Seventh Research Division and the Center for Information and Control,
School of Automation Science and Electrical Engineering,
Beihang University (BUAA), Beijing 100191, China
e-mail: annebuaa@hotmail.com

Y. Jia et al. (eds.), *Proceedings of 2016 Chinese Intelligent Systems Conference*, Lecture Notes in Electrical Engineering 405,
DOI 10.1007/978-981-10-2335-4_4

freedom, which is limited in vertical direction. Without the above limitation and high reliability, the suspension method [5–7] becomes an economical and practical method.

There are two kinds of suspension compensation methods. One is called passive suspension compensation [5] and the other one is active suspension compensation [6, 7]. The active suspension method performs much better than the passive one. The compensation accuracy of the suspend system depends on the structure and controller of the vertical direction subsystem greatly. Usually, the wire rope is one of the necessary parts of the vertical direction subsystem. However, when a force is opposite or greater than the object gravity functions on the object, the rope will be invalid. To solve this problem, a spring is combined with rope, but the compensation accuracy is affected by the rope reversing. In this paper, we develop a new means that combines a spring buffer unit with the transmission unit to avoid the shortage of the wire rope. Employing the Lagrange method we build the mathematical model and design a controller for the subsystem. Since the spring stiffness coefficient and subsystem friction can hardly measure accurately, there are uncertain parameters in the system. Due to the uncontrollability of the subject force, we consider it as disturbances. Then it is a system with uncertainties and disturbances.

The rest of this paper is structured as follows. Section 2 states the structure design of the subsystem. Section 3 presents the progress of building the model and the controller design. The simulation results are provided in Sect. 4, and concluding remarks are given in Sect. 5.

2 System Design

The installation is a servo system consisting of the transmission compensation module (CM), the buffer module (BM), and the control part (CP). The transmission compensation module includes the rack, the gear, and the fixing plate. The buffer module includes the linear bearing, the steel shaft, the compression spring, the bearing grooves and the bearing guide, the rolling bearings, the bearing retainer, and the connection block. The control part contains the servo motor, the tension sensor, and the relative displacement sensor. Figure 1 shows the connection relationship and information flow of the upright subsystem. The rolling bearings are tangent to the bearing grooves and fixed by the bearing retainer (Fig. 2). The linear bearing installs in the upper end of the bearing guide. The steel shaft passes through and mates the linear bearing and its upper end links to the lower end of the tension sensor. The tension sensor's upper end connects to the rack. The rack mates the gear and the motion of the gearing is controlled by the servo motor. The servo motor fixes on the fixing plate. The relative displacement attaches to the tension sensor. The bearing retainer fixes to the lower end of the steel raft and connects the bottom of the compress spring. The upper end of the compress spring attaches to the bearing guide. The object attaches to the bearing guide with the connection block. The object attaches to the bearing guide through the connection block.

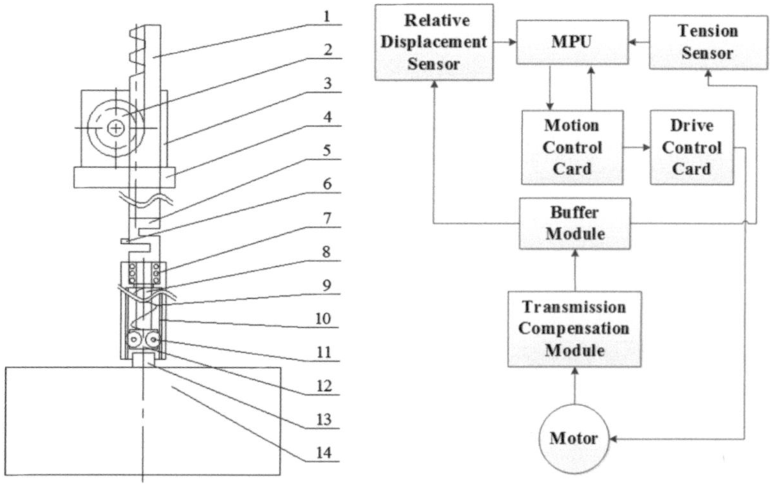

Fig. 1 Structure and information flow of the subsystem. *1* Rack, *2* gear, *3* servo motor, *4* fixing plate, *5* tension sensor, *6* relative displacement sensor, *7* linear bearing, *8* steel shaft, *9* compression spring, *10* bearing grooves and bearing guide, *11* rolling bearings, *12* bearing retainer, *13* connection block, *14* object

Fig. 2 Structure of rolling bearings and bearing retainer

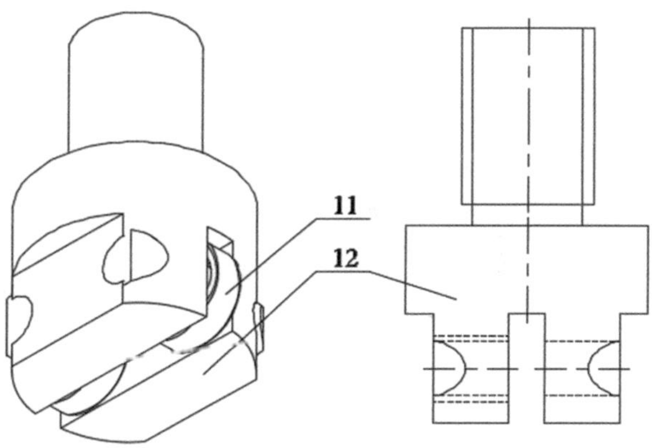

As described above, the force of the compress spring is equal to the gravity of the object (The weight of the connection structure is negligible relative to the object's) when the system is static or in uniform motion. The compress length of the compress spring will vary when the upright direction force functions on the object. It is our subject to keep the force or the length of the compress spring equal to the object gravity so the object motion only decided by the external force. The force of the compress spring can be controlled by the servo motor indirectly. The tension sensor is used to measure the spring force and the relative displacement is utilized to measure the variation length of the compress spring. The centroids of the rack and the object are in the same vertical line.

3 Model Building and Controller Design

The installation coordinate is shown in Fig. 3. The O_1, O_2, O_3 represent the centroid of the gear, the rack, and the object, respectively. Their mass marks m_1, m_2, M. Connect point O_2, O_3 and view the straight line in which the segment O_2O_3 lies as y-axis and the upward direction is positive. A straight line that goes through the point O_1 and is perpendicular to the y-axis serves as x-axis and its right direction is positive. The intersection of x-axis and the y-axis is the origin point O. Denote α as the rotation angle of the gear and the counterclockwise is positive. The gear radius is R. The length of the rack is h_0. O_2 and O coincide at the rack initial position. The compress spring is lightweight and its free length is l_0. Define g as the gravitational acceleration. The spring stiffness is k. $l_1 = \frac{Mg}{k}$. The displacement of the rack driven by the gear is $y = \alpha R$. The displacement of the object is $x = y - l$. In the above formula l is the spring deformation length and $l = 0$ when the fore of the spring is equal to the gravity of the object.

The system kinetic energy is

$$T = \frac{1}{2}J_1\dot{\alpha}^2 + \frac{1}{2}M\dot{x}^2 + \frac{1}{2}m_2\dot{y}^2$$
$$J_1 = \frac{1}{2}m_1R^2, \dot{x} = \dot{y} - \dot{l}, y = \alpha R$$

Choose xOy plane as the zero potential energy surface and the system potential energy is

$$V = \frac{1}{2}k(l + l_1)^2 + Mg\left(y - l_0 - l - \frac{1}{2}h_0\right) + m_2gy$$

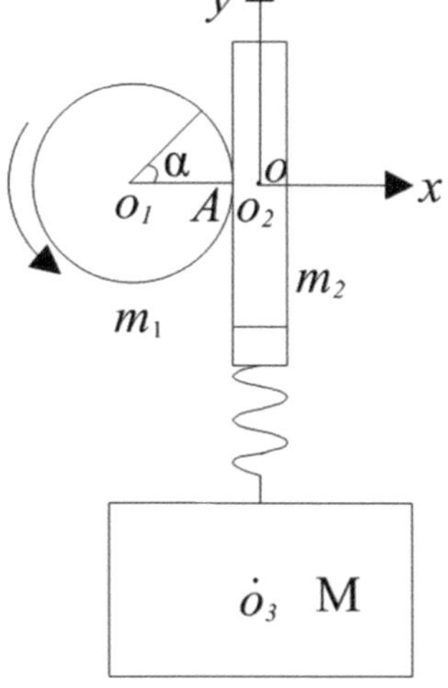

Fig. 3 Coordinate system

$$
\begin{aligned}
L &= T - V \\
&= \frac{1}{2} J_1 \dot{\alpha}^2 + \frac{1}{2} M \dot{x}^2 + \frac{1}{2} m_2 \dot{y}^2 - \frac{1}{2} k (l + l_1)^2 - Mg\left(y - l_0 - l - \frac{1}{2} h_0 \right) - m_2 g y
\end{aligned} \tag{1}
$$

Lagrange equation is

$$
\frac{d}{dt}\left(\frac{\partial L}{\partial \dot{q}}\right) - \frac{\partial L}{\partial q} = Q_j (j = 1, 2, \ldots\ldots) \tag{2}
$$

Choose α, l as system generalized coordinates. Substitute Eq. (1) into Eq. (2) and we gain

$$
\begin{cases} (J_1 + MR^2 + m_2 R^2)\ddot{\alpha} - MR^2 \ddot{l}\,) + (M + m_2) g R = C_g \tau \\ -MR\ddot{\alpha} + \ddot{l} + kl = F_l \end{cases} \tag{3}
$$

where τ is the motor driving torque, C_g is the transmission efficiency, and F_l is the object driving force.

Consider $\ddot{\alpha}, \ddot{l}$ as variables and solve Eq. (4). We get

$$
\begin{cases} \ddot{\alpha} = \frac{2}{m_1 R^2 + 2 m_2 R^2} \left(-C_l R \dot{l} - klR + C_g \tau - (M + m_2) g R + F_l R \right) \\ \ddot{l} = \frac{2}{m + 2m_2} \left(-\mu C_l \dot{l} - \mu k l + C_g \frac{\tau}{R} - (M + m_2) g + \mu F_l \right), \mu = 1 + \frac{m_1 + 2m_2}{2M} \end{cases} \tag{4}
$$

Our goal is to control l varying according to the given law and α is only the intermediate variables. Then the subsystem dynamic equation is

$$
(0.5 m_1 + m_2)\ddot{l} + \mu k l + \mu C_l \dot{l} = C_g \tau / \mathrm{R} - (M + m_2) g + \mu F_l \tag{5}
$$

Define

$$
\begin{aligned}
&\rho = \frac{1}{0.5 m_1 + m_2},\ k_1 = \rho \mu k,\ C_1 = \rho \mu C_l,\ u(t) = \rho F,\ d(t) = \rho \mu F_l, \\
&F = C_g \tau / R - (M + m_2) g,\ d(t) = \mu F_l
\end{aligned}
$$

$$
\ddot{l} + C_1 \dot{l} + k_1 l = u(t) + d(t) \tag{6}
$$

The initial value of l, $\dot{l}$ is $l(0) = \dot{l}(0) = 0$. Then the Laplace transform of Eq. (6) is

$$
s^2 L(s) + {}_1 C_1 s L(s) + k L(s) = U(s) + D(s) \tag{7}
$$

The open-loop transfer function is

$$G(s) = \frac{L(s)}{U(s)} = \frac{1}{s^2 + C_1 s + k_1} \tag{8}$$

We choose the PID controller

$$G_c(s) = k_P + \frac{k_I}{s} + k_D s \tag{9}$$

The closed-loop system characteristic function is

$$\Phi(s) = 1 + G(s)G_c(s) = \frac{s^3 + (k_D + C_1)s^2 + (k_P + k_1)s + k_I}{s(s^2 + C_1 s + k_1)}$$

Define

$$D(s) = s^3 + (k_D + C_1)s^2 + (k_P + k_1)s + k_I = a_0 s^3 + a_1 s^2 + a_2 s + a_3 \tag{10}$$

$$a_0 = 1,\ a_1 = k_D + C_1,\ a_2 = k_P + k_1,\ a_3 = k_I$$

Based on the Routh–Hurwitz stability criterion, the system stable conditions are

$$a_i > 0,\ (i = 0,\ 1 \ldots 3),\ a_1 a_2 - a_0 a_3 > 0$$

For this system the stable conditions are

$$k_D > -C_1, k_P + k_1 > 0, k_I > 0, (k_D + C_1)(k_P + k_1) > k_I \tag{11}$$

In order to improve the system dynamic performance, define

$$e = l - l^d,\ \dot{e} = \dot{l} - \dot{l}^d,\ \upsilon = \dot{l}^d - \lambda e,\ \dot{\upsilon} = \ddot{l}^d - \lambda \dot{e},$$
$$r = \dot{l} - \upsilon = \dot{e} + \lambda e, \dot{r} = \ddot{l} - \dot{\upsilon} = \ddot{e} + \lambda \dot{e}$$

where l^d is the desired value.

$$u(t) = -k_P r - k_I \int r - \rho_0 \tanh[(a + bt)r] \tag{12}$$

$\rho_0 > \|d(t)\|$, and $\rho_0 \tanh[(a + bt)r]$ is used to suppress interference $d(t)$.

The main advantage of Eq. (12) named PISS controller taking $e, \dot{e}$ into consideration at the same time can reduce overshoot. The simulation results prove this benefit perfectly.

4 Simulation Results

The parameters of the subsystem are

$$m_1 = 0.08\,\text{kg},\ m_2 = 0.64\,\text{kg},\ k = 1500,\ C_l = 0.05,\ C_g = 0.90,\ g = 10$$

We set $d(t) = 2\sin 2t$, $l^d = \cos t$. Figure 4 shows that both the PID controller and the PISS controller could track the desired signal l^d well. However, from Fig. 5 we can conclude that the PISS controller has a smaller overshoot and better performance of interference suppression. The steady-state tracking error of PID controller and PISS controller are less than 0.4 and 0.1 %.

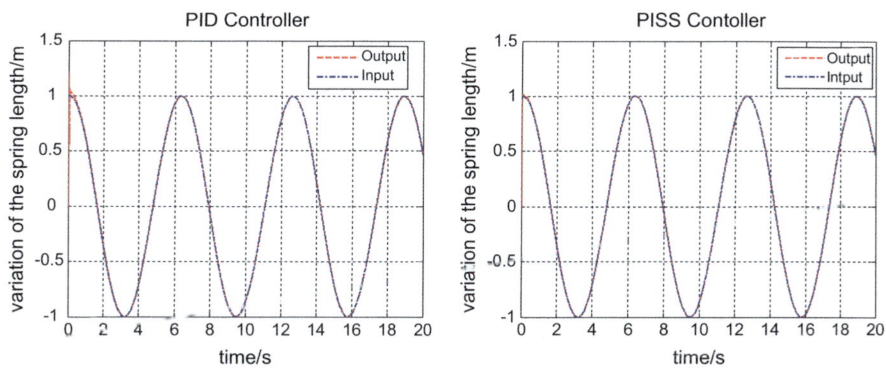

Fig. 4 The object input and output of the system

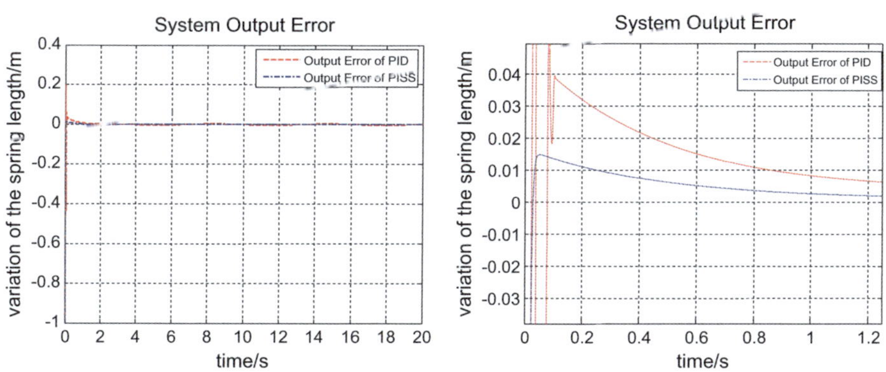

Fig. 5 The system output error

5 Conclusions

In this paper, a novel upright controllable force installation is proposed and then the mathematic model based on Lagrange equation is developed. Finally, a stable and high compensation accuracy controller is given. Then the system could provide different kinds of force to lunar rovers, satellites, aircrafts, etc. when they are tested and verified on the ground. However, the existence of the sensor errors and modeling errors will increase tracking error. Depth study should be conducted to find more practical controllers.

Acknowledgments This work was supported by the National Basic Research Program of China (973 Program: 2012CB821200, 2012CB821201), the NSFC (61134005, 61327807, 61521091, 61520106010), and the Fundamental Research Funds for the Central Universities (YWF-16-GJSYS-31,YWF-16-GJSYS-32).

References

1. Watanabe Y, Kengo A et al (1998) Microgravity experiments for a visual feedback control of a space robot capturing a target. In: Proceedings of the IEEE/RSJ international conference on intelligent robots and systems, vol 3, pp 1993–1998
2. Sawada H, Ui K, Mori M et al (2004) Micro-gravity experiment of a space robotic arm using parabolic flight. Adv Robot 18(3):247–267
3. Menon C, Busolo S, Cocuzza S et al (2007) Issues and solutions for testing free-flying robots. Acta Astronaut 60(12):957–965
4. Robertson A et al (1999) Spacecraft formation flying control design for the Orion mission. AIAA guidance, navigation, and control conference, pp 1562–1575
5. Kienholz DA, Crawley EF, Harvey TJ (1989) Very low frequency suspension systems for dynamic testing. In: Proceedings of the 30th structures, structural dynamics and materials, pp 327–336
6. White QC, Xu Y (1994) An active vertical-direction gravity compensation system. IEEE Trans Instrum Meas 43(6):786–792
7. Han O, Kienholz D, Janzen P et al (2010) Gravity-offloading system for large-displacement ground testing of spacecraft mechanisms. In: Proceedings of 40th aerospace mechanisms symposium, pp 119–132

Fading Unscented–Extended Kalman Filter for Multiple Targets Tracking with Symmetric Equations of Nonlinear Measurements

Cui Zhang, Yingmin Jia and Changqing Chen

Abstract This paper is devoted to the problem of multiple targets tracking based on symmetric equations of nonlinear measurements. We develop a nonlinear stochastic model with unknown random bias to provide a unified structure for the tracking systems with different types of symmetric measurement equations. Moreover, the fading unscented–extended Kalman filter (FUEF) is designed to deal with the strong nonlinearities by embedding the unscented transform into the extended Kalman filter and to conduct the effect of unknown bias by inserting the fading factor. The performance of the novel filter paired with two of symmetric measurement equations are illustrated and compared by the Monte Carlo simulation results.

Keywords Kalman filter · Multiple targets tracking · Nonlinear system · Random bias

1 Introduction

Multiple targets tracking (MTT) has always been an active research field for being widely applied in civilian and military areas including computer vision [1], vehicle environment perception [2], air traffic control [3], etc. The major challenge of MTT is due to the uncertainty in the association of measurements and targets.

C. Zhang (✉) · Y. Jia
The Seventh Research Division and the Center for Information and Control,
School of Automation Science and Electrical Engineering,
Beihang University (BUAA), 100191 Beijing, China
e-mail: zhangcui0521@163.com

Y. Jia
e-mail: ymjia@buaa.edu.cn

C. Chen
The Science and Technology on Space Intelligent Control Laboratory,
Beijing Institute of Control Engineering, 100190 Beijing, China

Y. Jia et al. (eds.), *Proceedings of 2016 Chinese Intelligent Systems Conference*, Lecture Notes in Electrical Engineering 405,
DOI 10.1007/978-981-10-2335-4_5

Many traditional multiple targets tracking approaches have been proposed. The nearest neighbor (NN) [4] is the easiest method to implement. The probability data association (PDA) [5] and the joint probabilistic data association (JPDA) [6] are the wildly used method. Multiple hypothesis tracking (MHT) [7] is also a well-known method. The robustness of NN is poor, PDA and JPDA suffer the restrictions of huge computation burden as the number of targets increasing, while MHT is hard to implement. There are also various studies which have made contributions to the implicit data association MTT technologies. One popular used method is the probabilistic hypothesis density (PHD) filter developed by using the random finite set theory [8]. However, PHD filter only gains the set of probably state estimations at each time instant, and needs further work to maintain track continuity.

This paper is about another considerable implicit data association approach called the symmetric measurement equation (SME). This method obviates the association uncertainty by introducing the symmetric transformations to the original measurements. It was firstly come up by Kamen in the early 1990s, [9] proposed two types of SME and the later work [10–13] applied different nonlinear techniques to deal with the nonlinearity caused by symmetric transformation. The Kernel-SME filter was presented in [14] which constructed a symmetric transformation by using a Gaussian mixture to match the measurements.

In most of the literature, the symmetric transformed method is used to handle the MTT with linear measurements. To the best of the our knowledge, the issue of SME based on nonlinear original measurements, which commonly occurs in practical situations such as navigational system and radar tracking, has been left open. It is noted that, the multiple dimensional symmetric transformation is a nonlinear transformation, the model of measurements after symmetric transformation would have strong nonlinearities and complex noise term. The well-known nonlinear filter, extend Kalman filter (EKF) [15], has inherent disadvantages such as the requirement of gentle nonlinearities and the calculation errors of Jacobian matrices, while the traditional unscented Kalman filter (UKF) [16] could not handle the complex noise term. Thus an easy implementing and computational efficient filter should be designed.

2 Problem Formulation

Focus on the problem of tracking multiple targets based on nonlinear measurements, where the association between the measurements and the targets is unavailable. The tracking system can be given by

$$x_{k+1}^i = \Phi_k^i x_k^i + \Gamma_k^i w_k^i \tag{1}$$

$$m_k^i = h_k^i(x_k^i) + v_k^i \qquad i = 1, \ldots, N \tag{2}$$

where the subscript k denotes the discrete sampling time, the superscript i represents the ith target and N is the number of targets. $x_k^i \in \mathbb{R}^n$ is the target state, $m_k^i \in \mathbb{R}$ is the measurement, h_k^i is the nonlinear observation function, Φ_k^i and Γ_k^i are the state transition matrix and the noise matrix with appropriate dimensions respectively. w_k^i is the process noise and v_k^i is the observation noise. The noise vectors w_k and v_k^i are assumed to be zero mean and

$$E\{w_k^i(w_\tau^i)^T\} = Q_k^i\delta_{k\tau}, \quad E\{v_k^i(v_\tau^i)^T\} = R_k^i\delta_{k\tau}, \quad E\{w_k^i(v_k^j)^T\} = 0, \quad i,j = 1,\dots,N$$

where the $\delta_{k\tau}$ denotes the Kronecker delta function, $Q_k^i > 0$ and $R_k^i > 0$ are the covariance matrices of the process noise and measurement noise, respectively.

In order to the estimate the states of multiple targets, we comprise all states of single target to a joint target state $\mathbf{x}_k = [(x_k^1)^T, \dots, (x_k^N)^T]^T$, where the superscript T denotes the transpose of a vector or a matrix. Typically in this paper, the bold-faced vector variables indicate the joint vector of overall target and the italics symbols represent individual vector elements of single target. Then the dynamic model of the tracking system (1) can be rewritten as an overall dynamic model

$$\mathbf{x}_{k+1} = \Phi_k\mathbf{x}_k + \Gamma_k\mathbf{w}_k \tag{3}$$

Then a "pseudomeasurement" could be generated by the symmetric functions of the original measurement data (2) as $s_k^j = g^j(m_k^1, \dots, m_k^N)(j = 1, \dots, N)$. It should be noted that the rearrangement of the original data m_i would not affect the s because of its symmetric structure. Thus the data association problem can be avoided without losing basic information. To ensure the original measurements m_i could be recovered uniquely from s_k^j, the dimension of symmetric transformation should equal to the number of targets.

As the model of the symmetric measurement does not remain the conventional structure for filtering, our aim is to evolve it to an approximate formulation as

$$s_k^j = g^j(x_k^1, \dots, x_k^N) + \varphi^j(v_k^1, \dots, v_k^N) \qquad j = 1, \dots, N \tag{4}$$

For simplicity of illustration, take example of three targets moving in one dimension to present the calculation of the approximation. Firstly, using the sum-of-products SME, we can obtain

$$\begin{aligned}
\mathbf{s}_k = \mathbf{g}_k(m_k^i) &= \begin{bmatrix} m_k^1 + m_k^2 + m_k^3 \\ m_k^1 m_k^2 + m_k^1 m_k^3 + m_k^2 m_k^3 \\ m_k^1 m_k^2 m_k^3 \end{bmatrix} \\
&= \begin{bmatrix} h_k^1 + v_k^1 + h_k^2 + v_k^2 + h_k^3 + v_k^3 \\ (h_k^1 + v_k^1)(h_k^2 + v_k^2) + (h_k^1 + v_k^1)(h_k^3 + v_k^3) + (h_k^2 + v_k^2)(h_k^3 + v_k^3) \\ (h_k^1 + v_k^1)(h_k^2 + v_k^2)(h_k^3 + v_k^3) \end{bmatrix} \\
&= \begin{bmatrix} h_k^1 + h_k^2 + h_k^3 \\ h_k^1 h_k^2 + h_k^1 h_k^3 + h_k^2 h_k^3 \\ h_k^1 h_k^2 h_k^3 \end{bmatrix} + \begin{bmatrix} v_k^1 + v_k^2 + v_k^3 \\ (h_k^2 + h_k^3) v_k^1 + (h_k^1 + h_k^3) v_k^2 + (h_k^1 + h_k^2) v_k^3 \\ h_k^1 h_k^2 v_k^3 + h_k^1 h_k^3 v_k^2 + h_k^2 h_k^3 v_k^1 \end{bmatrix} \\
&\quad + \begin{bmatrix} 0 \\ v_k^1 v_k^2 + v_k^1 v_k^3 + v_k^2 v_k^3 \\ h_k^3 v_k^1 v_k^2 + h_k^2 v_k^1 v_k^3 + h_k^1 v_k^2 v_k^3 + v_k^1 v_k^2 v_k^3 \end{bmatrix}
\end{aligned} \tag{5}$$

where the first term is a nonlinear function of target state, the second term is a nonlinear function of state and observation noise, while the third term is a function containing complicated noise, so we regard it as an unknown random bias. Next for the sum-of-powers SME, similar result can be calculated as

$$\begin{aligned}
\mathbf{s}_k = \mathbf{g}_k(m_k^i) &= \begin{bmatrix} m_k^1 + m_k^2 + m_k^3 \\ (m_k^1)^2 + (m_k^2)^2 + (m_k^3)^2 \\ (m_k^1)^3 + (m_k^2)^3 + (m_k^3)^3 \end{bmatrix} = \begin{bmatrix} h_k^1 + v_k^1 + h_k^2 + v_k^2 + h_k^3 + v_k^3 \\ (h_k^1 + v_k^1)^2 + (h_k^2 + v_k^2)^2 + (h_k^3 + v_k^3)^2 \\ (h_k^1 + v_k^1)^3 + (h_k^2 + v_k^2)^3 + (h_k^3 + v_k^3)^3 \end{bmatrix} \\
&= \begin{bmatrix} h_k^1 + h_k^2 + h_k^3 \\ (h_k^1)^2 + (h_k^2)^2 + (h_k^3)^2 \\ (h_k^1)^3 + (h_k^2)^3 + (h_k^3)^3 \end{bmatrix} + \begin{bmatrix} v_k^1 + v_k^2 + v_k^3 \\ 2h_k^1 v_k^1 + 2h_k^2 v_k^2 + 2h_k^3 v_k^3 \\ 3(h_k^1)^2 v_k^1 + v3(h_k^2)^2 v_k^2 + 3(h_k^3)^2 v_k^3 \end{bmatrix} \\
&\quad + \begin{bmatrix} 0 \\ (v_k^1)^2 + (v_k^2)^2 + (v_k^3)^2 \\ (v_k^1)^3 + (v_k^2)^3 + (v_k^3)^3 + 3h_k^1 (v_k^1)^2 + 3h_k^2 (v_k^2)^2 + 3h_k^3 (v_k^3)^2 \end{bmatrix}
\end{aligned} \tag{6}$$

also the first term is the nonlinear function containing only target state, the second term is the nonlinear function of target state and zero-mean Gaussian noise and the third term could be treated as unknown random bias whose mean is nonzero.

From above, the observation model of MTT system based on SME with nonlinear measurements can be reformed as a nonlinear model with unknown random bias:

$$\mathbf{s}_k = \mathbf{g}_k(\mathbf{x}_k) + \Psi_k(\mathbf{x}_k, \mathbf{v}_k) + \mathbf{b}_k \tag{7}$$

Thus the assignment problem of measurement-to-target is traded to the problem of dealing with nonlinearities and unknown random bias.

Remark 1 This technology evolving the SME-based tracking system with nonlinear measurements into the biased nonlinear system has several characteristics. On one hand, it can obviate the need of calculating the addition item as in [13] to ensure the mean of pseudomeasurement noise is zero. On the other hand, it provides a uniform structure for multiple targets tracking system based on different types of SME which makes it convenient and tractable for filtering. Also it would improve the tracking accuracy since the high-order terms of observation noise are altered into the unknown random bias to be estimated instead of ignored.

3 Fading Unscented–Extended Kalman Filter

Suppose that the state estimate and the estimate error covariance have been obtained at time $k-1$. The specific filtering algorithm for the nonlinear system (3) and (7) proceeds as follows.

3.1 State and Measurement Prediction

First of all, the prediction of target state and prediction error covariance can be calculated directly since the dynamic equation (3) is assumed in linear

$$\widehat{\mathbf{x}}_{k|k-1} = \Phi_{k-1}\widehat{\mathbf{x}}_{k-1|k-1} \tag{8}$$

$$\mathbf{P}_{k|k-1} = \Phi_{k-1}\mathbf{P}_{k-1|k-1}\Phi_{k-1}^T + \Gamma_{k-1}\mathbf{Q}_{k-1}\Gamma_{k-1}^T \tag{9}$$

where $\mathbf{Q}_{k-1} = diag[Q_{k-1}^1, \cdots, Q_{k-1}^N]$ is the covariance of noise $\mathbf{w}_{k-1}$.

Second, we should get the prediction of measurements based on the observation model (7). The nonlinear function Ψ in (7) can be approximated by the first order linearization as

$$\Psi_k(\mathbf{x}_k, \mathbf{v}_k) \approx \Psi_k(\widehat{\mathbf{x}}_{k|k-1}, 0) + \Psi_k^{\mathbf{x}}\widetilde{\mathbf{x}}_{k|k-1} + \Psi_k^{\mathbf{v}}\mathbf{v}_k \tag{10}$$

where

$$\widetilde{\mathbf{x}}_{k|k-1} = \mathbf{x}_k - \widehat{\mathbf{x}}_{k|k-1},\ \Psi_k^{\mathbf{x}} = \frac{\partial\Psi_k(\mathbf{x}_k, \mathbf{v}_k)}{\partial\mathbf{x}_k}|_{\mathbf{x}_k=\widehat{\mathbf{x}}_{k|k-1}, \mathbf{v}_k=0}, \tag{11}$$
$$\Psi_k^{\mathbf{v}} = \frac{\partial\Psi_k(\mathbf{x}_k, \mathbf{v}_k)}{\partial\mathbf{v}_k}|_{\mathbf{x}_k=\widehat{\mathbf{x}}_{k|k-1}, \mathbf{v}_k=0}$$

and the error covariance would be

$$\mathbf{RR}_k \approx \Psi_k^{\mathbf{x}}\mathbf{P}_{k|k-1}(\Psi_k^{\mathbf{x}})^T + \Psi_k^{\mathbf{v}}\mathbf{R}_k(\Psi_k^{\mathbf{v}})^T \tag{12}$$

where $\mathbf{R}_k = diag[R_k^1, \ldots, R_k^N]$ is the covariance of $\mathbf{v}_k$.

What could reduce the complexity of computation is that, by derivation and induction, we can get $\boldsymbol{\Psi}_k(\mathbf{x}_{k|k-1}, 0) = 0$ and $\boldsymbol{\Psi}_k^{\mathbf{x}} = 0$. Hence the predicted mean of $\boldsymbol{\Psi}_k$ would be zero and for the error covariance only $\boldsymbol{\Psi}_k^{\mathbf{v}}$, the Jacobian matrix whose elements contain the nonlinear function of state $h_k^i(\mathbf{x}_k)$, is still underestimate. Take advantage of the unscented transformation, the predicted sigma points of $\widehat{\mathbf{x}}_{k|k-1}$ can be calculated by

$$\xi_k^0 = \widehat{\mathbf{x}}_{k|k-1}, \tag{13}$$

$$\xi_k^l = \widehat{\mathbf{x}}_{k|k-1} + \left(\sqrt{(n+\lambda)\mathbf{P}_{k|k-1}}\right)_l, \qquad l = 1, \ldots, n \tag{14}$$

$$\xi_k^l = \widehat{\mathbf{x}}_{k|k-1} - \left(\sqrt{(n+\lambda)\mathbf{P}_{k|k-1}}\right)_{l-n}, \qquad l = n+1, \ldots, 2n \tag{15}$$

where the subscript l denotes the lth column of the matrix and n is the dimension of joint state. $\lambda = \alpha^2(n+\kappa) - n$ is a scaling factor, where α determines the dispersion of sigma points which is selected as 0.01 and κ is usually set as 0.

Then substituting the sigma points (13)–(15) to the functions h_k^i, we can obtain another sigma points set $\zeta_k^l = \left[h_k^1(\xi_k^l)^T \ \ldots \ h_k^N(\xi_k^l)^T\right]^T, (l = 0, \ldots, 2n)$. The expectation is captured as

$$\widehat{h}_{k|k-1}^i = \left(\sum_{l=0}^{2n} \omega_d^l \zeta_k^l\right)^i, \quad l = 0, \ldots, 2n, \ i = 1, \ldots, N \tag{16}$$

where the superscript i denotes the ith row of the matrix, and

$$\omega_d^0 = \frac{\lambda}{n+\lambda}; \quad \omega_d^l = \frac{1}{2(n+\lambda)}, \qquad l = 1, \ldots, 2n \tag{17}$$

Thus the Jacobian matrix $\boldsymbol{\Psi}_k^{\mathbf{v}}$ can be estimated by

$$\boldsymbol{\Psi}_k^{\mathbf{v}} \approx \boldsymbol{\Psi}_k^{\mathbf{v}}(\widehat{h}_{k|k-1}^i) \quad i = 1, \ldots, N \tag{18}$$

Now implementing the nonlinear symmetric function $\mathbf{g}_k$ in formulation (7) to these new sigma points, we can get the third sigma set for the predicted measurement

$$\delta_k^l = \mathbf{g}_k(\zeta_k^l), \quad l = 0, \ldots, 2n \tag{19}$$

Recall the results about the statistical characteristics of $\boldsymbol{\Psi}_k$, the predicted mean of symmetric measurements would be

$$\widehat{\mathbf{s}}_{k|k-1} = \sum_{l=0}^{2n} \omega_d^l \delta_k^l, \quad l = 0, \ldots, 2n \tag{20}$$

the predicted measurement covariance and the cross-correlation covariance are

$$\mathbf{P}_{\mathbf{s}_k} = \sum_{l=0}^{2n} \omega_c^l (\delta_k^l - \widehat{\mathbf{s}}_{k|k-1})(\delta_k^l - \widehat{\mathbf{s}}_{k|k-1})^T + \Psi_k^{\mathbf{v}} \mathbf{R}_k (\Psi_k^{\mathbf{v}})^T \tag{21}$$

$$\mathbf{P}_{\mathbf{x}_k \mathbf{s}_k} = \sum_{l=0}^{2n} \omega_c^l (\xi_k^l - \widehat{\mathbf{x}}_{k|k-1})(\delta_k^l - \widehat{\mathbf{s}}_{k|k-1})^T \tag{22}$$

where

$$\omega_c^0 = \frac{\lambda}{n+\lambda} + (1 - \alpha^2 + \beta), \quad \omega_c^l = \frac{1}{2(n+\lambda)}, l = 1, \ldots, 2n \tag{23}$$

where $\beta = 2$ is optimal under Gaussian distribution.

3.2 Adaptive Fading Factor

The random bias $\mathbf{b}_k$ in (7) with complex distribution is hard to be estimated. In addition to the unknown bias, the approximation of $\Psi^{\mathbf{v}}$ would also introduce the inaccuracies. To compensate these affections in measurement prediction, the fading factor μ is inserted in (21).

Assume that the actual measurement covariance is $\widetilde{\mathbf{C}}_k$ and has the equivalent relation that $\widetilde{\mathbf{C}}_k = \mu_k \mathbf{P}_{\mathbf{s}_k}$ where $\mathbf{P}_{\mathbf{s}_k}$ is calculated as mentioned above without considering the unknown bias.

Define the innovation vector as $\widetilde{\mathbf{s}}_k = \mathbf{s}_k - \widehat{\mathbf{s}}_{k|k-1}$. Making use of innovation sequence from previous times to now, $\widetilde{\mathbf{C}}_k$ can be estimated by

$$\widetilde{\mathbf{C}}_k = \frac{1}{L-1} \sum_{\tau=k+1-L}^{k} \widetilde{\mathbf{s}}_\tau \widetilde{\mathbf{s}}_\tau^T \tag{24}$$

where L is a window size.

For the sake of ensuring the optimality of the Kalman gain, the innovation sequence should be a white sequence and the condition that the autocovariance is zero has to be satisfied. So the orthogonality should be held

$$\mathbf{P}_{\mathbf{x}_k \mathbf{s}_k} - \widetilde{\mathbf{K}}_k \widetilde{\mathbf{C}}_k = 0 \tag{25}$$

where $\widetilde{\mathbf{K}}_k$ is the actual Kalman gain. To reduce computation complexity, the scalar variable μ_k can be approximated as

$$\mu_k = max\left\{1, \frac{1}{n} trace(\widetilde{\mathbf{C}}_k \mathbf{P}_{\mathbf{s}_k}^{-1})\right\} \tag{26}$$

3.3 State Update with Measurement

Based on the conclusions above, we can finally obtain the state estimation by updating the state prediction with received measurements

$$\widetilde{\mathbf{P}}_{\mathbf{s}_k} = \mu_k \mathbf{P}_{\mathbf{s}_k} \tag{27}$$

$$\widetilde{\mathbf{K}}_k = \mathbf{P}_{\mathbf{x}_k \mathbf{s}_k} \widetilde{\mathbf{P}}_{\mathbf{s}_k}^{-1} \tag{28}$$

$$\widehat{\mathbf{x}}_{k|k} = \widehat{\mathbf{x}}_{k|k-1} + \widetilde{\mathbf{K}}_k \widetilde{\mathbf{s}}_k \tag{29}$$

$$\mathbf{P}_{k|k} = \mathbf{P}_{k|k-1} - \widetilde{\mathbf{K}}_k \widetilde{\mathbf{P}}_{\mathbf{s}_k} \widetilde{\mathbf{K}}_k^T \tag{30}$$

Remark 2 It is remarkable that the fading factor μ_k can be adaptively adjusted in each iteration to compensate the influence of unknown random bias in observation model, as well as be obtained efficiently with tiny calculation burden. The modified gain matrix holds $\widetilde{\mathbf{K}}_k = \mathbf{K}_k / \mu_k$, which is accordance with the intuition that the gain should be reduced in the situation where the measurements model with unknown information is not reliable.

4 Simulation Results

The performance of the proposed FEUF paired with two kinds of SME are evaluated in this section. It is compared with the nearest neighbor method and conventional EKF via computer simulations.

4.1 Simulation Setup

Consider tracking three moving targets in $x - y$ plane simultaneously whose state $x_k^i (i = 1, 2, 3)$ is of the form $x_k^i = \left[x^i(k)\ \dot{x}^i(k)\ y^i(k)\ \dot{y}^i(k) \right]^T$ with position in meters and velocity in meters per second. The state of three targets evolves with the initial state $x_0^1 = \left[700\ {-8}\ 100\ 3 \right]^T$, $x_0^2 = \left[100\ 3\ 50\ 6 \right]^T$ and $x_0^3 = \left[300\ 0\ {-100}\ 9 \right]^T$ respectively, which are chosen to ensure the trajectories of three targets would intersect with one another.

The target dynamics are described by the wiener velocity model, so the matrices in (3) are given by:

$$\Phi_k^i = \begin{bmatrix} 1 & T & 0 & 0 \\ 0 & 1 & 0 & 0 \\ 0 & 0 & 1 & T \\ 0 & 0 & 0 & 1 \end{bmatrix}, \quad \Gamma_k^i = \begin{bmatrix} \frac{T^2}{2} & 0 \\ T & 0 \\ 0 & \frac{T^2}{2} \\ 0 & T \end{bmatrix} \tag{31}$$

where $T = 1$ s is the sampling period. w_k^i is a zero-mean white Gaussian with covariance $Q_k^i = 0.04I$. The nonlinear observations are chosen as the radar observation model generated by true range and angle

$$r_k^i = \sqrt{x^i(k)^2 + y^i(k)^2} + (v_r)_k^i, \quad \theta_k^i = arctan\frac{y^i(k)}{x^i(k)} + (v_\theta)_k^i \tag{32}$$

where the independent white Gaussian noise $(v_r)_k^i$ and $(v_\theta)_k^i$ are with covariance

$$R_k^1 = \begin{bmatrix} 10 & 0 \\ 0 & 10^{-4} \end{bmatrix}, \quad R_k^2 = \begin{bmatrix} 50 & 0 \\ 0 & 10^{-4} \end{bmatrix}, \quad R_k^3 = \begin{bmatrix} 25 & 0 \\ 0 & 10^{-4} \end{bmatrix} \tag{33}$$

We take the root mean square (RMS) errors as the measures of tracking performance. At the kth sampling period, RMS position error $RMS_p^i(k)$ and RMS velocity error $RMS_v^i(k)$ are respectively defined as

$$RMS_p^i(k) = \sqrt{\frac{1}{M}\sum_{j=1}^{M}(\eta_p^{i,j}(k))^2}, RMS_v^i(k) = \sqrt{\frac{1}{M}\sum_{j=1}^{M}(\eta_v^{i,j}(k))^2} \tag{34}$$

with the notation M denoting the number of Monte Carlo runs. $\eta_p^{i,j}(k)$ and $\eta_v^{i,j}(k)$ which stand for kth time position and velocity error of ith target from the jth run are given by

$$\eta_p^{i,j}(k) = \sqrt{(x_j^i(k) - \hat{x}_j^i(k))^2 + (y_j^i(k) - \hat{y}_j^i(k))^2} \tag{35}$$

$$\eta_v^{i,j}(k) = \sqrt{(\dot{x}_j^i(k) - \hat{\dot{x}}_j^i(k))^2 + (\dot{y}_j^i(k) - \hat{\dot{y}}_j^i(k))^2} \tag{36}$$

where $\hat{x}_j^i(k)$, $\hat{y}_j^i(k)$, $\hat{\dot{x}}_j^i(k)$ and $\hat{\dot{y}}_j^i(k)$ designate the estimate state of target.

4.2 *Effectiveness in Trajectories Maintaining*

To verify the effectiveness of designed filter in tracking multiple targets, we compare performance of the FEUF paired with two types of SME (sum-of-products and sum-of-powers) with that of the nearest neighbor method paired with UKF. The NN

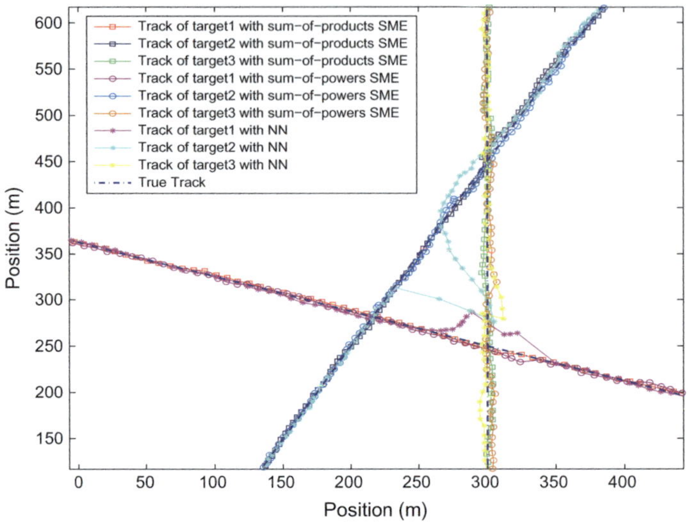

Fig. 1 Comparison of estimated position

algorithm chooses the observations whose distance to the predicted measurements are minimal as the associated observations to update the state estimation [4].

Figure 1 presents the true and estimated trajectories of the three moving targets by FEUF with sum-of-products SME, FEUF with sum-of-powers SME, and UKF with NN method. Although three filters all may fail in maintaining the correct track, the FEUF based on SME have higher percentages of successful tracking. In the successful tracking case, as shown in Fig. 1, the proposed filter can provide accurate tracking results almost all the time without data association even around the path cross points where the NN method always performs poorly. It can be demonstrated more directly by the RMS errors. Figure 2 shows the comparison of RMSE in position, the FEUF have much less error than NN method has when the trajectories are across. While the results of comparison of RMSE in velocity are similar and omitted.

More details are given in Table 1. The terms "ARMSE in position" and "ARMSE in velocity" are calculated by averaging the RMSE over the tracks of three targets only during the valid runs. The CPU time are the average of each run among total 100 runs. As can be seen that the proposed filter outperforms the NN method in maintaining the correct target association. The FEUF products has the best accuracy and FEUF powers is not as ideal as FEUF products may be affected by the complexity structure in the bias. It is expected that the FEUF with two types of SME have higher computational cost than UKF-NN due to the calculation of more sigma sets and adaptive fading factor.

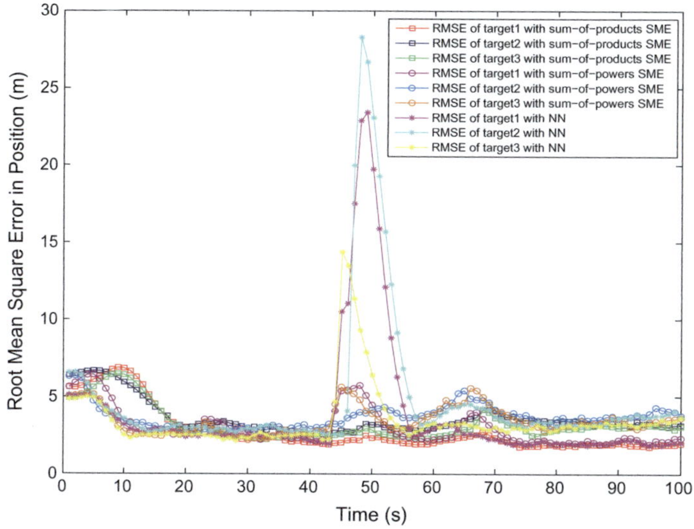

Fig. 2 Comparison of RMS position errors

Table 1 Comparison of tracking performance

	ARMSE in position	ARMSE in velocity	Association maintaining	CPU time
FEUF-products	3.2076	0.3055	99 %	0.2764
FEUF-powers	3.3815	0.4364	84 %	0.3012
UKF-NN	4.0783	0.5569	52 %	0.1608

4.3 *Effectiveness in Nonlinearities Handling*

Figures 3 and 4 shows the simulation results of applying SME with the developed filter and the EKF filter. As the RMS errors of EKF are instable while the errors of FEUF remain little, it can be observed that the EKF fails in dealing with both types of symmetric equations of nonlinear measurements and the FEUF performs well in handling strong nonlinearities.

Remark 3 It is noted that, when implemented the FEUF, the sum-of-products SME beats the sum-of-powers SME in accuracy of both state estimation and trail association with less computational burden. This method conquers the shortcoming of sum-of-products SME with the UKF in the previous studies of Leven [13], as well as the results are consistent with the conclusion in [13] that the performance of the SME approach is rest with the pairing of a specific SME implementation and an appropriate filter.

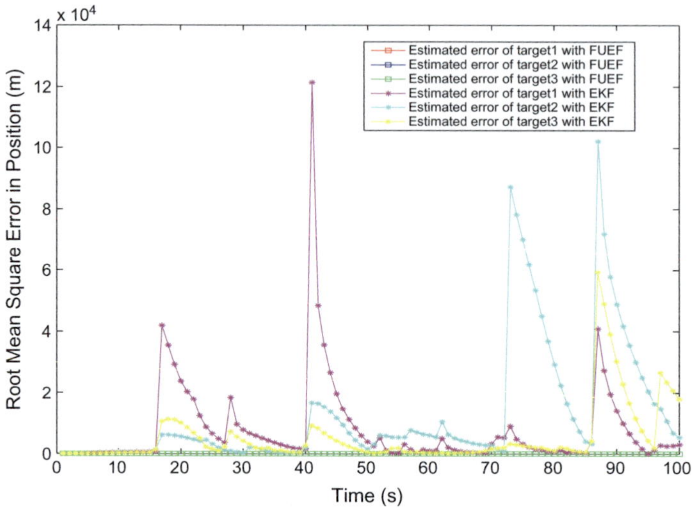

Fig. 3 Comparison of RMSE in position based on sum-of-products SME

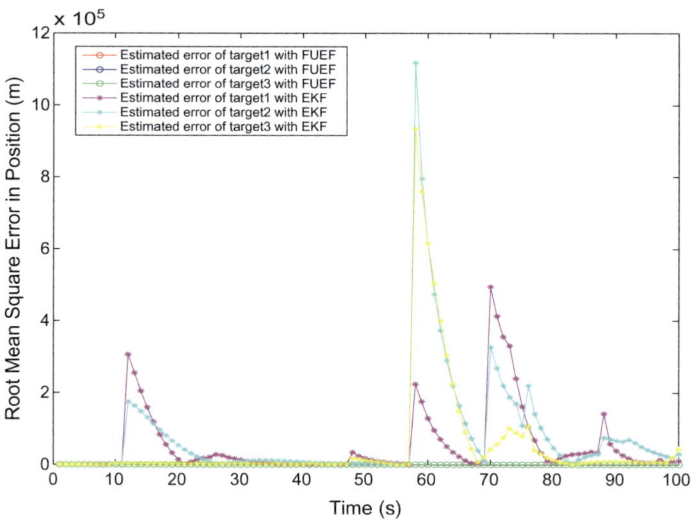

Fig. 4 Comparison of RMSE in position based on sum-of-powers SME

5 Conclusions

This paper presented a uniform structure model for multi-target tracking system with nonlinear observations implemented with different forms of symmetric measurement equations. Moreover, the prior calculation to approximate the noise covariance matrix of symmetric transformed measurement can be avoided. To resolve the obstacles brought by the unknown random noise in the new model and to deal with the high nonlinearities caused by the symmetric transformation of nonlinear measurements, the fading unscented–extended filter algorithm has been designed. The proposed filter was proved to track multiple moving targets successfully with some additional computation cost.

Acknowledgments This work was supported by the National Basic Research Program of China (973 Program: 2012CB821200, 2012CB821201), the NSFC (61134005, 61327807, 61521091, 61520106010, 61304232) and the Fundamental Research Funds for the Central Universities (YWF-16-GJSYS-31, YWF-16-GJSYS-32).

References

1. Cox I, Hingorani S (1996) An efficient implementation of Reids multiple hypothesis tracking algorithm and its evaluation for the purpose of visual tracking. IEEE Trans Pattern Anal Intell 18(2):138–150
2. Cristiano P, Nunes U (2006) A multi-target tracking and GMM-classifier for intelligent vehicles. In: IEEE conference on intelligent transportation systems, Toronto, pp 313–318
3. Kuchar JK, Yang LC (2000) A review of conflict detection and resolution modeling methods. IEEE Trans Intell Transp Syst 1(4), 179–189
4. Blackman S, Popoli R (1999) Design and analysis of modern tracking systems. Artech House, Norwell, MA
5. Bar-Shalom Y, Fortman TE (1988) Tracking and data association. Academic, Boston
6. Bar-Shalom Y (1978) Tracking methods in a multi-target environment. IEEE Transl Automat Control AC-23(4):618–626
7. Reid DB (1979) An algorithm for tracking multiple targets. IEEE Transl Automat Control AC-24(6):843–854
8. Vo BN, Ma WK (2006) The Gaussian mixture probability hypothesis density filter. IEEE Transl Signal Process 54(11):4091–4104
9. Kamen E (1992) Multiple target tracking based on symmetric measurement equations. IEEE Transl Automat Control 37(4):371–374
10. Kamen E, Sastry C (1993) Multiple target tracking using products of position measurements. IEEE Transl Aerosp Electron Syst 29(2):476–493
11. Sastry C, Kamen E (1993) SME filter approach to multiple target tracking with radar measurements. Proc Inst Elect Eng F 140(4):251–260
12. Muder D, O'Neil S, Drummond O (eds) The multi-dimensional SME filter for multitarget tracking. In: Proceedings of signal data processing small targets, vol 1954, pp 587–599
13. Leven WF, Lanterman AD (2009) Unscented Kalman filters for multiple target tracking with symmetric measurement equations. IEEE Transl Automat Control 54(2):370–375
14. Baum M, Hanebeck UD (2013) The Kernel-SME filter for multiple target tracking. In: 16th international conference on information fusion, pp 288–295
15. Simon D (2006) Optimal state estimation: Kalman, H infinity and nonlinear approaches. Wiley, New Jersey
16. Julier SJ, Uhlmann JK (2004) Unscented filtering and nonlinear estimation. Proc IEEE 92(3):401–422

RSS-Based Target Tracking with Unknown Path Loss Exponent

Jian Zhang, Wenling Li and Jian Sun

Abstract This paper studies the problem of target tracking by using the received signal strength (RSS) with unknown path loss exponent. The path loss exponent is estimated by using the least square approach and the unscented Kalman filter is used to address the nonlinear filtering problem. A numerical example and an experiment involving localization of a mobile robot are provided to demonstrate the effectiveness of the proposed algorithm.

Keywords Target tracking · Received signal strength · Path loss exponent

1 Introduction

In recent years, target tracking has received great attention due to its applications in civil and military areas. Although the GPS can generate accessible tracking accuracy in many situations, it cannot be used in indoor localization for multipath effect and signal missing. In wireless sensor networks (WSN), target tracking algorithms can be commonly classified into two categories: range-based and rang-free. The representative signal models include time of arrival (TOA) [1], time difference of arrival (TDOA) [2], angle of arrival (AOA) [3], and received signal strength (RSS) [4].

RSS-based target tracking has been widely used in WSN. This basic idea is that the transmitting node transmits a signal with fixed strength, and the receiving node calculate the distance with the measurement of received signal strength. Common propagation model mainly includes the path loss model [5], the multiwall model (MWM) [6], and the new empirical model (NEM) [7]. For example, it has been

J. Zhang (✉)
School of Mathematics and Systems Science, Beihang University, Beijing 100191, China
e-mail: zhangjian1991@126.com

W. Li · J. Sun
School of Automation Science and Electrical Engineering, Beihang University, Beijing 100191, China

Y. Jia et al. (eds.), *Proceedings of 2016 Chinese Intelligent Systems Conference*, Lecture Notes in Electrical Engineering 405,
DOI 10.1007/978-981-10-2335-4_6

shown that the RSS is susceptible to be disturbed by the environment factors [8], which means the unknown factors in engineering applications could largely reduce the performance of RSS. In [9], an interacting multiple model estimator has been proposed to estimate the path loss exponent and simulation results show that the tracking accuracy is better than the filter based on the jointly estimating the path loss exponent and the target state in an augmented form [9]. In [10], the target state and the path loss exponent are jointly estimated where the latter is derived by using the least square approach. The results are extended by using the weighted least square approach in [11].

In this paper, we attempt to study the target tracking by using the RSS with unknown path loss exponent. The filter is developed by combining with the multiple model approach and the least square technique. The unscented Kalman filter is used to address the nonlinear filtering problem. A numerical example and an experiment are provided to illustrate the effectiveness of the proposed algorithm.

The rest of this paper is organized as follows. The target tracking problem formulated is given in section. The tracking algorithm is presented in Sect. 3. In Sect. 4, simulation and experimental results are provided, followed by conclusion in Sect. 5.

2 System Model

We consider the following target tracking model:

$$\begin{aligned} x_k &= F_{k-1}x_{k-1} + G_{k-1}w_{k-1} \\ z_{k,s} &= h_s(x_k) + v_{k,s}, s = 1, 2, \ldots, N \end{aligned}$$

where $x_k = \left[p_{x,k}, p_{y,k}, v_{x,k}, v_{y,k}\right]^{\mathrm{T}}$ represents the target state vector at time instant k. $\left(p_{x,k}, p_{y,k}\right)$ and $\left(v_{x,k}, v_{y,k}\right)$ denote the position and velocity components at time instant k. F_{k-1} and G_{k-1} are known matrices with appropriate dimensions. The process noise w_{k-1} and the measurement noise $v_{k,s}$ are assumed to be zero-mean white Gaussian with covariance matrices Q_{k-1} and R_{k-1}, respectively. $z_{s,k}$ is the received power $S_{s,k}$ obtained by sth reference node.N is the number of the sensor nodes. $h_s(x_k)$ is the measurement function of the sth node, and the path loss model is given by

$$S_{s,k} = S_0 - 10n_{k,s}\log_{10}\frac{\sqrt{\left(p_{x,k} - r_{x,s}\right)^2 + \left(p_{y,k} - r_{y,s}\right)^2}}{d_0}$$

where d_0 denotes the reference distance. S_0 is the RSS at the reference distance d_0. $\left(r_{x,s}, r_{y,s}\right)$ is the position of the sth reference node. $n_{k,s}$ denotes the path loss exponent of the sth reference node.

The aim of the target tracking is to estimate the target state x_k by using the measurements $z_{k,s}$ up to time k.

3 Target Tracking

The unscented Kalman filter has been proposed by Julier and Uhlmann [12, 13]. As the UKF has been extensively studied, we omit the details for space considerations. A cycle of the UKF runs as follows.

Step 1. Generate sigma points

$$\chi_k(0) = \hat{x}_{k|k}$$
$$\chi_k(i) = \gamma\sqrt{P^i_{k|k}} + \hat{x}_{k|k}, i = 1, 2, \ldots, 2L$$

where $\gamma = \sqrt{L+\lambda}$, and $\lambda + L = 3$, $L = 4$. $P^i_{k|k}$ denotes the columns of the $\pm P_{k|k}$. The combined weights are defined as follows:

$$W_0^{\text{mean}} = W_0^{\text{cov}} = \frac{\lambda}{\lambda + L}$$
$$W_1^{\text{mean}} = W_1^{\text{cov}} = \frac{1}{2(\lambda + L)}, i = 1, 2, \ldots, 2L$$

Step 2. Predict the state estimate

$$\chi_{k+1}(i) = f(\chi_k(i), k)$$

The predicted estimate and its corresponding covariance are calculated by

$$\hat{x}_{k+1|k} = \sum_{i=0}^{2L} W_i^{\text{mean}} \chi_{k+1}(i)$$
$$P_{k+1|k} = \sum_{i=0}^{2L} W_i^{\text{cov}} \left(\chi_{k+1}(i) - \hat{x}_{k+1|k}\right)\left(\chi_{k+1}(i) - \hat{x}_{k+1|k}\right)^{\text{T}} + G_k Q_k G_k^{\text{T}}$$

The predicted observation is given by

$$\hat{y}_{k+1|k,s,j} = \sum_{i=0}^{2L} W_i^{\text{mean}} h_{s,j}(\chi_{k+1}(i), k+1), s = 1, 2, \ldots, N$$
$$P^{vv}_{k+1|k,j} = \sum_{i=0}^{2L} W_i^{\text{cov}} \left(z_{k+1,s,j}(i) - \hat{y}_{k+1|k,s,j}\right)\left(z_{k+1,s,j}(i) - \hat{y}_{k+1|k,s,j}\right)^{\text{T}} + R_k$$
$$P^{xy}_{k+1|k,j} = \sum_{i=0}^{2L} W_i^{\text{cov}} \left(\chi_{k+1}(i) - \hat{x}_{k+1|k}\right)\left(z_{k+1,s,j}(i) - \hat{y}_{k+1|k,s,j}\right)^T$$

Step 3. Update the state estimate
The updated estimate and its corresponding covariance are computed by

$$\hat{x}_{k|k-1,j} = \hat{x}_{k|k-1} + K_{k,j} v_{k,s,j}$$
$$P_{k|k,j} = P_{k|k-1,j} - K_{k,j} P^{vv}_{k|k-1,j} K_{k,j}$$

where the gain and innovation are given by

$$K_{k,j} = P^{xy}_{k+1|k,j} \left(P^{vv}_{k+1|k,j} \right)^{-1}$$
$$v_{k,s,j} \equiv z_{k,s} - \hat{y}^{m}_{k|k-1,s}$$

$\hat{y}_{k+1|k,s,j}$ denotes the predicted observation using parameter $n_{k,s,j}$.
Step 4. Fuse the state estimates

$$\hat{x}_{k|k} = \frac{\sum_j^m \hat{x}_{k|k-1,j} / w^e_{k,j}}{\sum_j^m 1/w^e_{k,j}}$$
$$P_{k|k} = \frac{\left. \sum_j^m P_{k|k,j} / w^e_{k,j} \right)^2}{\left(\sum_1^m 1/w^e_{k,j} \right)^2}$$

where

$$w^e_{k,j} = \mathrm{mean}\left(w_{k,s,j} \right)$$
$$w_{k,s,j} = \left\| z_{k,s} - \hat{y}_{k+1|k,s,j} \right\|$$

In reality, PLE is strongly affected by environment. The PLE of different nodes are interrelated in the same location at the same time, and they are different for the signal contamination around. Since the PLE seriously affect the positioning accuracy, adaptive gridding strategy and least squares method are used in this paper. The main idea of the adaptive gridding strategy is to move the value of the PLE to the most likely state. Three coupling parameter $n_{k,s,m}$ are set through $n_{k-1,s,m}$, which is given by:

$$n_{k,s,2} = \frac{\sum_j^3 n_{k-1,s,j} / w_{k-1,s,j}}{\sum_j^3 1/w_{k-1,s,j}}$$

$$n_{k,s,1} = n_{k,s,2} - d * \alpha$$
$$n_{k,s,3} = n_{k,s,2} + d * \alpha$$

where α denotes the scaling factor. d denotes the distance from the low state to the high state, and it is adjusted by the follow scheme:

$$d = \begin{cases} d + 0.02, & \text{if trace}\left(P^m_{k|k}\right) > \text{trace}\left(P^m_{k-1|k-1}\right) \\ d - 0.04, & \text{if trace}\left(P^m_{k|k}\right) < \text{trace}\left(P^m_{k-1|k-1}\right) \end{cases}$$

Finally, if d is bigger than 1, then $d = 1$. otherwise, the least square method is used to modify the PLE as follows

$$n_{k,s,1} = N_k$$
$$n_{k,s,2} = \frac{\sum_j^4 n_{k-1,s,j}/w_{k-1,s,j}}{\sum_j^4 1/w_{k-1,s,j}}$$

If $(\|n_{k,s} - N_k\| < 0.02)$, then $d = 0.02$, else $d = (n_{k,s} - N_k)$.

$$n_{k,s,3} = n_{k,s,2} + d * \alpha$$
$$n_{k,s,4} = n_{k,s,2} - d * \alpha$$

The least square estimate N_k is given by the follow scheme:

$$N_{k+1} = N_k + L_{k+1} * (\tilde{y}_{k+1} - S_0 - H_{k+1} * N_k)$$
$$L_{k+1} = A_k H_{k+1}^{\mathrm{T}} \left[H_{k+1} A_k H_{k+1}^{\mathrm{T}} + R^{-1}\right]^{-1}$$
$$A_{k+1} = [I - L_{k+1} H_{k+1}] A_k$$
$$H_k = \begin{bmatrix} -5\left(\log_{10}\left((p_{x,k} - r_{x,1})^2 + (p_{y,k} - r_{y,1})^2\right) + 2\right) \\ \vdots \\ -5\left(\log_{10}\left((p_{x,k} - r_{x,s})^2 + (p_{y,k} - r_{y,s})^2\right) + 2\right) \end{bmatrix}_{s \times 1}$$

where the priori estimate is $N_1 = A_1 H_1^{\mathrm{T}} R(\tilde{y}_1 - S_0)$, and the covariance estimate is $A_1 = \left(H_1 R H_1^{\mathrm{T}}\right)^{-1}$.

4 Simulation and Experimental Result

We consider a two-dimensional tracking scenario. Four sensors are located at (0, 10), (10, 0), (0, −10), (−10, 0) in meters. To describe the target motion, the nearly constant velocity is used as follow

$$x_k = \begin{bmatrix} 1 & 0 & T & 0 \\ 0 & 1 & 0 & T \\ 0 & 0 & 1 & 0 \\ 0 & 0 & 0 & 1 \end{bmatrix} x_{k-1} + \begin{bmatrix} T^2/2 & 0 \\ 0 & T^2/2 \\ T & 0 \\ 0 & T \end{bmatrix} w_{k-1}$$

where $T = 0.1$ s is the sampling interval. The process noise w_{k-1} is assumed to be zero-mean white Gaussian with $Q_{k-1} = 0.1^2 I_2$. The target starts at (5, 0) in meters, and its initial velocity is (0, 0.63) in m/s. the estimate of the initial state of the target is set to be $[6\ 6\ 0.3\ 0.3]^T$.

In simulations, we assume S_0 is close to -12 dBm when d_0 is 0.1 m. The PLE is set to be 2.3. The sensor's sampling error is set zero mean and variance of 1, and the sampled values are rounded according to the real situation. In the filter, $R = 1.4$, $n_{k,s,2}$ is set to be 2.4, d is set to be 0.8. The scaling factor α is set to be 0.3.

To evaluate the performance of the filter, 100 Monte Carlo runs is used to calculate the root mean square error (RMSE) in position in simulation. The performance of the proposed filter with joint estimate of least square method and adaptive gridding strategy (UKF-LS&AG) is compared with that of the augmented state unscented Kalman filter (UKF-AS). The filter with known PLE (UKF-K) serves as the baseline algorithm to evaluate the performance.

The performance comparison with the positioning averages of RMSE (ARMSE) is shown in Fig. 1. The result suggests that the UKF-LS&AG is better than UKF-AS. Although the UKF-LS&AG tracking accuracy show bad at the beginning of the filter, it is better than the UKF-AS.

When the PLE is chosen to be different value, the performance comparison with the averages of RMSE (ARMSE) is shown in Fig. 2. In simulations, the true value of PLE has been set from 2 to 2.8. The simulation result indicates that UKF-LS&AG is better than UKF-AS.

The performance of the UKF-LS&AG filter with respect to the scaling factor α is shown in Fig. 3. The factor α is taken from 0.1 to 0.9. The ARMSE in position suggests that the larger the scaling factor α is, the less the ARMSE in position is. Therefore, the scaling factor α is advised to take larger than 0.5 to improve

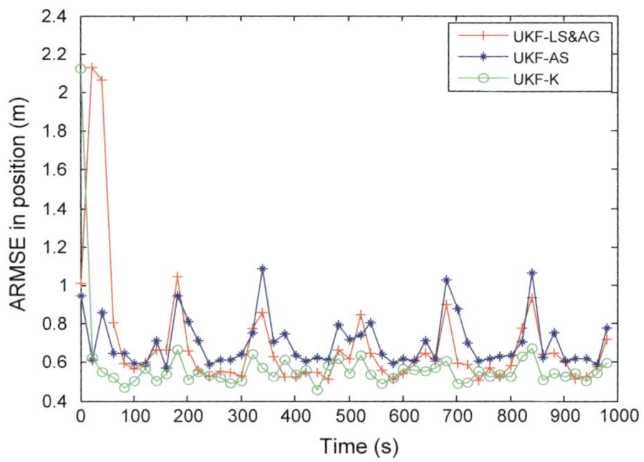

Fig. 1 Performance comparison with respect to RMSE in position

Fig. 2 Effect of the PLE $n_{k,s}$ on the estimation error

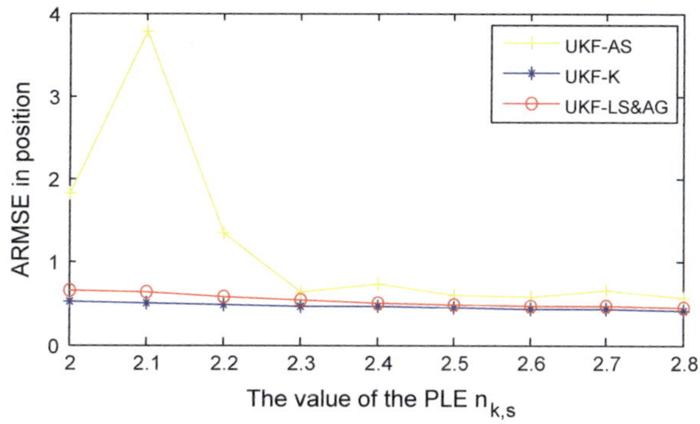

Fig. 3 Effect of the scaling factor α on the estimation error

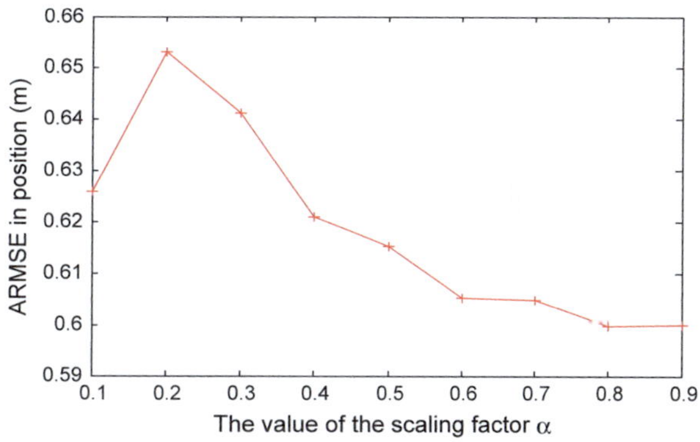

accuracy. In the simulations, the true measurement noise covariance is set to be $R_k = \mathrm{diag}([1\ 1\ 1\ 1])$, while the measurement noise covariance adopted in the filter is taken to be $\widetilde{R}_k = \gamma * diag([1\ 1\ 1\ 1])$. The ARESM in position with γ of a set of value is shown in Fig. 4. The simulation results indicate that UKF-LS&AG do well for mismatched noise covariance matrices. A notable feature is that while the measurement noise covariance is larger than the true covariance, ARESM in position of UKF-LS&AG may reduce.

We have done a RSS-based indoor positioning experiment with TI's CC2530 chip. Three nodes were placed at (1.04, 2.05), (1.54, 1.55) and (2.04, 2.05). The RSS is gotten from the chip and transmitted to the computer. The sampling interval is $T = 1$ s, and the process noise covariance is $Q_k = 0.00489I_2$. The target track round, whose locus is shown in Fig. 5, and the target starts at the position (3.1 2.1) in meters and initial velocity is (0 0.2) in m/s. The scaling factor α is chosen to be 0.5. The positioning using filter with the joint estimator of PLE is shown in Fig. 5 as well. The RMSE is calculated to be 0.5279 m. This indicates that the filter with the joint estimator of PLE can work well in realistic cases either.

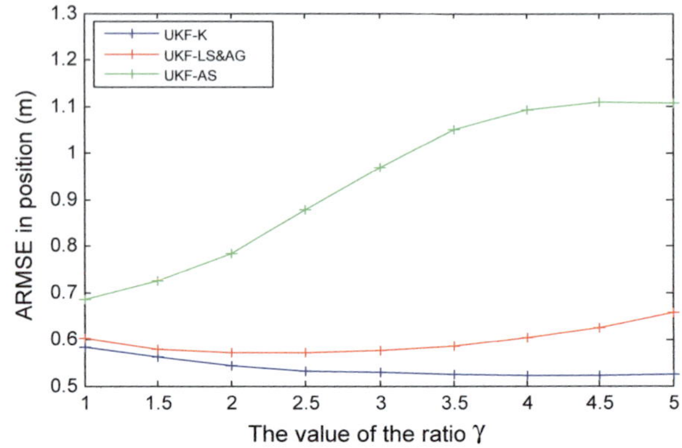

Fig. 4 Effect of the radio γ on the estimation error

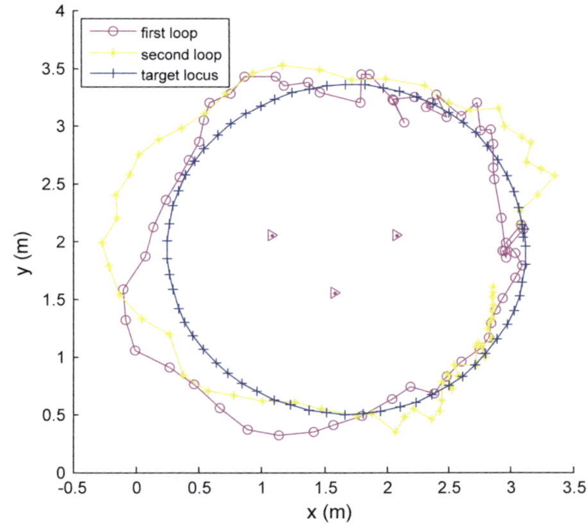

Fig. 5 Target locust and position estimation

5 Conclusion

The adaptive gridding strategy is used to estimate the path loss exponent and the least square method is used to modify the path loss exponent. The unscented Kalman filter is used to address the nonlinear filtering problem. Simulation and experiment results show the effectiveness and the applicability of the proposed algorithm.

Acknowledgments This work was supported by the National Basic Research Program of China (973 Program, 2012CB821200, 2012CB821201), the NSFC (61573031, 61134005, 61327807, 61520106010).

References

1. Güvenç İ, Chong CC (2009) A survey on TOA based wireless localization and NLOS mitigation techniques. IEEE Commun Surv Tutorials 11(3):107–124
2. Yang L, Ho KC (2009) An approximately efficient TDOA localization algorithm in closed-form for locating multiple disjoint sources with erroneous sensor positions. IEEE Trans Signal Process 57(12):4598–4615
3. Cong L, Zhuang W (2002) Hybrid TDOA/AOA mobile user location for wideband CDMA cellular systems. IEEE Trans Wireless Commun 1(3):439–447
4. Patwari N, Ash JN, Kyperountas S et al (2005) Locating the nodes: cooperative localization in wireless sensor networks. IEEE Signal Process Mag 22(4):54–69
5. Borrelli A, Monti C, Vari M et al (2004) Channel models for IEEE 802.11b indoor system design. In: 2004 IEEE international conference on communications, vol 6. IEEE, pp 3701–3705
6. Cheung KW, Sau HM, Murch RD (1998) A new empirical model for indoor propagation prediction. IEEE Trans Veh Technol 47(3):996–1001
7. Narzullaev A, Park Y, Yoo K et al (2011) A fast and accurate calibration algorithm for real-time locating systems based on the received signal strength indication. AEU Int J Electr Commun 65(4):305–311
8. Wu RH, Lee YH, Tseng HW et al (2008) Study of characteristics of RSSI signal. In: IEEE international conference on industrial technology, pp 104–111
9. Li W, Jia Y (2015) Distributed target tracking by time of arrival and received signal strength with unknown path loss exponent. IET Signal Proc 9(9):681–686
10. Li X (2006) RSS-based location estimation with unknown pathloss model. IEEE Trans Wireless Commun 5(12):3626–3633
11. Wang G, Chen H, Li Y et al (2012) On received-signal-strength based localization with unknown transmit power and path loss exponent. IEEE Wireless Commun Lett 1(5):536–539
12. Julier SJ, Uhlmann JK (1997) New extension of the Kalman filter to nonlinear systems. In: AeroSense'97. International society for optics and photonics, pp 182–193
13. Julier SJ, Uhlmann JK, Durrant-Whyte HF (1995) A new approach for filtering nonlinear systems. American control conference, proceedings of the 1995, vol 3. IEEE, pp 1628–1632

Further Analysis on Observability of Stochastic Periodic Systems with Application to Robust Control

Hongji Ma, Ting Hou and Jie Wang

Abstract This paper is concerned with a class of discrete-time stochastic systems with periodic coefficients and multiplicative noise. Above all, observability is studied by analyzing the unobservable subspace of concern dynamics. Further, invariant-subspace approach is applied to derive an operator-spectral criterion of observability, which improves the observability test presented by Ma et al. (Proceedings of 2016 American control conference, to appear) [1]. Based on the proposed observability criterion, an infinite-horizon stochastic periodic H_2/H_∞ control is obtained in the presence of (x, u, v)-dependent noise.

Keywords Stochastic periodic systems · Observability · Unobservable subspace · Spectral criterion · H_2/H_∞ control

1 Introduction

Discrete-time stochastic systems with multiplicative noise provide an ideal mathematical model for characterizing various dynamics subject to parametric perturbation caused by random environmental disturbance. This type of models have attracted considerable attention due to wide applications in engineering and economic fields, such as signal processing [2], networked control [3], portfolio optimization [4], etc. For example, an elaborate analysis has been presented in [5] for stabilization of stochastic discrete-time control systems. Reference [6] considered the linear quadratic control problem of discrete-time stochastic systems with control-dependent noise and indefinite cost weighting matrix. Besides, robust control theory

H. Ma (✉) · T. Hou · J. Wang
College of Mathematics and Systems Science, Shandong University of Science and Technology, Qingdao 266590, China
e-mail: ma_math@163.com

T. Hou
State Key Laboratory of Mining Disaster Prevention and Control Co-founded by Shandong Province and the Ministry of Science and Technology, Qingdao 266590, China

Y. Jia et al. (eds.), *Proceedings of 2016 Chinese Intelligent Systems Conference*, Lecture Notes in Electrical Engineering 405,
DOI 10.1007/978-981-10-2335-4_7

has been addressed for discrete-time stochastic systems with/without random jump parameters; see, e.g., [2, 7, 8] and the references therein.

Observability is an essential structural performance of dynamic systems, which plays a fundamental role in the analysis of linear systems. In the available literature, there have been two distinct ways for investigating the observability of stochastic linear systems. The first way is to introduce a cumulated energy function associated with the measurement output. From the relationship between the value of output energy function and the initial system state, a necessary and sufficient condition can be derived for examining the observability of linear stochastic systems with state-dependent noise [9] or Markov jump linear systems [10]. It is remarkable that this method generally does not lead to Hautus-type test (i.e., spectral criterion). Another line is the so-called operator-spectral method, whose idea is to convert the concerned linear stochastic system into a linear deterministic system by means of $\mathcal{H}$-representation formula, see [11, 12]. Recently, [13] has generalized the operator-spectral method to explore the observability and detectability of linear Markov jump systems with periodic coefficients. It is shown that the spectral criteria for observability and detectability of periodic Markov jump systems does not depend on a single monodromy operator, but on a sequence of positive evolution operators with the total being the period.

In this paper, we will present some further results on observability of discrete-time stochastic periodic systems. To this end, unobservable subspace is first studied for linear stochastic periodic systems. Some criteria of observability are developed based on the analysis of unobservable subspace. In addition, an invariant-subspace approach is employed to prove an operator-spectral criterion of observability, which improves the Hautus-type test [1] obtained by use of $\mathcal{H}$-representation technique. More specifically, the observability of concerned models can be verified by checking the proposed spectral criterion within the domain of real positive semi-definite matrices, instead of the set of all complex symmetric matrices. Further, by combining with a periodic version of stochastic bounded real lemma, the proposed observability criterion will be applied to settle an infinite-horizon H_2/H_∞ stochastic periodic control in the presence of (x, u, v)-dependent noise. It is shown that under the condition of observability, the solvability of considered H_2/H_∞ control problem is equivalent to that of a group of generalized periodic Riccati equations.

The rest of this paper is organized as follows. Section 2 gives some preliminaries about discrete-time stochastic periodic systems. Section 3 addresses the property of unobservable subspace and then presents an operator-spectral criterion for observability. Section 4 supplies an infinite-horizon H_2/H_∞ stochastic periodic control by using the proposed observability criterion. Finally, Sect. 5 will end the paper with a brief concluding remark.

Notations. $\mathcal{C}(\mathcal{R})$: the set of all complex (real) numbers; $\mathcal{R}^n$: n-dimensional complex (real) vector space with the usual Euclidean norm $\|\cdot\|$; $\mathcal{R}^{n\times m}$: the space of all $n \times m$ complex (real) matrices with the operator norm $\|\cdot\|_2$; $A > 0\,(\geq 0)$: A is a symmetric positive (semi-)definite matrix; S_{n+}: the set of real positive semi-definite matrices; A': the transpose of a matrix (vector) A; $A^{-1}(A^+)$: the inverse

(Moore-Penrose pseudoinverse) of a matrix A; I_n: the $n \times n$ identity matrix; $\mathbb{Z}_+ := \{0, 1, \ldots\}$ and $\mathbb{Z}_{1+} := \{1, 2, \ldots\}$.

2 Preliminaries

On a complete probability space $(\Omega, \mathscr{F}, P)$, we consider the following stochastic periodic systems:

$$x_{t+1} = A_t x_t + C_t x_t w_t,\ t \in \mathbb{Z}_+, \tag{1}$$

where $x_t \in \mathcal{R}^n$ represents the state, A_t and C_t are deterministic matrices. The initial state satisfies $E\|x_0\|^2 < \infty$ and for each $t \in \mathbb{Z}_+$, x_0 is independent of σ-algebra $\mathscr{G}_t = \sigma\{w_t | 0 \le k \le t\}$. Denote by $\mathscr{F}_t = \sigma(x_0) \vee \mathscr{G}_t$ the minimal σ-algebra consisting of $\sigma(x_0)$ and $\mathscr{G}_t$. Let $l^2(0, \infty; \mathcal{R}^m)$ be the space of $\mathcal{R}^m$-valued, nonanticipative square summable stochastic processes $\{y_{(t,\omega)} : \mathbb{Z}_+ \times \Omega \to \mathcal{R}^m\}$, which are $\mathscr{F}_t$-measurable and $\sum_{t=0}^{\infty} E\|y_t\|^2 < \infty$. It can be verified that $l^2(0, \infty; \mathcal{R}^m)$ is a real Hilbert space with the norm induced by the usual inner product: $\|y\|_{l^2(0,\infty;\mathcal{R}^m)} = (\sum_{t=0}^{\infty} E\|y_t\|^2)^{1/2}$.

The following assumptions are used throughout this paper.

Hypothesis 1 (i) The coefficients of considered systems are all real matrix-valued sequences with a common period $\theta \in \mathbb{Z}_{1+}$ and suitable dimensions, e.g., $A_{t+\theta} = A_t$, and $A_t \in \mathcal{R}^{n \times n}$; (ii) $\{w_t\}_{t \in \mathbb{Z}_+}$ is a sequence of independent random variables with $Ew_t = 0$ and $Ew_s w_t = \delta_{(t-s)}$ for $t, s \in \mathbb{Z}_+$, where $\delta_{(\cdot)}$ is a Kronecker function. Next, we give some basic concepts of stability. □

Definition 1 System (1) (or (**A, C**) for short) is called asymptotically mean square stable (AMSS) if

$$\lim_{t \to +\infty} E\|x_{(t;\xi,t_0)}\|^2 = 0$$

for all $\xi \in \mathcal{R}^n$ and $t_0 \ge 0$, where $x_{(t;\xi,t_0)}$ is the state of (1) corresponding to $x_{t_0} = \xi \in \mathcal{R}^n$. Moreover, the controlled stochastic periodic system (**A, B; C, D**):

$$\begin{cases} x_{t+1} = (A_t x_t + B_t u_t) + (C_t x_t + D_t u_t) w_t, \\ u_t \in l^2(0, \infty; \mathcal{R}^{n_u}),\ t \in \mathbb{Z}_+ \end{cases} \tag{2}$$

is called stochastically stabilizable if there exists a θ-periodic sequence $K_t \in \mathcal{R}^{n_u \times n}$ ($t \in \mathbb{Z}_+$) such that $u_t = K_t x_t$ guarantees the closed-loop system (**A + BK, C + DK**):

$$x_{t+1} = (A_t + B_t K_t) x_t + (C_t + D_t K_t) x_t w_t,\ t \in \mathbb{Z}_+$$

to be AMSS. In this case, K_t is called a stabilizing feedback gain. □

Let S_n be the set of all $n \times n$ symmetric matrices whose entries may be complex. Thus, S_n is a Hilbert space with the inner product

$$\langle X, Y\rangle = Tr(XY),\ X, Y \in \mathcal{S}_n. \tag{3}$$

Define a Lyapunov operator $\mathcal{L}_t : \mathcal{S}_n \to \mathcal{S}_n$ as follows:

$$\mathcal{L}_t(X) = A_t X A_t' + C_t X C_t',\ X \in \mathcal{S}_n,\ t \in \mathcal{Z}_+. \tag{4}$$

According to the inner product (3), the adjoint operator of $\mathcal{L}_t$ turns out to be

$$\mathcal{L}_t^*(X) = A_t' X A_t + C_t' X C_t,\ X \in \mathcal{S}_n,\ t \in \mathcal{Z}_+. \tag{5}$$

Moreover, the linear evolution operator of $\mathcal{L}_t$ is given by $\mathcal{T}_{t,s} = \mathcal{L}_{t-1}\mathcal{L}_{t-2}\dots\mathcal{L}_s$ for $t > s \geq 0$, and $\mathcal{T}_{t,t} = \mathcal{I}$ (the identity operator). It can be verified that $\mathcal{T}_{t,s}$ $(t \geq s)$ is a linear positive operator.

3 Observability

This section focuses on observability of the following stochastic periodic system with measurement equation:

$$\begin{cases} x_{t+1} = A_t x_t + C_t x_t w_t, \\ y_t = Q_t x_t,\ t \in \mathcal{Z}_+, \end{cases}$$

where $y_t \in \mathcal{R}^{n_y}$ is the measurement output and the other terms are the same as in (1).

Definition 2 $\xi \in \mathcal{R}^n$ is called an unobservable state of (6) at t_0 if the measurement output corresponding to $x_{t_0} = \xi$ satisfies that

$$y_t \equiv 0\ (a.s.),\ T \geq t \geq t_0,\ \forall T > t_0 \geq 0.$$

System (6), denoted by $(\mathbf{A}, \mathbf{C}|\mathbf{Q})$, is called observable at t_0 if there is no nonzero unobservable state at t_0. Further, $(\mathbf{A}, \mathbf{C}|\mathbf{Q})$ is (uniformly) observable if it is observable for all $t_0 \in \mathcal{Z}_+$. □

Let $\mathcal{N}_t$ be the set of all unobservable states of (6) at time t, then it is easy to verify that $\mathcal{N}_t$ constitutes a subspace of $\mathcal{R}^n$, i.e., the unobservable subspace of (6) at time t. Next, we present an algebraic characterization for the unobservable subspace of (6).

Lemma 1 *$\mathcal{N}_t = Ker(\mathbb{O}_t)$, where $Ker(\cdot)$ is the kernel of a matrix and*

$$\begin{aligned} \mathbb{O}_t = (&Q_t'\ A_t'Q_{t+1}'\ C_t'Q_{t+1}'\ A_t'A_{t+1}'Q_{t+2}' \\ &C_t'A_{t+1}'Q_{t+2}'\ A_t'C_{t+1}'Q_{t+2}'\ C_t'C_{t+1}'Q_{t+2}' \dots)'. \end{aligned} \tag{6}$$

Proof By Definition 1, an unobservable state ξ at t results in $y_t \equiv 0$ $(a.s.)$ on $[t, +\infty)$, which implies

$$Q_t\xi = 0,\ Q_{t+1}x_{t+1} = 0,\ Q_{t+2}x_{t+2} = 0, \dots,\ (a.s.).$$

According to the state equation of (6), we have

$$\begin{cases} Q_t\xi = 0, \\ Q_{t+1}(A_t\xi + C_t\xi w_t) = 0, \\ Q_{t+2}[A_{t+1}(A_t\xi + C_t\xi w_t) + C_{t+1}(A_t\xi + C_t\xi w_t)w_{t+1}] = 0, \\ \quad\vdots \end{cases} \tag{7}$$

in almost sure sense. Due to the randomness of $\{w_t\}_{t\in\mathbb{Z}_+}$, (7) holds iff

$$\begin{aligned} &Q_t\xi = 0,\ Q_{t+1}A_t\xi = 0,\ Q_{t+1}C_t\xi = 0, \\ &Q_{t+2}A_{t+1}A_t\xi = 0,\ Q_{t+2}A_{t+1}C_t\xi = 0, \\ &Q_{t+2}C_{t+1}A_t\xi = 0,\ Q_{t+2}C_{t+1}C_t\xi = 0, \dots, \end{aligned}$$

which means $\xi \in Ker(\mathbb{O}_t)$. Therefore, $\mathcal{N}_t \subseteq Ker(\mathbb{O}_t)$. Conversely, if $\xi \in Ker(\mathbb{O}_t)$, then the above analysis suggests that the output y_t of (6) corresponding to $x_t = \xi$ almost surely equals to zero over $[t, +\infty)$. So, $\xi \in \mathcal{N}_t$. The proof is ended. □

Remark 1 Let $\tilde{\mathbb{O}}_t = (Q_t'\ A_t'Q_{t+1}'\ A_t'A_{t+1}'Q_{t+2}' \dots)'$, then the following deterministic periodic system:

$$\begin{cases} x_{t+1} = A_t x_t, \\ y_t = Q_t x_t,\ t \in \mathbb{Z}_+, \end{cases} \tag{8}$$

or $(\mathbf{A}|\mathbf{Q})$ is (completely) observable at t iff $Ker(\tilde{\mathbb{O}}_t) = \{0\}$ ([14], p.127). By Lemma 1, it is clear that $Ker(\mathbb{O}_t) \subseteq Ker(\tilde{\mathbb{O}}_t)$. So, the observability of $(\mathbf{A}|\mathbf{Q})$ implies that of $(\mathbf{A}, \mathbf{C}|\mathbf{Q})$, but the converse is not true. □

As simple applications of Lemma 1, we can get some further properties of the unobservable subspace.

Proposition 1 (i) $\mathcal{N}_{t+\theta} = \mathcal{N}_t$. (ii) $(\mathbf{A}, \mathbf{C}|\mathbf{Q})$ *is observable iff* $\mathcal{N}_t = \{0\}$ *for* $t = 0, 1, \dots, \theta - 1$.

Proposition 2 *(i)* $\mathcal{N}_t$ *is an* $(\mathbf{A}, \mathbf{C})$*-invariant subspace of* $Ker(Q_t)$, *i.e.,*

$$A_t\mathcal{N}_t \subseteq \mathcal{N}_{t+1},\ C_t\mathcal{N}_t \subseteq \mathcal{N}_{t+1},\ \mathcal{N}_t \subseteq Ker(Q_t),\ \forall t \in \mathbb{Z}_+.$$

(ii) $\mathcal{N}_t$ *is the largest* $(\mathbf{A}, \mathbf{C})$*-invariant subspace of* $Ker(Q_t)$.

Proof Obviously, $\mathcal{N}_t \subseteq Ker(Q_t)$. For arbitrary $\xi \in \mathcal{N}_t$, it follows from Lemma 1 that

$$Q_{t+1}A_t\xi = 0, Q_{t+2}A_{t+1}A_t\xi = 0, Q_{t+2}C_{t+1}A_t\xi = 0, \dots,$$

which implies $A_t\xi \in Ker(\mathbb{O}_{t+1}) = \mathcal{N}_{t+1}$. Therefore, $A_t\mathcal{N}_t \subseteq \mathcal{N}_{t+1}$. Similarly, it can be shown that $C_t\mathcal{N}_t \subseteq \mathcal{N}_{t+1}$.

To prove (ii), let $\mathbb{S}_t$ be an $(\mathbf{A}, \mathbf{C})$-invariant subspace of $Ker(Q_t)$. Then, for arbitrary $\xi \in \mathbb{S}_t$, we have $A_t\xi \in \mathbb{S}_{t+1} \subseteq Ker(Q_{t+1})$, $C_t\xi \in \mathbb{S}_{t+1} \subseteq Ker(Q_{t+1})$, and $\xi \in Ker(Q_t)$. Therefore,

$$Q_t\xi = 0, Q_{t+1}A_t\xi = 0, Q_{t+1}C_t\xi = 0.$$

Further, by the invariant property, there are $A_{t+1}\mathbb{S}_{t+1} \subseteq \mathbb{S}_{t+2}$, $C_{t+1}\mathbb{S}_{t+1} \subseteq \mathbb{S}_{t+2}$, and $\mathbb{S}_{t+2} \subseteq Ker(Q_{t+2})$. These relations lead to

$$Q_{t+2}A_{t+1}A_t\xi = 0,\ Q_{t+2}A_{t+1}C_t\xi = 0,\ Q_{t+2}C_{t+1}A_t\xi = 0,\ Q_{t+2}C_{t+1}C_t\xi = 0.$$

Repeating the above procedure, it is shown that $\xi \in Ker(\mathbb{O}_t)$. By Lemma 1, $\xi \in \mathcal{N}_t$. Hence, $\mathbb{S}_t \subseteq \mathcal{N}_t$. □

Remark 2 From Propositions 1 and 2, we can conclude that $(\mathbf{A}, \mathbf{C}|\mathbf{Q})$ is observable at time t iff $Ker(Q_t)$ does not have nontrivial $(\mathbf{A}, \mathbf{C})$-invariant subspace. □

Next, we consider the influence of output-feedback controls on the observability.

Proposition 3 *If the open-loop stochastic periodic system*

$$\begin{cases} x_{t+1} = (A_tx_t + B_tu_t) + (C_tx_t + D_tu_t)w_t, \\ y_t = Q_tx_t,\ t \in \mathbb{Z}_+ \end{cases}$$

is observable at t, then by applying an output-feedback controller $u_t = v_t - K_ty_t$ (K_t may not be periodic here) to (9), the resulting system (9) remains observable at t:

$$\begin{cases} x_{t+1} = [(A_t - B_tK_tQ_t)x_t + B_tv_t] + [(C_t - D_tK_tQ_t)x_t + D_tv_t]w_t, \\ y_t = Q_tx_t,\ t \in \mathbb{Z}_+. \end{cases}$$

Proof If system (9) is not observable at t, then by Remark 2, $Ker(Q_t)$ has a nontrivial $(\mathbf{A} - \mathbf{BKQ}, \mathbf{C} - \mathbf{BKQ})$-invariant subspace $\mathbb{S}_t$, i.e.,

$$\begin{cases} (A_t - B_tK_tQ_t)\mathbb{S}_t \subseteq \mathbb{S}_{t+1},\ (C_t - D_tK_tQ_t)\mathbb{S}_t \subseteq \mathbb{S}_{t+1}, \\ \mathbb{S}_t \subseteq Ker(Q_t),\ \exists 0 \neq \xi \in \mathbb{S}_t, \end{cases} \tag{9}$$

which is further reduced to

$$A_t\mathbb{S}_t \subseteq \mathbb{S}_{t+1}, C_t\mathbb{S}_t \subseteq \mathbb{S}_{t+1}, \mathbb{S}_t \subseteq Ker(Q_t), \exists 0 \neq \xi \in \mathbb{S}_t.$$

So, $Ker(Q_t)$ has a nontrivial $(\mathbf{A}, \mathbf{C})$-invariant subspace. By Remark 2, this contradicts the observability of (9) at time t. The proof is ended. □

Now, we will employ the invariant-subspace method to show an operator-spectral criterion for detectability of $(\mathbf{A}, \mathbf{C}|\mathbf{Q})$.

Theorem 1 *$(A, C|Q)$ is observable at time t_0 iff there does not exist nonzero $X \in S_{n+}$ such that*

$$\begin{cases} \mathcal{T}_t^{\theta}(X) = \lambda X, \ \lambda \in C, \\ Q_j\mathcal{T}_{j,t_0}(X) = 0, \ j = t_0, t_0+1, \ldots, t_0+\theta-1. \end{cases} \tag{10}$$

Proof Necessity. Let us define $X_t = E(x_t x_t')$ and $Y_t = E(y_t y_t')$ associated with (6). It can be verified that $X_j = \mathcal{T}_{j,t}(X_t)$ and $Y_j = Q_j\mathcal{T}_{j,t}(X_t)Q_j'$ for $j \geq t$. Moreover, the following relationships are valid:

$$\begin{cases} x_t = 0\ (a.s.) \Leftrightarrow X_t = 0, \ t \in \mathbb{Z}_+, \\ y_t = 0\ (a.s.) \Leftrightarrow Y_t = 0 \Leftrightarrow Tr(Y_t) = 0, \ t \in \mathbb{Z}_+. \end{cases} \tag{11}$$

If the criterion is not satisfied at $t_0 \in \mathbb{Z}_+$, there exists some nonzero $X \in S_{n+}$ satisfying (10). It follows that

$$Tr(Y_j) = \langle \mathcal{T}_{j,t_0}(X), Q_j'Q_j \rangle = 0 \tag{12}$$

for $j = t_0, \ldots, t_0+\theta-1$. When $j \geq t_0+\theta$, there are $l \in \mathbb{Z}_+$ and $k \in [0, \theta-1]$ such that $j = t_0 + l\theta + k$, which together with (10) and the bi-periodicity of $\mathcal{T}_{j,t}$, yields that

$$\begin{aligned} Tr(Y_j) &= \langle \mathcal{T}_{j,t_0}(X), Q_j'Q_j \rangle \\ &= \langle \mathcal{T}_{t_0+l\theta+k,t_0}(X), Q'_{t_0+l\theta+k}Q_{t_0+l\theta+k} \rangle \\ &= \langle \mathcal{T}_{t_0+l\theta+k,t_0+l\theta}\mathcal{T}_{t_0+l\theta,t_0+(l-1)\theta} \cdots \mathcal{T}_{t_0+\theta,t_0}(X), Q'_{t_0+k}Q_{t_0+k} \rangle \\ &= \langle \mathcal{T}_{t_0+k,t_0}(\mathcal{T}_{t_0}^{\theta})^l(X), Q'_{t_0+k}Q_{t_0+k} \rangle \\ &= \lambda^l \langle \mathcal{T}_{t_0+k,t_0}(X), Q'_{t_0+k}Q_{t_0+k} \rangle = 0, \end{aligned}$$

which justifies that $Tr(Y_j) = 0$ for all $j \geq t_0$. As a result, we have $Y_t = Q_t\mathcal{T}_{t,t_0}(X)Q_t' = 0$ for all $t \geq t_0$. Note that $X \in S_{n+}$ can be expressed as $X = \sum_{s=1}^{n} x_s x_s'$ with $x_s \in \mathcal{R}^n$. Thus, $\mathcal{T}_{t,t_0}(X) = \sum_{s=1}^{n} \mathcal{T}_{t,t_0}(x_s x_s')$, where there is at least one $\bar{s} \in [1, n]$ satisfying $x_{\bar{s}} \neq 0$. On the other hand, because

$$0 = Y_t = Q_t\mathcal{T}_{t,t_0}(X)Q_t' \geq Q_t\mathcal{T}_{t,t_0}(x_{\bar{s}}x_{\bar{s}}')Q_t' \geq 0, \ \ t \geq t_0,$$

we conclude that the measurement output arising from $x_{t_0} = x_{\bar{s}} \neq 0$ fulfills $y_{(t;x_{\bar{s}},t_0)} = 0$ (a.s.) on the interval $[t_0, \infty)$. Hence, $(\mathbf{A}, \mathbf{C}|\mathbf{Q})$ is not observable at t_0.

Sufficiency. Suppose that the criterion holds at $t_0 \in \mathbb{Z}_+$. If $(\mathbf{A}, \mathbf{C}|\mathbf{Q})$ is unobservable at time t_0, let $x_{t_0} = x \neq 0 \in \mathcal{R}^n$ be an unobservable state and

$$X_t = E[x_{(t;x,t_0)}x'_{(t;x,t_0)}] = \mathcal{T}_{t,t_0}(X),$$
$$Y_t = E[y_{(t;x,t_0)}y'_{(t;x,t_0)}] = Q_t\mathcal{T}_{t,t_0}(X)Q'_t,$$

where $X = xx'$. Thus, $y_{t;t_0,x} = 0$ (a.s.) for $t \in [t_0, \infty)$, and further the necessity argument leads to that

$$\langle \mathcal{T}_{t,t_0}(X), Q'_tQ_t \rangle = 0, \quad t \geq t_0. \tag{13}$$

Let $\mathcal{X}_+^{t_0} (\subset S_{n_+})$ be the closed convex hull of all positive semi-definite matrices satisfying (13). Moreover, $\mathcal{X}^{t_0}$ denotes the minimal subspace generated by $\mathcal{X}_+$. Then, $\mathcal{X}_+^{t_0}$ is a closed solid pointed convex cone in $\mathcal{X}^{t_0}$. From (13), for any $X_0 \in \mathcal{X}_+^{t_0}$, it can be derived that

$$\begin{aligned} &\langle \mathcal{T}_{t,t_0}(\mathcal{T}_{t_0}^{\theta}(X_0)), Q'_tQ_t \rangle \\ &= \langle \mathcal{T}_{t+\theta,t_0+\theta}(\mathcal{T}_{t_0}^{\theta}(X_0)), Q'_tQ_t \rangle \\ &= \langle \mathcal{T}_{t,t_0}(X_0), Q'_{t+\theta}Q_{t+\theta} \rangle = 0, \quad \forall t \geq t_0, \end{aligned} \tag{14}$$

which means $\mathcal{T}_{t_0}^{\theta}(\mathcal{X}_+^{t_0}) \subset \mathcal{X}_+^{t_0}$. Consequently, $\mathcal{X}^{t_0}$ is an invariant subspace with respect to $\mathcal{T}_{t_0}^{\theta}$. By Krein–Rutman theorem [15], there exists a nonzero $X_1 \in \mathcal{X}_+^{t_0}$ with $\mathcal{T}_{t_0}^{\theta}(X_1) = \rho X_1$, where $\rho \geq 0$ is the spectral radius of $\mathcal{T}_{t_0}^{\theta}$ on $\mathcal{X}^{t_0}$. Since X_1 belongs to $\mathcal{X}_+^{t_0}$, it follows that X_1 also satisfies (13). This contradicts the assumption that (10) holds. So, $(\mathbf{A}, \mathbf{C}|\mathbf{Q})$ is observable. □

Remark 3 It is remarkable that in our previous study [1], $\mathcal{H}$-representation method is applied to present an operator-spectral test for observability, which requires to check (10) within the subspace of all complex symmetric matrices. In comparison, the spectral criterion given in Theorem 1 is restricted to the spectra associated with real positive semi-definite matrices, which is more convenient to verify. □

4 Application to H_2/H_∞ Control

In this section, we will employ the observability criterion to tackle the infinite-horizon H_2/H_∞ control problem for the following DTSPS:

$$\begin{cases} x_{t+1} = (A_t x_t + B_t u_t + F_t v_t) + (C_t x_t + D_t u_t + G_t v_t) w_t, \\ y_t = \begin{bmatrix} Q_t x_t \\ u_t \end{bmatrix}, \ t \in \mathcal{Z}_+. \end{cases} \tag{15}$$

More specifically, for a given disturbance attenuation level $\gamma > 0$, we aim to find a linear state-feedback controller $u_t^* \in l^2(0, \infty; \mathcal{R}^{n_u})$, if possible, such that

(i) For arbitrary $x_0 \in \mathcal{R}^n$, the state response of (15) corresponding to $v = 0$ and $u = u_t^*$ is AMSS.

(ii) $\|L_\infty^{u^*}\| := \sup\limits_{v \in l^2(0,\infty;\mathcal{R}^{n_v}),\ v \neq 0, x_0 = 0} \dfrac{(\sum_{t=0}^{\infty} E[\|Q_t x_t\|^2 + \|u_t^*\|^2])^{\frac{1}{2}}}{(\sum_{t=0}^{\infty} E\|v_t\|^2)^{\frac{1}{2}}} < \gamma.$

(iii) When the worst case disturbance v_t^*, if it exists, is imposed on (15), u_t^* minimizes the output energy: $J_2(u, v^*) := \sum_{t=0}^{\infty} E\|y_t\|^2$.

From a perspective of game theory, the solution (u^*, v^*) of H_2/H_∞ control problem establishes a Nash equilibrium strategy as follows:

$$J_1(u^*, v) \leq J_1(u^*, v^*), \ J_2(u^*, v^*) \leq J_2(u, v^*),$$

where $J_1(u, v) := \sum_{t=0}^{\infty} E[\|y_t\|^2 - \gamma^2 \|v_t\|^2]$.

First of all, we need the following lemma about the stabilizing solution of generalized periodic Riccati equation.

Lemma 2 *If system (2) is stochastically stabilizable and* $(\mathbf{A}, \mathbf{C}|\mathbf{M}^{1/2})$ *is observable, then the GPRE (16) has a* θ*-periodic stabilizing solution* $X_t > 0$.

$$\begin{cases} X_t = A_t X_{t+1} A_t + C_t' X_{t+1} C_t + M_t - V_t(X_{t+1}) W_t(X_{t+1})^{-1} V_t(X_{t+1})', \\ V_t(X_{t+1}) - A_t' X_{t+1} B_t + C_t' X_{t+1} D_t, \\ W_t(X_{t+1}) = R_t + B_t' X_{t+1} B_t + D_t' X_{t+1} D_t > 0, \\ R_t > 0, \ M_t \geq 0, \ t \in \mathcal{Z}_+, \end{cases} \tag{16}$$

Proof If system (2) is stochastically stabilizable, then the GPRE (16) admits a minimal solution $X_t^m \geq 0$ (Theorem 5.9, [16]). Moreover, the optimal cost of the free end-point LQR problem (no requirement of $\lim_{t \to +\infty} E\|x_t\|^2 = 0$}):

$$\begin{cases} J(\xi, t_0) = \min\limits_{u_t \in l^2(0,\infty;\mathcal{R}^u)} \sum_{t=t_0}^{\infty} E(x_t' M_t x_t + u_t' R_t u_t), \\ s.t. \ \ (2) \text{ with } x_{t_0} = \xi \in \mathcal{R}^n \end{cases} \tag{17}$$

is given by $J(\xi, t_0) = \xi' X^m_{t_0} \xi$ with the optimal control $u^*_t = -W_t(X^m_{t+1})^{-1} V_t(X^m_{t+1})' x_t$ (Theorem 6.1, [16]). If $(\mathbf{A}, \mathbf{C}|\mathbf{M}^{1/2})$ is observable, we infer that $X^m_t > 0$. Otherwise, there is $0 \neq \xi \in \mathcal{R}^n$ and $t_0 \in \mathbb{Z}_+$ such that $X^m_{t_0}\xi = 0$. It follows from (17) that

$$0 \leq \sum_{t=t_0}^{\infty} E(x_t' M_t x_t + u_t^{*\prime} R_t u_t^*) = J_2(\xi, t_0) = \xi' X^m_{t_0} \xi = 0,$$

which yields $M_t^{1/2} x_t \equiv 0$ (a.s.) for $T \geq t \geq t_0$ and $\forall T > t_0$. Thus, $x_{t_0} = \xi$ is a nonzero unobservable state, which contradicts the observability of $(\mathbf{A}, \mathbf{C}|\mathbf{M}^{1/2})$. So, there must be $X^m_t > 0$. To proceed, we substitute X^m_t into the GPRE (16) and rewrite it as follows:

$$\begin{cases} X^m_t = (A_t + B_t K^m_t)' X_{t+1} (A_t + B_t K^m_t) + (C_t + D_t K^m_t)' X_{t+1} (C_t \\ \qquad +D_t K^m_t) + M_t + K^{m\prime}_t R_t K^m_t, \\ K^m_t = -W_t(X^m_{t+1})^{-1} V_t(X^m_{t+1})', \ R_t > 0, \ M_t \geq 0, \ t \in \mathbb{Z}_+. \end{cases} \tag{18}$$

Moreover, we deduce that $(\mathbf{A} + \mathbf{B}\mathbf{K}^m, \mathbf{C} + \mathbf{D}\mathbf{K}^m | \mathbf{M} + \mathbf{K}^{m\prime}\mathbf{R}\mathbf{K}^m)$ is observable. Otherwise, by Theorem 1, there is a nonzero $X \in S_{n+}$ and $t \in \mathbb{Z}_+$ such that

$$\begin{cases} \check{\mathcal{T}}_t^{\theta}(X) = \lambda X, \ \lambda \in C, \\ (M_j + K^{m\prime}_j R_j K^m_j) \check{\mathcal{T}}_{j,t}(X) = 0, \ j = t, t+1, \ldots, t+\theta-1, \end{cases} \tag{19}$$

where $\check{\mathcal{T}}_{t,s} := \check{\mathcal{L}}_{t-1} \check{\mathcal{L}}_{t-2} \ldots \check{\mathcal{L}}_s$ $(\check{\mathcal{T}}_{t,t} = \mathcal{I})$ and $\check{\mathcal{L}}_t(X) := (A_t + B_t K^m_t) X (A_t + B_t K^m_t)'$. Since $R_t > 0$ and $M_t \geq 0$, the second equality of (19) implies that for $j = t, t+1, \ldots, t+\theta-1$,

$$M_j^{1/2} \check{\mathcal{T}}_{j,t}(X) = 0, \tag{20}$$

$$K^m_j \check{\mathcal{T}}_{j,t}(X) = 0. \tag{21}$$

Particularly, when $j = t$, it follows from (21) that $K^m_t X = 0$. As a result, we have $\check{\mathcal{L}}_t(X) = \mathcal{L}_t(X)$ and $\check{\mathcal{T}}_{t,s} = \mathcal{T}_{t,s}$. Now, combing (20) with the first equality of (19), we conclude that there exists a nonzero $X \in S_{n+}$ and $t \in \mathbb{Z}_+$ such that

$$\begin{cases} \mathcal{T}_t^{\theta}(X) = \lambda X, \ \lambda \in C, \\ M_j^{1/2} \mathcal{T}_{j,t}(X) = 0, \ j = t, t+1, \ldots, t+\theta-1. \end{cases} \tag{22}$$

According to Theorem 1, (22) implies that $(\mathbf{A}, \mathbf{C}|\mathbf{M}^{1/2})$ is not observable. This contradicts to the assumption. So, $(\mathbf{A} + \mathbf{B}\mathbf{K}^m, \mathbf{C} + \mathbf{D}\mathbf{K}^m | \ \mathbf{M} + \mathbf{K}^{m\prime}\mathbf{R}\mathbf{K}^m)$ must be observable.

Then, applying Theorem 3.2 (i) [1] to (18), we get that $(\mathbf{A}+\mathbf{B}\mathbf{K}^m, \mathbf{C}+\mathbf{D}\mathbf{K}^m)$ is AMSS. Hence, K_t^m (see (18)) is a stabilizing feedback gain, which proves X_t^m to be a stabilizing solution. The proof is ended. □

In what follows, a stochastic bounded real lemma will be shown for the perturbed stochastic periodic system:

$$\begin{cases} x_{t+1} = (A_t x_t + F_t v_t) + (C_t x_t + G_t v_t) w_t, \\ y_t = Q_t x_t, \ t \in \mathcal{Z}_+, \end{cases} \tag{23}$$

where $v_t \in l^2(0, \infty; \mathcal{R}^{n_v})$ is an exogenous random disturbance. If $(\mathbf{A}, \mathbf{C})$ is AMSS, a linear perturbation operator $\mathcal{L}_\infty : l^2(0, \infty; \mathcal{R}^{n_v}) \to l^2(0, \infty; \mathcal{R}^{n_y})$ can be defined as $\mathcal{L}_\infty(v_t) = Q_t x_t$. The norm of $\mathcal{L}_\infty$ is given by:

$$\|\mathcal{L}_\infty\| := \sup_{\substack{v \in l^2(0,\infty;\mathcal{R}^{n_v}), \\ v \neq 0, x_0 = 0}} \frac{\| \mathcal{L}_\infty(v) \|_{l^2(0,\infty;\mathcal{R}^{n_y})}}{\| v \|_{l^2(0,\infty;\mathcal{R}^{n_v})}}.$$

Lemma 3 *Given $\gamma > 0$, (23) is internally stable (i.e., AMSS when $v_t = 0$) and $\|\mathcal{L}_\infty\| < \gamma$, iff the following GPRE has a θ-periodic stabilizing solution $X_t \geq 0$:*

$$\begin{cases} X_t = A_t X_{t+1} A_t + C_t' X_{t+1} C_t + Q_t' Q_t - \bar{V}_t(X_{t+1}) \bar{W}_t(X_{t+1})^{-1} \bar{V}_t(X_{t+1})', \\ \bar{V}_t(X_{t+1}) = A_t' X_{t+1} F_t + C_t' X_{t+1} G_t, \\ \bar{W}_t(X_{t+1}) = F_t' X_{t+1} F_t + G_t' X_{t+1} G_t - \gamma^2 I_{n_v} < -\varepsilon_0 I_{n_v}, \\ \varepsilon_0 \in (0, \gamma^2 - \|\mathcal{L}_\infty\|^2), \ t \in \mathcal{Z}_+. \end{cases} \tag{24}$$

Proof In the model (5) of [8], we assume that Markov chain has a unique state and there is only one multiplicative noise (i.e., Markov jump parameter vanishes and $r = 1$). Then, by adapting the coefficient notions, it follows from Theorem 1 [8] that (23) is internally stable and $\|\mathcal{L}_\infty\| < \gamma$ iff the discrete-time Riccati equation (24) has a stabilizing solution $X_t \geq 0$. Since the coefficients of (24) are θ-periodic, by Theorem 5.5 [16], its stabilizing solution X_t must be θ-periodic. This completes the proof. □

By use of Lemma 3 and the spectral criterion of observability, we can devise the following state-feedback H_2/H_∞ controller.

Theorem 2 (i) *For (15), if the following coupled GPREs admit a group of θ-periodic solutions $(X_t^1, K_t^1; X_t^2, K_t^2) \in S_{n+} \times \mathcal{R}^{n \times n_v} \times S_{n+} \times \mathcal{R}^{n \times n_u}$:*

$$\begin{cases} X_t^1 = (A_t + B_t K_t^2)' X_t^1 (A_t + B_t K_t^2) + (C_t + D_t K_t^2)' X_t^1 (C_t + D_t K_t^2) \\ \quad +Q_t' Q_t + K_t^{2\prime} K_t^2 + K_t^3 H_t^{-1} K_t^{3\prime}, \\ H_t = \gamma^2 I_{n_v} - F_t' X_t^1 F_t - G_t' X_t^1 G_t > \varepsilon_0 I_{n_v},\ X_t^1 \geq 0, \\ \varepsilon_0 \in (0, \gamma^2 - \|\mathcal{L}_\infty^{u^*}\|^2), \end{cases} \tag{25}$$

$$K_t^1 = H_t^{-1} K_t^{3\prime}, \tag{26}$$

$$\begin{cases} X_t^2 = (A_t + F_t K_t^1)' X_t^2 (A_t + F_t K_t^1) + (C_t + G_t K_t^1)' X_t^2 (C_t + G_t K_t^1) \\ \quad +Q_t' Q_t - K_t^4 \bar{H}_t^{-1} K_t^{4\prime}, \\ \bar{H}_t = I_{n_u} + B_t' X_t^2 B_t + D_t' X_t^2 D_t,\ X_t^2 \geq 0, \end{cases} \tag{27}$$

$$K_t^2 = -\bar{H}_t^{-1} K_t^{4\prime}, \tag{28}$$

where

$$K_t^3 = (A_t + B_t K_t^2)' X_t^1 F_t + (C_t + D_t K_t^2)' X_t^1 G_t,$$
$$K_t^4 = (A_t + F_t K_t^1)' X_t^2 B_t + (C_t + G_t K_t^1)' X_t^2 D_t,$$

and $(\boldsymbol{A}, \boldsymbol{C}|\boldsymbol{Q})$, $(\boldsymbol{A} + \boldsymbol{F}\boldsymbol{K}^1, \boldsymbol{C} + \boldsymbol{G}\boldsymbol{K}^1|\boldsymbol{Q})$ *are both observable, then the desired* H_2/H_∞ *control and the worst case disturbance are given by* $(u_t^* = K_t^2 x_t, v_t^* = K_t^1 x_t)$.
(ii) *If the* H_2/H_∞ *control problem of (15) is solved by* $(u_t^* = K_t^2 x_t, v_t^* = K_t^1 x_t)$ *and* $(\boldsymbol{A} + \boldsymbol{F}\boldsymbol{K}^1, \boldsymbol{C} + \boldsymbol{G}\boldsymbol{K}^1|\boldsymbol{Q})$ *is detectable, then the coupled GPREs (25) and (27) have a* θ*-periodic quaternion solution* $(X_t^1, K_t^1; X_t^2, K_t^2) \in S_{n+} \times \mathcal{R}^{n \times n_v} \times S_{n+} \times \mathcal{R}^{n \times n_u}$. *Moreover,* (K_t^1, K_t^2) *must has the expression (26) and (28).*

Proof (i) Let us first show that (15) is stabilized by u^* when $v = 0$. Note that (25) and (27) can be rearranged as follows:

$$X_t^1 = (A_t + B_t K_t^2)' X_t^1 (A_t + B_t K_t^2) + (C_t + D_t K_t^2)' X_t^1 (C_t + D_t K_t^2) + \hat{Q}_t' \hat{Q}_t, \tag{29}$$

$$\begin{aligned} X_t^2 &= (A_t + B_t K_t^2 + F_t K_t^1)' X_t^2 (A_t + B_t K_t^2 + F_t K_t^1) + (C_t + D_t K_t^2 + G_t K_t^1)' \\ &\quad \cdot X_t^2 (C_t + D_t K_t^2 + G_t K_t^1) + \tilde{Q}_t' \tilde{Q}_t, \end{aligned} \tag{30}$$

where $\hat{Q}_t$ and $\tilde{Q}_t$ are given by

$$\hat{Q}_t = \begin{bmatrix} Q_t \\ K_t^2 \\ H_t^{-1/2} K_t^3 \end{bmatrix},\ \tilde{Q}_t = \begin{bmatrix} Q_t \\ K_t^2 \end{bmatrix}.$$

It can be asserted that $(\mathbf{A} + \mathbf{FK}^1 + \mathbf{BKBK}^2, \mathbf{C} + \mathbf{GK}^1 + \mathbf{BKDK}^2|\tilde{\mathbf{Q}})$ is observable. Otherwise, by Theorem 1, for some $t \in \mathbb{Z}_+$, there exists a nonzero $X \in S_{n+}$ such that

$$\mathcal{T}_{t+\theta,t}^1(X) = \lambda X,\ \lambda \in C, \tag{31}$$

$$\tilde{Q}_j \mathcal{T}_{j,t}^1(X) = 0,\ j = t, t+1, \ldots, t+\theta-1, \tag{32}$$

where $\mathcal{T}^1_{j,t} = \mathcal{L}^1_{j-1}\mathcal{L}^1_{j-1}\cdots\mathcal{L}^1_t$ ($\mathcal{T}^1_{t,t} = \mathcal{I}$), and

$$\mathcal{L}^1_t(X) = (A_t + B_tK^2_t + F_tK^1_t)X(A_t + B_tK^2_t + F_tK^1_t)' \\ +(C_t + D_tK^2_t + G_tK^1_t)X(C_t + D_tK^2_t + G_tK^1_t)'. \quad (33)$$

When $j = t$, (32) yields $\tilde{Q}_tX = 0$, which implies $K^2_tX = 0$. Substituting it into (31)–(33), we get that for the above $t \in \mathbb{Z}_+$, there exists a nonzero $X \in S_{n+}$ such that

$$\mathcal{T}^2_{t+\theta,t}(X) = \lambda X,\ \lambda \in C, \\ Q_j\mathcal{T}^2_{j,t}(X) = 0,\ j = t, t+1, \ldots, t+\theta-1,$$

where $\mathcal{T}^2_{j,t} = \mathcal{L}^2_{j-1}\mathcal{L}^2_{j-1}\cdots\mathcal{L}^2_t$ ($\mathcal{T}^2_{t,t} = \mathcal{I}$), and

$$\mathcal{L}^2_t(X) = (A_t + F_tK^1_t)X(A_t + F_tK^1_t)' + (C_t + G_tK^1_t)X(C_t + G_tK^1_t)'.$$

It follows from Theorem 1 that $(\mathbf{A}+\mathbf{FK}^1, \mathbf{C}+\mathbf{GK}^1|\mathbf{Q})$ is not observable at t, which contradicts the assumption. Thus, $(\mathbf{A}+\mathbf{FK}^1+\mathbf{BKBK}^2, \mathbf{C}+\mathbf{GK}^1+\mathbf{BKDK}^2|\tilde{\mathbf{Q}})$ is observable. Applying Lemma 2 to (31), we obtain that $(\mathbf{A}+\mathbf{FK}^1+\mathbf{BK}^2, \mathbf{C}+\mathbf{GK}^1+\mathbf{DK}^2)$ is AMSS, which implies $u^* \in l^2(0,\infty;\mathcal{R}^{n_u})$ and $v^* \in l^2(0,\infty;\mathcal{R}^{n_v})$. By the same procedure, we can show that $(\mathbf{A}+\mathbf{BK}^2, \mathbf{C}+\mathbf{DK}^2)$ is AMSS. So, (15) is stabilized by u^* when $v = 0$. In addition, because $(\mathbf{A}+\mathbf{FK}^1+\mathbf{BK}^2, \mathbf{C}+\mathbf{GK}^1+\mathbf{BKDK}^2)$ is AMSS, it is obvious that $(\mathbf{A}+\mathbf{FK}^1, \mathbf{B}; \mathbf{C}+\mathbf{GK}^1, \mathbf{D})$ is stochastically stabilizable. Recalling that $(\mathbf{A}+\mathbf{FK}^1, \mathbf{C}+\mathbf{GK}^1|\mathbf{Q})$ is observable, by Lemma 2, X^2_t is a stabilizing solution of (27).

Next, we will prove that $\|L^{u^*}_\infty\| < \gamma$. Substituting u^* into (15), the following system is got:

$$\begin{cases} x_{t+1} = [(A_t + B_tK^2_t)x_t + F_tv_t] + [(C_t + D_tK^2_t)x_t + G_tv_t]w_t, \\ y_t = \begin{bmatrix} Q_t \\ K^2_t \end{bmatrix} x_t,\ t \in \mathbb{Z}_+. \end{cases} \quad (34)$$

In the preceding analysis, $(\mathbf{A}+\mathbf{BK}^2, \mathbf{C}+\mathbf{DK}^2)$ is shown to be AMSS, which means X^1_t is a stabilizing solution of (27). As an immediate result of Lemma 3, we have $\|L^{u^*}_\infty\| < \gamma$.

There still needs to show that u^* minimizes the cost function $J_2(u, v^*)$. To this end, let us first prove that the worst case disturbance

$$v^* := \arg \min_{v\in l^2(0,\infty;\mathcal{R}^{n_v})} J_1(u^*, v) \quad (35)$$

is given by $K^1_tx_t$. As concerns (25) and (34), by completing square, we get

$$\begin{aligned} J_1(u^*, v) &= x_0' X_0^1 x_0 - \sum_{t=0}^{\infty} E\{[v(t) - H_t^{-1} K_t^{3\prime} x(t)]' H_t [v(t) - H_t^{-1} K_t^{3\prime} x(t)]\} \\ &\le x_0' X_0^1 x_0 = J_1(u^*, v^*), \end{aligned} \tag{36}$$

which indicates $v_t^* = K_t^1 x_t$ with $K_t^1 = H_t^{-1} K_t^{3\prime}$. Under the influence of v^*, (15) becomes

$$\begin{cases} x_{t+1} = [(A_t + F_t K_t^1) x_t + B_t u_t] + [(C_t + G_t K_t^1) x_t + D_t u_t] w_t, \\ y_t = \begin{bmatrix} Q_t x_t \\ u_t \end{bmatrix}, \ t \in \mathbb{Z}_+. \end{cases} \tag{37}$$

Thus, we have to prove that u^* is the optimal control to the following infinite-horizon LQ problem:

$$\begin{cases} \min\limits_{u \in l^2(0,\infty; R^{n_u})} J_2(u, v^*) = \sum\limits_{t=0}^{\infty} E\|y_t\|^2, \\ s.t. \qquad (72). \end{cases} \tag{38}$$

Because X_t^2 has been shown to be a stabilizing solution of (27), by using again the technique of completing square, we get

$$\begin{aligned} J_2(u, v^*) &= \sum_{t=0}^{\infty} E[\|Q_t x_t\|^2 + \|u_t\|^2] \\ &= x_0' X_0^2 x_0 + \sum_{t=0}^{\infty} E\{[u_t + \bar{H}_t^{-1} K_t^{4\prime} x_t]' \bar{H}_t [u_t + \bar{H}_t^{-1} K_t^{4\prime} x_t]'\} \\ &\ge x_0' X_0^2 x_0 = J_2(u^*, v^*), \end{aligned} \tag{39}$$

which validates the optimality of $u_t^* = K_t^2 x_t$, where $K_t^2 = -\bar{H}_t^{-1} K_t^{4\prime}$. This ends the proof of (i).

(ii) If $u_t^* = K_t^2 x_t$ is an H_2/H_∞ controller, then system (34) is internally stable and $\|L_\infty^{u^*}\| < \gamma$. On one hand, by Corollary 3.9 [16], the state of (34) satisfies $x_t \in l^2(0, \infty; R^n)$ for any $v_t \in l^2(0, \infty; R^{n_v})$. On the other hand, by Lemma 3, it is deduced that (25) has a θ-periodic stabilizing solution $X_t^1 \ge 0$. As shown in the proof of (i), by means of completing square, we can derive from (34) and (25) that (36) holds. So, the worst case disturbance is given by $v_t^* = H_t^{-1} K_t^{3\prime} x_t$. Since $X_t^1 \ge 0$ is a stabilizing solution of (25), $(\mathbf{A} + \mathbf{F}\mathbf{K}^1 + \mathbf{B}\mathbf{K}^2, \mathbf{C} + \mathbf{G}\mathbf{K}^1 + \mathbf{D}\mathbf{K}^2)$ is AMSS. So, $(\mathbf{A} + \mathbf{F}\mathbf{K}^1, \mathbf{B}; \mathbf{C} + \mathbf{G}\mathbf{K}^1, \mathbf{D})$ is stochastically stabilizable. Bearing in mind that $(\mathbf{A} + \mathbf{F}\mathbf{K}^1, \mathbf{C} + \mathbf{G}\mathbf{K}^1 | \mathbf{Q})$ is observable, by Lemma 2, we conclude that (27) has a θ-periodic stabilizing solution $X_t^2 \ge 0$. For (27) and (37), by completing square, we arrive at (39), which implies that the optimal control of the LQ problem (38) is given by $\hat{u}_t = -\bar{H}_t^{-1} K_t^{4'} x_t$. Recalling the formulation (iii) of H_2/H_∞ controller, there holds $u_t^* = \hat{u}_t$, i.e., $K_t^2 = -\bar{H}_t^{-1} K_t^{4'}$. The proof is completed. □

Remark 4 As pointed out in [8], it is impossible to solve the infinite-horizon time-varying H_2/H_∞ control problem with (x, u, v)-dependent noise under the assumption of stochastic detectability, the dual concept of stochastic stabilizability. With the aid of (uniform) detectability criterion, the main result of this section will produce an infinite-horizon periodic H_2/H_∞ controller with (x, u, v)-dependent noise, which also answers the question raised by [8] within the periodic framework.

5 Conclusion

In this paper, we have studied the observability of stochastic periodic systems. An improved operator-spectral criterion has been derived in terms of the invariant-subspace method. As an application of the proposed observability criterion, the infinite-horizon stochastic periodic H_2/H_∞ control problem has been settled in the presence of all state-, control- and disturbance-dependent noises. Next, we intend to extend the operator-spectrum analytical approach to study the observability of linear stochastic systems with general time-varying coefficients, which merits a further study.

Acknowledgments This work was supported by the National Natural Science Foundation of China (No.61304074), the Research Fund for the Taishan Scholar Project of Shandong Province, the SDUST Research Fund (No.2014JQJH103), Scientific Research Foundation of Shandong University of Science and Technology for Recruited Talents (No. 2016RCJJ031) and the Shandong Joint Innovative Center for Safe and Effective Mining Technology and Equipment of Coal Resources.

References

1. Ma H, Hou T, Zhang W Stability and structural properties of stochastic periodic systems: an operator-spectral approach. In: Proceedings of 2016 American control conference, to appear
2. Gershon E, Shaked U, Yaesh I (2005) H_∞ Control and estimation of state-multiplicative linear systems. springer, London
3. Stoica AM (2010) Mixed H_2 and H_∞ performance analysis of networked control systems with fading communication channels. In: Proceedings of 9th European conference of control, pp 218–222
4. Yin G, Zhou XY (2004) Markowitz's mean-variance portfolio selection with regime switching: From discrete-time models to their continuous-time limits. IEEE Trans Autom Control 49:349–360
5. Morozan T (1983) Stabilization of stochastic discrete-time control systems. Stoch Anal Appl 1:89–116
6. Ait Rami M, Chen X, Zhou XY (2002) Discrete-time indefinite LQ control with state and control dependent noise. J Global Optim 23:245–265
7. El Bouhtouri A, Hinrichsen D, Pritchard AJ (1999) H_∞-type control for discrete-time stochastic systems. Int J Robust Nonlinear Cntrol 9:923–948

8. Ma H, Zhang W, Hou T (2012) Infinite horizon H_2/H_∞ control for discrete-time time-varying Markov jump systems with multiplicative noise. Automatica 48:1447–1454
9. Li ZY, Wang Y, Zhou B, Duan GR (2009) Detectability and observability of discrete-time stochastic systems and their applications. Automatica 45:1340–1346
10. Shen L, Sun J, Wu Q (2013) Observability and detectability of discrete-time stochastic systems with Markovian jump. Syst Control Lett 62:37–42
11. Ni Y, Zhang W, Fang H (2010) On the observability and detectability of linear stochastic systems with Markov jumps and multiplicative noise. J Syst Sci Complexity 23:102–115
12. Zhang W, Chen BS (2004) On stabilizability and exact observability of stochastic systems with their applications. Automatica 40:87–94
13. Hou T, Ma H, Zhang W (2016) Spectral tests for observability and detectability of periodic Markov jump systems with nonhomogeneous Markov chain. Automatica 63:175–181
14. Bittanti S, Colaneri P (2009) Periodic systems: filtering and control. Springer, London
15. Schneider H (1965) Positive operator and an inertia theorem. Numerische Mathematik 7:11–17
16. Dragan V, Morozan T, Stoica AM (2010) Mathematical methods in robust control of discrete-time linear stochastic systems. Springer, New York

Development of a Simulation Platform for Spacecraft Omni-directional Rendezvous

Shihao Sun, Hao Li, Yingmin Jia and Changqing Chen

Abstract A novel semi-physical simulation platform that can be used for verification on spacecraft omni-directional rendezvous is presented. The platform is comprised of a 6-DOFs motion module and a 3-DOFs motion module, which provides the capability for reconstructing the motion of spacecraft rendezvous on the ground. The mechanical structure and the dynamics model of the platform are given at first. Then the reference trajectory for the platform's motion module is developed based on the similarity theory and a tracking controller is designed. Finally, the effectiveness of the simulation platform is demonstrated by the numerical simulation of flying-by rendezvous.

Keywords Omni-directional rendezvous · Semi-physical simulation · Motion simulator · Similarity theory · Tracking control

1 Introduction

Omni-directional rendezvous of spacecraft is an essential technology for in-orbit service [1], e.g., capturing and repairing a tumbling failed spacecraft [2]. To accomplish these challenge missions, the server spacecraft need to determine the relative position and attitude [3] of failed spacecraft through flying around and inspection the target at first. Then rendezvous and docking with the target through relative position and attitude control [4]. Meanwhile, early ground-based testing of the Guidance, Navigation, and Control (GNC) algorithms is necessary to improve the reliability, and reduce risks and costs for these missions. Currently, many kinds of the ground

S. Sun (✉) · H. Li · Y. Jia
The Seventh Research Division and the Center for Information and Control,
School of Automation Science and Electrical Engineering,
Beihang University (BUAA), Beijing 100191, China
e-mail: jxcrssh@126.com

C. Chen
The Science and Technology on Space Intelligent Control Laboratory,
Beijing Institute of Control Engineering, Beijing 100190, China

Y. Jia et al. (eds.), *Proceedings of 2016 Chinese Intelligent Systems Conference*, Lecture Notes in Electrical Engineering 405,
DOI 10.1007/978-981-10-2335-4_8

simulation platforms have been designed for spacecraft rendezvous verification, and these platforms can be divided into two groups by physical and semi-physical simulation.

The air-bearing platform, like the platform designed in SPHERES project [5], can offset the gravity of the tested object through gas buoyancy, which provide the similar micro-gravity environment in space for validation and verification various space missions such as spacecraft formation and flying-by rendezvous. This platform belongs to physical simulation which means that it can be used to test the real dynamics response. However, it can simulate only 5-DOFs motion, and it is very difficult to simulate the orbit motion through control the jet thrust implement on the platform. These problems with air-bearing platform is limited to use in the omni-directional rendezvous verification.

The mechanical servo platform, like EPOS designed by German Aerospace Center [6], can reconstruct the motion of spacecraft through a hybrid method where the motion of spacecraft is calculated by its dynamical models and this calculated motion is realized by the simulation platform tracking. This platform belongs to semi-physical simulation which means that the motion of spacecraft in space is not realized by physical method, but the real sensors and on-board microprocessor can be incorporated with the simulation loops. Semi-physical simulation is a reliable and cheaper way for the reasons that we have precise orbit-attitude dynamics and the critical components such as sensors, processing unit and GNC algorithms are included. However, There are two limitations on the semi-physical simulation platform: size of experimental area and duration of experiment. For instance, in the mission about rendezvous with a tumbling failed spacecraft [7], the starting relative distance is about 100–200 m to avoid collisions, and the duration of multi-circle flying-by rendezvous is long for attitude and orbit synchronous. Moreover, the existing semi-physical simulation platform can only simulate uni-directional rendezvous e.g. v-bar direction. To the best of our knowledge, there are no platforms suitable for omni-directional rendezvous.

Motivated by the above discussions, in this paper, a novel semi-physical simulation platform is presented for verification on spacecraft omni-directional rendezvous. This novel platform decompose the 3-DOFs translational motion of server spacecraft into cylindrical coordinate system by circumferential, radial and vertical motion module. Meanwhile, combining the similarity theory into simulation system [8, 9] in which the time scale, linear velocities and distances are scaled to meet with the demand of the ground simulation.

The rest of this study is organized as follows. In Sect. 2, an overview of the simulation platform is given, with a detailed descriptions of the mechanical structure and the dynamics model. The tracking control design are given in Sect. 3 where the reference trajectory is calculated by the dynamic similitude model.Simulation results and conclusion are given in Sects. 4 and 5, respectively.

2 Overview of the Platform

2.1 System Description

The simulation platform for spacecraft omni-directional rendezvous is a semi-physical simulator (also named as hardware-in-the-loop simulator). The simulation concept of the platform is dynamical emulation and kinematical equivalence [10], which makes full use of both precise spacecraft orbit and attitude dynamics and real sensors, processing unit hardware and GNC algorithms. Figure 1-left shows the simulation rig layout. The rig is a 9-DOFs motion system, which consists of a center fixed bed, two three-axis turntables, a circumferential motion module, a radial motion module, and a vertical motion module. The two turntables simulate the attitude motion of target and chaser spacecraft respectively, and the three motion modules simulate the relative orbit motion between target and chaser spacecraft. As such it is suitable for omni-directional rendezvous.

This simulation platform is a similarity theory based semi-physical simulation system, which consists of machinery, motor drive and control, and mathematical computation. The data flow in the simulation platform is as follows (Fig. 1-right): ①The reference trajectory calculated by Spacecraft Similitude Dynamics System. ②Within facility controller, Platform Mechanical System track the reference trajectory. ③The platform motion states are provided by the sensors system. ④The scaled motion states are feedback to the Spacecraft Rendezvous Control System via similarity transformation. ⑤The control commands of orbit and attitude are calculated by rendezvous control algorithms and then transfer to the Spacecraft Similitude Dynamics System via similarity inverse transformation. As such the closed-loop simulation achieves the equivalent rendezvous motion between the ground simulation platform and the spacecraft.

2.2 System Models

The structure and link coordinate frames of the chaser part in the simulation platform is shown in Fig. 2. Note that frame O_0 is the base coordinate fixed to the ground,

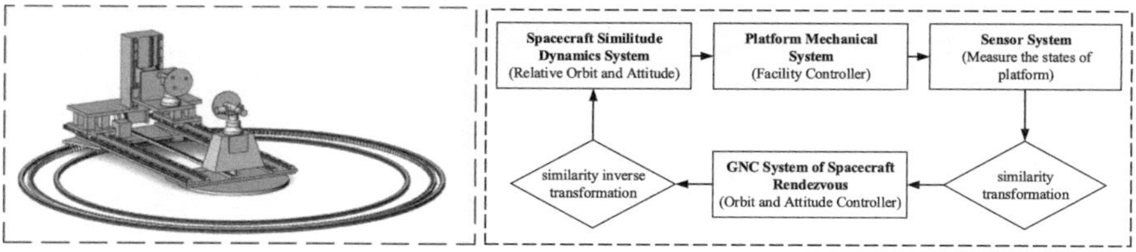

Fig. 1 *Left* Schematic diagram; *right* block diagram

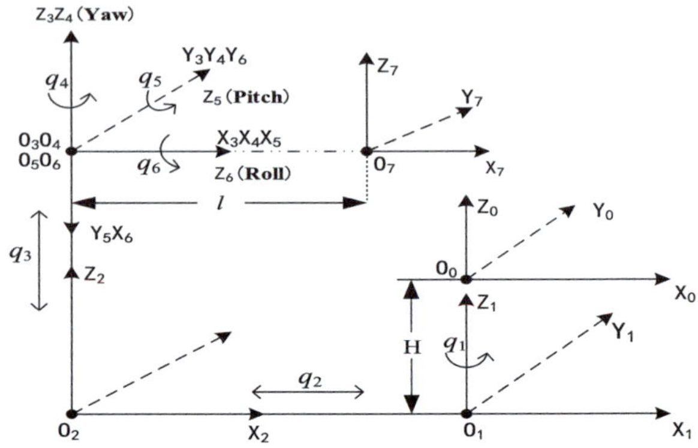

Fig. 2 Coordinate systems of platform

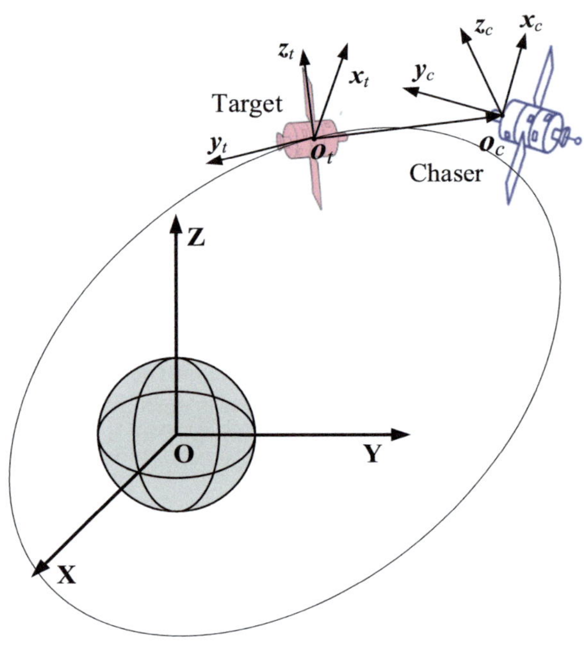

Fig. 3 Coordinate systems of spacecraft

which is corresponding to orbital coordinate O_t of the target spacecraft shown in Fig. 3. This frame can be regarded as the Hill coordinate of the platform. Moreover, frames O_1, O_2, O_3, O_4, O_5, and O_6 are set coincident with the three motion modules and the three-axis turntable, respectively. The frame O_7 is set coincident with the docking panel of the chaser simulator, which is corresponding to frames O_c. According to these definitions, The transformation matrix ${}^{i}\boldsymbol{T}_{i-1}$, which connect the frame O_i to the frame O_{i-1} is obtained by

$$ {}^{0}\boldsymbol{T}_1 = \begin{pmatrix} cq_1 & -sq_1 & 0 & 0 \\ sq_1 & cq_1 & 0 & 0 \\ 0 & 0 & 1 & H \\ 0 & 0 & 0 & 1 \end{pmatrix} \quad {}^{1}\boldsymbol{T}_2 = \begin{pmatrix} 1 & 0 & 0 & q_2 \\ 0 & 1 & 0 & 0 \\ 0 & 0 & 1 & 0 \\ 0 & 0 & 0 & 1 \end{pmatrix} \quad {}^{2}\boldsymbol{T}_3 = \begin{pmatrix} 1 & 0 & 0 & 0 \\ 0 & 1 & 0 & 0 \\ 0 & 0 & 1 & q_3 \\ 0 & 0 & 0 & 1 \end{pmatrix} \quad {}^{3}\boldsymbol{T}_4 = \begin{pmatrix} cq_4 & -sq_4 & 0 & 0 \\ sq_4 & cq_4 & 0 & 0 \\ 0 & 0 & 1 & 0 \\ 0 & 0 & 0 & 1 \end{pmatrix} $$

$$ {}^{4}\boldsymbol{T}_5 = \begin{pmatrix} cq_5 & -sq_5 & 0 & 0 \\ 0 & 0 & 1 & 0 \\ -sq_5 & -cq_5 & 0 & 0 \\ 0 & 0 & 0 & 1 \end{pmatrix} \quad {}^{5}\boldsymbol{T}_6 = \begin{pmatrix} 0 & 0 & 1 & 0 \\ cq_6 & -sq_6 & 0 & 0 \\ sq_6 & cq_6 & 0 & 0 \\ 0 & 0 & 0 & 1 \end{pmatrix} \quad {}^{6}\boldsymbol{T}_7 = \begin{pmatrix} 0 & 0 & -1 & 0 \\ 0 & 1 & 0 & 0 \\ 1 & 0 & 0 & l \\ 0 & 0 & 0 & 1 \end{pmatrix} $$

The docking panel center position $\boldsymbol{x}_p = \begin{bmatrix} x_{ct} & y_{ct} & z_{ct} \end{bmatrix}^T$ and the orientation cosine matrix $\boldsymbol{R} = \begin{bmatrix} \boldsymbol{R}_1 & \boldsymbol{R}_2 & \boldsymbol{R}_3 \end{bmatrix}$ relative to the base frame O_0 can be calculated by implementing the products of matrix ${}^{i}\boldsymbol{T}_{i-1}$:

$$ \begin{bmatrix} \boldsymbol{x}_p \\ \boldsymbol{R}_1 \\ \boldsymbol{R}_2 \\ \boldsymbol{R}_3 \end{bmatrix} = \begin{bmatrix} lcq_5\mathrm{c}(q_1 + q_4) + q_2 cq_1 \\ lcq_5\mathrm{s}(q_1 + q_4) + q_2 sq_1 \\ q_3 - lsq_5 \\ cq_5\mathrm{c}(q_1 + q_4) \\ cq_5\mathrm{s}(q_1 + q_4) \\ -sq_5 \\ sq_5 sq_6 \mathrm{c}(q_1 + q_4) - cq_6\mathrm{s}(q_1 + q_4) \\ sq_5 sq_6 \mathrm{s}(q_1 + q_4) + cq_6\mathrm{c}(q_1 + q_4) \\ cq_5 sq_6 \\ cq_6 sq_5 \mathrm{c}(q_1 + q_4) + sq_6\mathrm{s}(q_1 + q_4) \\ cq_6 sq_5 \mathrm{s}(q_1 + q_4) - cq_6\mathrm{c}(q_1 + q_4) \\ cq_6 cq_5 \end{bmatrix} \tag{1} $$

By differential equation (1), the kinematic equations which show the relationship between the joint velocities and the docking panel linear and angular velocity are derived:

$$ \begin{bmatrix} \boldsymbol{v} \\ \boldsymbol{\omega} \end{bmatrix} = \boldsymbol{J}\dot{\boldsymbol{q}} \quad \boldsymbol{J} = \begin{pmatrix} -q_2 sq_1 - lc(q_5)s(q_1 + q_4) & cq_1 & 0 & -lcq_5 s(q_1 + q_4) & -lsq_5 c(q_1 + q_4) & 0 \\ q_2 cq_1 + lcq_5 c(q_1 + q_4) & sq_1 & 0 & lcq_5 c(q_1 + q_4) & -lsq_5 s(q_1 + q_4) & 0 \\ 0 & 0 & 1 & 0 & -l\cos q_5 & 0 \\ 0 & 0 & 0 & 0 & -\mathrm{s}(q_1 + q_4) & cq_5\mathrm{c}(q_1 + q_4) \\ 0 & 0 & 0 & 0 & \mathrm{c}(q_1 + q_4) & cq_5\mathrm{s}(q_1 + q_4) \\ 1 & 0 & 0 & 1 & 0 & -sq_5 \end{pmatrix} \tag{2} $$

where $\boldsymbol{v}$ and $\boldsymbol{\omega}$ are the linear velocity and angular velocity of the chaser docking panel relative to the base frame O_0. $\dot{\boldsymbol{q}} = [\dot{q}_1, \dot{q}_2, \ldots, \dot{q}_6]^T$ is the generalized velocity vector of system motion module. $\boldsymbol{J}$ is the mapping matrix termed geometric Jacobian. (abridged notation: $cq_i = \cos q_i$, $sq_i = \sin q_i$).

With Lagrange formulation, the dynamical equations of the chaser part simulator can be derived by

$$ \boldsymbol{H}(\boldsymbol{q})\ddot{\boldsymbol{q}} + \boldsymbol{C}(\boldsymbol{q}, \dot{\boldsymbol{q}})\dot{\boldsymbol{q}} + \boldsymbol{G}(\boldsymbol{q}) = \boldsymbol{\tau} \tag{3} $$

where $\boldsymbol{\tau}$ is actuation torques or force, and

$$\boldsymbol{H}(\boldsymbol{q}) = [h_{jk}]_{6\times 6} \quad h_{jk} = \sum_{i=\max(j,k)}^{6} tr(\frac{\partial^0 \boldsymbol{T}_i}{\partial q_j} \boldsymbol{I}_i \frac{\partial (^0\boldsymbol{T}_i)^{\mathrm{T}}}{\partial q_k}) \quad j,k = 1,\ldots,6$$

$$\boldsymbol{C}(\boldsymbol{q},\dot{\boldsymbol{q}}) = [c_{jk}]_{6\times 6} \quad c_{jk} = \sum_{i=1}^{6} \frac{1}{2}(\frac{\partial h_{jk}}{\partial q_i} + \frac{\partial h_{ji}}{\partial q_k} - \frac{\partial h_{ki}}{\partial q_j})\dot{q}_i \quad j,k = 1,\ldots,6$$

$$\boldsymbol{G}(\boldsymbol{q}) = [g_j]_{6\times 1} \quad g_j = -\sum_{i=1}^{6} m_i[0\ 0\ -g\ 0]\frac{\partial^0 \boldsymbol{T}_i}{\partial q_j} {}^i\tilde{r}_{Ci} \quad j = 1,\ldots,6$$

For the target part simulator, its kinematics and dynamics model are consistent with three-axis simulation turntable, which is omitted here.

3 Controller Design

3.1 Reference Trajectory Generated by Similitude Dynamics

In this part, we will scale the time and distance in the semi-physical simulation to meet with limiting conditions on the size of experimental area and duration of experiment. With dimension analysis in similarity theory [9, 11], the similitude dynamics of attitude and relative orbit can be given by

$$\begin{gathered}
\frac{\boldsymbol{I}}{\lambda_I}\dot{\boldsymbol{\omega}}^r + \boldsymbol{\omega}^r \times \boldsymbol{I}\boldsymbol{\omega}^r = \frac{\boldsymbol{M}}{\lambda_M} \\
\begin{bmatrix} \dot{\psi}^r \\ \dot{\varphi}^r \\ \dot{\theta}^r \end{bmatrix} = \boldsymbol{T}\boldsymbol{\omega}^r = \begin{bmatrix} 0 & \frac{\sin\theta^r}{\cos\varphi^r} & \frac{\cos\theta^r}{\cos\varphi^r} \\ 0 & \cos\theta^r & -\sin\theta^r \\ 1 & \tan\varphi^r\sin\theta^r & \tan\varphi^r\cos\theta^r \end{bmatrix}\boldsymbol{\omega}^r \\
\psi_0^r = \Psi_0, \varphi^r{}_0 = \Phi_0, \theta_0^r = \Theta_0, \boldsymbol{\omega}^r{}_0 = \boldsymbol{\Omega}_0/\lambda_T \\
\begin{cases}
\ddot{x}_{ct}^r = \frac{2\omega}{\lambda_T}\dot{y}_{ct}^r - \frac{k\omega^{3/2}}{\lambda_T^2}x_{ct}^r + \frac{\omega^2}{\lambda_T^2}x_{ct}^r + \frac{\dot{\omega}}{\lambda_T^2}y_{ct}^r + a_x\frac{\lambda_L}{\lambda_T^2} \\
\ddot{y}_{ct}^r = -\frac{2\omega}{\lambda_T}\dot{x}_{ct}^r + \frac{2k\omega^{3/2}}{\lambda_T^2}y_{ct}^r + \frac{\omega^2}{\lambda_T^2}y_{ct}^r - \frac{\dot{\omega}}{\lambda_T^2}x_{ct}^r + a_y\frac{\lambda_L}{\lambda_T^2} \\
\ddot{z}_{ct}^r = \frac{k\omega^{3/2}}{\lambda_T^2}z_{ct}^r + a_z\frac{\lambda_L}{\lambda_T^2}
\end{cases} \\
x_{ct0}^r = \lambda_L X_{ct0}, y_{ct0}^r = \lambda_L Y_{ct0}, z_{ct0}^r = \lambda_L Z_{ct0} \\
\dot{x}_{ct0}^r = \frac{\lambda_L}{\lambda_T}\dot{X}_{ct0}, \dot{y}_{ct0}^r = \frac{\lambda_L}{\lambda_T}\dot{Y}_{ct0}, \dot{z}_{ct0}^r = \frac{\lambda_L}{\lambda_T}\dot{Z}_{ct0}
\end{gathered} \tag{4}$$

where λ_L, λ_T and λ_m are the scalar factors of the length, time, and quality dimension, respectively. $\lambda_{\boldsymbol{M}} = \frac{\lambda_T}{\lambda_m \lambda_L^2}$ and $\lambda_{\boldsymbol{I}} = \frac{1}{\lambda_m \lambda_L^2}$ are the scalar factors of torque and inertia matrix. $\boldsymbol{I}$ is the inertia matrix of spacecraft. $\boldsymbol{M}$ is the attitude actuation torque of spacecraft. $k = \mu^{-2} p^{-3}$ is the relative motion constant of spacecraft rendezvous. μ is the gravitational constant. p, ω and $\dot{\omega}$ are the semi-latus rectum, angular velocity and angular accelerate of target spacecraft orbit, respectively. a_x, a_y and a_z are the three axis orbit thrust acceleration of chaser spacecraft. $\boldsymbol{\Psi}_0$, $\boldsymbol{\Phi}_0$ $\boldsymbol{\Theta}_0$ and $\boldsymbol{\Omega}_0$ are the initial attitude angle and angular velocity of spacecraft. X_{ct0}, Y_{ct0}, Z_{ct0}, $\dot{X}_{ct0}$, $\dot{Y}_{ct0}$, and $\dot{Z}_{ct0}$ are the initial relative positions between chaser and target spacecraft. ψ^r, φ^r, θ^r and $\boldsymbol{\omega}^r$ are the reference Z-Y-X Euler angles and angular velocity of the simulator. x_{ct}^r, y_{ct}^r, and z_{ct}^r are the reference position between chaser and target simulator in the base frame.

Remark 1 Note that in the semi-physical simulation, the propagation of dynamics are only by numerical calculation. Thus the dynamical parameters such as inertia matrix $\boldsymbol{I}$ and mass m are numerical values, which are irrelevant to the real inertia or mass. As such we can set the scalar factors of mass λ_m equal to 1.

3.2 *Derivation of the Tracking Controller*

In this part, the tracking controller will be developed using the resolved acceleration control method [12].

From Eq. (4), the reference position vector and the reference attitude cosine matrix of the chaser simulator relative to the base frame O_0 can be obtained:

$$\boldsymbol{x}_p{}^r = \begin{bmatrix} x_{ct}^r \\ y_{ct}^r \\ z_{ct}^r \end{bmatrix}$$

$$\boldsymbol{R}^r = \begin{bmatrix} \boldsymbol{R}_1{}^r & \boldsymbol{R}_2{}^r & \boldsymbol{R}_3{}^r \end{bmatrix} = \begin{bmatrix} \mathrm{c}\psi^r \mathrm{c}\varphi^r & -\mathrm{s}\psi^r \mathrm{c}\theta^r + \mathrm{c}\psi^r \mathrm{s}\varphi^r \mathrm{s}\theta^r & \mathrm{s}\psi^r \mathrm{s}\theta^r + \mathrm{c}\psi^r \mathrm{s}\varphi^r \mathrm{c}\theta^r \\ \mathrm{s}\psi^r \mathrm{c}\varphi^r & \mathrm{c}\psi^r \mathrm{c}\theta^r + \mathrm{s}\psi^r \mathrm{s}\varphi^r \mathrm{s}\theta^r & -\mathrm{c}\psi^r \mathrm{s}\theta^r + \mathrm{s}\psi^r \mathrm{s}\varphi^r \mathrm{c}\theta^r \\ -\mathrm{s}\varphi^r & \mathrm{c}\varphi^r \mathrm{s}\theta^r & \mathrm{c}\varphi^r \mathrm{c}\theta^r \end{bmatrix} \tag{5}$$

From the Euler's formula, we can obtain

$$\boldsymbol{R}^r = \boldsymbol{R}_k(\alpha) \boldsymbol{R} \tag{6}$$

where $\boldsymbol{R}_k(\alpha) = \cos \alpha \boldsymbol{I} + (1 - \cos \alpha) \boldsymbol{k} \boldsymbol{k}^{\mathrm{T}} + \sin \alpha \boldsymbol{S}(\boldsymbol{k})$, $\boldsymbol{I}$ is unit matrix, $\boldsymbol{S}(\boldsymbol{k})$ is the cross product skew-symmetric matrix corresponding to $\boldsymbol{k}$.

The position error $\boldsymbol{e}_p$ and orient error $\boldsymbol{e}_o$ are defined as

$$\boldsymbol{e}_p = \boldsymbol{x}_p{}^r - \boldsymbol{x}_p \quad \boldsymbol{e}_o = \boldsymbol{k} \sin \alpha \tag{7}$$

Then it is easier to validate that $\boldsymbol{e}_o = \frac{1}{2}\left(\boldsymbol{R}_1 \times \boldsymbol{R}_1{}^r + \boldsymbol{R}_2 \times \boldsymbol{R}_2{}^r + \boldsymbol{R}_3 \times \boldsymbol{R}_3{}^r\right)$, and $\dot{\boldsymbol{e}}_o \approx \boldsymbol{\omega}^r - \boldsymbol{\omega}, \ddot{\boldsymbol{e}}_o \approx \dot{\boldsymbol{\omega}}^r - \dot{\boldsymbol{\omega}}$.

Theorem 1 *Consider the definition of position error and orient error, If the tracking controller is designed as in Eq. (8), and the control gains constant satisfy* $k_p > 0$ *and* $k_d > 0$*, then the asymptotical convergence of the tracking error can be guaranteed.*

$$\boldsymbol{\tau} = \boldsymbol{H}(\boldsymbol{q})\ddot{\boldsymbol{q}}^* + \boldsymbol{C}(\boldsymbol{q}, \dot{\boldsymbol{q}})\dot{\boldsymbol{q}} + \boldsymbol{G}(\boldsymbol{q}) \tag{8}$$

where $\ddot{\boldsymbol{q}}^* = J^{-1}\left\{\begin{bmatrix} \ddot{\boldsymbol{x}}_p{}^r \\ \dot{\boldsymbol{\omega}}^r \end{bmatrix} + k_d \begin{bmatrix} \dot{\boldsymbol{x}}_p{}^r \\ \boldsymbol{\omega}^r \end{bmatrix} + k_p \begin{bmatrix} \boldsymbol{e}_p \\ \boldsymbol{e}_o \end{bmatrix} - \dot{J}\boldsymbol{q}\right\} - k_d\dot{\boldsymbol{q}}$

Proof Substituting Eq. (8) into the dynamics equation (3) yields the following:

$$\boldsymbol{H}(\boldsymbol{q})\ddot{\boldsymbol{q}} + \boldsymbol{C}(\boldsymbol{q}, \dot{\boldsymbol{q}})\dot{\boldsymbol{q}} + \boldsymbol{G}(\boldsymbol{q}) = \boldsymbol{H}(\boldsymbol{q})\ddot{\boldsymbol{q}}^* + \boldsymbol{C}(\boldsymbol{q}, \dot{\boldsymbol{q}})\dot{\boldsymbol{q}} + \boldsymbol{G}(\boldsymbol{q}) \tag{9}$$

Then, multiply both sides by $\boldsymbol{JH}^{-1}$, we can obtain the following:

$$\boldsymbol{J}\ddot{\boldsymbol{q}} = \begin{bmatrix} \ddot{\boldsymbol{x}}_p{}^r \\ \dot{\boldsymbol{\omega}}^r \end{bmatrix} + k_d \begin{bmatrix} \dot{\boldsymbol{x}}_p{}^r \\ \boldsymbol{\omega}^r \end{bmatrix} + k_p \begin{bmatrix} \boldsymbol{e}_p \\ \boldsymbol{e}_o \end{bmatrix} - \dot{\boldsymbol{J}}\dot{\boldsymbol{q}} - k_d\boldsymbol{J}\dot{\boldsymbol{q}} \tag{10}$$

According to the kinematic equation (2):

$$\boldsymbol{J}\dot{\boldsymbol{q}} = \begin{bmatrix} \boldsymbol{v} \\ \boldsymbol{\omega} \end{bmatrix} = \begin{bmatrix} \dot{\boldsymbol{x}}_p \\ \boldsymbol{\omega} \end{bmatrix} \tag{11}$$

Take the derivative of Eq. (11),

$$\boldsymbol{J}\ddot{\boldsymbol{q}} + \dot{\boldsymbol{J}}\dot{\boldsymbol{q}} = \begin{bmatrix} \ddot{\boldsymbol{x}}_p \\ \dot{\boldsymbol{\omega}} \end{bmatrix} \tag{12}$$

Then, substituting the above equation into Eq. (10) gives

$$\begin{bmatrix} \ddot{\boldsymbol{x}}_p \\ \dot{\boldsymbol{\omega}} \end{bmatrix} = \begin{bmatrix} \ddot{\boldsymbol{x}}_p{}^r \\ \dot{\boldsymbol{\omega}}^r \end{bmatrix} + k_d \begin{bmatrix} \dot{\boldsymbol{x}}_p{}^r \\ \boldsymbol{\omega}^r \end{bmatrix} + k_p \begin{bmatrix} \boldsymbol{e}_p \\ \boldsymbol{e}_o \end{bmatrix} - k_d \begin{bmatrix} \dot{\boldsymbol{x}}_p \\ \boldsymbol{\omega} \end{bmatrix} \tag{13}$$

Rewriting the above equation

$$\begin{bmatrix} \ddot{\boldsymbol{x}}_p^r - \ddot{\boldsymbol{x}}_p \\ \dot{\boldsymbol{\omega}}^r - \dot{\boldsymbol{\omega}} \end{bmatrix} + k_d \begin{bmatrix} \dot{\boldsymbol{x}}_p^r - \dot{\boldsymbol{x}}_p \\ \boldsymbol{\omega}^r - \boldsymbol{\omega} \end{bmatrix} + k_p \begin{bmatrix} \boldsymbol{e}_p \\ \boldsymbol{e}_o \end{bmatrix} = 0 \tag{14}$$

Base on the fact that $\dot{\boldsymbol{e}}_o \approx \boldsymbol{\omega}^r - \boldsymbol{\omega}$, $\ddot{\boldsymbol{e}}_o \approx \dot{\boldsymbol{\omega}}^r - \dot{\boldsymbol{\omega}}$, $\dot{\boldsymbol{e}}_p = \dot{\boldsymbol{x}}_p{}^r - \dot{\boldsymbol{x}}_p$, $\ddot{\boldsymbol{e}}_p = \ddot{\boldsymbol{x}}_p{}^r - \ddot{\boldsymbol{x}}_p$, and denote that $\tilde{e} = \left[\boldsymbol{e}_p\ \boldsymbol{e}_o\right]^{\mathrm{T}}$, we can obtain:

$$\ddot{\tilde{e}} + k_d\dot{\tilde{e}} + k_p\tilde{e} = 0 \tag{15}$$

Because of $k_d > 0$ and $k_p > 0$, it is obvious that $\tilde{\boldsymbol{e}} \to 0, \dot{\tilde{\boldsymbol{e}}} \to 0$.

Remark 2 Constraints due to the conditions for the approximate equalities to hold, we need set the initial states of simulator consistent with the reference states. For the real experiment, we can plan a trajectory from static states to the initial states and track this trajectory first; Moreover, because Eq. (9) contains the inverse of Jacobian matrix, the reference trajectory should avoid to pass by the system singularity point.

Remark 3 The controller designed in this section is based on the perfect model, when applied to the real physical system, we can adopt the robust controller such as sliding-mode control [13, 14] to deal with uncertainty and disturbance.

4 Simulation Results

In this section, an example of flying-by rendezvous is provided to illustrate the effectiveness of the simulation platform. In this example, the orbit and attitude parameters of the target spacecraft are listed as follow, the orbit semi-major axis $a = 6907900$ m, the eccentricity $e = 0.2$, initial true anomaly is 0 rad, three Euler attitude angles are 0 rad. The initial relative parameters between chaser and target spacecraft are listed as follow, the initial relative position, velocity, attitude angle and attitude angle velocity are $X_{ct0} = -100$ m, $Y_{ct0} = -100$ m, $Z_{ct0} = 10$ m, $\dot{X}_{ct0} = 1$ m/s, $\dot{Y}_{ct0} = 1$ m/s, $\dot{Z}_{ct0} = -1$ m/s, $\Psi_0 = 0.6$ rad, $\Phi_0 = 0.8$ rad, $\Theta_0 = 0.7$ rad, $\Omega_0 = (0.5, 0.3, 0.5)$ rad/s. Moreover, the scalar factors of fundamental dimensions are assumed as $\lambda_L = 1/10$, $\lambda_T = 1/10$, $\lambda_m = 1/1$.

In this process of flying-by rendezvous, the desired position trajectory of the chaser relative to the target in the orbit plane of the target is set as helical line:

$$\begin{aligned} X_d &= 140\sqrt{2}\exp(-0.008t)\cos(0.04t + 5\pi/4) \\ Y_d &= 140\sqrt{2}\exp(-0.008t)\sin(0.04t + 5\pi/4) \end{aligned} \tag{16}$$

Furthermore, for simplicity, the orbital and attitude tracking controller for rendezvous are designed uses feedback linearization method:

$$\begin{cases} a_x = k\omega^{3/2}X_{ct} - \omega^2 X_{ct} - 2\omega\dot{Y}_{ct} - \dot{\omega}Y_{ct} + \ddot{X}_d - (X_{ct} - X_d) - (\dot{X}_{ct} - \dot{X}_d) \\ a_y = -2k\omega^{3/2}Y_{ct} - \omega^2 Y_{ct} + 2\omega\dot{X}_{ct} + \dot{\omega}X_{ct} + \ddot{Y}_d - (Y_{ct} - Y_d) - (\dot{Y}_{ct} - \dot{Y}_d) \\ a_z = -k\omega^{3/2}Z_{ct} - 0.0001Z_{ct} - 0.02\dot{Z}_{ct} \\ \boldsymbol{M} = \boldsymbol{I}\boldsymbol{T}^{-1}(-\dot{\boldsymbol{T}}\boldsymbol{T}^{-1}\begin{bmatrix}\dot{\Psi}\\ \dot{\Phi}\\ \dot{\Theta}\end{bmatrix} + \boldsymbol{T}\boldsymbol{I}^{-1}((\boldsymbol{T}^{-1}\begin{bmatrix}\dot{\Psi}\\ \dot{\Phi}\\ \dot{\Theta}\end{bmatrix}) \times (\boldsymbol{I}\boldsymbol{T}^{-1}\begin{bmatrix}\dot{\Psi}\\ \dot{\Phi}\\ \dot{\Theta}\end{bmatrix})) - \boldsymbol{K}_p\begin{bmatrix}\Psi\\ \Phi\\ \Theta\end{bmatrix} - \boldsymbol{K}_v\begin{bmatrix}\dot{\Psi}\\ \dot{\Phi}\\ \dot{\Theta}\end{bmatrix}) \end{cases} \tag{17}$$

where $\boldsymbol{T} = \begin{bmatrix} 0 & \frac{\sin(\Theta)}{\cos(\Phi)} & \frac{\cos(\Theta)}{\cos(\Phi)} \\ 0 & \cos(\Theta) & -\sin(\Theta) \\ 1 & \tan(\Phi)\sin(\Theta) & \tan(\Phi)\cos(\Theta) \end{bmatrix}$, $\boldsymbol{K}_p = \mathrm{diag}(\,0.01\;\;0.01\;\;0.02\,)$, $\boldsymbol{K}_v = \mathrm{diag}(\,1\;\;1\;\;2\,)$.

The parameters of controller (8) are set as $k_p = 20$, $k_d = 50$. Next, we simulate the dynamic behavior of simulation platform and spacecraft using

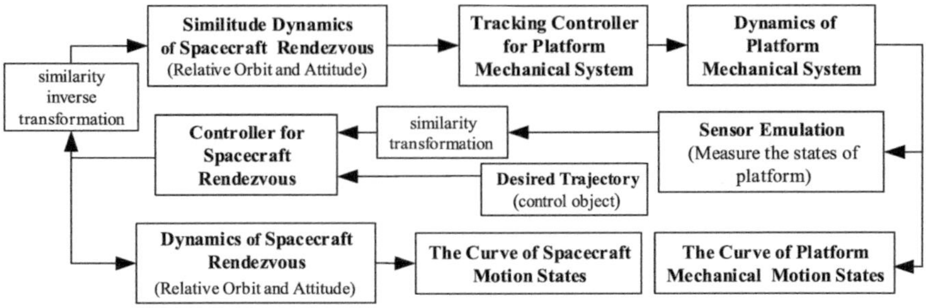

Fig. 4 Block diagram of simulation

Fig. 5 Trajectory of chaser simulator

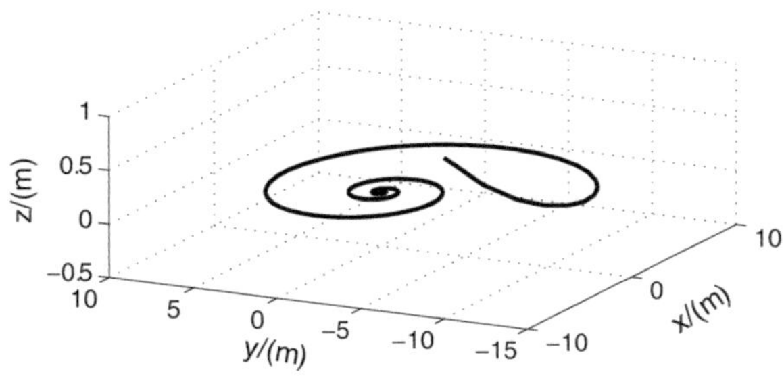

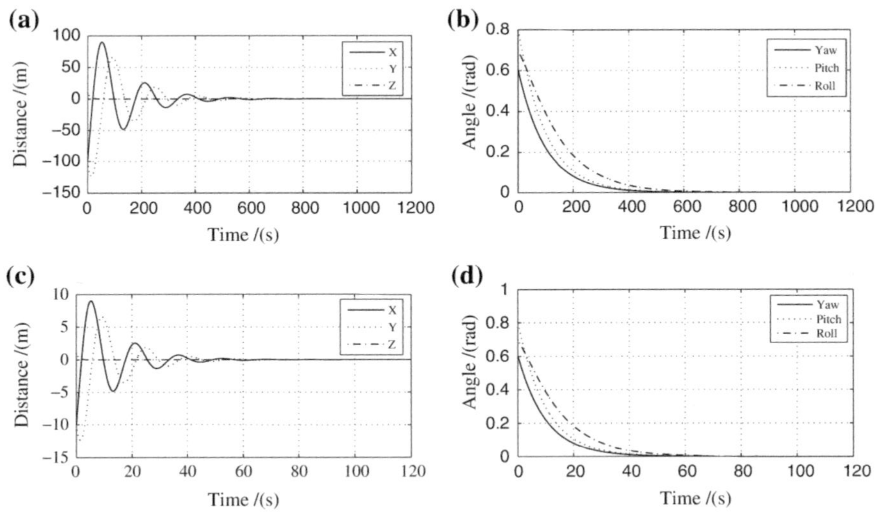

Fig. 6 Curve of relative position and attitude

MATLAB/Simulink. The block diagram of simulation is shown in Fig. 4, and the position response trajectory of the chaser simulator under controller (8) are depicted in Fig. 5. It can be seen from Fig. 5 that the motion of chaser simulator realize the fly-by rendezvous in the small size of experimental area (20 m × 20 m × 2 m). The relative position and attitude of spacecraft and simulator are depicted in Fig. 6. As we set $\lambda_L = 1/10$ and $\lambda_T = 1/10$, the duration and distance of real rendezvous are ten times as much as simulation rendezvous, which meets the expectation.

Remark 4 The integrator module is modified to changeable integration step, which is necessary to simulate two different time scales simultaneously.

5 Conclusions

This paper presents a 9-DOFs semi-physical simulation platform for verification of spacecraft omni-directional rendezvous. The mechanical structure, design principle, system models, and tracking controller design of the platform were discussed. Moreover, a similarity theory based simulation methodology had been proposed to meet with limiting conditions on the size of experimental area and duration of the experiment. An illustrative example about fly-by rendezvous has shown the effectiveness of this simulation platform. As future works, the uncertainty and disturbance in the experimental environment will be considered, and experiment with the prototypes of the platform to illustrate the effectiveness of the platform and the test method.

Acknowledgments This work was supported by the National Basic Research Program of China (973 Program: 2012CB821200, 2012CB821201), the NSFC (61134005, 61327807, 61521091, 61520106010,61304232), and the Fundamental Research Funds for the Central Universities (YWF-16-GJSYS-31,YWF-16 GJSYS-32).

References

1. Fehse W (2005) Automated rendezvous and docking of spacecraft. Cambridge University Press, Cambridge
2. Inaba N, Oda M, Asano M (2006) Rescuing a stranded satellite in space experimental robotic capture of non-cooperative satellites. Trans Jpn Soc Aeronaut spaceences 48(162):213–220
3. Kim SG, Crassidis JL, Cheng Y et al (2007) Kalman filtering for relative spacecraft attitude and position estimation. J Guidance Control Dyn 30(1):133–143
4. Zhao Lin, Jia Yingmin (2014) Multi-objective output feedback control for autonomous spacecraft rendezvous. J Franklin Inst Eng Appl Math 351(5):2804–2821
5. Nolet S, Kong E, Miller DW (2004) Autonomous docking algorithm development and experimentation using the SPHERES testbed. Proc SPIE—Int Soc Opt Eng
6. Benninghoff H, Rems F, Boge T (2014) Development and hardware-in-the-loop test of a guidance, navigation and control system for on-orbit servicing. Acta Astronautica 102(1):67–80
7. Persson S, et al. (2006) PRISMA—An autonomous formation flying mission. In: Small satellite systems and services-The 4S symposium, Sardinia, Italy, ESA, pp 25–29
8. Guglieri G, Maroglio F, Pellegrino P et al (2014) Design and development of guidance navigation and control algorithms for spacecraft rendezvous and docking experimentation. Acta Astronautica 94(1):395–408
9. Sun Shi-Hao, Zhao Lin, Jia Ying-Min (2014) Similitude design method for motion reconstruction of space cooperative vehicles. J Astronaut 35(7):802–810
10. Xu W, Liang B, Xu Y et al (2007) A ground experiment system of free-floating robot for capturing space target. J Intell Robot Syst 38(2):187–208

11. Kline SJ (1986) Similitude and approximation theory. Springer, New York, pp 76–103
12. Luh JYS, Walker MW, Paul RPC (1980) Resolved-acceleration control of mechanical manipulators. IEEE Trans Autom Control 25(3):468–474
13. Zhao Lin, Jia Ying-Min (2015) Finite-time attitude tracking control for a rigid spacecraft using time-varying terminal sliding mode techniques. Int J Control 88(6):1–13
14. Zhao Lin, Jia Ying-Min (2014) Decentralized adaptive attitude synchronization control for spacecraft formation using nonsingular fast terminal sliding mode. Nonlinear Dyn 78(4): 2279–2294

Stabilization of Perturbed Linear Systems by an Event-Triggered Robust H_∞ Controller

Hao Jiang, Yingmin Jia and Changqing Chen

Abstract Event-triggered control systems have been increasingly studied as an alternative to the time-triggered sample-data system, mainly for its advantage of reducing the resource utilization. In this paper, we propose a robust H_∞ controller for a linear perturbed model and implement it in an event-triggered feedback scheme, to achieve the L_2 stability of the system. A sufficient condition guaranteeing the L_2 stability of the system is provided in the form of matrix inequality. The optimization of this inequality condition is settled by eliminating the nonlinear components, and an adjustable parameter is brought in to reduce the conservatism. Illustrative examples are given to show this controller's ability at enlarging sample time and disturbance attenuation.

Keywords Event-triggered control · L_2 stability · H_∞ control

1 Introduction

1.1 Event-Triggered Control

Feedback control laws are mainly implemented in digital platforms since microprocessors offer many advantages over analog platforms [1]. The principal problem is to determine how frequently to execute the control task so that the desired system

H. Jiang (✉) · Y. Jia
The Seventh Research Division and the Center for Information and Control, School of Automation Science and Electrical Engineering, Beihang University (BUAA), Beijing 100191, China
e-mail: 031jianghao@gmail.com

Y. Jia
e-mail: ymjia@buaa.edu.cn

C. Chen
The Science and Technology on Space Intelligent Control Laboratory, Beijing Institute of Control Engineering, Beijing 100190, China

Y. Jia et al. (eds.), *Proceedings of 2016 Chinese Intelligent Systems Conference*, Lecture Notes in Electrical Engineering 405,
DOI 10.1007/978-981-10-2335-4_9

performance is achieved. The traditional way is to execute the control task consisting of sampling the state, computing the control law, and updating the actuator at equidistant sampling time intervals. It is the most common implementation so far mainly because of the existence of a well-established system theory for sample-data system which makes it easy to analyze the difference between the customary continuous design and the digital implementation. However, it seems unnatural to update the control signal in a periodic fashion regardless of the current state of the system, especially for nonlinear systems in which something significant could happen to the state rapidly. Moreover, the control task being executed at the constant frequency which is chosen based on the worst situation also leads to a waste of resource utilization, such as communication bandwidth and processor.

The event-triggered control system, in which the control task execution is triggered by an "event condition," is an alternative to the time-triggered feedback system. The event-triggered control can lead to a remarkable reduction of the system resource, especially when the plant states barely change, so it is widely used in the networked systems ([2, 3]). Another important reason why event-triggered control is worthy of researching is that it can improve the overall system performance in certain cases, because the control input is transmitted to the plant exactly when the system is in an unexpected condition, that is, the state information is utilized in a more effective way. Some research on the theoretical analysis of event-triggered control is shown in [4]. A state-feedback approach was applied to event-triggered control in [5].

1.2 Literature and Contribution of This Paper

In this paper, we consider a robust H_∞ controller implemented in event-triggered feedback strategy for linear perturbed systems, to achieve the closed-loop system's L_2 stability. Similar research on event-triggered robust H_∞ control was done in few papers. Wang and Lemmon [6] studied a self-triggered control for linear perturbed system in which the magnitude of the disturbance is bounded by a linear function of the norm of the system state. Wang and Lemmon [7] relaxes this assumption to disturbance whose induce L_2 norm is finite. Recent works also drive event-triggered control into perturbed systems meeting with other problems, such like time-delay [8], observer-based control [9], and simple nonlinear components [10]. In the above papers, they either design the controller simply making the closed-loop system matrix Hurwitz ([8, 9]), or design a robust H_∞ controller in a continuous feedback scheme, and achieve the L_2 stability of the system by adjusting the conservatism of the event condition ([6, 7]). In other words, most of these controllers designed based on continuous feedback scheme are not the optimal ones in the sense of reducing the sampling frequency, which is the way we design the controller in this paper. Some previous work was presented in [11]. In this paper, we improve the conservatism of the matrix inequality condition proposed in [11] by introducing in an adjustable parameter α. Eliminate the nonlinear components in the inequality, we proposed an LMI (linear matrix inequality) that also guarantees the system's L_2 stability. By solving the optimization problem of the LMI, we can design an optimal (suboptimal) controller in sense of requiring the loosest event condition given the demanded maximum L_2 gain, or obtaining the minimum L_2 gain given the demanded event condition. Moreover, the conservatism of this LMI can be reduced by adjusting the value of parameter α.

2 Problem Statement

In this paper, we consider a perturbed linear time-invariant system given as

$$\begin{cases} \dot{x} = Ax + B_1 w + B_2 u \\ z = Cx + D_1 w + D_2 u \end{cases} \tag{1}$$

where $x \in \mathbb{R}^n$ represents the system state, with initial state $x(0) = x_0$. $u \in \mathbb{R}^m$ and $z \in \mathbb{R}^q$ are the control input and output signals respectively. $w \in L_2[0, \infty)$ is the exogenous disturbance. The system matrices A, B_1, B_2, C, D_1, D_2 are real matrices of appropriate dimensions.

System (1) is stabilized by a state-feedback controller

$$u(t) = Kx(t), t \geq 0. \tag{2}$$

The feedback is implemented not by equal-distant time sampling, but in an event-triggered (self-triggered) strategy. And the sampling of the state, calculating and updating the control input are triggered by an event condition.

We define a sampling time sequence $\{t_i\}_{i=0}^{\infty}$ to record the sample points. The control input stays constant between two sample points,

$$u(t) = Kx(t_i), \ t \in [t_i, t_{i+1}) \tag{3}$$

Then the dynamics of the closed-loop system between t_i and t_{i+1} can be expressed by

$$\begin{aligned} \dot{x} &= Ax(t) + B_1 w(t) + B_2 Kx(t_i) \\ &= Ax(t) + B_1 w(t) + B_2 Kx(t) - B_2 Kx(t) + B_2 Kx(t_i) \\ &= (A + B_2 K)x(t) + B_1 w(t) + B_2 Ke(t) \end{aligned}$$

where the state error e is defined as

$$e(t) = x(t_i) - x(t), \ t \in (t_i, t_{i+1})$$

Thus we can rewrite system (1) as

$$\begin{cases} \dot{x} = (A + B_2K)x + B_1w + B_2Ke \\ z = (C + D_2K)x + D_1w + D_2Ke \end{cases} \tag{4}$$

The event condition triggering the sampling of state is set to be

$$\|e(t)\|^2 < \sigma\|x(t)\|^2, \ \sigma > 0. \tag{5}$$

This event condition is also used in [12], and it can be considered as a restriction of the relative error between the current state $x(t)$ and the last sampled state $x(t_i)$, which means if this error gets too large, we can no longer use the current control input $Kx(t_i)$ to control the plant.

Consider the appearance of the exogenous disturbance, we invoke L_2 stability to describe the stability of the closed-loop system. The main aim of this paper is to design an optimal controller K in sense of requiring the loosest event condition given the demanded maximum L_2 gain, or obtaining the minimum L_2 gain given the demanded event condition.

3 Design of the Controller and Event Condition

3.1 L_2 Stability Condition

In this section, we will investigate a sufficient condition in the form of matrix inequality guaranteeing the required L_2 stability of the system. The main result in this section is given in the following theorem.

Theorem 1 *Consider system* (4). *For arbitrarily given* $\alpha > 0$, *if there exist a positive definite symmetric matrix* $X \in \mathbb{R}^{n\times n}$, *a matrix* $K \in \mathbb{R}^{m\times n}$ *and a positive scalar* $\sigma > 0$, *satisfying the following inequality*

$$\begin{bmatrix} (A+B_2K)^T X + X(A+B_2K) + \alpha\sigma I & XB_2K & XB_1 & (C+D_2K)^T \\ K^T B_2^T X & -\alpha I & 0 & K^T D_2^T \\ B_1^T X & 0 & -\gamma^2 I & D_1^T \\ C + D_2K & D_2K & D_1 & -I \end{bmatrix} < 0 \tag{6}$$

Then we have, system (4) *under the event condition* (5) *is* L_2 *stable and the input-output* L_2 *gain is smaller than* γ, *i.e.*,

$$\|z\|_2 < \gamma \|w\|_2.$$

Proof Define $V(x) = x^T Xx$, where X is one of the positive definite solutions of (6), and set $w = 0$. We consider the time derivative of $V(x)$ along the solution of (4) within one triggering interval, i.e., within $t \in [t_i, t_{i+1})$

$$\begin{aligned}\dot{V}(t) &= 2x^T X\left(\left(A + B_2K\right)x + B_2Ke\right)\\ &= x^T\left(X\left(A+B_2K\right)+\left(A+B_2K\right)^T X\right)x+2(\frac{1}{\sqrt{\alpha}}x^T)XB_2K(\sqrt{\alpha}e)\\ &\le x^T\left(X\left(A+B_2K\right)+\left(A+B_2K\right)^T X+\frac{1}{\alpha}XB_2KK^TB_2^TX\right)x+\alpha e^Te\end{aligned}$$

Since we have $\|e(t)\|^2 < \sigma\|x(t)\|^2$, $\dot{V}$ can be reduced to

$$\dot{V} \le x^T(X\left(A+B_2K\right)+\left(A+B_2K\right)^T X+\frac{1}{\alpha}XB_2KK^TB_2^TX+\alpha\sigma I)x$$

We can see from (6) that

$$\begin{bmatrix}\left(A + B_2K\right)^T X + X\left(A + B_2K\right) + \alpha\sigma I & XB_2K\\ K^TB_2^TX & -\alpha I\end{bmatrix} < 0, \tag{7}$$

by the Schur complement formula. Furthermore (7) is equivalent to

$$X\left(A + B_2K\right) + \left(A + B_2K\right)^T X + \frac{1}{\sigma}XB_2KK^T B_2^TX + \alpha\sigma I < 0,$$

which indicates that $\dot{V}(t) < 0$ for every triggering interval $t \in [t_i, t_{i+1})$. Also notice that $w \in L_2[0, \infty)$, then we have

$$V(t) \to 0, \text{when } t \to \infty. \tag{8}$$

Then we define

$$J_T = \int_0^T \|z\|^2 \mathrm{d}t - \gamma^2 \int_0^T \|w\|^2 \mathrm{d}t,$$

and we have

$$\begin{aligned}J_T &= \int_0^T \left[z^Tz - \gamma^2w^Tw\right]\mathrm{d}t = \int_0^T \left[z^Tz - \gamma^2w^Tw + \dot{V}(x)\right]\mathrm{d}t - V(x(T))\\ &= \int_0^T \left[z^Tz-\gamma^2w^Tw+2x^TX\left(\left(A+B_2K\right)x+B_1w+B_2Ke\right)\right]\mathrm{d}t - V(x(T))\\ &< \int_0^T \begin{bmatrix}x\\ e\\ w\end{bmatrix}^T\left(\begin{bmatrix}\left(C + D_2K\right)^T\\ \left(D_2K\right)^T\\ D_1^T\end{bmatrix}\left[C + D_2K \;\; D_2K \;\; D_1\right]\right.\end{aligned}$$

$$+\begin{bmatrix} X(A+B_2K)+(A+B_2K)^T X & XB_2K & XB_1 \\ B_2^T X & 0 & 0 \\ B_1^T X & 0 & -\gamma^2 I \end{bmatrix}\Bigg)\begin{bmatrix} x \\ e \\ w \end{bmatrix} \mathrm{d}t + \int_0^T (\alpha\sigma x^T x - \alpha e^T e)\mathrm{d}t - V(x(T))$$

$$= \int_0^T \begin{bmatrix} x \\ e \\ w \end{bmatrix}^T \left(\begin{bmatrix} (C+D_2K)^T \\ (D_2K)^T \\ D_1^T \end{bmatrix} \begin{bmatrix} C+D_2K & D_2K & D_1 \end{bmatrix} \right.$$

$$\left. + \begin{bmatrix} X(A+B_2K)+(A+B_2K)^T X+\alpha\sigma I & XB_2K & XB_1 \\ B_2^T X & -\alpha I & 0 \\ B_1^T X & 0 & -\gamma^2 I \end{bmatrix} \right) \begin{bmatrix} x \\ e \\ w \end{bmatrix} \mathrm{d}t - V(x(T))$$

by Schur complement formula, (6) is equivalent to

$$\begin{bmatrix} (C+D_2K)^T \\ (D_2K)^T \\ D_1^T \end{bmatrix} \begin{bmatrix} C+D_2K & D_2K & D_1 \end{bmatrix} + \begin{bmatrix} X(A+B_2K)+(A+B_2K)^T X+\alpha\sigma I & XB_2K & XB_1 \\ B_2^T X & -\alpha I & 0 \\ B_1^T X & 0 & -\gamma^2 I \end{bmatrix} < 0$$

So we get

$$\int_0^T \left[z^T z - \gamma^2 w^T w + \dot{V}(x) \right] \mathrm{d}t < 0$$

that is,

$$V(T) + \int_0^T \|z\|^2 \mathrm{d}t < \gamma^2 \int_0^T \|w\|^2 \mathrm{d}t$$

Let $T \to \infty$, we have $\|z\|_2^2 < \gamma^2 \|w\|_2^2$, or equivalently, $\|z\|_2 < \gamma \|w\|_2$. □

3.2 Design of Controller and Event Condition

Theorem 1 provides a sufficient condition guaranteeing the finite-gain L_2 stability of system (4). In a traditional H_∞ control problem, it is important not only to be able to find out that the system is finite-gain L_2 stable, but also to calculate the L_2 gain or an upper bound on it. However, in our event-triggered control system, because of the existence of the state error e, the matrix inequality (6) is not a linear matrix inequality, for it contains nonlinear components about matrix K and symmetric matrix X. This means we cannot get the optimal controller directly by solving inequality (6). In the following we will make some transformations on (6) and turn it into a solvable LMI.

First, we pre- and post-multiply the left side of (6) by a diagonal matrix $diag(X^{-1}, X^{-1}, I, I)$, we get an equivalent form of (6)

$$\begin{bmatrix} X^{-1}(A+B_2K)^T+(A+B_2K)X^{-1}+\alpha\sigma X^{-1}X^{-1} & B_2KX^{-1} & B_1 & X^{-1}(C+D_2K)^T \\ X^{-1}K^TB_2^T & -\alpha X^{-1}X^{-1} & 0 & X^{-1}K^TD_2^T \\ B_1^T & 0 & -\gamma^2 I & D_1^T \\ (C+D_2K)X^{-1} & D_2KX^{-1} & D_1 & -I \end{bmatrix} < 0 \quad (9)$$

Notice that

$$-X^{-1}X^{-1} \leq -X^{-1} - X^{-1} + I \quad (10)$$

and apply the Schur complement formula to (9), we can see that (6) holds if

$$\begin{bmatrix} X^{-1}(A+B_2K)^T+(A+B_2K)X^{-1} & B_2KX^{-1} & B_1 & X^{-1}(C+D_2K)^T & X^{-1} \\ X^{-1}K^TB_2^T & -\alpha(2X^{-1}-I) & 0 & X^{-1}K^TD_2^T & 0 \\ B_1^T & 0 & -\gamma^2 I & D_1^T & 0 \\ (C+D_2K)X^{-1} & D_2KX^{-1} & D_1 & -I & 0 \\ X^{-1} & 0 & 0 & 0 & -\frac{1}{\alpha\sigma} \end{bmatrix} < 0 \quad (11)$$

holds, and (11) can be considered as an LMI by defining

$$P = X^{-1},\ Q = KX^{-1},\ \xi = \frac{1}{\sigma}.$$

Given the demanded upper bound of the L_2 gain γ, and set a proper value of α, the controller (3) can be designed by solving the following optimization problem

$$\min \xi \quad (12)$$

$$s.t. \begin{cases} \begin{bmatrix} AP+B_2Q+(AP+B_2Q)^T & B_2Q & B_1 & (CP+D_2Q)^T & P \\ Q^TB_2^T & -\alpha(2P-I) & 0 & Q^TD_2^T & 0 \\ B_1^T & 0 & -\gamma^2 I & D_1^T & 0 \\ CP+D_2Q & D_2Q & D_1 & -I & 0 \\ P & 0 & 0 & 0 & -\frac{1}{\alpha}\xi \end{bmatrix} < 0 \\ X > 0 \\ \xi > 0 \end{cases}$$

We denote the optimum solution of (12) as $(P_{opt}, Q_{opt}, \xi_{\min})$, and define

$$\sigma_{\max} = \frac{1}{\xi_{\min}}$$

then a suboptimal controller K_{opt} of the system is

$$K_{opt} = Q_{opt}P_{opt}^{-1} \quad (13)$$

and the corresponding event condition is

$$\|e(t)\|^2 < \sigma_{\max}\|x(t)\|^2 \quad (14)$$

Similarly, if given a demanded event condition σ, we could also achieve a suboptimal controller K_{opt} and an upper bound of the system's L_2 gain γ_{opt} by solving an optimization problem of (11).

Remark 1 In Theorem 1, the inequality condition (6) can always guarantee the L_2 stability of system (4) for an arbitrary value of α. And when solving the optimization problem (12) to design the controller, α can be set as any value.

Actually α can be considered as a parameter which can adjust the conservatism we bring in when we achieve the LMI condition in (12). This conservatism consists of conservatism brought in by inequality (6) and the process when we use inequality (10) to transform (6) into an LMI. Especially for the inequality (10), it is commonly used to transform matrix inequalities with nonlinear components into LMI, but it usually brings in much conservatism at the same time.

We denote these two parts of conservatism as

$$J_1(\alpha) = (\sqrt{\alpha\sigma}I - \frac{1}{\sqrt{\alpha}}K^T B_2^T X)^T (\sqrt{\alpha\sigma}I - \frac{1}{\sqrt{\alpha}}K^T B_2^T X)$$

$$J_2(\alpha) = \frac{1}{\alpha} X B_2 K((2X - XX)^{-1} - I)K^T B_2^T X$$

and we could see that these two parts are both affected by the value of α. That allows us to adjust the parameter α into a suitable value to minify the overall conservatism of the LMI in (12)

$$J(\alpha) = J_1(\alpha) + J_2(\alpha)$$

and let the optimization process (12) give the largest value of $\sigma_{\max}$. The detailed way of obtaining the best value of α will be demonstrated in the simulation section.

4 Simulation

In this section, an illustrative example is given to show the efficiency of the theoretical result.

We consider a second-order perturbed linear system

$$\begin{cases} \dot{x} = \begin{bmatrix} 5 & 10 \\ 1 & 0 \end{bmatrix} x + \begin{bmatrix} 1 \\ 1 \end{bmatrix} w + \begin{bmatrix} 0 \\ 1 \end{bmatrix} u \\ z = \begin{bmatrix} 1 & 0 \\ 0 & 1 \end{bmatrix} x \\ x(0) = [-1\ 2]^T \end{cases}$$

The external disturbance w is given as

$$w(t) = w = \sin(8\pi t),\ t \in [0.5, 1]$$

and the demanded upper bound of the L_2 gain of the controlled system is set to be $\gamma = 1$.

Giving α different values and solve the optimization problem (12), we get the relationship between the optimum solution $\bar{\sigma}_{\max}$ and α as shown in Fig. 1. From Fig. 1 we could get $\alpha = 45$ brings out the maximum value of $\bar{\sigma}_{\max} = 0.042$. And the corresponding optimal controller is

$$K_{opt} = [-81, -56] \tag{15}$$

This process shows how to find a proper value of α in order to get the largest possible $\bar{\sigma}_{\max}$. During this process we also find that the norm of the controller K obtained from the optimization problem (13) shows a similarly positive correlation with α, as shown in Fig. 2. This relationship can be used to choose the value of α if the system has control input constraints.

The dynamics of the closed loop and the sample time are shown in Figs. 3 and 4. We can see that the robust H_∞ controller (15) not only guarantees the trajectory of the closed-loop system converging rapidly, but also obtains a good capacity of disturbance attenuation.

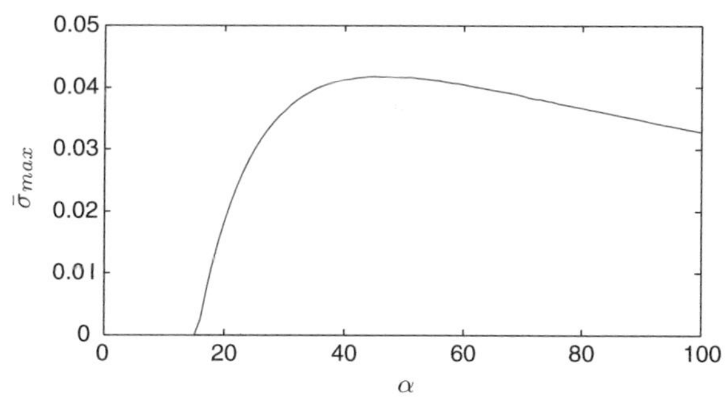

Fig. 1 Relationship between $\bar{\sigma}_{\max}$ and α

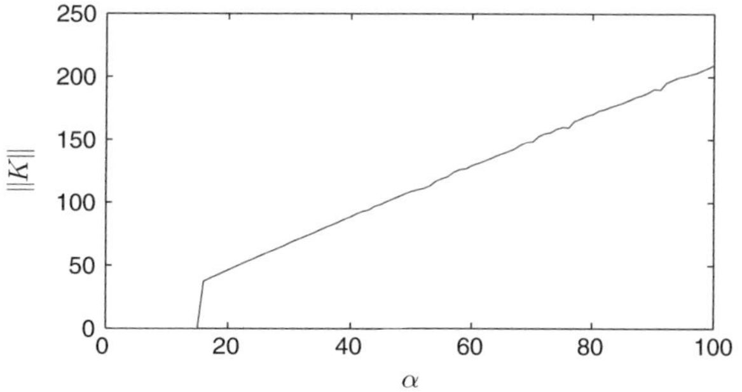

Fig. 2 Relationship between $\|K\|$ and α

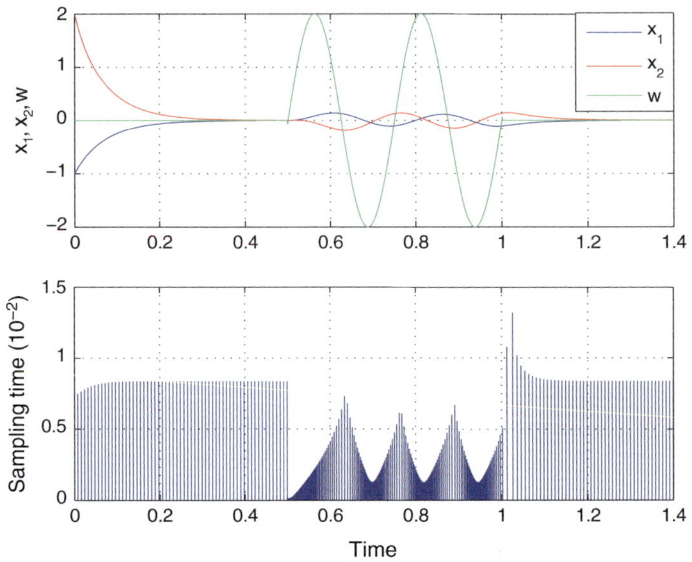

Fig. 3 The state responses of the system

Fig. 4 Disturbance attenuation of the system

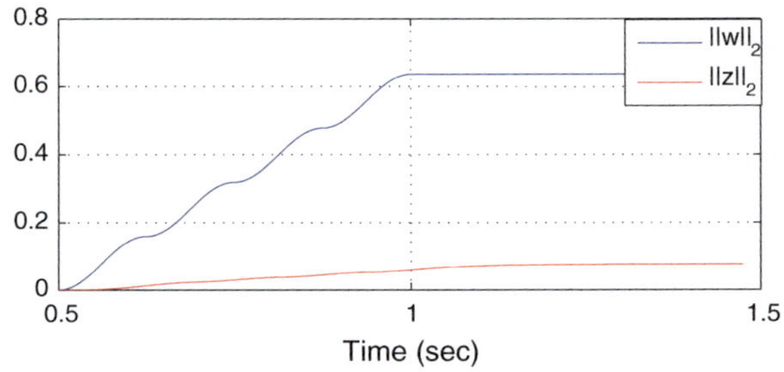

5 Conclusions and Future Work

In this paper, we introduce robust H_∞ theory into event-triggered control mechanism to reduce the resource utility implementing feedback. We investigate a state-feedback H_∞ control based on the event-triggered scheme for the perturbed linear system in a general form, a sufficient condition which guarantees the L_2 stability of the system is given in the form of Matrix Inequality. And the optimization of this inequality condition is settled by eliminating the nonlinear components, and an adjustable parameter is brought in to reduce the conservatism. The illustrative example shows that the closed-loop system has a good capability on disturbance attenuation, while the triggering frequency of the self-triggering scheme still has room for improvements.

Following this paper, we will improve the optimization program of the matrix inequality by applying some cone optimization methods for further reduction of the conservatism of the stability condition. And other absolute-error-based event condition will be studied instead of the relative-error-based event condition we apply in this paper, to improve the triggering efficiency of the event/self-triggered control system.

Acknowledgments This work was supported by the National Basic Research Program of China (973 Program: 2012CB821200, 2012CB821201), the NSFC (61134005, 61327807, 61521091, 61520106010) and the Fundamental Research Funds for the Central Universities (YWF-16-GjSYS-31, YWF-16-GJSYS-32).

References

1. Anta A, Tabuada P (2010) To sample or not to sample: self-triggered control for nonlinear systems. IEEE Trans Autom Control 55(9):2030–2042
2. Zhang X, Han Q (2014) Event-triggered dynamic output feedback control for networked control systems. IET Control Theor Appl 8(4):226–234
3. Hu S, Yin X, Zhang Y, Tian EG (2012) Event-triggered guaranteed cost control for uncertain discrete-time networked control systems with time-varying transmission delays. IET Control Theor Appl 6(18):2793–2804
4. Heemles WPMH, Sandee JH, Van Den Bosch PPJ (2008) Analysis of event-based controllers for linear systems. Int J Control 81(4):571–590
5. Lunze J, Lehmann D (2010) A state-feedback approach to event-based control. Automatica 46(1):211–215
6. Wang X, Lemmon M (2009) Self-triggered feedback control systems with finite-gain L_2 stability. IEEE Trans Autom Control 54(3):452–467
7. Wang X, Lemmon M (2010) Self-triggering under state-independent disturbances. IEEE Trans Autom Control 55(6):1494–1500
8. Fiter C, Hetel L, Perruquetti W, Richard J-P (2012) State-dependent sampling for perturbed time-delay systems. In: Proceedings 51st IEEE conference on decision and control, Maui, Hawaii, USA, December 2012, pp. 2358–2363
9. Almeida J, Silvestre C, Pascoal AM (2012) Observer based self-triggered control of linear plants with unknown disturbances. In: Proceedings of the (2012) American control conference Montréal, Canada 2012, pp 5688–5693
10. Tiberi U, Johansson KH (2012) A simple self-triggered sampler for nonlinear systems. In: IFAC Conference on Analysis and Design of Hybrid Systems, Eindhoven, the Netherlands, 2012, pp 76–81
11. Jiang H, Zhou H, Jia Y, Du J (2013) Event-triggered robust H_∞ control for linear systems with disturbance. In: Proceedings of 32nd Chinese control conference, China, 2013, pp 2102–2107
12. Tabuada P (2007) Event-triggered real-time scheduling of stabilizing control tasks. IEEE Trans Autom Control 52(9):1680–1685

Consensus of Linear Multi-agent Systems with Persistent Disturbances

Shaoyan Guo, Lipo Mo and Tingting Pan

Abstract This paper focuses on the consensus problem of continuous-time multi-agent systems with persistent disturbances. A distributed protocol is designed, which consists of two parts, one is the traditional control protocol, the other one is the estimation of disturbances. Then, using the method of matrix analysis, the sufficient conditions for achieving consensus of the closed-loop systems are found out. Finally, simulations are provided to demonstrate the effectiveness of the proposed algorithm.

Keywords Multi-agent system · Consensus · Disturbances · Control protocol

1 Introduction

Multi-agent systems have the characteristic of autonomy, distribution, and coordination, and have the ability of self-organization, learning, and reasoning. Multi-agent systems are efficient to deal with the practical systems, such as the formation flight of the UAV, multi-robot systems, and so on [1, 2]. More and more attentions have been paid on cooperative control of multi-agent systems in recent years.

The consensus problem of multi-agent systems is one of the most fundamental issues. Starting from the Viçsek model [3], a broad spectrum of scholars are much more kindly to study the consensus problems of multi-agent [6] systems with different characteristics. For example, the consensus problems of discrete-time were investigated in [4, 5]. For the continuous-time multi-agent systems, consensus problems were discussed in [6, 7]. It is shown that the consensus of first-order systems can be achieved if and only if the network topology contains a directed spanning tree. And then these results were extended to stochastic switching systems [6], some average consensus conditions were obtained. All of these results were given for the first-order multi-agent systems. In practical systems, the control objects may be accelerated velocity rather than velocity and the methods can not be applied to second-order

S. Guo · L. Mo (✉) · T. Pan
School of Science, Beijing Technology and Business University, Beijing 100048, China
e-mail: molipo@th.btbu.edu.cn

Y. Jia et al. (eds.), *Proceedings of 2016 Chinese Intelligent Systems Conference*, Lecture Notes in Electrical Engineering 405,
DOI 10.1007/978-981-10-2335-4_10

systems straightforward, so it is meaningful to investigate the consensus problems of second-order multi-agent systems. In [2], it shows that the second-order systems might not achieve consensus even if the network topology has a directed spanning tree. And a necessary and sufficient condition was given for the consensus of second-order systems with directed topologies. Recently, the consensus problems of linear multi-agent systems were also considered. In [8], it was proved that the consensus can be reached if and only if all of the nonzero eigenvalues of the Laplacian matrix lie in the stable regions.

In practical systems, it is inevitable that the system can be affected by external disturbances, so it is important to discuss the consensus problem of the multi-agent systems under disturbance. In [9–11], the H_∞ is used to solve the consensus problem under disturbance. To attenuate the communication noises, a distributed stochastic approximation type protocol is also adapted. Using probability limit theory and algebraic graph theory, consensus conditions for this kind of protocols are obtained [12]. In [13], a new controller is proposed to solve the consensus problem of the multi-agent systems under unknown persistent disturbances. In [14], The stochastic consensus problem of linear multi-input multi-output (MIMO) multi-agent systems (MASs) with communication noises and Markovian switching topologies is studied by designing consensus protocol. In [15], the consensus problem of second-order discrete-time multi-agent systems with white noise disturbance under Markovian switching topologies is discussed. And for more consensus problems of the multi-agent systems under disturbance, refer [16–18]. However, to the best of our knowledge, the consensus problem of the linear multi-agent systems with constant persistent disturbances have not been discussed, this paper we focus on this problem. The stochastic consensus problem of linear multi-input multi-output (MIMO) multi-agent systems (MASs) with communication noises and Markovian switching topologies

The main contribution of this paper is that sufficient conditions were obtained for the consensus of linear multi-agent systems with persistent disturbances. Based on the graph theory and matrix theory, the consensus protocol was designed and the consensus state was also obtained. Comparing with the literature, the result herein is more simple and general, and it is easy to verify in practical engineering systems.

2 Preliminaries

An undirected graph $\mathcal{G}$ is defined by a set $V_{\mathcal{G}} = \{1, \ldots, N\}$ of nodes and a set $E = \mathcal{E}_{\mathcal{G}} \times \mathcal{E}_{\mathcal{G}}$ of edges. If $(i,j) \in \mathcal{E}_{\mathcal{G}}$, then the node i and j are neighbors and the neighboring relation is indicated with $i \sim j$. The neighborhood $N_i \subseteq V$ is denoted the set $\{v_j \in V | (i,j) \in E\}$, then the degree of a node is given by the number of its neighbors. Let d_i be the degree of node i, then the degree matrix of a graph $\mathcal{G}$, $\mathcal{D} \in \mathbb{R}^{n\times n}$, is given by $\mathcal{D} = diag\{d_1, d_2, \ldots, d_N\}$, the adjacency matrix of a graph $\mathcal{G}$, $\mathcal{A} \in R^{n\times n}$, is given by $\mathcal{A} = [a_{ij}]$, if $(i,j) \in \mathcal{E}_{\mathcal{G}}, a_{ij} = 1$, otherwise $a_{ij} = 0$. And the Laplacian matrix is given by $\mathcal{L} = \mathcal{D} - \mathcal{A}$. By the definition of Laplace matrix, we can obtain the spectrum of the Laplacian matrix for a connected, undirected graph can be ordered as $0 = \lambda_1 < \lambda_2 \leq \cdots \leq \lambda_N$. And 1_N is the eigenvector belongs to the zero eigenvalue λ_1, and $\mathcal{L}1_N = 0_N$ where 1_N denote the $N \times 1$ vector of all ones.

Lemma 1 *Let A, B, C, D are constant matrices with proper dimensions. Then*
$A \otimes (B + C) = A \otimes B + A \otimes C,$
$(A + B) \otimes (C + D) = A \otimes C + B \otimes D,$
where $\otimes$ represents the Kronecker product.

Lemma 2 *[19] For partitioned matrix $X = \begin{pmatrix} X_{11} & X_{12} \\ X_{12}^T & X_{22} \end{pmatrix}$, the following inequalities are equivalent:*

(a) $X > 0$;

(b) $X_{11} - X_{12}X_{22}^{-1}X_{12}^T > 0$ *and* $X_{22} > 0$;

(c) $X_{22} - X_{12}^T X_{11}^{-1} X_{12} > 0$ *and* $X_{11} > 0$.

Lemma 3 *Consider two symmetric matrices $A \in \mathbb{R}^{n\times n}$ and $B \in \mathbb{R}^{n\times n}$. If all eigenvalues of A are no more than 0, and all eigenvalues of B are less than 0, then all eigenvalues of $A + B$ are less than 0.*

Proof Because all eigenvalues of A are no more than 0 and all eigenvalues of B are less than 0, there exists a nonzero vector $x = (x_1, x_2, \ldots, x_n)$, such that $x^T Ax \leq 0, x^T Bx < 0$. Then

$$x^T Ax + x^T Bx = x^T (A + B)x < 0,$$

so all eigenvalues of $A + B$ are less than 0.

3 System Model

Consider the multi-agent systems consisting of N agents. The dynamic of i-th agent is represented by

$$\dot{x}_i(t) = Ax_i(t) + B[u_i(t) + w_i], \qquad x_i(0) = x_{i0}, \qquad i = 1, 2, \ldots, N, \tag{1}$$

where $x_i(t) \in \mathbb{R}^n$, $u_i(t) \in \mathbb{R}^q$, $w_i \in \mathbb{R}^q$ represent the state, control input, and persistent disturbances of i-th agent, respectively, $A \in \mathbb{R}^{n\times n}, B \in \mathbb{R}^{n\times q}$ are system matrices. To discuss the consensus problem of the multi-agent system (1), we propose the following control protocol for agent i

$$u_i(t) = K \sum_{j\in N_i} a_{ij}(x_j(t) - x_i(t)) - \hat{w}_i(t), \tag{2}$$

where $K \in \mathbb{R}^{q\times n}$ is the control gain, and $\hat{w}_i(t) \in \mathbb{R}^q$ is the estimation of w_i, the dynamic equations of which are as follows:

$$\dot{\hat{w}}_i = F \sum_{j\in N_i} a_{ij}[(x_i(t) - x_j(t)) - (\hat{x}_i(t) - \hat{x}_j(t))], \tag{3}$$

where $F \in \mathbb{R}^{q\times n}$ is a constant matrix which will be determined, $\hat{x}_i(t) \in \mathbb{R}^n$ is the estimation of the state of the agent i, the dynamic equations of which are as follows:

$$\dot{\hat{x}}_i(t) = Ax_i(t) + BK \sum_{j\in N_i} a_{ij}[\hat{x}_j(t) - \hat{x}_i(t)] + M \sum_{i=1}^{N} a_{ij}[(\hat{x}_j(t) - \hat{x}_i(t) - (x_j(t) - x_i(t))], \tag{4}$$

where $M \in \mathbb{R}^{n\times q}$ is also a constant matrix which will be determined. Under the control protocol (2), system (1) can be rewritten as

$$\dot{x}_i(t) = Ax_i(t) + B[K \sum_{j\in N_i} a_{ij}(x_j(t) - x_i(t) - \hat{w}_i(t) + w_i]. \tag{5}$$

Then the consensus problem of system (1) can be transferred into the stability problem of system (5).

4 Stability Analysis

Let $\tilde{x}(t) = x(t) - \hat{x}(t)$ and $\tilde{w}(t) = \hat{w}(t) - w$ be the state estimation error and the disturbance estimation error, respectively. According to (4) and (5), we can get

$$\dot{\tilde{x}}(t) = (I_N \otimes A - \mathcal{L} \otimes BK + \mathcal{L} \otimes M)\tilde{x}(t) - (I_N \otimes B)\tilde{w}(t), \tag{6}$$

$$\tilde{w}(t) = \hat{w}(t) - w, \tag{7}$$

Denote $e(t) = [x^T(t), \tilde{x}^T(t), \tilde{w}^T(t)]^T$, then according to (5), (6) and (7), we can get

$$\dot{e}(t) = A_o e(t), \tag{8}$$

where

$$A_0 = \begin{pmatrix} I_N \otimes A - \mathcal{L} \otimes BK & 0 & -I_N \otimes B \\ 0 & I_N \otimes A - \mathcal{L} \otimes BK + \mathcal{L} \otimes M & -I_N \otimes B \\ 0 & \mathcal{L} \otimes F & 0 \end{pmatrix}.$$

Then the consensus problem of the system (1) transfers the stability problem of the system (8). The system matrix A_0 plays an important role in the stability analysis. Now we analyze this matrix.

For the Laplacian matrix of undirected graph, there exists a matrix so that

$$U^T \mathcal{L} U = \mathrm{diag}(\lambda_1, \cdots, \lambda_N), \tag{9}$$

where $\lambda_i, i = 1, 2, \ldots, N$, are the eigenvalues of $\mathcal{L}$. let $\bar{e}(t) = U_0 e(t)$, where $U_0 = \begin{pmatrix} U \otimes I_n & & \\ & U \otimes I_n & \\ & & U \otimes I_q \end{pmatrix}$. By the orthogonal transformation, we can obtain $\dot{\bar{e}}(t) = A_1 \bar{e}(t)$, where

$$A_1 = \begin{pmatrix} I_N \otimes A - U^T \mathcal{L} U \otimes BK & 0 & -I_N \otimes B \\ 0 & I_N \otimes A - \mathcal{L} \otimes BK + U^T \mathcal{L} U \otimes M & -I_N \otimes B \\ 0 & U^T \mathcal{L} U \otimes F & 0 \end{pmatrix}.$$

Since the eigenvalues of a matrix are not affected by exchanging the row and corresponding column of a matrix simultaneously, A_1 can be transferred to a block diagonal, $\bar{A}_1 = diag(A_{11}, \ldots, A_{1N})$

where $A_{1i} = \begin{pmatrix} A - \lambda_i BK & 0 & -B \\ 0 & A + \lambda_i M - \lambda_i BK & -B \\ 0 & \lambda_i F & 0 \end{pmatrix}$, $i = 1, 2, \ldots, N$.

Theorem 1 *Consider system (1), the control protocol solves the consensus problem if there exist a positive-definite matrix P, and* $\mu_1 > 0, \mu_2 > 0, \mu_3 > 0$, *such that*

$$A^T P_1 + P_1 A - 2P_1 BB^T P_1 < 0, \tag{10}$$

$$B^T P_2 + P_2 B > 0, \tag{11}$$

$$-2\lambda_{max}\mu_1 I_N + 3P_1 B[B^T P_2 + P_2 B]^{-1} B^T P_1 < 0, \tag{12}$$

$$-2\lambda_{max}\mu_2 I_N + 3A^T P_2[B^T P_2 + P_2 B]^{-1} P_2^T A < 0, \tag{13}$$

$$-2\lambda_{max}\mu_3 I_N + 3\lambda_{min}^2 P_1 BB^T P_2[B^T P_2 + P_2^T B] P_2 BB^T P_1 < 0, \tag{14}$$

where $\mu_1 + \mu_2 + \mu_3 = \mu$, $K = \tau B^T P_1, \tau = 1$, $\begin{pmatrix} M \\ F \end{pmatrix} = -\mu P^{-1} \begin{pmatrix} I_n \\ 0 \end{pmatrix}$ *and* $P = \begin{pmatrix} P_1 & P_2 \\ P_2^T & P_3 \end{pmatrix}$, P_1, P_2, P_3 *have appropriate dimensions.*

Proof Noted the form of A_{1i}, we analyze the two block matrices $E = \begin{pmatrix} A - \lambda_i BK \end{pmatrix}$ and $G = \begin{pmatrix} A + \lambda_i M - \lambda_i BK & -B \\ \lambda_i F & 0 \end{pmatrix}$.

For matrix E, by taking $K = \iota B^T P_1$ $\iota > (\frac{1}{\lambda_i})$, then $(A - \lambda_i BK)^T P_1 + P_1(A - \lambda_i BK) = A^T P_1 + P_1 A - 2\lambda_i \iota P_1 BB^T P_1 < A^T P_1 + P_1 A - 2P_1 BB^T P_1 < 0$, so $A - \lambda_i BK$ is Hurwitz stable.

For matrix G, let

$$G = \begin{pmatrix} A - \lambda_i BK & 0 \\ 0 & 0 \end{pmatrix} + \begin{pmatrix} 0 & -B \\ 0 & 0 \end{pmatrix} + \begin{pmatrix} \lambda_i M & 0 \\ \lambda_i F & 0 \end{pmatrix}.$$

By taking $\begin{pmatrix} M \\ F \end{pmatrix} = -\mu P^{-1} \begin{pmatrix} I_n \\ 0 \end{pmatrix}$, we can obtain

$$\begin{pmatrix} \lambda_i M & 0 \\ \lambda_i F & 0 \end{pmatrix}^T P + P \begin{pmatrix} \lambda_i M & 0 \\ \lambda_i F & 0 \end{pmatrix} = \begin{pmatrix} -2\lambda_i \mu I_n & 0 \\ 0 & 0 \end{pmatrix}, \tag{15}$$

then

$$\begin{pmatrix} 0 & -B \\ 0 & 0 \end{pmatrix}^T P + \begin{pmatrix} 0 & -B \\ 0 & 0 \end{pmatrix} P = \begin{pmatrix} 0 & -P_1 B \\ -B^T P_1 & -B^T P_2 - P_2^T B \end{pmatrix}, \tag{16}$$

and

$$\begin{aligned} &\begin{pmatrix} A - \lambda_i BK & 0 \\ 0 & 0 \end{pmatrix}^T P + P \begin{pmatrix} A - \lambda_i BK & 0 \\ 0 & 0 \end{pmatrix} \\ &= \begin{pmatrix} A^T P_1 + P_1 A - 2\lambda_i \tau P_1 BB^T P_1 & A^T P_2 - \lambda_i \tau P_1 BB^T P_2 \\ P_2^T A - \lambda_i \tau P_2^T BB^T P_1 & 0 \end{pmatrix}. \end{aligned}$$

Denote $\mu = \mu_1 + \mu_2 + \mu_3$ and make a sum of the three matrices, we can get

$$A_{1i}^T P + P A_{1i} = M_1 + M_2 + M_3 + M_4, \tag{17}$$

where

$$M_1 = \begin{pmatrix} A^T P_1 + P_1 A - 2\lambda_i \tau P_1 BB^T P_1 & 0 \\ 0 & 0 \end{pmatrix},$$

$$M_2 = \begin{pmatrix} -2\lambda_i \mu_1 I_N & A^T P_2 \\ P_2^T A & -\frac{1}{3}(B^T P_2 + P_2 B) \end{pmatrix},$$

and

$$M_3 = \begin{pmatrix} -2\lambda_i \mu_2 I_N & -P_1 B \\ -B^T P_1 & -\frac{1}{3}(B^T P_2 + P_2 B) \end{pmatrix},$$

$$M_4 = \begin{pmatrix} -2\lambda_i \mu_3 I_N & -\lambda_i \tau P_1 BB^T P_2 \\ -\lambda_i \tau P_2^T BB^T P_1 & -\frac{1}{3}(B^T P_2 + P_2 B) \end{pmatrix},$$

Since(10)–(14) hold, according to Lemma 2, $M_1 \leq 0$ $M_2 < 0$, $M_3 < 0$, $M_4 < 0$, then according to Lemma 3, we have

$$G^T P + PG = M_1 + M_2 + M_3 + M_4 < 0.$$

So matrix $A_{1i}, i = 2, 3, \ldots, N$ is Hurwitz stable. According to *Theorem* 1 the consensus problem can be solved.

Remark 1 According to Theorem 1, we not only solved the consensus problem but also got the consensus state. Since $A_{1i}, i = 2, 3, \ldots, N$ are Hurwitz stable, $\tilde{x}_i, i = 2, 3, \ldots, N$ are asymptotically stable. Now we consider the first block of $\bar{A}_1$, since $\lambda = 0$, we can get

$$\begin{pmatrix} \dot{\bar{x}}_1(t) \\ \dot{\bar{\bar{x}}}_1(t) \\ \dot{\bar{\bar{w}}}_1(t) \end{pmatrix} = \begin{pmatrix} A & 0 & -B \\ 0 & A & -B \\ 0 & 0 & 0 \end{pmatrix} \begin{pmatrix} \bar{x}_1(t) \\ \bar{\bar{x}}_1(t) \\ \bar{\bar{w}}_1(t) \end{pmatrix}, \tag{18}$$

By solving the differential equations, we obtain

$$\bar{\bar{w}}_1 = \bar{\bar{w}}_1(t_0),$$

$$lim_{t\to\infty}\bar{x}_1(t) = e^{At}\bar{x}_1(t_0) + \int_0^\infty e^{A(t-\tau)}B\bar{\bar{w}}(t_0)dt, \tag{19}$$

denote $U = \begin{pmatrix} 1_N & U_1, \end{pmatrix}$, and according to the non singular transformation, we can get

$$x(t) = \{\begin{pmatrix} 1_N & U_1 \end{pmatrix} \otimes I_N\}\bar{x}(t) = \frac{1}{\sqrt{N}}(1_N \otimes I_N)\bar{x}_1(t), \tag{20}$$

So

$$x_1(t_0) = \frac{1}{\sqrt{N}}(1_N^T \otimes I_N)x(t_0) = \frac{1}{\sqrt{N}}\sum_{i=1}^{N} x_i(t_0), \tag{21}$$

Remark 2 Comparing with [14], the result we get in this paper are more simple and more general. For system (8) we just use the Lyapunov stability criterion to get the result, the generality of the result make it more meaningful.

Fig. 1 Network topology

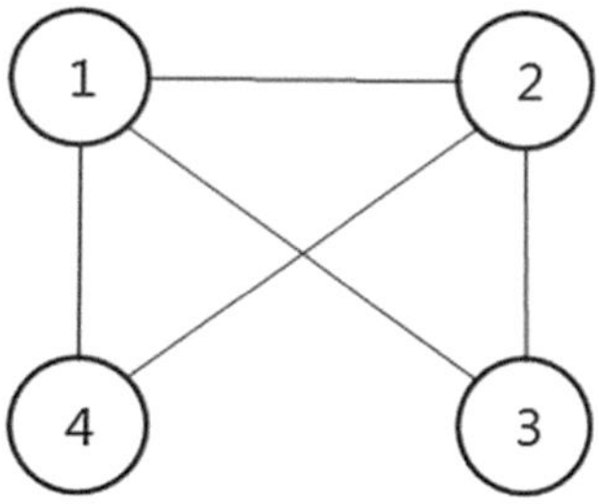

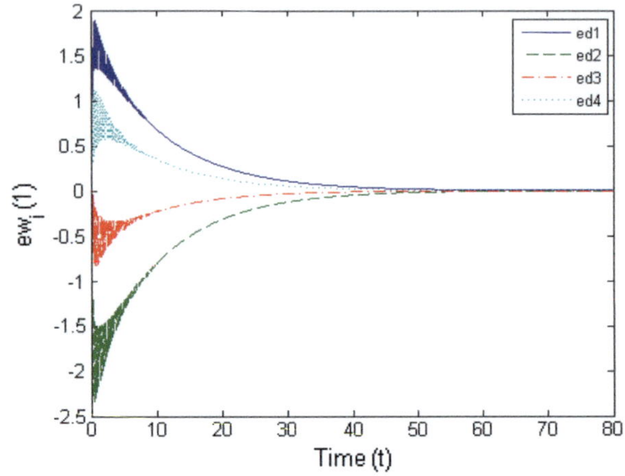

Fig. 2 The error of states' estimation

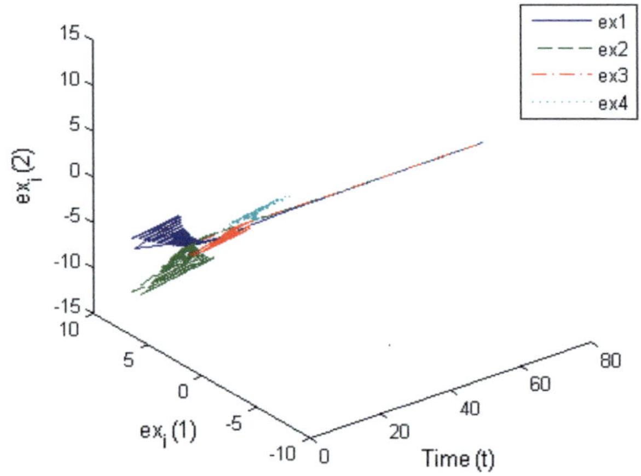

Fig. 3 The error of disturbance' estimation

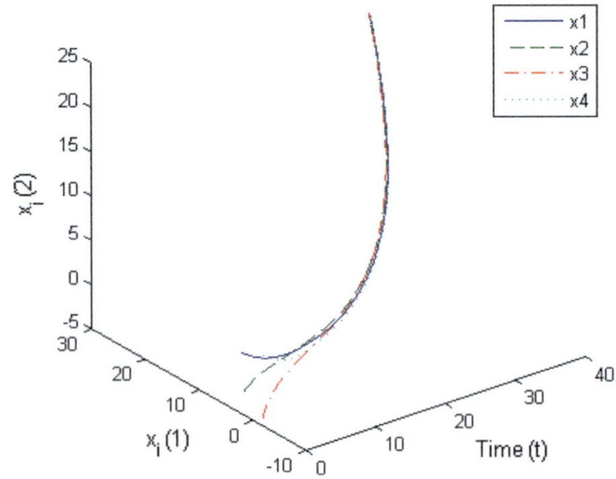

Fig. 4 The states of agents

5 Simulation Example

In this section, a simulation example is provided to validate the effectiveness of our algorithm. Consider a network of four agents, the system matrices are $A = \begin{pmatrix} 0 & 1 \\ 1 & 0 \end{pmatrix}$, $B = \begin{pmatrix} 1 \\ 1 \end{pmatrix}$, and the topological structure is shown in Fig. 1.

So the Laplacian matrix can be determined as

$$L = \begin{pmatrix} 3 & -1 & -1 & -1 \\ -1 & 3 & -1 & -1 \\ -1 & -1 & 2 & 0 \\ -1 & -1 & 0 & 2 \end{pmatrix}, \text{ with eigenvalues } 0, 2, 4, 4.$$

For simplicity,we choose $\tau = 1$, and we get the solution P, M, F as follows

$$P = \begin{pmatrix} 0.5 & 0.3 & 0.2 \\ 0.3 & 0.2 & 0.2 \\ 0.2 & 0.2 & 0.4 \end{pmatrix}, M = \begin{pmatrix} -26.25 & 17.5 \\ 17.5 & 8.75 \end{pmatrix}, F = \begin{pmatrix} 0.8 & 0.7 \end{pmatrix}.$$

The error of states' and disturbances' estimation are shown in Figs. 2 and 3 respectively, and the consensus state is shown in Fig. 4.

6 Conclusions

In this paper, we addressed the consensus problems of multi-agent systems when dynamics of agents are perturbed by constant persistent disturbances. We derived a sufficient condition for achieving consensus of multi-agent system with constant persistent disturbances. Specifically, it is shown that the consensus state converges to the mean states of all agents [20]. The future work will focus on the consensus problems of high-order systems with Markov switching topologies.

Acknowledgments This work was supported by National Natural Science Foundation of China (61304155) and Beijing Municipal Government Foundation for Talents (2012D005 003000005).

References

1. Fax JA, Murray RM (2004) Information flow and coorperative control of vehicle formations. IEEE Trans Autom Control 49(4):1465–1476
2. Ren W, Atkins E (2007) Distributed multi-vehicle coordinated control via local information exchange. Int J Robust Nonlinear Control 17(10–11):1002–1033
3. Vicsek T, Czirok A, Benjacob E (1995) Novel type of phase transition in a system of self-driven particles. Phys Rev Lett 75(6):1226–1229
4. Czirok A, Barabasi AL, Vicsek T (1997) Collective motion of self-propelled particles: kinetic phase transition in one dimension. J Mental Health Training Educ Pract 82(1):15–22

5. Jadbabaie A, Lin J, Morse AS (2003) Coordination of groups of mobile autonomous agents using nearest neighbor rules. IEEE Trans Autom Control 48(6):985–1001
6. Olfati-Saber R, Murray RM (2004) Consensus problems in networks of agents with switching topology and time-delays. IEEE Trans Autom Control 49(9):1520–1533
7. Ren W, Beard RW (2005) Consensus seeking in multiagent systems under dynamically changing interaction topologies. IEEE Trans Autom Control 50(5):655–661
8. Yu W, Chen G, Cao M (2010) Some necessary and suffcient conditions for second-order consensus in multi-agent dynamical systems. Automatica 46(6):1089–1095
9. Yu W, Chen G, Ren W (2011) Distributed higher order consensus protocols in multi-agent dynamical systems. Circuits Syst I Regular Pap IEEE Trans 58(8):1924–1932
10. Liu Y, Jia Y (2010) Consensus problem of high-order multi-agent systems with external disturbances: an H analysis approach. Int J Robust Nonlinear Control 20(14):1579–1953
11. Mo Lp, Zhou Yj, Zhou Hb (2012) L-two-L-infinity consensus in uncertain high-order multi-agent systems with a leader. Control Theor Appl 29(9):1125–1131
12. Lin P (2008) JIa Y, Li L Distributed robust H-infinity consensus control in directed networks of agents with time-delay. Syst Control Lett 57(8):643–653
13. Li T, Zhang JF (2010) Consensus conditions of multi-agent systems with time-varying topologies and stochastic communication noises. IEEE Trans Autom Control 55(9):2043–2057
14. Yucelen T, Egerstedt M (2012) Control of multi-agent systems under persistent disturbances. In: Proceedings of the American control conference, pp 5264–5269
15. Wang Y, Cheng L, Ren W, Hou ZG (2015) Seeking consensus in networks of linear agents: communication noises and markovian switching topologies. IEEE Trans Autom Control 60(5):1374–1379
16. Pan Tt, Mo Lp, Cao Xb (2015) Consensus of discrete-time multi-agent systems with white noise disturbance. IFAC-PapersOnLine 48(28):202–205
17. Chen Y, Dong H, Lu J, Sun X (2015) Robust consensus of nonlinear multiagent systems with switching topology and bounded noises. IEEEXplore 2(1)
18. Cao W, Zhang J, Ren W (2015) LeaderCfollower consensus of linear multi-agent systems with unknown external disturbances. Syst Control Lett 82:64–70
19. Mei SW, Shen TL, Liu KZ (2008) Modern robust control theory and application. Tsinghua University press, p 89
20. Godsil C, Royle GF (2001) Algebraic graph theory. Springer
21. Yang H, Zou H, Liu H, Han F, Zhang S (2015) Consensus of fractional multi-agent systems with disturbances. World Congr Intell Control Autom 2015:1154–1158

Cluster Synchronization in Complex Dynamical Networks with Linear Generalized Synchronization in Each Community

Yan Liu, Zhengquan Yang, Jiezhi Wang, Qing Zhang and Zengqiang Chen

Abstract This paper concentrates on the cluster synchronization such that each community achieves linear-generalized synchronization in complex networks. Here two control schemes are introduced to realize cluster synchronization. One is to adjust the coupling strength of edges which connect different groups automatically, and to use the pinning control simultaneously. The other scheme is to adjust the coupling strength of edges in each group and to select only one node in each group to control. Finally, we present some simulations to demonstrate the correction of our conclusion.

Keywords Complex networks · Cluster synchronization · Linear generalized synchronization · Adaptive control

1 Introduction

In recent years, complex networks have drawn more attention in virtue of its potential application in various fields [1–4], such as social networks, food webs and communication networks. Many research works were interested in global synchronization, complete synchronization, lag synchronization, and projective synchronization [5–10]. Meanwhile, there are many control methods have been used to achieve synchronization in complex networks, such as pinning control [11–13], adaptive control [13] and impulsive control [14, 15]. For more literature focusing on the synchronization problem we refer the reader(s) to [16–18]. Cluster synchronization is a particular synchronization phenomena which has been studied to be more momentous in brain science [19] and communication engineering [20]. This phe-

Y. Liu (✉) · Z. Yang · J. Wang · Q. Zhang · Z. Chen
College of Science, Civil Aviation University of China, Tianjin 300300, China
e-mail: yanliucauc@163.com

Z. Chen
College of Computer and Control Engineering, Nankai University,
Tianjin 300071, China

Y. Jia et al. (eds.), *Proceedings of 2016 Chinese Intelligent Systems Conference*, Lecture Notes in Electrical Engineering 405,
DOI 10.1007/978-981-10-2335-4_11

nomena indicates that nodes synchronize with each other in each community, but the nodes in distinct communities have no synchronization. [21] considered cluster synchronization in connected chaotic networks. Cluster synchronization of linearly coupled complex network has been studied under pinning control in [22]. In fact, a great amount of complex networks contain distinct types of nodes [23] in real life. In [24], by the adaptive scheme adjusting each edge only according to its local information, then networks satisfy cluster synchronization. Wu and Lu [25, 26] proposed an control scheme with varying the coupling strength of every node only due to its neighborhood information. Lately, some research works have considered some particular cases in cluster synchronization. For instance, cluster mixed synchronization and cluster projective synchronization are proposed in [27, 28], respectively. In [29], consider cluster synchronization in networks with time-varying and distributed delay coupling with adaptive control.

Generalized synchronization (GS) means that stats of the response system synchronize that of the drive system through a nonlinear smooth functional mapping. Linear GS which is a special kind of GS has been studied in [9, 10].

In this paper, we study cluster synchronization such that each groups achieves linear GS respectively. To reach the aim, two useful coupling schemes are introduced by adjusting the coupling strength of some edges and selecting some nodes in each groups to control.

This paper is outlined as follows. In Sect. 2, here are some preliminaries introduced. Section 3 presents two effective control methods which need both the adaptive control and pinning control. In Sect. 4, we give a few examples to indicate the validness of our presented controllers. The conclusion is given in last section.

2 Preliminaries

This paper discusses a complex network which has l communities and N dynamical nodes. If node i belongs to the kth community, then we denote $\phi_i = k$.

Now, we describe a controlled complex network

$$\dot{x}_i(t) = f_{\phi_i}(x_i(t)) + \varepsilon \sum_{j=1}^{N} a_{ij}(t)x_j(t) + u_i(t), \quad i = 1, 2, \ldots, N, \tag{1}$$

where $x_i(t) = (x_{i1}(t), \ldots, x_{in}(t))^T \in R^n$ is the state of the ith node, $f_{\phi_i} : R^n \to R^n$ is the local dynamics of nodes in the ϕ_ith community. For any nodes i and j, if $\phi_i \neq \phi_j$, i.e., nodes i and j are in different communities, then $f_{\phi_i} \neq f_{\phi_j}$. ε is the coupling strength and $A = (a_{ij})_{N\times N}$ is the coupling configuration matrix, if i and j connect, then $a_{ij} = a_{ji} > 0$; otherwise, $a_{ij} = 0$; $a_{ii} = -\sum_{j=1, j\neq i}^{N} a_{ij}$, $i = 1, 2, \ldots, N$.

Definition 1 Assume that $s_{\phi_i}(t) \in R^n$ is any smooth dynamics for $i = 1, 2, \ldots, N$. The controlled network (1) is said to be a cluster synchronization if

$$\lim_{t\to\infty} \|x_i(t) - s_{\phi_i}(t)\| = 0, \quad i = 1, 2, \ldots, N,$$

and

$$\lim_{t\to\infty} \|x_i(t) - x_j(t)\| \neq 0, \quad \phi_i \neq \phi_j,$$

where $s_1(t), \ldots, s_l(t)$ are distinct matrices.

Define the error variables as $e_i(t) = x_i(t) - s_{\phi_i}(t)$ for $i = 1, 2, \ldots, N$. Let $s(t) \in R^n$ be any smooth dynamic, and $s(t)$ may be a chaotic orbit, a periodic orbit, or even an equilibrium point. This paper discusses

$$s_k(t) = P_k s(t) + Q_k, \qquad k = 1, 2, \ldots, l, \tag{2}$$

where P_k and Q_k are matrices such that $s_1(t), \ldots, s_l(t)$ are different matrices. Hence

$$\lim_{t\to\infty} \|x_i(t) - s_{\phi_i}(t)\| = \lim_{t\to\infty} \|x_i(t) - (P_{\phi_i} s(t) + Q_{\phi_i})\| = 0 \tag{3}$$

i.e., each community achieves linear GS, respectively.

Assumption 1 For $g(z)$, there is a positive constant L such that

$$\|g(z_1) - g(z_2)\| \leq L\|z_1 - z_2\|$$

for any z_1, z_2.

3 Main Results

In the following, we present a scheme to make the network (1) and the dynamic (2) realize cluster synchronization. By the definition of coupling matrix A, we know that

$$\varepsilon \sum_{j=1}^{N} a_{ij}(t) s_{\phi_j}(t) = 0, \quad i \in V_{\phi_i} \backslash \bar{V}_{\phi_i}, \tag{4}$$

where V_{ϕ_i} represent all the nodes in the ϕ_ith group and $\bar{V}_{\phi_i}$ which is the subset of V_{ϕ_i} consists of nodes with at least a neighbor in other groups. E represents all the edges in networks, and E_1 only contains the edges whose ends are in different communities. Due to the above conditions, a controller is introduced as

$$u_i = \begin{cases} \dot{s}_{\phi_i} - f_{\phi_i}(s_{\phi_i}) - \varepsilon \sum_{j=1}^{N} a_{ij} s_{\phi_j} - g_i e_i, \; i \in \bar{V}_{\phi_i} \\ \dot{s}_{\phi_i} - f_{\phi_i}(s_{\phi_i}), \qquad i \in V_{\phi_i} \backslash \bar{V}_{\phi_i} \end{cases} \tag{5}$$

$$\dot{a}_{ij} = d \sum_{(i,j)\in E_1} \|e_i - e_j\|, \tag{6}$$

$$\dot{g}_i = k_i \|e_i\|^2, \quad i \in \bar{V}_{\phi_i}, \tag{7}$$

where the constants d and k_i are positive and k_i is the control gain. In fact, in order to reach cluster synchronization, it need not only adjust the coupling strength of edges which connect different groups but also to control those nodes which have direct connections to other groups.

Thus, because of the above control law, the error variable can be given as

$$\dot{e}_i = \begin{cases} f_{\phi_i}(x_i) - f_{\phi_i}(s_{\phi_i}) - \varepsilon \sum_{j=1}^{N} a_{ij}e_j - g_i e_i, \; i \in \bar{V}_{\phi_i} \\ f_{\phi_i}(x_i) - f_{\phi_i}(s_{\phi_i}) - \varepsilon \sum_{j=1}^{N} a_{ij}e_j, \qquad i \in V_{\phi_i} \backslash \bar{V}_{\phi_i}. \end{cases} \tag{8}$$

Theorem 1 *Assume that A1 holds. Then, the network (1) and the state (2) can realize cluster synchronization with the controllers (5), (6), and (7).*

Proof Construct the Lyapunov function:

$$V = \frac{1}{2}\sum_{i=1}^{N} e_i^T e_i + \frac{1}{2d}(\bar{a}_{ij} - a_{ij})^2 + \frac{1}{2}\sum_{i\in\bar{V}_{\phi_i}} \frac{1}{k_i}(\bar{g}_i - g_i)^2, \tag{9}$$

where $\bar{a}_{ij}$ is larger than the corresponding edge coupling strength a_{ij}, and $\bar{g}_i$ is a positive constant to be determined.

Let $\tilde{e}_k(t) = (e_{1k}, e_{2k}, \ldots, e_{Nk})^T$, $G = diag(\bar{g}_1, \ldots, \bar{g}_N)$ with $\bar{g}_i = 0$ for $i \in V_{\phi_i} \backslash \bar{V}_{\phi_i}$. Hence

$$\begin{aligned}
\dot{V} &= \sum_{i=1}^{N} e_i^T \dot{e}_i + \frac{1}{d}(\bar{a}_{ij} - a_{ij})(-\dot{a}_{ij}) + \sum_{i\in\bar{V}_{\phi_i}} \frac{1}{k_i}(\bar{g}_i - g_i)(-\dot{g}_i) \\
&= \sum_{i=1}^{N} e_i^T [f_{\phi_i}(x_i) - f_{\phi_i}(s_{\phi_i}) - \varepsilon \sum_{j=1}^{N} a_{ij}e_j] - \sum_{i\in\bar{V}_{\phi_i}} e_i^T g_i e_i - \sum_{(i,j)\in E_1} (\bar{a}_{ij} - a_{ij})\|e_i - e_j\| \\
&\quad - \sum_{i\in\bar{V}_{\phi_i}} (\bar{g}_i - g_i)\|e_i\|^2 \\
&= \sum_{i=1}^{N} e_i^T [f_{\phi_i}(x_i) - f_{\phi_i}(s_{\phi_i})] - \varepsilon \sum_{i=1}^{N}\sum_{j=1}^{N} e_i^T a_{ij} e_j - \sum_{(i,j)\in E_1} (\bar{a}_{ij} - a_{ij})\|e_i - e_j\| \\
&\quad - \sum_{i\in\bar{V}_{\phi_i}} \bar{g}_i \|e_i\|^2 \\
&\leq \sum_{i=1}^{N} e_i^T L e_i - \varepsilon \sum_{k=1}^{n} \tilde{e}_k^T A \tilde{e}_k - \sum_{(i,j)\in E_1} (\bar{a}_{ij} - a_{ij})\|e_i - e_j\| - \sum_{k=1}^{n} \tilde{e}_k^T G \tilde{e}_k
\end{aligned}$$

$$= \sum_{k=1}^{n} \tilde{e}_k^T L\tilde{e}_k - \varepsilon \sum_{k=1}^{n} \tilde{e}_k^T A\tilde{e}_k - \sum_{(i,j)\in E_1} (\bar{a}_{ij} - a_{ij})\|e_i - e_j\| - \sum_{k=1}^{n} \tilde{e}_k^T G\tilde{e}_k$$

$$= \sum_{k=1}^{n} \tilde{e}_k^T (LI_N + \varepsilon A - G)\tilde{e}_k - \sum_{(i,j)\in E_1} (\bar{a}_{ij} - a_{ij})\|e_i - e_j\|, \tag{10}$$

where denote by I_N an N-dimension unit matrix.

Since the eigenvalues of the matrix A satisfy $\lambda_N(A) \leq \cdots \leq \lambda_2(A) < \lambda_1(A) = 0$, it is obvious that one can select suitable constants $\bar{g}_i$ to make $LI_N + \varepsilon A - G$ be a negative definite matrix. Furthermore, the value $\bar{a}_{ij}$ is larger than the corresponding edge coupling strength a_{ij}. Hence, we have

$$\dot{V} \leq \sum_{k=1}^{n} \tilde{e}_k^T (LI_N + \varepsilon A - G)\tilde{e}_k < 0, \tag{11}$$

and $\lim_{t\to\infty} \|e_i\| = 0$, i.e., the network (1) and the state (2) realize cluster synchronization. □

Similar to Theorem 1, in order to make the network (1) and the dynamic (2) reach cluster synchronization, we design another scheme as the following:

$$\dot{a}_{ij} = d \sum_{(i,j)\in E\setminus E_1} \|e_i - e_j\|, \tag{12}$$

for any $d > 0$.

Now let $J = \{m_1, m_2, \dots, m_l\}$, where m_k represents a node in $\bar{V}_k$ for $k = 1, \dots, l$.

$$\dot{g}_i = k_i \|e_i\|^2, \quad i \in J \ \ and \ \ g_i = 0, \ \ otherwise, \tag{13}$$

for any constant $k_i > 0$.

Theorem 2 *Suppose A1 holds. Then, the network (1) and the state (2) can reach cluster synchronization with the controllers (5), (12) and (13).*

Proof Similar to the proof of Theorem 1, so it is omitted here. □

4 Simulation

In this section, we will give some simulations to verify the correction of the proposed control schemes.

In the following, the complex dynamic network consists of 13 nodes and 3 communities. The size of three communities are $N_1 = 4$, $N_2 = 4$ and $N_3 = 5$. The local state of three communities are described by two Lü systems and a Lorenz system. The two Lü systems are stated as follows:

$$\dot{x} = f_1(x) = \begin{pmatrix} \rho(x_2 - x_1) \\ -x_1x_3 + \nu_i x_2 \\ x_1x_2 - \mu x_3 \end{pmatrix}, \tag{14}$$

where $\rho = 36$, $\mu = 3$, $\nu_1 = 20$ and $\nu_2 = 25$.

The Lorenz system is drawn as

$$\dot{x} = f_3(x) = \begin{pmatrix} a(x_2 - x_1) \\ -x_1x_3 - x_2 - bx_1 \\ x_1x_2 - cx_3 \end{pmatrix}, \tag{15}$$

where $a = 10$, $b = 28$ and $c = \frac{8}{3}$. Evidently, Assumption 1 is satisfied.

Choose the Rössler system as the $s(t)$ of (2). Then the synchronized aims

$$s_k = P_k y(t) + Q_k, \quad k = 1, 2, 3 \tag{16}$$

where $P_1 = diag(1, -1, 1)$, $P_2 = diag(1, 1, -1)$, $P_3 = diag(-1, 1, 1)$, $Q_1 = [0, 1, 1]^T$, $Q_2 = [1, 0, 1]^T$ and $Q_3 = [1, 1, 0]^T$. The Rössler system can be described by

$$\dot{y} = \begin{pmatrix} -y_2 - y_3 \\ y_1 + \alpha y_2 \\ \beta + y_3(y_1 - \gamma) \end{pmatrix}, \tag{17}$$

where $\alpha = 0.2$, $\beta = 0.2$ and $\gamma = 5.7$.

The topology of the network is given as follows:

$$\begin{pmatrix}
-3 & 1 & 1 & 0 & 1 & 0 & 0 & 0 & 0 & 0 & 0 & 0 & 0 \\
1 & -3 & 1 & 1 & 0 & 0 & 0 & 0 & 0 & 0 & 0 & 0 & 0 \\
1 & 1 & -3 & 1 & 0 & 0 & 0 & 0 & 0 & 0 & 0 & 0 & 0 \\
0 & 1 & 1 & -5 & 0 & 1 & 0 & 0 & 1 & 1 & 0 & 0 & 0 \\
1 & 0 & 0 & 0 & -4 & 1 & 1 & 1 & 0 & 0 & 0 & 0 & 0 \\
0 & 0 & 0 & 1 & 1 & -4 & 1 & 1 & 0 & 0 & 0 & 0 & 0 \\
0 & 0 & 0 & 0 & 1 & 1 & -5 & 1 & 1 & 0 & 0 & 0 & 1 \\
0 & 0 & 0 & 0 & 1 & 1 & 1 & -3 & 0 & 0 & 0 & 0 & 0 \\
0 & 0 & 0 & 1 & 0 & 0 & 1 & 0 & -5 & 1 & 1 & 0 & 1 \\
0 & 0 & 0 & 1 & 0 & 0 & 0 & 0 & 1 & -5 & 1 & 1 & 1 \\
0 & 0 & 0 & 0 & 0 & 0 & 0 & 0 & 1 & 1 & -3 & 1 & 0 \\
0 & 0 & 0 & 0 & 0 & 0 & 0 & 0 & 0 & 1 & 1 & -3 & 1 \\
0 & 0 & 0 & 0 & 0 & 0 & 1 & 0 & 1 & 1 & 0 & 1 & -4
\end{pmatrix}$$

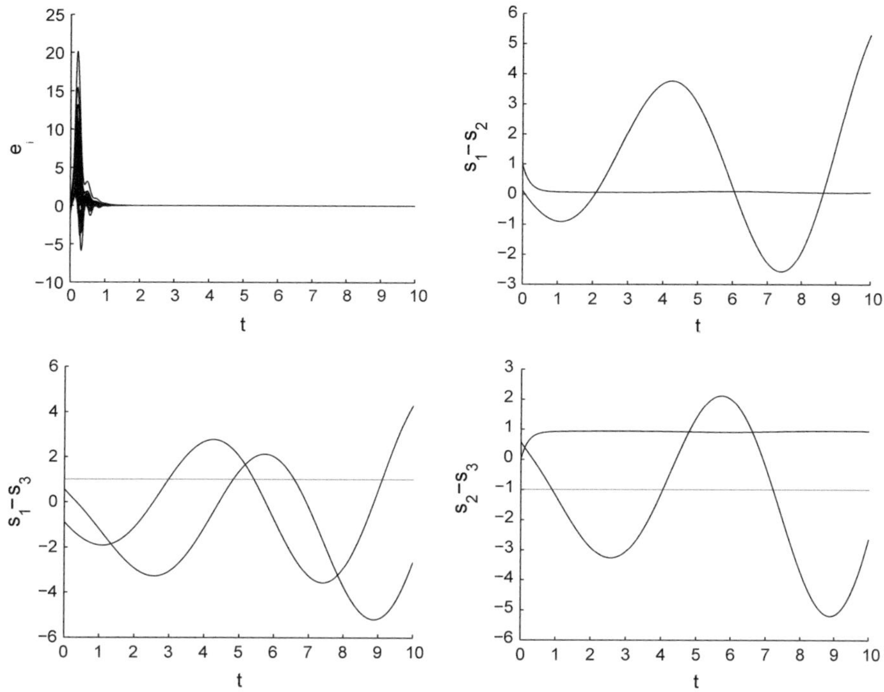

Fig. 1 The evolution of synchronization errors

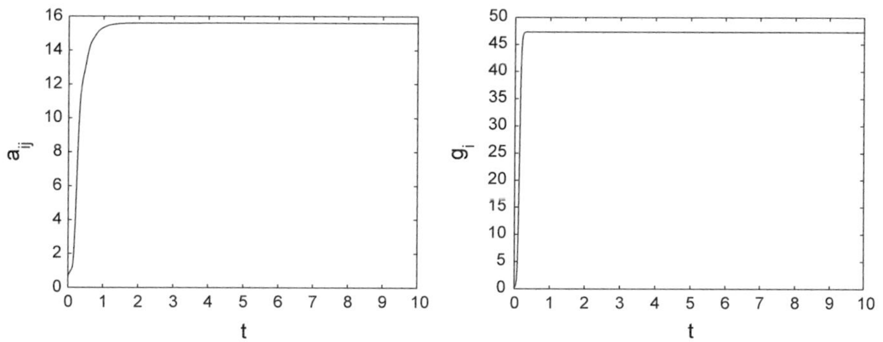

Fig. 2 The coupling strength a_{ij} and the feedback gain g_i

Example 1 In the first example, by Theorem 1, the network can be synchronized to the state of (9) using the controllers (3), (4), and (5). Now, the initial values of state vectors X_0, Y_0, $g_i(0)$, $i = 1, 4, 5, 6, 7, 9, 10, 13$ are chosen in (0, 1) randomly. Furthermore, select $\varepsilon = 20$, $d = 0.8$ and $k_i = 10$, $i = 1, 4, 5, 6, 7, 9, 10, 13$. From Figs. 1 and 2, we find that the coupling strength of edges a_{ij} and the feedback gain g_i are tended to positive constants when the network realizes cluster synchronization.

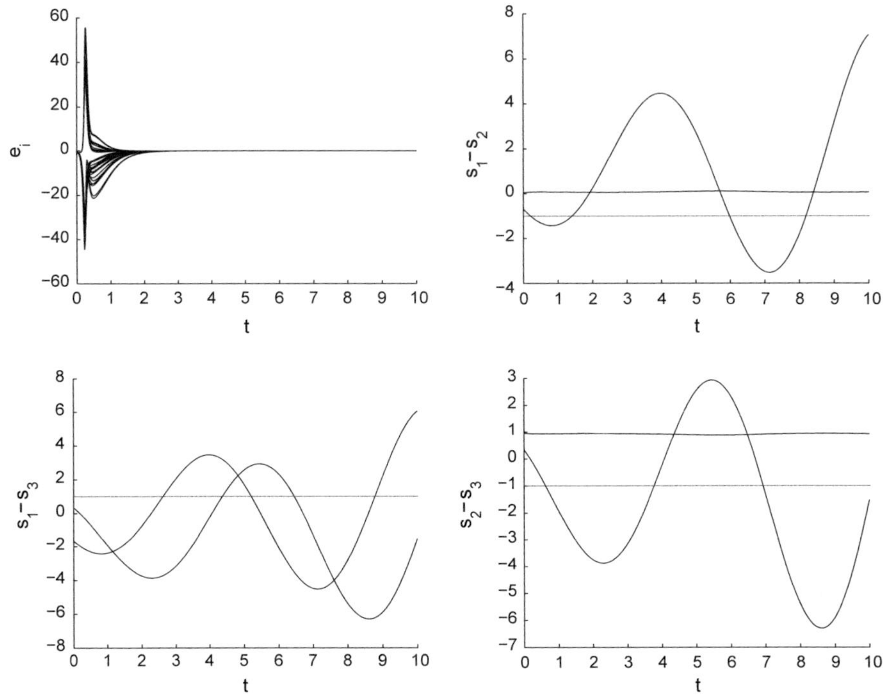

Fig. 3 The evolution of synchronization errors

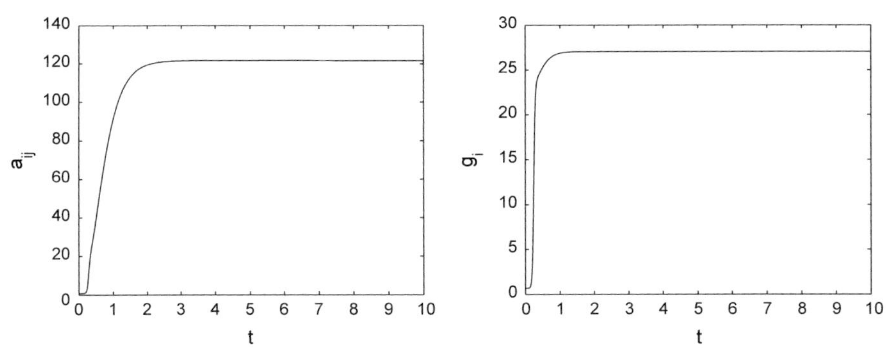

Fig. 4 The coupling strength a_{ij} and the feedback gain g_i

Example 2 In the second example, by Theorem 2, the network can be synchronized to the state of (9) using the controllers (3), (7), and (8). Here, the original values of state vectors X_0, Y_0, $g_i(0)$, $i = 4, 7, 13$ are chosen in (0, 1) randomly. In addition, choose $\varepsilon = 30$, $d = 0.8$ and $k_i = 0.1$, $i = 4, 7, 13$. From Figs. 3 and 4, we learn that the coupling strength of edges a_{ij} and the feedback gain g_i are tended to positive constants while cluster synchronization achieves.

5 Conclusion

In this paper, cluster synchronization has been investigated by presenting two adaptive control schemes. Both of the control methods need to select some edges and nodes to control simultaneously. One is to use the pinning control method and also adjust the coupling strength of some edges which connect different groups. Selecting only one node in each group to control and adjusting the coupling strength of edges in each group is another control scheme.

Acknowledgments This work was supported by Basic Research Projects of High Education (ZXH2012K002, ZXH2010D011), National Natural Science Foundation of China (61174094, 61039001), and National Natural Science Foundation of Tianjin (14JCYBJC18700).

References

1. Stogatz SH (2001) Exploring complex networks. Nature 410:268–276
2. Wang X, Chen G (2003) Complex network: small-world, scale-free and beyond. IEEE Circuits Syst Mag 3:6–20
3. Boccaletti S, Latora V, Moreno Y, Chavezf M, Hwanga DU (2006) Complex networks: structure and dynamics. Phys Reports-review Sect Phys Lett 424:175–308
4. Arenas A, Diaz-Guilera A, Kurths J, Moreno Y, Zhou C (2008) Synchronization in complex networks. Phys Reports-review Sect Phys Lett 469:93–153
5. Cao J, Chen G, Li P (2008) Global synchronization in an array of delayed neural networks with hybrid coupling. IEEE Trans Syst Man Cybern-Part B: Cybern 38(2):488–498
6. Yu W, Cao J, Lv J (2008) Global synchronization of linearly hybrid coupled networks with time-varying delay. SIAM J Appl Dyn Syst 7(1):108–133
7. Li P, Zhang Y, Zhang L (2006) Global synchronization of a class of delayed complex networks. Chaos Solitons Fractals 30:903–908
8. Gao H, Lam J, Chen G (2006) New criteria for synchronization stability of general complex dynamical networks with coupling delays. Phys Lett A 360:263–273
9. Sun M, Zeng CH, Tian L (2010) Linear generalized synchronization between two complex networks. Commun Nonlinear Sci Numer Simulat 15:2162–2167
10. Yang ZQ, Zhang Q, Chen ZQ (2012) Adaptive linear generalized synchronization between two nonidentical networks. Commun Nonlinear Sci Numer Simul 17:2628–2636
11. Li X, Wang X, Chen G (2004) Pinning a complex dynamical network to its equilibrium. IEEE Trans Circuits Syst I-Regular Pap 51:2074–2087
12. Chen T, Liu X, Lu W (2007) Pinning complex networks by a single controller. IEEE Trans Circuits Syst I-Regular Pap 54:1317–1326
13. Zhou J, Lu J, Lü J (2008) Pinning adaptive synchronization of a general complex dynamical network. Automatica 44:996–1003
14. Sun W, Hu T, Chen Z, Chen S, Xiao L (2011) Implusive synchronization of a general nonlinear coupled complex network. Commun Nonlinear Sci Numer Simul 16:4501–4507
15. Li C, Yu W, Huang T (2014) Impulsive synchronization schemes of stochastic complex networks with switching topology: average time approach. Neural Netw 54:85–94
16. Yu W, Chen G, Lü J, Kuths J (2013) Synchronization via pinning control on general complex networks. SIAM J Control Optim 51(2):1395–1416
17. Cao Y, Yu W, Ren W, Chen G (2013) An overview of recent progress in the study of distributed multi-agent coordination. IEEE Trans Ind Inf 9(1):427–438

18. Hu H, Yu W, Xuan Q, Yu L, Xie G (2013) Consensus for second-order agent dynamics with velocity estimators via pinning control. IET Control Theory Appl 7(9):1196–1205
19. Schnitzler A, Gross J (2005) Normal and pathological oscillatory communication in the brain. Nature Rev Neurosci 6:285–286
20. Rulkov NF (1996) Images of synchronized chaos: experiments with circuits. Chaos 6:262–279
21. Ma Z, Liu Z, Zhang G (2006) A new method to realize cluster synchronization in connected chaotic networks. Chaos 16(2) (023103)
22. Wu W, Zhou W, Chen T (2009) Cluster synchronization of linearly coupled complex networks under pinning control. IEEE Trans Circuits Syst I-Regular Pap 56:829–839
23. Hu CH, Yu J, Jiang H, Teng ZH (2011) Synchronization of complex community networks with nonindentical nodes and adaptive coupling strength. Phys Lett A 375 873–879
24. Lu X, Qin B (2009) Adaptive cluster synchronization in complex dynamical networks. Phys Lett A 373:3650–3658
25. Wu X, Lu H (2011) Cluster synchronization in the adaptive complex dynamical networks via a novel appoach. Phys Lett A 375:1559–1565
26. Wu ZH (2012) Comment on cluster synchronization in the adaptive complex dynamical networks via a novel appoach. Phys Lett A 376:1176–1177
27. Wu ZH, Fu X (2012) Cluster mixed synchronization via pinning control and adaptive coupling strength in community networks with nonidentical nodes. Commun Nonlinear Sci Numer Simul 376:1628–1636
28. Wu ZH, Fu X (2012) Cluster projective synchronization between community networks with nonidentical nodes. Physica A 391:6190–6198
29. Li K, Zhou J, Yu W, Small M, Fu X (2014) Adaptive cluster synchronization in networks with time-varying and distributed delay couplings. Appl Math Modell 38(4):1300–1314

Output Feedback Stabilization of Stochastic Non-holonomic Mobile Robots

Wenli Feng, Hongyu Wei, Hongmei Zhang and Dongkai Zhang

Abstract We discussed the output feedback stabilization of stochastic non-holonomic mobile robots. Output feedback controllers are given with backstepping method. So, the original closed-loop system can be stabilized in probability based on provided switching control strategy. In the end, we give an example to explain these results.

Keywords Stochastic non-holonomic mobile robots · Output feedback stabilization · Backstepping technique

1 Introduction

Since backstepping method was first introduced [1], stochastic control became one of the active areas of nonlinear control, especially the output feedback control [2–7]. With the development of stochastic systems, there were some results of stabilization for stochastic non-holonomic systems [8–11]. Two classes of stabilization were considered: state feedback [12–15] and output feedback [16–18]. One of the practical non-holonomic systems is non-holonomic mobile robots, whose state feedback stabilization was discussed by backstepping method [9, 19–21]. Output feedback control is studied when robots satisfy strict triangle structure because only its linear velocity is related to stochastic disturbance [17, 18]. However, if linear and angular velocities of non-holonomic mobile robots are related to stochastic disturbances, the systems do not satisfy strict triangle structure. To our knowledge,

W. Feng (✉) · H. Zhang · D. Zhang
School of Mathematics and Information Science, Shijiazhuang University, Hebei 050035, China
e-mail: fwl1028@126.com

H. Wei
Management Center of China Post Online College, Shijiazhuang Posts and Telecommunications Technical College, Hebei 050021, China

Y. Jia et al. (eds.), *Proceedings of 2016 Chinese Intelligent Systems Conference*, Lecture Notes in Electrical Engineering 405,
DOI 10.1007/978-981-10-2335-4_12

this problem has not been discussed in existed literatures. The purpose of paper is to discuss the output feedback stabilizing controllers of stochastic non-holonomic mobile robots.

2 Preliminaries and Problem Formulation

2.1 Preliminaries

Let stochastic nonlinear system as follows:

$$\mathrm{d}x = f(x)\mathrm{d}t + g(x)\mathrm{d}B, x(t_0) \in \mathbb{R}^n, \tag{1}$$

where $x \in \mathbb{R}^n$ is the state, $f: \mathbb{R}^n \to \mathbb{R}^n$ and $g: \mathbb{R}^n \to \mathbb{R}^n$ are Borel measurable functions and satisfy locally Lipschitz in x, $B \in \mathbb{R}^r$ is a r-dimensional independent standard Wiener process defined on the complete probability space $(\Omega, \mathcal{F}, P)$.

Definition 1 [22] For any given $V(x) \in \mathcal{C}^2$, the differential operator $\mathcal{L}$ of (1) can be defined as follows:

$$\mathcal{L}V(x) = \frac{\partial V(x)}{\partial x} f(x) + \frac{1}{2} \mathrm{Tr}\left\{ g^T(x) \frac{\partial^2 V}{\partial x^2} g(x) \right\}.$$

Definition 2 [23] The equilibrium $x = 0$ of system (1) is

- globally stable in probability if there exists a class $\mathcal{K}$ function $\gamma(\cdot)$ for $\forall \varepsilon > 0$, such that

$$P\{|x(t)| < \gamma(|x(t_0)|)\} \geq 1 - \varepsilon, \quad \forall t \geq 0, x(t_0) \in \mathbb{R}^n \backslash \{0\}.$$

- globally asymptotically stable in probability if it satisfies the above condition and

$$P\left\{ \lim_{t \to \infty} |x(t)| = 0 \right\} = 1, \quad \forall x(t_0) \in \mathbb{R}^n.$$

Lemma 1 [23] *For system (1), if there exist a $\mathcal{C}^2$ function $V(x)$, class $\mathcal{K}_\infty$ function $\alpha_1(\cdot)$ and $\alpha_2(\cdot)$, constants $c_1 > 0, c_2 > 0$, and a nonnegative function $W(x)$ such that*

$$\alpha_1(\|x\|) \leq V(x) \leq \alpha_2(\|x\|),$$
$$\mathcal{L}V = \frac{\partial V(x)}{\partial x} f(x) + \frac{1}{2} Tr\left\{ g^T(x) \frac{\partial^2 V(x)}{\partial x^2} g(x) \right\} \leq -c_1 W(x) + c_2,$$

then

(i) *an almost surely unique solution is existed on* $[t_0, \infty)$ *for each* $x(t_0) \in \mathbb{R}^n$,
(ii) *when* $W(x)$ is *continuous and* $c_2 = 0, f(0) = 0, g(0) = 0$, *the equilibrium* $x = 0$ *be globally stable in probability and*

$$P\left\{\lim_{t\to\infty} W(x) = 0\right\} = 1, \text{for } \forall x(t_0) \in \mathbb{R}^n.$$

2.2 Problem Formulation

The mobile robot (2) was discussed in [24]:

$$\begin{cases} \ddot{\theta} = w, \\ \dot{x}_c = v \cos\theta, \\ \dot{y}_c = v \sin\theta, \end{cases} \tag{2}$$

where v and w are control inputs which denote linear and angular velocity, respectively. Let

$$\begin{cases} w(\theta) = w_1(\theta) + w_2(\theta)\dot{B}(t), \\ v(\theta) = v_1(x, y, \theta) + v_2(x, y, 0)\dot{B}(t), \end{cases} \tag{3}$$

where $B(t)$ denote a Brownian motion and $\dot{B}(t)$ is the derivative of $B(t)$. With the similar method in [17, 18], we have the output model.

$$\mathrm{d}x_0 = u_0 \mathrm{d}t + g_0 \mathrm{d}B, \tag{4.1}$$

$$\left.\begin{array}{l} \mathrm{d}x_1 = x_2 u_0 \mathrm{d}t + f_1 \mathrm{d}t + g_1 \mathrm{d}B \\ \mathrm{d}x_2 = u\mathrm{d} \ + f_2 \mathrm{d}t + g_2 \mathrm{d}B \end{array}\right\}, \tag{4.2}$$

$$y = [x_0, x_1], \tag{4.3}$$

where $g_0 = w_2$, $f_1 = -\frac{1}{2}x_1 w_2^2 + v_2 w_2$, $f_2 = -\left\{x_1 u_0 + \frac{1}{2}x_2 w_2^2\right\}$, $g_2 = v_2 - x_1 w_2$ and $g_1 = x_2 w_2$.

Remark 1 The system (4.1)–(4.3) is not strict triangle structure.

3 Main Results

In the following parts, Assumptions 1–2 are necessary.

Assumption 1 For smooth function $w_2(\theta)$, one has $w_2(\theta) = m_1\theta$, where m_1 is a known nonnegative constant.

Assumption 2 For smooth function $v_2(x_c, y_c, \theta)$, one has

$$|v_2(x_c, y_c, \theta)| \le m_2(|x_1| + |x_2|),$$

where m_2 is a known nonnegative constant.

3.1 The First State Stabilization

We choose u_0 in subsystem (4.1) as follows:

$$u_0 = -\eta_0 x_0, \ \eta_0 = \lambda + \frac{1}{2} m_1^2 \tag{5}$$

where $\lambda > 0$.

The Lyapunov function can be chosen as

$$V_0(x_0) = \frac{1}{2} x_0^2. \tag{6}$$

By (4.1) and (5), one has

$$\mathcal{L} V_0 \le -\lambda x_0^2. \tag{7}$$

Theorem 1 *If Assumption 1 holds, choosing $m_1 > 0$, $\lambda > 0$, controller u_0 as (5), respectively, then*

(i) *an almost surely unique solution of (4.1) and (5) is existed on $[t_0, \infty)$ for $\forall x_0(t_0)$;*
(ii) *the equilibrium $x_0 = 0$ of (4.1) and (5) is globally asymptotically stable in probability.*

Remark 2 From Lemma 2.3 ([25], p 93), for initial condition $x_0(t_0) \neq 0$, the solutions of (4.1) and (5) never reach the zero.

3.2 Other States Stabilization

We introduce the following transformation:

$$z_1 = \frac{x_1}{u_0}, z_2 = x_2. \tag{8}$$

So, subsystem (4.2) with (8) can be transformed into

$$\begin{cases} \mathrm{d}z_1 = z_2\mathrm{d}t + \eta_0 z_1\mathrm{d}t + \phi_1\mathrm{d}t + \psi_1\mathrm{d}B, \\ \mathrm{d}z_2 = u\,\mathrm{d}t + \phi_2\mathrm{d}t + \psi_2\mathrm{d}B, \end{cases} \tag{9}$$

where

$$\phi_1 = \frac{f_1}{u_0} + z_1\frac{g_0^2}{x_0^2} + \eta_0 z_1\frac{g_0 g_1}{u_0}, \psi_1 = \frac{g_1}{u_0} - z_1\frac{g_0}{x_0}, \phi_2 = f_2, \psi_2 = g_2.$$

The following observer is needed in the processor of design of the output feedback controller.

$$\begin{cases} \dot{\hat{z}}_1 = \hat{z}_2 + \eta_0\hat{z}_1 + La_1(y_1u_0 - \hat{z}_1), \\ \dot{\hat{z}}_2 = u + L^2a_2(y_1u_0 - \hat{z}_1), \end{cases}$$

where $L \geq 1$ is a gain parameter, $a_1, a_2 > 0$ are real numbers, and $p(s) = s^2 + a_1s + a_2$ is Hurwitz.

Letting the state estimation errors

$$\xi_1 = z_1 - \hat{z}_1, \xi_2 = \frac{z_2 - \hat{z}_2}{L}, \tag{10}$$

one has error dynamics as follows:

$$\mathrm{d}\xi = (LA + B)\xi\mathrm{d}t + \Phi\mathrm{d}t + \Psi\mathrm{d}B, \tag{11}$$

where

$$A = \begin{bmatrix} -a_1 & 1 \\ -a_2 & 0 \end{bmatrix}, B = \begin{bmatrix} \eta_0 & 0 \\ 0 & 0 \end{bmatrix}, \Phi = \begin{bmatrix} \phi_1 \\ \frac{\phi_2}{L} \end{bmatrix}, \Psi = \begin{bmatrix} \psi_1 \\ \frac{\psi_2}{L} \end{bmatrix}.$$

Obviously, A is Hurwitz. So, a positive definite matrix P exists and satisfy $A^TP + PA = -I$.

Therefore, the following entire systems can be given:

$$\begin{cases} \mathrm{d}\xi = (LA + B)\xi\mathrm{d}t + \Phi\mathrm{d}t + \Psi\mathrm{d}B, \\ \dot{\hat{z}}_1 = \hat{z}_2 + \eta_0\hat{z}_1 + La_1(y_1u_0 - \hat{z}_1), \\ \dot{\hat{z}}_2 = u + L^2a_2(y_1u_0 - \hat{z}_1). \end{cases} \tag{12}$$

By Assumptions 1–2 and Theorem 1, Proposition 1 holds.

Proposition 1 *There exist constants* τ_1, τ_2, *and*

$$\begin{cases} |\Phi(x_0, z)|^2 \leq 2\left(|z_1| + \frac{|z_2|}{L}\right)^2\tau_1, \\ |\Psi(x_0, z)|^2 \leq 2\left(|z_1| + \frac{|z_2|}{L}\right)^2\tau_2, a.s. \end{cases} \tag{13}$$

where $z = [z_1, z_2]^T$.

We need define some variables which can invoke backstepping technique.

$$\varepsilon_1 = \hat{z}_1, \varepsilon_2 = \hat{z}_2 - \alpha_1(\hat{z}_1), \tag{14}$$

where α_1 is the virtual smooth controller.

Step 1 Choose Lyapunov function as $V_1 = \delta\xi^T P\xi + \frac{1}{2}\varepsilon_1^2$, where $\delta > 0$ is a parameter. From It $\hat{o}$ differentiation formula, we have

$$\begin{aligned}\mathcal{L}V_1 = & -\left\{\delta\left(L - 2\|P\|^2 - \|B\|^2\right) - 4C_e\right\}\|\xi\|^2 + 4C_e\left(\left|\hat{z}_1^2\right| + \frac{\left|\hat{z}_2^2\right|}{L^2}\right)\\ & + \hat{z}_1(\hat{z}_2 + \eta_0\hat{z}_1 + La_1\xi_1),\end{aligned}$$

where $C_e = 2\delta\left(\tau_1^2 + \lambda_{\max}(P)\tau_2\right)^2$.

By simple calculation, the following inequalities hold:

$$4C_e\frac{\left|\hat{z}_2^2\right|}{L^2} \le 8C_e\frac{\varepsilon_2^2 + \alpha_1^2}{L^2}, \hat{z}_1\hat{z}_2 \le 8C_e\frac{\varepsilon_2^2 + \alpha_1^2}{L^2}, La_1\hat{z}_1\xi_1 \le \frac{La_1|\hat{z}_1|^2}{4} + L\xi_1^2.$$

So, one can obtain

$$\begin{aligned}\mathcal{L}V_1 = & -\left\{\delta\left(L - 2\|P\|^2 - \|B\|^2\right) - 4C_e - L\right\}\|\xi\|^2 + \hat{z}_1\left\{\alpha_1 + 4LC_e\hat{z}_1 + \frac{La_1^2}{4}\hat{z}_1\right\}\\ & + \frac{8C_e\varepsilon_2^2}{L^2} + \frac{8C_e\alpha_1^2}{L^2} + \frac{1}{4L}\varepsilon_2^2.\end{aligned} \tag{15}$$

Therefore

$$\alpha_1 = -Lb_1\varepsilon_1,\ b_1 = 4 + 4C_e + \frac{a_1^2}{4}. \tag{16}$$

By (15)–(16), one has

$$\begin{aligned}\mathcal{L}V_1 = & -\left\{\delta\left(L - 2\|P\|^2 - \|B\|^2\right) - 4C_e - L\right\}\|\xi\|^2 + \left\{\frac{8C_e}{L^2} + \frac{1}{4L}\right\}\varepsilon_2^2\\ & -\left\{3L - 8C_e b_1^2\right\}\varepsilon_1^2.\end{aligned} \tag{17}$$

Step 2 From (14) and (16), letting

$$\varepsilon_2 = \hat{z}_2 + Lb_1\hat{z}_1, \tag{18}$$

we have

$$\mathrm{d}\,\varepsilon_2 = \{u + \hat{z}_2 + \eta_0\hat{z}_1 + (L^2a_2 + La_1)\xi_1\}\mathrm{d}t. \tag{19}$$

Defining the quadratic Lyapunov candidate function $V_2 = V_1 + \frac{1}{2}\varepsilon_2^2$, from It ô differentiation formula, one has

$$\begin{aligned}\mathcal{L}V_2 = &-\left\{\delta\left(L - 2\|P\|^2 - \|B\|^2\right) - 4C_e - L\right\}\|\xi\|^2 + \left\{\frac{8C_e}{L^2} + \frac{1}{4L}\right\}\varepsilon_2^2 \\ &-\{3L - 8C_eb_1^2\}\varepsilon_1^2 + \varepsilon_2\{u + \hat{z}_2 + \eta_0\hat{z}_1 + (L^2a_2 + La_1)\xi_1\}.\end{aligned}$$

By simple calculation, we have

$$\begin{aligned}&\varepsilon_2\hat{z}_2 \le L\varepsilon_2^2 + L\varepsilon_2^2 + \frac{b_1^2}{4}\varepsilon_1^2, \eta_0\varepsilon_2\varepsilon_1 \le L\varepsilon_2^2 + \frac{\eta_0^2}{4L}\varepsilon_1^2, \\ &L(La_2 + a_1)^2\varepsilon_2\xi_1 \le \frac{L(La_2 + a_1)^2}{4L}\varepsilon_2^2 + L\xi_1^2.\end{aligned}$$

So,

$$\begin{aligned}\mathcal{L}V_2 = &-\left\{\delta\left(L - 2\|P\|^2 - \|B\|^2\right) - 4C_e - 2L\right\}\|\xi\|^2 - \left(3L - 8C_eb_1^2 - \frac{\eta_0^2}{4L} - \frac{b_1^2}{4}\right)\varepsilon_1^2 \\ &+ \varepsilon_2\left\{u + \left(2L + L^2 + \frac{L(La_2 + a_1)^2}{4L} + \frac{8C_e}{L^2} + \frac{1}{4L}\right)\varepsilon_2\right\}.\end{aligned} \tag{20}$$

Choosing the virtual control law u

$$\begin{cases} u = -Lb_2\varepsilon_2, \\ b_2 = 3 + L + \frac{(La_2 + a_1)^2}{4L} + \frac{8C_e}{L^3} + \frac{1}{4L^2}, \end{cases} \tag{21}$$

and substituting (21) into (20), one has

$$\begin{aligned}\mathcal{L}V_2 = &-\left\{\delta\left(L - 2\|P\|^2 - \|B\|^2\right) - 4C_e - 2L\right\}\|\xi\|^2 \\ &-\left(3L - 8C_eb_1^2 - \frac{\eta_0^2}{4L} - \frac{b_1^2}{4}\right)\varepsilon_1^2 - L\varepsilon_2^2.\end{aligned} \tag{22}$$

From above analysis and simple proof, one has the following result.

Theorem 2 *If Assumptions 1–2 hold, choosing proper constants* $\lambda > 0, m_1 > 0$, *and* $m_2 > 0$, *we have*

(i) *the systems (12) and (21) have an almost surely unique solution on* $[t_0, \infty)$ *for* $\forall z(t_0)$,
(ii) *the equilibrium* $z = 0$ *of this system is globally stable in probability.*

4 Switching Control Design

In the above section, $x_0(t_0) \neq 0$ is discussed. But, when initial condition $x_0(t_0) = 0$, with the similar results in Sect. 5 in [9], one gets the following switching control procedure.

Theorem 3 *If Assumptions 1–2 hold, the following switching method can be applied to (4)*:

(i) *When initial condition belongs to*

$$\{(\theta(t_0), x_c(t_0), y_c(t_0)) \in \mathbb{R}^3 | \theta(t_0) \neq 0\},$$

u_0 *and* u *chosen as (5) and (21), respectively,*
(ii) *When initial condition belongs to*

$$\{(\theta(t_0), x_c(t_0), y_c(t_0)) \in \mathbb{R}^3 | \theta(t_0) = 0\}$$

for $t \in [t_0, t_s)$, *we choose* $u_0 = -u_0^* \neq 0$ *and* $u = u^*$; *for* $t \in [t_s, \infty)$ *and switch* u_0 *and* u *into (5) and (21) at the time* $t = t_s^*$.

So, let the closed-loop system be stable in probability.

5 An Example

Let $w_2 = 0.1\theta$ and $v_2 = 0.1(x_c \cos\theta + y_c \sin\theta)$, $\eta_0 = 2$, $a_1 = 0.75$, $a_2 = 1.25$, $\delta = 2.1$, $L = 25$ and initial conditions $\theta(0) = 3$, $x_c(0) = -0.289$, $y_c(0) = -0.17$ in system (4). Figure 1 illustrates responses of states in Eqs. (4.1–4.3), (5), and (21). Figure 2 gives responses of u_0 and u.

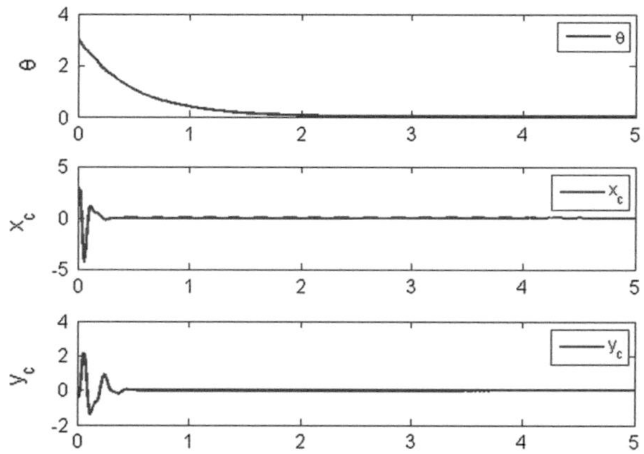

Fig. 1 Responses of θ, x_c, and y_c

Fig. 2 Responses of u_0 and u

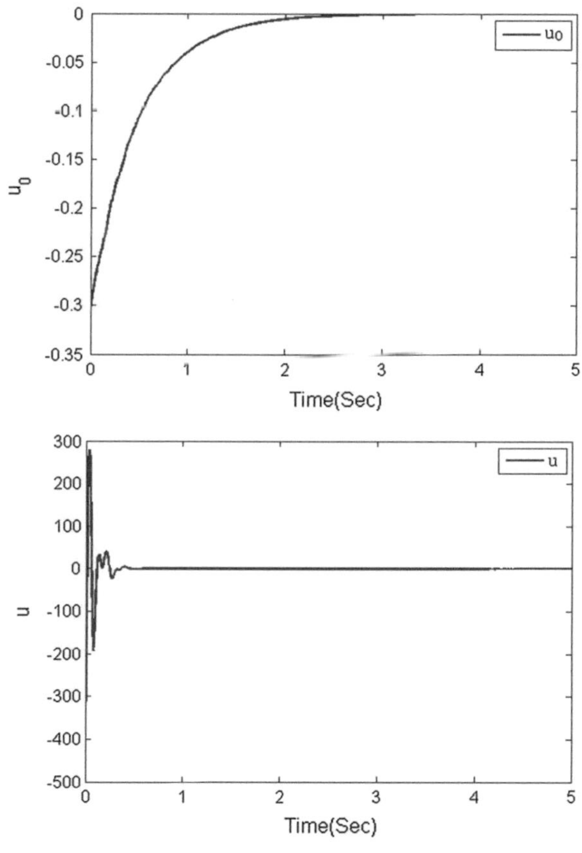

6 Conclusions

In this paper, when linear velocity and angular velocity of non-holonomic mobile robots are all related to Brownian motion, we design output feedback controllers based on backstepping method.

Acknowledgment This paper is partially supported by the National Natural Science Foundation (no. 61503262) and Natural Science Foundation of Hebei Province of China (no. A2014106035).

References

1. Pan Z, Baçar T (1999) Backstepping controller design for nonlinear stochastic systems under a risk-sensitive cost criterion. SIAM J Control Optim 37(3):957–995
2. Deng H, Krstić M (1999) Output-feedback stochastic nonlinear stabilization. IEEE Trans Autom Control 44(2):328–333
3. Li W, Liu X, Zhang S (2012) Further results on adaptive state-feedback stabilization for stochastic high-order nonlinear systems. Automatica 48(8):1667–1675
4. Liu L, Xie X (2013) State feedback stabilization for stochastic feedforward nonlinear systems with time-varying delay. Automatica 49(4):936–942
5. Xie X, Liu L (2013) A homogeneous domination approach to state feedback of stochastic high-order nonlinear systems with time-varying delay. IEEE Trans Autom Control 58 (2):494–499
6. Li F, Liu Y (2015) Global stabilization via time-varying output-feedback for stochastic nonlinear systems with unknown growth rate. Syst Control Lett 77:69–79
7. Deng H, Krstić M (2000) Output-feedback stabilization of stochastic nonlinear systems driven by noise of unknown covariance. Syst Control Lett 39(3):173–182
8. Liu S, Krstić M (2010) Stochastic source seeking for nonholonomic unicycle. Automatica 46 (9):1443–1453
9. Wu Z, Liu Y (2012) Stochastic stabilization of nonholonomic mobile robot with heading-angle-dependent disturbance. Math Probl Eng 2012:1–17
10. Du Q, Wang C, Wang G, Zhang D (2015) State-feedback stabilization for stochastic high-order nonholonomic systems with Markovian switching. Nonlinear Anal: Hybrid Syst 18:1–14
11. Do K (2015) Global inverse optimal stabilization of stochastic nonholonomic systems. Syst Control Lett 75:41–55
12. Wang J, Gao Q, Li H (2006) Adaptive robust control of nonholonomic systems with stochastic disturbances. Sci China Ser F: Inf Sci 49(2):189–207
13. Zhao Y, Yu J, Wu Y (2011) State-feedback stabilization for a class of more general high order stochastic nonholonomic systems. Int J Adapt Control Sig Process 25(8):687–706
14. Zhang D, Wang C, Chen H, Yang F, Du J (2013) Adaptive stabilization of stochastic non-holonomic systems with nonhomogeneous uncertainties. Trans Inst Meas Control 35 (5):648–663
15. Gao F, Yuan F (2013) Adaptive stabilization of stochastic nonholonomic systems with nonlinear parameterization. Appl Math Comput 219(16):8676–8686
16. Zheng X, Wu Y (2010) Output feedback stabilization of stochastic nonholonomic systems. In: Proceedings of 8th conference on world congress on intelligent control and automation, Jinan, China. pp 2091–2096
17. Liu Y, Wu Y (2011) Output feedback control for stochastic nonholonomic systems with growth rate restriction. Asian J Control 13(1):177–185

18. Zhang D, Wang C, Wei G, Chen H (2014) Output feedback stabilization for stochastic nonholonomic systems with nonlinear drifts and Markovian switching. Asian J Control 16 (6):1679–1692
19. Shang Y, Meng H (2012) Exponential stabilization of nonholonomic mobile robots subject to stochastic disturbance. J Inf Comput Sci 9(9):2635–2642
20. Feng W, Sun Q, Cao Z, Zhang D, Chen H (2013) Adaptive state-feedback stabilization for stochastic nonholonomic mobile robots with unknown parameters. Discrete Dyn Nat Soc 2013:1–9
21. Zhang D, Wang C, Wei G, Zhang H, Chen H (2014) State-feedback stabilization for stochastic nonholonomic mobile robots with uncertain visual servoing parameters. Int J Syst Sci 45(7): 1451–1460
22. Krstić M, Deng H (1998) Stability of nonlinear uncertain systems. Springer Publishing
23. Deng H, Krstić M, Williams R (2001) Stabilization of stochastic nonlinear driven by noise of unknown covariance. IEEE Trans Automatic Control 46(8):1237–1253
24. Campion G, Bastin G, D'Andréa-Novel B (1996) Structural properties and classification of kinematic and dynamic models of wheeled mobile robots. IEEE Trans Robot Autom 12 (1):47–62
25. Mao X (1997) Stochastic differential equations and their applications. Horwood Publishing, Chichester

Research on Grasp Force Control of Apple-Picking Robot Based on Improved Impedance Control

Wei Tang, Wei Ji, Xiangli Meng, Bo Xu, Dean Zhao and Shihong Ding

Abstract In order to solve the problem during the force control process of apple-picking robot the control effects are affected by robot dynamic parameters, contact environment uncertainties, and the noise interference of force sensors. This paper presents an improved impedance control algorithm. The algorithm only considers the direction of grasping apples, avoiding the complexity of multi-DOF manipulator impedance control; meanwhile impedance parameters are adjusted by RLS on time to improve the real-time control and the robustness to the force interference. Simulation results show the control algorithm is effective in force tracking and the output of force is steady; it can reduce the grasp damage remarkably. The research can provide a reference for the apple-picking robot compliant grasp.

Keywords Picking robot · Impedance control · Recursive least square · Forgetting factor

1 Introduction

The compliance of robot plays a significant role during the grasp process of apple-picking robot under the complex and particular work environment as improper force will lead the failure of grasping apple [1]. To achieve compliant grasp, lots of experts and scholars put forward many effective control methods. Force/position control and impedance control are two main compliant force control methods. The basic idea of position/force control is to decompose the task into corresponding two orthogonal subspaces of position and contact force in compliant coordinate space; then computes the position and force control, respectively; finally, the computation result is converted into joint space and combined into a unified joint control torque to drive the end-effector to achieve the compliant behavior.

W. Tang · W. Ji (✉) · X. Meng · B. Xu · D. Zhao · S. Ding
School of Electrical and Information Engineering,
Jiangsu University, Zhenjiang 212013, China
e-mail: jwhxb@163.com

Y. Jia et al. (eds.), *Proceedings of 2016 Chinese Intelligent Systems Conference*, Lecture Notes in Electrical Engineering 405,
DOI 10.1007/978-981-10-2335-4_13

However, when the state transforms from free state to contact state, the system will be unstable due to impact force especially the stiffness coefficient of the contact environment is large [2]. Impedance control does not control desired force and position directly, and it regulates the dynamic relationship between the force and position to realize the compliant control. However, the control accuracy relies on the choice of parameters of impedance controller, and actually the parameters are difficult to obtain accurate value [3]. This paper presents an improved impedance control algorithm based on the features and requirements of apple-picking robot and the end-effector designed by our group, to adjust the impedance parameters online real time with recursive least square (RLS). The effectiveness of the proposed method is verified by simulation experiments. This method also provides a theoretical basis to achieve the compliance of apple-picking robot.

2 The Strategy of Impedance Control

When the end-effector grasps the apple, there is a dynamic relationship between them. The characteristics of the relationship is determined by target impedance parameters M_t, B_t, K_t, and the dynamic behavior can be set with these parameters shown as follows:

$$F_e - F_r = M_t(X - X_r) + B_t(\dot{X} - \dot{X}_r) + K_t(X - X_r) \tag{1}$$

where M_t, B_t, K_t are the target inertia matrix, damping matrix, stiffness and matrix; $X, \dot{X}, X, X_r, \dot{X}_r, X_r$ are the actual position, velocity, acceleration and reference position, velocity, and acceleration of end-effector, respectively. F_e and F_r are the actual grasp force and reference force, respectively. Before grasping apple, the contact force is 0, so the trajectory tracking control can be realized. While the end-effector contacts the apple, the correction of reference trajectory is generated by impedance controller automatically according to the deviations from actual force and desired force, to adjust the grasp force between end-effector and apple, to achieve the dynamic grasp finally.

A second-order model is formed when contact occurs, and the model contains the impedance characteristics of end-effector and apple. The model is shown in Fig. 1.

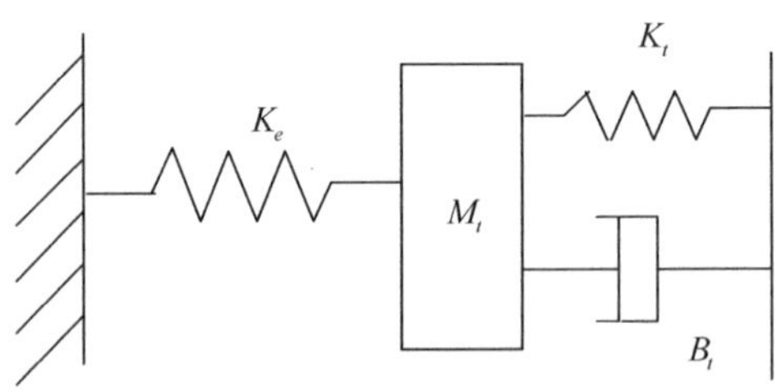

Fig. 1 Contact impedance model between finger and environment

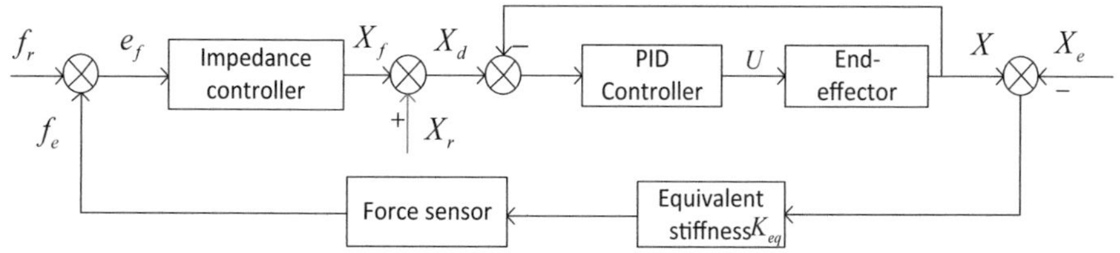

Fig. 2 Impedance control system based on position

In the actual apple-picking robot control system, the value of acceleration of end-effector is calculated by the operation of twice differentials to the position signal [4]. Taking the situation of 1-D environment impedance into consideration, the force tracking impedance controller can be designed as

$$M_t(\ddot{x} - \ddot{x}_r) + B_t(\dot{x} - \dot{x}_r) + K_t(x - x_r) = f_e - f_r \tag{2}$$

In the frequency domain, the impedance relationship can be described as

$$X_f(s) = \frac{E(s)}{M_t s^2 + B_t s + K_t} \tag{3}$$

The above Eq. (3) is like a second-order low-pass filter as M_t, B_t, K_t are diagonal matrices. It filters each input E, obtaining the value of position correction, and then the correction adds with the reference position produced by trajectory planning, so the input of position controller shown can be described as

$$X_d = X_r + X_f \tag{4}$$

The impedance controller based on Eq. (2) is shown in Fig. 2 [5].

f_e is the actual grasp force, f_r is the reference force, e_f is the force error, X is the actual grasp position, X_r is the reference position, and X_e is the initial position.

3 Error Analysis of Impedance Control

Taking the signal degree of freedom into consideration, the contact model is simplified as a linear spring system:

$$x = f_e / k_e + x_e \tag{5}$$

The reference position usually is a constant value; therefore, $\ddot{x}_r = \dot{x}_r = 0$, and the equation is substituted into the impedance control Eq. (2), getting the differential equation of force error as

$$m_d\ddot{e} + b_d\dot{e} + (k_d + k_e)e = m_d\ddot{f}_d + b_d\dot{f}_d + k_d f_d - k_d k_e(x_r - x_e) \tag{6}$$

where f_r is constant, and the force error is

$$e_{ss} = \frac{k_d}{k_d + k_e}[f_d + k_e(x_e - x_r)] = k_{eq}\left(\frac{f_d}{k_e} + x_e - x_r\right) \tag{7}$$

where k_{eq} is the equivalent stiffness of apple and target impedance controller:

$$k_{eq} = \left(k_d^{-1} + k_e^{-1}\right)^{-1} = \frac{k_d k_e}{k_d + k_e} \tag{8}$$

The stable contact force is

$$f_{ss} = f_r - e_{ss} = k_{eq}\left(\frac{f_r}{k_d} + x_r - x_e\right) \tag{9}$$

The above equations show that grasp force is the function of equivalent stiffness and desired position, which only knows the exact reference position and equivalent stiffness in advance, so that the actual force can track the reference force on time to realize the error-free force tracking. In fact, the information of environment is difficult to obtain exactly, in order to improve the precision of grasp, applying method of online adjustment to adjust the target stiffness through adjusting the equivalent stiffness. In the actual control system, the stiffness is always obtained by empirical data, as it represents a physical property of apple, and the target stiffness has a vital influence on the effect of force control [6]. Currently, the parameters of impedance controller are fixed; obviously, it cannot solve the problems caused by the robot system uncertainty and the noise caused by force sensor, so it is necessary to adjust the parameters of impedance control online to adapt the varied and complex working environment of apple-picking robot.

4 Target Stiffness Online Adjustment

The impedance controller contains three parameters: they are stiffness, damping, and inertia. Stiffness is the most important parameter, as the value shows the changes of stiffness between end-effector and apple [7], so it is necessary to adjust it to meet the system requirement. The principle of stiffness parameter adjustment is to make the system in the critical damping or damping state, so as to ensure the stability of the system. The recursive least square method (RLS) has become a widely used method for online identification because of simple principle, fast convergence, and easy to understand and program. Due to the parameters, this system is time-varying, so the RLS with forgetting factor is introduced to identify the impedance parameter [8, 9]. As the stiffness parameter k_t cannot be measured

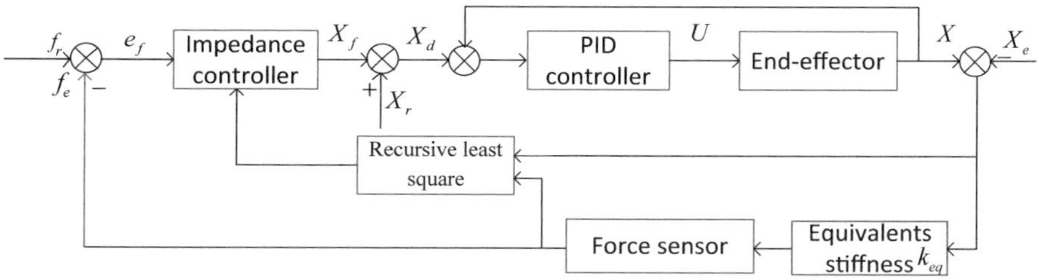

Fig. 3 The impedance controller based on RLS

directly, this paper measures the equivalent stiffness k_{eq}, and then calculates the k_t with Eq. (8); meanwhile, the RLS algorithm is used to carry out the iterative calculation to get the stiffness parameters that can meet requirement of the force for grasping apple. The impedance controller based on RLS is shown in Fig. 3.

The RLS algorithm is proposed to calculate the stiffness of the second contact model for an ideal result. In general, the contact model is always simplified as a linear spring: $y = k_{eq}x$, where y is the contact force measured by force sensor, and x is the displacement which can be calculated by the angular displacement and robot dynamic equations. After getting a set of data, the k_t can be calculated with the following equations:

$$\hat{k}_{eq,t} = \left(x_t^T x_t\right)^{-1} x_t^T y_t (t = 1, 2 \cdots) \tag{10}$$

The iterative formula of k_t based on RLS is

$$P_{t+1} = P_t / (C_t + x_{t+1} P_t x_{t+1}) \tag{11}$$

$$\hat{K}_{eq,t+1} = \hat{K}_{eq,t} + \theta_{t+1}(y_{t+1} - x_{t+1}\theta_t) \tag{12}$$

$$\theta_{t+1} = p_{t+1} x_{t+1} \tag{13}$$

$$k_{t+1} = \frac{k_{eq,t+1} \cdot k_e}{k_e - k_{eq,t+1}} \tag{14}$$

where $P_1 = 1$; $\hat{K}_{s,t} = 0$; C_t is the forgetting factor. The choice of C_t is very important because improper value will cause the covariance matrix P_t become inaccurate. In the time-varying system, it is considered a reasonable method to create a function about error $\hat{e}_t$ to calculate C_t. While following the common RLS algorithm, the forgetting factor is set as a function of error:

$$C_t = 1 - a_1[\arctan(a_2(|\hat{e}_t| - a_3))/\pi + 1/2] \tag{15}$$

where $\hat{e}_t = y_t - \hat{K}_{eq,t}x_t$; a_1, a_2, a_3 are tunable parameters. Adapting the value of C_t according to the error $\hat{e}_t$, when C_t is small, the value of K_t converges quickly but is sensitive to interference. On the contrary, the value of K_t converges slower but has a strong anti-jamming capability, and the value of a_1, a_2, a_3 can be selected according to Ref. [9]. In every sampling period of the system, taking average value of force sensor as the measured value of the contact force, when the error between measured value and reference value is around 2 %, the K_t is considered stable and the iterative process is finished.

5 Simulation Experiment

The parameters of end-effector designed by our group are shown as follows [10]: the finger of end-effector is driven by small steering gear HS-225MG; the signals of force and position are collected by data acquisition card; then the controller drives the motor which drives the finger to realize the non-damage grasp; the motor armature resistance is $R = 1.5\,\Omega$; the motor torque constant is $K_t = 0.9$; the motor EMF constant is $K_b = 0.7$; the motor inertia is $J = 0.1\,\mathrm{kg \cdot m^2}$, according to the mechanical characteristics of the DC motor; and the relationship between the voltage and the output angle is obtained:

$$AU_c = \dot{\theta}_L + B\dot{\theta}_L + \xi_o \tag{16}$$

where $\dot{\theta}_L$ is the load angle, the value of ξ_o is related to the steering gear load and reduction ration, and A 、B are constant related with steering gear:

$$A = \frac{k_t \cdot k_s}{nRJ}, B = \left(\frac{D}{J} + \frac{k_t k_b}{RJ}\right), \zeta_o = \frac{M_{coul} + \dfrac{M_n}{n}}{Jn} \tag{17}$$

The transfer function of input voltage and output angle based on motor parameter is shown as

$$g(s) = \frac{6}{s^2 + 8s} \tag{18}$$

The parameters of impedance controller are set as follows according to Ref. [5, 11] $m_d = 50\,\mathrm{N/m}$, $b_d = 200\,\mathrm{N/m}$, $k_e = 15000\,\mathrm{N/m}$, $k_d = 4000\,\mathrm{N/m}$, $k_{eq} = 10909.09\,\mathrm{N/m}$ $x_r = 0.006\,\mathrm{m}$ $x_e = 0.004\,\mathrm{m}$. The parameters of forgetting factor in RLS algorithm are set as follows in Ref. [8]: $a_1 = 0.4, a_2 = 0.5, a_3 = 1$, the simulation model of impedance control block diagram is shown in Fig. 4.

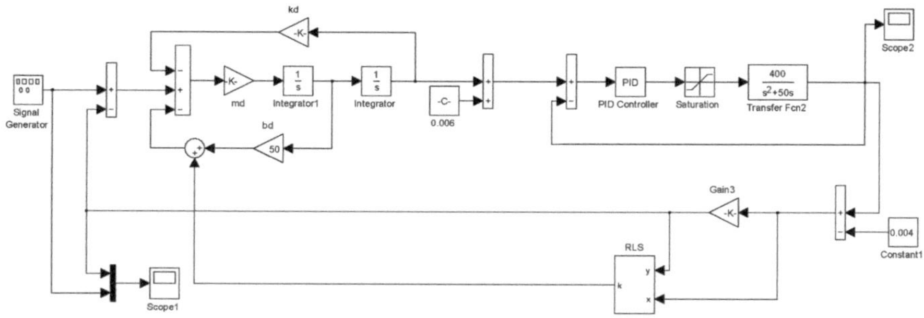

Fig. 4 The simulation of impedance control

6 Step Force Tracking Experiment

According to experimental data, the grasp force is designed as 8.1 N, and the simulation time is set as 5 s [12]. The results are shown in Figs. 5 and 6.

The simulation results are shown in the same condition: the improved impedance controller has a shorter adjustment time, smaller force error, and better force track performance than original impedance controller.

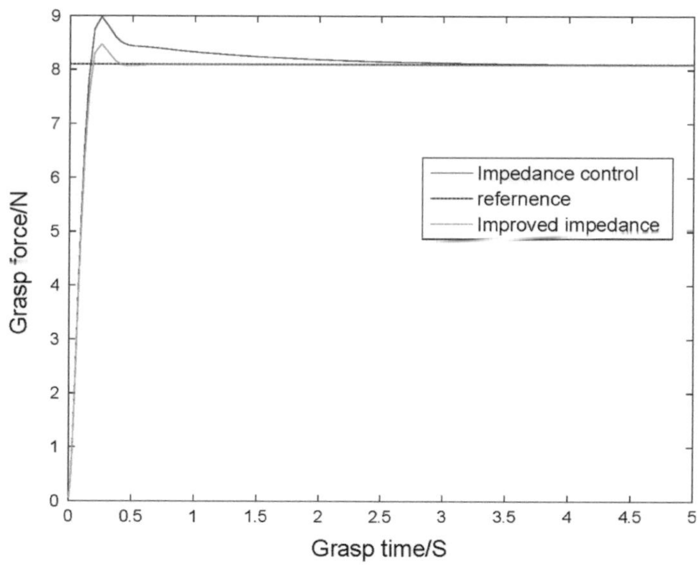

Fig. 5 Step force tracking

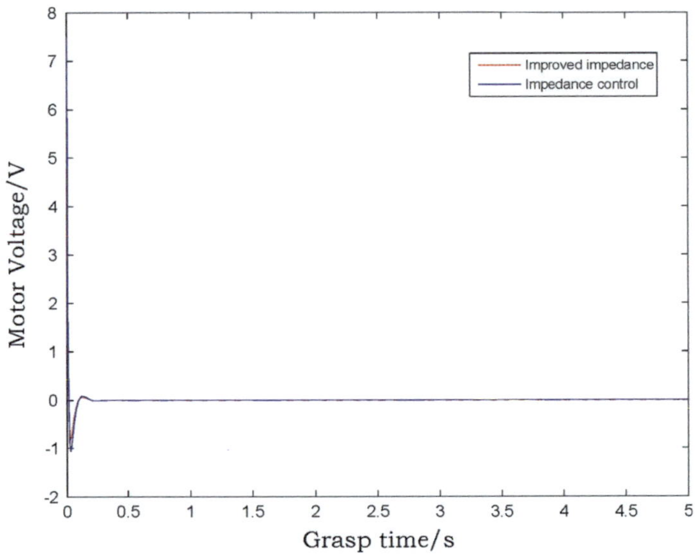

Fig. 6 Motor voltage under step force tracking

7 Sine Force Tracking Experiment

The input force is disturbed by the environment during the process of picking apple, so the sinusoidal force tracking experiment is carried out. The sinusoidal force is 2.5 N, and frequency is 0.5 Hz. The simulation results are shown in Figs. 7 and 8.

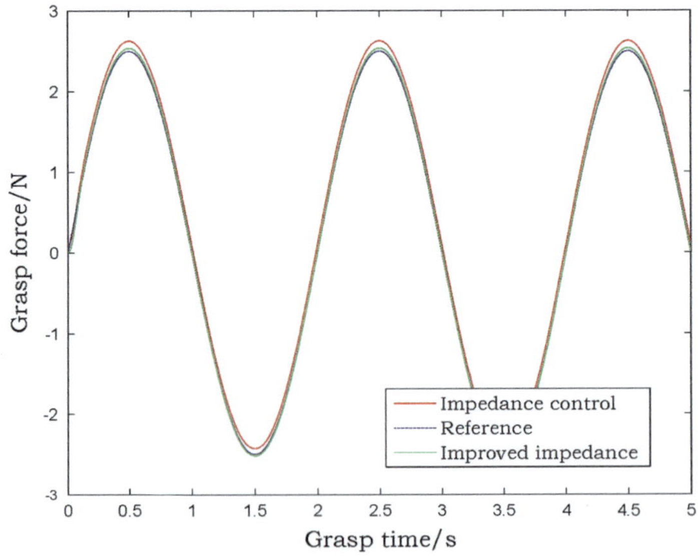

Fig. 7 Sine force tracking

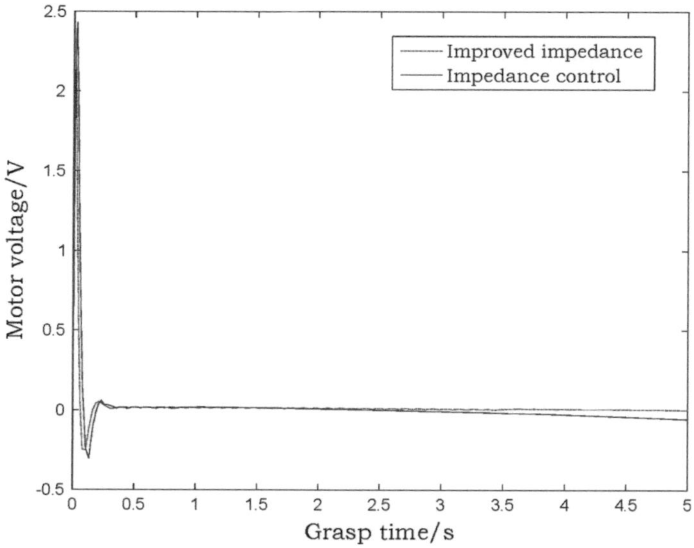

Fig. 8 Motor voltages under sine force tracking

The results show that the improved impedance control has a smaller overshoot, short state time, and better track performance to dynamic force than impedance control.

8 Conclusion

In order to avoid damaging apple, the grasp force is controlled strictly during the process of grasping apple. This paper aimed at the problem of the impedance control: due to the existence of the uncertainties of environment and random disturbance, and the parameters of impedance control are fixed, so it cannot meet the grasp requirement. This paper presents a novel impedance control based on the RLS to overcome the shortcoming of traditional method, and the simulation result shows that this method is effective and feasible. This method can provide a reference to compliant control of end-effector.

Acknowledgments This work was supported in part by the National Natural Science Foundation of China (NSFC) Grant 31571571, the Natural Science Foundation of Jiangsu Province Grant BK20150530, the Professional Research Foundation for Advanced Talents of Jiangsu University Grant 14JDG077, the Research Fund for the Doctoral Program of Higher Education of China under Grant 20133227110024, and a Project Funded by the Priority Academic Program Development of JiangSu Higher Education Institutions (PAPD).

References

1. Wu ZY, Hu XC et al (2003) Research progress and problems of agricultural robot. Trans CSAE 19(1):20–24
2. Jung S, Hsia TC (2010) Reference compensation technique of neural force tracking impedance control for robot manipulators. In: 2010 8th world congress on intelligent control and automation (WCICA). IEEE, pp 650–655
3. Villani L, Natale C, Siciliano B et al (2000) An experimental study of adaptive force/position control algorithms for an industrial robot. IEEE Trans Control Syst Technol 8(5):777–786
4. Ting Z, Li J, Hong L (2012) A novel grasping force control strategy for multi-fingered prosthetic hand. J Cent South Univ 19:1537–1542
5. Wang X, Xiao Y, Bi S et al (2015) Design of test platform for robot flexible grasping and grasping force tracking impedance control. Trans CSAE 31(1):58–63
6. Erickson D, Weber M, Sharf I (2003) Contact stiffness and damping estimation for robotic systems. Int J Robot Res 22(1):41–57
7. Jung S, Hsia TC (2007) Force tracking impedance control of robot manipulators for environment with damping. In: 33rd annual conference of the IEEE industrial electronics society, IECON 2007. IEEE, pp 2742–2747
8. Wang Z, Peer A, Buss M (2009) Fast online impedance estimation for robot control. In: IEEE international conference on mechatronics, ICM 2009. IEEE, pp 1–6
9. Li Z, Tang X, Xiong S et al (2012) Application of Kalman active observers in robot force control. J Huazhong Univ Sci Technol (Nature Science Edition) 40(2)
10. Ji W, Luo D, Li J et al (2014) Compliance grasp force control for end-effector of fruit-vegetable picking robot. Trans CSAE 30(9):19–26
11. Ma X Research on compliant control of parallel manipulator based on adaptive impedance control. Harbin Institute of Technology, Harbin
12. Luo D The research of compliance control for end-effector of fruit picking robot. Jiangsu University, Zhenjiang

Adaptive Neural Network Control for a Class of Nonlinear Systems

Chao Yang, Yingmin Jia and Changqing Chen

Abstract An adaptive neural network control scheme is developed for perturbed nonlinear systems with unknown functions. To avoid the curse of dimensionality, dynamic surface-control (DSC) technique is introduced in the progress of controller design. Moreover, the problem of singularity is solved in estimation of the unknown functions by designing a novel strategy of estimation. It is shown that the DSC-based controller can ensure semi-global uniform ultimate bounded of the closed-loop system, and the tracking error can be arbitrarily small with appropriate design parameters. A simulation example is used to demonstrate the validness of the proposed algorithm.

Keywords Adaptive neural control · Nonlinear systems · Dynamic surface-control

1 Introduction

As an effective method in the field of nonlinear system control, backstepping technique has been widely concerned in recent years. For a class of strict feedback

---

This work was supported by the National Basic Research Program of China (973 Program: 2012CB821200, 2012CB821201), the NSFC (61134005, 61327807, 61521091, 61520106010, 61304232), and the Fundamental Research Funds for the Central Universities (YWF-16-GJSYS-31, YWF-16-GJSYS-32).

C. Yang (✉) · Y. Jia
School of Automation Science and Electrical Engineering, The Seventh Research Division and the Center for Information and Control, Beihang University (BUAA), Beijing 100191, China
e-mail: yangchao_2011@sina.cn

Y. Jia
e-mail: ymjia@buaa.edu.cn

C. Chen
The Science and Technology on Space Intelligent Control Laboratory, Beijing Institute of Control Engineering, Beijing 100190, China

Y. Jia et al. (eds.), *Proceedings of 2016 Chinese Intelligent Systems Conference*, Lecture Notes in Electrical Engineering 405,
DOI 10.1007/978-981-10-2335-4_14

systems, [1–4] apply backstepping technique to the design of adaptive law. Based on the integral-type Lyapunov function, [5] proposes an adaptive neural network control strategy, but the integral calculation is complicated and constrains the application of the method in practice. In [1], the control of the strict feedback system with unknown virtual control gains is solved by the backstepping method. In [6], by combing backstepping technique with neural network, a robust control method is used for the induction motor. For nonlinear systems with unknown virtual control function but bounded derivative, [7] uses the backstepping method to design a simple structure, which relaxes the requirement of calculation of integral function. However, with the increase of the order of the plant, the computational complexity of the conventional backstepping method has increased dramatically. For a class of SISO nonlinear systems, the dynamic surface-control method is first introduced in [8], which introduces a first order filter in each step of the backstepping design to overcome the defect of traditional method. In [9], the adaptive neural network and DSC are combined to broaden application of the method.

On the basis of above results, this paper focuses on a class uncertain nonlinear systems with external disturbances. By presenting a DSC-based adaptive neural network algorithm, the conventional high computational complexity problem in backstepping technique is solved. The designed controller is simple in structure and easy to be realized in engineering.

The paper is presented as follows. Section 2 describes the nonlinear systems to be considered and some necessary assumptions are shown. The DSC-based adaptive controller is designed in Sect. 3. Section 4 gives the theoretical analysis of system stability. A simulation example is given to illustrate the effectiveness of the proposed algorithm in Sect. 5.The final Sect. 6 concludes our work.

2 Problem Formulation and Preliminaries

Consider the following nonlinear system:

$$\begin{cases} \dot{x}_i &= f_i(\bar{x}_i) + g_i x_{i+1} + \Delta_i(t), 1 \leq i \leq n-1, \\ \dot{x}_n &= f_n(x) + g_n u + \Delta_n(t), \\ y &= x_1 \end{cases} \tag{1}$$

where $x = [x_1, x_2, \dots, x_n]^T \in R^n, u \in R$, and $y \in R$ are state variable, system input, and system output, respectively, $\bar{x}_i = [x_1, x_2, \dots, x_i]^T \in R^i$; $f_i(.) : R^i \to R, i = 1, 2, \dots, n$ are unknown smooth nonlinear functions, g_i are unknown constants and $\Delta_i(.)$, $i = 1, 2, \dots, n$, are the external disturbances and satisfy $|\Delta_i(t)| \leq D_i$, with D_i being a constant.

The object is to design the adaptive neural network controller for the nonlinear system (1), on the basis of backstepping method and the DSC technique, guarantee all the closed-loop signals are semi-globally uniformly bounded, and the tracking error can be arbitrarily small. To design the controller, following assumptions are made:

Assumption 1 The reference signal $y_d(t)$ is smooth bounded and has a two-order continuous bounded derivative $\ddot{y}_d$.

Assumption 2 The control gains g_i, $1 \leq i \leq n$ are unknown constants, but the sign of g_i is known, and there exist positive constants $\bar{g}_i \geq \underline{g}_i > 0$, such that

$$\bar{g}_i \geq g_i \geq \underline{g}_i \tag{2}$$

Without loss of generality, suppose that $g_i > 0$.

The RBF neural network (RBFNN) can be used to approximate any continuous nonlinear function. An n-input-single-output RBFNNs with N neutrons in the middle layer are constructed as follows:

$$y = \theta^T \xi(Z) \tag{3}$$

where $Z \in R^n$ is the input vector, y is the output of the networks, $\theta \in R^N$ is the adjustable weight vector, $\xi(Z) = [\xi_1(Z), \xi_2(Z), \ldots, \xi_l(Z)]^T$ is nonlinear function vector with $\xi_i(Z)$ being Gaussian function, that is

$$\xi_i(Z) = \exp\left[-\frac{(Z - u_i)^T(Z - u_i)}{b_i^2}\right], i = 1, 2, \ldots, l \tag{4}$$

where $u_i = [u_{i1}, u_{i2}, \ldots, u_{iq}]^T$ is the center and b_i is the width of the $i - th$ Gaussian function. Thus, any continuous nonlinear function$f(x)$ can be appropriated as

$$f(Z) = \theta^{*T}\xi(Z) + \delta^*, Z \in \Omega \tag{5}$$

where θ^* is the ideal weight vector and defined as

$$\theta^* := arg \min_{\theta \in \bar{R}^l}\left\{\sup_{Z \in \Omega} |f(Z) - \theta^T \xi(Z)|\right\}$$

and the approximation error δ^* satisfies $|\delta^*| \leq \delta_M$.

3 Adaptive Dynamic Surface-Control Design

In this section, an DSC-based adaptive NN control method will be proposed. As always in the backstepping approach, the whole method contains n steps.

Step 1: Define $e_1 = x_1 - y_d$, the time derivative of e_1 is given by

$$\dot{e}_1 = f_1(x_1) + g_1 x_2 + \Delta_1 - \dot{y}_d \tag{6}$$

define

$$\phi_1 = \frac{1}{g_1}(f_1(x_1) - \dot{y}_d) = \theta_1^{*T} \xi_1(Z_1) + \delta_1^* \tag{7}$$

where $Z_1 = [x_1, \dot{y}_d]^T$

The visual control law is designed as

$$\alpha_2 = -(c_1 + \frac{1}{\gamma_1^2})e_1 - \hat{\theta}_1^T \xi_1(Z_1) \tag{8}$$

where c_1, γ_1 are positive design parameters, $\hat{\theta}_1$ is the estimation of θ_1^*, the adaptive law is

$$\dot{\hat{\theta}}_1 = \Gamma_1(\xi_1(Z_1)e_1 - \sigma_1 \hat{\theta}_1) \tag{9}$$

where $\Gamma_1 = \Gamma_1^T$ is a positive definite matrix, σ_1 is positive parameter to be designed.

To avoid the exploration of computation, the DSC method is considered. Let α_2 pass through the first-order filter β_2 with time constant τ_2. Then one has

$$\tau_2 \dot{\beta}_2 + \beta_2 = \alpha_2, \beta_2(0) = \alpha_2(0) \tag{10}$$

Define $e_2 = x_2 - \beta_2, \eta_2 = \beta_2 - \alpha_2 = \beta_2 + (c_1 + \frac{1}{\gamma_1^2})e_1 + \hat{\theta}_1^T \xi_1(Z_1)$, then

$$\dot{e}_1 = g_1(e_2 - (c_1 + \frac{1}{\gamma_1^2})e_1 - \hat{\theta}_1^T \xi_1(Z_1) + \delta_1^* + \eta_2) + \Delta_1 \tag{11}$$

The derivation of filtering error η can be calculated as

$$\begin{aligned} \dot{\eta}_2 = \dot{\beta}_2 - \dot{\alpha}_2 &= -\frac{\eta_2}{\tau_2} - \frac{\partial \alpha_2}{\partial e_1}\dot{e}_1 - \frac{\partial \alpha_2}{\partial x_1}\dot{x}_1 - \frac{\partial \alpha_2}{\partial y_d}\dot{y}_d - \frac{\partial \alpha_2}{\partial y_d}\ddot{y}_d - \frac{\partial \alpha_2}{\partial \hat{\theta}_1}\dot{\hat{\theta}}_1 \\ &= -\frac{\eta_2}{\tau_2} + B_2(e_1, e_2, \eta_2, \hat{\theta}_1, y_d, \dot{y}_d, \ddot{y}_d) \end{aligned} \tag{12}$$

where, $B_2(\cdot)$ is a continuous bounded function. (Details can be seen in [8]).

Step i ($2 \le i \le n-1$): For $\dot{x}_i = f_i(\bar{x}_i) + g_i(\bar{x}_i)x_{i+1} + \Delta_i(t)$, define

$$\phi_i(Z_i) = \frac{1}{g_i}(f_i(\bar{x}_i) - \dot{\beta}_i) = \theta_i^{*T} \xi_i(Z_i) + \delta_i^* \tag{13}$$

where $Z_i = [\bar{x}_i, \dot{\beta}_i]^T$

The visual control law is designed as

$$\alpha_i = -(c_i + \frac{1}{\gamma_i^2})e_i - \hat{\theta}_i^T \xi_i(Z_i) \tag{14}$$

where c_i, γ_i are positive design parameters, e_i will be given below, $\hat{\theta}_i$ is the estimation of θ_i^*, and the adaptive law is

$$\dot{\hat{\theta}}_i = \Gamma_i[\xi_i(Z_i)e_i - \sigma_i\hat{\theta}_i] \tag{15}$$

where $\Gamma_i = \Gamma_i^T$ is a positive definite matrix, σ_i is positive parameter to be designed.

To avoid the exploration of computation, the DSC method is considered. Let α_i pass through the first-order filter β_i with time constant τ_i. That is,

$$\tau_i\dot{\beta}_i + \beta_i = \alpha_i, \beta_i(0) = \alpha_i(0) \tag{16}$$

Define $e_i = x_i - \beta_i, \eta_i = \beta_i - \alpha_i = \beta_i + (c_i + \frac{1}{\gamma_i^2})e_1 + \hat{\theta}_i^T \xi_i(Z_i)$, then

$$\dot{e}_i = g_i(e_i - (c_i + \frac{1}{\gamma_i^2})e_i - \hat{\theta}_i^T \xi_i(Z_i) + \delta_i^* + \eta_i) + \Delta_i \tag{17}$$

The derivation of filtering error η_i is

$$\begin{aligned}\dot{\eta}_i = \dot{\beta}_i - \dot{\alpha}_i &= -\frac{\eta_i}{\tau_i} - \frac{\partial \alpha_i}{\partial e_i}\dot{e}_i - \frac{\partial \alpha_i}{\partial x_i}\dot{x}_i - \frac{\partial \alpha_i}{\partial y_d}\dot{y}_d - \frac{\partial \alpha_i}{\partial y_d}\ddot{y}_d - \frac{\partial \alpha_i}{\partial \hat{\theta}_i}\dot{\hat{\theta}}_i \\ &= -\frac{\eta_i}{\tau_i} + B_2(e_1, \ldots, e_i, \eta_2, \hat{\theta}_i, y_d, \dot{y}_d, \ddot{y}_d)\end{aligned} \tag{18}$$

where, $B_i(\cdot)$ is also a continuous bounded function.

Step n: Define

$$\phi_n(Z_n) = \frac{1}{g_n}(f_n(\bar{x}_n) - \dot{\beta}_n) = \theta_n^{*T} \xi_n(Z_n) + \delta_n^* \tag{19}$$

where $Z_n = [\bar{x}_n, \dot{\beta}_n]^T$

The system control input u is designed as:

$$u = -(c_n + \frac{1}{\gamma_n^2})e_n - \hat{\theta}_n^T \xi_n(Z_n) \tag{20}$$

where c_n, γ_n are positive design parameters, $e_n = x_n - \beta_n$, $\hat{\theta}_n$ is the estimation of θ_n^*, and the adaptive law is

$$\dot{\hat{\theta}}_n = \Gamma_i[\xi_n(Z_n)e_n - \sigma_n\hat{\theta}_n] \tag{21}$$

where $\Gamma_n = \Gamma_n^T$ is a positive definite matrix, σ_n is positive design parameter.

4 Stability Analysis

Next, the dynamic performance of the system will be discussed. The main result is shown in the following theorem.

Theorem *Consider nonlinear system (1) with Assumptions 1, 2. Under the control law (21), the virtual control (8), (14) and the adaptive law (9), (15), the closed-loop system is semi-globally uniform ultimate bounded, and the tracking error can be arbitrarily small with appropriate designed parameters.*

Proof The Lyapunov function candidate is chosen as

$$V = \frac{1}{2}(\sum_{i=1}^{n}\frac{e_i^2}{g_i} + \sum_{i=1}^{n}\tilde{\theta}_i^T\Gamma_i^{-1}\tilde{\theta}_i + \sum_{i=2}^{n}\eta_i^2) \tag{22}$$

The time derivative of V is given by

$$\begin{aligned}\dot{V} = &\sum_{i=1}^{n}(-(c_i + \frac{1}{\gamma_i^2})e_i^2 + e_i\delta_i^*) + \sum_{i=1}^{n-1}(e_ie_{i+1} + e_i\eta_i + |\eta_{i+1}B_{i+1}| \\ &- \frac{\eta_{i+1}^2}{\tau_{i+1}}) - \sum_{i=1}^{n}(\tilde{\theta}_i^T(\xi_i(Z_i)e_i + \Gamma_i^{-1}\dot{\tilde{\theta}}_i) + \frac{e_i\Delta_i}{g_i})\end{aligned} \tag{23}$$

Note that,

$$|e_i\Delta_i| \le \frac{1}{\gamma_i^2} + \frac{\gamma_i^2\Delta_i^2}{4} \le \frac{1}{\gamma_i^2} + \frac{\gamma_i^2D_i^2}{4} \tag{24}$$

$$e_ie_{i+1} \le e_i^2 + \frac{e_{i+1}^2}{4} \tag{25}$$

$$e_i\eta_{i+1} \le e_i^2 + \frac{\eta_{i+1}^2}{4} \tag{26}$$

$$e_i\delta_i^* \le e_i^2 + \frac{\delta_i^{*2}}{4} \tag{27}$$

Design parameters as follows $c_{11} = 3 + \alpha_0 - c^*_{10}, c_{i1} = 3.25 + \alpha_0 - c^*_{i0}, c_{n1} = 1.75 + \alpha_0 - c^*_{n0}$, where α_0 is a positive constant. Then the Eq. (25) can be rewritten as

$$\dot{V} \leq \sum_{i=1}^{n}(-\alpha_0 e_i^2 - \frac{\sigma_i}{2\lambda_m ax(\Gamma_i^{-1})}\dot{\tilde{\theta}}_i^T \Gamma_i^{-1}\tilde{\theta}_i) + \frac{\eta_i^{*2}}{4} + \frac{\sigma_i}{2}\theta_i^{*2} + \sum_{i=1}^{n-1}(\frac{\eta_{i+1}^2}{4} - \frac{\eta_{i+1}^2}{\tau_{i+1}} + |\eta_{i+1}B_{i+1}|) + \sum_{i=1}^{n}\frac{\gamma_i^2 D_i^2}{4g_i^2} \quad (28)$$

Let $\frac{1}{\tau_{i+1}} = 0.25 + 0.5M_{i+1}^2, e_M = 0.25\delta_m^2 + 0.5\sigma_i\theta_M^2$, where $|\delta_i^*| \leq \delta_m, |\theta_i^*| \leq \theta_M, |B_{i+1}| < M_{i+1}$, then one has

$$\dot{V} \leq \sum_{i=1}^{n}(-\alpha_0 e_i^2 - \frac{\sigma_i}{2\lambda_{max}(\Gamma_i^{-1})}\dot{\tilde{\theta}}_i^T \Gamma_i^{-1}\tilde{\theta}_i) + ne_M + 0.5n - \sum_{i=1}^{n-1}\alpha_0 \underline{g}_i \eta_{i+1}^2 + \sum_{i=1}^{n}\frac{\gamma_i^2 D_i^2}{4\underline{g}_i^2} \quad (29)$$

Design $\alpha_0 \geq 0.5r/\underline{g}_i, \sigma_i \leq r\lambda_{max}(\Gamma_i^{-1})$ and $Q = ne_M + 0.5n + \sum_{i=1}^{n}\frac{\gamma_i^2 D_i^2}{4\underline{g}_i^2}$ with r being a positive constant. Then it comes to

$$\dot{V} \leq -\frac{1}{2}(\sum_{i=1}^{n}\frac{re_i^2}{g_i(\bar{x}_i)} + \sum_{i=1}^{n} r\tilde{\theta}_i^T \Gamma_i^{-1}\tilde{\theta}_i + \sum_{i=2}^{n} r\eta_i^2) + Q \leq -rV + Q \quad (30)$$

which means

$$V \leq \frac{Q}{r} + (V(0) - \frac{Q}{r})e^{-rt} \quad (31)$$

At this point, the theorem is proved.

5 Simulation Example

To illustrate the effectiveness of the proposed approach, an example is considered in this section. Consider the following nonlinear system:

$$\begin{cases} \dot{x}_1 = f_1(x_1) + g_1 x_2 + d_1 \\ \dot{x}_2 = f_2(\bar{x}_2) + x_3 + d_2 \\ \dot{x}_3 = f_3(\bar{x}_3) + g_2 u + d_3 \\ y = x_1 \end{cases} \quad (32)$$

where the actual value of the unknown functions are $f_1 = -0.05cos(x_1) + 0.1, f_2 = 0.05cos(x_1) - x_2, f_3 = 0.1x_2 - 0.03x_3, g_1 = 1, g_2 = 1,$ and the disturbances $d_1 = 0.01sin(2t), d_2 = 0.1cos(2t), d_3 = 0.04sin(t)cos(2t)$

In the proposed method, the ideal tracking trace is $y_d = 5sin(t)$. Parameters are designed as follows. $\sigma_i = 0.05, \delta_M = 0.05, r = 0.1, c_{11} = 1.5, c_{21} = 2.0, c_{31} = 1.0,$ $\gamma_i = 0.1, \Gamma_i = diag20, \tau_i = 0.03, \underline{g}_i = 1$, and the initial conditions as $x_0 = [0.5, 0, 0]$ and $\theta_i(0) = 0$.The first NN contains 7 nodes with centers evenly located in [−3, 3] and the widths are 2. The second NN contains 9 nodes with centers evenly spaced in [−9, 9] and the widths are 4. The third neural network contains 5 nodes with centers evenly located in [−5, 5] and the widths are 2.

The simulation results are given in figures below. From these figures, it is clearly shown that under the proposed control law, all the closed-loop signals are semi-globally bounded (Fig. 1a, b).

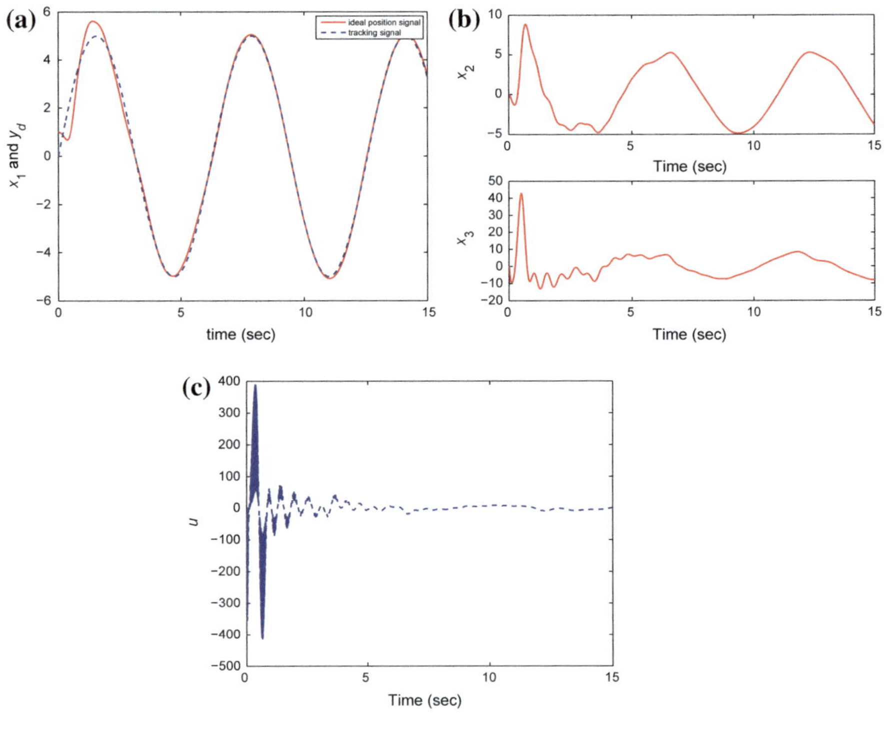

Fig. 1 **a** x_1 and y_d. **b** x_2, x_3. **c** u

6 Conclusions

Based on DSC method, an adaptive neural network control algorithm is designed for a class of uncertain nonlinear systems with external disturbances. In this scheme, the problem of singularity is solved in estimation of the unknown functions by designing a novel strategy of estimation. The designed controller is simple in structure and easy to be realized in engineering. The proposed method also avoids the curse of dimensionality in traditional backstepping method. Simulation examples demonstrate the effectiveness of the proposed algorithm.

References

1. Krstic M, Kanellakopoulos I, Kokotovic PV (1995) Nonlinear and adaptive control design. Wiley, New York
2. Zhou J, Wen C, Zhang Y (2004) Adaptive backstepping control of a class of uncertain nonlinear systems with unknown backlash-like hysteresis. IEEE Trans Autom Control 49(10):1751–1757
3. Sun L, Tong S, Liu Y (2011) Adaptive backstepping sliding mode H∞ control of static var compensator. IEEE Trans Control Syst Technol 19(5):1178–1185
4. Yu J, Chen B, Yu H, Gao J (2011) Adaptive fuzzy tracking control for the chaotic permanent magnet synchronous motor drive system via backstepping. Nonlinear Anal Real World Appl 12(1):671–681
5. Zhang T, Ge SS (2000) Adaptive neural network control for strict-feedback nonlinear systems using backstepping design. Automatica 36(12):1835–1846
6. Kwan CM, Lewis FL (2000) Robust backstepping control of induction motors using neural networks. IEEE Trans Neural Netw 11(5):1178–1187
7. Ge SS, Wang C (2002) Adaptive NN control of uncertain nonlinear pure-feedback systems. Automatica 38(4):671–682
8. Swaroop D, Hedrick JK, Yip PP, Gerdes JC (2000) Direct adaptive NN control for a class of nonlinear systems. IEEE Trans Neutral Netw 45(10):1893–1899
9. Ge SS, Wang C (2002) Direct adaptive NN control for a class of nonlinear systems. IEEE Trans Neutral Netw 13(1):214–221

Robust Coupling-Observer-Based Linear Quadratic Regulator for Air-Breathing Hypersonic Vehicles with Flexible Dynamics and Parameter Uncertainties

Na Wang, Lin Zhao, Chong Lin and Yumei Ma

Abstract This paper studies the anti-disturbance control problem for air-breathing hypersonic vehicles (AHVs) with flexible dynamics and parameter uncertainties. A novel anti-disturbance control method is presented, which includes a robust coupling observer (RCO) and a linear quadratic regulator (LQR). The compensator is designed to reject the disturbance generated by rigid-flexible couplings (RFCs). The LQR is presented to track desired trajectories. Finally, simulation results show that the control performance can be improved by using the RCO-based LQR compared with the traditional linear quadratic regulator.

Keywords Robust coupling observer · Flexible air-breathing hypersonic vehicle · Uncertainties · Linear quadratic regulator

1 Introduction

Air-breathing hypersonic vehicles (AHVs) present a cost-efficient way to make access to space routine. It is very important and difficult to design a flight control law for AHVs, especially for flexible AHVs (FAHVs) with uncertainties. Many control methods have been proposed for AHVs [1–23] in the past years. An FAHV model has been given in [1]. Then several effective controllers are presented for this nonlinear model [21–23]. But the nonlinear control methods presented in [22, 23] are proposed base on FAHV models without RFCs. It has been shown that RFCs may degrade the flight control performance of FAHVs (see [1, 23]). Therefore the RFCs are investigated and transformed into some forces and moment terms in [23]. Large amount of calculations may be caused by online adaptive laws. Thus, an RCO-based compensator is designed to reject the flexible effects to deal with the problem in

N. Wang (✉) · L. Zhao · C. Lin · Y. Ma
College of Automation and Electrical Engineering, Qingdao University, Qingdao 266071, Shandong Province, China
e-mail: wangnaflcon@126.com

Y. Jia et al. (eds.), *Proceedings of 2016 Chinese Intelligent Systems Conference*, Lecture Notes in Electrical Engineering 405,
DOI 10.1007/978-981-10-2335-4_15

[23]. Then a novel nonlinear composite hierarchical control method is constructed to achieve tracking objectives and compensate the disturbances produced by RFCs.

Recently, the disturbance-observer-based control strategy has attracted considerable attention. In most cases, principle of disturbance observer design in the time domain is similar to one of the state observers presented in [24], which can be concluded as follows: with suitable selection of disturbance observer gain, the dynamics of disturbance observer must be faster than the ones of the actual system, so that the estimate of disturbance can approach the actual disturbance as quickly as possible. In [25], it has been shown that this kind of composite hierarchical anti-disturbance control scheme can be applied to solve many control problems with multiple disturbances. Moreover, it has been applied effectively in many practical systems (see e.g. [7, 20, 26–30]). For example, a disturbance-observer-based dynamic inversion controller was presented for missile systems in [30], and several effective controllers based on disturbance observer have been proposed for the AHV model in [7, 29, 30]. But the work for the nonlinear FAHV model with RFCs using composite control strategy is still insufficient. For example, a coupling observer based nonlinear controller was presented for FAHV in [31], but the parameter uncertainties were not considered in that paper. So an RCO-based sliding mode controller was proposed for FAHVs in [32]. However, there are still many control methods which can be combined with the RCO. And different problem need be solved for different control method. So a novel anti-disturbance controller is proposed, which includes an RCO-based compensator and a linear quadratic regulator.

The rest paper is organized as follows. In Sect. 2, the FAHV model is introduced and the control objective is defined. In Sect. 3, a novel composite hierarchical control strategy is proposed. Simulation results are shown in Sect. 4, and conclusions are offered in Sect. 5.

2 Problem Formulation

In [1], a nonlinear FAHV model for the longitudinal dynamics, as developed by Bolender and Doman, is described as

$$\dot{h} = V \sin \gamma \tag{1}$$

$$\dot{V} = \frac{T \cos \alpha - D}{m} - g \sin \gamma \tag{2}$$

$$\dot{\alpha} = -\frac{T \sin \alpha + L}{mV} + Q + \frac{g}{V} \cos \gamma \tag{3}$$

$$\dot{\gamma} = \frac{T \sin \alpha + L}{mV} - \frac{g}{V} \cos \gamma \tag{4}$$

Table 1 Nomenclature

h	Altitude	L	Lift
γ	Flight path angle	N_i	Generalized elastic force
V	Velocity	D	Drag
θ	Pitch angle, $\theta = \alpha - \gamma$	M	Pitching moment
α	Angle of attack	T	Thrust
Q	Pitch rate, $Q = \dot{\theta}$	m	Vehicle mass
η_i	Generalized elastic coordinate	I_{yy}	Moment of inertia
ζ_i	Damping ratio for η_i	Φ	Fuel equivalence ratio
$\tilde{\psi}_i$	Inertial coupling parameter	g	Acceleration due to gravity
ω_i	Natural frequency for η_i	δ_e	Elevator deflection
S	Reference area	$\bar{c}$	Mean aerodynamic chord
z_T	Thrust moment arm	q	Dynamic pressure, $q = (\rho V^2)/2$
ρ	Air density, $\rho = \rho_0 \exp(-(h - h_0)/h_s)$	h_s^{-1}	Air density decay rate
ρ_0	Air density at trimmed cruise condition	h_0	Altitude at trimmed cruise condition

$$I_{yy}\dot{Q} = M + \tilde{\psi}_1\ddot{\eta}_1 + \tilde{\psi}_2\ddot{\eta}_2 \tag{5}$$

$$\ddot{\eta}_i = -2\zeta_i\omega_i\dot{\eta}_i - \omega_i^2\eta_i + N_i + \tilde{\psi}_i\dot{Q}, \ i = 1,2 \tag{6}$$

where $x_r = [h\ V\ \alpha\ \gamma\ Q]^T$ is rigid-body state, $u = [\delta_e\ \Phi]^T$ is control input and flexible mode η_i, and the output is $y = [V\ \gamma]^T$. The nomenclature is given in Table 1. Coefficients of forces and moment are subject to uncertainties in this paper. So $L = L^0 + \Delta L$, $D = D^0 + \Delta D$, $T = T^0 + \Delta T$, $N_i = N_i^0 + \Delta N_i$, $M = M^0 + \Delta M$. Where the approximations of the forces and moment see [21].

3 RCO-Based LQR Design

3.1 Control System Model

The control system model with multiple disturbances is developed from model (1)–(6), which is given by

$$\dot{V} = f_1(x_r, \Phi) + d_1 \tag{7}$$

$$\dot{\alpha} = f_2(x_r, \Phi) + d_2 \tag{8}$$

$$\dot{\gamma} = f_3(x_r, \Phi) + d_3 \tag{9}$$

$$\dot{Q} = f_4(x_r, \Phi) + g_4(x_r, \Phi)\delta_e + d_4 \tag{10}$$

$$\dot{x}_f = F_f(x_r, \Phi) + A_f x_f + B_f(x_r)\delta_e + E_f d_f \tag{11}$$

where flexible state $x_f = [\eta_1 \ \ \eta_2 \ \ \dot{\eta}_1 \ \ \dot{\eta}_2]^T$, d_1, d_2, d_3 and d_f are disturbances generated by the uncertain parameters, and $d_4 = d_u + d_c$, d_u is composite disturbance generated by uncertain parameters, while d_c is generated by the RCOs and $d_c = C_f x_f$. The concrete expressions of $f_1(x_r, \Phi)$, $f_2(x_r, \Phi)$, $f_3(x_r, \Phi)$, $f_4(x_r, \Phi)$, $g_4(x_r, \Phi)$, d_1, d_2, d_3, d_u, A_f, $B_f(x_r)$, C_f, E_f, $F_f(x_r, \Phi)$ and d_f are given in [32].

3.2 RCO Design

From (10) and (11), an RCO can be designed as follows

$$\begin{cases} \hat{d}_c = C_f \hat{x}_f \\ \hat{x}_f = v_f + L_f Q \\ \dot{v}_f = (A_f - L_f C_f)(v_f + L_f Q) + B_f(x_r)\delta_e \\ \qquad + F_f(x_r, \Phi) - L_f(f_4(x_r, \Phi) + g_4(x_r, \Phi)\delta_e) \end{cases} \tag{12}$$

where $\hat{d}_c$ is estimate of d_c, $\hat{x}_f$ is estimate of x_f, v_f is the internal state, and we need design the observer gain $L_f \in \mathcal{R}^{4\times 1}$.

Define $e_f = x_f - \hat{x}_f$ and $e_c = d_c - \hat{d}_c$, then $e_c = C_f e_f$. Using the RCO (12), the differential equation for e_f is described as

$$\begin{cases} \dot{e}_f = \bar{A}_f e_f + \bar{E}_f \bar{d}_f \\ z_f = C_f e_f \end{cases} \tag{13}$$

where $\bar{A}_f = A_f - L_f C_f$, $\bar{E}_f = [-\beta_f L_f \ \ E_f]$, $\bar{d}_f = [\bar{d}_u \ \ d_f]^T$, $\bar{d}_u = \beta_f^{-1} d_u$, tuning parameter β_f can ensure that $\bar{d}_u$, d_{f1} and d_{f2} have the same order of magnitude to obtain favorable observer gain L_f, and z_f is the reference output, so $z_f = e_c$.

Theorem 1 *Consider system (13), if there exist matrices $X_f \in \mathcal{R}^{4\times 4}$ and $R_f \in \mathcal{R}^{4\times 1}$ such that*

$$\begin{bmatrix} sym(X_f A_f - R_f C_f) & -R_f & X_f E_f & C_f^T \\ \star & -\gamma_f & 0 & 0 \\ \star & \star & -\gamma_f I & 0 \\ \star & \star & \star & -\gamma_f \end{bmatrix} < 0 \tag{14}$$

then by selecting $L_f = X_f^{-1} R_f$, the system (13) is asymptotically stable when the disturbance $\bar{d}_f$ is ignored and $||z_f||_2 \leq \gamma_f ||\bar{d}_f||_2$ is satisfied, where matrix X_f is positive, and γ_f represents the level of disturbance attenuation and $\gamma_f > 0$.

Proof Consider the Lyapunov function $W_f = e_f^T P_f e_f$, where matrix $P_f \in \mathcal{R}^{4\times4}$ is positive. To prove the stability and robustness of system (13), we choose

$$J_f(t) = \int_0^t [z_f^T(\tau)z_f(\tau) - \gamma_f^2 \bar{d}_f(\tau)^T \bar{d}_f(\tau) + \dot{W}_f(\tau)]dt.$$

Defining $H_f = z_f^T z_f - \gamma_f^2 \bar{d}_f^T \bar{d}_f + \dot{W}_f$, then it is shown that

$$H_f = \begin{bmatrix} e_f \\ \bar{d}_u \\ d_f \end{bmatrix}^T \begin{bmatrix} P_f\bar{A}_f + \bar{A}_f^T P_f + C_f^T C_f & -P_f\beta_f L_f & P_f E_f \\ \star & -\gamma_f^2 & 0 \\ \star & 0 & -\gamma_f^2 I \end{bmatrix} \begin{bmatrix} e_f \\ \bar{d}_u \\ d_f \end{bmatrix}$$

Substituting $R_f = X_f L_f$ into (14), and pre- and post-multiplying (14) by $diag\{\gamma_f^{\frac{1}{2}} I, \gamma_f^{\frac{1}{2}}, \gamma_f^{\frac{1}{2}} I, \gamma_f^{-\frac{1}{2}}\}$, and using the Schur Complement Property, we can obtain

$$\begin{bmatrix} sym(P_f A_f - P_f L_f C_f) + C_f^T C_f & -P_f\beta_f L_f & P_f E_f \\ \star & -\gamma_f^2 & 0 \\ \star & \star & -\gamma_f^2 I \end{bmatrix} < 0 \tag{15}$$

If the matrix inequality (15) holds, it can be seen that $H_f < 0$. Finally, $J_f(t) < 0$ and $||z_f||_2 \le \gamma_f ||\bar{d}_f||_2$ hold under the zero initial condition. If LMI (15) holds, we can obtain $P_f\bar{A}_f + \bar{A}_f^T P_f < 0$. *End*

3.3 RCO-Based LQR Design

At first, an actuator model is described as

$$\ddot{\Phi} = -2\zeta\omega\dot{\Phi} - \omega^2\Phi + \omega^2\Phi_c \tag{16}$$

Then a model can be developed from (7)–(10). Selecting $\bar{u} = [\delta_e \;\; \Phi_c]^T$ as the new input. According to (7)–(10) and (16), we can obtain

$$\begin{aligned} \begin{bmatrix} V^{(3)} \\ \gamma^{(3)} \end{bmatrix} &= \begin{bmatrix} f_V(x_r, \Phi) \\ f_\gamma(x_r, \Phi) \end{bmatrix} + \begin{bmatrix} g_{V1}(x_r, \Phi) & g_{V2}(x_r, \Phi) \\ g_{\gamma1}(x_r, \Phi) & g_{\gamma2}(x_r, \Phi) \end{bmatrix} \bar{u} \\ &+ \begin{bmatrix} g_{V3}(x_r, \Phi) \\ g_{\gamma3}(x_r, \Phi) \end{bmatrix} d_c + \begin{bmatrix} d_V \\ d_\gamma \end{bmatrix} \end{aligned} \tag{17}$$

where control input $\bar{u}$ and disturbance d_c appear explicitly, nonlinear functions $f_V(x_r, \Phi)$, $f_\gamma(x_r, \Phi)$, $g_{Vj}(x_r, \Phi)$ and $g_{\gamma j}(x_r, \Phi)$, $j = 1, 2, 3$ are sufficiently smooth with respect to x_r and Φ, disturbances d_V and d_γ are functions of disturbances d_u, d_1, d_3 and their derivatives. Nonlinear functions $f_V(x_r, \Phi)$, $f_\gamma(x_r, \Phi)$, $g_{Vj}(x_r, \Phi)$, $g_{\gamma j}(x_r, \Phi)$, $j = 1, 2, 3$ and disturbances d_V, d_γ see [32].

When the estimation value of d_c is developed, the RCO-based LQR (RCOBLQR) can be obtained

$$\bar{u} = -G^{-1}(x_r, \Phi)(u_1 + u_2 + u_3) \tag{18}$$

where

$$G(x_r, \Phi) = \begin{bmatrix} g_{V1}(x_r, \Phi) & g_{V2}(x_r, \Phi) \\ g_{\gamma 1}(x_r, \Phi) & g_{\gamma 2}(x_r, \Phi) \end{bmatrix}, \; u_1 = \begin{bmatrix} f_V(x_r, \Phi) \\ f_\gamma(x_r, \Phi) \end{bmatrix},$$

$$u_2 = \begin{bmatrix} -V_d^{(3)} + K_V \bar{e}_V(t) \\ -\gamma_d^{(3)} + K_\gamma \bar{e}_\gamma(t) \end{bmatrix}, \; u_3 = \begin{bmatrix} g_{V3}(x_r, \Phi)\hat{d}_c \\ g_{\gamma 3}(x_r, \Phi)\hat{d}_c \end{bmatrix}.$$

in which $\hat{d}_c$ is the disturbance estimation value obtained by the RCO (12), the velocity tracking error is $e_V(t) = V - V_d$, the flight-path angle tracking error is $e_\gamma(t) = \gamma - \gamma_d$, $\bar{e}_V(t) = [\int_0^t e_V(\tau)d\tau \;\; e_V(t) \;\; \dot{e}_V(t) \;\; \ddot{e}_V(t)]^T$, $\bar{e}_\gamma(t) = [\int_0^t e_\gamma(\tau)d\tau \;\; e_\gamma(t) \;\; \dot{e}_\gamma(t) \;\; \ddot{e}_\gamma(t)]^T$, the control gains $K_V \in \mathcal{R}^{1\times 4}$ and $K_\gamma \in \mathcal{R}^{1\times 4}$ are to be obtained by solving following two Riccati equations

$$P_1 A_0 + A_0^T P_1 - P_1 B_0 R_V^{-1} B_0^T P_1 + Q_V = 0 \tag{19}$$

$$P_2 A_0 + A_0^T P_2 - P_2 B_0 R_\gamma^{-1} B_0^T P_2 + Q_\gamma = 0 \tag{20}$$

where

$$A_0 = \begin{bmatrix} 0\,1\,0\,0 \\ 0\,0\,1\,0 \\ 0\,0\,0\,1 \\ 0\,0\,0\,0 \end{bmatrix}, \; B_0 = \begin{bmatrix} 0 \\ 0 \\ 0 \\ 1 \end{bmatrix},$$

and $P_1 \in \mathcal{R}^{4\times 4}$, $P_2 \in \mathcal{R}^{4\times 4}$ and $P_3 \in \mathcal{R}^{4\times 4}$ are positive definite matrices, the weights $Q_V \in \mathcal{R}^{4\times 4}$, $Q_\gamma \in \mathcal{R}^{4\times 4}$, $R_V \in \mathcal{R}$ and $R_\gamma \in \mathcal{R}$, satisfying

$$Q_V \geq 0, \;\; Q_\gamma \geq 0$$

$$0 < R_f, \;\; 0 < R_V \leq \beta_V^{-2}, \;\; 0 < R_\gamma \leq \beta_\gamma^{-2}$$

then by selecting $K_V = -R_V^{-1} B_0^T P_2$ and $K_\gamma = -R_\gamma^{-1} B_0^T P_3$, the velocity tracking error $e_V(t)$ and the flight-path angle tracking error $e_\gamma(t)$ can converge to zero for FAHV without uncertainties in [31]. When the estimation value $\hat{d}_c = 0$, that is, there is no RCO, the RCO-based LQR (18) reduces to the dynamic-inversion-based LQR (DIBLQR).

4 Simulations

In this section, the uncertainties $\Delta C_L^{(\cdot)}$, $\Delta C_D^{(\cdot)}$, $\Delta C_M^{(\cdot)}$, $\Delta C_T^{(\cdot)}$, $\Delta C_{T\Phi}^{(\cdot)}$ and $\Delta C_{N_i}^{(\cdot)}$ are all chosen as the random signal that remains within [−15 %, 15 %].

In the simulation, maneuver task is acceleration at constant dynamic pressure ($V_d(t) = 2500V_d^0(t)$, $\gamma_d(t) = arcsin(2h_s\dot{V}_d(t)/V_d^2(t))$) at time $t = 200$ s (details see [22]). We set weights $Q_V = diag\{1, 10, 1, 1\}$, $Q_\gamma = I_{4\times4}$, $R_V = 1$, $R_\gamma = 0.1$, and parameters $\beta_V = 1$, $\beta_\gamma = 1$. Then we have $K_V = -[3.4849\ \ 5.5722\ \ 4.5983\ \ 1]$, $K_\gamma = -[5.4399\ \ 9.7962\ \ 8.4827\ \ 3.1623]$ by solving the two Riccati equations (19)

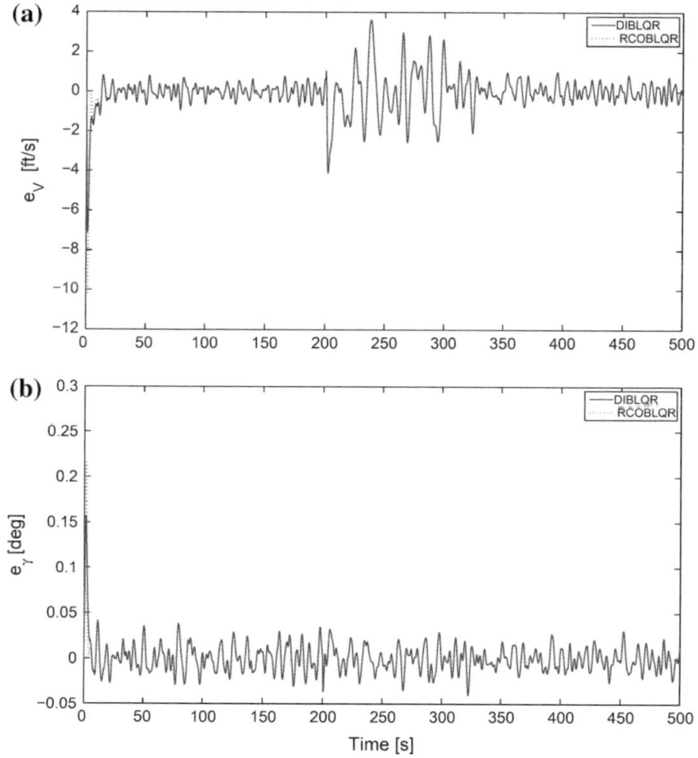

Fig. 1 **a** Velocity tracking error $e_V(t)$, **b** Flight-path angle tracking error $e_\gamma(t)$

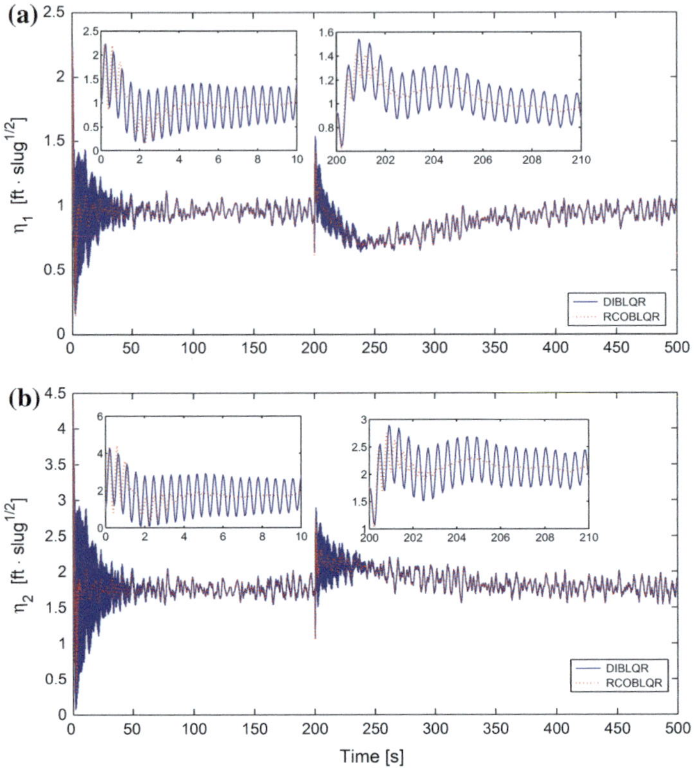

Fig. 2 **a** Flexible mode $\eta_1(t)$, **b** Flexible mode $\eta_2(t)$

and (20). And we obtain $L_f = [2.1616 \quad -4.7157 \quad -0.6872 \quad -8.4052]^T$ according to Theorem 1. In addition, the gains of DIBLQR are chosen as $\bar{K}_V = K_V$ and $\bar{K}_\gamma = K_\gamma$.

The simulation results of $e_V(t)$, $e_\gamma(t)$, $\eta_1(t)$ and $\eta_2(t)$ using RCOBLQR and DIBLQR are shown in Figs. 1 and 2. It can be observed from Fig. 1 that the tracking errors of RCOBLQR and DIBLQR are similar because of $\bar{K}_V = K_V$ and $\bar{K}_\gamma = K_\gamma$. However, the high-frequency oscillations of flexible modes can be alleviated using RCOBLQR, as shown in Fig. 2. So performance degradation or large oscillations of $\eta_1(t)$ and $\eta_2(t)$ may be caused under DIBLQR, while the RCOBLQR can alleviate these oscillations.

5 Conclusions

In this paper, the flexible effects produced by RCOs are considered as unknown interferences. A novel nonlinear controller which includes an RCO-based feedforward compensator and a dynamic-inversion-based LQR is presented, where the compensator is constructed to reject the flexible effects on pitch rate and the LQR is designed to guarantee velocity and flight-path angle track the desired signals. Simulation results show that performance degradation or large oscillations of flexible modes may be caused using the traditional dynamic-inversion-based LQR without considering the RFCs, while the RCO-based LQR can alleviate these oscillations.

References

1. Bolender MA, Doman DB (2007) Nonlinear longitudinal dynamical model of an air-breathing hypersonic vehicle. J Spacecraft Rockets 44(2):374–387
2. Schmidt D (1992) Dynamics and control of hypersonic aeropropulsive/aeroelastic vehicles. AIAA paper 1992-4326
3. Chavez FR, Schmidt DK (1994) Analytical aeropropulsive/aeroelastic hypersonic-vehicle model with dynamic analysis. J Guidance Control Dyn 17(6):1308–1319
4. Chavez FR, Schmidt DK (1999) Uncertainty modeling for multivariable-control robustness analysis of elastic high-speed vehicles. J Guidance Control Dyn 22(1):87–95
5. Williams T, Bolender M, Doman D (2006) An aerothermal flexible mode analysis of a hypersonic vehicle. AIAA paper 2006-6647
6. Gregory IM, Chowdhry RS, McMinn JD, Shaughnessy JD (1994) Hypersonic vehicle model and control law development using H_∞ and μ synthesis. NASA TM-4562
7. Chen M, Jiang CS, Wu QX (2011) Disturbance-observer-based robust flight control for hypersonic vehicles using neural networks. Adv Sci Lett 4(5):1771–1775
8. Groves KP, Sigthorsson DO, Serrani A, Yurkovich S, Bolender MA, Doman DB (2005) Reference command tracking for a linearized model of an air-breathing hypersonic vehicle. AIAA paper 2005-6144
9. Groves KP, Serrani A, Yurkovich S, Bolender MA, Doman DB (2006) Anti-windup control for an air-breathing hypersonic vehicle model. AIAA paper 2006-6557
10. Sigthorsson DO, Serrani A, Yurkovich S, Bolender MA, Doman DB (2006) Tracking control for an overactuated hypersonic air-breathing vehicle with steady state constraints. AIAA paper 2006-6558
11. Sigthorsson D, Jankovsky P, Serrani A, Yurkovich S, Bolender M, Doman D (2008) Robust linear output feedback control of an airbreathing hypersonic vehicle. J Guidance Control Dyn 31(4):1052–1066
12. Hu X, Wu L, Hu C, Gao H (2012) Adaptive sliding mode tracking control for a flexible air-breathing hypersonic vehicle. J Franklin Inst Eng Appl Math 349(2):559–577
13. Marrison C, Stengel R (1998) Design of robust control systems for a hypersonic aircraft. J Guidance Control Dyn 21(1):58–63
14. Wang Q, Stengel R (2000) Robust nonlinear control of a hypersonic aircraft. J Guidance Control Dyn 23(4):577–585
15. Xu H, Mirmirani M, Ioannou P (2004) Adaptive sliding mode control design for a hypersonic flight vehicle. J Guidance Control Dyn 27(5):829–838
16. Butt WA, Yan L, Kendrick AS (2010) Robust adaptive dynamic surface control of a hypersonic flight vehicle. In: Proceedings of the 49th IEEE conference on decision control. Atlanta, Georgia, U.S.A., pp 3632–3637
17. Huang H,Zhang Z (2015) Characteristic model-based H2/H robust adaptive control during the re-entry of hypersonic cruise vehicles. Sci China Inf Sci 58(1):012202(21)
18. Lan XJ, Wang YJ, Liu L (2015) Dynamic decoupling tracking control for the polytopic LPV model of hypersonic vehicle. Sci China Inf Sci 58(9):092203(14)
19. Wang P, Tang GJ, Wu J (2015) Sliding mode decoupling control of a generic hypersonic vehicle based on parametric commands. Sci China Inf Sci 58(5):052202(14)

20. Li SH, Sun HB, Sun CY (2012) Composite controller design for an air-breathing hypersonic vehicle. Proc Inst Mech Eng Part I: J Syst Control Eng 69(7):595–611
21. Parker JT, Serrani A, Yurkovich S, Bolender MA, Doman DB (2007) Control-oriented modeling of an air-breathing hypersonic vehicle. J Guidance Control Dyn 30(3):856–869
22. Fiorentini L, Serrani A (2012) Adaptive restricted trajectory tracking for a non-minimum phase hypersonic vehicle model. Automatica 48(7):1248–1261
23. Fiorentini L, Serrani A, Bolender M, Doman D (2009) Nonlinear robust adaptive control of flexible air-breathing hypersonic vehicles. J Guidance Control Dyn 32(2):401–416
24. Driels M (1996) Linear control systems engineering. McGraw-Hill, New York
25. Guo L, Cao S (2013) Anti-disturbance control for systems with multiple disturbances. CRC Press, Boca Raton
26. Guo L, Cao SY (2014) Anti-disturbance control theory for systems with multiple disturbances: a survey. ISA Trans 53(4):846–849
27. Guo L, Chen WH (2005) Disturbance attenuation and rejection for systems with nonlinearity via DOBC approach. Int J Robust Nonlinear Control 15(3):109–125
28. Guo L, Wen XY (2011) Hierarchical anti-disturbance adaptive control for non-linear systems with composite disturbances and applications to missile systems. Trans Inst Meas Control 33(8):942–956
29. Wen XJ, Guo L (2010) Composite disturbance-observer-based control and H_∞ control for complex continuous models. Int J Robust Nonlinear Control 20(1):106–118
30. Chen WH (2003) Nonlinear disturbance observer-enhanced dynamic inversion control of missiles. J Guidance Control Dyn 26(1):161–166
31. Wang N, Wu HN, Guo L (2014) Coupling-observer-based nonlinear control for flexible air-breathing hypersonic vehicles. Nonlinear Dyn 78(3):2141–2159
32. Wang N, Li WS, Guo L, Han HL (2014) Robust coupling-observer-based sliding mode control for flexible air-breathing hypersonic vehicles. In: Proceedings of 2014 IEEE Chinese guidance, navigation and control conference. Yantai, China, pp 1415–1422

An Intelligent Surveillance System for Crowed Abnormal Detection

Xin Tan, Chao Zhang, Chubin Zhuang and Hongpeng Yin

Abstract In this paper, an intelligent surveillance system for emergency detection is designed for campus. The framework of the system includes following steps. First, the foreground extraction is used with background difference and three-frame-difference methods to get the moving objects. Second, based on canny operator, an edge detection is employed to eliminate the effect of brightness. Third, to enhance the accuracy of the results at non-vertical visual angle, the image is divided into limited layers to detect separately after edge detection, based on pixel features. In this way, the crowd density information can be obtained. Experimental results show the effectiveness of the designed system.

Keywords Intelligent surveillance · Foreground extraction · Canny operator · Pixel features · Detect separately

1 Introduction

Nowadays, the security issues in campus is hot, since stampede happened sometimes. On 26th September, 2014, a stampede occurred in a primary school of Kunming, Yunnan province. Six students died and twenty-six students injured. It is a great potential danger threatening the safety of students. So it is urgent to protect the students against stampede dangers.

There are some limitations in traditional manual monitoring, including heavy labor cost and high false alarm rate. In order to improve the efficiency and reduce the probability of public safety accident in campus, intelligent security system is necessary.

Computer vision technology is widely used in the field of surveillance in public places. Fruin et al. [1] proposed that the dangerous threshold of population density is about 1 p/m^2, when the population density in the monitored area becomes higher

X. Tan · C. Zhang · C. Zhuang · H. Yin (✉)
School of Automation, Chongqing University, Chongqing 400044, China
e-mail: yinhongpeng@gmail.com

Y. Jia et al. (eds.), *Proceedings of 2016 Chinese Intelligent Systems Conference*, Lecture Notes in Electrical Engineering 405,
DOI 10.1007/978-981-10-2335-4_16

than this value, this area may be at risk status. This theorem explains the relationship between population density and the probability of abnormal events. Terada et al. [2] use stereo vision to calculate people density, improving the detection accuracy. But due to the use of multiple cameras, the expense also increases. Davies [3] and Chow [4], etc. use the image pixel features to calculate people density, but for monitor with inclination angle, the results of detection is not good. On this basis, this paper presents a method with low-cost and high-detection accuracy, which can effectively improve the performance of the actual monitoring system.

In this paper, first, the process of video image processing is described, combining three-frame-difference method with background subtraction method, the pixels of people in the videos can be well extracted. Based on the pixel feature, the population density can be easily calculated. Second, a method was proposed for monitor with inclination angle. Calculating the image of video layer-by-layer, a more accurate result could be got by giving different weight to different layer. The intelligent surveillance system can apply in the high density situation in campus. The system can make contributions to protect the students with relatively low cost. Meanwhile, it can detect the crowed density and predict the abnormal situation with higher accuracy.

2 The Framework of the Intelligent Surveillance System

The topology of an intelligent surveillance system for crowed abnormal detection includes several parts: cameras on spot, router, network video decoder, network center, control center, and the administration client.

Figure 1 shows the topology of the system. Different areas, including the key areas and the general areas, are monitored by different cameras. Meanwhile, the camera can connect with the network by the router. In one part, the videos are returned by the network to be saved to refer to it sometimes. In another part, the videos are processed by the video decoder and return to the control center. The control center can detect the crowed abnormal density and motions to make an early-warning, which can be presented in the administration client. In this way, it's convenient for the managers to supervise and maintain.

The algorithm process [5] of the crowed abnormal detection is showed as Fig. 2. First, background difference and three-frame-difference methods are mixed used to extract the foreground image. In this way, the background difference method can offer the relatively intact foreground image, meanwhile, it can keep the good dynamic property by the three-frame-difference method. After that, the canny operator is employed to make edge detection, in order to reduce the effect of skin color, brightness, and the colour of clothes on the detection accuracy. Then, based on the distance between center of camera with different areas and the different viewing angle, the detection areas can be divided into several layers and detect separately. So that the effect of distance on the accuracy can be eliminated. Based on the method of pixel features, the crowed density can be obtained in different

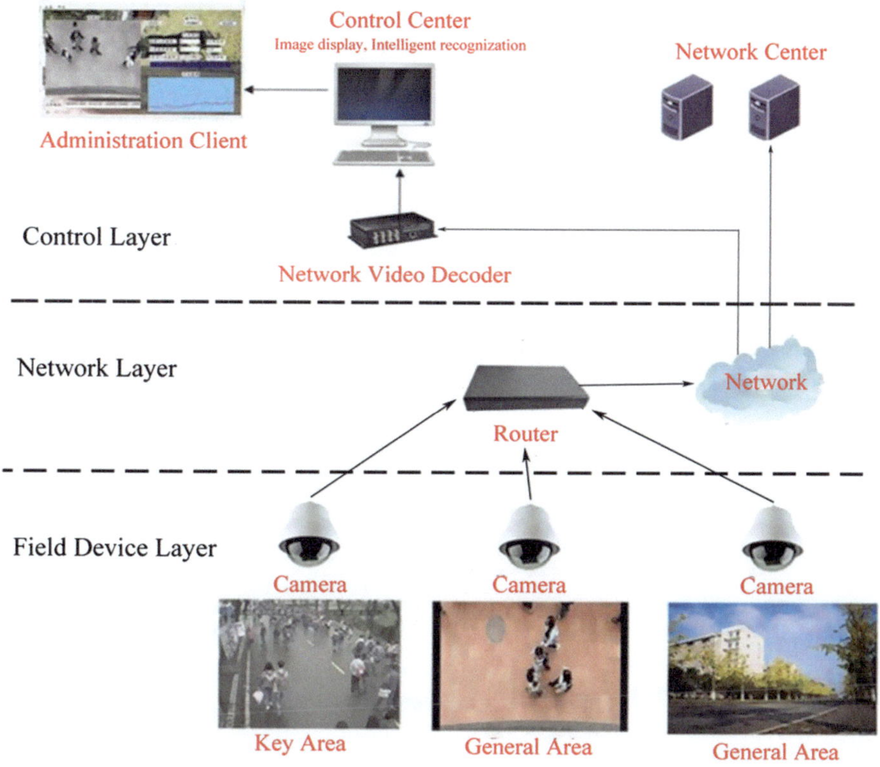

Fig. 1 The topology of the system

areas. The texture features are ingnored bacause it's complicated to account [6]. Finally, the whole crowed density information can be got by weighted mean.

3 The Detection of Crowd Abnormal Density

3.1 Foreground Extraction with Background Difference and Three-Frame-Difference Methods

In the intelligent surveillance system, it is the beginning and key stage to make the foreground extraction, which is to obtain moving object and process it. So that accurate moving object is significant to improve the accuracy of the detection and maintain system stability.

The background difference method [7] is one of the commonest method to make the foreground extraction. Compared with the background, the pixel of the motion

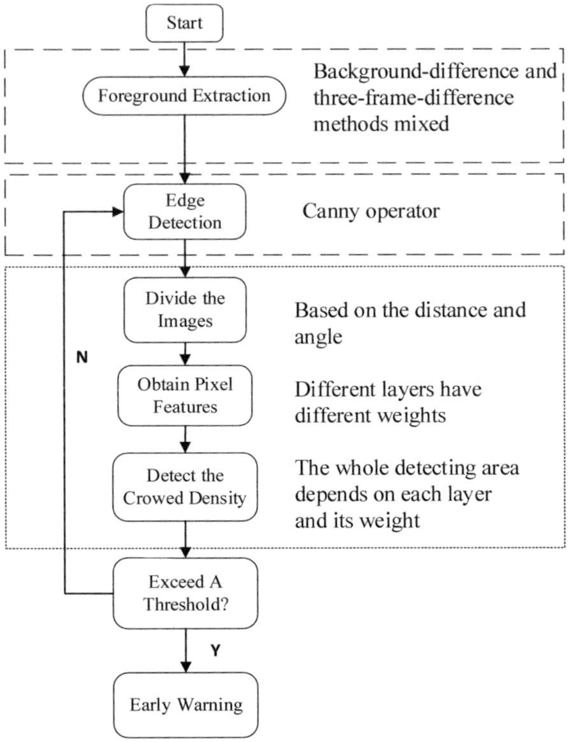

Fig. 2 The algorithm of the detecting

is quite different. Based on the difference, subtraction is made between current image and background to have the image of the moving object.

Assume that the background is $B(x, y)$, it is saved at start and $C(x, y)$ is the image in current time. So that the moving object $D(x, y)$ is:

$$D(x, y) = C(x, y) - B(x, y) \tag{1}$$

It is obvious that the background difference method has its own advantage, which is easy to calculate. What is more, the moving object is much more intact. However, the intelligent surveillance system is applied in the spot which is large-scale scene and crowded. It means that the brightness is not stable and crowed always overlap each other. If only background difference method is employed, the image of moving object has a lot of noise, resulting in the inaccuracy of detecting for crowed density.

Three-frame-difference method [8] is useful to solve the problem mentioned, one of the frame-difference method essentially. But differently, three-frame-difference method avoids the more or less detection in nonuniform velocity situation. The three-frame-difference method is shown as following.

STEP 1 The $(k+1)$ frame's gray value of the initial video subtracts the k frame's gray value, and the absolute value of the result is used to have $F_1(x,y)$. Then a proper threshold value $T_i^{'}$ is chosen to have the binary image $\Delta F_1(x,y)$.

STEP 2 Then the $(k+2)$ frame's gray value of the initial video subtracts the $(k+1)$ frame's, and the absolute value of the result to is used have $F_2(x,y)$. Then the same threshold value $T_i^{'}$ is chosen to have the binary image $\Delta F_2(x,y)$.

STEP 3 Make the binary image $\Delta F_1(x,y)$ and $\Delta F_2(x,y)$ collation operation to obtain the result by three-frame-difference method $\Delta F(x,y)$

$$\Delta F(x,y) = \Delta F_1(x,y) \cap \Delta F_2(x,y) \tag{2}$$

Compared with the background difference method, the three-frame-difference method takes good dynamic properties. But if only three-frame-difference method used, as Fig. 3 shows, the moving object is not intact leading to a smaller result. Based on Lin [9], the paper proposes a mixed method with background difference and three-frame-difference. So that the intact moving object can be obtained and good dynamic properties can be kept with filtering and noise reduction. The following equation shows the way to mix them.

$$D(x,y) \cup \Delta F(x,y) \tag{3}$$

The result of background difference and three-frame-difference method mixed can be shown as follow.

According to the Fig. 3 it can be seen that the intact moving object can be obtained by background difference method. The three-frame-difference method can keep the favorable dynamic character. So it is an acceptable method to get the moving object.

Fig. 3 Different methods of foreground extraction

3.2 An Edge Detection Based on Canny Operator

Generally, the edge of image exists between different areas for its different gray value. There is a visible difference between areas, which is the edge. Essentially, the edge of image means the saltus of gray value. In the intelligent surveillance system, objective condition is complicated in public. For example, the brightness is not stable and crowed overlap seriously. So the only foreground extraction results in inaccuracy. Particularly, in reality everyone wears clothes in different colors also leads to big errors. To reduce the effect, the proper takes edge detection to reflect the features by objects' contour.

Canny operator, an effectual way, takes Gaussian filter to have the gradients of image and makes use of first-order derivative to detect the edge. The local maximum of gradients can be obtained also [10]. The detailed steps are following.

STEP 1 Before the edge detection, the 2D Gaussian function is used to make smoothing filtering. In the 2D Gaussian function, δ is a parameter of Gaussian filter, it is used to control smoothness. A proper parameter of Gaussian filter can keep a favorable accuracy of edge, also a high signal-to-noise ratio. The paper gives $\delta = 1$ based on the experience.

STEP 2 Image G is obtained by Gaussian filter. After that, the amplitude image P consisting of gradient values and direction image θ consisting of gradient directions, can be got by gradient computation on G. So that the partial derivative in (i,j) is $P_x(i,j)$ in x direction and $P_y(i,j)$ in y direction.

STEP 3 Obtain the value and direction of gradient in (i,j) by partial derivative.

$$\begin{aligned} S(i,j) &= \sqrt{P_x^2(i,j) + P_y^2(i,j)} \\ \theta(i,j) &= \arctan\frac{P_x(i,j)}{P_y(i,j)} \end{aligned} \tag{4}$$

STEP 4 Make a certain point as the center to have interpolation method around it in gradient direction. Then choose the maximum point to use non-maximum suppression to eliminate a part of non-peripheral points. In this way, the new edge can be obtained.

STEP 5 Set threshold value T_H and T_L to have the double threshold detection. A pixel can be regarded as the edge, if its value is bigger than T_H. Oppositely, a pixel would not be called the edge, if its value is smaller than T_L. Based on the connectivity of edge, the point, which pixel value is between two thresholds, can be called edge if its proximal point is the edge point. Or it can't be.

After the edge detection, the effect of brightness instability and any other problems has receded or eliminated. So the image can be used for crowed density detection.

3.3 Detect the Crowed Density Separately

Based on the pixel features method, crowed density detection is easy to calculate and takes simple technique. But the accuracy does not meet the requirement. In order to enlarge the detection area and improve the use ratio of surveillance system, cameras always work at the non-vertical visual angle. However, the angle leads to inaccuracy for the distance difference. The paper proposes a method to detect separately to eliminate the effect of distance and angle. Based on the different distance and angle between the camera and detecting area, the whole area is divided into several levels. It can be illuminated by the following figure.

As Fig. 4 shows, the angle must satisfy that

$$0 < \alpha_i < 90^{\circ},$$

where α_i is the angle of detecting area and camera. The number of layers are divided by the accuracy of the system. Too few layers cannot satisfy the accuracy but too many layers are hard to calculate.

The most important is to decide the weights a_i in different layers, which depends on the distance and angle with camera. In the paper, the a_i is defined as:

$$a_i = \frac{H \tan \alpha_i}{L}, \tag{5}$$

where the L is the length of the detection area and the H is the height of the camera.

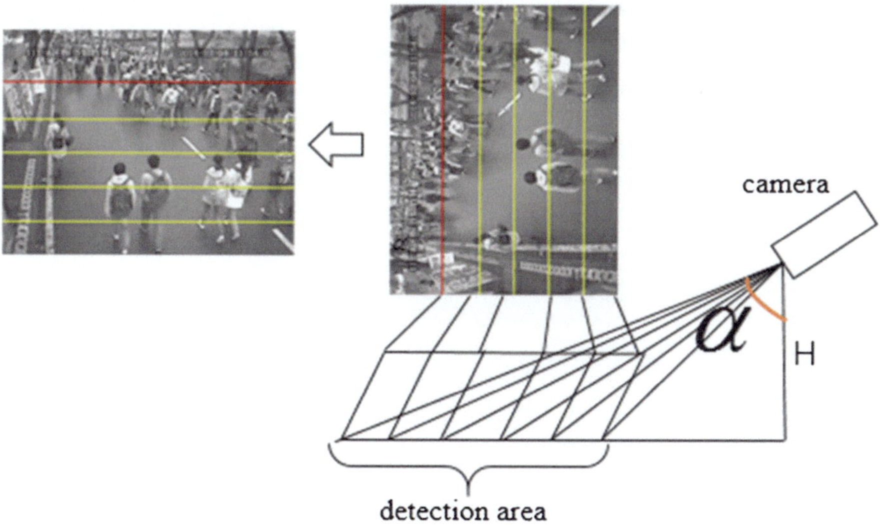

Fig. 4 Detect the crowed density separately

In different layers, the number of edge pixels is different, which is supposed T_{i0}. Compared with threshold number of edge pixels T_i, the detecting area density ρ_i can be obtained.

$$\rho_i = \frac{T_{i0}}{T_i} \tag{6}$$

Finally, it is easy to have the whole crowed density by weighted average.

$$\rho = \sum_{i=1}^{5} a_i \times \rho_i (0 < a_i < 1, a_{i+1} > a_i), \tag{7}$$

where a_i is the weight of different areas, decided by the angle and distance between camera and detecting area.

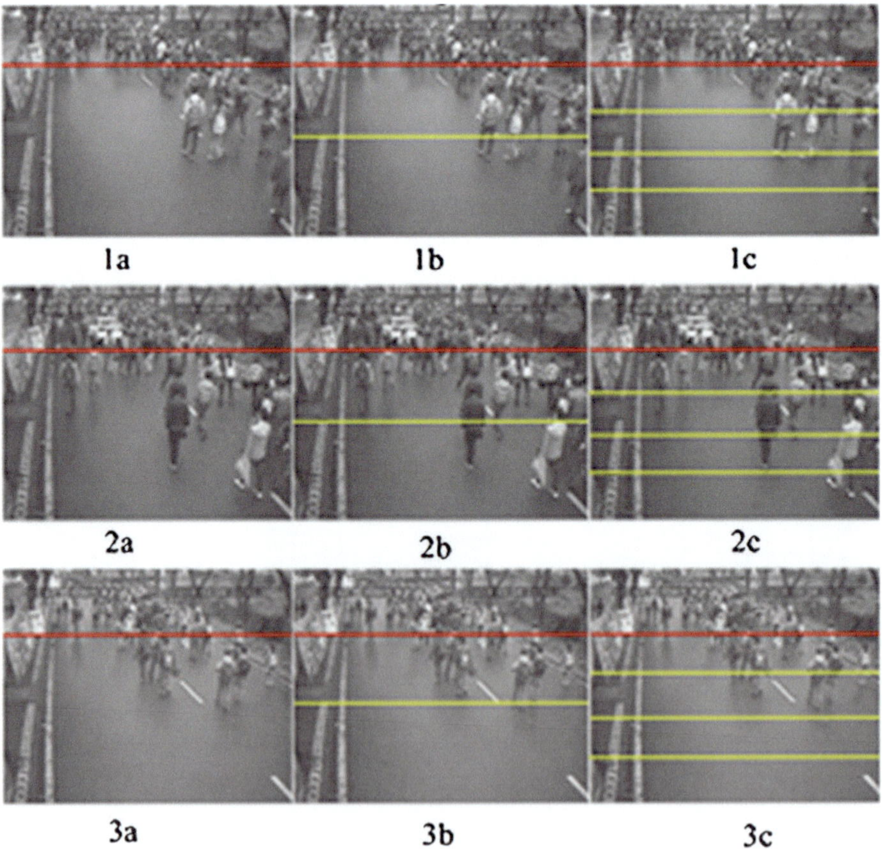

Fig. 5 Detection for low density situation

4 Experiment and Results

To illustrate the effectiveness of the proposed approach, an experiment of crowed density abnormal detection is conducted on the playground. Based on the Microsoft Visual Studio 2010, the experiment can be finished.

Figure 5 shows the detection for low density situation.

Among them, Fig. 5-1a, 2a and 3a are different low density situations on different time. Correspondingly, Fig. 5-1b, 2b and 3b are divided into two layers to

Table 1 Detection results for low density situation

Low density	Reference density p/m^2	0 layers p/m^2	2 layers p/m^2	4 layers p/m^2
Situation 1	0.08	0.12	0.12	0.10
Situation 2	0.13	0.15	0.13	0.11
Situation 3	0.09	0.14	0.16	0.12

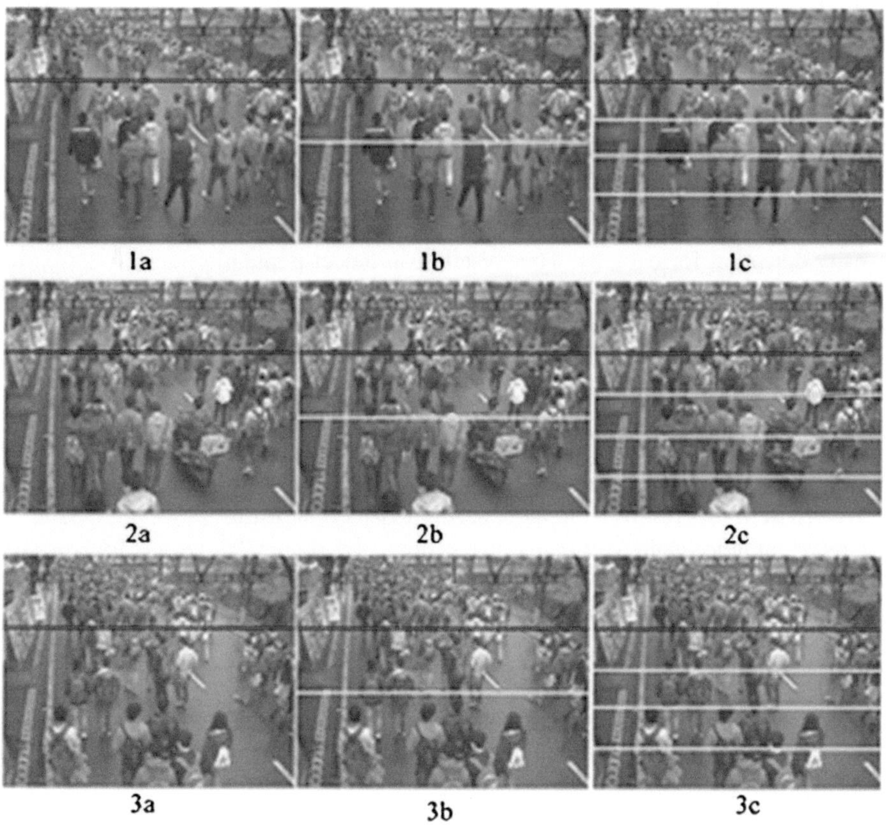

Fig. 6 Detection for high density situation

Table 2 Detection results for high density situation

High density	Reference density p/m^2	0 layers p/m^2	2 layers p/m^2	4 layers p/m^2
Situation 1	0.25	0.35	0.31	0.24
Situation 2	0.38	0.49	0.41	0.37
Situation 3	0.28	0.34	0.32	0.29

detect. Meanwhile, Fig. 5-1c, 2c and 3c are divided into four layers to detect. The following table shows the detecting results (Table 1).

In the low density situation, dividing images into several layers can improve the accuracy, however, it is not obviously that more layers lead to more accurate.

Figure 6 shows the detection for high density situation.

Among them, Fig. 6-1a, 2a and 3a are different low density situations on different time. Correspondingly, Fig. 6-1b, 2b and 3b are divided into two layers to detect. Meanwhile, Fig. 6-1c, 2c and 3c are divided into four layers to detect. The following table shows the detecting results (Table 2).

In the high density situation, more layers are efficient to detect the situation more accurately. Averagely, the error date decreases by 25 % with four layers.

5 Conclusion and Discussion

In this paper, an intelligent surveillance system is designed for crowed abnormal density detection for campus. This system can detect population density based on pixel features. Importantly, the system can detect the area layer-by-layer to improve the accuracy. In the high density situation, the error date decreases by 25 % with four layers. But averagely, the error date only decreases by 8 % with four layers and there is no difference in different layers. In addition, the system also provides a warning mechanism based on the detection results to help enhance the efficiency of public safety management. The experiments illustrate that this system has a good performance, like high accuracy of detection and rapidity of warning mechanism.

The more layers result in the more accurate detection results, however, leading to the more complex calculation. In this system, the proper layers are four or five according to the experiment. But it is not a universal choice in any other situations. A method to find the proper layers for every situation in needed in the further time.

Acknowledgments We would like to thank the supports by National Natural Science Foundation of China (61374135, 61203321), Chongqing Nature Science Foundation for Fundamental science and frontier technologies (cstc2015jcyjB0569), China Central Universities Foundation (106112016CDJZR175511) and the National College Students' Innovative Experiment Project (201510611063).

References

1. Fruin JJ (1984) Crowd dynamics and auditorium management. Auditorium News 22(5):4
2. Terada K, Yoshida D, Oe S, Yanmaguchi J (1990) A counting method of the number of passing people using a stereo camera. In: IECON-Proceedings
3. Davies AC, Yin JH, Velastin SA (1995) Crowd monitoring using image processing. Electron Commun Eng 7(1):37–47
4. Chow TWS, Yam JYF, Cho SY (1999) Fast training algorithm for feed forward neural networks: application to crowd estimation at underground stations. Artif Intell Eng 13:301–307
5. Yue X The algorithm of crowd density detection in intelligent monitoring system
6. Marana AN, Velastin SA, Costa LF et al (1998) Automatic estimation of crowd density using texture. Saf Sci 28(3):165–175
7. Han HZ, Wang ZL, Liu JW et al (2003) Adaptive background modeling with shadow suppression. In: The IEEE 6th international conference on intelligent transportation systems 1:720–724
8. Gupte S, Masound O, Martin R et al (2002) Detection and classification for vehicles. IEEE Trans Intell Transp Syst 3(1):37–47
9. Jiayi L, Zhezhou Y, Jian Z, Anna M, Yefeng C (2008) A new moving detection method based on background differencing and coterminous frames differencing 29(4):1–4
10. Canny JA (1986) Computational approach to edge detection. IEEE Trans Pattern Anal Mach Intell 8(1):679–697

The Active Disturbance Rejection Control with a Square-Root Amplifier for Non-minimum Phase System

Tong Wu, Weicun Zhang and Weidong Li

Abstract This paper adopts the active disturbance rejection control (ADRC) technology to control the non-minimum phase system (NMP). A new amplifier called square-root amplifier (SQF) is added in the ADRC to reduce the undershoot and the settling times of the step response. Simulation results are given to verify the effectiveness of the proposed scheme.

Keywords Non-minimum phase system · Active disturbance rejection control · Square-root amplifier

1 Introduction

Non-minimum phase system is a kind of system with poles or zeros in the right-half plane (RHP). It is well known that the terrible characteristic is caused by the unstable zeros or the zeros in the right-half plane (RHP) [1]. For NMP system, the undershoot in the step response is always a headache, which means the high-frequency gain and low-frequency gain of the system have opposite signs [2]. To deal with the NMP system, the undershoot is to be reduced along with the system to be stabilized and the overshoot as well as the settling time to be reduced.

Active Disturbance Rejection Control (ADRC) technique as an emerging technique has been widely used in solving various types of control problems [3]. And the applications of this technology are mostly confined to minimum phase system [2], while there are lots of non-minimum phase (NMP) system in the real world, such as flexible manipulators, vessels, aircrafts [4], and hydraulic turbines.

T. Wu · W. Zhang (✉)
School of Automation and Electrical Engineering, University of Science and Technology, Beijing, China
e-mail: weicunzhang@ustb.edu.cn

W. Li
Southwest Institute of Clean Energy and Electric Power System, Dali University, Dali 671003, Yunnan, China

Y. Jia et al. (eds.), *Proceedings of 2016 Chinese Intelligent Systems Conference*, Lecture Notes in Electrical Engineering 405,
DOI 10.1007/978-981-10-2335-4_17

The non-minimum phase behavior is generally determined by the system's dynamic characteristics. In the real world we cannot just neglect the NMP phenomenon. The new control strategy, Active Disturbance Rejection Control, will be used to control the NMP system in the following parts.

To achieve a better performance of NMP control systems, a new kind of amplifier called square-root amplifier (SQF) is added in the ADRC technique. The past researches have shown that the SQF can be used to improve the system's rapidity without leading to a big overshoot [5], which means the rapidity of a system can be improved by SQF without strengthening the control force. Also the NMP system is wanted a smooth control signal which is usually coupled with a longer settling time. Maybe the SQF will bring us some good luck.

2 Active Disturbance Rejection Control Technique

ADRC technique is a new control technique using extended state observer (ESO) to estimate the total disturbance and cancel it with a feedback compensation strategy. It makes the controller design much easier, as the plant is approximately reduced to a double integrator using ESO [6].

2.1 Extended State Observer (ESO)

The extended state observer is a talent reform of the observer, of which the extended state is used to estimate the internal and external disturbance or uncertainty. The good estimation is followed by a compensation employing the feedback technique to cancel the total disturbance in the input. Considering a second-order plant like (1), its extended state observer is usually designed as (2) [7]

$$\begin{aligned} \dot{x}_1 &= x_2 \\ \dot{x}_2 &= f(x_1, x_2) + f_d + bu \end{aligned} \tag{1}$$

$$\begin{aligned} \dot{z}_1 &= \dot{z}_2 - \beta_1(z_1 - y) \\ \dot{z}_2 &= \dot{z}_3 + bu - \beta_2(z_1 - y) \\ \dot{z}_3 &= -\beta_3(z_1 - y) \end{aligned} \tag{2}$$

The states in the ESO can be a good estimation for the states of the plant to be controlled and the total disturbance if the observer estimation error is small enough to be neglected. Of course the states in the observer can be used in the control algorithm in the place of the real states of the plant to be controlled. Also the extended state is feedback to the input part to cancel the total disturbance reshaping the original plant into a double integrator [8], which means the plant to be controlled is always an approximate double integrator when the real model is a second-order plant. The following part of controller design benefits a lot from this approximation.

$$\begin{cases} \dot{x}_1 = x_2 \\ \dot{x}_2 = f(x_1, x_2) + f_d + b\frac{u_0 - z_3}{b} \end{cases} \tag{3}$$

$\Rightarrow$

$$\begin{cases} \dot{x}_1 = x_2 \\ \dot{x}_2 = u_0 \end{cases} \tag{4}$$

2.2 Parameterization of Extended State Observer

The observer gain vector, $[\beta_1, \beta_2, \beta_3]$, is usually obtained using the pole placement technique. Assigning all eigenvalues at ω_o as (5) is a simple way to get a good speed at which the observer tracks the states while coupling with a minimum sensitivity to the sensor noises [9].

$$\lambda(s) = s^3 + \beta_1 s^2 + \beta_2 s + \beta_3 = (s + \omega_o)^3 \tag{5}$$

That is

$$\beta_1 = 3\omega_o, \beta_2 = 3\omega_o^2, \beta_3 = \omega_o^3 \tag{6}$$

2.3 The Extended State Observer for NMP System

Consider a NMP system (7),

$$\begin{aligned} \dot{x}_1 &= x_2 \\ \dot{x}_2 &= f(x_1, x_2) + f_d + b_0\dot{u} + bu \end{aligned} \tag{7}$$

where the value of b_0 is negative. Assuming the NMP system is a minimum phase system as (8), using low-frequency gain b in the ESO design, taking the derivative of its input signal as a part of the total disturbance, the corresponding extended state observer can be designed as (2).

$$\ddot{y} = f(y, \dot{y}, u) + bu \tag{8}$$

2.4 Controller Design

The PD controller has many good performances such as accelerating the system's response rate and increasing the stability of the system. But usually a PD controller is not a good choice for a system because there always exist some noise in the output. The noise is amplified when we try to get the derivative of the output. However, the noise is not mixed into the derivative part in ADRC for the derivative of the output is obtained from z_2, the estimation of the output's derivative component. In the ADRC control system the PD controller can be securely designed as (9) to achieve the control requirements for a NMP system.

$$\begin{aligned} u_0 &= k_p(r - z_1) - k_d z_2 \\ u &= \frac{u_0 - z_3}{b} \end{aligned} \tag{9}$$

3 Square-Root Amplifier (SQF)

Square-root amplifier is a new nonlinear amplifier proposed to improve the quality indicators of optimal control systems [5]. It is generally used in improving the rapidity of the system without introducing big overshoot. The SQF is used for ADRC design-making the control law as (10)

$$\begin{aligned} u_0 &= \pm\sqrt{\left|k_p(r - (\pm\sqrt{|z_1|})) - k_d z_2\right|} \\ u &= \frac{u_0 - z_3}{b} \end{aligned} \tag{10}$$

A square-root amplifier like (11) has the characteristic that it gives a much larger output when the value of the input u is $0 < |u| < 1$, a much less output when the value of the input u is $|u| > 1$. The decrease and increase of the input value are much more significant as $|u|$ is far away from 1, while they are less significant as $|u|$ is close to 1.

$$y = \pm\sqrt{|u|} = \begin{cases} \to \pm 1, 0 < |u| < 1 \\ \to \pm 1, \ |u| > 1 \end{cases} \tag{11}$$

4 Simulation

4.1 Hydraulic Turbine 1

The second-order hydraulic turbine model is written as (12) [10]

$$G(s) = \frac{5(1-0.8s)}{(1+0.4s)(1+4.8s)} \tag{12}$$

PID, ADRC, ADRC with SQF methods are applied to control the plant, with tracking-differentiator (TD) used in ADRC and ADRC with SQF [10]. The transfer function of tracking-differentiator is (13)

$$G_{td} = \frac{\omega^2}{s^2+2\zeta\omega+\omega^2}\left[\frac{\alpha\omega(n-2)}{s+\alpha\omega(n-2)}\right]^{n-2} \tag{13}$$

where $\omega = 1.5$, $\zeta = 1$, $n = 2$, $\alpha = 5$.

The step response of control systems are shown in Fig. 1. The drawing of partial enlargement for Fig. 1 is shown in Fig. 2. Under the same overshoot of the step response, the ADRC with SQF has the minimum settling time along with a small undershoot.

4.2 Hydraulic Turbine 2

The third-order hydraulic turbine model is written as (14) [11]

$$\frac{1}{1+0.2s}\frac{1-0.8s}{1+0.5\times 0.8s}\frac{1}{0.2+0.96s} \tag{14}$$

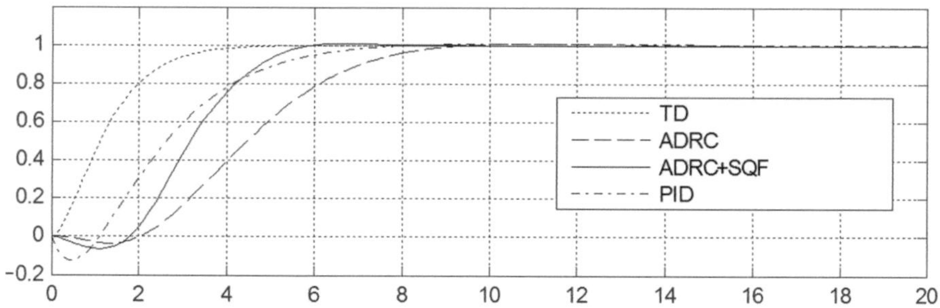

Fig. 1 Step response of the hydraulic turbine 1 control systems

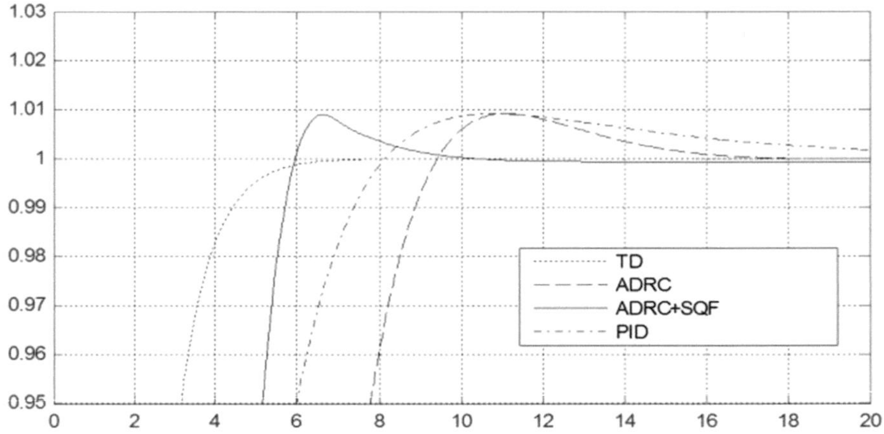

Fig. 2 Partial enlargement for Fig. 1

ADRC, ADRC with SQF methods are applied to control the plant, with tracking-differentiator (TD) used in ADRC and ADRC with SQF [10]. The transfer function of tracking-differentiator is the same with Eq. (13); where $\omega = 1.5$, $\zeta = 1$, $n = 2$, $\alpha = 5$.

The step response of control systems are shown in Fig. 3. The drawing of partial enlargement for Fig. 3 is shown in Fig. 4. Under the same overshoot of the step response, the ADRC with SQF has the minimum settling time along with a small undershoot.

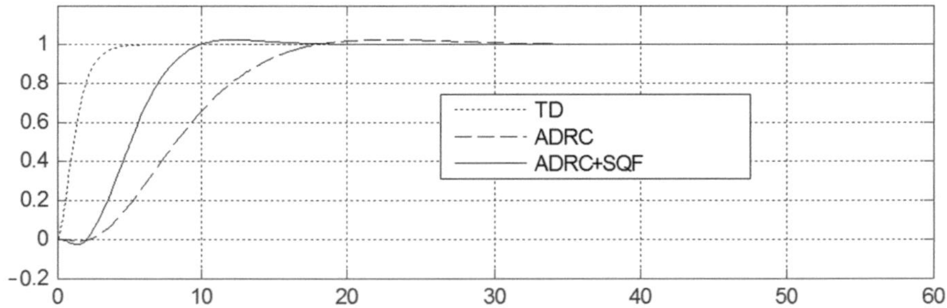

Fig. 3 Step response of the hydraulic turbine 2 control systems

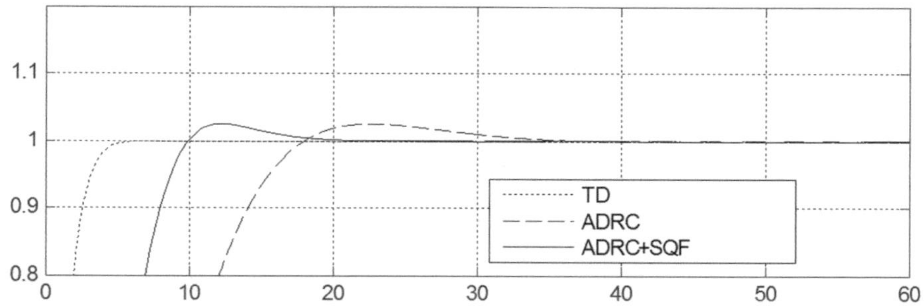

Fig. 4 Partial enlargement for Fig. 3

5 Conclusion

In this paper, a new amplifier is added in the ADRC to improve the control effect for the NMP phase system. Simulation results have shown that the step response of the system controlled by ADRC with SQF amplifier has the minimum settling time compared with regular ADRC and PID, along with a small undershoot.

Acknowledgments The author would like to thank the anonymous reviewers for their constructive and insightful comments for further improving the quality of this work. This work was supported by the Major State Basic Research Development Program (973 Program) (No. 2012CB821200) and National Science Foundation of China (No. 61520106010).

References

1. Ling W, Wenfeng L, Dazhong Z (2003) Optimal design of controllers for non-minimum phase system. Acta Automatica Sinica 29(1):135–141
2. Shen Z, Zhiqiang G (2010) Active disturbance rejection control for non-minimum phase systems active disturbance rejection control for non-minimum phase systems. In: Proceedings of the 29th Chinese control conference, pp 6066–6070, 29–31 July 2010
3. Congzhi H, Zhiqiang G (2013) On transfer function representation and frequency response of linear active disturbance rejection control. In: Proceedings of the 32nd Chinese control conference, pp 72–77, 26–28 July 2013
4. Shanwei S, Bo Z, Jinwu X, Yan L (2015) A survey on the control of non-minimum phase systems. Acta Automatica Sinica 41(1):9–20
5. Guobo X, Nianyin Z (2014) Square-root amplifier and its application in the optimal control systems. Electric Drive 44(4):46–50
6. Qing Z, Linda Q, Zhiqiang G (2012) On validation of extended state observer through analysis and experimentation. J Dyn Syst Meas Control 134(2):024505(6 pages)
7. Xing C, Donghai L, Zhiqiang G, Chuangfeng W (2011) Tuning method for second-order active disturbance rejection control. In: Proceedings of the 30th Chinese control conference. Yantai, China, 22–24 July 2011, pp 6322–6327
8. Jingqing H (2008) Active disturbance rejection control technique—the technique for estimating and compensating the uncertainties (in Chinese). National Defense Industry Press, Beijng, Chap. 6

9. Zhiqiang G Scaling and bandwidth-parameterization based controller tuning. In: Proceedings of the American control conference, pp 4989–4996
10. Xiang L, Donghai L, Xuezhi J, Xuejiao H (2001) Simulation study on auto-disturbance-rejection control for unstable system and non-minimum phase system. Control Decis 16(4):420–424
11. Zhen H, Yuping L, Zheng C-J (2006) A novel control method for non-minimum phase systems. Inf Control 35(5):560–563

Human Action Recognition Based on Multifeature Fusion

Shasha Zhang, Weicun Zhang and Yunluo Li

Abstract A human action recognition method based on histogram of oriented gradient (HOG) of motion history image (MHI) and Speeded Up Robust Features (SURF) is presented in this paper. The method overcomes the shortcomings of the algorithm based only on HOG of motion history image. The experiment results show that the new method improves recognition rate and has a promising performance.

Keywords Histogram of oriented gradient (HOG) · Motion histogram image (MHI) · Speeded Up Robust Features (SURF) · Action recognition

1 Introduction

In recent years, human action recognition based on video has been attracting more and more attention, and many applications are reported, such as video surveillance in public places, the natural human–computer interaction, elderly and healthcare analysis. Many effective recognition algorithms for human action recognition have been proposed. Ke et al. [1] commented the current video-based human motion recognition process, which includes major algorithm of action recognition, the action recognition system, and application environment of action recognition technology. Aggarwal et al. [2] compared the advantages and disadvantages of various advanced action recognition algorithms and the applicable range of various algorithms. Xu et al. [3] summarized the features representation of action and the latest developments of action recognition methods.

The most common method of human action recognition is feature extraction from a sequence of images, following by classification. Actually, the process of feature extraction is the representation of the original image sequence which is

S. Zhang · W. Zhang (✉) · Y. Li
School of Automation and Electrical Engineering, University of Science and Technology Beijing, Beijing 100083, China
e-mail: weicunzhang@ustb.edu.cn

Y. Jia et al. (eds.), *Proceedings of 2016 Chinese Intelligent Systems Conference*, Lecture Notes in Electrical Engineering 405,
DOI 10.1007/978-981-10-2335-4_18

conducive to the movement classification. At present, the common methods of image sequence representation are divided into two kinds, i.e., global representation and local representation [4]. Global representation encodes the human region of interest (ROI) as a whole, and the main global representation includes silhouette, contour, optical flow, and so on. Bobick and Davis [5] proposed the motion energy image (MEI) and the motion history image (MHI). Ali and Shah [6] extracted a series of motion features from optical flow diagram. In local representation, an image sequence is represented as a local descriptor or local block, then the statistic character of which is regarded as the final human action descriptors. Local block is extracted through intensive sampling or spatiotemporal interest points corresponding to location of interest action. Williams et al. [7] put forward Hessian detector. Hessian temporal points of interest are searched through maximizing the row-column of 3D Hessian matrix. Laptev et al. [8] expanded Harris 2D detector into Harris 3D detector. Up to now, many efforts have been devoted to describe the spatiotemporal interest points. Laptev [9] put forwarded HOG/HOF descriptor to describe spatial and temporal points of interest.

This paper thus proposes a novel method which combines the histogram of oriented gradient (HOG) of the MHI with SURF to enhance the action recognition rate. The block diagram of the proposed method is shown in Fig. 1. First, we calculate a motion history image from the frame difference of an action video and the HOG feature from the image. Second, the action occurrence region is obtained by integral projection, and the interest points of action occurrence region are obtained. And these interest points are represented by SURF descriptor. Third, the HOG feature of MHI and the information of interest points of action occurrence region are combined. Finally, the combined feature is fed into support vector machine (SVM) classifier for recognizing human action types.

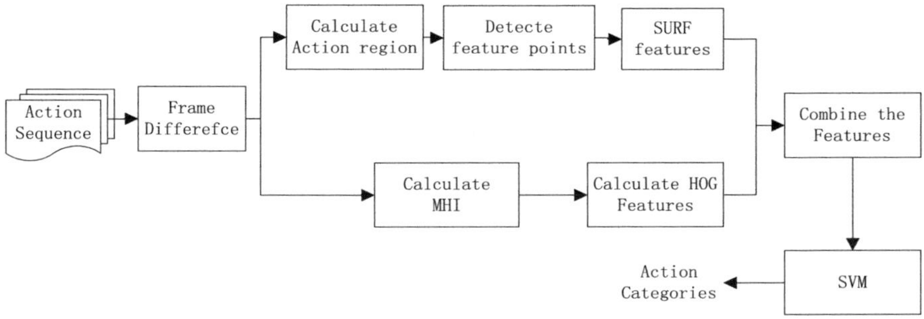

Fig. 1 The block diagram of the proposed method

2 Feature Extraction Based on MHI-HOG

Human actions include information of time and space, and motion histogram image (MHI) reflects not only the spatial position of action but also the order of movement. MHI as a global feature extraction can be a good representation of human action. Compared with other feature description method, histogram of oriented gradients (HOG) features has an optical invariance in the image, which is good to overcome the MHI shortcomings. Huang et al. [10] proposed a method called MHI-HOG [11], in which they extracted the HOG feature of MHI to recognize human action. Therefore, this method can improve the accuracy of the classification.

2.1 *Motion History Image (MHI)*

MHI is mainly used to describe the movement trend of action and has a good description of temporal information of action. The MHI of each frame in a given video can be obtained through the following steps. (a) Extract a binary image sequence $DI_i(x, y)$ which indicates the regions of motion. $DI_i(x, y)$ can be obtained by frame difference. (b) Based on $DI_i(x, y)$, MHI is defined as:

$$H_\tau(x, y, t) = \left\{ \begin{array}{ll} \tau & DI_i(x, y) \geq T \\ \mathrm{Max}(0, H_\tau(x, y, t-1) - 1) & DI_i(x, y) < T \end{array} \right\}, \quad (1)$$

where τ is duration of the temporal extent of a movement. Figure 2 shows an example of the motion history image when a person is jacking.

Fig. 2 Motion history image

2.2 Histogram of Oriented Gradient (HOG) Feature of MHI

Although the MHI includes temporal information, it is too sensitive to background noise and can only be used for relatively simple action. And the information of MHI is likely to be covered by the previous information. What is more, when silhouette is incomplete or fuzzy, the correct rate of classification action will be significantly reduced. In computer vision and image processing, HOG is a feature descriptor used for object detection. The descriptor is obtained by calculating gradient histogram of image in local area. HOG is represented by an edge (gradient) of the structural features, and it can suppress the influence of the translation and rotation to some extent caused by the quantization of the spatial position and orientation. The method of dividing image into blocks makes the relationship between the local image pixels characterized better. In order to attack the drawbacks, we can calculate the HOG feature of MHI. The method does not need to extract the silhouette images. Specific method following five steps:

(a) The MHI is normalized.
(b) We use Eqs. (2) and (3) to calculate the gradient of each pixel point.

$$G_x(x,y) = H(x+1,y) - H(x-1,y) \tag{2}$$

$$G_y(x,y) = H(x,y+1) - H(x,y-1) \tag{3}$$

where $G_x(x,y)$, $G_y(x,y)$, $H(x,y)$ represents horizontal gradient, vertical gradient, and pixel value, respectively. Calculate the orientation and magnitude of the gradient of each pixel as follow:

$$G(x,y) = \sqrt{G_x(x,y)^2 + G_y(x,y)^2} \tag{4}$$

$$\alpha(x,y) = \tan^{-1}\left(\frac{G_y(x,y)}{G_x(x,y)}\right) \tag{5}$$

(c) Construct gradient orientation histograms for each cell

The image is divided into a number of cells, and then the number of gradient histogram for each cell is counted.

(d) Collect HOG features

Combining feature vectors of all cells within a block, HOG features of the block will be got. In order to ensure speed, the MHI of each frame is divided into 4 * 2 cells, and the gradient orientation of each cell is quantized into nine angles. The HOG features of MHI are obtained by collecting the feature of all the cells, and feature dimension of each frame image is 288. The extracted features will be used for subsequent classification.

3 SURF Feature Points Detection

In machine vision, the feature points are widely used, which is because that it can show the characteristics variation of the location in space and time, and it is also widely used to describe the action in the video. The method, SURF feature point detection, has the good rapidity and efficiency. So, it is more popular than others in recent years. SURF (Speeded Up Robust Features) was first proposed by Bay [12] in 2006. SURF is obtained by improving SIFT algorithm. Thus, standard SURF can run three times faster than SIFT in generally, and it has the stability in multiple photos. The most important feature of SURF is that the Haar features and the concept of the integral image are used, which greatly reduces the running time of the program.

The strongest feature points can be detected very well by SURF. However, video background noise and body without movement can produce a lot of useless feature points, which will increase the amount of calculation. In this paper, a method, combining integral projection with SURF, is presented. That is

Step 1: The frame differential method and morphology are used to get the binary image of the video sequence;

Step 2: Based on the binary image in step 1, the motion area can be obtained by integral projection, as shown in Fig. 3.

Fig. 3 Result of motion area

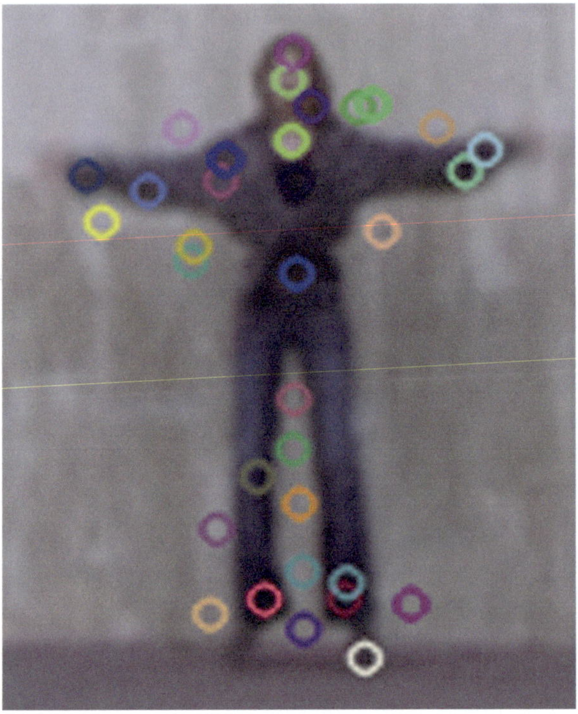

Fig. 4 Detection of feature points

Step 3: The feature points of motion area are got by SURF, as shown in Fig. 4.

Figure 4 shows the examples of our interest point detection results (circle points). It is evident that our detected interest points are much less compared to those detected by the whole image.

4 Action Representation

4.1 SURF Descriptor

Although we have got the feature points of the motion region in Sect. 3, the feature points using SURF feature descriptors should be expressed before delivering them to SVM. The method of constructing SURF descriptor is shown as following. First, feature point is set to the center, and the main direction of the feature points is set to the x-axis direction; then, a square area of 4 * 4 should be selected; next, the values of Haar wavelet characteristic response can be obtained in the horizontal and vertical direction (Fig. 5).

Where $\sum dx$ is the sum of Haar wavelet feature value in horizontal direction, $\sum |dx|$ is the sum of Haar wavelet feature absolute value in horizontal

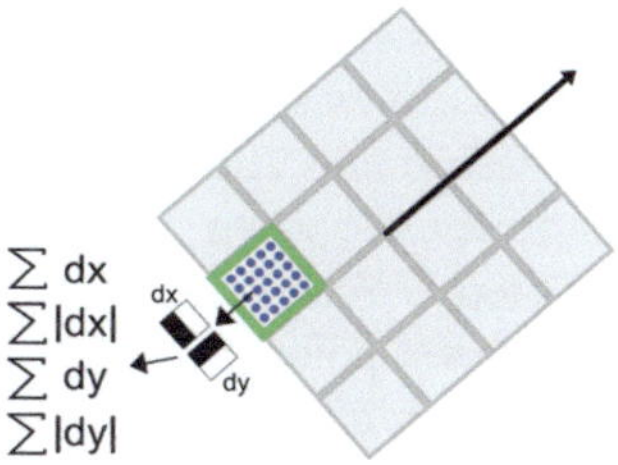

Fig. 5 The process of calculating the SURF descriptor

direction, $\sum dy$ is the sum of Haar wavelet feature value in vertical direction, $\sum |dy|$ is the sum of Haar wavelet feature absolute in value vertical direction. So each feature point can be represented by a $4 * 4 * 4 = 64$ dimensional vector.

4.2 Motion Features

In Sect. 4.1, the SURF descriptor of each interest point is a 64 dimensional vector. If the number of interest points in each frame is N, the feature dimension of each is $N * 64$. The dimension of feature is so high that it cannot reasonably combine the HOG features obtained by Sect. 2. Therefore, the dimension of features of interest points should be reduced before the feature fusion, shown in Fig. 6. The feature fusion is obtained by combining the reduced features with HOG features. The process is:

Step 1. Single frame dimension reduction. Principle component analysis is used to perform longitudinal dimension reduction for SURF descriptor extracted from

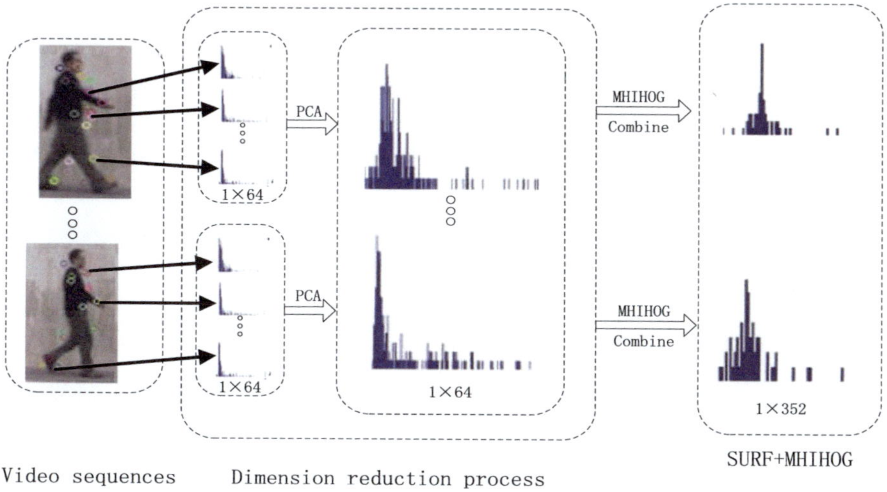

Fig. 6 Description of motion features

interest points in the same frame. It means that $N * 64$ dimension features can be reduced to 64 dimensions by gathering principal components of all the descriptors for each frame. Although part of information will be lost, the loss of the information is acceptable when N is chosen as 60 in our experiment.

Step 2. Feature combination. Combine the 64 dimension spatiotemporal features of interest points and the HOG feature of MHI in each frame; finally, we get 352 dimension features for each frame (SURF + MHI-HOG).

5 Analysis of Results

In this section, experiments are performed on the Weizmann dataset using the combined spatiotemporal feature. The performance of the proposed algorithm is illustrated in this section by comparing with the recent reports associated with the related features.

SVM [13] (Support Vector Machine) is a novel small sample learning method with intuitive geometric interpretation and good generalization ability. In the case of limited training data set, Corinna [14] has shown that SVM has a better classification performance than BP, KNN, and Bayes. The samples of human action recognition are generally relatively small, so SVM is selected as the data classifier.

The testing video used in the experiments is the standard Weizmann movement dataset as shown in Fig. 7, which includes 9 people performing 10 movements—bending, jacking, jumping, pjumping, running, siding, skipping, walking, waving1, waving2.

The "leave-one-out" cross validation is used to evaluate the recognition rate of action classifier in our experiment. One actions of each actor is used in turn as test samples, and the rest of all the actions are used as the training samples. The test process continues until all the actions have been completely tested. The experimental results are shown in Tables 1 and 2.

According to the results, the method based on MHI-HOG + SURF achieves a higher accuracy rate than the method based on MHI-HOG. However, the method proposed in this paper shows a relatively low recognition rates when the action is occurred at a faster frequency, like running and siding.

Fig. 7 Weizmann samples

Table 1 Confusion matrix of MHI-HOG

	Bend	Jack	Jump	Pjump	Run	Side	Skip	Walk	Wave1	Wave2
Bend	1	0	0	0	0	0	0	0	0	0
Jack	0.02	0.89	0.02	0	0	0	0.05	0	0.02	0
Jump	0.08	0	0.71	0.06	0	0.05	0.08	0.08	0	0
Pjump	0.05	0.05	0	0.86	0	0	0	0	0.04	0
Run	0	0	0	0	0.64	0	0.16	0.2	0	0
Side	0	0	0.1	0	0	0.65	0.15	0.1	0	0
Skip	0	0	0	0	0	0.03	0.79	0.18	0	0
Walk	0	0	0	0	0.08	0.06	0	0.86	0	0
Wave1	0	0	0	0	0	0	0	0	0.84	0.16
Wave2	0	0	0	0	0	0	0	0	0.18	0.82

Table 2 Confusion matrix of MHI-HOG + SURF

	Bend	Jack	Jump	Pjump	Run	Side	Skip	Walk	Wave1	Wave2
Bend	1	0	0	0	0	0	0	0	0	0
Jack	0.02	0.96	0	0	0	0	0	0	0.02	0
Jump	0.06	0	0.73	0.06	0	0	0.07	0.07	0	0
Pjump	0.01	0	0	0.96	0	0	0	0	0.03	0
Run	0	0	0	0	0.72	0	0.14	0.14	0	0
Side	0	0	0.1	0	0	0.76	0.1	0.04	0	0
Skip	0	0	0	0	0	0	0.82	0.18	0	0
Walk	0	0	0	0	0.08	0	0	0.92	0	0
Wave1	0	0	0	0	0	0	0	0	0.98	0.02
Wave2	0	0	0	0	0	0	0	0	0.04	0.96

6 Conclusions and Future Work

In this paper, a method for human action recognition based on MHI-HOG + SURF features is presented. The experiment results show that this method has a better performance than that one based the HOG of MHI. High recognition rate has been achieved for static movement or actions at a slower frequency in our experiments. However, dynamic movements or actions at a faster frequency are too difficult to be recognized by this method. Therefore, the future research will be focused on dynamic movements.

Acknowledgments The author would like to thank the anonymous reviewers for their constructive and insightful comments for further improving the quality of this work. This work was supported by the Major State Basic Research Development Program (973 Program) (No. 2012CB821200), National Key Technologies R&D Program (No. 2013BAB02B07), and National Science Foundation of China (No. 61520106010).

References

1. Ke, S-R, Thuc HLU, Lee, Y-J et al (2013) A review on video-based human activity recognition. Computers 2(2):88–131
2. Aggarwal JK, Ryoo MS (2011) Human activity analysis: a review. ACM Comput Surv 43 (3):1–47
3. Qinjun X, Zhenyang W (2014) Research progress on activity recognition in video. J Electron Meas Instrum 4:4–28
4. Poppe R (2010) A survey on vision-based human action recognition. Image Vis Comput 28 (6):976–990
5. Bobick Aaron F, Davis James W (2001) The recognition of human movement using temporal templates. IEEE Trans Pattern Anal Mach Intell (PAMI) 23(3):257–267
6. Saad A, Mubarak S (2010) Human action recognition in videos using kinematic features and multiple instance learning. IEEE Trans Pattern Anal Mach Intell (PAMI) 32(2): 288–303
7. Willems G, Tinne T, Gool Lic J (2008) An efficient dense and scale-invariant spatio-temporal interest point detector. In: Proceedings of the ECCV, pp 650–663
8. Ivan L, Tony L (2003) Space-time interest points. In: Proceedings of the IEEE international conference on computer vision, pp 432–439
9. Ivan L, Marszalek M, Schmid C et al (2008) Learning realistic human actions from movies. In Proceedings of CVPR, pp 1–8
10. Thi TH, Cheng L, Zhang J, Wang L, Satoh S (2012) Integrating local action elements for action analysis. Comput Vis Image Underst 116(3):378–395
11. Huang C-P, Hsieh C-H, Lai K-T, Huang W-Y (2011) Human action recognition using histogram of oriented gradient of motion history image. Int Conf Instrum Meas Comput Commun Control 353–356
12. Bay H, Ess A, Tuytelaars T et al (2008) SURF: speeded up robust features. Comput Vis Image Underst 110(3):346–359
13. Chang C-C, Lin C-J (2011) Libsvm: a library for support vector machines. Acm Trans Intell Syst Technol 2(3):389–396
14. Cortes C, Vapnik V (1995) Support-vector networks. Mach Learn 20(3):273–297

Adaptive Blocks-Based Target Tracking Method Fusing Color Histogram and SURF Features

Ruilin Cao, Qing Li, Weicun Zhang, Zhao Pei and Yichang Liu

Abstract In order to solve the object tracking problem under occlusion, this paper proposes an adaptive weighted tracking algorithm based on color features.The target is divided into pieces according to the features of color in this algorithm. SURF features are used to regulate the existing tracking results to improve tracking accuracy. In the tracking process, we can judge the occlusion by the degree of matching, and adaptively adjust the weight of each block to achieve accurately tracking. Simulation results show the effectiveness of the proposed algorithm.

Keywords Adaptive weighted · Object tracking · Color histogram · SURF features

1 Introduction

Moving object tracking, integrating computer vision, and pattern recognition technology, is widely used in intelligent control, human–computer interaction, and vehicle navigation, and other fields [1–3]. In the real environment, the deformation and occlusion problems of tracking target may lead to inaccurate result, even losses of the tracking target. Therefore, how to effectively deal with target occlusion and deformation is the difficulty [4, 5]. There are many block-target tracking research results based on detection of occlusion have been reported. Reference [6] divided the target template into several subblock evenly. It matched with color histogram and did an exhaustive search to determine the target position in the current frame by weighted votes, which is a good solution to the problem of occlusion and performance changing. But the calculation amount of exhaustive search is very large. Reference [7] divided the template binary image into blocks through the horizontal

R. Cao · Q. Li · W. Zhang (✉) · Z. Pei · Y. Liu
School of Automation and Electrical Engineering, University of Science and Technology Beijing, Beijing 100083, China
e-mail: weicunzhang@ustb.edu.cn

Y. Jia et al. (eds.), *Proceedings of 2016 Chinese Intelligent Systems Conference*, Lecture Notes in Electrical Engineering 405,
DOI 10.1007/978-981-10-2335-4_19

and vertical projection, characterized by the color histogram. However, there will be a drift when the subblock is similar to background color. Hossain et al. [8] used the local invariant features of SIFT (Scale Invariant Feature Transform) to achieve the target tracking, due to the local region feature invariance of the light, scale, and rotation invariance. But SIFT feature matching computes complex, which cannot meet the real-time tracking requirements.

To solve the above problems, this paper presents an adaptive block color histogram fusing SURF target tracking method. We will divide target into blocks with color projection. Each subblock using the color histogram and SURF characterization, not only can judge occlusion, also overcome similar color target interference. SURF can also correct the tracking window.

2 Color-Based Adaptive Segmentation

The occluded part of the target includes a little information of target, while unobstructed part saves a lot of target information. By combining block information, target can be effectively tracked. The fixed number of block was unable to adapt the diversity of the target. The color distribution of the target tends to have high similarity in the local small scale. So dividing the target directly from color distribution can be considered. First, the image block should be distinguished from the selected feature space. The excessively similarity between the image blocks will lead the tracker to the similar image blocks, which cannot reflect the true trajectory of the target. Second, too many blocks will affect the real-time computing of the image. And at the same time, too few blocks will reduce the ability of anti-occluding. In this paper, we will use the method based on adaptive blocks to improve the tracking ability of anti-occluding.

In the paper, $A(M,N)$ is the target rectangle area, M is the target height and $N$is the width, $C(i,j)$ is the color value of the target point (i,j). Extract of the vertical projection of the target $C^H=\left\{C_j^h=\frac{1}{N}\sum_{i=1}^{N}C(i,j), j=1,2,\ldots,M\right\}$. Select the threshold Tc of vertical projection and judge the projection image: if $|C_{j+1}^h-C_j^h|<Tc$, then C_{j+1}^h and C_j^h will be divided into the same block. Assume that the vertical direction projection is divided into K subblocks, then each subblock will be projected, respectively in horizontal $C^V(k)=\left\{C_i^v=\frac{1}{M}\sum_{j=1}^{M}C(i,j),\right.$ $\left.i=1,2,\ldots,N\right\}$. The threshold for each subblock is determined in the same way. If $|C_{i+1}^v-C_i^v|<Tc$, C_{i+1}^v and C_i^v will be included into the same block.

In order to enhance the target representation capability, we can refer to the block method in Ref. [6] and [9]. Then each subblock is divided into two blocks in left and right or up and down, consisting the ultimate target subblock sets $P=\{P^{2l}\}$. Each subblock is expressed $P_j=\{dx,dy,w,h\}$. (dx,dy) is the offset for the current block from the center. w is the block width and h is the height. According to the result of adaptive block, a target model,$P^T = \{p_i^T, w_{p_i}\}$, $i = 1, 2, \ldots, k$, can be built. k is the number of subblock, p_i^T is the ith block histogram. w_{p_i} is the corresponding subblock weight.

3 Improved Target Segmented Tracking Fusing SURF Feature

3.1 The Adaptive Block Fusing Based on Color Histogram

The purpose of tracking is to find the target in the current frame image. The target template has a few blocks, so we can get the color histogram. And we also get the candidate region. We can define a similarity measure function.

$$\hat{\rho}(y) = \rho(\hat{p}^{(k)}, \hat{q}^{(k)}) = \sum_{n=1}^{m} \sqrt{p_u(y) q_u}, \tag{1}$$

where $\{\hat{q}^{(k)}\}_{k=1,\ldots,K}$ is histogram of the template, and $\{\hat{q}^{(k)}\}$ is histogram of the template, k is the number of blocks. The formula (1) uses Bhattacharyya coefficient expresses similarity between $\{\hat{q}^{(k)}\}$ and $\hat{p}^{(k)}$.

There are some blocks, merging them with adaptive weights can make a good tracking. So a merging frame is defined.

$$\hat{\rho}(y) = \lambda^{(1)} \rho^{(1)} + \cdots + \lambda^{(k)} \rho^{(k)} + \cdots + \lambda^{(K)} \rho^{(K)} = \sum_{k=1}^{K} \lambda^{(k)} \rho^{(k)}, \tag{2}$$

where $\lambda^{(k)}$ is the weight of the block. The higher weight value $\lambda^{(k)}$, the more important block. The weights are satisfied with $\sum_{k=1}^{K} \lambda^{(k)} = 1$. The formula (2) is the similarity measure function between the template and candidate target.

Reference [7] put forwarded the weight sum into the mean shift, and provided a target tracking frame. The position of target is given by formula (3) and (4).

$$\hat{y} = \frac{\sum_{i=1}^{N} \sum_{j=1}^{n(i)} \lambda_i \omega_i^j (y_j^j - \Delta y_i) g\left(\| \frac{y_{i0} - y_i^j}{h_i} \|^2 \right)}{\sum_{i=1}^{N} \sum_{j=1}^{n(i)} \lambda_i \omega_i^j g\left(\| \frac{y_{i0} - y_i^j}{h_i} \|^2 \right)}, \tag{3}$$

$$\omega_i^j = \sum_{u=1}^{M} \delta\left[b(y_i^j) - u \right] \sqrt{\frac{q_u^i}{p_u^i(y_{i0})}}, \tag{4}$$

where $n(i)$ is the number of pixels of the subblock. y_i^j is the coordinate; y_{i0} is the position center of each subblock in last frame. Δy_i is the center offset relative to the overall target; h_i is the bandwidth of kernel function; q_u^i is probability distribution of the template block i; $p_u^i(y_{i0})$ is probability distribution target of subblock i in the center of the candidate region.

3.2 The Extraction and Matching of SURF Feature Points

The SURF algorithm has made many improvements based on SIFT. By using the approximate image of box filtering in integration, the speed of feature extraction is greatly improved. And also it has the features of rotation invariant and scale invariant. In this paper, the target can be effectively described and the location performance can be better achieved by using target tracking.

Through extracting the SURF feature points of each image block from the target in the initial frame, the information on the scale of the image block is obtained. Then, all the image blocks are matched with feature points. The paper uses approximate nearest neighborhood (ANN) algorithm. Through SURF, feature points sets S_1 for template A and S_2 for current frame B, are collected. Calculate the Euclidean distance between each point p_{1i} in S_1 and all points in S_2. We can get the smallest distance d_1 and second smallest distance d_2, and the corresponding feature points p_{2i} and p_{2i}' in S_2. If $d_1 \leq \alpha d_2$ (α is the ratio of minimum distance and second smallest distance, and it is 0.65 in this paper), we would conclude that p_{2i} and p_{1i} are matched. Otherwise that point would be discarded. Meanwhile, we will conclude that there is no point matched with p_{1i} in S_2.

3.3 Image Block Weights Update and Occlusion Judging

During tracking, the target may be segmented, the occluded block cannot match SURF feature points, but unobstructed image block will play a significant role in the process of matching and tracking. Therefore, the subblock of the target role is not the same. We should configure for each subblock weights based on the color similarity and SURF feature matching number. In the trace, the subblock makes a larger contribution, then the weight become lager. Conversely, the subblock has a smaller weight. Adaptive weights are updated as follows:

(1) If color similarity between the ith subblock and corresponding subtemplate block is greater than a given threshold, and SURF feature matching number is

the same as above. So its weight $\lambda_{it} = \lambda_{it-1} + \beta\lambda_{it-1}$ should be increased, where β is a positive integer, λ_{it-1} is the weight at time $t-1$.

(2) If color similarity between the ith subblocks and corresponding subtemplate block is less than a given threshold, but SURF feature matching number is greater than the given threshold, subblock color maybe change. Keep the weight unchanged. We think the light maybe changed.

(3) If color similarity between the ith subblocks and corresponding subtemplate block is greater than a given threshold, but SURF feature matching number is less than the given threshold. We conclude that the position of the subblock is offset. Then the weight of the subblock reduces.

(4) If color similarity between the ith subblocks and corresponding subtemplate block and SURF feature matching number are less than a given threshold, so it may be segmented. Then the subblock weight is assigned a smaller weight.

(5) For weight of each subblock, normalization is done.

3.4 Algorithm Steps

This article proposes an adaptive tracking algorithm based on color block, it also combines the SURF feature matching. Figure 1 is a flowchart of the tracking algorithm. Tracking algorithm description is given as follow.

Step 1 initialization phase

(1) According to the above method, the target template is divided into N blocks.
(2) Calculate the center position and scale information for each subblock.
(3) Establish a histogram vector for each subblock and calculate the SURF feature points.
(4) Initialize each subblock weight: $\lambda_i = 1/N$, where N is the total number of subblock.

Step 2 tracking phase

(1) The initial target position of each frame is the previous target position. Calculate the subblock color similarity between candidate target and the template, and the SURF feature matching numbers.
(2) According to the color similarity and the SURF feature matching numbers from (1), the occlusion can be judged.
(3) By matching the feature point sets, a region of interest and a back projection can be obtained.
(4) The tracking window is returned by using Cam Shift tracking algorithm.
(5) Update each image block weight, and go on the next frame tracking.

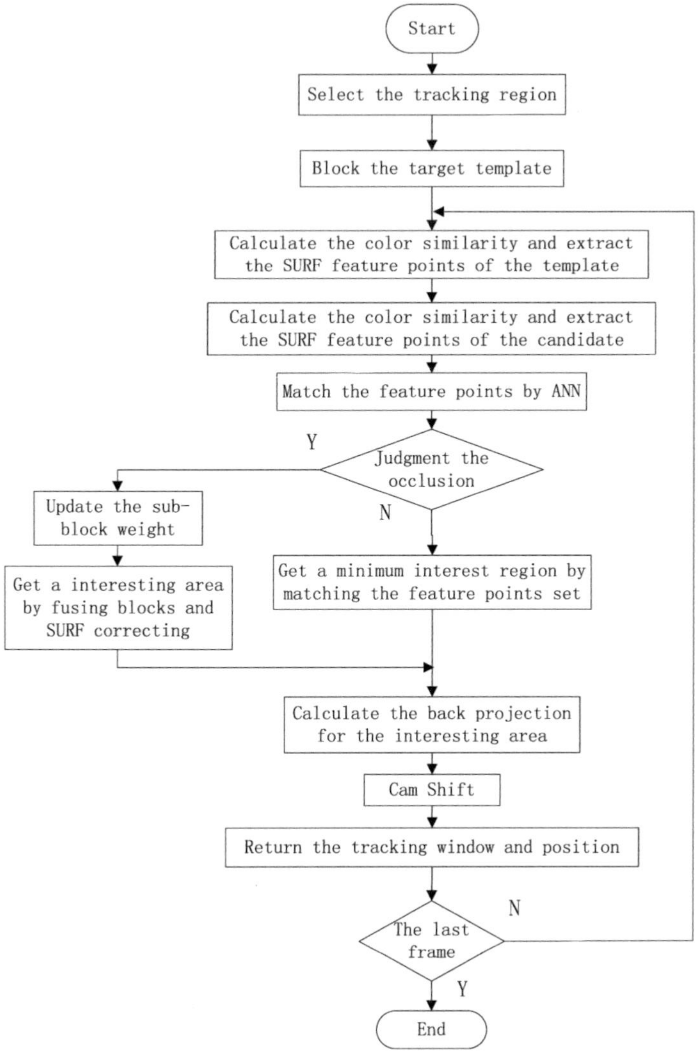

Fig. 1 Tracking flow chart

4 Experiment Results

To test the effectiveness of this algorithm, this paper carries out some experiments for multiple situations. Case 1 is to test the algorithm in scale adaption. And case 2 is to test the algorithm in occlusion.

Case 1: Scale adaption in tracking with Cam Shift.

The Fig. 2 shows the tracking result. The four images are four frames from the video. The background of video is simple and the target is not occluded.

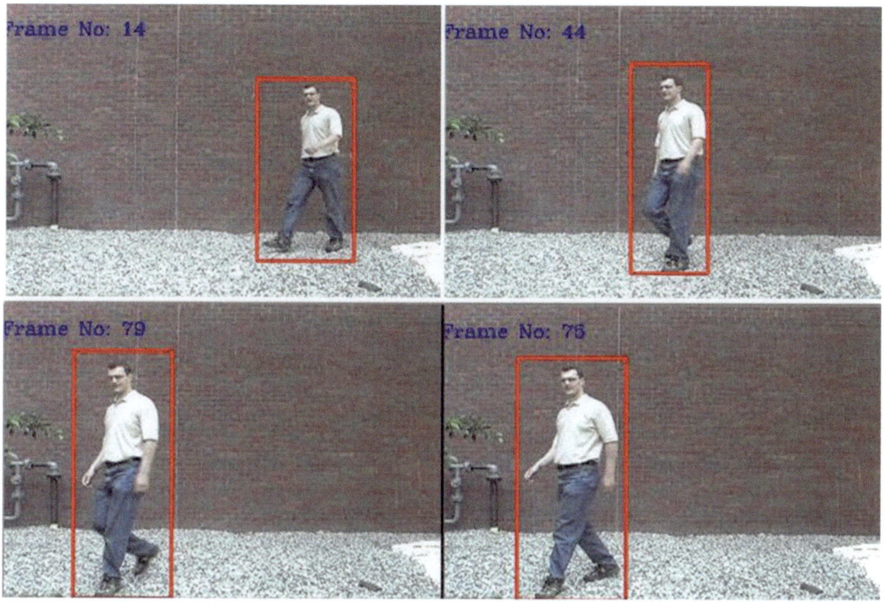

Fig. 2 Tracking result in scale change

We can see the scale of the person in video is varying, but the tracking effect is still good, which proves the algorithm has scale adaptation.

Case 2:

The following video is in order to test the algorithm tracking effect in occlusion situations. The target in the course of the video sometimes is segmented, but not completely segmented. But the algorithm still achieves a good tracking. After the occlusion, the target is still kept a well tracking (Fig. 3).

Fig. 3 Tracking result in occlusion

5 Conclusions and Future Work

Based on the target segment tracking of Cam shift, this paper puts forward the fusion of SURF matching. The block method based on the color is adaptive, which is good for tracking. Through judging occlusion and adjusting block weight, the problem of tracking drift in the occlusion is improved. Experimental results show that the proposed algorithm can keep good tracking performance in the occlusion and object scale change scenarios. However, there are some problems in this algorithm. When the template is extracted, it also contains the background. Then the SURF matching and the tracking effect will be influenced. The next we also need to improve the algorithm to reduce the impact of background on the template.

Acknowledgments The author would like to thank the anonymous reviewers for their constructive and insightful comments for further improving the quality of this work. This work was supported by the Major State Basic Research Development Program (973 Program) (No. 2012CB821200), National Key Technologies R&D Program (No. 2013BAB02B07), and National Science Foundation of China (No. 61520106010).

References

1. Bakhtari A, Mackay M, Benhabib B (2009) Active-vision for the autonomous surveillance of dynamic, multi-object environments. J Intell Robot Syst 54(4):567–593
2. Vadakkepat P, Lim P, Desilva L et al (2008) Multimodal approach to human-face detection and tracking. IEEE Trans Ind Electron 55(3):1385–1392
3. Matei B, Sawhney H, Samarasekera S (2011) Vehicle tracking across non-overlapping cameras using joint kinematic and appearance features. In: The 24th IEEE conference on computer vision and pattern recognition. Spring, Colorado, pp 3465–3472
4. Kalal Z, Mikolajczyk K, Matas J (2012) Tracking-learning-detection. IEEE Trans Pattern Anal Mach Intell 34(7):1409–1422
5. Yang H, Shao L, Zheng F et al (2012) Recent advances and trends in visual tracking: a review. Neurocomputing 74(18):3823–3831
6. Adam A, Rivlin E, Shimshoni I (2006) Robust fragments-based tracking using the integral histogram. In: IEEE Computer society conference on computer vision and pattern recognition. IEEE, New York, pp 798–805
7. Fang J, Yang J, Liu H (2011) Efficient and robust fragments-based multiple kernels tracking. Int J Electron Commun 65(11):915–923
8. Hossain K, Oh C, Lee C-W et al (2012) Multi-part SIFT feature based particle filter for rotating object tracking. In: Informatics, electronics and vision (ICIEV) 2012, pp 1016–1020
9. Erdem E, Dubuisson S, Bloch I (2012) Fragments based tracking with adaptive cue integration. Comput Vis Image Underst 116(7):827–841

Study on Camera Calibration for Binocular Stereovision Based on Matlab

Haibo Liu, Yujie Dong and Fuzhong Wang

Abstract Camera calibration is designed to build a 3D word coordinates and the corresponding relationship between the 2D image coordinates. It is the foundation of binocular stereo measurement system, and the calibration precision also affects the measuring precision of whole system eventually. This paper proposes an improved plane calibration method based on studies of existing calibration method. First of all, the method used camera pinhole model for calculating initial interior and exterior parameters, then the radial and tangent distortion were introduced, and the nonlinear optimizing method was used to solve distortion coefficients. Finally, a simulation example is provided to prove the effectiveness of the conclusion.

Keywords Camera calibration · Binocular stereo vision · Camera model · Matlab

1 Introduction

Camera calibration is the indispensable premise and foundation of binocular stereo vision system, accurate to calculate interior and exterior parameters of the camera, and the radial and tangential distortion is not only can improve the accuracy of 3D restore directly, and can lay good foundation for the subsequent feature matching, which is one of the hot issues in the field of stereo vision research [1, 2].

Camera calibration is to determine the mapping relationship between the world coordinate system for three-dimensional objects and the camera image coordinate, including calibration for interior and exterior the geometric and optical parameters

---

H. Liu (✉) · F. Wang
School of Electrical Engineering and Automation, Henan Polytechnic University, Jiaozuo 454003, China
e-mail: liuhaibo09@hpu.edu.cn

Y. Dong
Wanfang College of Science & Technology, Henan Polytechnic University, Jiaozuo 454003, China

Y. Jia et al. (eds.), *Proceedings of 2016 Chinese Intelligent Systems Conference*, Lecture Notes in Electrical Engineering 405,
DOI 10.1007/978-981-10-2335-4_20

of camera imaging system, and the relative position between the two cameras calibration. Existing camera calibration techniques can be roughly divided into the traditional methods of camera calibration and camera self-calibration method. Traditional calibration method need to use the machined calibration block, through the establishment of 3D coordinate of known points on the calibration block and its corresponding image point to calculate interior and exterior parameters for the camera [3]. This method can achieve high accuracy, but time-consuming for calibration process. Camera self-calibration does not need calibration, calibrating directly rely on the relationship between the corresponding points of many images. The Flexibility of method is strong, which is mainly used in place of accuracy is not high. Two-step method proposed by Zhang Zhengyou is a compromise between self-calibration and traditional calibration methods; it is to avoid the high equipment requirement and complicated faults of the traditional methods, and it is of high accuracy and better robustness than self-calibration, therefore has been widely applied. But in the case of a wide-angle lens distortion larger, the correction results deviation is larger. To this end, this paper puts forward an improved plane calibration method.

2 The Camera Imaging Model

One of the basic tasks of binocular stereo vision is to calculate the geometry information of target in the three-dimensional space from the goal of image information obtained by the camera, and rebuild and locate the target. The interrelation between the three-dimensional geometry position of point on the space target surface and the corresponding points in the image is determined by the geometrical model of camera imaging. The camera imaging model is divided into linear model and nonlinear model.

3 Linear Model

Linear model is refers to the pinhole imaging model, as shown in Fig. 1. p is projection point on the image from space point P, which is intersection point with OP and image plane. Projecting from three-dimensional space on a two-dimensional image plane involves four coordinate conversions [4].

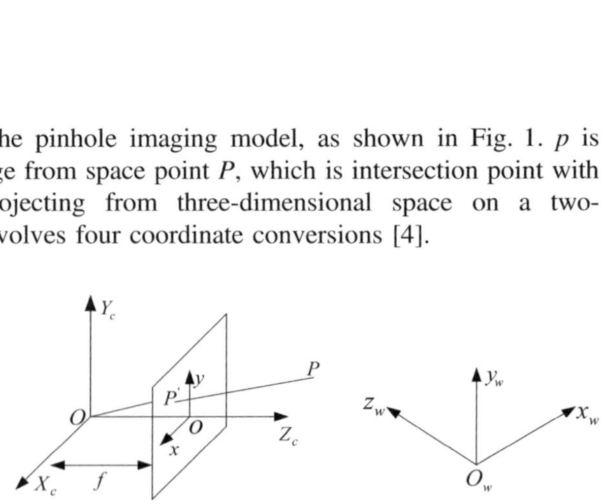

Fig. 1 Pinhole model of camera

(1) The world coordinate system X_w-Y_w-Z_w: the real coordinate system is the absolute coordinates of the objective world, and can describe the location of the cameras or any other object.
(2) The camera coordinate system X_c-Y_c-Z_c: the camera light heart O_c is the origin, X_c, Y_c axis is respectively parallel to x and y, Z_c is camera optical axis, O_cO is the camera focal length.
(3) The image plane coordinate system x- y: the origin O is intersection point for optical axis and the image plane, x, y is respectively parallel to u and v, in mm for unit.
(4) Computer image plane coordinate system u-v: each digital image is expressed as M × N matrix inside a computer, (u, v) is said as rows and columns of pixels in the matrix, the origin of coordinates is on the upper left corner for image, in Pixels for unit.

Camera calibration is to solve the conversion between computer image coordinates (u, v) and the world coordinate system (X_w, Y_w, Z_w). According to the principle of pinhole imaging, the conversion relation is as follow.

$$S\begin{bmatrix} \mathrm{u} \\ v \\ 1 \end{bmatrix} = F[R \quad T]\begin{bmatrix} x_w \\ y_w \\ z_w \\ 1 \end{bmatrix} = M\begin{bmatrix} x_w \\ y_w \\ z_w \\ 1 \end{bmatrix} \tag{1}$$

On the type, S is scale factor, F is the interior parameter matrices, $[R\ T]$ is the camera exterior parameter matrix, R and T is respectively rotation and translation components for the camera coordinate system relative to the world coordinate system. F, R, and T is defined respectively as follow.

$$F = \begin{bmatrix} f_x & 0 & C_x \\ 0 & f_y & C_y \\ 0 & 0 & 1 \end{bmatrix}, R = \begin{bmatrix} r_{1,1} & r_{1,2} & r_{1,3} \\ r_{2,1} & r_{2,2} & r_{2,3} \\ r_{3,1} & r_{3,2} & r_{3,3} \end{bmatrix}, T = \begin{bmatrix} r_{1,4} \\ r_{2,4} \\ r_{3,4} \end{bmatrix} \tag{2}$$

On the type, f_x, f_y is denote respectively the focal length in the direction of x, y; C_x, C_y is denote respectively the computer image coordinates of intersection point for the main optical axis and image plane. The task of camera calibration is obtained interior and exterior parameter matrix.

4 Nonlinear Model

The ideal camera model is pinhole model, its object and image is to meet similar triangle relations, while the actual camera does not meet this condition [5]. Because the camera optical system has processing error and assembling error, there are radial distortion and tangential two main error of optical distortion between real and ideal imaging. The radial distortion is symmetric about main optical axis of camera lens, but tangential distortion is not. Although there is actually issue such as no central and thin mirror image aberration, but too much distortion parameters not only can not improve the calibration accuracy, it will make the calibration result is unstable, so generally consider only radial distortion and tangential distortion. Sets the physical coordinates of ideal imaging point to (x_u, y_u), (x_d, y_d) is the physical coordinates after the distortion, the relationship is as follow.

$$\begin{cases} x_d = x_u + \delta_x(x_u, y_u) \\ y_d = x_u + \delta_y(x_u, y_u) \end{cases} \tag{3}$$

On the type, $\delta_x(x_u, y_u)$, $\delta_y(x_u, y_u)$ is denote respectively the amount of distortion in the direction of x, y, the expression is:

$$\begin{cases} \delta_x(x_u, y_u) = x_u(k_1\rho^2 + k_2\rho^4) + p_1(3x_u^2 + y_u^2) + 2p_2x_uy_u \\ \delta_y(x_u, y_u) = y_u(k_1\rho^2 + k_2\rho^4) + 2p_1x_uy_u + p_2(x_u^2 + 3y_u^2) \end{cases} \tag{4}$$

On the type, $\rho = \sqrt{x_u^2 + y_u^2}$, k_1 and k_2 is respectively the first-and second-order radial distortion coefficient; p_1 and p_2 is respectively the first-and second-order tangential distortion coefficient.

The linear model and nonlinear model has its own advantages and disadvantages. Linear model computing speed is fast, but because regardless of the distortion, the accuracy is not high. Nonlinear model accuracy is higher, but a large number of nonlinear equations are solved, computing speed is slow [6]. This paper obtained the initial internal and external parameters matrix of camera through the linear model, then the radial and tangential distortion are introduced, the distortion parameters are get through a nonlinear optimization algorithm.

5 Improved Method of Camera Calibration

Calibration algorithm is divided into two steps, one is to calculate interior and exterior parameters matrix of camera through improved plane calibration method; the second is calculate structural parameters of binocular vision system by using the internal and external parameter matrix of camera [7].

According to plane calibration method, the template plane is seen as X_w-Y_w plane of the world coordinate system, type 1 can be simplified as follow.

$$S\begin{bmatrix} \mathrm{u} \\ \mathrm{v} \\ 1 \end{bmatrix} = \begin{bmatrix} f_xr_{1,1} & f_xr_{1,2} & f_xr_{1,4} \\ f_yr_{2,1} & f_yr_{2,2} & f_yr_{2,4} \\ r_{3,1} & r_{3,2} & r_{3,4} \end{bmatrix}\begin{bmatrix} x_w \\ y_w \\ 1 \end{bmatrix} = H\begin{bmatrix} x_w \\ y_w \\ 1 \end{bmatrix} \tag{5}$$

On the type, H is called single matrix. To calculate the internal and external parameters is converted into calculating H. According to the principle of the linear imaging model, it can get two linear equation correspond to each pair of the known world coordinates and computer image coordinates, H is obtained when there are enough corresponding points, then decomposed to get the internal and external parameters. Optimization process is divided into two steps on the whole.

(1) Regardless local optimization of the distortion

Regardless of distortion is ignored radial distortion and tangential distortion of camera calibration in the process of optimization. Local optimization refers to the selection of corresponding points, not adopt all point in the image, but considering the due to ignore the influence of distortion, so only choose less distortion point near the image center as a corresponding points to optimize, thus reducing the error, make the optimized PARAMETERS in the first step as accurate as possible. When selecting points are near the image center, due to its distortion volume is small, so cannot consider distortion, evaluation criterion for optimization model is established according to the ideal perspective model.

$$\min = \sum_{i=1}^{n}\sum_{j=1}^{l_i}\left\| m'_{i,j} - m\left(F, R_i, T_i, m_{i,j}\right)\right\| \tag{6}$$

On the type, $m'_{i,j}$ is image coordinates of j point in image i, $m_{\mathrm{i,\ j}}$ is for the corresponding world coordinates; l_{i} is the number of point for image ii. Constraint equation is that R_{i} should satisfy the six orthogonally equation. The purpose for optimization model is put F, R_{i} and T as the initial value of optimizing function, the optimization algorithm is Levenberg–Marquardt for its optimization.

(2) To obtain precise camera model parameters, it should consider all of world coordinate points and the corresponding image points, and establish the optimization model to optimize all internal and external parameters. Levenberg–Marquardt algorithm is adopted to establish the optimization model.

$$\min = \sum_{i=1}^{n}\sum_{j=1}^{l_i}\left\| m'_{i,j} - m\left(k_1, k_2, p_1, p_2, F, R_i, T_i, m_{i,j}\right)\right\| \tag{7}$$

The distortion coefficient k_1, k_2, p_1, p_2 is very small, so the initial value can be set to 0. The initial value of model is parameters values and the initial distortion coefficient for the first step to calculate. Internal parameter matrix, external parameter matrix, and distortion coefficient are obtained by two-step optimization, which is as a result of the camera calibration.

In a binocular vision system, it should get interior and exterior parameters of the camera and know the relative positions between the two cameras, structure parameters of the system are rotation matrix R and translation vector T.

Exterior parameters of the left and right camera in binocular vision system are respectively R_l, T_l and R_r, T_r, R_l and T_l is the relative position of the world coordinate system and the left camera; R_r, T_r is the relative position of the world coordinate system and the right camera; x coordinate for any point in space is respectively x_w, x_l, x_r in the world coordinate system, the left and right camera coordinate system.

$$\begin{cases} x_l = R_l x_w + T_l \\ x_r = R_r x_w + T_r \end{cases} \tag{8}$$

Therefore, the relative geometric relationship between two cameras is as follow.

$$\begin{cases} R = R_r R_l^{-1} \\ T = T_r - R_r R_l^{-1} T_l \end{cases} \tag{9}$$

6 Simulation Experiment

The checkerboard is used as 2D plane targets in experiment, selected the checkerboard vertices as feature point to calibrate camera parameters, which is as shown in Fig. 2. Assume that the internal parameters of camera are constant; put the 2D plane targets in some position in the measurement volume. In the measurement volume, the 2D plane targets can move free, do not need to know the parameters of the movement. The CCD camera shoot 28 groups plane target images for different position and angle. Considering the orthogonally of rotation matrix, camera parameters are obtained. In the process of calibration, assuming that the internal parameters of camera is constant whether the camera shoot targets from which angle, only the external parameters changed.

In this binocular vision measurement system, the internal and external parameters of two cameras are:

Fig. 2 2D plane targets

(1) Internal parameters for left camera

$$f_{cl} = [594.22, 591.74], C_{cl} = [154.33, 254.46],$$
$$k_{1l} = -0.14, k_{2l} = 0.26, p_{1l} = -3.1 \times 10^{-3}, p_{2l} = -8.7 \times 10^{-4}.$$

(2) Internal parameters for right camera

$$f_{cr} = [3675.41, 3670.17], C_{cr} = [806.61, 612.92],$$
$$k_{1r} = -0.17, k_{2r} = 0.88, p_{1r} = -1.72 \times 10^{-3}, p_{2r} = 1.13 \times 10^{-3}.$$

According the calculated camera parameters, it reconstructed the location relationship between the camera and different location targets, which is as shown in Fig. 3.

The internal and external parameters of two CCD cameras are processed by stereo calibration toolbox in MATLAB. According to the relationship between the camera coordinate system and world coordinate system, stereo calibration is made to get the location relationship for two CCD camera coordinate systems, which is as shown in Fig. 4. The calibration results of the binocular stereo vision measurement system are as follows.

$$R = [-2.9 \times 10^{-2}, 0.38, -1.17 \times 10^{-3}],$$
$$T = [-200.34, 100.35, 150.75].$$

The standard model of stage is placed in measuring volume of binocular vision measurement system, the left of the model is composed of multiple parallel plane; the spacing between adjacent plane are 1.00 mm, the right is a standard cant. A 3D reconstruction result of the stage model is shown in Fig. 5, using the least squares fitting plane equation, the distance between the adjacent parallel planes is 0.99 mm by calculating.

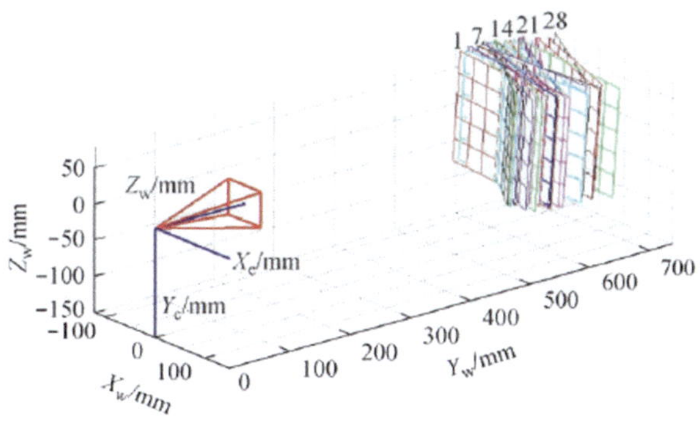

Fig. 3 Location relationship between the camera and targets

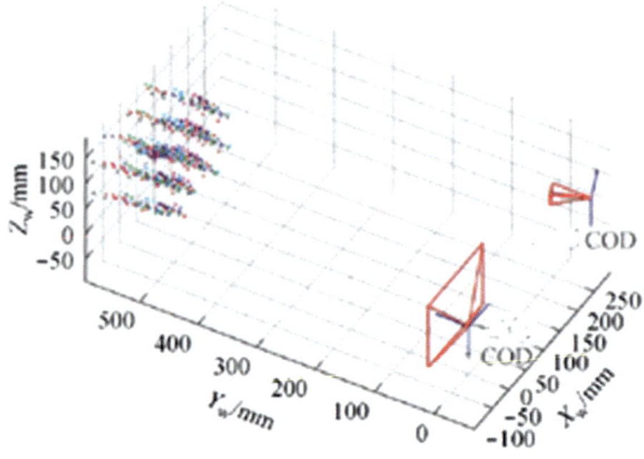

Fig. 4 Stereo calibration results

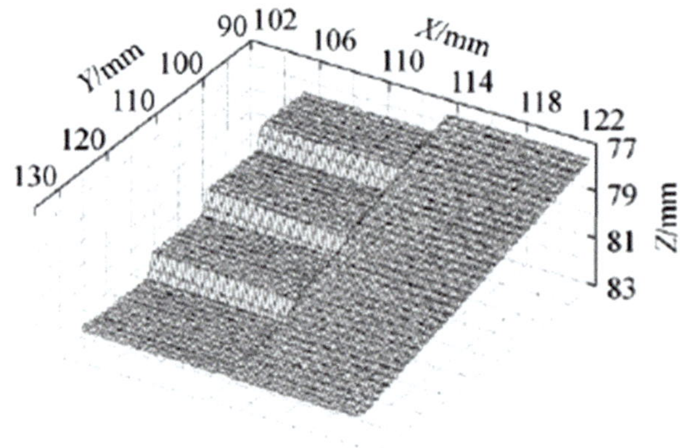

Fig. 5 Measurement results of stage model

7 Conclusion

The camera calibration is one of the key technology for binocular vision measurement system, this paper used camera pinhole model for calculating initial interior and exterior parameters, then the radial and tangent distortion were introduced, and the nonlinear optimizing method was used to solve distortion coefficients. The interior and exterior parameters of CCD cameras are processed by stereo calibration toolbox in MATLAB, so that it can get the relationship poison of two cameras. Finally, a standard model of stage is measured, and its three-dimensional shape is restored. The experimental results verify the accuracy and stability of this method.

References

1. Armstrong M, Zisserman A (1996) Self-calibration from image triplets. In: Proceedings of European conference on computer vision, pp 4–12
2. Zhang Z (1999) Flexible camera calibration by viewing a plane from unknown orientations. In: Proceedings of 7th international conference on computer vision, pp 666–673
3. Zhang ZY (2000) A flexible new technique for camera calibration. IEEE Trans Pattern Anal Mach Intell 22(11):1330–1334
4. Song L, Lu L (2007) High precision camera calibration in vision measurement. Opt Laser Technol 39(5):1413–1420
5. Wu F, Li H (2000) A new camera self-calibration method based on active vision system. Chin J Comput 23(11):1130–1139
6. Liu Y, Kong J (2011) Nonlinear camera calibration method of binocular stereo vision system. Appl Res Comput 28(9):3398–3400
7. Liu J, Yuan S (2008) Research on camera calibration in binocular stereo vision. Comput Eng Appl 44(6):237–243

The Structure Shaping of ITAE Optimal Control System Base on PSO

Yuzhen Zhang, Qing Li, Weicun Zhang, Yichang Liu and Zhao Pei

Abstract A method, structure shaping, based on the standard form of ITAE optimal control, is presented in the paper. First, structure shaping is related to adjusting the factors of local feedback and open-loop amplification; second, the PSO is used to obtain the optimal result of structure shaping. The results show that after structure shaping, tracking response, stability and robustness have been greatly improved, which possess a high practical value in engineering.

Keywords Standard form · ITAE optimal control · PSO · Structure shaping

1 Introduction

The standard form of closed-loop system transfer function, satisfying ITAE optimal control, was pointed out by Dunstan Graham and R. C. Lathrop in the 1950s [1, 2]. After a period of silence, it was gradually applied in engineering [3]. Actually, to the best of the authors' knowledge, Ref. [4] showed the ITAE optimal regulation law in SCR direct current drives systems. Reference [5] provided the ITAE optimal control digital servo system of type III. Twice optimum control for time delay system was referred to [6, 7]. In order to design the twice optimum control for a kind of second-order systems with deadtime, the new standard forms of ITAE optimum transfer function were given in [8] with the better performance index than before. The application proved that the ITAE optimal control law can be suitable for application in engineering with good performance.

On the basis of ITAE optimal control system, in order to strengthen the system damping in the middle band, the value of local feedback should be increased, improving stability and robustness, but followed by slowing down the rising speed inevitably [9]. Correspondingly, the open-loop amplification factor also needs to be

Y. Zhang · Q. Li · W. Zhang (✉) · Y. Liu · Z. Pei
School of Automation and Electrical Engineering, University of Science and Technology Beijing, Beijing 100083, China
e-mail: weicunzhang@ustb.edu.cn

Y. Jia et al. (eds.), *Proceedings of 2016 Chinese Intelligent Systems Conference*, Lecture Notes in Electrical Engineering 405,
DOI 10.1007/978-981-10-2335-4_21

increased to ensure the smoother and faster system response process. This approach is called structure shaping [9].

Particle Swarm Optimization (PSO) is a kind of evolutionary computation technology, proposed in 1995 by Dr. Eberhart and Dr. Kennedy [10]. It is originated from research on bird flock preying behavior [10, 11]. The basic idea of particle swarm optimization algorithm is to find the optimal solution, through collaboration and information sharing between individuals in the group. The advantages of PSO are simple and easy to achieve without many parameters to be adjusted.

In the paper, in order to complete the structure shaping with the optimal factor, the PSO is introduced. Based on the ITAE optimal standard form, the PSO is used to obtain the result of structure shaping.

2 Particle Swarm Optimization

The basic idea of PSO can be described as following. Consider the search space $S \in R^n$ and the particle swarm $M \in R^m$ consisting of m particles, the position of each particle $X_i = (X_{i1}, X_{i2}, \ldots, X_{in})$ represents a candidate solution for one question. The quality of these solutions is determined by the value of the fitness function. Fitness function relates to the objective function of the specific problem. Particle jumps in the search space S at the speed of $V_i = (V_{i1}, V_{i2}, \ldots, V_{in})$. In the process, the best position of every particle can be expressed as $P_i = (P_{i1}, P_{i2}, \ldots, P_{in})$. Individual particles record themselves the optimum results P_i in the search process, meanwhile recording the global optimum result $P_g = (P_{g1}, P_{g2}, \ldots, P_{gn})$. Then, individual particles update X_i and V_i based on the P_i and P_g. Through iteration searching, the optimal solution can be found finally.

PSO has the following advantages: a. realized easily; b. relatively few parameters; c. fast convergence.

2.1 PSO Process

Initialization. Generating m particles' random position vectors $x_i(t)$, $i = 1, 2, \ldots, m$, in the search space, then the velocity vector $v_i(t) = \left[v_i^1, v_i^2, \ldots, v_i^n\right]$, $i = 1, 2, \ldots, m$ can be obtain by the position. Set the maximum value of iterations $T_{\max}$ and accelerated constant C_1, C_2.

The fitness function $F(x_i)$ is calculated, thus the quality of the particle fitness can be evaluated.

For each particle, its current fitness function value is compared with its previous best fitness value *pbest*. If the current value is superior to the original value, the original value will be replaced by the current value. The position of the particle P_i consists with its fitness value.

Similar to the previous step, the global optimal fitness value *gbest* of particle swarm also can be updated.

Fig. 1 PSO flow chart

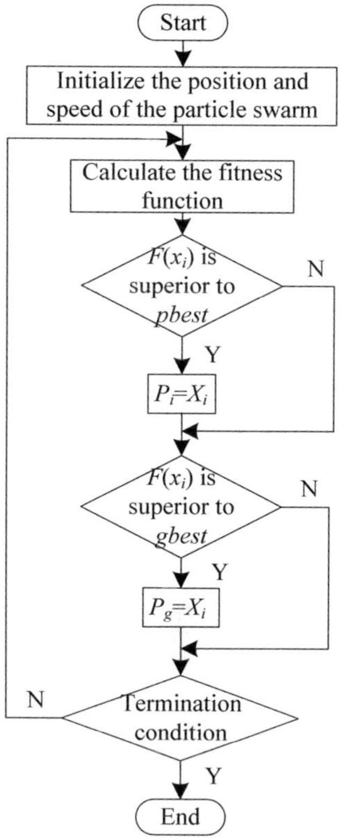

Each particle's speed and position is updated, then, the cycle continues execution until the termination condition is satisfied.

From the above, the PSO is completed, as shown in Fig. 1, thus we got the optimal solution for the problem.

2.2 PSO Design

Consider the search space $S \in R^n$ and the particle swarm $M \in R^m$, the position of each particle is given by

$$x_i(t+1) = x_i(t) + v_i(t+1) \tag{1}$$

Further, define the velocity of each particle, that is,

$$v_i(t+1) = \omega(t)v_i(t) + C_1(t)r_1(p_i(t) - x_i(t)) + C_2(t)r_2(p_g(t) - x_i(t)) \tag{2}$$

where inertia weight ω controls the impact of the velocity at last moment on the current velocity. It balances the global search ability and the local search ability of the algorithm. In this paper, inertia weight is set as variable, decreasing gradually. Thus, the exploration and exploitation is better than the constant one.

Learning factor $C_1(t)$ indicates the cognitive ability of particles themselves experience, adjusting step length forward the best position. Learning factor $C_2(t)$ represents the cognitive ability of the social experience, regulating step length to the global best particle position. Further, we set the learning factors as variable, those are,

$$C_1(t) = c_{1f} - \frac{2 * t}{T_{\max}} \tag{3}$$

$$C_2(t) = c_{2f} + \frac{2 * t}{T_{\max}} \tag{4}$$

where c_{1f} and c_{2f} are constant.

r_1 and r_2 are both random number, which is to allow particles can go forward with equal probability of acceleration. Thus particles fly to the best position themselves and the global best position. The iteration, $t = 1, 2, 3, \ldots$, is satisfied with $t \leq T_{\max}$.

In addition, the maximum velocity $V_{\max}$ limits particle's biggest displacement distance. If $V_{\max}$ is too large, the optimal solution may be skipped. On the contrary, the result is a local optimal solution.

3 Structure Shaping

For a system, there are several optimum control laws. Through comparative analysis, the ITAE optimum control law is more practical and selective than others [12]. However, the system with good dynamic property the performance can be improved further by structure shaping. As structure shaping is an engineering term, practical examples can show the conclusion obviously. In order to achieve a more pronounced effect, the standard form of transfer function, satisfying ITAE optimal control, is used to implement structure shaping.

Through analysis and discussion, on the basis of ITAE optimal control system, in order to improve stability and robustness of the system, the system damping in the middle band should be strengthened. It requires that feedback coefficient can be introduced. Meanwhile, most controller parameters are unchanged. So, the value of local feedback coefficient is increased. But it slows down the rising speed inevitably. Correspondingly, the open-loop amplification factor is also need to increase to ensure the smoother and faster system response process. This approach is called structure shaping. For more details, the reader is referred to [9].

From the above, we can know that the standard form of closed-loop system transfer function, satisfying ITAE optimal control has good generality. For the transfer function, structure shaping is completed by adjusting the molecules and the denominator coefficients of the transfer function. From this point of view, some simulation experiments can be done. Meanwhile, the results can show the superiority of the system performance after structure shaping obviously. The simulation results are presented in the next section.

4 Simulation Results

Consider the transfer function standard form of ITAE optimal control system, which is presented in Ref. [12]. In the author's opinion, the generic form of controlled object after structure shaping can be shown as

$$G(s) = \frac{b}{s^n + a_1 * s^{n-1} + \cdots + a_{n-1}s + a_n} \tag{5}$$

where $a_1, a_2, \ldots, a_n, b$ are the adjustable factors.

Based on MATLAB/Simulink platform, PSO is accomplished by M-file, and the simulation control model is established. Appropriate fitness function is used iteratively to search the best transfer function form after structure shaping.

It is worth noting that the integral bound of ITAE is zero to the limited time, rather than zero to infinity. Besides, the system performance index J is not the ITAE optimum control law simply, but the overshoot σ is considered. That is

$$J = (1 + a * \sigma) * \int_0^{\tau} \tau |e(\tau)| \mathrm{d}\tau \tag{6}$$

where a is weighting coefficient, $e(\tau)$ is the error.

Case 1
The second-order ITAE optimal control system is shown in Fig. 2a. On this occasion, there are two factors should be adjusted. The transfer function after structure shaping is also presented in Fig. 2a. The reference signal and the output signal $y(t)$ are shown in Fig. 2b.

From the Fig. 2b, after structure shaping, response speed has been greatly improved. To some extent, system stability has improved.

Case 2
The third-order ITAE optimal control system is shown in Fig. 3a. According to the generic form, as formula (5) shown, there are three factors should be adjusted. The transfer function after structure shaping is also presented in Fig. 3a. The reference signal and the output signal $y(t)$ are shown in Fig. 3b.

As shown in Fig. 3b, after structure shaping, response speed has been greatly improved obviously. However, comparing to the Case 1, the effect of the performance improvement is decreased slightly. To some extent, system stability has improved.

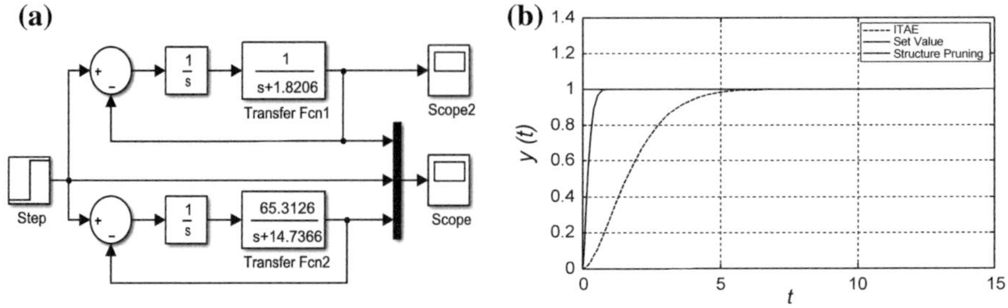

Fig. 2 s-order system structure and its response. **a** Contrasting system structure of Case 1. **b** Set value and output $y(t)$ of Case 1

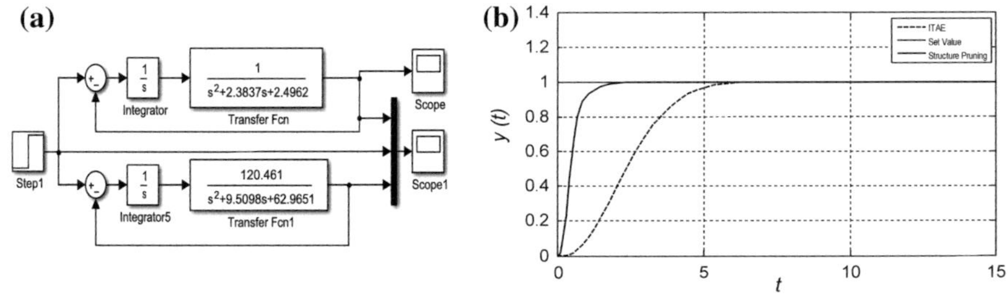

Fig. 3 Third-order system structure and its response. **a** Contrasting system structure of Case 2. **b** Set value and output $y(t)$ of Case 2

Case 3

Similar to the above, the fourth-order ITAE optimal control system is shown in Fig. 4a. Then, there are four factors should be adjusted. The transfer function after structure shaping is also presented in Fig. 4a. The reference signal and the output signal $y(t)$ are shown in Fig. 4b.

Similar to the above, system performance has been greatly improved obviously. However, compare to the Case 1, the effect of the performance improvement is decreased slightly.

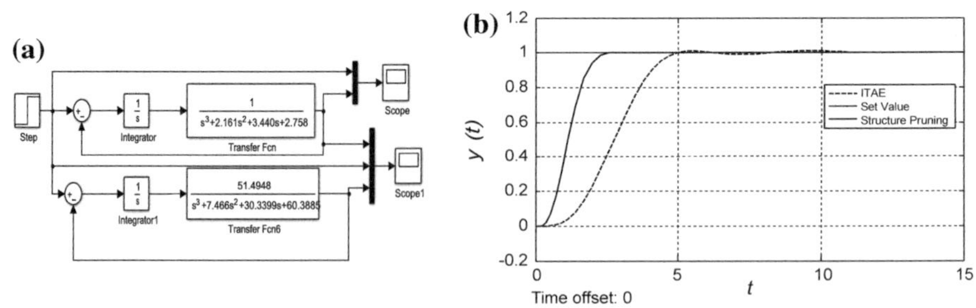

Fig. 4 Fourth-order system structure and its response. **a** Contrasting system structure of Case 3. **b** Set value and output $y(t)$ of Case 2

In summary, as shown in Figs. 2, 3 and 4, after structure shaping, tracking response has improved. However, through the comparison, we also can find that, to some extent, the effect is decreased as the increase of system order.

5 Conclusions and Future Work

With the help of PSO, the parameter tuning issue on the structure shaping has been addressed with the simple methods, satisfying ITAE optimal control law. The results show that after structure shaping, tracking response and stability of the systems have improved.

However, in the author's opinion, the fitness function could be more appropriate and accurate. To a certain extent, structure shaping possesses a high practical value in engineering. Thus, the future research will be focused on the practical application in motor control.

Acknowledgments The author would like to thank the anonymous reviewers for their constructive and insightful comments for further improving the quality of this work. This work was supported by the Major State Basic Research Development Program (973 Program) (No. 2012CB821200), National Key Technologies R&D Program (No. 2013BAB02B07) and National Science Foundation of China (No. 61520106010).

References

1. Graham D, Lathrop RC (1953) The synthesis of "Optimum" transient response: criteria and standard forms. AIEE Trans 72:273–288
2. Schultz WC, Rideout VC (1961) Control system performance measures: past, present and future. IRE Trans Autom Control 2:22–35
3. Xiang G (1977) Linear feedback ITAE optimum control system. Electr Drive 2:16–40
4. Lei X (1983) The ITAE optimal regulation law in SCR direct current drive systems. Acta Automatica Sinica 9(3):229–232
5. Chen M, Wu Y, Jiang Q (1993) ITAE optimal control digital servo system of type III. Acta Automatica Sinica 19(2):177–182
6. Xiang G, Yang Y, Yang Q (1995) Twice optimum control for a kind of first order systems with dead-time. Inf Control 24(4):208–214
7. Yang Q, Yang Y, Xiang G (1995) ITAE twice optimal control for a kind of second order systems with dead-time. Chin Control Conf (10):475–480
8. Yang Y, Xiang G (1997) The new standard forms of ITAE optimum transfer function. Inf Control 26(4):259–265
9. Yu Y (2015) Twice optimal control for a class of unstable process with time delay and its structure pruning. Guangdong University of Technology, Guang Zhou
10. Kennedy J, Eberhart R (1995) Particle swarm optimization. In: Proceedings of IEEE international conference on neural networks, vol 4, pp 1942–1948
11. Liu B (2010) Particle swarm optimization algorithm and its engineering application. Publishing House of Electronics Industry, Beijing
12. Xiang G (2010) Optimum control of system with deadtime. China Electric Power Press, Beijing

Vehicle Queue Detection Method Based on Aerial Video Image Processing

Haiyang Yu, Yawen Hu and Hongyu Guo

Abstract Vehicle queue length is one of the important traffic parameters in intelligent traffic management system. High-altitude video monitoring avoids environmental object barrier and has advantages of dynamic video monitoring like larger view range, multi-angle and high precision, at the same time, it provides technical support for the vehicle queue length detection at the intersection. In order to detect the vehicle queue length in real time and apply it to the management of intelligent traffic system, a new algorithm based on video vehicle queue length detection is proposed in this paper. First, the frame difference method is used to construct the background of the image so that the background modeling error of moving objects could be reduced. On this basis, relief operation is taken to the image background and the current frame image in order to avoid the impact of light changes on the algorithm. Finally, the two-value image is analyzed to obtain the real queue length. The experimental results show that the improved method is simple to achieve and it can obtain a more accurate queue length.

Keywords Intelligent transportation · Vehicle queue length · Frame difference method · High-altitude video image · Two-value image analysis

---

H. Yu (✉)
Key Laboratory of Urban ITS Technology Optimization and Integration Ministry of Public Security, Hefei 230088, People's Republic of China
e-mail: hyyu@buaa.edu.cn

H. Yu · Y. Hu
Beijing Key Laboratory for Cooperative Vehicle Infrastructure Systems and Safety Control, School of Transportation Science and Engineering, Beihang University, 37 Xueyuan Road, Haidian District, Beijing 100191, China
e-mail: huyawen@buaa.edu.cn

H. Guo
Anhui Keli Information Industry Co. Ltd., Hefei 230088, China
e-mail: guohongyu1995@163.com

Y. Jia et al. (eds.), *Proceedings of 2016 Chinese Intelligent Systems Conference*, Lecture Notes in Electrical Engineering 405,
DOI 10.1007/978-981-10-2335-4_22

1 Introduction

In recent years, the increasing number of vehicles makes the city intersection congestion increasingly more serious. It also makes the intelligent transportation system (Intelligent transportation system, ITS) gradually become one of the inevitable trends of traffic management in the future. Intelligent transportation system utilizes effective combination with advanced information technology, computer technology, sensor technology, communication technology, and control technology to carry out an accurate, efficient, and scientific management system for the transportation.

Vehicle queue length is one of the important traffic parameters in intelligent traffic management system [2]. The traditional way of loop detector is easy to count, but it is difficult to detect the specific length of the vehicle. Most of the domestic traffic intersections use the timing control of traffic signal lamp as the fixed time, there are also part of the intersections using piecewise control which means according to the preestimated traffic peak, the allocated time of traffic lights is designed in advance. However, the way of fixed time allocation is difficult to adapt to the random variation of traffic flow. Intelligent traffic signal machine command system should distribute the time of traffic signal lamp accurately in real time according to the change of vehicle queue length in intersection [3], and satisfy the road passing guarantee to the greatest extent, which can greatly improve the intersection congestion phenomenon. High-altitude video monitoring avoid environmental object barrier and has advantages of dynamic video monitoring like larger view range, multi-angle, and high precision, at the same time, it provides technical support for the vehicle queue length detection at the intersection. With the rapid development of computer vision technology, the application of image processing technology in the field of intelligent transportation systems is more and more widely used. At present, the methods of vehicle queue length automatic extraction commonly used based on image processing are: (1) the optical flow method: the method is built on the assumption that the background gray scale changes smaller, and the time and space of the image are analyzed in optical flow field to segment the moving object boundaries, but the algorithm has complex realization, a great mount of computation, and it is difficult to compute in real time. (2) frame difference method: the method usually calculate the difference image of the front and back frames directly and segment the moving object region by setting the threshold, but when the vehicle speed at the intersection is slow or stop, it will lead to the failure of the method; (3) background method: this method is the difference between the current frame and background image to extract the vehicle, which is susceptible to illumination changes and shadows [1]; (4) object extraction method: such as the application of edge detection algorithm to segment the number of vehicles in the image, and the effect of background information of the image is great on the algorithm, which leads to the increasing of its implementation complexity, and the calculation error of queue length caused by vehicle mutual occlusion is large [4].

The real time, accurate, and effective vehicle queue length automatic extraction method is very few now, this is because the complex changes in urban traffic environment seriously affect the effectiveness and real time of the algorithm. In order to detect the length of the vehicle accurately and in real time and be applied to the intelligent transportation system management. A new algorithm based on video vehicle queue length detection is proposed in this paper. The algorithm is divided into three parts. At first, the sample information on time series is trained for a period of time, and the frame difference accumulation method is used to establish a clean image background. Then the current frame image and the background template are pretreated by relief, and the motion template is extracted by the image background method. Finally, the two-value target template obtained is carried on logical operation to extract the relevant parameters of the queue length for mathematical statistical modeling analysis.

2 Background Building

The frame difference detection algorithm has the strong robustness to the light variation in traffic environment, which can quickly extract the movement area. Therefore, this paper uses the frame difference method to construct the background image. The thought of constructing the image background by frame difference method is that the gray scale of background pixel points changes relatively slowly, and the gray scale of moving regions in front and back frames will have more obvious changes. Through two frames subtraction, the absolute value image of brightness difference is obtained, and moving object region is extracted by setting the threshold segmentation. We assume that $\mathbf{I}(x, y, t)$ represents the current frame at time t, and $\mathbf{I}(x, y, t-1)$ represents the front frame at time $t-1$. The relation operation of frame difference image and the two-value template image between the two adjacent frames is as follows:

$$\mathbf{B}(x, y, t) = \begin{cases} \alpha \times \mathbf{I}(x, y, t) + (1-\alpha) \times \mathbf{I}(x, y, t-1) & |\mathbf{I}(x, y, t) - \mathbf{I}(x, y, t-1)| > T \\ (1-\alpha) \times \mathbf{I}(x, y, t) + \alpha \times \mathbf{I}(x, y, t-1) & |\mathbf{I}(x, y, t) - \mathbf{I}(x, y, t-1)| \le T \end{cases} \quad (1)$$

In the formula, $\alpha \in (0, 1)$ is the weight parameter which should be taken the smaller value. T is a predetermined threshold. $\mathbf{B}(x, y, t)$ is the pixel value of image background at time t.

From the formula (1), we can know that the image $\mathbf{B}(x, y, t)$ fully utilizes the information of the front and back frames image, focus on conserving the background information which is the part of the front and back frames whose change is smaller, and suppress vehicle movement area which is the part of the frame difference that has the larger change. By calculating the background image sequence $\mathbf{B}(x, y, t)$ in the time [0, T], the mean value of the cumulative value in time (t[0, T])

Fig. 1 Interested lane extraction

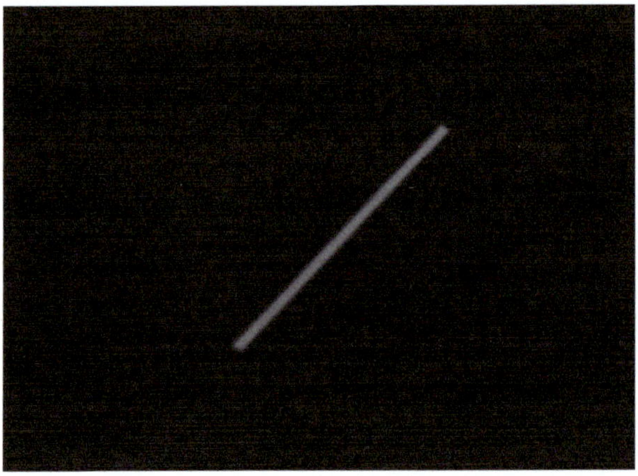

Fig. 2 Frame difference background image

is calculated to express the image background $\mathbf{BG}(x, y)$. This paper adopts the method of extracting image interest to improve the running speed of the algorithm and avoid the influence of illumination change on the model. Figure 1 is the interest lane extraction, Fig. 2 is the image background obtained by frame difference method.

3 Moving Segmentation

In this paper, moving target extraction is divided into two processes. First, relief operation is taken to the image background and the current frame image in order to reduce the influence of illumination and shadow on the algorithm. Finally, background subtraction method is used on the preprocessing image to separate the motion region, and the template image is analyzed so that the real queue detection results can be obtained.

3.1 Image Relief Operation

The complex changes of traffic environment such as light and shadow have a great influence on the detection model, and the literature [6], and so on give the illumination robustness detection algorithm and the method of reducing the shadow effect. But these methods are only designed for monitoring high-definition image at close quarters. For the high-altitude monitoring images, light effect is more obvious, which greatly reduces the effectiveness of the detection model based on color information. In order to avoid complex calculation, this paper uses simple image relief operation, and the convolution is as follows:

$$\mathbf{BG}_{\text{new}}(x,y) = \int\int_{\Omega} K \times \mathbf{BG}(x,y)\mathrm{d}x\mathrm{d}y + 128 \tag{2}$$

In the formula, Ω is the whole image region, $K = \begin{bmatrix} 1 & 0 \\ 0 & -1 \end{bmatrix}$ is the convolution kernel.

Figure 3 is a relief image of the current frame, Fig. 4 is the background relief image. After analyzing and processing the image we can know that, the effect of relief image treatment is to strengthen the object boundary, which greatly increases the part of interest border and weakens the influence of illumination on image, and the ideal treatment effect is achieved.

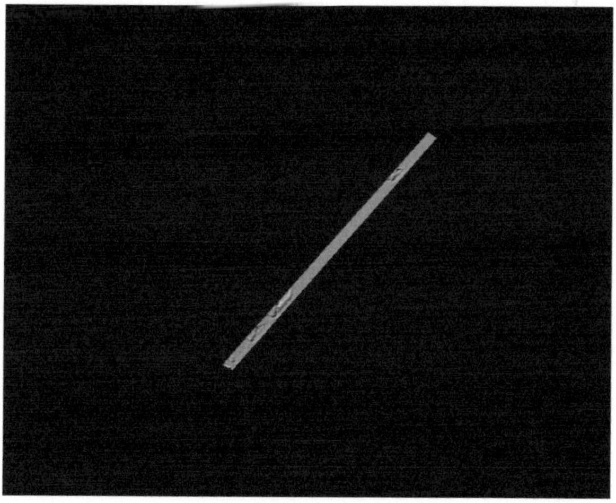

Fig. 3 Relief image of the current frame

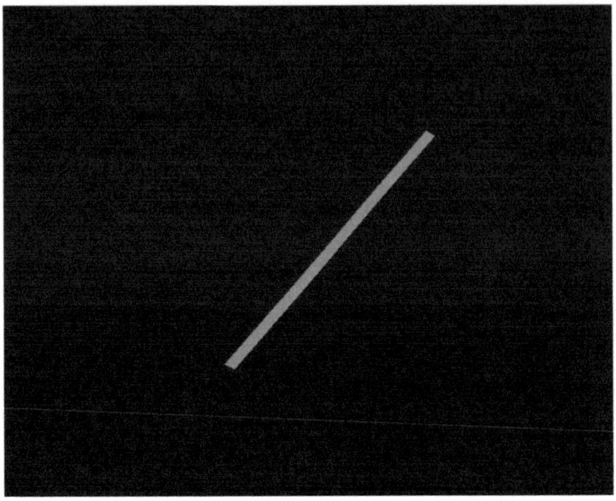

Fig. 4 background relief image

3.2 Threshold Segmentation

After image relief operation, the light and shadow or other external environment influence are greatly reduced. In this paper, the background subtraction method is used to extract the moving object, and the difference image **Diff** between the current frame and the background image is obtained.

$$\mathbf{Diff} = |\mathbf{I}_{\text{new}} - \mathbf{BG}_{\text{new}}| \tag{3}$$

The difference image **Diff** is carried on Otsu threshold segmentation to achieve automatic target segmentation and avoid the weakness of artificial selected threshold. We assume that the image has L gray levels, the pixel number of gray scale i is n_i, the total number of pixels is N, the occurrence probability of each gray value is $p_i = n_i/N$. And assume that there is a threshold T that segments the image into two regions which are the background class $A = (0, 1, 2, \ldots, T)$ and target class $B = (T, T+1, T+2, \ldots, L-1)$. The each occurrence probability of the two classes is,

$$p_A = \sum_{i=0}^{T} p_i, \quad p_B = \sum_{i=T+1}^{L-1} p_i \tag{4}$$

The calculation of gray mean values of A and B are,

$$\omega_A = \sum_{i=0}^{T} ip_i/p_A, \quad \omega_B = \sum_{i=T+1}^{L-1} ip_i/p_B \tag{5}$$

The whole image gray mean value is further get

$$\omega_0 = p_A\omega_A + p_B\omega_B = \sum_{i=0}^{L-1} ip_i \tag{6}$$

So the inter-class variance between two regions of A and B can be obtained,

$$\sigma^2 = p_A(\omega_A - \omega_0)^2 + p_B(\omega_B - \omega_0)^2 \tag{7}$$

Because the larger the inter class variance, the bigger the gray level difference of two classes. The optimal threshold T_{best} is obtained by the maximum type (7),

$$T_{\text{best}} = \underset{0 \le T \le L-1}{\arg\max}\; p_A(\omega_A - \omega_0)^2 + p_B(\omega_B - \omega_0)^2 \tag{8}$$

The difference image **Diff** is segmented by threshold T_{best} to obtain the two-value image $\mathbf{O}_{\text{bject}}$,

$$\mathbf{O}_{\text{bject}}(x,y) = \begin{cases} 1 & \mathbf{Diff}(x,y) > T_{\text{best}} \\ 0 & \mathbf{Diff}(x,y) < T_{\text{best}} \end{cases} \tag{10}$$

4 Queue Detection

Motion detection two-value template is an important analysis object of the following vehicle detection or other tasks [5]. Two-value template takes the noise of the foreground region in the case of keeping the foreground information, this paper hopes that the information of the image can get the minimum loss and achieve the purpose of removing noise.

The first goal of this paper is to eliminate the noise points in the foreground region, and to eliminate the boundary points of the foreground object by using the erosion operation, which makes the boundary interior shrink, and at the same time, the object smaller than the structural elements can be removed. In practice, the whole object needs to be obtained, not only the independent pixels after erosion. Finally, the foreground boundary points are expanded by expansion, which is effective to fill the void of object after the image segmentation. After processing the two-value template filling such as the void of the moving object, the boundary and noise of the moving object are smoothed, which lay the foundation for the follow-up target analysis. Figure 5 is the threshold segmentation image, Fig. 6 is

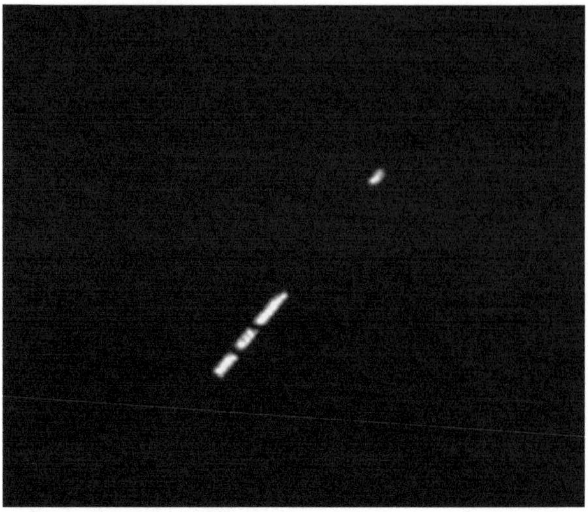

Fig. 5 Threshold segmentation image

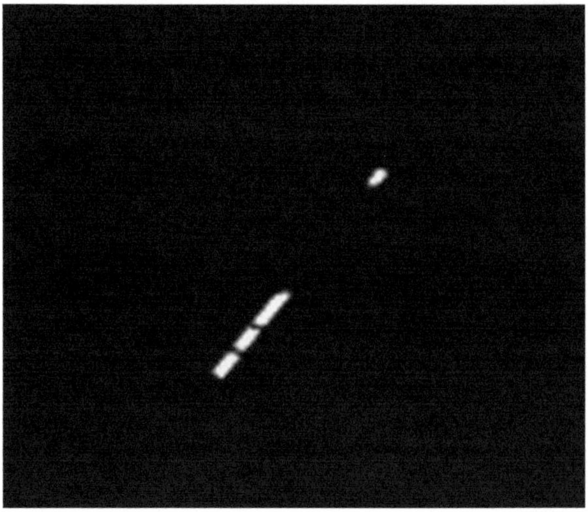

Fig. 6 Image after corrosion expansion

the image after corrosion and expansion, from them we can see that the processed image better reflects the moving target.

Figure 7 is the setting of temple center line image of interest lane. From the geometric characteristics of the lane at intersection we can know that, lane geometry is approximated as rectangular. And in this paper, the length of temple center line is considered approximate to the length of the vehicle lane. Figure 8 is the intersection

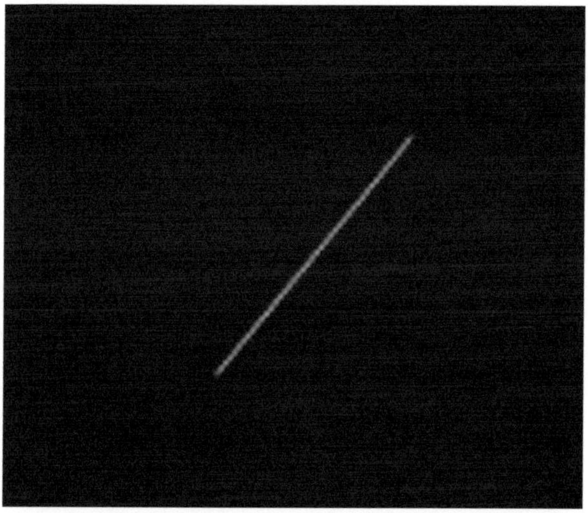

Fig. 7 Interested template center line

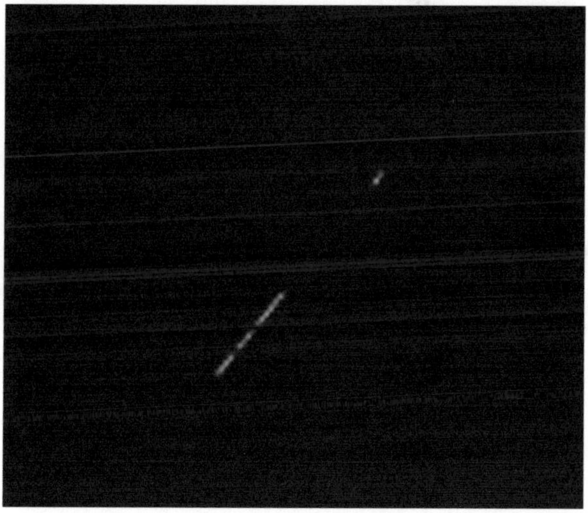

Fig. 8 Queue length extraction

of the temple center line image of the lane and the image after processing, and the virtual queue length can be obtained which is the length between the first goal and the last one. In this paper, we set the length of the real queue which is the sum value of the length of the moving target in Fig. 8.

5 Experimental Results and Analysis

In this paper, the experimental test hardware are P4/3.0 GHz CPU,1 GB memory, WinXP system, and VC development environment. In order to further verify the results of this algorithm, this paper uses the following three parameters to illustrate the performance of this algorithm.

S1 = real detection length/interest template length;
S2 = real detection length/virtual length ratio;
S3 = error length/interest template length, length of error = Virtual length ratio − real detection length.

Figure 9 is the queue detection results of intersection obtained by the method in this paper. Figure 9a–d are the images of 500 frames test results in total in different

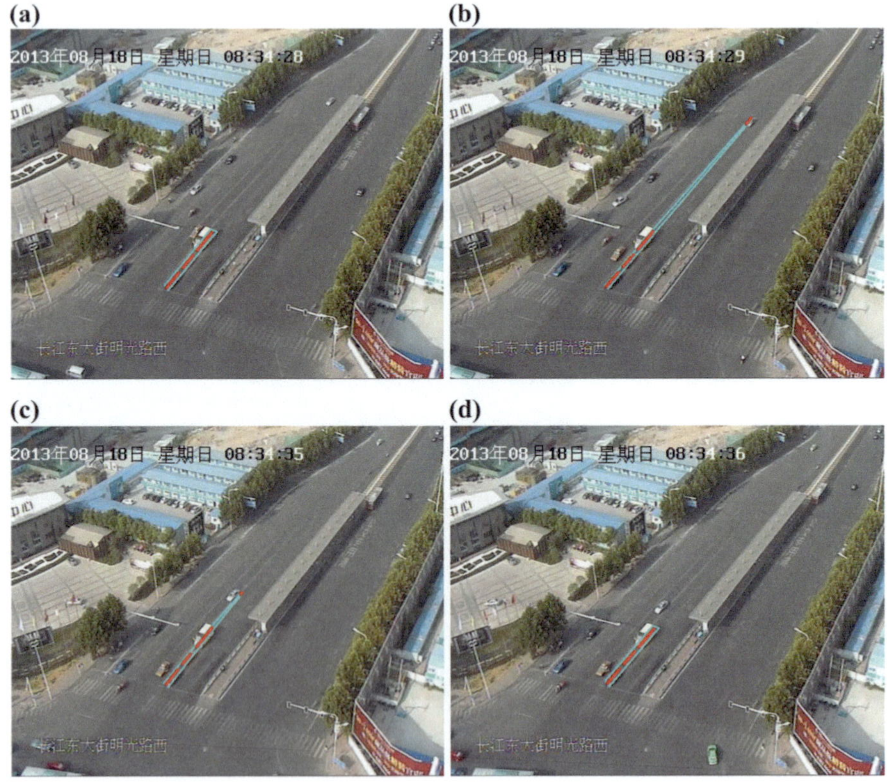

Fig. 9 Test results of the method in this paper

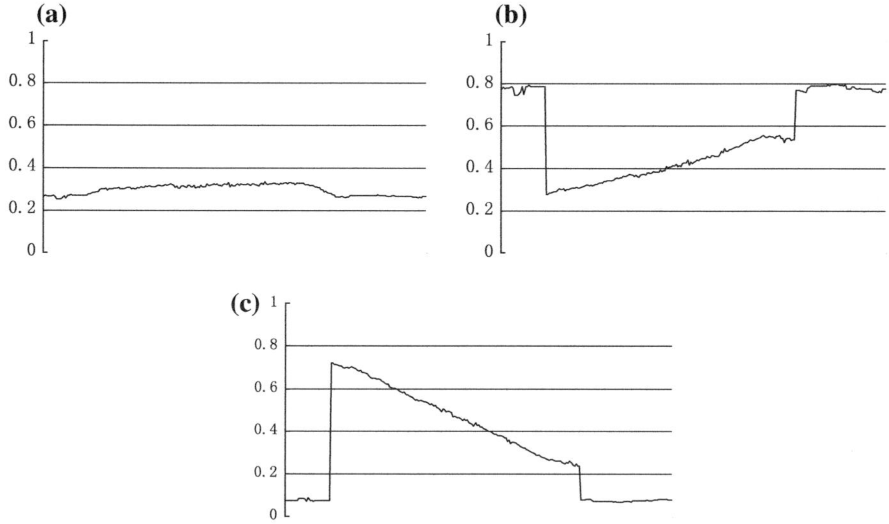

Fig. 10 Variation curves of parameter s1, s2, and s3 in test results. **a** Variation curve of parameter s1, **b** Variation curve of parameter s2, **c** Variation curve of parameter s3

period. From them, the red line represents the actual queue length, the blue solid box represents virtual queue length. Figure 10 is about the variation curves of parameter S1, S2, and S3 in detection results of this paper. The analysis of Figs. 7 and 8 shows that, when the new target enters in interest lane, as is shown in Fig. 10, vehicle queue length increases at first, the curve of parameter S1 begins to grow and then tends to be stable, the curve of parameter S2 decreases sharply at first and then increases steadily, the curve of parameter S3 first increases sharply and then gradually decreases. This is because the virtual queue length increases first and then decreases. When the new target drives out of interest lane, the curves of parameters S1, S2, and S3 return to the start state.

Figure 11 is the queue detection results of morning peak obtained by the method in this paper. Figure 11a–d are the images of 500 frames test results in total in different period. Figure 12 is about the variation curves of parameter S1, S2, and S3 in morning peak queue. The analysis of Figs. 11 and 12 shows that, when the new vehicle enters in target region, the real queue length increases constantly, the difference value of virtual queue length and the real queue length continues to

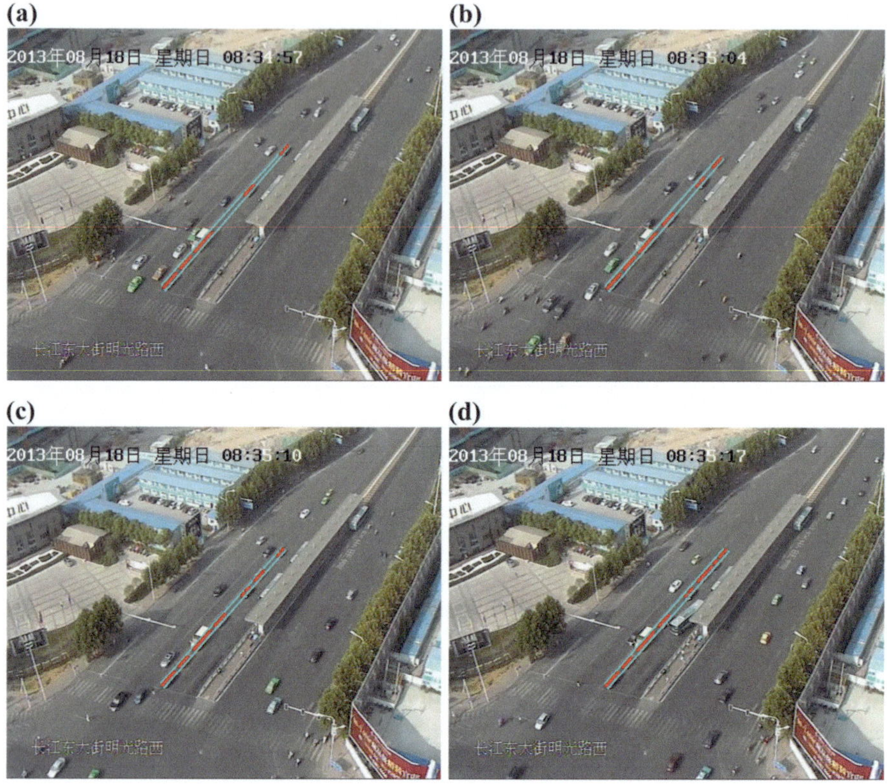

Fig. 11 Test results of queue in morning peak

decrease. When another new target enters in the target region, the real queue length also increases constantly, but the difference value of virtual queue length and the real queue length begins to increase first and then decrease. So the curve of parameter S1 increases constantly, the curve of parameter S2 increases first and then decreases and finally increases again, the curve of parameter S3 decreases first and then increases and finally decreases again.

Figure 13 is the queue detection results of evening peak obtained by the method in this paper. Figure 13a–d are the images of 500 frames test results in total in different period. Figure 14 is about the variation curves of parameter S1, S2 and S3

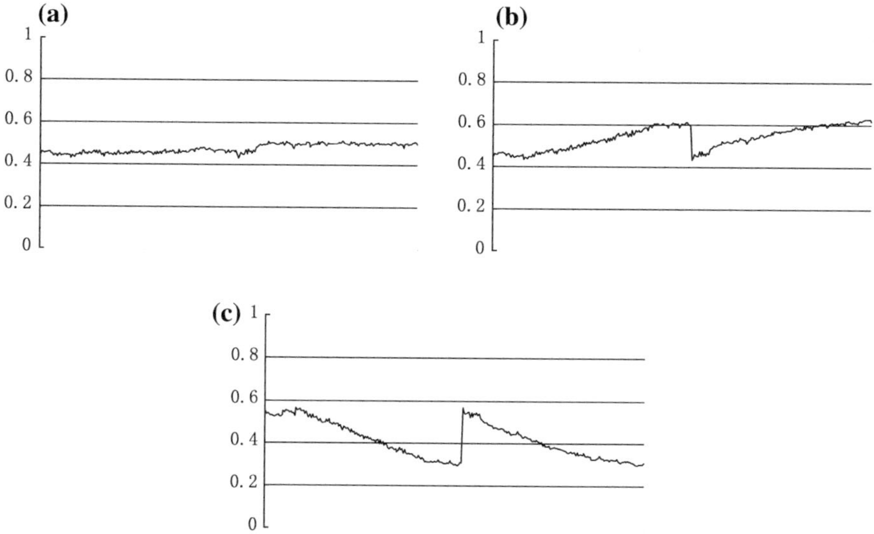

Fig. 12 Variation curves of parameter s1, s2, and s3 in morning peak queue. **a** Variation curve of parameter s1, **b** Variation curve of parameter s2, **c** Variation curve of parameter s3

in evening peak queue. The analysis of Figs. 13 and 14 shows that, when the new vehicle enters in target region constantly, the real queue length increases, the difference value of virtual queue length and the real queue length increases first and then decreases constantly which means the oscillation phenomenon. So the curve of parameter S1 increases constantly, the curve of parameter S2 shows the oscillation phenomenon that decreases first and then increases constantly, the curve of parameter S3 firstly increases first and then decreases and performs the repeated alternation.

6 Conclusions

A new algorithm based on video vehicle queue length detection is proposed in this paper. The algorithm uses the frame difference method to extract the background of the image, and reduces the influence of motion blur and noise or other factors on the background modeling. The image preprocessing is carried on the relief operation, which reduces the impact of light change on the model, thus greatly improves the

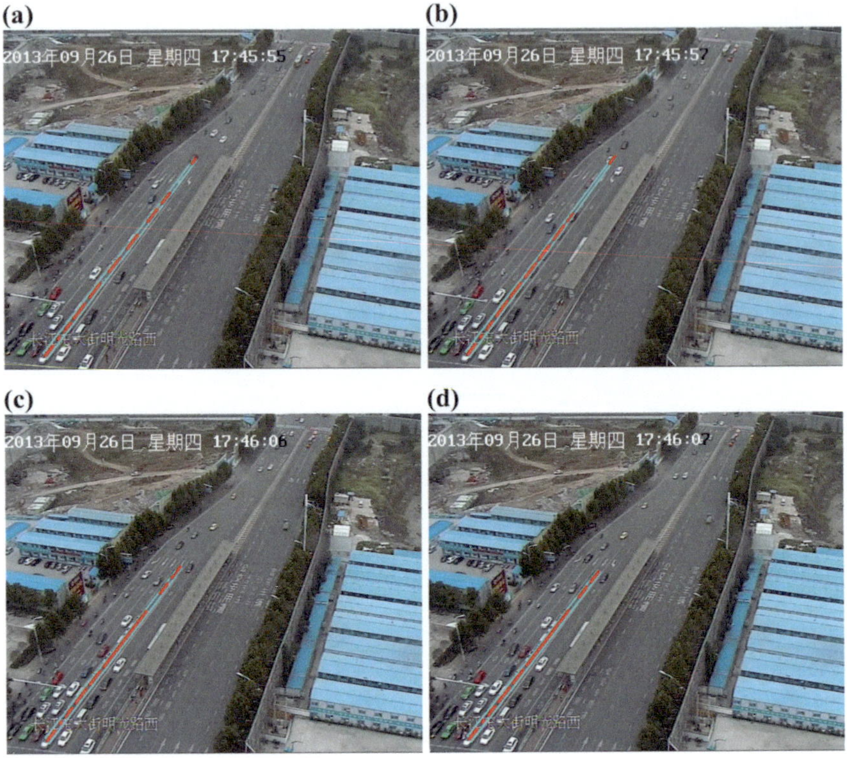

Fig. 13 Test results of queue in evening peak

accuracy of the algorithm. Finally, the real queue length is extracted by two-value image analysis. By analyzing the experimental detection results in the morning and evening peak, we can see that the improved method is simple to achieve and has good practicability and accuracy.

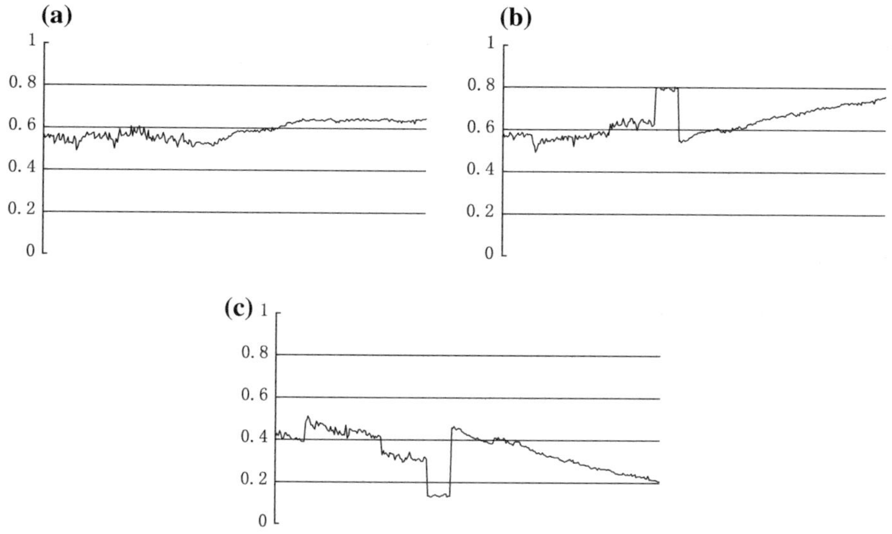

Fig. 14 Variation curves of parameter s1, s2, and s3 in evening peak queue. **a** Variation curve of parameter s1, **b** Variation curve of parameter s2, **c** Variation curve of parameter s3

Acknowledgments This research was funded by the Key Laboratory of Urban ITS Technology Optimization and Integration Ministry of Public Security People's Republic of China in Hefei and the National Science and Technology Support Program of China under Grant #2014BAG03B02. The views are those of the authors alone.

References

1. Chiranjeevi P, Sengupta S (2012) Spatially correlated background subtraction, based on adaptive background maintenance. J Vis Commun Image Represent 23(6):948–957
2. Asakuraa Y, Kusakabea T, Long NX et al (2015) Incident detection methods using probe vehicles with on-board GPS equipment. Transp Res Procedia 6:17–27
3. Shen Z, Miao C, Zhang F (2014) Vehicle queue length detection method at intersection based on vision. Comput Eng 40(4):218–222
4. Ashwini B, Yuvaraju BN (2015) Application of edge detection for vehicle detection in traffic surveillance system. IEEE Int J Comput Appl 115(15):8
5. Desai D, Somani S (2014) Instinctive traffic control and vehicle detection techniques. Eng Res 5 (1):2192–2195
6. Yua X, Shia Z (2015) Vehicle detection in remote sensing imagery based on salient information and local shape feature. Optik - Int J Light Electron Opt 126(20):2485–2490

The Design and Application of a Manipulator's Motion Controller for Changing CNC Machine Tools

Wenhao Tong, Weicun Zhang and Weidong Li

Abstract Computer numerical control (CNC) machine is one of the most important processing means in the manufacturing industry right now. The paper presents and designs a motion controller of manipulator which is used to change the tools for the CNC machine. First, the paper describes the necessity and significance of automatic tools change for CNC machine. Then the details about the hardware and function of the motion controller of manipulator for tools change are introduced. In the end, the paper analyses the data clearly about the whole system when the manipulator is working.

Keywords Manipulator · Computer numerical control machine tools · Tools change · Motion controller

1 Introduction

The first CNC machine in the world was invented at Massachusetts Institute of Technology in 1952 [1]. With the widely utilizing of CNC machine in machinery manufacturing industry, especially in automotive industry, aerospace and military [2], computer numerical control technology are developing rapidly in hardware and software [3]. In 1956, the Fujitsu invented the Numerical Control Turret Punch Press (NCT). The company IBM of America invented AP machining center at the same time. The company KT of America had invented the Auto Tool Change (ATC) in 1958. And our country started researching the CNC in 1958 [4].

W. Tong · W. Zhang (✉) · W. Li
School of Automation and Electrical Engineering, University of Science and Technology Beijing, Beijing 100083, China
e-mail: weicunzhang@ustb.edu.cn

W. Tong · W. Li
Southwest Institute of Clean Energy and Electric Power System, Dali University, Dali, 671003 Yunnan, China

Y. Jia et al. (eds.), *Proceedings of 2016 Chinese Intelligent Systems Conference*, Lecture Notes in Electrical Engineering 405,
DOI 10.1007/978-981-10-2335-4_23

With the rapid development of modern industry, various machines' automation and intelligence having been improved rapidly [5]. And it promotes the usage rate of the CNC machine. Tools are the one of the most important part of the CNC machine [6], and it is frail. In addition, the lifetime of the tools has serious impact on reliability of CNC machine [7]. When the CNC machine is running, the condition of machining and the hardness of the work are always changing [8]. And this could cause low accuracy of the work. In order to ensure high accuracy of the work and save machining time, the CNC machine needs to change tools timely. Because of the consumption of the tools, it is frequently needed to change tools.

The manipulator of the CNC machine is very important for tools change. In this paper, the manipulator has four free degrees. This manipulator can carry two tools at the same time. And one is the old tool, another one is the new tool. So, it can install the new tool after taking out the old tool.

The paper mainly talks about how to design a motion controller for manipulator of CNC machine which is used to change tools. And the system mainly contains communication technology and motor driving technology.

2 System Description

The structure block diagram of the motion controller for manipulator of CNC machine is described in Fig. 1.

Fig. 1 The structure of the system

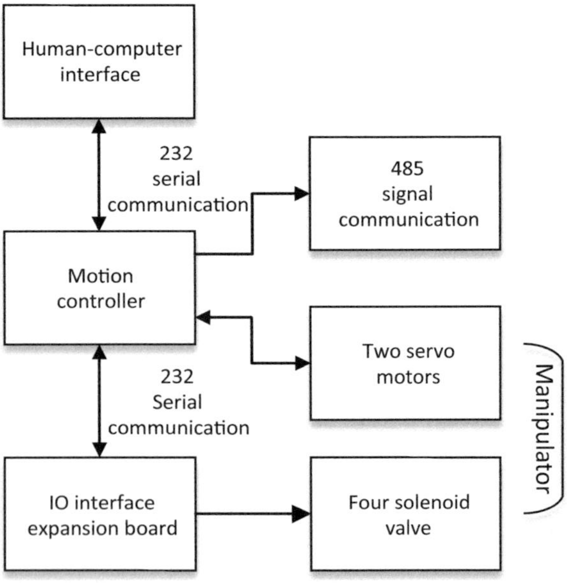

(1) Human–computer interface: workers enter the instructions into PC. PC sends these instructions to motion controller through serial port. The instructions contain some information, such as speed, the number of pulse, direction of the motion, the command of tools changing, and the command of solenoid valve.
(2) Motion controller: the core chip of the motion controller is TMS320F28335, which produced by Texas Instrument. The chip can manage a large number of data. The clock frequency of the chip is up to 150 MHz. Its single precision floating point operation unit with IEEE-754 standard. It has 16 × 16 and 32 × 32 multiplication accumulation operation. It has Harvard bus architecture with rapid interrupt response and processing. The code has high efficiency. Therefore, this motion controller could realize more difficult algorithm and logic calculation in software. And it also prepares for the motion controller's upgrade. The work of the controller is analyzing the order from the PC. On one hand, it sends the control commands to the driver, and receives the feedback message at the same time which means the position of the two servo motors. And it also represents the position of the manipulator. On the other hand, it sends the solenoid valve control command to the IO interface expansion board, and these commands are used to control the motion of the manipulator with four free degree. And at the same time, the controller reads the position of each free degree feedback from the IO interface expansion board, which can avoid the fault when the manipulator is running.
(3) IO interface expansion board: the board receives the instrument from controller to control the solenoid valves directly. And it sends the position of the solenoid valves to controller timely.
(4) 232 serial's communication: this communication mode is full duplex, with high speed and good stability. It has far transmission distance, and it is used in the industrial field with bad environment.
(5) AC servo motor: the Panasonic AC servo motor is accurate. And its task is to move the manipulator installed on the CNC. In order to adjust the position of the manipulator, it also needs to send the position information of the manipulator to the controller.
(6) Manipulator: there are four solenoid valves. The first one is used to rotate the manipulator, and the second one moves the manipulator up and down. The third one is linked to the finger 1, and the fourth one is linked to the finger 2. And the two fingers' functions are taking tools and changing tools.

An interface is retained, which can achieve with 485 communication protocol. And it can do some software test.

Fig. 2 The physical map of the manipulator

3 Mechanism of the Manipulator

As we can see in Fig. 2, at the top of the picture, there are two black axes driven by two servo motors. And the two axes could move the manipulator to the destination. The manipulator is installed below the axes, with four free degrees that are already described above. Because the degree of the manipulator is controlled by the solenoid valve, each degree of the manipulator could only realize two actions. The axis of rotation can only be rotated 180°. The vertical axis can only be moved up and down in a fixed distance. And the finger 1 and 2 can only open and close.

4 Hardware Circuit Design

The controller is the heart of the system. And it must have these functions: input differential signal; output differential signal; two path of 232 serial communication. The paper chooses TMS320F28335 as the core chip. For this controller, it mainly

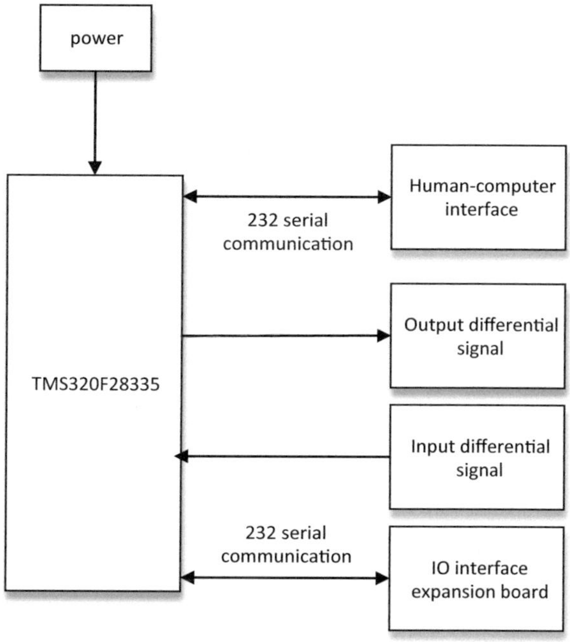

Fig. 3 The functional block diagram of the control board

contains the power module, communication module, and input/output signal module. The functional block diagram of the control board is described in Fig. 3.

(1) The power supply is 24 V. 24 V through the chip LM2596 could be changed to 5 V. And the core chip needs 3.3 and 1.9 V power supply. 5 V through the chip PS767D301 could be changed to 3.3 and 1.9 V.
(2) Serial communication module. This module uses a chip MAX3232 which can achieve two path of serial communication.
(3) The output differential signal is 5 V, So that the controller uses the chip MC3487 which can output 5 V differential signal. And the chip can transform the unilateral 5 V TTL signal into bilateral 5 V differential signal.
(4) The input differential signal needs to be transformed into TTL signal. Therefore, the paper chooses chip 6N137 which can transform 5 V differential signal into 5 V TTL signal. Then the core chip can receive the signal.
(5) There is a very important circuit besides all above. The core chip can only receive the signal with 3.3 V. So the TTL signal with 5 V must be changed, or the signal with 5 V will damage the core chip [9]. The paper choses chip SN74LVC4345. This chip can achieve two-way transformation.

5 Software Process

The controller is waiting for the data from the PC. If PC sent data to the controller, the serial port interrupt will be triggered. Then the core chip will save these data to its register. Then the controller will analyze and open these data. Because part of them- motor code will be used to be sent to the servo driver. Then the driver will follow the order to drive the motors. And at the same time, servo driver will send the position of the motors timely to the controller. And the rest data will be sent to IO interface expansion board. Then the board could control the solenoid valve independently and logically. And it means the action of manipulator could act correctly. And besides, the controller will send the message that contains manipulator's position to PC to be saved and read (Fig. 4).

Fig. 4 Software process

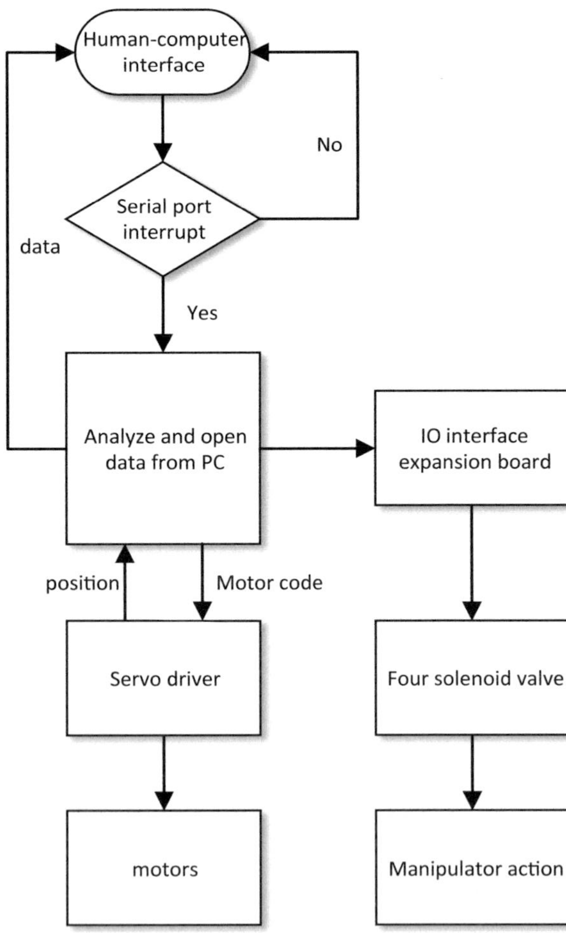

6 Test Results

There is a group of data. It represents the action that the manipulator is taking tools. The computer gets these data every 0.5 s. The unit in the table of the paper is millimeter.

Because there is no difference between taking tools and changing tools in trace, I just collect the data of taking tools. The only difference between them is that the action of changing tools has one more action the rotation axis will rotate.

7 Analysis of Experimental Results

In this part, the paper shows the error line between the standard values with the actual values.

We can see that the manipulator's trace from the Table 1: first, the manipulator moves in x axis. Then it moves in y axis. When the y axis is static which means that the motor is static, there are errors. It could because of mechanical vibration. And it is the same as x axis. Besides, when the motor is running, there are also some errors. It is reasonable because of the error are negligible.

They some error that cannot be able to show through table or figure. For example, if the finger 1 or 2 be abraded, it will cause a failure of changing or taking tools. Generally speaking, the error is controlled in a good range.

Besides all above, there are some points that need to be promoted. For example, it is short of the check of the solenoid valve. And the controller does not have the limitation of the motor. So my next research is about perfecting the controller and making it more reliable (Figs. 5 and 6).

Table 1 Taking tools

Position	Standard	(0, 0)	(200, 0)	(400, 0)	(600, 0)
	Actual	(0, 0)	(200.003, 0.001)	(400.001, 0.005)	(599.997, 0.001)
	Standard	(800, 0)	(800, 100)	(800, 200)	(800, 300)
	Actual	(800.002, 0.002)	(800.001, 99.996)	(800.001, 200.001)	(800.002, 299.999)
	Standard	(800, 400)	(800, 500)	(800, 600)	
	Actual	(800.001, 400.000)	(800.001, 500.001)	(800.001, 600.000)	

Fig. 5 The error of x axis

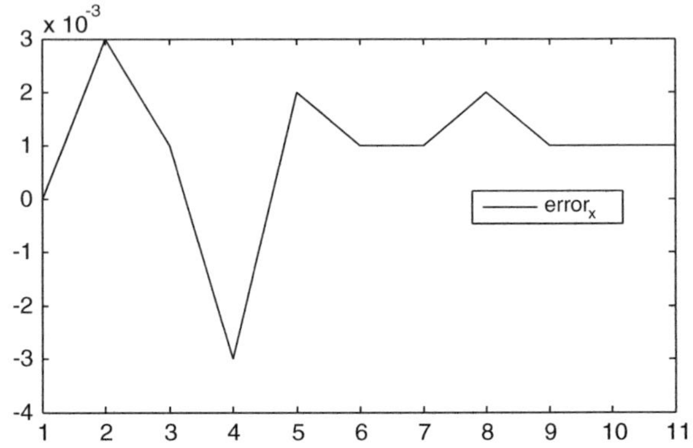

Fig. 6 The error of y axis

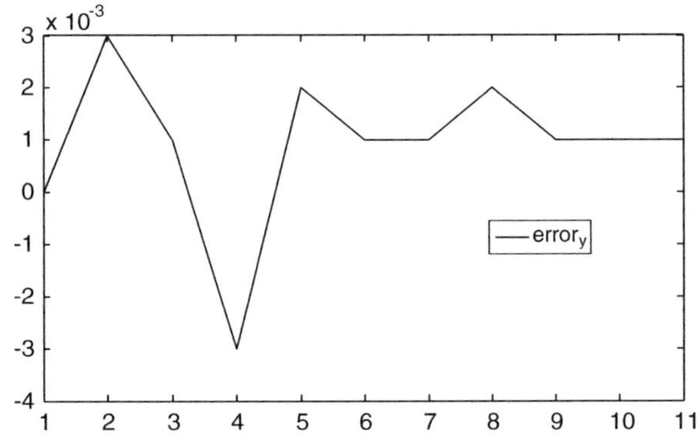

Acknowledgments The author would like to thank the anonymous reviewers for their constructive and insightful comments for further improving the quality of this work. This work was supported by the Major State Basic Research Development Program (973 Program) (No. 2012CB821200), National Key Technologies R&D Program (No. 2013BAB02B07) and National Science Foundation of China (No. 61520106010)

References

1. Wang Z (2014) The research of the auto tools change based on computer numerical control machine tools. Technol Dev Enterprise 33(3)
2. Qian D, Ji W, Du S et al (2016) Research of tool reliability and lifetime distribution model based on statistical analysis. Modern Manuf Eng (2)
3. Huang P, Yang D (2016) A design of reliable protection model based on computer numerical control machine tools. Light Ind Sci Technol (4)
4. Yang Z, Chen C, Chen F et al (2013) Progress in the research of reliability technology of machine tools. J Mech Eng 49(20)
5. Zhang G, Liu J, Wang G (2012) Fuzzy reliability allocation of CNC machine tools based on task. Comput Integr Manuf Syst, 18(4)

6. Tang K (2012) Development situation and trend of NC machine tool in China. Mach Tool Hydraulics 40(5)
7. Li N, Lu X, Han P et al (2012) Study on the reliability of CNC and machine tools and key function units. Modular Machine Tool & Automatic Manufacturing Technique (11)
8. Wang W, Yang J, Yao X et al (2012) Synthesis modeling and real-time compensation of geometric error and thermal error for CNC machine tools. J Mech Eng 48(7)
9. Li J (2012) Study of the measuring method dust for concentration based on DSP. Shandong University of Science and Technology

Adaptive Terminal Sliding Mode Control for Servo Systems with Nonlinear Compensation

Tianyi Zeng, Xuemei Ren, Wei Zhao and Shubo Wang

Abstract To achieve high accurate position tracking of servo systems with backlash and friction nonlinearity, a terminal sliding mode controller with adaptive compensation is proposed in this paper. The nonlinear backlash model is converted into the linear expression in order to simplify the system for control-oriented design. The presented controller consists two parts, which are position tracking controller and nonlinear compensator. A novel terminal sliding mode controller is proposed by adopting a terminal sliding mode manifold, while making sure the control system could reach the sliding surface and converge to equilibrium point in finite time. The adaptive compensator is used to compensate the error caused by linearization and friction including static friction and viscous dissipation. Simulation results verify the reliability and effectiveness of the proposed method.

Keywords Servo system · Terminal sliding mode control · Adaptive control · Nonlinear compensation

1 Introduction

Servo systems have been widely used in the fields of military and industry applications. To satisfy the industry requirement, servo systems should be qualified for high-tracking speed and high-accuracy performance. Furthermore, compared with other control systems, nonlinearity of gear backlash and friction always existed in servo system control problem.

Terminal sliding mode control proposed by Venkataraman [1] can converge to equilibrium in finite time. Feng [2] developed a kind of hybrid terminal sliding mode observer in order to achieve better performance in rotor position and speed estimation of permanent-magnet synchronous motor control system. A new form of

---

T. Zeng · X. Ren (✉) · W. Zhao · S. Wang
School of Automation, Beijing Instituted of Technology, Beijing 100081, China
e-mail: xmren@bit.edu.cn

Y. Jia et al. (eds.), *Proceedings of 2016 Chinese Intelligent Systems Conference*, Lecture Notes in Electrical Engineering 405,
DOI 10.1007/978-981-10-2335-4_24

terminal sliding mode is used in [3] for rigid robotic manipulators, which can achieve faster and high-precision tracking performance. In [4], terminal sliding mode control and its homogeneity are researched and used in design proposal for general finite time switching manifolds. Yan [5] designed a new terminal sliding mode controller for MEMS gyroscopes, which can make the control system reach sliding surface in finite time. In [6], a novel approach of adaptive fuzzy hierarchical terminal sliding mode controller is designed to make the error converge to zero in finite time. Obviously a variety of terminal sliding mode control methods are adopted in servo system control, especially for position tracking problems, in order to achieve higher control precision.

The system performance is affected significantly by the nonlinearity of gear backlash and friction. Adaptive compensate methods are widely used as the treatment for nonlinear characteristic. In [7], to estimate the optimal bound of the lumped uncertainty, a fuzzy inference mechanism with center adaptation of the membership functions was proposed. To achieve high precision and robustness of rocket launcher servo system, Wu [8] adopted Chebyshev neural network to identify the system's Jacobian information. In [9], a novel adaptive law was designed which can provide fast adaptation and chattering reduction in the vicinity of sliding manifold. Bartolini [10] proposed an adaptation strategy with uncertainty compensation which can adjust the control magnitude online. To deal with a class of uncertain nonlinear MIMO systems, Li [11] adopted an adaptive tuning law to estimate the unknown system uncertainties which are bounded.

In this paper, we study the servo system with nonlinearity of backlash and friction. The terminal sliding mode control is used to achieve position tracking, where the adaptive method is introduced to compensate the bounded errors caused by linearization and uncertain parameters of friction model.

The rest of the paper is organized as follows. In Sect. 2, the servo system and its linearized form are described; the state space model is also given in this section. In Sect. 3, an adaptive terminal sliding mode controller is designed to guarantee tracking performance and deal with the nonlinearity. In Sect. 4, the stability and convergence of the closed-loop system based on the control method proposed are analyzed. Finally, the simulation results are given in Sect. 5 to illustrate the effectiveness of the proposed control method.

2 Problem Formulation

The motor driving servo system can be described as follows:

$$\begin{aligned} &J_m\ddot{\theta} + b_m\dot{\theta} = u - T_m \\ &J_l\ddot{\theta}_l + f_l(\dot{\theta}_l) = T_l \end{aligned}, \tag{1}$$

where θ and θ_l are the angle position of motor and load, respectively; J_m and J_l are the moments of inertia of motor and load, respectively; T_m and T_l are the input and output of the gear box, respectively; $f_l(\dot{\theta}_l)$ is the friction torque.

The backlash nonlinearity of gears can be usually defined as follows:

$$T_l(t) = \begin{cases} n(T_m - \alpha), & \dot{T}_m > 0 \\ n(T_m + \alpha), & \dot{T}_m < 0 \\ T_l(t_-), & \text{otherwise} \end{cases} \tag{2}$$

where α is the backlash distance; n is gear ratio.

The backlash nonlinearity model (2) function can usually be transformed as follows [12]:

$$T_l(t) = nT_m(t) + d, \tag{3}$$

where n is gear ratio; d is a nonlinear function on T_m and bounded as $|d| \le \xi$, ξ is a positive constant, which can be handled in the following research.

To simplify the controller design, the system (1) can be replaced by the following form according to [13]:

$$\left(n^2 J_m + J_l\right)\dot{\theta}_l + n^2 b_m \dot{\theta}_l = nu + f(\dot{\theta}_l) + d \tag{4}$$

The original friction models are discontinuous or piecewise continuous which may cause the problems of deriving smooth control actions. Furthermore, the friction model parameters are difficult to be identified. A novel continuously friction model which is also differentiable is adopted in this paper, where the friction torque $f(\ddot{\theta})$ can be expressed as follows:

$$f(\dot{\theta}_l) = A_1(\tanh(\beta_1\dot{\theta}_l) - \tanh(\beta_2\dot{\theta}_l)) + A_2\tanh(\beta_3\dot{\theta}_l) + A_3\dot{\theta}_l, \tag{5}$$

where $A_1, A_2, A_3, \beta_1, \beta_2$ and β_3 are positive parameters. A_1 and A_2 stand for the static friction coefficients; the Stribeck effect can be described by $\tanh(\beta_1\dot{\theta}) - \tanh(\beta_2\dot{\theta})$; $A_2\tanh(\beta_3\dot{\theta})$ is the Coulomb friction; $A_3\theta$ stands for the viscous dissipation. Further information can be referred to [14]. It is different from general friction models, (5) is a continuously and differentiable form which can be controlled more flexible.

The state variables x_1 and x_2 are defined as

$$\begin{cases} x_1 = \theta_l \\ x_2 = \dot{\theta}_l \end{cases} \tag{6}$$

Then the state space equation is expressed as

$$\begin{cases} \dot{x}_1 = x_2 \\ \dot{x}_2 = \frac{1}{n^2 J_m + J_l}(nu + f(x_2) - b_m n^2 x_2 + d) \end{cases} \tag{7}$$

Consequently, the purpose of this paper is to control the system output tracking a desired reference signal with satisfactory transient performance under the nonlinearity of backlash and friction.

3 Controller Design

To guarantee the tracking output performance, a terminal sliding mode controller with adaptive compensation is designed. The tracking error is defined as

$$e(t) = y(t) - y_d(t). \tag{8}$$

Then

$$\dot{e} = x_2 - \dot{y}_d. \tag{9}$$

$$\ddot{e} = \dot{x}_2 - \ddot{y}_d. \tag{10}$$

The sliding mode surface is defined as

$$s = \dot{e} + c_1 e + c_2 \lambda(e), \tag{11}$$

where $c_1, c_2 > 0$ are constants, $\lambda(e)$ is defined as

$$\lambda(e) = \begin{cases} e^{\frac{p}{q}}, & \bar{s} = 0 \text{ or } \bar{s} \neq 0, |e| \geq \varepsilon \\ \tau_1 e + \tau_2 \operatorname{sgn}(e) e^2, & \bar{s} \neq 0, |e| \leq \varepsilon \end{cases}, \tag{12}$$

where $p > q$ are positive odd integers, $\bar{s} = \dot{e} + c_1 e + c_2 e^{\frac{p}{q}}$, ε is a small positive constant, $\tau_1 = \left(2 - \frac{p}{q}\right)\varepsilon^{\frac{p}{q}-1}$, $\tau_2 = \left(\frac{p}{q} - 1\right)\varepsilon^{\frac{p}{q}-2}$. Taking the derivative of $\lambda(e)$ along with (12) yields

$$\lambda(e) = \begin{cases} \frac{p}{q} e^{\frac{p}{q}-1}\dot{e}, & \bar{s} = 0 \text{ or } \bar{s} \neq 0, |e| \geq \varepsilon \\ \tau_1 \dot{e} + 2\tau_2 \operatorname{sgn}(e) e\dot{e}, & \bar{s} \neq 0, |e| \leq \varepsilon \end{cases} \tag{13}$$

Then the following equation can be obtained:

$$\dot{s} = \ddot{e} + c_1 \dot{e} + c_2 \lambda(e), \tag{14}$$

where (14) is a continuous function.

The controller u_s can be designed as follows:

$$u_s = [\ddot{y}_d - c_1\dot{e} - c_2\lambda(e) - k_1 s]\left(n^2 J_m + J_l\right) + b_m n^2 x_2 \tag{15}$$

An adaptive strategy is adopted in order to deal with the unknown parameters of friction model (5), and compensate the error caused by linearization.

The adaptive friction compensator is designed as

$$u_f = -\left(\widehat{A}_1 f_1 + \widehat{A}_2 f_2 + \widehat{A}_3 f_3\right), \tag{16}$$

where $f_1 = \tanh(\beta_1\dot{\theta}_l) - \tanh(\beta_2\dot{\theta}_l)$, $f_2 = \tanh(\beta_3\dot{\theta}_l)$, $f_3 = \dot{\theta}_l$; $\widehat{A}_1, \widehat{A}_2$ and $\widehat{A}_3$ are the estimates of the uncertain parameters A_1, A_2 and A_3, respectively.

The adaptive error compensator is designed as

$$u_n = -\xi \operatorname{sgn}(\Lambda), \tag{17}$$

where $\hat{\xi}$ is the estimation of the uncertain bound ξ; $\Lambda = (1 + k_2 k_3)\delta_1 + (k_2 + k_3)\delta_2 + \delta_3$, $\delta_1 = \int (\theta - \theta_r)dt$, $\delta_2 = \theta - \theta_r$, $\delta_3 = \theta - \theta_r$. k_2 and k_3 are positive constant.

The adaptive update law is given as follows:

$$\dot{\hat{\psi}} = -\gamma(\phi s + k\hat{\psi}) \tag{18}$$

where $\dot{\hat{\psi}} = \left[\dot{\hat{A}}_1 \quad \dot{\hat{A}}_2 \quad \dot{\hat{A}}_3 \quad \dot{\hat{\xi}}\right]^T$, $\gamma = [\gamma_1 \quad \gamma_2 \quad \gamma_3 \quad \gamma_4]$, $\gamma_1, \gamma_2, \gamma_3, \gamma_4$ and k are adaptive tuning parameters, $\phi = [f_1 \quad f_2 \quad f_3 \quad \operatorname{sgn}(\Lambda)]^{\mathrm{T}}$.

According to (20), (15), (16), and (17), the Eq. (14) can be rewritten as

$$\dot{s} = -k_1 s - \frac{1}{n^2 J_m + J_l}\tilde{\psi}^T\phi, \tag{19}$$

where $\tilde{\psi} = \psi - \hat{\psi}$ is the weight error.

Consequently, the controller can be expressed as follows:

$$u = u_s + u_n + u_f, \tag{20}$$

where u_s, u_n and u_f are controlling quantities of sliding mode controller, nonlinearity compensator, and friction compensator, respectively.

The proposed controller consists of a sliding mode controller for position tracking, adaptive nonlinearity compensators for the bounded error of linearization, and nonlinear friction function (5). The structure of the controller is shown in Fig. 1.

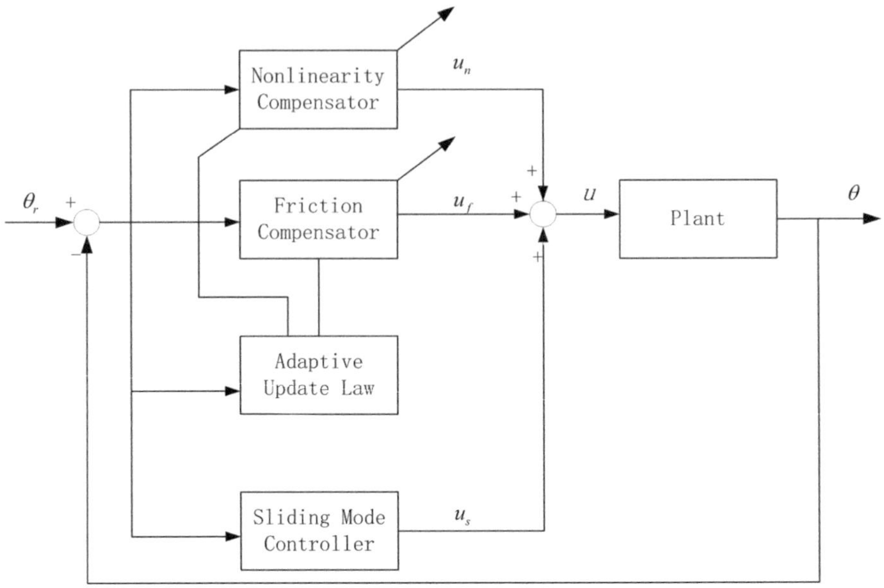

Fig. 1 Controller structure of servo system

4 Stability Analysis

Theorem 1 *For system* (7), *the control u is designed as* (20) *and the adaptive update law* (18) *is adopted, the weight error* $\tilde{\psi}$, *the sliding mode manifold s, and the tracking error e are uniformly ultimately bounded.*

Proof The Lyapunov function is chosen as

$$V = \frac{1}{2}s^2 + \frac{1}{2}\tilde{\psi}^T\Gamma\tilde{\psi}, \tag{21}$$

where $\Gamma = \frac{1}{n^2 J_m + J_l}\gamma^{-1}$. Taking the derivative of V along with (21) yields

$$\begin{aligned} \dot{V} &= s\dot{s} - \tilde{\psi}^T\Gamma^{-1}\hat{\dot{\psi}} \\ &= -k_1 s^2 - \frac{1}{J_m - nJ_l}\tilde{\psi}^T\phi s + \tilde{\psi}^T\Gamma^{-1}(\phi s + k\hat{\psi}) \\ &= -k_1 s^2 + k\tilde{\psi}^T\hat{\psi}. \end{aligned} \tag{22}$$

Consider the inequality as follow:

$$\begin{aligned}\tilde{\psi}^T\hat{\psi} &= \tilde{\psi}^T(\psi - \tilde{\psi}) = \langle \tilde{\psi}, \psi \rangle - \|\tilde{\psi}\|_F^2 \\ &\le \|\psi\|_F\|\tilde{\psi}\|_F - \|\tilde{\psi}\|_F^2 \le \|\tilde{\psi}\|_F\psi_N - \|\tilde{\psi}\|_F^2 \\ &= -\left(\|\tilde{\psi}\|_F - \frac{1}{2}\psi_N\right)^2 + \frac{1}{4}\psi_N^2\end{aligned} \tag{23}$$

where $\|\tilde{\psi}\|_F^2 = tr(\tilde{\psi}^T\tilde{\psi})$ is the Frobenius norm, ψ_N is positive constant and $\|\psi\| \le \psi_N$.

Then (22) can be rewritten as

$$\dot{V} = -k_1s^2 + k\tilde{\psi}^T\hat{\psi} \le -k_1s^2 - k\left[\left(\|\tilde{\psi}\|_F - \frac{1}{2}\psi_N\right)^2 - \frac{1}{4}\psi_N^2\right]. \tag{24}$$

To make the system stable, $\dot{V} < 0$ is needed, which means

$$\|\tilde{\psi}\|_F \ge \psi_N. \tag{25}$$

$$|s| \ge \frac{\psi_N^2}{4k_1}. \tag{26}$$

Under the conditions above, the closed-loop system is uniformly ultimately bounded according to Lyapunov theorem. Furthermore, $\tilde{\psi}$ and s are bounded, which means $\tilde{\psi}$ and s converge to a compact set near zero. So the tracking error e can be guaranteed bounded according to (11).

5 Simulation Results

In this section, the controller designed above will be tested to demonstrate the effectiveness. The physical parameters of the system are listed in Table 1.

The parameters of the controller are designed as $k_1 = 9$, $k_2 = 0.9$, $k_3 = 0.2$, $c_1 = 50$, $c_2 = 55$, $p = 7$, $q = 3$, $\gamma_1 = \gamma_2 = \gamma_3 = \gamma_4 = 0.5$, $k = 20$. The initial states are set as $\theta = \theta = 0$. The initial conditions of adaptive parameters are adopted as $\hat{A}_1(0) = 0.2$, $\hat{A}_2(0) = 0.5$, $\hat{A}_3(0) = 0.01$, $\hat{\xi}(0) = 0.6$.

Table 1 System parameters

Parameter	Value	Unit
J_m	0.185	$kg \cdot m^2$
J_l	0.028	$kg \cdot m^2$
b_m	1.2	Nm · s/rad
n	7/60	/

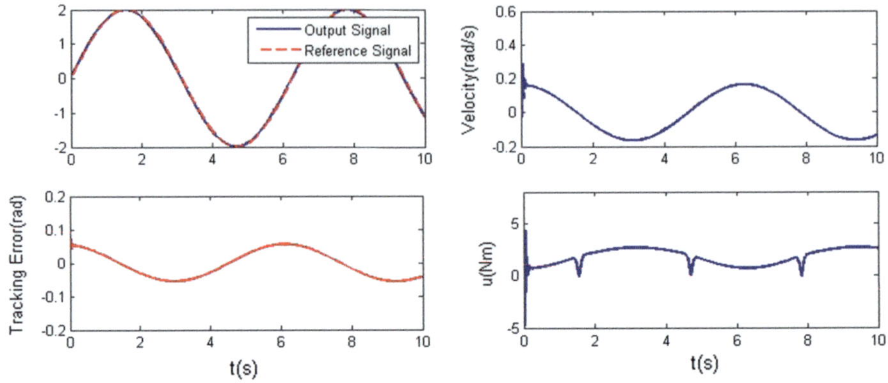

Fig. 2 Tracking profiles for $y_d = 2\sin(t)$ rad

The reference signal is adopted as $y_d = 2\sin(t)$ rad. The simulation results are shown as follows.

It can be seen from the simulation results in Fig. 2 that the proposed controller can track the reference signal with limited tracking error. The amplitude of control torque is in a reasonable range.

Change the reference signal as $y_d = 3$ rad, the simulation results are depicted as follows.

It can be seen in Fig. 3 that the controller can track the reference signal. However, the amplitude of control torque can reach as high as 150 Nm, which is unrealistic in real physical systems. Make the amplitude limited below 35 Nm, the simulation results are illustrated as follows

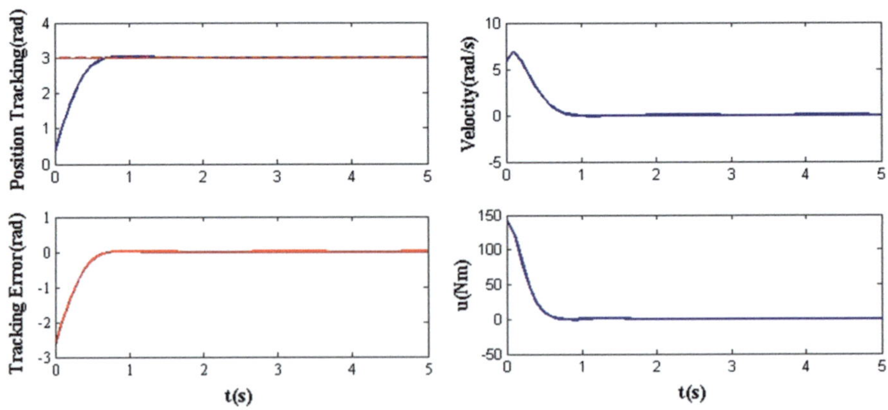

Fig. 3 Tracking profiles for $y_d = 3$ rad

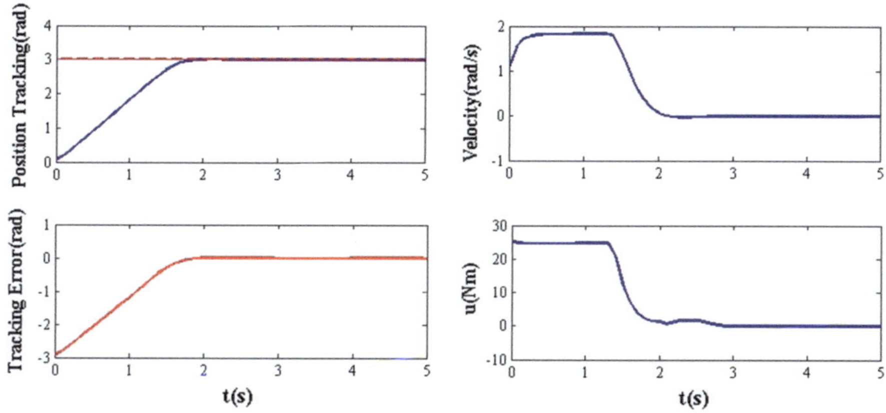

Fig. 4 Tracking profiles for $y_d = 3$ rad with torque limitation

Figure 4 shows the position tracking response when $y_d = 3$ and the control torque limitation is 35 Nm. The controller can still track the reference signal, but the rising time is relative long.

6 Conclusions

An adaptive terminal sliding mode control algorithm was proposed for servo systems with backlash nonlinearity and uncertain friction. The system with backlash nonlinearity was simplified for controller design, and a kind of continuously differentiable friction model was adopted to describe the friction dynamics. An adaptive terminal sliding mode controller was designed to guarantee tracking performance and deal with the nonlinearity of backlash and friction. The stability and convergence of the closed-loop system based on the control method proposed were analyzed. The simulation results illustrated the control performance of servo systems.

Acknowledgments This work is supported by National Natural Science Foundation of China (No. 61433003, 61273150 and 61321002).

References

1. Venkataraman ST, Gulati S (1991) Terminal sliding modes: a new approach to nonlinear control systems. In: Proceedings of 5th International Conference Advanced Robotics, pp 443–448
2. Feng Y, Zheng J, Yu X, Truong NV (2009) Hybrid terminal sliding-mode observer design method for a permanent-magnet synchronous motor control system. IEEE Trans Ind Electron 56(9):3424–3431

3. Yu S, Yu X, Shirinzadeh B, Man Z (2005) Continuous finite-time control for robotic manipulators with terminal sliding mode. Automatica 41(11):1957–1964
4. Yu X, Man Z (2002) Fast terminal sliding mode control design for nonlinear dynamic systems. IEEE Trans Circuits Syst I Fundam Theory Appl 49(2):261–264
5. Yan W, Hou S, Fang Y et al (2016) Robust adaptive nonsingular terminal sliding mode control of MEMS gyroscope using fuzzy-neural-network compensator. Int J Mach Learn Cybern, 1–13
6. Vaghei Y (2016) Trajectory tracking of under-actuated nonlinear dynamic robots: adaptive fuzzy hierarchical terminal sliding-mode control. J Artif Intell Data Min, 2322–5211
7. Lin FJ, Chiu SL (1998) Adaptive fuzzy sliding-mode control for PM synchronous servo motor drives. IEEE Proc Control Theory Appl 145(1):63–72
8. Wu YF, Ma DW, Le GG (2014) Adaptive sliding mode control of servo system. Appl Mech Mater 556–562:2466–2469
9. Baek J, Jin M, Han S (2016) A new adaptive sliding mode control scheme for application to robot manipulators. IEEE Trans Ind Electron, 1–1
10. Bartolini G, Levant A, Pisano A et al (2016) Adaptive second-order sliding mode control with uncertainty compensation. Int J Control, 1–19
11. Li Peng, Ma Jianjun, Zheng Zhiqiang (2016) Robust adaptive sliding mode control for uncertain nonlinear MIMO system with guaranteed steady state tracking error bounds. J Franklin Inst 353(2):303–321
12. Su CY, Stepanenko Y, Svoboda J, Leung TP (2000) Robust adaptive control of a class of nonlinear systems with unknown backlash-like hysteresis. IEEE Trans Autom Control 45(12):2427–2432
13. Jiao X, Zhang J, Shen T (2014) An adaptive servo control strategy for automotive electronic throttle and experimental validation. IEEE Trans Ind Electron 61(11):6275–6284
14. Makkar C, Dixon WE, Sawyer WG. Hu G (2005) A new continuously differentiable friction model for control systems design. In: Proceedings IEEE/ASME International Conference Advanced Intelligent Mechatronics, pp 600–605

Two-Stage Recursive Least Squares Parameter Identification for Cascade Systems with Dead Zone

Linwei Li, Xuemei Ren, Wei Zhao and Minlin Wang

Abstract In this paper, two-stage recursive least squares algorithm (TS-RLS) is investigated for parameter identification of cascade systems with dead zone. In order to estimate the slopes and endpoints of the dead zone, switching functions are presented to reconstruct the expression of dead zone. All the parameters of linear subsystems and dead zone are separated by using the key term separation principle, which is applied to convert the cascade systems into a quasilinear model. The proposed identification algorithm not only estimates all the parameters of the cascade systems, but also reduces the computation cost of identification process. The result of the simulation illustrates the flexibility and efficiency of proposed identification algorithm.

Keywords Parameter identification · Dead zone · Two-stage algorithm · Cascade systems

1 Introduction

The cascade systems are block-oriented nonlinear system as shown in Fig. 1, which consists of three parts, $f(\cdot)$ denotes the dead zone nonlinear function, L_1 and L_2 represent the two linear dynamic blocks. It is noted that this model can approximate general industrial processes,. e.g., DC servo motor, electrical devices, electrically stimulated muscle, copper convert slag, and x-y positioning tables [1]. So, the parameters estimation of cascade model is a hotspot problem.

There are some methods to estimate the parameters of the cascade system with dead zone [2–5]. Wang [6] proposed the corresponding noise estimation to replace the unmeasurable noise term, and an extended stochastic gradient identification on basis of the over-parameterization method was derived for cascade systems. Dong [7] handled the backlash by using the subgradient, and reported a recursive identification

L. Li · X. Ren (✉) · W. Zhao · M. Wang
School of Automation, Beijing Institute of Technology, Beijing 100081, China
e-mail: xmren@bit.edu.cn

Y. Jia et al. (eds.), *Proceedings of 2016 Chinese Intelligent Systems Conference*, Lecture Notes in Electrical Engineering 405,
DOI 10.1007/978-981-10-2335-4_25

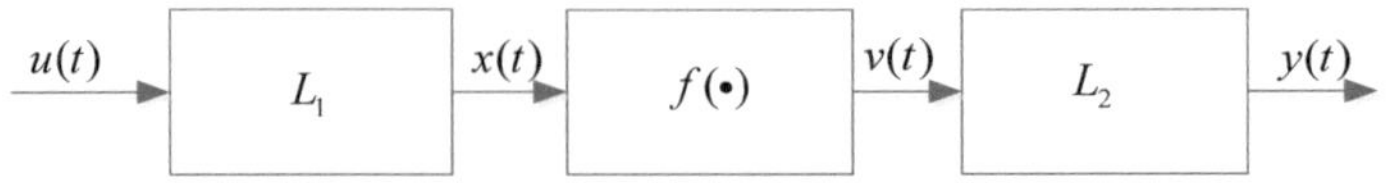

Fig. 1 The structure of cascade systems

for cascade systems with backlash-like hysteresis. Li [8] applied Newton iteration for Hammerstein nonlinear controlled autoregressive moving average systems. Vörös [9] provided decomposition technique to handle the effect of backlash, and presented iterative parameter identification for three block cascade model. Gomez [10] presented the subspace-based identification algorithm, in which two steps were applied to estimate the parameters of the multivariable Hammerstein and Wiener model. However, the identification systems of above literatures are mainly Hammerstein or Hammerstein-Wiener model. The identification algorithm has a large of computation cost.

In this paper, a recursive approach for the cascade systems with dead zone is proposed. Inspired by [11], some switching functions are applied to reconstruct the expression of dead zone, the parameters of linear subsystems and dead zone are separated by using the key term separation principle, which is applied to convert the cascade systems into a special model. In order to decrease the computation cost of identification algorithm, a TS-RLS algorithm is developed.

The paper is organized as follows. Section 2 describes dead zone model and linear subsystem. A two-stage identification algorithm is derived based on the interactive technology in Sect. 3. Section 4 provides a simulation to validate the effectiveness of the proposed algorithm. Finally, conclusions are offered in Sect. 5.

2 Model of Cascade Systems with Dead Zone

It is well known that the cascade systems consist of the two linear subsystems around a nonlinear block. On the one hand, the dead zone is described based on the switching function. On the other hand, the linear system is transformed into an output equation.

2.1 Dead Zone Model

Consider a dead zone model in Fig. 2, where $x(t)$ and $v(t)$ are the deadzone input and output, m_l and m_r are slope. b_l and b_r are left endpoint and right endpoint. The dead zone mathematical model is described by

Fig. 2 Dead zone

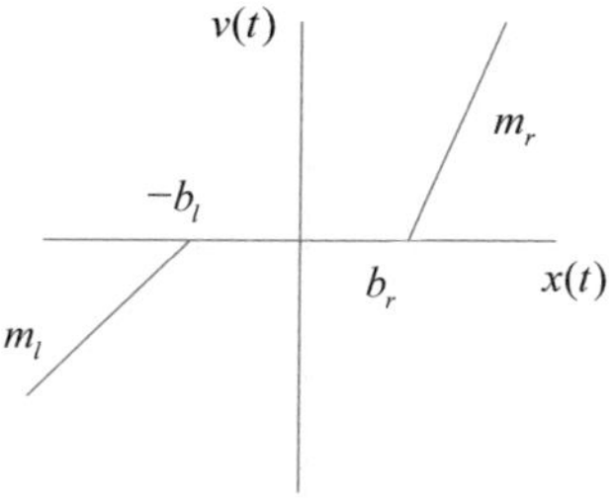

$$v(t)=\begin{cases} m_l(x(t)+b_l) & if\ x(t)\le -b_l \\ 0 & if\ -b_l<x(t)<b_r \\ m_r(x(t)-b_r) & if\ x(t)\ge b_r \end{cases} \tag{1}$$

According to formula (1), it can be seen that the dead zone model is unsuitable for parameter identification of the dead zone. Thus, it is necessary that dead zone model is transformed into an easy analysis of mathematical expression. This expression is linear or quasilinear. In the previous work [12], the switching function can be defined as

$$g(s)=\begin{cases} 0 & if\ s>0 \\ 1 & if\ s\le 0 \end{cases} \tag{2}$$

In order to describe piecewise characteristics of (1), it is desirable to define the left side of dead zone

$$f_1(t)=g[x(t)-b_l] \tag{3}$$

Define the right side of dead zone

$$f_2(t)=g[b_r-x(t)] \tag{4}$$

The middle section of the dead zone can be described by $f_1(t)$ and $f_2(t)$. Now the dead zone (1) can be rewritten as

$$v(t)=m_lx(t)f_1(t)+m_lb_lf_1(t)+m_rx(t)f_2(t)-m_rb_rf_2(t) \tag{5}$$

whose the input–output characteristic (5) is equal to function (1).

2.2 Cascade Systems with Dead Zone

In Fig. 1, the cascade systems consist of three parts, where dead zone nonlinear lies between the linear subsystem L_1 and linear subsystem L_2.

The linear subsystem L_1 can be expressed as

$$B(z^{-1})x(t) = A(z^{-1})u(t), \tag{6}$$

where $u(t)$ and $x(t)$ are input and output of the linear subsystem L_1, respectively. $B(z^{-1})$ and $A(z^{-1})$ are polynomial in the shift operator $z^{-1}(z^{-1}y(t) = y(t-1))$, and defined as

$$A(z^{-1}) = a_1 z^{-1} + a_2 z^{-2} + \cdots + a_{n_a} z^{-n_a} \tag{7}$$

$$B(z^{-1}) = 1 + b_1 z^{-1} + b_2 z^{-2} + \cdots + a_{n_b} z^{-n_b} \tag{8}$$

Assume that the orders n_a, n_b are known, the parameters $a_1, a_2, \ldots, a_{n_a}$, $b_1, b_2, \ldots, b_{n_b}$ are unknown and $u(t) = 0, x(t) = 0$ for $t \leq 0$.

The linear subsystem L_2 can be described by

$$D(z^{-1})y(t) = C(z^{-1})v(t), \tag{9}$$

where $v(t)$, $y(t)$ are input and output of the linear subsystem L_2, respectively. $D(z^{-1})$ and $C(z^{-1})$ are polynomial, and given by

$$C(z^{-1}) = c_1 z^{-1} + c_2 z^{-2} + \cdots + c_{n_c} z^{-n_c} \tag{10}$$

$$D(z^{-1}) = 1 + d_1 z^{-1} + d_2 z^{-2} + \cdots + d_{n_d} z^{-n_d} \tag{11}$$

The degrees n_c, n_d are known, the parameters $c_1, c_2, \ldots, c_{n_c}$, $d_1, d_2, \ldots, d_{n_d}$ are unknown.$v(t) = 0, y(t) = 0$ for $t \leq 0$.

The formulas (5), (6), and (9) can be rewritten as

$$x(t) = \sum_{i=1}^{n_a} a_i u(t-i) - \sum_{j=1}^{n_b} b_j x(t-j) \tag{12}$$

$$v(t) = m_l x(t) f_1(t) + m_l b_l f_1(t) + m_r x(t) f_2(t) - m_r b_r f_2(t) \tag{13}$$

$$y(t) = \sum_{m=1}^{n_c} c_m v(t-m) - \sum_{n=1}^{n_d} d_n y(t-n) + d(t), \tag{14}$$

where $d(t)$ is a white noise with zero mean. The output model of the whole system is obtained by a direct substitution of (12), (13) into (14), but it will lead to a complex output equation. Thus, the key term separation principle is introduced to simplify output equation and separate the parameters of nonlinear part and linear subsystem.

By half-substituting (12) into (13), $v(t)$ is rewritten as

$$\begin{aligned} v(t) &= m_l f_1(t) \sum_{i=1}^{n_a} a_i u(t-i) - m_l f_1(t) \sum_{j=1}^{n_b} b_j x(t-j) \\ &\quad + m_l b_l f_1(t) + m_r x(t) f_2(t) - m_r b_r f_2(t) \end{aligned} \tag{15}$$

Then, half-substituting (15) into (14), it yields

$$\begin{aligned} y(t) &= m_l c_1 f_1(t-1) \sum_{i=1}^{n_a} a_i u(t-i-1) - m_l c_1 f_1(t-1) \sum_{j=1}^{n_b} b_j x(t-j-1) \\ &\quad + m_l b_l c_1 f_1(t-1) + m_r c_1 x(t-1) f_2(t-1) \\ &\quad - m_r b_r c_1 f_2(t-1) + \sum_{m=2}^{n_c} c_m v(t-m) - \sum_{n=1}^{n_d} d_n y(t-n) + d(t) \end{aligned} \tag{16}$$

Therefore, the particular form of the cascade systems is constructed by (12), (13), (14), (15), and (16). Even though the output Eq. (16) is nonlinear in variables, all the model parameters are separated.

3 The Identification Method

In the previous section, $x(t)$ and $v(t)$ are not available and must be estimated. So, the TS-RLS [13] is considered. This algorithm is able to estimate the parameters of the output expression (16) during the identification process. The output Eq. (16) can be arranged as

$$\begin{aligned} y(t) &= m_l c_1 f_1(t-1) a_1 u(t-2) + m_l c_1 f_1(t-1) a_2 u(t-3) + \cdots \\ &\quad + m_l c_1 f_1(t-1) a_{n_a} u(t-n_a-1) - m_l c_1 f_1(t-1) b_1 x(t-2) \\ &\quad - m_l c_1 f_1(t-1) b_2 x(t-3) - \cdots - m_l c_1 f_1(t-1) b_{n_b} x(t-n_b-1) \\ &\quad + m_l b_l c_1 f_1(t-1) + m_r c_1 x(t-1) f_2(t-1) - m_r b_r c_1 f_2(t-1) \\ &\quad + c_2 v(t-2) + c_3 v(t-3) + \cdots + c_{n_c} v(t-n_c) - d_1 y(t-1) \\ &\quad - d_2 y(t-2) - \cdots - d_{n_d} y(t-n_d) + d(t) \end{aligned} \tag{17}$$

Form the above-mentioned results, a recursive least squares(RLS) is applied to estimate all the model parameters, which would lead to a large calculated cost. Therefore, we can decompose the Eq. (17) into two subsystems shown as

$$\begin{aligned} y(t) &= \varphi_a^{\mathrm{T}}(t)\theta_a + \varphi_b^{\mathrm{T}}(t)\theta_b + d(t) \\ &= \varphi^{\mathrm{T}}(t)\theta + d(t) \end{aligned} \tag{18}$$

Define

$$\varphi(t) = [\varphi_a(t), \varphi_b(t)]^{\mathrm{T}}, \theta = [\theta_a, \theta_b]^{\mathrm{T}},$$

where

$$\begin{aligned}\varphi_a(t) = & [f_1(t-1)u(t-2), f_1(t-1)u(t-3), \ldots, f_1(t-1)u(t-n_a-1), \\ & -f_1(t-1)x(t-2), -f_1(t-1)x(t-3), \ldots, f_1(t-1)x(t-n_b-1), \\ & f_1(t-1), f_2(t-1)x(t-1), -f_2(t-1)]^{\mathrm{T}} \\ \theta_a = & [m_l c_1 a_1, m_l c_1 a_2, \ldots, m_l c_1 a_{n_a}, m_l c_1 b_1, m_l c_1 b_2, \ldots, \\ & m_l c_1 b_{n_b}, m_l b_l c_1, m_r c_1, m_r b_r c_1]^{\mathrm{T}} \\ \varphi_b(t) = & [v(t-2), v(t-3), \ldots, v(t-n_c), -y(t-1), -y(t-2), \ldots, -y(t-n_d)]^{\mathrm{T}} \\ \theta_b = & [c_1, c_2, c_3, \ldots, c_{n_c}, d_1, d_2, \ldots, d_{n_d}]^{\mathrm{T}},\end{aligned}$$

where $\varphi_a(t), \varphi_b(t)$ include the unknown variables $x(t), v(t)$. Hence, defining $\hat{x}(t), \hat{v}(t)$ as the estimate of $x(t), v(t)$. Similarly, $\hat{\theta}_a(t-1), \hat{\theta}_b(t-1)$ are the estimate of $\theta_a(t), \theta_b(t)$. Thus, the TS-RLS estimation algorithm can be summarized as follows:

$$\hat{\theta}_a(t) = \hat{\theta}_a(t-1) + L_a(t)\left[y(t) - \hat{\varphi}^{\mathrm{T}}(t)\hat{\theta}_a(t-1)\right] \tag{19}$$

$$L_a(t) = P_a(t-1)\hat{\varphi}_a(t)\left[1 + \hat{\varphi}_a^{\mathrm{T}}(t)P_a(t-1)\hat{\varphi}_a(t)\right]^{-1} \tag{20}$$

$$P_a(t) = \left(I - L_a(t)\hat{\varphi}_a^{\mathrm{T}}(t)\right)P_a(t-1), P_a(0) = p_0 I, \tag{21}$$

where

$$\begin{aligned}\hat{\varphi}_a(t) = & [f_1(t-1)u(t-2), f_1(t-1)u(t-3), \ldots, f_1(t-1)u(t-n_a-1), \\ & -f_1(t-1)\hat{x}(t-2), -f_1(t-1)\hat{x}(t-3), \ldots, \\ & f_1(t-1)\hat{x}(t-n_b-1), f_1(t-1), f_2(t-1)\hat{x}(t-1), -f_2(t-1)]^{\mathrm{T}}\end{aligned} \tag{22}$$

$$\hat{x}(t) = \sum_{i=1}^{n_a} a_i u(t-i) - \sum_{j=1}^{n_b} b_j \hat{x}(t-j) \tag{23}$$

Similar algorithm is used in the parameters estimation of θ_b, it yields

$$\hat{\theta}_b(t) = \hat{\theta}_b(t-1) + L_b(t)[y(t) - \hat{\varphi}^{\mathrm{T}}(t)\hat{\theta}_b(t-1)] \tag{24}$$

$$L_b(t) = P_b(t-1)\hat{\varphi}_b(t)\left[1 + \hat{\varphi}_b^{\mathrm{T}}(t)P_b(t-1)\hat{\varphi}_b(t)\right]^{-1} \tag{25}$$

$$P_b(t) = \left[I - L_b(t)\hat{\varphi}_b^{\mathrm{T}}(t)\right]P_b(t-1), P_b(0) = p_0 I \tag{26}$$

where

$$\hat{\varphi}_b(t) = [\hat{v}(t-2), \hat{v}(t-3), \ldots, \hat{v}(t-n_c), -y(t-1), -y(t-2), \ldots, -y(t-n_d)]^{\mathrm{T}} \tag{27}$$

$$\hat{v}(t) = m_l\hat{x}(t)f_1(t) + m_l b_l f_1(t) + m_r\hat{x}(t)f_2(t) - m_r b_r f_2(t) \tag{28}$$

The steps of the TS-RLS algorithm is summarized as follows:

Step 1: Let $t = 1$, set the initial values $\theta_a(0), \theta_b(0)$ as small positive vectors. $p_0 = 10^6$, $\hat{v}(t) = 0$, $\hat{x}(t) = 0$, $y(t) = 0$, for $t \le 0$.
Step 2: Collect the input data $u(t)$ and the output data $y(t)$, Compute $\hat{\varphi}_a(t)$ by (22), $\hat{\varphi}_b(t)$ by (27), $\hat{\varphi}(t)$ by (18).
Step 3: Compute $L_a(t)$ by (20), $L_b(t)$ by (25), $P_a(t)$ by (21), $P_b(t)$ by (26).
Step 4: Update $\theta_a(t)$ by (19), $\theta_b(t)$ by (24).
Step 5: Increase t by 1 and go to Step 2.

4 The Simulation

In this section, an example is given to validate the proposed algorithm. Without loss of generality, the coefficient $a_1 = 1, c_1 = 1$ is set. Consider the following cascade systems, where dead zone model parameters are given by $m_l = 0.8$,
$m_r = 0.8, b_l = 0.2, b_r = 0.2$ and the two linear subsytems are described by

$$L_1{:}\, x(t) = u(t-1) + 0.3u(t-2) - 0.4x(t-1) - 0.64x(t-2)$$
$$L_2{:}\, y(t) = v(t-1) + 0.5v(t-2) - 0.45y(t-1) - 0.4y(t-2)$$

In simulation, $u(t)$ is zero mean and unit variance random, and the noise $d(t)$ is a white noise with zero mean and variance $\sigma^2 = 0.1^2$. Signal noise ratio SNR = 15.14 is added to the output. Data length is 3000. The initial parameters of the proposed algorithm are show as, $\hat{\theta}_b(0) = [0.001,\ 0.001,\ 0.001]^{\mathrm{T}}$, $p_0 = 10^6$

$$\hat{\theta}_b(0) = [0.001,\ 0.001, 0.001]^{\mathrm{T}}.$$

The results of parameters identification are shown in Figs. 3, 4 and 5. Figure 3 is the parameters estimation curve of linear subsystem L_1. The parameters identification curve of dead zone is shown in Fig. 4. The parameter estimation curve of linear subsystem L_2 is displayed in Fig. 5. The parameter estimation and corresponding errors ($\delta = \|\theta(t) - \theta\| / \|\theta\|$) are shown in Tables 1 and 2. The computation cost of the TS-RLS and RLS at each recursive identification is shown in Table 3,

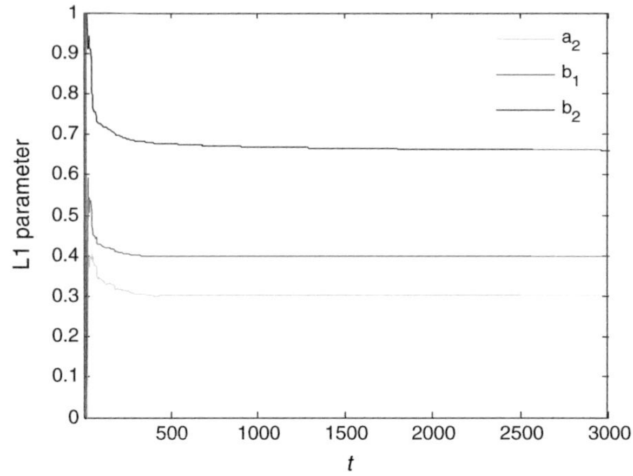

Fig. 3 Parameter estimates of L_1

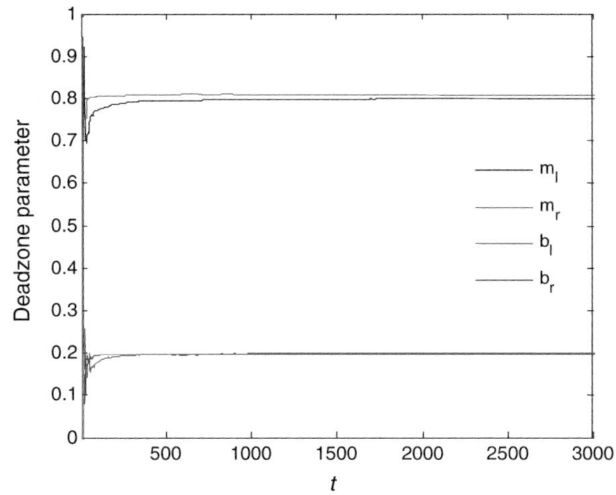

Fig. 4 Parameter estimates of dead zone

where the numbers in the brackets represent the computation cost for $n_a = 1, n_b = 2, n_c = 1$ and $n_d = 2$.

From Figs. 3, 4 and 5, Tables 1, 2 and 3, it can be seen that the parameters estimation values can converge to their true values by the TS-RLS algorithm and estimation errors become small with the increasing of data length. Although the parameters estimation accuracy of the TS-RLS is less than that the RLS, the RLS needs more computation cost than the TS-RLS. These results show that the TS-RLS algorithm is effective for estimating the parameters of the cascade systems.

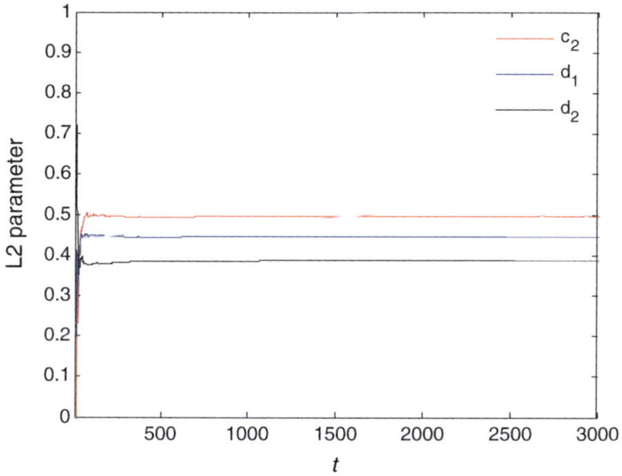

Fig. 5 Parameter estimation of L_2

Table 1 Estimation results of the TS-RLS

Algorithms	t	a_2	b_1	b_2	c_2	d_1	d_2
TS-RLS	500	0.30204	0.39952	0.67656	0.49276	0.44394	0.38388
	1000	0.30258	0.39947	0.66878	0.49458	0.44478	0.38585
	2000	0.30182	0.39923	0.66294	0.49620	0.44550	0.38743
	3000	0.30195	0.39948	0.66075	0.49708	0.44591	0.38837
RLS	500	0.29998	0.40000	0.64006	0.50002	0.44998	0.39996
	1000	0.30004	0.39994	0.64005	0.50002	0.45003	0.40000
	2000	0.30003	0.39996	0.64005	0.49995	0.44998	0.40001
	3000	0.30002	0.39997	0.64004	0.49995	0.44999	0.40001
	Ture values	0.30000	0.40000	0.64000	0.50000	0.45000	0.40000

Table 2 Estimation results of the TS-RLS

	t	b_r	b_l	m_r	m_l	δ (%)
TS-RLS	500	0.19627	0.19740	0.80973	0.79374	2.64706
	1000	0.19713	0.19888	0.80968	0.79671	2.14065
	2000	0.19772	0.19945	0.80830	0.79860	1.74237
	3000	0.19831	0.19947	0.80784	0.79904	1.58416
RLS Ture values	500	0.19991	0.20002	0.80004	0.80004	0.11488
	1000	0.19991	0.20001	0.79996	0.80004	0.00891
	2000	0.19997	0.20003	0.79998	0.80000	0.00625
	3000	0.19997	0.20000	0.79997	0.79998	0.00565
		0.20000	0.20000	0.80000	0.80000	

Table 3 The computation cost of the TS-RLS and RLS $(n = n_a + n_b + n_c + n_d)$

Algorithms	Number of multiplications	Number of additions
RLS	$2n^2 + 4n$ (96)	$2n^2 + 2n$ (84)
TS-RLS	$2(n_a + n_b)^2 + 4(n_a + n_b) + \ldots + 2(n_c + n_d)^2 + 4(n_c + n_d)$ (60)	$2(n_a + n_b)^2 + 2(n_a + n_b) + \ldots + 2(n_c + n_d)^2 + 2(n_c + n_d)$ (48)

5 Conclusions

This paper proposes the two stage recursive parameter identification algorithm to identify the cascade systems with dead zone. The cascade systems are transformed into a special form by using the key term separation principle and interactive estimation method. The proposed identification algorithm not only estimates all the parameters of the cascade systems, but also reduces the computation cost of identification process. Simulation results illustrate the efficiency of the proposed algorithm.

Acknowledgments This work is supported by National Natural Science Foundation of China (No. 61433003,61273150 and 61321002).

References

1. Ibrir S, Xie W-F, Su C-Y (2007) Adaptive tracking of nonlinear systems with non-symmetric deadzone input. Automatica 43(3):522–530
2. Crama P, Schoukens S (2005) Computing an initial estimate of a Wiener-Hammerstein system with a random phase multisine excitation. IEEE Trans Instrum Meas 54(1):117–122
3. Boutayeb M, Darouach M (1995) Recursive identification method for MISO Wiener-Hammerstein model. IEEE Trans Autom Control 40(2):287–291
4. Wang DQ, Ding F, Chu YY (2013) Data filtering based recursive least square algorithm for Hammerstein systems using the key-term separation principle. Inf Sci 222(10):203–212
5. Vörös J (2014) Iterative identification of nonlinear dynamic systems with output backlash using three-block cascade models. J Franklin Inst 79(3):5455–5466
6. Wang DQ, Ding F (2008) Extended stochastic gradient identification algorithms for Hammerstein Wiener ARMAX systems. Comput Math Appl 56(12):3157–3164
7. Dong RL, Tan YH, Chen H, Xie YQ (2012) Non-smooth recursive identification of sandwich systems with backlash-like hysteresis, pp 3800–3844
8. Li JH (2013) Parameter estimation for Hammerstein CARARMA systems based on the Newton iteration. Appl Math Lett 26(1):91–96
9. Vörös J (2014) Iterative identification of nonlinear dynamic systems with output backlash using three block cascade models. J Franklin Inst 79(3):5455–5466
10. Gomez JC, Baeyens E (2005) Subspace-based identification algorithms for Hammerstein and Wiener Models, 11(2): 127–136
11. Vörös J (2010) Modeling and identification of systems with backlash. Automatica 46(2):369–374

12. Vörös J (2014) Identification of nonlinear dynamic systems with input saturation and output backlash using three-block cascade models. J Franklin Inst 351(12):5455–5466
13. Duan HH, Jia J, Ding RF (2012) Two-stage recursive least squares parameter estimation algorithm for output error models. Math Comput Model 55(3–4):1151–1159

Robust Tracking Control for Flexible Space End Effector

Yi Li, Xiaodong Zhao and Yingmin Jia

Abstract In this paper, a robust tracking control strategy is proposed for flexible space end effector. First, a flexible end effector is designed for the purpose of decreasing the rigid collision caused by the contact between space manipulator with the target. Permanent magnet spherical motor (PMSM) combining yaw, pitch, and roll motions in a single joint is employed to track the trajectory of the end effector. Then, the dynamic model of PMSM rotor is built according to the second Lagrange equation and the Cardan angle coordinate transformation. Finally, based on computed torque method, a robust control strategy is presented to reject the external disturbance. The simulation results illustrate the strong robustness of proposed control scheme.

Keywords Robust control · Flexible end effector · PMSM · Computed torque method

1 Introduction

With the development of space explorations, the success of capturing the target is the key of on-orbit service techniques, such as space repairing, rescuing, and refueling failed satellites. In [1–3], space manipulator end effector was employed to capture the target, which would cause violent rigid collision. The ROGER system was used as an end effector to haul defunct satellites or space debris into grave orbit [4]. However, this design is only suitable for capturing small targets by large space system, and the process of capturing is uncontrollable.

As a brand-new motor, spherical motor has special three-dimensional structure, thus it's difficult to establish appropriate dynamic model and design a high-

Y. Li (✉) · X. Zhao · Y. Jia
The Seventh Research Division and the Center for Information and Control, School of Automation Science and Electrical Engineering, Beihang University (BUAA), Beijing 100191, China
e-mail: liyi309@163.com

Y. Jia et al. (eds.), *Proceedings of 2016 Chinese Intelligent Systems Conference*, Lecture Notes in Electrical Engineering 405,
DOI 10.1007/978-981-10-2335-4_26

performance trajectory tracking controller. Lee et al. [5] presented the effects of a variable-reluctance spherical motor (VRSM). A robust back-stepping controller was developed to compensate for uncertainties account for imperfect modeling and intentional computational simplification. In order to restrain the effects of the external disturbances and parameter variations, an adaptive back-stepping sliding mode controller was investigated in [6], but the chattering of variable structure controller is a problem. With the development of intelligent control theory, [7] modeled the dynamic of PMSM and put forward a dynamic decoupling control algorithm of the motor using fuzzy controllers (FCs) and a neural network identifier (NNI), this algorithm improved the static and dynamic performances of the control systems with strong robustness to uncertainties, however, it was limited to the specific engineering application because of its structure and the complexity of calculation.

In this paper, a flexible space end effector is designed for decreasing rigid collision, and the process of capturing is simple and controllable. In addition, PMSM is used to track the trajectory of the end effector, a robust control strategy based on computed torque method is proposed. This method can effectively eliminate the influences of inter-axis nonlinear coupling and has strong robustness.

2 Problem Formulation

2.1 Description of Flexible Space End Effector

The flexible space end effector includes gripper, ball-joint, PMSM and spring-damper systems as shown in Fig. 1. Since the mass of the mechanism is much smaller than the target's, it would be bounced off when collided with the target. The ball-joint and spring-damper system are used for buffering impact force and turning collision into contact. Meanwhile, the end effector can realize yaw, pitch, and roll motions by controlling PMSM.

2.2 Dynamics Modeling

The development of the methodology described in this paper is based on following basic assumptions:

1. Non-contact position detection device will be employed in the effector, so the dynamic model of rotor can be used here.
2. The motion state of the rotor is fully observed.
3. The effector is in microgravity environment, and the gravity is neglected.

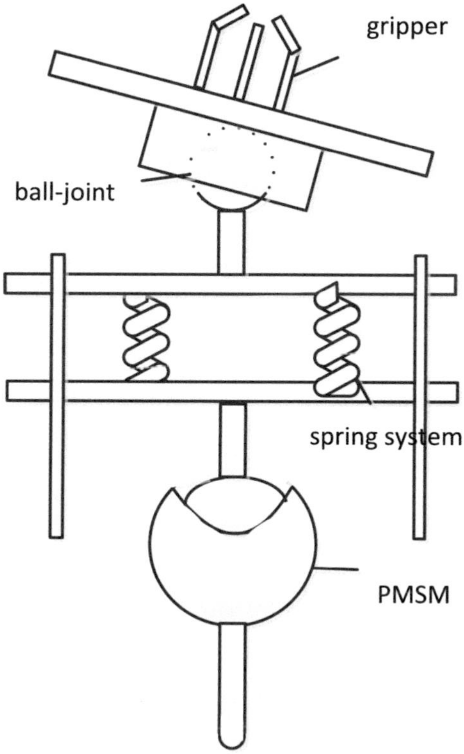

Fig. 1 Flexible space end effector

Actually, three-dimensional motion of the rotor in PMSM can be equivalent to a single rigid successively rotating about three axes of the sphere center coordinate frame.

Define the stator coordinate frame XYZ and the rotor coordinate frame xyz (inertial principal axis). Based on the Cardan angle (α, β, γ) coordinate transformation (showed in Fig. 2) and Lagrangian method, the PMSM rotor dynamic model is built as

$$H(q)(\ddot{q}) + C(q, \dot{q})(\dot{q}) = \tau + \omega \tag{1}$$

where $q = (\alpha, \beta, \gamma)^{\mathrm{T}}$, $\dot{q} = (\dot{\alpha}, \dot{\beta}, \dot{\gamma})^{\mathrm{T}}$ is the corresponding angular velocity vector, $H(q) \in R^{3\times3}$ is a symmetric positive definite inertia matrix, $C(q, \dot{q}) \in R^{3\times3}$ includes the nonlinear Coriolis and centrifugal forces acting on the system, $\tau = (\tau_\alpha, \tau_\beta, \tau_\gamma)^{\mathrm{T}}$ is the control torque, and ω is the external-uncertain disturbance. Moreover,

$$H(q) = \begin{bmatrix} J_1 \cos^2 \beta + J_2 \sin^2 \beta & 0 & J_2 \sin \beta \\ 0 & J_1 & 0 \\ J_2 \sin \beta & 0 & J_2 \end{bmatrix} \tag{2}$$

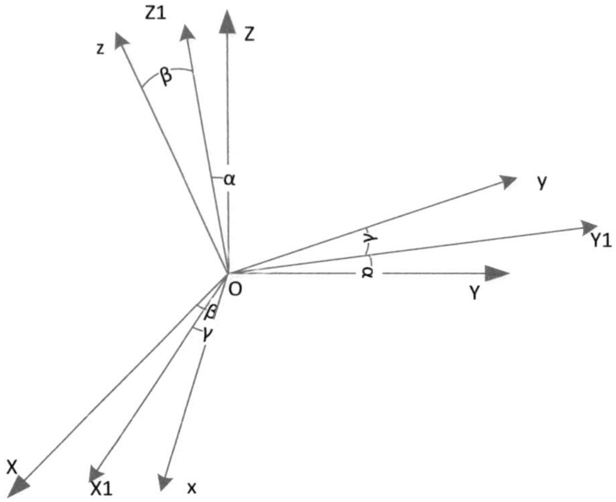

Fig. 2 Coordinate transformation based on Cardan angle

$$C(q,\dot{q}) = \begin{bmatrix} (J_2 - J_1)S_\beta C_\beta \dot{\beta} & (J_2 - J_1)S_\beta C_\beta \dot{\alpha} + \frac{1}{2}J_2 C_\beta \dot{\gamma} & \frac{1}{2}J_2 C_\beta \dot{\beta} \\ (J_1 - J_2)S_\beta C_\beta \dot{\alpha} - \frac{1}{2}J_2 C_\beta \dot{\gamma} & 0 & -\frac{1}{2}J_2 C_\beta \dot{\alpha} \\ \frac{1}{2}J_2 C_\beta \dot{\beta} & \frac{1}{2}J_2 C_\beta \dot{\alpha} & 0 \end{bmatrix} \tag{3}$$

where J_1 is the moment of inertia of x-y axis, and J_2 is the moment of inertia of z axis. The dynamic model bears the following properties.

Property 1 *The inertia matrix $H(q)$ is symmetric and uniformly positive definite and satisfies the following inequalities*

$$\underline{\lambda} I \leq H(q) \leq \overline{\lambda} I \tag{4}$$

Property 2 *The external-uncertain disturbance is norm-bounded as*

$$\|\omega\| \leq \rho \tag{5}$$

where $\underline{\lambda}, \overline{\lambda}, \rho$ are positive constants.

3 Robust Control Synthesis

Two steps are considered in the development of the robust tracking control scheme. First, a nominal feedback controller is designed to make the nominal system (without external disturbance) track the desired reference trajectory with an exponential convergence rate. Then, a robust compensator is designed to reduce the effects of the external disturbance so that the output tracking error of the closed loop system with external disturbance converges to zero.

Let q_d represent the desired trajectory that the rotor of the PMSM must follow and the output tracking error can be defined as

$$e = q_d - q \tag{6}$$

The corresponding tracking error velocity and acceleration are

$$\begin{aligned} \dot{e} &= \dot{q}_d - \dot{q} \\ \ddot{e} &= \ddot{q}_d - \ddot{q} \end{aligned} \tag{7}$$

where $\dot{q}_d, \ddot{q}_d$ are the velocity and acceleration of the desired reference trajectory, respectively

Based on the computed torque method, the desired torque for the nominal system has the following form

$$\tau_T = C(q, \dot{q})\dot{q} + H(q)u \tag{8}$$

Substituting Eq. (8) into the nominal system and with the property 1, we obtain a LTI system as

$$\ddot{q} = u \tag{9}$$

Application of the bias PD control, the control vector u is given as

$$\begin{aligned} u &= \ddot{q}_d + K_d(\dot{q}_d - \dot{q}) + K_p(q_d - q) \\ &= \ddot{q}_d + K_d\dot{e} + K_p e \end{aligned} \tag{10}$$

where $K_d, K_p \in R^{3\times3}$ are the positive definite matrices. Substituting Eq. (10) into (8) leads to

$$\tau_T = C(q, \dot{q})\dot{q} + H(q)(\ddot{q}_d + K_d\dot{e} + K_p e) \tag{11}$$

To account for the external disturbance in the sphere motor dynamics, the robust controller takes the following form

$$\tau = \underbrace{C(q, \dot{q})\dot{q} + H(q)(\ddot{q}_d + K_d\dot{e} + K_p e)}_{\text{tracking control}} + \underbrace{H(q)v}_{\text{robust compensator}} \tag{12}$$

Thus, using expressions (1), (6), (7) and (12), we get the error dynamic equation

$$\ddot{e} = -K_d\dot{e} - K_p e - H(q)^{-1}\omega - v \tag{13}$$

which can be written as

$$\dot{\zeta} = \begin{bmatrix} 0 & I \\ -K_p & -K_d \end{bmatrix} \zeta + \begin{bmatrix} 0 \\ -I \end{bmatrix} (\eta + v) \\ =: A\zeta + B(\eta + v) \tag{14}$$

where $\zeta = \begin{bmatrix} e \\ \dot{e} \end{bmatrix}, \eta = H(q)^{-1}\omega, A \in R^{6\times6}, B \in R^{6\times3}$. Due to the inequalities (4) and (5), η is also norm-bounded:

$$\|\eta\| = \|H(q)^{-1}\omega\| \le \overline{\rho} \tag{15}$$

In the proof of our main results, we need the following lemma:

Lemma 1 (Mei [8]) *Consider the nonlinear system*

$$\dot{x} = f(x, t) \tag{16}$$

if a continuously differentiable positive definite $V(x, t)$ exists and satisfies the following hypotheses

$$\sigma_1 \|x\|^2 \le V(x, t) \le \sigma_2 \|x\|^2 \tag{17}$$

$$\dot{V}(x, t) \le -\sigma_3 \|x\|^2 + \varepsilon exp(-\theta t) \tag{18}$$

where $\sigma_1, \sigma_2, \sigma_3, \varepsilon, \theta$ are the given positive constants. Then the state of Eq. (16) converges to zero with an exponential convergence rate. Moreover, the exponential convergence can be expressed by the following inequality

$$\|x\| \le \begin{cases} \frac{1}{\sigma_1} V(x(t_0)) exp(-\frac{\sigma_3}{\sigma_2}(t - t_0)) + \phi & \theta = \frac{\sigma_3}{\sigma_2} \\ \frac{1}{\sigma_1} V(x(t_0)) exp(-\frac{\sigma_3}{\sigma_2}(t - t_0)) + \psi & \theta \ne \frac{\sigma_3}{\sigma_2} \end{cases} \tag{19}$$

where $\phi = \frac{\varepsilon}{\sigma_1}(t - t_0)) exp(-\frac{\sigma_3}{\sigma_2} t)$

$\psi = \frac{\varepsilon exp(-\theta t_0)}{\sigma_1(\frac{\sigma_3}{\sigma_2} - \theta)} (exp(-\theta(t - t_0) - exp(-\frac{\sigma_3}{\sigma_2}(t - t_0)))$.

Theorem 1 *The tracking error $e(t)$ for the end effector system (1) with external disturbance converges to zero when $t \to \infty$ if the control input is designed such that*

$$\tau = \tau_T + H(q)v \tag{20}$$

where τ_T is the nominal feedback controller given in expression (11) and v is the variable structure compensator given as

$$v = -\frac{\overline{\rho}^2 z^T}{\|z\|\overline{\rho} + \epsilon exp(-\delta t)} \tag{21}$$

where $z = \zeta^T PB$, ϵ, δ *are the given positive constants,* $P \in R^{6\times 6}$ *is the positive definite solution of Eq. (25).*

Proof Define a Lyapunov function

$$V = \zeta^T P\zeta > 0 \tag{22}$$

It obviously satisfies that

$$\lambda_1 \|\zeta\|^2 \le V(\zeta, t) \le \lambda_2 \|\zeta\|^2 \tag{23}$$

where λ_1, λ_2 are the minimum and maximum eigenvalue of the matrix P. The time derivative of the Lyapunov function along the trajectories of the error dynamics is given by

$$\dot{V} = \zeta^T (A^T P + PA)\zeta + 2\zeta^T PB(\eta + v) \tag{24}$$

For a negative definite matrix A,

$$A^T P + PA = -Q \tag{25}$$

where Q is a symmetric positive definite matrix.

Substituting Eqs. (21) and (25) into Eq. (24), we have

$$\begin{aligned}
\dot{V} &= -\zeta^T Q\zeta + 2\zeta^T PB(\eta - \frac{\overline{\rho}^2 B^T P\zeta}{\|\zeta^T PB\|\overline{\rho} + \epsilon exp(-\delta t)}) \\
&\le -\zeta^T Q\zeta + 2(\|\zeta^T PB\| \|\eta\| - \frac{\overline{\rho}^2 \|\zeta^T PB\|^2}{\|\zeta^T PB\|\overline{\rho} + \epsilon exp(-\delta t)}) \\
&\le -\zeta^T Q\zeta + 2(\|\zeta^T PB\|\overline{\rho} - \frac{\overline{\rho}^2 \|\zeta^T PB\|^2}{\|\zeta^T PB\|\overline{\rho} + \epsilon exp(-\delta t)}) \\
&= -\zeta^T Q\zeta + \frac{2\|\zeta^T PB\|\overline{\rho}\epsilon exp(-\delta t)}{\|\zeta^T PB\|\overline{\rho} + \epsilon exp(-\delta t)} \\
&\le -\lambda \|\zeta\|^2 + 2\epsilon exp(-\delta t)
\end{aligned} \tag{26}$$

where λ is the minimum eigenvalue of matrix Q. In terms of Lemma 1, the tracking error $e(t)$ converges to zero with an exponential convergence rate when $t \to \infty$. this completes the proof.

4 Illustrative Example

In this section, we provide an example to illustrate the effectiveness of the controller design method proposed in this paper. Here, we consider Cardan angle ($\sin(0.05\pi t)$, $\cos(0.05\pi t)$, $\frac{\pi t}{50}$) as the desired reference trajectory. According to the analysis software

ADAMS, we obtain that $J_1 = 1.3682\,\text{kg.m}^2$, $J_2 = 1.3469\,\text{kg.m}^2$. Let the initial position $q(0) = (0.2, 0.3, 0.1)^T$, the corresponding initial angle velocity $\dot{q}(0) = (0, 0, 0)^T$ Assume that the external disturbance $\omega = (0.5\sin(0.2\pi t), 0.5\sin(0.2\pi t), 0.5\sin(0.2\pi t))^T$. We choose the parameters of the controller $K_p = diag(10, 20, 30)$, $K_d = diag(8, 35, 12.5)$, $Q = 5I_{6\times6}, \overline{\rho} = 0.8$, $\epsilon = 3$, $\delta = 0.05$ otherwise.

Figure 3 demonstrates that the Cardan angle α, β and γ track the reference trajectory well with the effect of the robust controller.

Figure 4 shows the control torque of robust controller. Note that the fluctuation is not chattering caused by the proposed controller as the control algorithm (1) includes periodic term $\ddot{q}_d$.

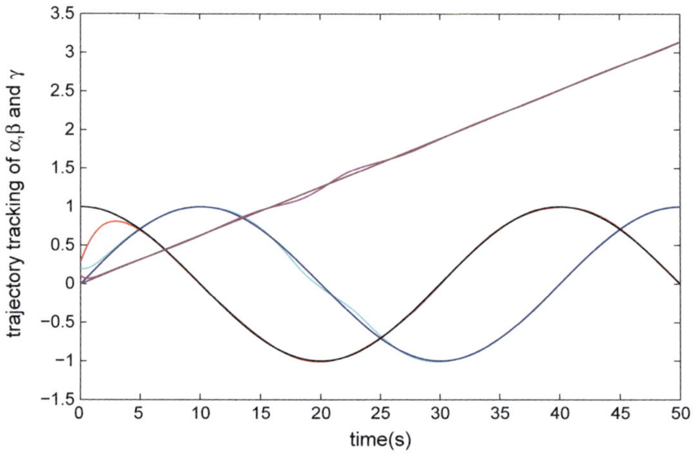

Fig. 3 Position tracking of Cardan angle α, β and γ

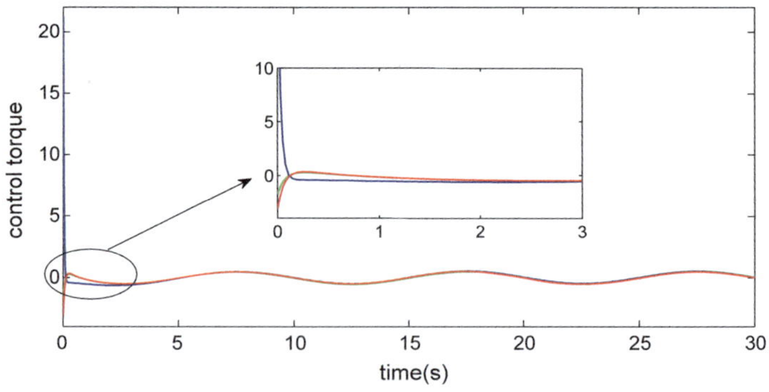

Fig. 4 Control torque of robust controller

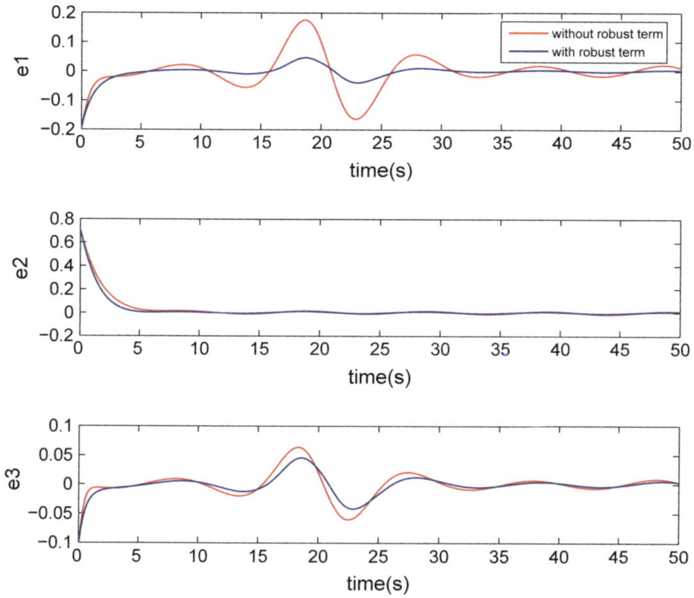

Fig. 5 Tracking error with and without robust term

Figure 5 compares the performance of the controller with and without robust term. the result shows that the robust controller is apparently better in suppressing external disturbance.

5 Conclusion

In this paper, a flexible space end effector is designed to decrease the collision when capturing space target, based on the computed torque method, a robust control law is presented for tracking the trajectory of the effector. An illustrative example is provided to demonstrate that the control algorithm is robust to bounded external disturbance.

Acknowledgments This work was supported by the National Basic Research Program of China (973 Program: 2012CB821200, 2012CB821201), the NSFC (61327807, 61521091, 61520106010, 61304232) and the Fundamental Research Funds for the Central Universities (YMF-16-GJSYS-31,YMF-16-GJSYS-32).

References

1. Bosse AB, Barnds WJ, Brown MA et al (2010) SUMO: spacecraft for the universal modification of orbits, defense and security. Int Soc Opt Photonics 33(3):707–723
2. Creamer G (2005) The SUMO/FREND project: technology development for autonomous grapple of geosynchronous satellites (AAS 07–015). Adv Astronaut Sci 50(4):511–515
3. Zimpfer D, Kachmar P, Tuohy S (2005) Autonomous Rendezvous, Capture and In-space Assembly: Past, Present and Future. In: Proceedings of 1st Space Exploration Conference: Continuing the Voyage of Discovery, vol 1. Orlando, Florida, USA, pp 234–245
4. Kassebom M, Koebel D, Tobehn C et al (2003) Roger-an advanced solution for a geostationary service satellite. In: Proceedings of International Astronautical Congress
5. Lee KM, Sosseh R (2002) Effects of fixture dynamics on back-stepping control of a VR spherical motor. In: 7th International Conference on Control, Automation, Robotics and Vision, 2002. ICARCV 2002, vol 1. IEEE, pp 384–389
6. Guo X, Wang Q, Li G et al (2014) Adaptive backstepping sliding mode control in permanent magnet spherical motor. J Nanjing Univ Aeronaut Astronaut 46(1):59–64
7. Xia C, Guo C, Shi T (2010) A neural-network-identifier and fuzzy-controller-based algorithm for dynamic decoupling control of permanent-magnet spherical motor. IEEE Trans Ind Electron 57(8):2868–2878
8. Mei S (2003) Modern Robust Control Theory and Application. Tsinghua Press, Beijing
9. Lee KM, Kwan CK (1991) Design concept development of a spherical stepper for robotic applications. IEEE Trans Robot Autom 7(1):175–181
10. Lee KM, Pei J, Roth R (1994) Kinematic analysis of a three-degrees-of-freedom spherical wrist actuator. Mechatronics 4(6):581–605
11. Wang Q, Li Z, Chen L et al (2005) Kinematic analysis and simulation of permanent magnet spherical stepper motor. In: Acta Simulata Systematica Sinica, vol 9, no 56
12. Huo W (2005) Robot Dynamics and Control. Higher Education Press, Beijing
13. Wang S (1998) Three Degrees-of-freedom Motor and Control. Huazhong University of Science and Technology Press, Wuhan

Robust Control for Elliptical Orbit Spacecraft Rendezvous Using Implicit Lyapunov Function

Xiwen Tian, Mingdong Hou, Yingmin Jia and Changqing Chen

Abstract In this paper, a robust control scheme using implicit Lyapunov function is proposed to solve the elliptical orbit spacecraft rendezvous problem in the presence of parameter uncertainties and external disturbances. First, an implicit Lyapunov function is constructed that is to be used in control design, and the finite-time convergence is guaranteed by some linear matrix inequalities depending on variable parameter and the implicit Lyapunov function. Then, an analytical feasible solution of these inequalities is provided for practical implementation. Simulation example is given to illustrate the effectiveness of the proposed method.

Keywords Robust control · Implicit lyapunov function · Spacecraft rendezvous

1 Introduction

Spacecraft rendezvous is an essential operational technology in many space missions, such as assembling, resupplying, exchanging of crew, repairing and docking [1]. In close range rendezvous which starts at a range of only a few hundreds of meters or even less, the navigation accuracy and rapidity requirements are increased. How-

X. Tian (✉) · M. Hou · Y. Jia
The Seventh Research Division and the Center for Information and Control, School of Automation Science and Electrical Engineering, Beihang University (BUAA), Beijing 100191, China
e-mail: tianxiwen123@163.com

M. Hou
e-mail: 603483879@qq.com

Y. Jia
e-mail: ymjia@buaa.edu.cn

C. Chen
The Science and Technology on Space Intelligent Control Laboratory, Beijing Institute of Control Engineering, Beijing 100190, China
e-mail: changqingchen@hotmail.com

Y. Jia et al. (eds.), *Proceedings of 2016 Chinese Intelligent Systems Conference*, Lecture Notes in Electrical Engineering 405,
DOI 10.1007/978-981-10-2335-4_27

ever, in general, precise measurements of orbital parameters and perturbations are hard to achieve. Therefore, robust control laws rejecting parameter uncertainties and external disturbances are desirable.

Variable structure controls (VSC) were proposed for the purpose of strong robustness and fast convergency, the most famous of which is sliding mode control (SMC). SMC constrains the motion of the system to a designed manifold where the system states slide towards the origin, and its robustness with respect to uncertainties and disturbances is guaranteed by high-frequency switching control [2]. Nevertheless, the consequent chattering phenomenon occurring in the actuator becomes the major problem. High order sliding mode (HOSM), first provided by A. Levant [3], attenuates the effect of chattering phenomenon and achieves finite-time stability mathematically. Plenty of researches based on HOSM were introduced, such as chattering analysis [4], Lyapunov stability [5] and extensions [6–8]. Despite of these excellent theoretical results, there are still some design problems in engineering. As Utkin [9] pointed out, HOSM is not always better than conventional sliding model control at chattering attenuation, and finite-time convergence is lost at the presence of unmodeled dynamics.

Soft VSC is a concept of intentionally precluding sliding modes to achieve continuous control and shorter settling time [10], which was first realized employing implicit Lyapunov function (ILF) by Adamy [11]. A Lyapunov function is defined by an implicit equation in ILF method to construct a robust control strategy, and the parameter tuning is based on linear matrix inequalities (LMIs). Using ILF, the finite-time and fixed-time stability were developed [12], the actuator saturation problem was solved [13], and the differentiator [14] and application for time delay systems [15] were also studied. However, the control laws designed for LTI systems in above literatures are not suitable for elliptical orbit spacecraft rendezvous described by LPV system [16], which motivates our present study.

In this paper, we put forward a robust control strategy using ILF for elliptical orbit spacecraft rendezvous in the presence of parameter uncertainties and external disturbances. Some linear parameter-variable matrix inequalities are given to guarantee the finite-time stabilization, and the existence of feasible solution of these inequalities is confirmed.

The rest of this paper is organized as follows: Sect. 2 summarizes the spacecraft rendezvous dynamics and some preliminaries about ILF. Section 3 presents the robust control strategy using implicit Lyapunov function. Sections 4 and 5 demonstrates the simulation and conclusion, respectively.

2 Problem Formulation

In this section, the dynamics of elliptical orbit spacecraft rendezvous is introduced, and some preliminaries used for control design are provided. The bold font is denoted as vectors or matrices, such as $\boldsymbol{a}$ or $\boldsymbol{A}$.

Fig. 1 The spacecraft rendezvous system

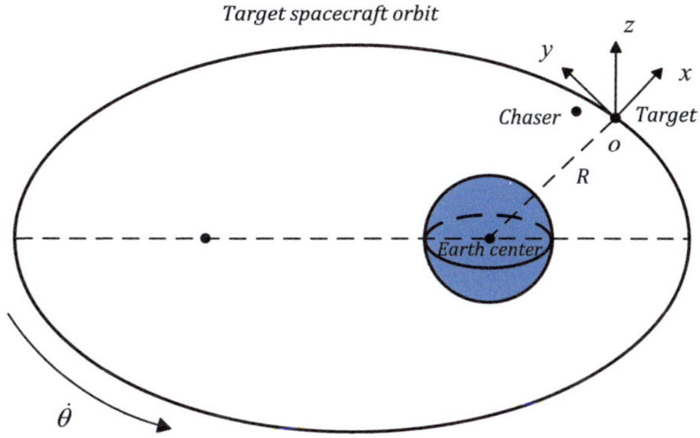

2.1 Elliptical Orbit Spacecraft Rendezvous

The spacecraft rendezvous system is illustrated in Fig. 1. The target spacecraft is assumed to move along an elliptical orbit, and the orbital coordinate frame is right-handed Cartesian coordinate with the origin attached to the centroid of the target, where the x-axis is along the vector from the earth center to the centroid of the target, the y-axis is along the target orbit circumference and the z-axis completes the right-handed frame. If the chaser and the target are close enough, the proximity relative motion can be described by the following linear parameter-variable equations [16]:

$$\begin{cases} \ddot{x} = \dot{\theta}^2 x + \ddot{\theta} y + 2\dot{\theta}\dot{x} + \frac{2\mu}{R^3} x + \frac{1}{m}(f_x + f_{dx}) \\ \ddot{y} = -\ddot{\theta} x + \dot{\theta}^2 y - 2\dot{\theta}\dot{y} - \frac{\mu}{R^3} x + \frac{1}{m}(f_y + f_{dy}) \\ \ddot{z} = -\frac{\mu}{R^3} z + + \frac{1}{m}(f_z + f_{dz}) \end{cases} \tag{1}$$

where x, y and z are the components of the relative position, μ is the gravitational constant of the Earth, θ is the true anomaly of the target, $\dot{\theta} = \frac{n(1+e\cos\theta)}{(1-e^2)^{3/2}}$, $n = \sqrt{\frac{\mu}{a^3}}$ is the mean motion of the target, a is the semimajor axis of the target orbit, e is the eccentricity, $R = \frac{a(1-e^2)}{1+e\cos\theta}$ is the orbital radius of the target, m is the mass of the chaser, $f_i (i = x, y, z)$ is the ith component of the control input force and f_{di} is the ith component of the external disturbances.

Let $\boldsymbol{r} = [x, y, z]^T$, $\boldsymbol{v} = [\dot{x}, \dot{y}, \dot{z}]^T$, $\boldsymbol{f} = [f_x, f_y, f_z]^T$, and $\boldsymbol{f}_d = [f_{dx}, f_{dy}, f_{dz}]^T$. In view of inaccuracy measurement of θ, (1) can be rewritten as

$$\begin{cases} \dot{\boldsymbol{r}} = \boldsymbol{v} \\ \dot{\boldsymbol{v}} = [\boldsymbol{A}_1 + \Delta\boldsymbol{A}_1]\boldsymbol{r} + [\boldsymbol{A}_2 + \Delta\boldsymbol{A}_2]\boldsymbol{v} + \frac{1}{m}(\boldsymbol{f} + \boldsymbol{f}_d) \end{cases} \tag{2}$$

where $\boldsymbol{A}_1 = \begin{bmatrix} \dot{\theta}^2 + \frac{2\mu}{R^3} & \ddot{\theta} & 0 \\ -\ddot{\theta} & \dot{\theta}^2 - \frac{\mu}{R^3} & 0 \\ 0 & 0 & -\frac{\mu}{R^3} \end{bmatrix}$, $\boldsymbol{A}_2 = \begin{bmatrix} 0 & 2\dot{\theta} & 0 \\ -2\dot{\theta} & 0 & 0 \\ 0 & 0 & 0 \end{bmatrix}$, $\boldsymbol{\Delta A}_1$ and $\boldsymbol{\Delta A}_2$ are the lumped uncertainties, and the norm-bounded condition is assumed as follow.

Assumption 1 The uncertain matrices $\boldsymbol{\Delta A}_1$, $\boldsymbol{\Delta A}_2$ and the external disturbances are norm-bounded as

$$\|\boldsymbol{\Delta A}_1\|_2 \leq c_1, \quad \|\boldsymbol{\Delta A}_2\|_2 \leq c_2, \quad \|\boldsymbol{f}_d\|_2 \leq c_3 \tag{3}$$

where c_1, c_2, and c_3 are known positive constant.

2.2 Preliminaries

Consider the following system:

$$\dot{\boldsymbol{x}} = \boldsymbol{f}(\boldsymbol{x}), \quad \boldsymbol{x}(t_0) = \boldsymbol{x}_0 \tag{4}$$

where $\boldsymbol{x} \in \mathbb{R}^n$ is the system state, $\boldsymbol{f} : \mathbb{R}^n \to \mathbb{R}^n$ is a continuous function, and the origin is an equilibrium of (4).

Lemma 1 *[12] If there exists a continuous function $Q(V, \boldsymbol{x}) : \mathbb{R}_+ \times \mathbb{R}^n \to \mathbb{R}$, and the following conditions are satisfied*

(C1) for any $\boldsymbol{x} \in \mathbb{R}^n \backslash \{\boldsymbol{0}\}$, Q is continuously differentiable, and there exist $V \in \mathbb{R}_+$ such that $Q(V, \boldsymbol{x}) = 0$;
(C2) let $\Omega = \{(V, \boldsymbol{x}) \in \mathbb{R}_+ \times \mathbb{R}^n : Q(V, \boldsymbol{x}) = 0\}$ and

$$\lim_{\substack{\boldsymbol{x}\to\boldsymbol{0} \\ (V,\boldsymbol{x})\in\Omega}} V = 0^+, \quad \lim_{\substack{V\to 0^+ \\ (V,\boldsymbol{x})\in\Omega}} \|\boldsymbol{x}\| = 0, \quad \lim_{\substack{\|\boldsymbol{x}\|\to+\infty \\ (V,\boldsymbol{x})\in\Omega}} V = +\infty; \tag{5}$$

(C3) $\frac{\partial Q(V,\boldsymbol{x})}{\partial V} < 0$ for any $V \in \mathbb{R}_+$ and $\boldsymbol{x} \in \mathbb{R}^n$;
(C4) $\frac{\partial Q(V,\boldsymbol{x})}{\partial \boldsymbol{x}} \boldsymbol{f}(\boldsymbol{x}) < 0$ for any $(V, \boldsymbol{x}) \in \Omega$;

then the origin of system (4) is globally uniformly asymptotically stable, and an implicit Lyapunov function is determined by the equation $Q(V, \boldsymbol{x}) = 0$ uniquely.

Lemma 2 *[17] Let $\boldsymbol{x}$ be the solution of (4). If there exists a continuous function $V : \mathbb{R}^n \to \mathbb{R}$, and $\frac{dV(\boldsymbol{x})}{dt} \leq -aV^c(\boldsymbol{x})$ for all $\boldsymbol{x} \in \mathbb{R}^n$ and some $a > 0$ and $c \in [0, 1)$. Then the origin of system (4) is globally finite-time stable, and the settling time satisfies $T(\boldsymbol{x}_0) \leq \frac{V^{1-c}(\boldsymbol{x}_0)}{a(1-c)}$.*

Theorem 1 *If there exists a continuous function* $Q(V,\boldsymbol{x}) : \mathbb{R}_+ \times \mathbb{R}^n \to \mathbb{R}$ *satisfies the conditions (C1–C3) and*

(C5) $\frac{\partial Q(V,x)}{\partial x}\boldsymbol{f}(t,\boldsymbol{x}) \le aV^c \frac{\partial Q(V,x)}{\partial V}$ *for all* $\boldsymbol{x} \in \mathbb{R}^n$ *and some* $a > 0$ *and* $c \in [0,1)$

then the origin of system (4) is globally finite-time stable, and the settling time satisfies $T(\boldsymbol{x}_0) \le \frac{V^{1-c}(\boldsymbol{x}_0)}{a(1-c)}$.

Proof The derivative of $Q(V,\boldsymbol{x}) = 0$ is

$$\frac{\partial Q(V,\boldsymbol{x})}{\partial \boldsymbol{x}}\boldsymbol{f}(\boldsymbol{x}) + \frac{\partial Q(V,\boldsymbol{x})}{\partial V}\frac{\mathrm{d}V}{\mathrm{dt}} = 0 \tag{6}$$

and condition (C3) and (C5) imply $\frac{dV}{dt} \le -aV^c$. From Lemma 2, the finite-time stability is proven.

3 Main Results

3.1 Robust Control Design Using Implicit Lyapunov Function

For elliptical orbit spacecraft rendezvous dynamics (2), the continuous function $Q(V,\boldsymbol{x})$ is chosen as

$$Q(V,\boldsymbol{x}) := \boldsymbol{x}^T\boldsymbol{D}(V^{-1})\boldsymbol{P}_1^{-1}\boldsymbol{D}(V^{-1})\boldsymbol{x} - 1 \tag{7}$$

where $V \in \mathbb{R}_+$, $\boldsymbol{x} = \left[\boldsymbol{r}^T, \boldsymbol{v}^T\right]^T$, $\boldsymbol{P}_1 \in \mathbb{R}^{6\times 6}$ is a constant positive definite matrix, $\boldsymbol{D}(V^{-1}) = \begin{bmatrix} V^{-r_1}\boldsymbol{I}_{3\times 3} & \boldsymbol{0}_{3\times 3} \\ \boldsymbol{0}_{3\times 3} & V^{-r_2}\boldsymbol{I}_{3\times 3} \end{bmatrix}$, and $r_1 = r_2 + g$, $r_2 > g$, $0 < g < 1$.

The continuous control law using ILF has the form

$$\boldsymbol{f} = mV^{r_2-g}\boldsymbol{P}_2(\theta, V)\boldsymbol{P}_1^{-1}\boldsymbol{D}(V^{-1})\boldsymbol{x} \tag{8}$$

where $\boldsymbol{P}_2(\theta, V) \in \mathbb{R}^{3\times 6}$ is a matrix depending on the variable parameter θ and implicit Lyapunov function V.

Theorem 2 *Denote that* $\boldsymbol{H}_\mu = \begin{bmatrix} -r_1\boldsymbol{I}_{3\times 3} & \boldsymbol{0}_{3\times 3} \\ \boldsymbol{0}_{3\times 3} & -r_2\boldsymbol{I}_{3\times 3} \end{bmatrix}$. *Given the spacecraft rendezvous system (2) satisfying Assumption 1, if the control law (8) is adopted and the linear matrix inequalities*

$$\begin{cases} \boldsymbol{AP}_1 + \boldsymbol{P}_1\boldsymbol{A}^T + \boldsymbol{BP}_2 + \boldsymbol{P}_2^T\boldsymbol{B}^T + (\alpha_1 + \alpha_2)\boldsymbol{P}_1 + \beta\boldsymbol{I}_{3\times3} \leq 0 \\ \boldsymbol{P}_1\Delta\boldsymbol{A}^T + \Delta\boldsymbol{AP}_1 \leq \alpha_2\boldsymbol{P}_1 \\ -\gamma\boldsymbol{P}_1 \leq \boldsymbol{P}_1\boldsymbol{H}_\mu + \boldsymbol{H}_\mu\boldsymbol{P}_1 < 0, \quad \boldsymbol{P}_1 > 0 \end{cases} \tag{9}$$

are feasible for some $\alpha_1 > \beta > 0$, $\alpha_2 > 0$, $\gamma > 0$, $\boldsymbol{P}_1 = \boldsymbol{P}_1^T \in \mathbb{R}^{6\times6}$ *and* $\boldsymbol{P}_2(\theta, V) \in \mathbb{R}^{3\times6}$, *where*

$$\boldsymbol{A} = \begin{bmatrix} \boldsymbol{0}_{3\times3} & \boldsymbol{I}_{3\times3} \\ V^{2g}\boldsymbol{A}_1 & V^g\boldsymbol{A}_2 \end{bmatrix}, \boldsymbol{B} = \begin{bmatrix} \boldsymbol{0}_{3\times3} \\ \boldsymbol{I}_{3\times3} \end{bmatrix}, \Delta\boldsymbol{A} = \begin{bmatrix} \boldsymbol{0}_{3\times3} & \boldsymbol{0}_{3\times3} \\ V^{2g}\Delta\boldsymbol{A}_1 & V^g\Delta\boldsymbol{A}_2 \end{bmatrix} \tag{10}$$

and V *is determined by* $Q(V, \boldsymbol{x}) = 0$ *with the external disturbances* $\boldsymbol{f}_d$ *satisfying the following condition*

$$\boldsymbol{f}_d^T\boldsymbol{f}_d \leq m^2\beta^2V^{2r_2-2g} \tag{11}$$

then the system state $\boldsymbol{x}$ *will converge to the desired terminal relative states* $\boldsymbol{x}_f = [\boldsymbol{r}_f^T, \boldsymbol{v}_f^T]^T$ *in finite time and the settling time* $T(\boldsymbol{x}) \leq \frac{\gamma V^g(\boldsymbol{x}_0)}{g(\alpha_1-\beta)}$.

Proof The smoothness, positive definiteness and radial unboundedness of V can be easily verified. The third inequalities of (9) guarantee the uniqueness of V, since $-\gamma V^{-1} = -\gamma V^{-1}\boldsymbol{x}^T\boldsymbol{D}(V^{-1})\boldsymbol{P}_1^{-1}\boldsymbol{D}(V^{-1})\boldsymbol{x} \leq V^{-1}\boldsymbol{x}^T\boldsymbol{D}(V^{-1})(\boldsymbol{H}_\mu\boldsymbol{P}_1^{-1} + \boldsymbol{P}_1^{-1}\boldsymbol{H}_\mu)\boldsymbol{D}(V^{-1})$ $\boldsymbol{x} = \frac{\partial Q}{\partial V} < 0$. Let $\xi_1 = \boldsymbol{D}(V^{-1})\boldsymbol{x}$ and $\xi_2 = \frac{1}{m}\boldsymbol{D}(V^{-1})\boldsymbol{Bf}_d$. It can be obtained that

$$\frac{\partial Q}{\partial \boldsymbol{x}}\dot{\boldsymbol{x}} = \begin{bmatrix} \xi_1 \\ \xi_2 \end{bmatrix}^T \boldsymbol{M} \begin{bmatrix} \xi_1 \\ \xi_2 \end{bmatrix} + \beta^{-1}V^g\xi_2^T\xi_2 - \alpha_1V^{-g}\xi_1^T\boldsymbol{P}_1^{-1}\xi_1 \tag{12}$$

where

$$\begin{aligned} \boldsymbol{M} &= \begin{bmatrix} V^{-g}\boldsymbol{N} & \boldsymbol{P}_1^{-1} \\ \boldsymbol{P}_1^{-1} & -\beta^{-1}V^gI_{3\times3} \end{bmatrix} \\ \boldsymbol{N} &= \boldsymbol{P}_1^{-1}(\boldsymbol{A} + \Delta\boldsymbol{A} + \boldsymbol{BP}_2\boldsymbol{P}_1^{-1}) + (\boldsymbol{A} + \Delta\boldsymbol{A} + \boldsymbol{BP}_2\boldsymbol{P}_1^{-1})^T\boldsymbol{P}_1^{-1} + \alpha_1\boldsymbol{P}_1^{-1} \end{aligned} \tag{13}$$

Applying the Schur complement to the first two matrix inequalities of (9), it has $\boldsymbol{M} \leq 0$. Taking into account that $\xi_1^T\boldsymbol{P}_1^{-1}\xi_1 = 1$ and condition (11), we have $\frac{\partial Q}{\partial \boldsymbol{x}}\dot{\boldsymbol{x}} \leq -(\alpha_1 - \beta)V^{-g} \leq \frac{\alpha_1-\beta}{\gamma}V^{1-g}\frac{\partial Q}{\partial V}$. The finite-time stability is proven according to Theorem 1.

Remark 1 In [12], by tuning the control parameters, the solution of matrix inequalities can be achieved off-line. However, in this paper, the matrix inequalities (9) needs to be solved on-line due to variable parameter θ and implicit Lyapunov function V that is also solved on-line, which may lead to unfeasible solution at some points. The existence of feasible solution will be discussed later.

3.2 Practical Implementation

LMI technique is an efficient method for linear matrix inequalities, however, whether the solution of (9) is feasible at every point should be considered. Here, we provide an analytical solution of these inequalities.

Theorem 3 *If there exist some* $\alpha_1 > \beta > 0$, $\alpha_2 > 0$, $\gamma > 0$, $\sigma_1 > 0$ *and* $\sigma_2 > 0$ *such that*

$$\begin{array}{l} \sigma_1 < \frac{2}{\alpha_1+\alpha_2}, \sigma_2 > \frac{g^2}{4r_1r_2}, \gamma > 2r_1 \\ \sigma_2\alpha_2^2 - 2(2\sigma_1\delta_1 + (2+\sigma_2)\delta_2)\alpha_2 - (\sigma_1\delta_1 + \delta_2)^2 \geq 0 \\ (\gamma - 2r_1)(\gamma - 2r_2) \geq \frac{g^2}{\sigma_2} \end{array} \tag{14}$$

where $\delta_1 = c_1 V_x^{2g}(\boldsymbol{x}_0)$ *and* $\delta_2 = c_2 V_x^g(\boldsymbol{x}_0)$, *then*

$$\boldsymbol{P}_1 = \begin{bmatrix} \boldsymbol{P}_{11} & \boldsymbol{P}_{12} \\ \boldsymbol{P}_{12} & \boldsymbol{P}_{13} \end{bmatrix}, \boldsymbol{P}_2 = \begin{bmatrix} \boldsymbol{P}_{21} & \boldsymbol{P}_{22} \end{bmatrix} \tag{15}$$

is a feasible solution of (9), where

$$\begin{array}{l} \boldsymbol{P}_{11} = -\sigma_1 \boldsymbol{P}_{12}, \boldsymbol{P}_{13} = -\frac{1+\sigma_2}{\sigma_1}\boldsymbol{P}_{12} \\ \boldsymbol{P}_{12} = diag\{p_{12}, p_{12}, p_{12}\}, p_{12} = -\frac{\beta}{2-(\alpha_1+\alpha_2)\sigma_1} \\ \boldsymbol{P}_{21} = -\boldsymbol{P}_{13} - \hat{\boldsymbol{A}}_1\boldsymbol{P}_{11} - \hat{\boldsymbol{A}}_2\boldsymbol{P}_{12} - (\alpha_1+\alpha_2)\boldsymbol{P}_{12} \\ \boldsymbol{P}_{22} = -\frac{1}{2}(\hat{\boldsymbol{A}}_1\boldsymbol{P}_{12} + \boldsymbol{P}_{12}\hat{\boldsymbol{A}}_1^T + \hat{\boldsymbol{A}}_2\boldsymbol{P}_{13} + \boldsymbol{P}_{13}\hat{\boldsymbol{A}}_2^T + (\alpha_1+\alpha_2)\boldsymbol{P}_{13} + \beta I_{3\times3}) \\ \hat{\boldsymbol{A}}_1 = V^{2g}\boldsymbol{A}_1, \hat{\boldsymbol{A}}_2 = V^g\boldsymbol{A}_2 \end{array} \tag{16}$$

Proof It can be verified that (15) is a solution of the following equation

$$\boldsymbol{A}\boldsymbol{P}_1 + \boldsymbol{P}_1\boldsymbol{A}^T + \boldsymbol{B}\boldsymbol{P}_2 + \boldsymbol{P}_2^T\boldsymbol{B}^T + (\alpha_1+\alpha_2)\boldsymbol{P}_1 + \beta\boldsymbol{I}_{3\times3} = 0 \tag{17}$$

Let $\Delta_1 = V^{2g}\Delta\boldsymbol{A}_1$ and $\Delta_2 = V^g\Delta\boldsymbol{A}_2$. Since $\dot{V} < 0$, they are bounded by

$$\|\Delta_1\|_2 \leq \delta_1, \quad \|\Delta_2\|_2 \leq \delta_2 \tag{18}$$

where $\delta_1 = c_1 V^{2g}(\boldsymbol{x}_0)$ and $\delta_2 = c_2 V^g(\boldsymbol{x}_0)$.

Dividing The second and second matrix inequalities of (9) are

$$\begin{array}{c} \begin{bmatrix} -\alpha_2\boldsymbol{P}_{11} & -\alpha_2\boldsymbol{P}_{12} + \boldsymbol{P}_{11}\Delta_1^T + \boldsymbol{P}_{12}\Delta_2^T \\ \star & -\alpha_2\boldsymbol{P}_{13} + \Delta_1\boldsymbol{P}_{12} + \boldsymbol{P}_{12}\Delta_1^T + \Delta_2\boldsymbol{P}_{13} + \boldsymbol{P}_{13}\Delta_2^T \end{bmatrix} \leq 0 \\ \begin{bmatrix} -2r_1\boldsymbol{P}_{11} & -(r_1+r_2)\boldsymbol{P}_{12} \\ \star & -2r_2\boldsymbol{P}_{13} \end{bmatrix} < 0 \\ \begin{bmatrix} -(\gamma-2r_1)\boldsymbol{P}_{11} & (-\gamma+r_1+r_2)\boldsymbol{P}_{12} \\ \star & -(\gamma-2r_2)\boldsymbol{P}_{13} \end{bmatrix} \leq 0 \end{array} \tag{19}$$

Applying Schur complement to (19), it has

$$\begin{cases} -\alpha_2 \boldsymbol{P}_{11} \le 0 \\ -\alpha_2 \boldsymbol{P}_{13} + \Delta_1 \boldsymbol{P}_{12} + \boldsymbol{P}_{12}\Delta_1^T + \Delta_2 \boldsymbol{P}_{13} + \boldsymbol{P}_{13}\Delta_2^T + \\ \quad (-\alpha_2 \boldsymbol{P}_{12} + \Delta_1 \boldsymbol{P}_{11} + \Delta_2 \boldsymbol{P}_{12})\frac{1}{\alpha_2}\boldsymbol{P}_{11}^{-1}(-\alpha_2 \boldsymbol{P}_{12} + \boldsymbol{P}_{11}\Delta_1^T + \boldsymbol{P}_{12}\Delta_2^T) \le 0 \\ -2r_1 \boldsymbol{P}_{11} < 0 \\ -2r_2 \boldsymbol{P}_{13} + \frac{(r_1+r_2)^2}{2r_1}\boldsymbol{P}_{12}\boldsymbol{P}_{11}^{-1}\boldsymbol{P}_{12} < 0 \\ -(\gamma - 2r_1)\boldsymbol{P}_{11} \le 0 \\ -(\gamma - 2r_2)\boldsymbol{P}_{13} + \frac{(\gamma-r_1-r_2)^2}{\gamma-2r_1}\boldsymbol{P}_{12}\boldsymbol{P}_{11}^{-1}\boldsymbol{P}_{12} \le 0 \end{cases} \tag{20}$$

Due to the special structure of matrix $\boldsymbol{P}_1$, (20) can be simplified as

$$\begin{cases} p_{12} \le 0 \\ p_{12}[\frac{\alpha_2(1+\sigma_2)}{\sigma_1} I_{3\times3} + \Delta_1 + \Delta_1^T - \frac{1+\sigma_2}{\sigma_1}\Delta_2 - \frac{1+\sigma_2}{\sigma_1}\Delta_2^T - \\ \quad \frac{1}{\alpha_2\sigma_1}(\alpha_2 I_{3\times3} + \sigma_1\Delta_1^T - \Delta_2^T)(\alpha_2 I_{3\times3} + \sigma_1\Delta_1^T - \Delta_2^T)] \le 0 \\ 2r_1 r_2(1+\sigma_2) - (r_1+r_2)^2 > 0 \\ -\gamma + 2r_1 \le 0 \\ (\gamma - 2r_1)(\gamma - 2r_2)(1+\sigma_2) - (\gamma - r_1 - r_2)^2 \ge 0 \end{cases} \tag{21}$$

which can be guaranteed by the scalar inequalities (14) and the norm-bounded condition (18). The proof is completed.

4 Simulation

In this section, simulation example is presented to illustrate the effectiveness of the control strategy using ILF. Consider the case that the target spacecraft is traveling along an elliptical orbit with semimajor axis $a = 24616$ km, eccentricity $e = 0.73074$, and the initial true anomaly of target is $\theta_0 = 15°$. The mass of chaser is $m = 300$ kg, the initial relative position $\boldsymbol{r}(0) = [-600, -400, 200]^T$ and velocity $\boldsymbol{v}(0) = [3.4, -4.8, 5.3]^T$. The uncertainties of the true anomaly θ is supposed to be 10 %, and external disturbances signal is $\boldsymbol{f}_d = [20\sin(100t), 20\sin(100t), 20\sin(100t)]^T$. The control parameters are $g = 0.5$, $r_2 = 0.8$, $\gamma = 5$, $\alpha_1 = 0.6$, $\alpha_2 = 0.2$, $\beta = 0.3$, $\sigma_1 = 1.8$, $\sigma_2 = 0.5$. To avoid singularity at $V = 0$, we set a minimum value of $V = V_{\min} = 0.1$, which allows a linear continuous control near the origin and reduces the chattering.

Figures 2 and 3 show the relative states response trajectories of closed-loop system, all of which converge fast. The control force is presented in Fig. 4, in which the chatting during the last tens of seconds is caused by periodical external disturbance we assumed. The value of implicit Lyapunov function is demonstrated in Fig. 5, which keeps decreasing until the minimum value we set.

Fig. 2 The relative position r

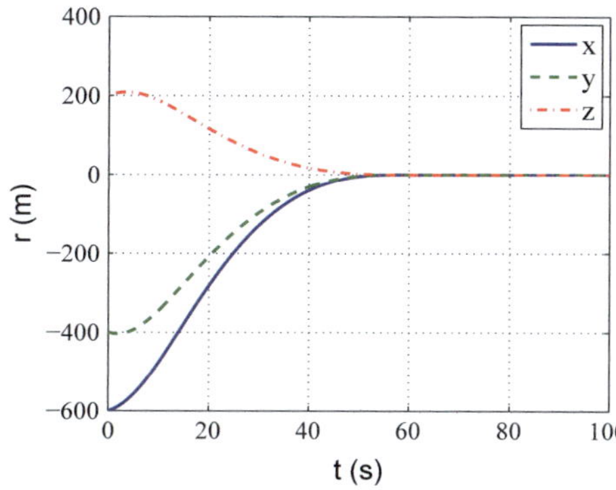

Fig. 3 The relative velocity v

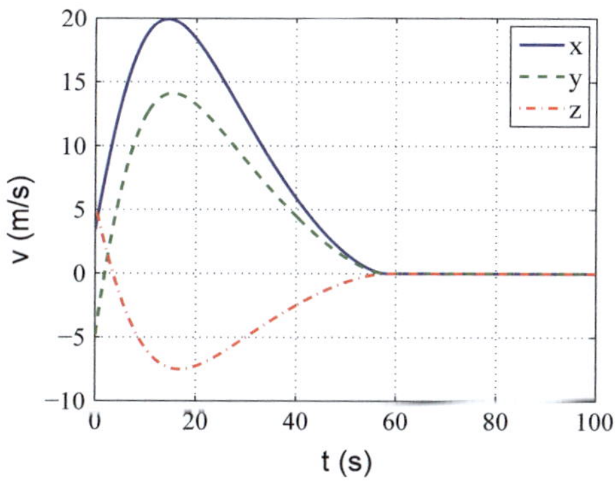

Fig. 4 The control force f

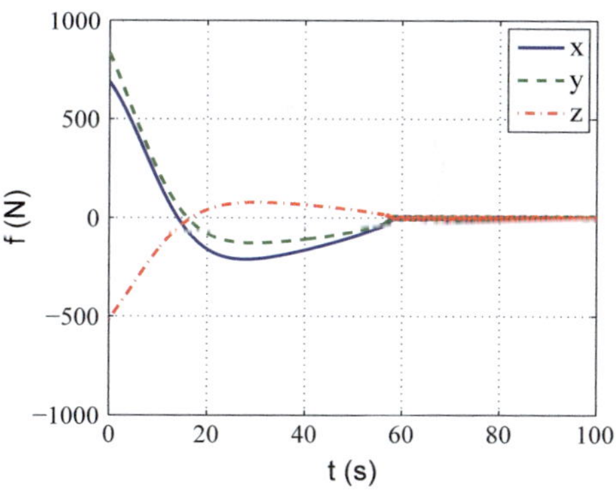

The simulation results above show that the robust control strategy using implicit Lyapunov function proposed in Sect. 3 is available and efficient, even in the presence of parameter uncertainties and external disturbance.

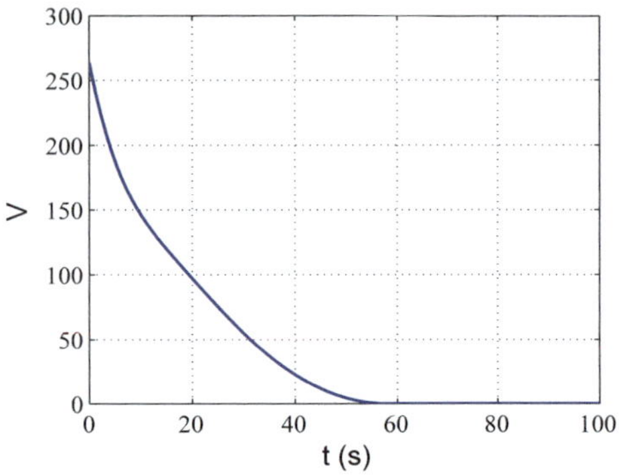

Fig. 5 The implicit Lyapunov function V

5 Conclusion

This paper proposes a robust control strategy for elliptical orbit spacecraft rendezvous system involving parameter uncertainties and external disturbances. The finite-time stability is guaranteed due to some linear parameter-variable matrix inequalities which need to be solved on-line. For practical implementation, an analytical solution of these inequalities is provided to satisfy the feasibility and reduce the on-line computation. Despite the strong robustness and fast convergence, the obtained control scheme using ILF still has some disadvantages, such as on-line computation of the ILF and singularity near the origin caused by finite numerical precision, which will be tackled in the further research.

Acknowledgments This work was supported by the National Basic Research Program of China (973 Program: 2012CB821200, 2012CB821201), the NSFC (61327807, 61521091, 61520106010, 61304232) and the Fundamental Research Funds for the Central Universities (YWF-16-GJSYS-31, YWF-16-GJSYS-32).

References

1. Fehse W (2003) Automated rendezvous and docking of spacecraft. Cambridge University Press, New York
2. Khalil HK (2002) Nonlinear system, 3rd edn. Prentice Hall, Upper Saddle River
3. Levant A (1993) Sliding order and sliding accuracy in sliding mode control. Int J Control 58(6):1247–1263
4. Bartolini G, Ferrara A, Usai E (1998) Chattering avoidance by second order sliding mode control. IEEE Trans Autom Control 43(2):241–246
5. Moreno JA (2011) Lyapunov approach to analysis and design of second order sliding mode algorithm. Slidng Modes after the First Decade of the 21st Century, Lecture Notes in Control and Information. Science 412:113–149
6. Levant A (2003) Higher-order sliding modes, differentiation and output-feedback control. Int J Control 76(9–10):924–941

7. Levant A (2005) Homogeneity approach to high-order sliding mode design. Automatica 41(5):823–830
8. Chalanga A, Kamal S, Fridman L, Bandyopadhyay B, Moreno JA (2014) How to implement super-twisting controller based on sliding mode observer?. In: 13th IEEE workshop on variable structure systems, pp 1–6
9. Utkin V (2016) Discussion aspects of high-order sliding mode control. IEEE Trans Autom Control 61(3):829–833
10. Adamy J, Flemming A (2004) Soft variable-structure controls: a survey. Automatica 40(11):1821–1844
11. Adamy J (1991) Strukturvariable regelungen mittels impliziter Ljapunov-funktionen. Ph.D. dissertation, University of Dortmund, Germany, VDI-Verlag
12. Polyakov A, Efimov D, Perruquetti W (2015) Finite-time and fixed-time stabilization: implicit Lyapunov function approach. Automatica 51:332–340
13. Lens H, Adamy J, Domont-Yankulova D (2011) A fast nonlinear control method for linear systems with input saturation. Automatica 47(4):857–860
14. Polyakov A, Efimov D, Perruquetti W (2014) Homogeneous differentiator design using implicit Lyapunov function method. In: European control conference, pp 288–293
15. Polyakov A, Efimov D, Perruquetti W, Richard JP (2015) Implicit Lyapunov-Krasovski functionals for stability analysis and control design of time-delay systems. In: IEEE transactions on automatic control, institute of electrical and electronics engineers, pp 1–6
16. Yamanaka K, Ankerson F (2002) New state transition matrix for relative motion on an arbitrary ellipcial orbit. J Guidance Control Dyn 25(1):60–66
17. Polyakov A, Fridman L (2014) Stability notions and Lyapunov functions for sliding mode control systems. J Franklin Inst 351(4):1831–1865

Dynamical Behaviors in Coupled FitzHugh-Nagumo Neural Systems with Time Delays

Yuan Zhang, Lan Xiang and Jin Zhou

Abstract It is observed that neuron encodes and integrates information employing a variety of complex dynamical behavior, such as spiking, bursting, periodicity, quasi-periodicity, and chaos. Time delay is an inevitable factor in the signal transmission between neurons, and neural system may lose its stability even for very small delay. In this paper, a model of coupled FitzHugh-Nagumo (FHN) neural system with two different delays is formulated, and its nonlinear dynamic behaviors such as stability, bifurcations, and chaos are then studied. It is shown that time delays can affect the stability of equilibrium states, and thereby lead to Hopf bifurcation and oscillation behavior. Moreover, some complex dynamics including quasi-periodic solutions and chaos are numerically demonstrated. Subsequently, numerical examples illustrate the effectiveness and feasibility of the theoretical results.

Keywords FitzHugh-Nagumo neural system · Delay · Hopf bifurcation · Quasi-periodic solution · Chaos

1 Introduction

Nonlinear phenomena appear in a wide variety of scientific applications such as plasma physics, solid state physics, optical fibers, biology, fluid dynamics, and chemical kinetics. In recent years, there has been an increase in activity and interest to explore various types of neural systems, such as Hopeld/Cohen-Grossberg neuron networks [1–4]. In 1952, Hodgkin and Huxley proposed a mathematical (HH) model that approximates the electrical characteristics of excitable cells [5]. The HH model

Y. Zhang · J. Zhou (✉)
Shanghai Institute of Applied Mathematics and Mechanics, Shanghai University, 200072 Shanghai, China
e-mail: jzhou@shu.edu.cn

L. Xiang
Department of Physics, School of Science, Shanghai University, 200444 Shanghai, China

Y. Jia et al. (eds.), *Proceedings of 2016 Chinese Intelligent Systems Conference*, Lecture Notes in Electrical Engineering 405,
DOI 10.1007/978-981-10-2335-4_28

is strongly nonlinear and complex, and its solution can be only obtained by numerical simulations. In 1961, Richard FitzHugh [6] developed an analog computer to study impulse propagation, which led him to simplify the HH model to a simpler one by modifying the van der Pol model. This model turned out to be similar to the model used by Nagumo et al. [7] to study nerve conduction, which became the famous FitzHugh-Nagumo (FHN) model. It is well known that the FHN model can well capture the general dynamics of the HH model and plays a key role in many theoretical studies [8–13]. As a result, the analysis and control of FHN neural systems have received much attention in recent years. For example, complete topological and qualitative investigation of the FHN equation with a cubic nonlinearity has been addressed by Bautin [9] and rich variety of nonlinear phenomena are observed. The dynamics of the coupled FHN neurons with symmetric and nonsymmetric coupling was considered in [10–12], in which many bifurcation behaviors for equilibrium point and limit cycle are investigated. For bursting in a set of coupled neuronal oscillators, Bekbolat Medetov et al. [13] have analyzed a coupled FHN model. An understanding of the mechanisms behind dynamical behaviors for FHN neural systems is the subject of intensive research for the last several years.

It has been observed that time delay is significant to the study of neural networks, since in real situation, the transmission of information from one neuron to another is not instantaneous. Moreover, time delay can induce instability and the occurrence of periodic oscillations, quasi-periodic solution, and even leads to multistability and chaotic motion [14–17]. Therefore, it is crucial to investigate neural networks subject to delays. Recently, dynamics of delayed FHN neural networks have received a lot of attentions [14, 15, 18–20]. For instance, Li et al. [14] introduced a time lag into a FHN neural system and investigated the effects of time delay on Hopf bifurcation and Bogdanov–Takens bifurcation. Yao and Tu [18] studied the stability switches and Hopf bifurcation in a coupled FHN neural system with multiple delays. Buric and Todorovic [19] studied the Hopf bifurcation (inverse and direct) and fold bifurcation of limit cycle in the delay-coupling FHN neurons. Regarding the sum of two delays as a parameter, Fan and Hong [20] investigated the stability and local Hopf bifurcation in the synaptically coupled nonidentical FHN model.

It should be noted that most of these previous delayed FHN neural models have tended to focus on stability and Hopf bifurcation analysis by normal form method and the center manifold theorem. To the best of our knowledge, Hopf bifurcation and chaos in delayed FHN neural systems have received little attention. Motivated by the above discussion, in this paper, we developed a delayed FHN neural model to explore how time delays affect the dynamic behavior quantitatively. The delayed model is described by the following rate equations

$$\begin{cases} \dot{x_1} = x_1 - \frac{x_1^3}{3} - y_1 + \gamma_1 x_2(t - \tau_1), \\ \dot{y_1} = \varepsilon_1(x_1 + a_1), \\ \dot{x_2} = x_2 - \frac{x_2^3}{3} - y_2 + \gamma_2 x_1(t - \tau_2), \\ \dot{y_2} = \varepsilon_2(x_2 + a_2), \end{cases} \tag{1}$$

where x_i and y_i, $i = 1, 2$, represent the voltage across the cell membrane, and the recovery state of the resting membrane of a neuron, respectively. a_1, a_2, ε_1, ε_2, γ_1 and γ_2 are parameters, and γ_1 and γ_2 are the coupling strengths between the network elements. The small positive value of $\varepsilon_i (i = 1, 2)$ ensures the separation of dynamical timescales between x_i and y_i, and the value of the excitability parameter a_i decides whether the cell is in the excitatory (stable equilibrium, $|a_i| > 1$) or in the oscillatory ($|a_i| < 1$) state. $\tau_{1,2} > 0$ represent the time delays in signal transmission between the neurons.

In the following, we shall discuss the stability, local Hopf bifurcation, and chaos of system (1) continuously. By using stability theory and Hopf bifurcation techniques, the sufficient conditions for the oscillation of the neural network are derived. Particularly, when choosing the sum of delays as bifurcation parameter, the delayed FHN system can exhibit sustained oscillations, quasi-periodic solutions and chaotic behaviors. This study might be helpful to the comprehension of FHN neurons and other neural network systems.

This paper is organized as follows. In Sect. 2, we provide some basic lemmas and obtain the main results on the stability and Hopf bifurcation in the system (1). Numerical simulations are carried out to illustrate the theoretical prediction and to explore the complex dynamics including chaos in Sect. 3. Conclusions are finally drawn in Sect. 4.

2 Local Stability and Hopf Bifurcation

Oscillation is ubiquitous in biological nervous system. It plays a major role in motor, sensory, and even cognitive function. Hopf bifurcations, including supercritical and subcritical Hopf bifurcations, are the key principle of designing biochemical oscillators. In this section, based on Hopf bifurcation theory, we investigate the effect of the time delays on Hopf bifurcation of system (1).

Obviously, the system (1) has a unique equilibrium state at

$$x_{1*} = -a_1, \qquad y_{1*} = \frac{a_1^3}{3} - a_1 - \gamma_1 a_2,$$
$$x_{2*} = -a_2, \qquad y_{2*} = \frac{a_2^3}{3} - a_2 - \gamma_2 a_1.$$

Let

$$\overline{x_1}(t) = x_1(t) - x_{1*}, \quad \overline{y_1}(t) = y_1(t) - y_{1*},$$
$$\overline{x_2}(t) = x_2(t) - x_{2*}, \quad \overline{y_2}(t) = y_2(t) - y_{2*}.$$

And still denote $\overline{x_1}(t), \overline{y_1}(t), \overline{x_2}(t), \overline{y_2}(t)$ by $x_1(t), y_1(t), x_2(t), y_2(t)$, then the system (1) is equivalent to the following system:

$$\begin{cases} \dot{x_1} = Mx_1(t) - y_1(t) + \gamma_1 x_2(t-\tau_1) \\ \qquad - x_{1*}x_1^2(t) - \frac{x_1^3(t)}{3}, \\ \dot{y_1} = \varepsilon_1 x_1(t), \\ \dot{x_2} = Qx_2(t) - y_2(t) + \gamma_2 x_1(t-\tau_2) \\ \qquad - x_{2*}x_2^2(t) - \frac{x_2^3(t)}{3}, \\ \dot{y_2} = \varepsilon_2 x_2(t), \end{cases} \tag{2}$$

where

$$M = 1 - x_{1*}^2, \qquad Q = 1 - x_{2*}^2.$$

Then the unique equilibrium $(x_{1*}, y_{1*}, x_{2*}, y_{2*})$ of system (1) is transformed into the zero equilibrium $(0,0,0,0)$ of system (2). It is easy to see that characteristic equation of the linearization system of system (2) at the zero equilibrium $(0,0,0,0)$ is

$$\lambda^4 + B_1\lambda^3 + B_2\lambda^2 + B_3\lambda + B_4 + A_1\lambda^2 e^{-\lambda\tau} = 0, \tag{3}$$

where

$$B_1 = -M - Q, \quad B_2 = MQ + \varepsilon_1 + \varepsilon_2,$$
$$B_2 = -Q\varepsilon_1 - M\varepsilon_2, \quad B_4 = \varepsilon_1\varepsilon_2,$$
$$A_1 = -\gamma_1\gamma_2, \quad \tau = \tau_1 + \tau_2 .$$

In order to investigate the distribution of roots of the transcendental equation (3), we introduce the following result from Ruan and Wei [21].

Lemma 2.1 *For the transcendental equation*

$$\begin{aligned} &p(\lambda, e^{-\lambda\tau_1}, \ldots, e^{-\lambda\tau_m}) \\ &= \lambda^n + p_1^{(0)}\lambda^{n-1} + \cdots + p_{n-1}^{(0)}\lambda + p_n^{(0)} \\ &+ [p_1^{(1)}\lambda^{n-1} + \cdots + p_{n-1}^{(1)}\lambda + p_n^{(1)}]e^{-\lambda\tau_1} + \cdots \\ &+ [p_1^{(m)}\lambda^{n-1} + \cdots + p_{n-1}^{(m)}\lambda + p_n^{(m)}]e^{-\lambda\tau_m} \\ &= 0, \end{aligned}$$

as $(\tau_1, \tau_2, \ldots, \tau_m)$ vary, the sum of orders of the zeros of $p(\lambda, e^{-\lambda\tau_1}, \ldots, e^{-\lambda\tau_m})$ in the open right half plane can change, and only a zero appears on or crosses the imaginary axis.

Next, we consider the local stability of the interior equilibrium $(x_{1*}, y_{1*}, x_{2*}, y_{2*})$ and the existence of Hopf bifurcation. It is well known that the stability of the zero equilibrium $(0,0,0,0)$ of system (2) is determined by the real parts of the roots of Eq. (3). If all roots of Eq. (3) locate the left-half complex plane, then the zero equilibrium $(0,0,0,0)$ of system (2) is asymptotically stable. If Eq. (3) has a root with positive real part, then the zero solution is unstable. Therefore, to study the stability of the zero equilibrium $(0, 0, 0, 0)$ of system (2), an important problem is to investigate the distribution of roots in the complex plane of the characteristic equation (3).

When $\tau = 0$, the characteristic equation (3) becomes

$$\lambda^4 + B_1\lambda^3 + (B_2 + A_1)\lambda^2 + B_3\lambda + B_4 = 0. \tag{4}$$

For equation (4), according to the Routh–Hurwitz criterion, we have the following result.

Lemma 2.2 *When* $\tau = 0$, A_1 *and* $B_k (k = 1, 2, 3, 4)$ *satisfy the condition:*

$$\begin{aligned}(H1) \quad & B_1 > 0, B_4 > 0, \quad B_1(B_2 + A_1) - B_3 > 0,\\ & B_1(B_2 + A_1)B_3 - B_3^2 - B_4B_1^2 > 0,\end{aligned}$$

then all roots of equation (3) have negative real parts, and hence the zero equilibrium $(0, 0, 0, 0)$ *of system (2) is asymptotically stable.*

Next, we consider the effects of a positive delay τ on the stability of the zero equilibrium $(0, 0, 0, 0)$ of system (2). Since the roots of the characteristic equation (3) depend continuously on τ, a change of τ must lead to change of the roots of Eq. (3). If there is a critical value of τ such that a certain root of Eq. (3) has zero real part, then at these critical values the stability of the zero equilibrium $(0, 0, 0, 0)$ of system (2) will switch, and under certain conditions a family of small amplitude periodic solutions can bifurcate from the zero equilibrium $(0, 0, 0, 0)$; that is, a Hopf bifurcation occurs at the zero equilibrium $(0, 0, 0, 0)$.

Considering the transcendental equation (3), clearly, $i\omega(\omega > 0)$ is a root of the Eq. (3) if and only if ω satisfies the following equation:

$$\begin{aligned}&\omega^4 - B_1 i\omega^3 - B_2\omega^2 + B_3 i\omega + B_4\\ &-A_1\omega^2\cos(\omega\tau) - i\sin(\omega\tau) = 0.\end{aligned} \tag{5}$$

Separating the real and imaginary parts, we have

$$\begin{cases}\omega^4 - B_2\omega^2 + B_4 = A_1\omega^2\cos(\omega\tau),\\ B_1\omega^3 - B_3\omega = A_1\omega^2\sin(\omega\tau).\end{cases} \tag{6}$$

Adding up the squares of both equations of (6), we obtain

$$\omega^8 + B_{11}\omega^6 + B_{12}\omega^4 + B_{13}\omega^2 + B_{14} = 0, \tag{7}$$

where

$$\begin{aligned}B_{11} = B_1^2 - 2B_2, \quad B_{12} = -A_1^2 + B_2^2 - 2B_1B_3 + 2B_4,\\ B_{13} = B_3^2 - 2B_2B_4, \quad B_{14} = B_4^2.\end{aligned}$$

Let $z = \omega^2$, then Eq. (7) can be denoted simply as the following equation:

$$z^4 + B_{11}z^3 + B_{12}z^2 + B_{13}z + B_{14} = 0. \tag{8}$$

In order to clearly analyze the existence and distribution of roots of Eq. (8), we denote

$$h(z) = z^4 + B_{11}z^3 + B_{12}z^2 + B_{13}z + B_{14}, \tag{9}$$

from Eq. (9), we can get

$$h'(z) = 4z^3 + 3B_{11}z^2 + 2B_{12}z + B_{13}. \tag{10}$$

Let $y = z + \frac{B_{11}}{4}$, then equation $4z^3 + 3B_{11}z^2 + 2B_{12}z + B_{13} = 0$ becomes

$$y^3 + p_1 y + q_1 = 0, \tag{11}$$

where

$$p_1 = \tfrac{B_{12}}{2} - \tfrac{3}{16}B_{11}^2, \quad q_1 = \tfrac{B_{11}^3}{32} - \tfrac{B_{11}B_{12}}{8} + \tfrac{B_{13}}{4}.$$

Define

$$\Delta = (\frac{q_1}{2})^2 + (\frac{p_1}{3})^3, \quad \varepsilon = \frac{-1 + i\sqrt{3}}{2},$$

$$y_1 = \sqrt[3]{-\frac{q_1}{2} + \sqrt{\Delta}} + \sqrt[3]{-\frac{q_1}{2} - \sqrt{\Delta}},$$

$$y_2 = \sqrt[3]{-\frac{q_1}{2} + \sqrt{\Delta}}\,\varepsilon + \sqrt[3]{-\frac{q_1}{2} - \sqrt{\Delta}}\,\varepsilon^2,$$

$$y_3 = \sqrt[3]{-\frac{q_1}{2} + \sqrt{\Delta}}\,\varepsilon^2 + \sqrt[3]{-\frac{q_1}{2} - \sqrt{\Delta}}\,\varepsilon.$$

Denote $z_i = y_i - \frac{B_{11}}{4} (i = 1, 2, 3)$. Then we can get the following lemma.

Lemma 2.3 *(I) If* $\Delta \geq 0$*, then (8) has positive roots if and only if* $z_1 > 0$ *and* $h(z_1) < 0$*;*
(II) If $\Delta < 0$*, then (8) has positive roots if and only if there exists at least one* z^* $\in \{z_1, z_2, z_3\}$ *such that* $z^* > 0$ *and* $h(z^*) \leq 0$.

Suppose that (8) has four positive roots, denoted by $z_k^* (k = 1, 2, 3, 4)$. Then (7) has four positive roots $\omega_k = \sqrt{z_k^*} (k = 1, 2, 3, 4)$.

Let

$$\tau_k^j = \frac{1}{\omega_k} \arccos\left\{\frac{\omega_k^4 - B_2\omega_k^2 + B_4}{\omega_k^2 A_1}\right\} + \frac{2j\pi}{\omega_k}, \tag{12}$$
$$k = 1, 2, 3, 4; j = 1, 2, 3, \ldots.$$

Therefore, $\pm i\omega_k$ is a pair of purely imaginary roots of (3) with $\tau = \tau_k^j$, and define $\tau_0 = min_{k\in\{1,2,3,4\}}\{\tau_k^0\}$, $\omega_0 = \omega_{k_0}$.

According to the Hopf bifurcation theorem, we need to verify the transversal condition, differentiating (3) with respect to τ, and noticing that λ is a function of τ, we can obtain

$$(\frac{d\lambda}{d\tau})^{-1} = -\frac{-2B_4 - B_3\lambda + B_1\lambda^3 + 2\lambda^4}{\lambda^2(B_4 + B_3\lambda + B_2\lambda^2 + B_1\lambda^3 + \lambda^4)} - \frac{\tau}{\lambda}.$$

Hence, a direct calculation shows that

$$Re(\frac{d\lambda}{d\tau})^{-1}_{\lambda=i\omega_0} > 0.$$

Therefore, the transversal condition holds and a Hopf bifurcation occurs at $\tau = \tau_0$. Then, we have the following results.

Theorem 2.1 *For system (1), assume that (H1) is satisfied. There exists a positive number τ_0 such that the equilibrium $(x_{1*}, y_{1*}, x_{2*}, y_{2*})$ is locally asymptotically stable when $\tau \in [0, \tau_0)$ and unstable when $\tau > \tau_0$. Furthermore, Eq. (1) undergoes a Hopf bifurcation at $(x_{1*}, y_{1*}, x_{2*}, y_{2*})$ when $\tau = \tau_0$.*

This theorem provides a sufficient time delay to generate oscillations in the neural network. It is shown that the total delay time of τ_1 and τ_2 is the key determinant of the oscillations. Moreover, the values of the other parameters provided that the conditions for sustained oscillation are satisfied.

3 Numerical Simulations

In this section, we will give numerical simulations to show Hopf bifurcation and to explore the possibility of quasi-periodic solutions and chaotic behavior in system (1). Here, we consider system (1) with the coefficients $a_1 = 0.96684, a_2 = 1.5, \varepsilon_1 = \varepsilon_2 = 0.1, \gamma_1 = 2, \gamma_2 = -0.62$. That is

$$\begin{cases} \dot{x_1} = x_1 - \frac{x_1^3}{3} - y_1 + 2 \times x_2(t - \tau_1), \\ \dot{y_1} = 0.1 \times (x_1 + 0.96684), \\ \dot{x_2} = x_2 - \frac{x_2^3}{3} - y_2 - 0.62 \times x_1(t - \tau_2), \\ \dot{y_2} = 0.1 \times (x_2 + 1.5). \end{cases} \tag{13}$$

Under the above parameter values, system (13) has an unique equilibrium (−0.96684, −3.66558, −1.5, 0.22444). Moreover, it is easy to check that conditions (H1) is satisfied.

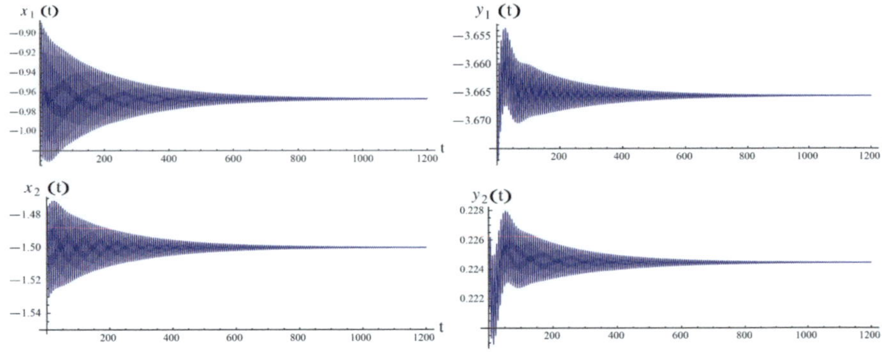

Fig. 1 The numerical approximations of system (13) when $(\tau_1 + \tau_2) = 0.96 < \tau_0$. The equilibrium $(x_{1*}, y_{1*}, x_{2*}, y_{2*})$ is asymptotically stable

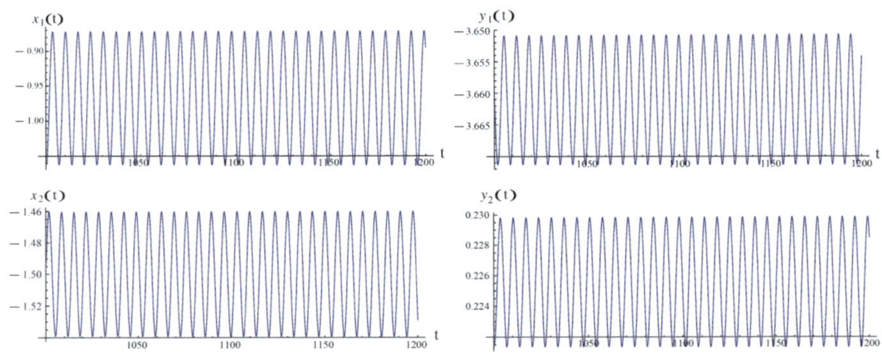

Fig. 2 The numerical approximations of system (13) when $(\tau_1 + \tau_2) = 0.98 > \tau_0$. The equilibrium $(x_{1*}, y_{1*}, x_{2*}, y_{2*})$ is unstable and a stable periodic solution bifurcates from $(x_{1*}, y_{1*}, x_{2*}, y_{2*})$

Without the effects of delays $(\tau_1 = \tau_2 = 0)$, the processes of transmission of information from one neuron to another are instantaneous events. According to the Routh–Hurwitz criterion, it indicates that the unique equilibrium $(-0.9668, -3.66558, -1.5, 0.22444)$ is asymptotically stable.

Under the regulation of delays, the time courses of x_1, y_1, x_2, and y_2 corresponding to the parameter $\tau = \tau_1 + \tau_2$ are shown in Figs. 1, 2, 3, 4, 5, 6 and 7. By means of the software Mathematica 8.0, we can obtain $\omega_0 = 0.932513$ and $\tau_0 = 0.974151$. From Fig. 1, it can be seen that the equilibrium point (–0.9668, –3.66558, –1.5, 0.22444) of system (13) is asymptotically stable when $0 < \tau < \tau_0$. As τ increases and passes through the critical delay τ_0, system (13) loses its stability and a Hopf bifurcation occurs, i.e., a family of periodic solutions bifurcate from the equilibrium as shown in Fig. 2, which leads to a series of sustained oscillations of x_1, y_1, x_2 and y_2.

Figures 3 and 4 respectively show that there exist stable quasi-periodic solutions and multi-humped periodic solutions in this neural system (13). Moreover, As τ continues to increase, we can find that chaotic behaviors occur. From Figs. 5 and 6, it

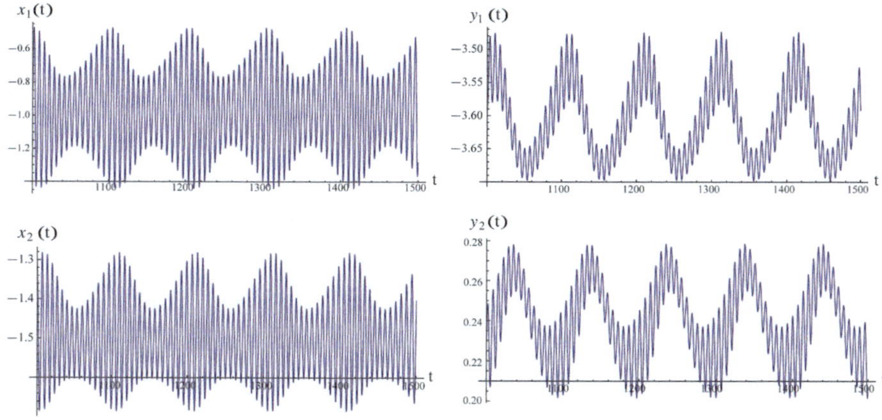

Fig. 3 Quasi-periodic solutions of system (13) with τ = 1.06

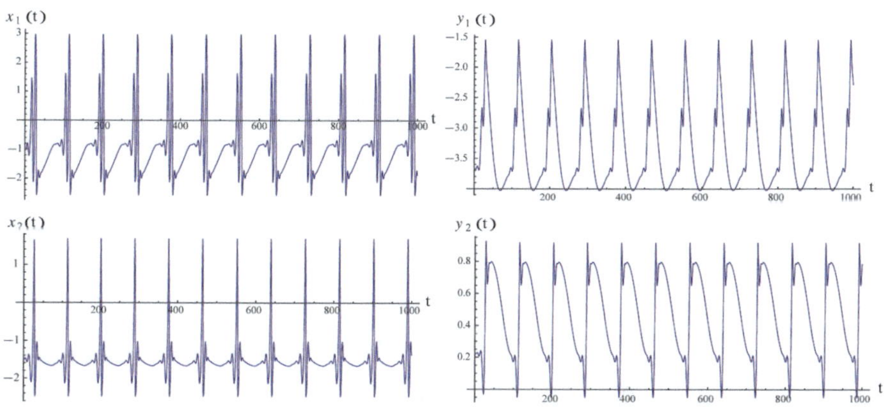

Fig. 4 Multi-humped periodic solutions of system (13) with τ = 2.8

can be seen that the trajectory of dynamical behavior is irregular and complicated, which is in consistence with the chaotic behaviors.

The unique character of chaotic dynamics may be seen most clearly by hyper sensitivity to initial conditions. That is, a small difference in initial conditions may lead to significantly different dynamic behaviors. For verifying the chaos, we perturbed the initial conditions to test the sensitivity of the system (13). Figure 7 shows a comparison of the solutions for the system (13) with initial conditions (–0.9668, –3.66557, –1.5501, 0.2244) and (–0.9668, –3.66557, –1.55, 0.2244). Near the initial time the two solutions appear to be the same, but as time increases there is a marked difference between the solutions supporting that the system (13) behavior is chaotic. According to the above analysis, these numerical results indicate that the FHN neuron system can exhibit the chaotic behaviors.

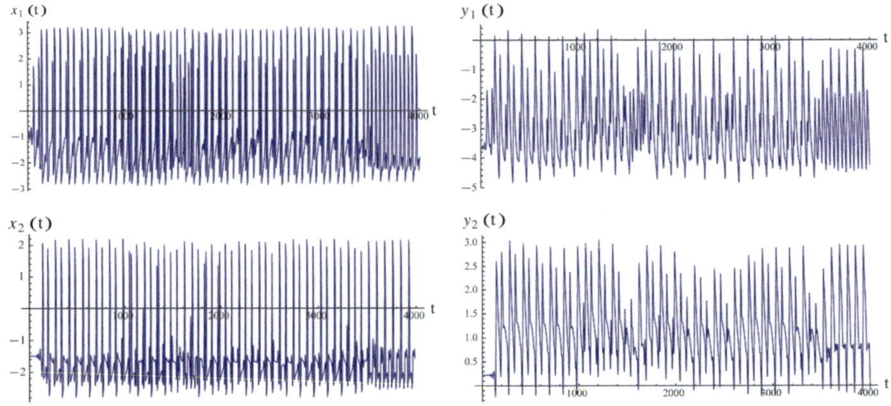

Fig. 5 Chaotic behavior of system (13) with $\tau = 69.5$

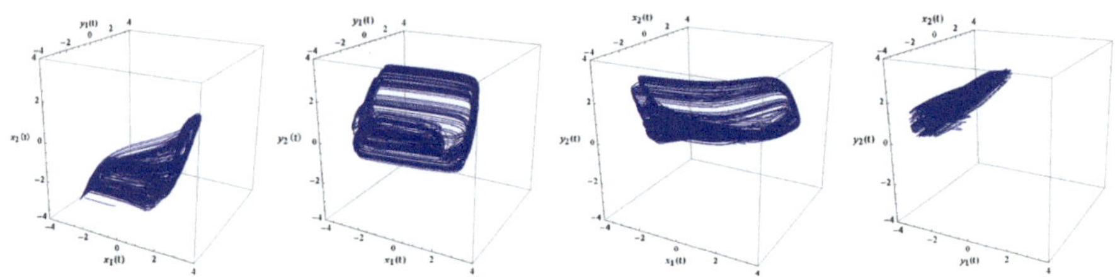

Fig. 6 Chaotic attractor of system (13) with $\tau = 69.5$

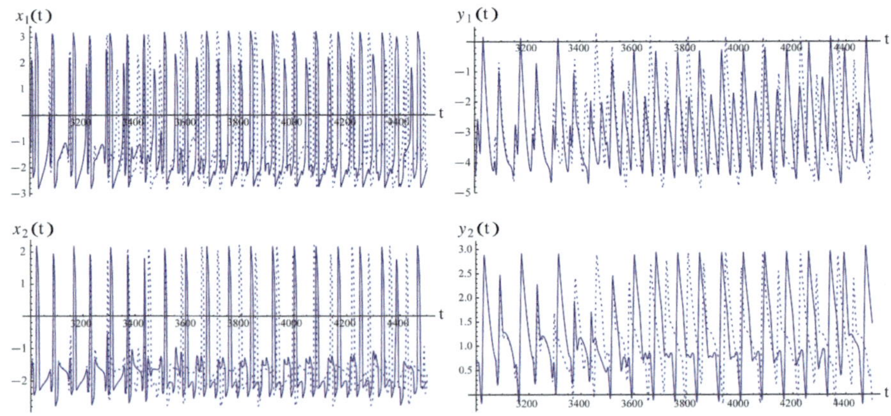

Fig. 7 Sensitive dependence on initial conditions: *solid line* and *dotted line* respectively represent the trajectories of system (13) ($\tau = 69.5$) with initial conditions $(-0.9668, -3.66557, -1.55, 0.2244)$ and $(-0.9668, -3.66557, -1.5501, 0.2244)$

4 Conclusions

In order to describe the intracellular process more precisely, we introduce time delays on a FHN neural system to account for the processes of transmission of information and explore how the time delays in the neural network can affect the dynamic behavior quantitatively. It is shown that (from Figs. 1, 2, 3, 4, 5 and 6), with the increase of time delay parameter, the attractor of system experience equilibrium point, periodic solution, quasi-periodic solution, multi-humped periodic solution and chaos successively. It turns out that time delays play an important role in determining the dynamic behavior quantitatively in the neural network. Since the transmission of information from one neuron to another is not instantaneous, considerable time delays are ubiquitous in all the neural processes. This study might provide some clues or helps to the fields of neuroscience and biomedicine.

Acknowledgments The authors wish to thank the editor and the reviewers for their insightful and constructive comments, which improved this paper significantly. This work is supported by the National Science Foundation of China (No. 11272191).

References

1. Hopfield JJ (1982) Neural networks and physical systems with emergent collective computational properties. Proc Nat Acad Sci 79:2554–2558
2. Hopfield JJ (1984) Neuronswith graded response have collective computational properties like those of two state neurons. Proc Nat Acad Sci 81:3088–3092
3. Hopfield JJ, Tank DW (1986) Computing with neural circuits: a model. Science 233:625–633
4. Hodgkin AL, Huxley AF (1952) A quantitative description of membrane current and its application to conduction and excitation in nerve. J Physiol 117(4):500–544
5. Hodgkin AL, Huxley AF (1952) A quantitative description of membrane and its application to conduction and excitation in nerve. J Physiol 117:500–44
6. FitzHugh R (1961) Impulses and physiological state in theoretical models of nerve membrane. Biophys J 1:445–466
7. Nagumo J, Arimoto S, Yoshizawa S (1962) An active pulse transmission line simulating nerve axon. Proc IRE 50:2061–2070
8. Kim S, Park SH, Ryu CS (1997) Multistability in coupled oscillator systems with time delay. Phys Rev Lett 79:2911–2914
9. Bautin AN (1975) Qualitative investigation of a particular nonlinear system. J Appl Math Mech 39(4):606–615
10. Ueta T, Miyazaki H, Kousaka T, Kawakami H (2004) Bifurcation and chaos in coupled BVP oscillators. Int J Bifurcat Chaos Appl Sci Eng 14(4):1305–1324
11. Ueta T, Kawakami H (2003) Bifurcation in asymmetrically coupled BVP oscillators. Int J Bifurcat Chaos Appl Sci Eng 13(5):1319–1327
12. Tsuji S, Ueta T, Kawakami H (2007) Bifurcation analysis of current coupled BVP oscillators. Int J Bifurcat Chaos Appl Sci Eng 17(3):837–850
13. Medetov B, Wei RC (2015) Numerically induced bursting in a set of coupled neuronal oscillators. Commun Nonlinear Sci Numer Simulat 20:1090–1098
14. Li YQ, Jiang WH (2011) Hopf and Bogdanov-Takens bifurcations in a coupled FitzHugh-Nagumo neural system with delay. Nonlinear Dyn 65:161–173

15. Jia JY, Liu HH, Xu CL, Yan F (2015) Dynamic effects of time delay on a coupled FitzHugh-Nagumo neural system. Alexandria Eng J 54:241–250
16. Xu X, Hu HY, Wang HL (2006) Stability switches, Hopf bifurcation and chaos of a neuron model with delay-dependent parameters. Phys Lett A 354:126–136
17. Lu HT (2002) Chaotic attractors in delayed neural networks. Phys Lett A 298:109–116
18. Yao SW, Tu HN (2014) Stability Switches and Hopf Bifurcation in a Coupled FitzHugh-Nagumo Neural System with Multiple Delays. Hindawi Publishing Corporation Abstract and Applied Analysis Vol 2014. Article ID 874701, 13 pp
19. Buric N, Todorovic D (2003) Dynamics of FitzHugh-Nagumo excitable systems with delayed coupling. Phys Rev E 67(Article ID 066222)
20. Fan D, Hong L (2010) Hopf bifurcation analysis in a synaptically coupled FHN neuron model with delays. Commun Nonlinear Sci Numer Simul 15(7):1873–1886
21. Ruan SG, Wei JJ (2003) On the zeros of transcendental functionals with applications to stability of delay differential equations with two delays. Dyn Contin Discrete Impuls Syst Ser A Math Anal 10:863–874

A Novel Algorithm Based on Avoid Determining Noise Threshold in DENCLUE

Ke Zhang, Yingzhi Xiong, Lei Huang and Yi Chai

Abstract This paper focuses on density-based clustering analysis. The determination of noise threshold set in DENCLUE is studied via analyzing several typical density-based clustering methods. An improved algorithm which does not use the noise threshold in DENCLUE is proposed based on the estimation of points in inner cluster. Compared to the original DENCLUE, smaller silhouette coefficients can be obtained from the proposed algorithm via experimental verification. Meanwhile, the noise in data sets can also be verified well in our method, which can be viewed as an improvement for applicability and performance of DENCLUE.

Keywords Density-based clustering · DENCLUE · Noise threshold · Estimation of points in inner cluster

1 Introduction

Clustering analysis is one of the most significant techniques in data mining area, which are formed by grouping data that have maximum similarity with other objects within the clusters, and minimum similarity with objects in other group [1]. At present, clustering analysis, a very hot research direction, is widely used in various areas of science, theoretical research, and engineering practice.

Existing mainstream clustering methods can be divided into five types: the clustering method based on partitioning, hierarchical-based clustering, grid-based

K. Zhang (✉)
School of Automation, Chongqing University, Chongqing 400044, China
e-mail: smeta@163.com

Y. Xiong · L. Huang · Y. Chai
State Key Laboratory of Power Transmission Equipment and System Security, Chongqing University, Chongqing 400030, China

Y. Chai
New Technology, College of Automation, Chongqing University, Chongqing 400030, China

Y. Jia et al. (eds.), *Proceedings of 2016 Chinese Intelligent Systems Conference*, Lecture Notes in Electrical Engineering 405,
DOI 10.1007/978-981-10-2335-4_29

clustering, model-based clustering and density-based clustering method [2, 3]. Different from other types of methods, for density-based clustering, first, different objects are divided by their distances in metric space. Second, different clusters are distinguished by density of the distribution of data sets. In this kind of algorithm, a dense area is regarded as a potential clusters. Data objects are all found by seeking for density connected region, making the clustering results complete enough.

Generally, advantages of density-based clustering are that it can discover not only globular clusters with similar size and density or non-convex kind of cluster, but also the clustering result can be more objective to reflect the distribution of data. Considering the actual situation, arbitrary shape of clusters is very strong. Dense sample points are discovered and classified to different categories by filtering out the area of low density in density-based clustering, making the shape of the cluster is not completely affected by data object in the metric space distance [4]. The degree of density of sample points be considered as the benchmark of clustering, and we can classify them as relevant clusters as long as a point of area of the density is greater than a certain threshold.

DENCLUE is such an efficient density-based clustering. Individual data objects are influenced by using kernel density function, and modeled by influence of point set and the total density. Such a concentration of thinking helps to find the data distribution of densely populated areas. Also, variable density data, irregular shape cluster, density difference, and the common problems in clustering can be solved well with DENCLUE. Since it is divided into clusters according to the density attractor, it needs to find the density of each object corresponding to the step attractor. Large amount of calculation of the estimation of each density attractor kernel density (commonly Gaussian kernel) will be caused. Thus, computing speed is its disadvantage.

To solve this issue, Hinneburg and Gabriel [5] proposed a method to control slow convergence speed to avoid the problem it can not reach local maximum point by introducing a Gaussian kernel as new mountain climbing algorithm, at the cost of the small precision, realizing the adaptive step length. To improve the similarity between objects, in Suganya and Nagarajan [6] proposed a DENCLUE method based on fuzzy theory. A Chang et al. [7] proposed a strong robust DENCLUE method without noise threshold via niching genetic algorithm. However, cluster centers can be identified and data objects can not be divided into clusters. In [8], Chonghui Guo optimized DENCLUE density function through introducing nichePSO. This method can solve the problem in [7]. In Sree [9], aimed at the problems of network information similarity measure, presented a kind of sequence similarity measure method, combined with network access sequence and the page information, verified the result whether DENCLUE can cluster the network information better through the experiments. In [10], according to online data and real-time fault diagnosis requirements, Zhu Liang solved poor robustness problem of DENCLUE for different data by substituting distance coefficients with similarity coefficients, which is simple and efficient in implementing and is particularly suitable for online. In [11], for optimization problem of smoothing parameter, Yan Jun made a probability density function can more accurately reflect the clustering center and the probability density change trend and got good effect of clustering by embedding optimal window algorithm into DENCLUE. In general, in order to improve the efficiency of DENCLUE, we want to be able to get accurate rapid method for determining of the preset parameters.

This paper proposes an algorithm to avoid determining noise threshold in DENCLUE. Noise threshold in DENCLUE is replaced by controlling convergence speed and inner clusters points estimation. First, by controlling the rate of convergence to complete the initial clustering, not enough dense collection (noise) can be filtered out by the inner cluster points estimation. Second, estimation of noise threshold can be avoided based on the inner cluster points analysis. Our method can avoid the comparison of kernel density estimation and noise threshold, and manually adjusting parameters, reducing the clustering time. Through the experiment, silhouette coefficients in our method is more compact than the original DENCLUE approach, noise points near the cluster are easily identified.

This paper is organized as follow. In Sect. 2, existing problems of original DENCLUE algorithm is given out. In Sect. 3, our method is introduced in detail. Several experiments are conducted for the original DENCLUE algorithms and the proposed method in Sect. 4. Discussions are summarized in Sect. 5.

2 Existing Problems in DENCLUE

DENCLUE method introduces the probability and statistical problems of non-parametric density estimation and describes the influence degree of each object to other objects through the influence function with a kernel function. Each kernel density estimation of objects is the sum of the object under the influence of all other objects. Such density measure makes the object located in the densely populated area near the center to obtain higher values. And every data object can reflect the location and the density contribution of surrounding area to the whole data set. So, in the case of unknown data distribution (the actual data sets is often a random sampling of whole data sets), this evaluation method, based on grid clustering, can objectively reflect the distribution of space object [12, 13].

DENCLUE, however, density attractor for every point has to be calculated. The kernel density estimation also has to be computed with their density attractor; this makes the computational complexity exponentially with the increase of data objects. Quantity of clustering speed will be slow especially for large data sets. At the same time, three parameters of DENCLUE need to be set: controlling convergence speed δ, smoothing parameter h, noise threshold ξ, in which δ effects seeking step length of density attractor, noise threshold will decide if density attractor can form a cluster center, and is a key of whether two clusters can be combined to a cluster. If the noise threshold is too large, some points within the cluster also can be considered as noise points, can also cause cluster that should be the same class is incorrectly divided into two categories. On the contrary, noise can be mistaken for the point inside the cluster, if the noise threshold is too small, this

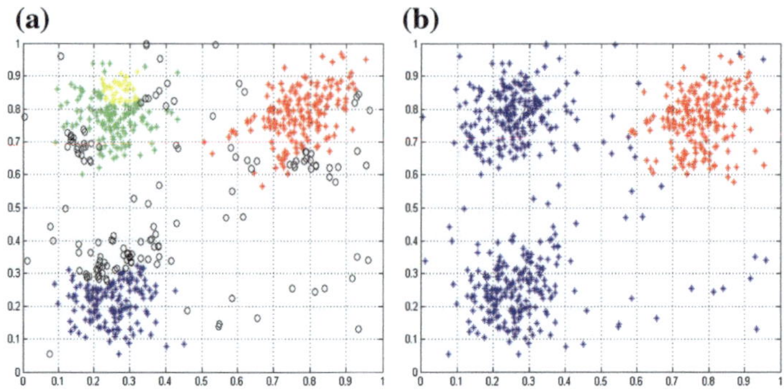

Fig. 1 The influence of different noise threshold for clustering results. **a** noise threshold is too large; **b** noise threshold is too small

also may cause the clusters, which were not of the same class, be classified into one category.

Shown in Fig. 1, data set has a typical Gaussian distribution characteristic (different color presents different cluster, black hollow point represents noise). When noise threshold is too large, there exists points in each cluster which are considered as noise points, cluster in top left of the regional is divided into two clusters. When noise threshold is too small, noise points in the regional can not be identified, and two different clusters in left side is divided into the same cluster.

According to the definition of DENCLUE, it is not hard to find that noise threshold is constrained by smoothing parameter, and controlling convergence speed should be adjusted when it changes. Therefore, the effectiveness of the clustering results will be directly affected by deviation of parameters setting. Especially in the application of large data sets, while lack of rich and powerful prior knowledge, artificial parameter deviation will make clustering process longer, and even lead to can not achieve results.

To solve this problem, we modify the DENCLUE. The core insight is to avoid frequency artificial adjustment of noise threshold. This method only need to focus on the choice of controlling convergence speed, and judgment of dense area is determined by the distribution of data sets. Another benefit is that, reducing the computational complexity and saving the running time of algorithm by abandoning the noise threshold and shunning comparison between kernel density estimation and noise threshold.

3 Methodology to Avoid Determining Noise Threshold

In three artificial parameters of DENCLUE, steep degree of kernel density estimation function is affected by smoothing parameter h. The smaller h is, the more steep the estimation function is, and vice versa [5] (see Fig. 2). In the case of

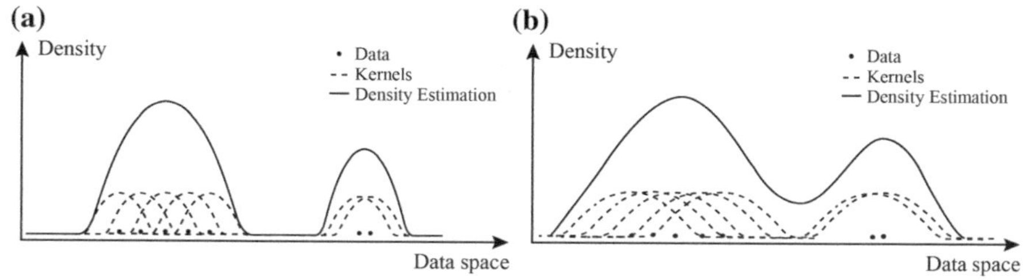

Fig. 2 Kernel density estimation curve with different smoothing parameters. **a** Smaller h; **b** bigger h

smaller h, if the value ξ is larger, the same cluster is divided into several smaller clusters; On the contrary, different clusters can not be effectively distinguished. That is, in the original DENCLUE method, the size of the ξ is constrained by smoothing parameter h, this increases the difficulty of parameters setting.

On the other hand, Noise threshold ξ and controlling convergence speed δ have remarkable influences on the clustering results. δ is step of seeking for density attractor, in which bigger one may miss density attractor and smaller one may lead too much steps, making the method time consuming. The function of noise threshold ξ mainly reflects in two aspects:

(1) Distinguish the noise points in data sets. The point is considered to be noise point if kernel density estimation of density attractor in each point of data sets is less than ξ.
(2) Judge whether two clusters existed density attractor can be combined into a cluster. If there is a path between the two density attractor, kernel density estimate is greater than or equal to ξ, then these two clusters should be regarded as the same cluster.

In original DENCLUE, comparison between noise threshold and the kernel density estimation is essential. It has high computational complexity when judging density attractor of each point on the path is greater than or equal to the noise threshold. But given such a situation: each cluster is attracted by its density attractor, and also has interaction among different density attractor, the farther the density attractor are, the less attractive each other does. To address this problem, the author of DENCLUE proposed a method to judge whether two clusters can be combined to one cluster by analyzing controlling the rate of convergence [5]. Setting a small δ as cluster distance threshold, two density attractor of the corresponding cluster can be combined into one cluster if the distance between the two density attractor is less than 2δ. But unfortunately, how to set an appropriate δ value is still unknown in the literature.

According to this insight, controlling convergence speed is adopted to judge whether different clusters should be combined into the same cluster instead of noise threshold value. As for no noise data, let ξ to 0 directly, then it can get good clustering effect by choosing an appropriate δ value. The noise threshold can be used to identify the noise points for data sets existing potential noise. Therefore, if there exists a condition which can replace noise threshold value method to judge whether the object is noise, we can avoid determining noise threshold for DENCLUE. Noise is, by definition, random error or variance of measured variables, and has significant difference with data objects within the cluster [14]. Typically, the noise points can be recognized as outliers, and it does not belong to any cluster. Even some noise points can be divided into a collection, the number of them should also be very few (its density contribution is significantly small to the whole data sets). That is, let $\xi = 0$ in DENCLUE, noise is divided in each sets, in which the number of objects should be no greater than the minimum quantity of the objects number in clusters. Therefore, noise identification can be converted into solving problem of determining objects number within the cluster.

Estimation of cluster points is a part of the problem of cluster evaluation, the appropriate points within the cluster can control the appropriate granularity clustering analysis [1]. Existing method including elbow method, approach based on variance in cluster and inflection point of cluster curve, information theory and cross validation. For the case of dense distribution of noise points in the space, inflection point or cross-validation-based method performances better. Information theory based approach is better for actual data sets. When noise points distributed evenly and for general far smaller than the number of objects in clusters, experience based method can be applied. For data sets with n points, the number of cluster is about $\sqrt{\frac{n}{2}}$, each cluster has $\sqrt{2n}$ points if the expectation is known.

Based on the above analysis, two function of noise threshold in the DENCLUE method can be replaced within the cluster point estimate and controlling the rate of convergence, respectively. The steps of our approach are as follows.

(1) Calculate density attractor of each point, and density attractor of x is

$$x^0 = x \tag{1}$$

$$x^{i+1} = x^i + \delta \frac{\nabla \hat{f}(x^i)}{|\nabla \hat{f}(x^i)|} \tag{2}$$

where δ is parameter controlling convergence speed, $\nabla \hat{f}(x^i)$ is the gradient of $\hat{f}(x^i)$, we have

$$\nabla \hat{f}(x) = \frac{1}{h^{d+2} n \sum_{i=1}^{n} K\left(\frac{x - x_i}{h}\right)(x_i - x)} \tag{3}$$

In the process of mountain climbing algorithm, if $\hat{f}_h(x^{i+1}) \leq \hat{f}_h(x^i)$, x^i is local maximum point, which is density attractor of x. And then, stop climbing process.

(2) A collection is divided according to the each point corresponding to density attractor. Then calculate the distance between the density attractor. Two density attractor can be merged as a collection if the distance is less than 2δ, and process stops when all set can not be merged.
(3) By using points in cluster to estimate whether the number of each set is greater than $\sqrt{2n}$, if it does, the sets is considered as a cluster; if not, all points in the collection are considered as noise points.

4 Simulation and Discussion

This section verifies the effectiveness of the proposed method via different data sets, and clustering evaluation method is used to compare the original DENCLUE and our method according to the clustering quality. In order to ensure the unity of the results, in this experiment, smoothing parameter h both in original DENCLUE and proposed method is set to 5. Parameter that needs to be debugged in the original DENCLUE is ξ and δ, and parameter in our method is δ.

Clustering evaluation is an inner method, which is suitable for the lack of benchmark data. The evaluation focuses on separation between clusters and compactness within a cluster, in which silhouette coefficient is viewed as a metric [15, 16]. The definitions of this evaluation method are as follows. Suppose there are n objects in data sets **D**, **D** is divided into k clusters: C_1, C_2, …, C_k. $a(x)$ presents average distance between x and the other objects in the cluster. $b(x)$ presents minimum average distance between x and points which does not belong to its own cluster. The compact degree of clusters is represented by silhouette coefficients via the following equation.

$$s(x) = \frac{b(x) - a(x)}{\max\{a(x), b(x)\}} \tag{4}$$

$s(x)$ is mean value of silhouette coefficients for all objects. The smaller $a(x)$ is, the more compact cluster is, and vice is established. The bigger $b(x)$ is, the more detached between x and other clusters, and vice versa. s(x) has the same effect with $a(x)$. The quality of clustering can be reflected by the three parameters.

Data sets "data_x_4xhelix_noise_824" are a two-dimensional space which contains [0, 1] distribution and random noise, and contains four independent Archimedes spiral data. This data set has the typical characteristics of flow pattern and it is difficult to distinguish by using the traditional clustering.

Clustering results through debugging value of ξ and δ repeatedly in the original DENCLUE (shown in Fig. 3). Experiments found that changing the value many times cannot accurately identify noise (increases from 0 to 0.025, the clustering results are the same, and with the bigger one, points in cluster will be recognized as noise). The reason for this is that noise points are widely distributed near clusters,

Fig. 3 Clustering result by using original DENCLUE on data sets data_x_4xhelix_noise_824

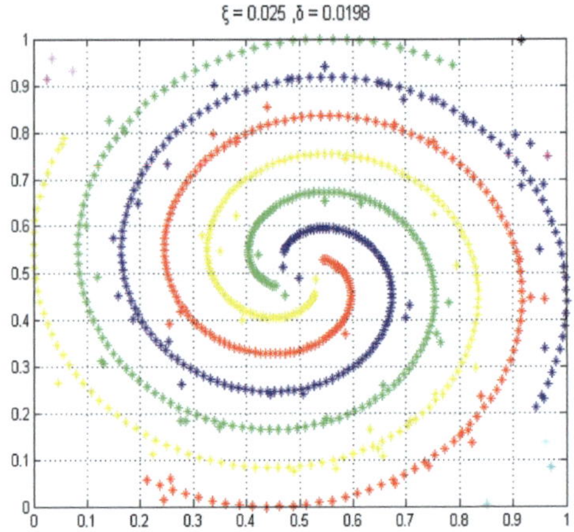

kernel density estimation of noise points, and points within clusters is very close, even both vary widely. At the same time, the noise threshold is constrained by smoothing parameter, and the value of rate of controlling convergence needs to be adjusted when noise threshold changes making original DENCLUE method of parameter adjustment very difficult.

Our result is shown in Fig. 4 (different δ). When the δ value is small, part of the point in cluster will be mistaken for noise (when the $\delta = 0.0129$, points in four parts at the end of clusters are treated as noise; when the $\delta = 0.0131$, points that mistaken for noise in four parts at the end of clusters are reducing). Increasing the δ value, we can get better clustering results (when the $\delta = 0.0136$, all points within the cluster can be classified correctly). That is to say, the performance of our approach is determined by δ entirely. For the noise points near the four clusters, in the clustering processing, they will be divided into the four clusters, respectively; therefore they can not be recognized by point estimates in clusters. For those far from clusters, we divide a collection for them separately in our method. Compared to the original DENCLUE, our method identifies noise more efficiently.

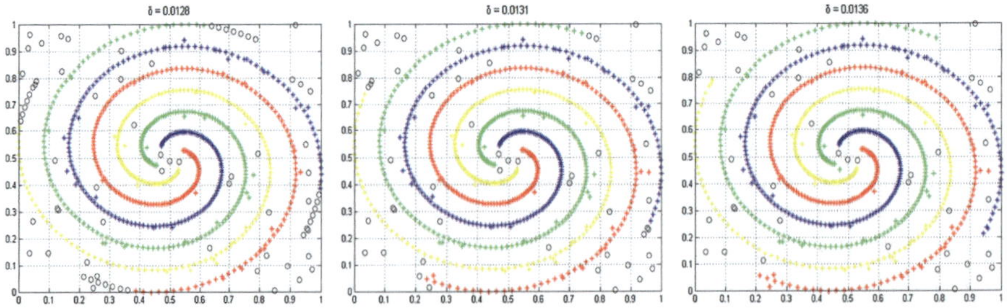

Fig. 4 Clustering result by using the proposed method on data_x_4xhelix_noise_824 with different controlling convergence speed

The clustering performance of original DENCLUE and our method are compared by silhouette coefficient s ($s = -0.0664$ in proposed method, $s = -0.07959$ in original DENCLUE). While distance of points in cluster is greater than minimum distance between points in clusters and other points in clusters, this will make silhouette coefficient negative and points in clusters incompact, which is determined by data distribution, and the silhouette coefficient in proposed approach is bigger than original DENCLUE, which demonstrates that a more compact clustering result will be obtained in the method.

In addition, to evaluate the proposed method, two-dimensional data sets with different characteristics (deformation, flow pattern, embedded, division of linear, density differences) are used. The results of original DENCLUE are shown in upper of Fig. 5 and the results of proposed method are shown in below of Fig. 5. For three data sets from left to right, ξ value in original DENCLUE is set as 0.0792, 0.0790, 0.0791, δ is set as 0.0150, 0.0269, 0.0514, respectively, while δ is set as 0.0149, 0.0149, 0.0507 in our method. According to the experimental result, high error rate of noise identification is occurred in original DENCLUE. A large amount of points will be recognized as noise points if it comes to an inappropriate ξ. Although noise points near cluster can not be fully identified in our method, the performance in three kinds of data sets is obvious better than original DENCLUE. A detailed comparison of their clustering performance with silhouette coefficient is shown in Table 1. Moreover, for the first and third data sets, the silhouette coefficient in our approach is bigger than original DENCLUE, leading to a better clustering result. The clustering result intuitively reflects our clustering effect, which is better. For the

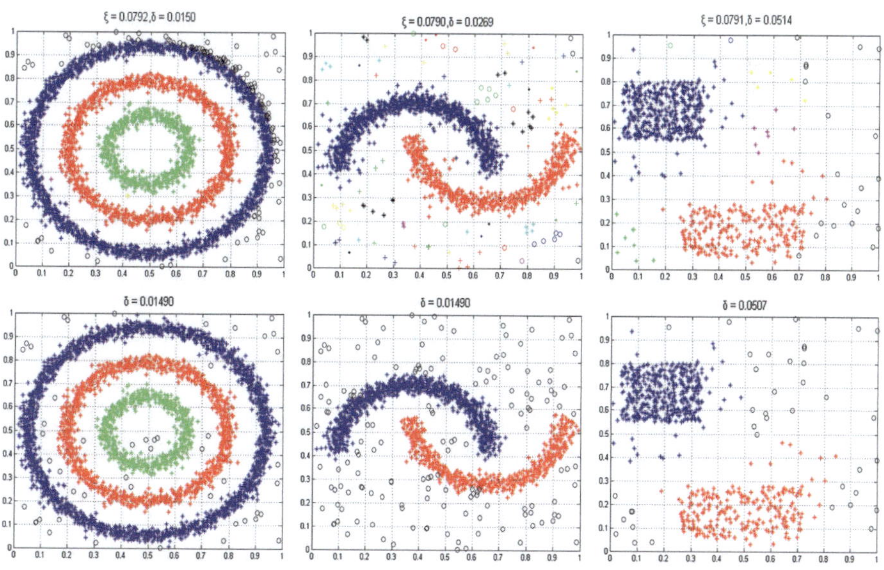

Fig. 5 Clustering result compared by the proposed method and original DENCLUE on different data sets

Table 1 Page Margins Silhouette coefficient of the proposed method and original DENCLUE on three different data sets

Method	Data sets	Silhouette coefficient
Proposed method	Three circles	−0.1803
	Two moons	0.3302
	Linear segmentation	0.6974
DENCLUE	Three circles	−0.2863
	Two moons	0.3313
	Linear segmentation	0.6158

second data sets, although, the silhouette coefficient in our approach is smaller than original DENCLUE, the reason is that most of the noise points were divided into a category in original DENCLUE. When calculating the silhouette coefficients, this kind of cluster contains only one object of noise point, it is very compact from the silhouette coefficients criterion. But in fact, these noise points were not correctly divided. It is obvious that the proposed method is better than original DENCLUE, considering the practical significance of the clustering.

5 Conclusion and Future Direction

The performance of the original DENCLUE is restricted because many parameters need to be set manually. A novel algorithm to avoid determining noise threshold has been proposed. We only consider its controlling convergence speed. The method does not calculate kernel density estimation, and focus on the setting of distance threshold δ in clustering processing. Considering clustering time and computational complexity, the proposed method is better than the original DENCLUE. A better result can be achieved by the experimental verification. Especially for the case that noise points close to the cluster, performance of identification of these points is better than the original DENCLUE. On the other hand, more compact clustering results are obtained via silhouette coefficient criterion, which also reflects the actual distribution of the data. The future directions are as follows:

(1) How to set controlling convergence speed that needs to be set in proposed method? Whether there exists an adaptive algorithm? For large amount of data, running time is too long.
(2) Effectiveness for data sets with variation intensely and very closed points has not yet been verified.
(3) Improve the accuracy of noise identification for the noise points closed to clusters that hard to be identified.
(4) High computational complexity of DENCLUE largely due to computational complexity of seeking density attractor and kernel density estimation. How to

find density attractor quickly and select an appropriate kernel function to improve its clustering performance are fundamental problems that need to be solved.

Acknowledgments This work is supported by National Natural Science Foundation of China under Grant 61203084 and 61374135, Basic Science and Advanced Technology Research Projects of Chongqing under Grant cstc2015jcyjA0480, and Chongqing University Postgraduates Innovation Project (CYB15051).

References

1. Han J, Kamber M, Pei J (2011) Data mining: concepts and techniques, 3rd edn. Morgan Kaufmann, San Francisco
2. Padhy N, Dr. Mishra P, Panigrahi R (2012) Summay of data mining. The survey of data mining applications and feature scope. Int J Comput Sci Eng Informa 2(3):43–58
3. Aggarwal CC, Reddy CK (2013) Data clustering: algorithm and applications. Chapman & Hall/CRC
4. Ware VS, Bharathi HN (2013) Study of density based algorithms. Int J Comput Appl 69 (26):1–4
5. Hinneburg A, Gabriel H-H (2007) DENCLUE 2.0: fast clustering based on kernel density estimation. In: International symposium on advances in intelligent data analysis. vol 4723, pp 70–80
6. Suganya M, Nagarajan S (2015) Message passing in clusters using fuzzy density based clustering. Indian J Sci Technol. 8(16):1–6
7. Chang DX, Zhang XD, Zheng CW, Zhang DM (2010) A robust dynamic niching genetic algorithm with niche migration for automatic clustering problem. Pattern Recogn 43 (4):1346–1360
8. Guo C, Zang Y (2012) Clustering algorithm based on density function and nichePSO. J Syst Eng Electron 23(3):445–452
9. Sree KS (2014) SSM-DENCLUE: enhanced approach for clustering of sequential data: experiments and test cases. Int J Comput Appl 96:7–13
10. Liang Z, D L, Fei H, Yifei T, Yanqiang Yuan (2015) Fault disgnosis of belt weigher using the improved DENCLUE and SVM. Harbin Gongye Daxue Xuebao/J Harbin Institute Technol 47 (7):122–128
11. Yan J, Yuan H, Shu X, Zhong S (2009) Optimal clustering algorithm for crime spatial aggregation states analysis. J Tsinghua Univ 49(2)
12. Yu X, Yu X (2010) On unsupervised clustering algorithm based on distance and density. Comput Appl Softw 27(7)
13. Hinneburg A, Keim DA (1998) An efficient approach to clustering in large multimedia databases with noise. In: Proceedings of the 4th international conference on knowledge discovery and data mining. AAAI Press, New York, pp 58–65
14. Aggarwal CC (2013) Outlier analysis. Data mining. pp 237–263
15. Zhou K, Yang S, Ding S, Luo H (2014) On cluster validation. Syst Eng-Theory Pract 34(9)
16. Liu Y, Li Z, Xiong H, Sen W (2013) Understanding and enhancement of internal clustering validation measures. IEEE Trans Syst Man Cybern Part B Cybern A Publ IEEE Syst Man Cybern Soc 43(3):982–994

Reactive Power Predictive Compensation Strategy for Heavy DC Hoist

CaiXia Gao, FuZhong Wang and ZiYi Fu

Abstract To deal with reactive power disturbances in power grid introduced by speed and load changes of heavy DC hoist in operation, a reactive power predictive compensation strategy is proposed. After analysis of reactive power absorbed by the DC hoist in different operation stages, the work principle and the reactive power compensation flowchart and modification module are presented with reactive power compensation regulation based on fuzzy control. Finally, experiment results are given to demonstrate the effectiveness and robustness of the proposed strategy in reducing external and internal disturbances, with the power factor ranging from 0.95 to 0.98.

Keywords DC hoist • Reactive power predictive compensation • Reactive power compensation flowchart • Fuzzy control

1 Introduction

Heavy DC hoist controlled by rectifier has been widely used in mining industry for its wide speed range, big torque at start, braking, and at a low speed, smooth operation, and high efficiency. However, in operation of the heavy DC hoist, the constant change of speed and load fluctuations may cause reactive power disturbances, severely threatening the secure operation of power grid. Thus reactive power compensation is needed for heavy DC hoist. There are many choices of such device, such as static capacitor, static VAR compensator (SVC), static VAR generator (SVG), magnetic controlled dynamic reactive power compensation device, etc. [1–5]. Unfortunately, all these compensation devices begin to work after the measurement of changes of reactive power. That means there is at least 20 ms delay time which will cause over-compensation or total compensation, over-voltage, or

C. Gao (✉) · F. Wang · Z.Y. Fu
School of Electrical Engineering and Automation, Henan Polytechnic University, 454000 Jiaozuo, China
e-mail: 254636920@qq.com

Y. Jia et al. (eds.), *Proceedings of 2016 Chinese Intelligent Systems Conference*, Lecture Notes in Electrical Engineering 405,
DOI 10.1007/978-981-10-2335-4_30

even security issues when the hoist transfers from one operation stage to another operation stage.

Therefore, this paper proposes a reactive power predictive compensation strategy in which the operation chart is firstly preorganized according to experiments; then in operation of the DC hoist, the compensation equipment automatically selects the operation chart and to compensate in advance (50 ms) before the transfer of operation stages, i.e., the significant changes of working speed and reactive power requirement, in order to avoid over-compensation or total compensation.

2 Reactive Power Analysis of the DC Hoist in Different Operation Stages

The DC hoist works according to the operation chart as shown in Fig. 1. In different stages, reactive power and power factor from power grid are different. Table 1 shows the power requirements of primary side of the rectifier transformer of a DC hoist (2100 kw) at different stages [1].

From Table 1, we see that at the main acceleration stage, although the speed is rather low, the reactive power impulse is the highest, and the apparent power taken from the grid is also the highest; at the constant speed stage, the elevating speed is the highest, the active power maintains at the rated value, and the reactive power is significantly reduced, the apparent power is much lower than that at main acceleration stage; at the deceleration stage, the active and reactive power become further lower owing to the reduction of current; at the creeping stage, the active and apparent power increase a little; at the braking stage, the electromotor current increases, but its speed reduces quickly, the reactive and active power taken from the grid are relatively small.

By analyzing Fig. 1 and Table 1, in operation of the hoist, the reactive power requirements introduced by speed changes may cause significant impulse to power grid. Thus reactive power compensation equipment is needed. The regular choices

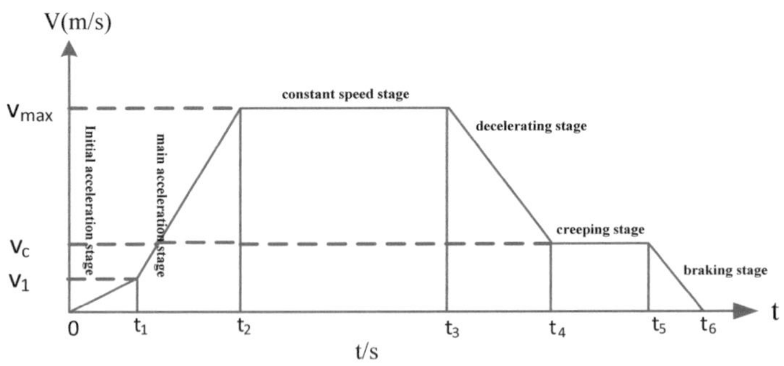

Fig. 1 The DC hoist operation chart

Table 1 The power requirements of primary side of the rectifier transformer of a DC hoist (2100 kw) at different stages

Power	Initial acceleration stage	Main acceleration stage	Constant speed stage	Decelerating stage	Creeping stage	Braking stage
Reactive power (kvar)	3170	6880	2027	1120	1390	1980
Active power (kw)	945	2720	2140	315	825	550
Apparent power (kva)	3310	7396	2950	1160	2950	2053

include static capacitor, static VAR compensator (SVC), static VAR generator (SVG), magnetic controlled dynamic reactive power compensation device, etc. All these compensation devices begin to work after the measurement of changes of reactive power. That means there is a delay time which will cause over-compensation or total compensation, over-voltage, when the hoist transfers from one operation stage to another operation stage. For example, if we want to increase the power factor to 0.95, then the reactive power requirements are 5986 kvar at the main acceleration stage, and 1324 kvar at the constant speed stage. In real operation, when the hoist transfer from main acceleration stage to constant speed stage, due to the delay time of reactive power compensation, the compensation value cannot be increased from 5986 to 1324 kvar instantly, thus over-compensation will be induced. After the delay time, reactive power compensation will transits from over-compensation to under-compensation via total compensation. Therefore, in the above-mentioned transition process, we will see over-voltage, or even security issue. Similarly, the same situation also exits when the hoist transfer from constant speed stage to decelerating stage. Consequently, predictive compensation is needed to avoid over-compensation and total compensation before the operation stage transits.

3 The Basic Principle of the Reactive Power Prediction Compensation Scheme for DC Hoist

To achieve the reactive power prediction compensation for DC hoist, this paper designs the reactive power predictive compensation device for DC hoist shown in Fig. 2, which is composed by reactive power compensation operation diagram presupposition and correction module, reactive power compensation operation diagram memory module, reactive power compensation operation diagram selection module, reactive power compensation adjustment module, and TCR thyristor

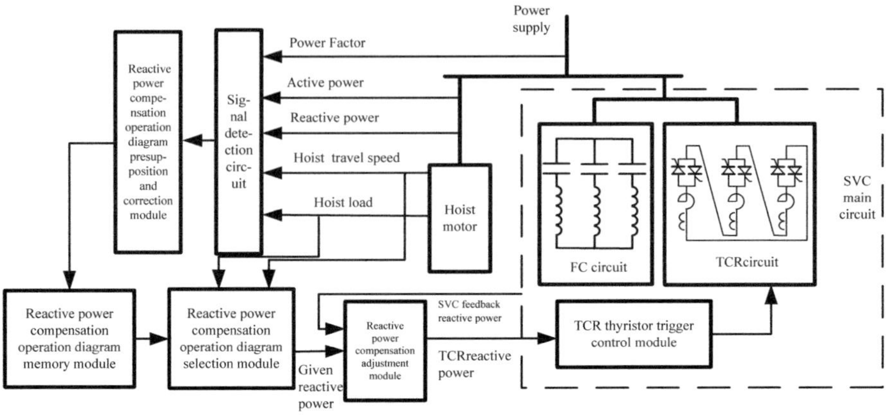

Fig. 2 DC hoist reactive power predictive compensation device

Fig. 3 Reactive power compensation operation diagram for full load

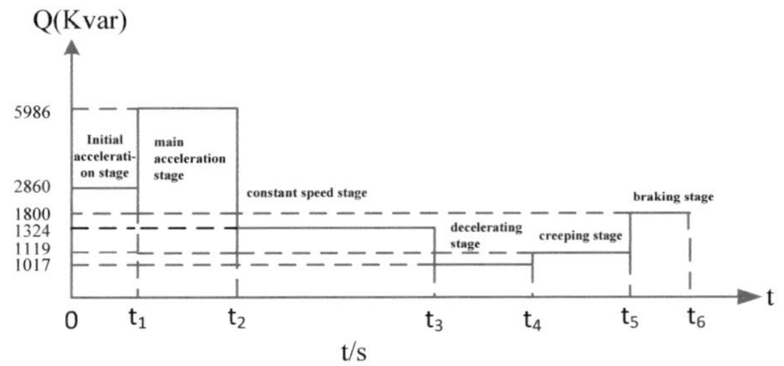

trigger control module; and its working principle is as follows: First, to preset reactive power compensation operation diagram according to the experiment and store the generated reactive power compensation operation diagram into the memory module of reactive power compensation; during the lifting operation, the selection module of reactive power compensation operation diagram selects the corresponding reactive power compensation operation diagram according to the measured load and running direction, and then according to the selected reactive power compensation operation diagram and SVC feedback reactive power, the reactive power compensation adjustment module realize the adjustment of reactive power through thyristor trigger control unit to change thyristor conduction angle in the TCR branch. Simultaneously, detect reactive power required by the hoist and power factor of the power-in after the compensation, analyze the compensation effect, to modify the original reactive compensation operation diagram, and deposit the modified one into the memory module of reactive power compensation operation diagram, for later use (Fig. 3).

3.1 The Design of Presupposition and Correction Module About Reactive Compensation Operation Diagram

The function of "presupposition and correction module of reactive compensation capacity operation diagram" is to setup the reactive compensation operation diagram and to correct original reactive compensation operation diagram online.

(1) The presupposition of reactive power compensation

According to Fig. 1 and Table 1, the reactive power of direct current lifter in all working stages is related to the speed and also related to the load of lifter. However, no matter how much the load is, the lifter runs basically in accordance with the fixed velocity diagram. Therefore, through online experiment, we can make actual measurement on the reactive power of the lifter in different working stages when the lifter raises or lowers the material with all kinds of loads from full load to light load and no load. In order to avoid over-compensation phenomenon, we can adopt reactive power control, use the power factor as constraints, determine reactive compensation capacity when the lifter raises or lowers with every file of load and different speed according to formula one, produce reactive compensation running chart(2n pieces), and deposit them in memory module of the reactive power compensation capacity.

$$0.95 \le \mathrm{COS}\phi = \frac{P_{ca}}{\sqrt{P_{ca}^2 + (Q_{ca} - Q_c)^2}} \le 0.98 \tag{1}$$

In formula (1), P_{ca} represents active power of the lifter when it works, Q_{ca} represents reactive power of the lifter when it works, and Q_c represents reactive compensation capacity. The power factor COS Φ takes 0.95 when the lifter is light load, and takes 0.98 when the lifter is overloaded.

(2) The online correction method of reactive power compensation

According to the active power and reactive power of the lifter from the online real-time detection, we can correct corresponding reactive power compensation operation diagram according formula one, and deposit it in the memory module of reactive power compensation operation diagram for next use.

3.2 The Design for Selection of Reactive Power Compensation Diagram

According to the direction of load and running direction, the reactive power compensation selection diagram selects the matching reactive power compensation diagram from the memory module and outputs the reactive power compensation diagram into the reactive power compensation controller module as a given value of reactive power compensation of reactive power regulator module. Specific methods are as follows:

In the operation of hoist, according to the actual load of hoist and the formula (2), the selection of reactive power compensation diagram determines which file number the load should belong to. Finally, according to the running direction, the reactive power compensation diagram selects the corresponding reactive power compensation diagram.

$$n_i = \left[\frac{W_i}{W_N} \times 30 + 0.5 \right] \tag{2}$$

In the formula (2), W_i—the load of the hoist operation; W_N—the rated load of the hoist.

3.3 Reactive Power Compensation Regulation Based on Fuzzy Control

Reactive power compensation regulation module based on fuzzy control is adjusting ahead the conduction angle of TCR thyristor triggering control module according to the reactive power given by the selected reactive power compensation flowchart and reactive SVC device, in order to change the reactive power, QC, output by SVC. The specific approach is as follows: During the operation, reactive power compensation regulation device calculates accurately the operating time of the hoist. When the hoist, in the operation period, switches from the higher needed reactive power compensation to the lower (like the switch, presented in the chart 3, from the main acceleration period to the isokinetic one or from the isokinetic to the deceleration period), switches ahead the conduction angle of the thyristor in the branch of TCR according to the needed reactive power compensation in the next operation period 50 ms earlier than the swift, thus changing the reactive power, QC, output by SVC in case that the operation of the hoist will enter the excessive compensation and complete compensation in the next operating period. The reactive power compensation regulation device resorts to the fuzzy control calculations.

Fuzzy control calculations are presented in Fig. 4: Qr stands for the given reactive power compensation by the reactive power compensation flowchart, the reactive power by the reactive power compensation device. Fuzzy control device chooses the compensation reactive power deviation and the changing rate, *ec*, as the input variables. In addition, it chooses the output reactive power, Q_{tcr}, needed by TCR branch of SVC as the output variables. The reactive power of the branch of TCR, Q_{tcr}, controls the units by the trigger of the Crystal Gate, realized by adjusting the thyristor angle α.

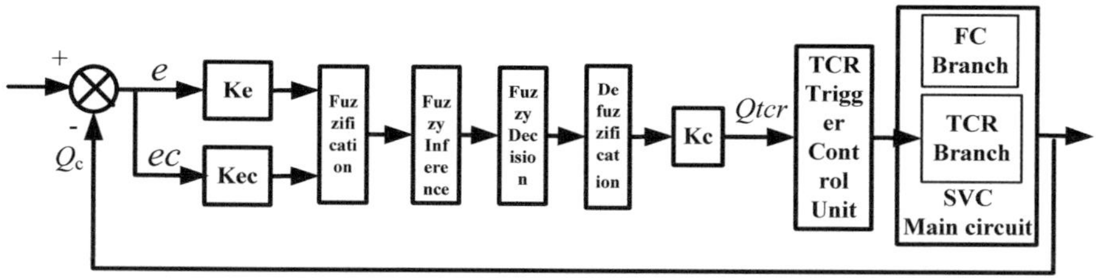

Fig. 4 The fuzzy controller structure

(1) Input and output variables fuzzification: we will set the change range of the reactive power Q_{tcr}, which is desired by the reactive deviation rate e and its change rate *ec* and TCR branch, as [–7 and interval +7], which is a continuous variation, to make it discrete, and to form a discrete set with 15 integers, that is {–7, –6, –5, –4, –3, –2, –1, 0, 1, 2, 3, 4, 5, 6, 7}.

We will set the variation range of *e*, *ec*, and Q_{tcr} as $[Q_m, Q_0]$, $[\triangle Q_m, \triangle Q_0]$, $[Q_{tcrm}, Q_{tcr0}]$, respectively. The three scale factors shown in [6] are $K_e = 14/(Q_m - Q_0)$, $K_{ec} = 14/(\triangle Q_m - \triangle Q_0)$, $Kc = (Q_{tcrm} - Q_{tcr0})/14$, respectively.

The language of the three input and output fuzzy is defined as {negative big (NB), negative middle (NM), negative small (NS), zero (Zo), positive small (PS), positive middle (PM), and positive big (PB)}, which belongs to membership function and adopts triangle.

(2) Fuzzy Control Rules: fuzzy control rules can be described using the following fuzzy statement:

IF A1 is NB AND A2 is NB THEN B is NM.

(3) Fuzzy Reasoning: The Mamdani reasoning is adopted. Namely: Known fuzzy relation matrix contains the relationship $A \rightarrow B$, for the given A', $A' \in U$, then the conclusion is $B' \in V$ and B' is [6].

$$B^{'} = \underset{u \in U}{\mathrm{SUP}}\left\{A^{'}(u)\,\Lambda(A(u)\Lambda\, B(u)\right\} \tag{3}$$

(4) Defuzzification: Resolving the fuzzy of the controlled elements with the gravity center method to get the reactive power Q_{tcr} that is needed to be output by the TCR branch of SVC. It is calculated as follows [6]:

$$Q_{tcr} = \frac{\int \mu_B(y) y dy}{\int \mu_B(y) dy} \tag{4}$$

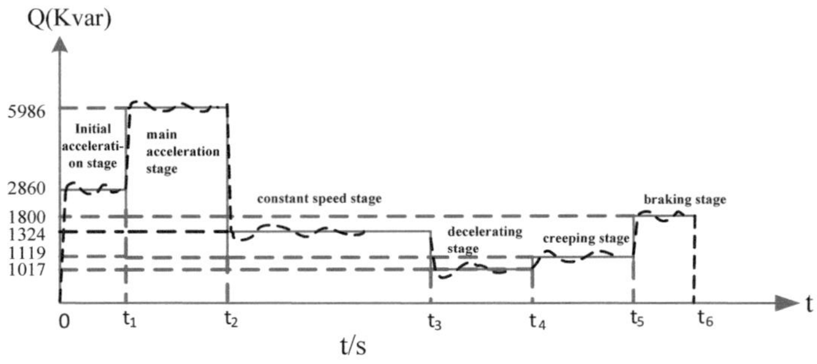

Fig. 5 The SVC reactive power diagram

Fig. 6 The power factor waveform diagram

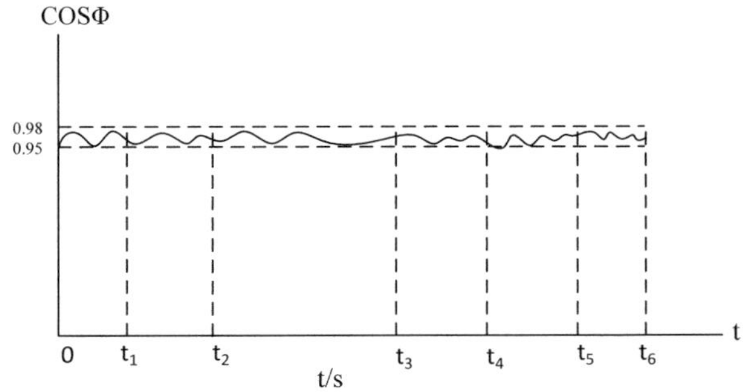

4 Experiments

This experiment takes a coal mine 2100 kW DC hoist as the research object, whose rectifier transformer once side voltage is 6 kV, II times side voltage is 0.78 kV. Furthermore, this experiment uses idle work ahead forecast compensation device to compensate for the idle work of elevator, testing SVC output of reactive power and power factor when the hoist at full load.

Figure 5 SVC output reactive power and power factor Fig. 6 wave forms.

From Fig. 5, we can see, when front forecasting compensation device of DC hoist with idle work is used, reactive power output by SVC can quickly compensate for the idle work needed by the elevator, and the track effect is good, and making the idle work output by grid side maintain at a lower level; As is shown in Fig. 6, when front forecasting compensation device of DC hoist with idle work is used, the power factor of grid line climb quickly, with a changing range between 0.95 and 0.98, to effectively improve the power factor of the power supply system of mine hoisting.

5 Conclusion

In this paper, a new method of compensation for reactive power prediction of DC hoist is presented, here are the following features:

(1) According to the load and running direction of the hoist (lifting or lowering), we determine the 2n kinds of reactive power compensation operation diagrams, and carry out reactive power compensation control according to the reactive power compensation operation diagram.
(2) Realize the reactive power predictive compensation. When the hoist, in the operation period, switches from the higher needed reactive power compensation to the lower (like the switch, presented in the Fig. 3, from the main acceleration period to the isokinetic one or from the isokinetic to the deceleration period), switches ahead the conduction angle of the thyristor in the branch of TCR according to the needed reactive power compensation in the next operation period 50 ms earlier than the swift, thus changing the reactive power, QC, output by SVC in case that the operation of the hoist will enter the excessive compensation and complete compensation in the next operating period.
(3) The control effect is good. The regulation of reactive power adopts the fuzzy control algorithm to realize rapid and accurate reactive power compensation, and has good robustness, which can effectively reduce the impact of internal and external interference on system stability. Ensure that the power factor to fluctuate within the range of 0.95–0.98.

References

1. Wang Y-M, Ai Y-L (2009) Research of scheme of dynamic reactive power compensation of mine hoist. Ind Mine Autom 6:36–38
2. Liu Y-P (2007) The reactive power compensation for impact load. North China Electric Power University
3. Song J-B, Wu Y-H (2014) Application of TCT SVC in the dynamic reactive power compensation of mine hoist. Power Capacitor React Power Compensation 8:32–35
4. Ji F, Khan MM, Chen C (2005) Static var compensator based on rolling synchronous symmetrical component method for unbalance three-phase system. In: IEEE international conference on industrial technology, pp 621–626
5. Liu Y-T (2009) Application of MSVC dynamic var compensation device in highvoltage coal mines. Coal Mine Machinery 9:216–218
6. Liu W-Y, Han J-G (2013) Research on the wind turbine fuzzy control method based on the Mamdani fuzzy inference. Renew En Resour 2: 48–51

Sensorless Vector Control of PMSM Based on Improved Sliding Mode Observer

Fangqiang Mu, Bo Xu, Guoding Shi, Wei Ji and Shihong Ding

Abstract This paper proposes a sensorless control algorithm for the permanent magnet synchronous motor (PMSM) based on a new sliding mode observer (SMO), which substitutes a sigmoid function for the signum function. The stability of proposed SMO is analyzed using the Lyapunov stability theorem. Adaptive law of switching gain is regulated with the rotor speed to expand the functional range of SMO. A back-electromotive force (back-EMF) observer is designed to eliminate the high frequency components of estimated back-EMF. An input-normalized phase-locked loop (PLL) is adopted to extract the rotor position and speed for compensating the phase lag resulted from the filter. The simulation results illustrate the validity of the analytical approach and the efficiency of the new sensorless control algorithm for PMSM based on the new SMO.

Keywords Permanent-magnet synchronous motor (PMSM) · Sliding mode observer (SMO) · Phase-locked loop (PLL) · Sensorless · Sigmoid function

1 Introduction

Permanent magnet synchronous motors (PMSMs) with their small size, high efficiency, large power density, and good dynamic performance are widely applied to high-performance servo and drive areas. Effective vector control of the PMSM requires accurate position information of rotor, since it is directly connected to the produced torque. The rotor position can be measured with an optical encoder, a resolver or with Hall sensors. However, such sensors installed on the rotor shaft increase the inserted noise, the motor size, and the overall cost of the drive system. To eliminate the need of position sensors, several sensorless control strategies have been developed.

F. Mu · B. Xu · G. Shi · W. Ji (✉) · S. Ding
School of Electrical and Information Engineering, Jiangsu University,
212013 Zhenjiang, China
e-mail: jwhxb@163.com

Y. Jia et al. (eds.), *Proceedings of 2016 Chinese Intelligent Systems Conference*, Lecture Notes in Electrical Engineering 405,
DOI 10.1007/978-981-10-2335-4_31

In most sensorless PMSM control methods, the rotor position is estimated using measured electrical quantities. According to the effectiveness of rotor position estimation for PMSM in different speed ranges, sensorless control methods are divided into two major categories: saliency-based methods and mode-based methods. The former is valid in very low or zero speed operation, and the latter is suitable for the rotor position estimation in medium speed and high speed operation. The mode-based methods are mainly based on the model of the PMSM include the direct calculation method using the stator terminal voltage and current [1, 2], extended Kalman filter (EKF)-based methods [3, 4], model reference adaptive system (MRAS)-based methods [5], and SMO-based methods [6–10].

Compared with other methods, SMO-based methods have advantages such as robustness for motor parameters, rapid responses, and less computational complexity. For an actual SMO, due to the time and space delay and dead zone, making the existence of chattering in the SMO is inevitable [11]. To eliminate the chattering, a boundary layer was used to solve the chattering problem and improve the precision of rotor position estimation [12]. In [13], a SMO based on two-stage filter for the estimation of rotor position and speed was proposed. The piecewise linear compensation was used for the rotor position estimate error compensation. The boundary layer methods and the filtering methods both are available to eliminate the chattering, but when a high control precision required, the two methods cannot meet the requirements for use in the SMO. In [14], a sigmoid function was used for the SMO as a switching function. The observer has fast responses and the chattering, which happens at the observer using the signum function, has been reduced significantly.

This paper proposes a new sensorless control algorithm for PMSM based on a new SMO which uses a sigmoid function as the switching function. It is proved that the proposed SMO is globally asymptotically stable based on the Lyapunov stability theory. Adaptive law of switching gain is regulated with the rotor speed to expand the functional range of SMO. The back-EMF observer, which has the structure of an extended Kalman filter, is designed to make the estimated back-EMF waveforms smoother. An input-normalized PLL is adopted to extract the rotor position and speed from the estimated back-EMF to eliminate the additional compensator for the rotor position. The superiority of proposed algorithm is proved through the comparison of simulation results.

2 Conventional SMO

Suppose that the distribution of air-gap magnetic field is sinusoidal, eddy current losses and hysteresis losses are negligible, and the induced EMF is sinusoidal. The model of PMSM in the stationary reference frame is

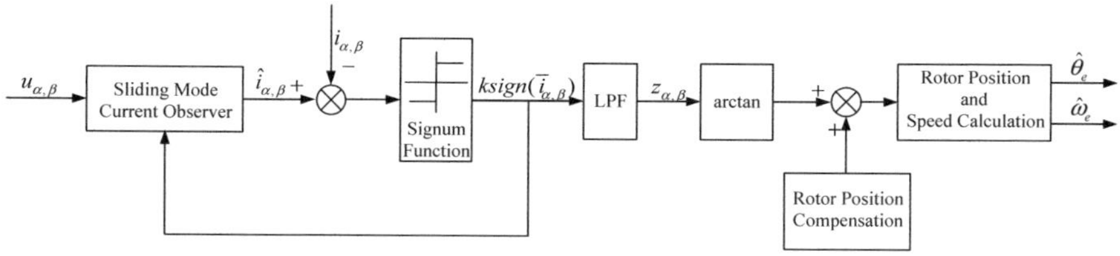

Fig. 1 Block diagram of traditional SMO

$$\begin{cases} \frac{\mathrm{d}i_\alpha}{\mathrm{dt}} = -\frac{R_S}{L_S} i_\alpha + \frac{u_\alpha}{L_S} - \frac{e_\alpha}{L_S} \\ \frac{\mathrm{d}i_\beta}{\mathrm{dt}} = -\frac{R_S}{L_S} i_\beta + \frac{u_\beta}{L_S} - \frac{e_\beta}{L_S} \end{cases} \quad (1)$$

where $i_{\alpha,\beta}$, $e_{\alpha,\beta}$, and $u_{\alpha,\beta}$ represent the stator current, back-EMF, and voltage for each phase in the stationary reference frame, respectively. R_S and L_S represent the stator resistance and inductance, respectively.

The back-EMF for each phase can be represented in the stationary reference frame as

$$\begin{cases} e_\alpha = -\omega_e \psi_f \sin\theta_e \\ e_\beta = \omega_e \psi_f \cos\theta_e \end{cases} \quad (2)$$

where ψ_f, ω_e, and θ_e represent the rotor permanent magnet flux linkage through the stator windings, the electric rotor speed, and the rotor position, respectively.

Suppose that the speed changes slowly ($\dot{\omega}_e \approx 0$) during the one estimation period. The dynamic behavior of back-EMF can be obtained by differentiation of (2):

$$\begin{cases} \frac{\mathrm{d}e_\alpha}{\mathrm{dt}} = -\omega_e e_\beta \\ \frac{\mathrm{d}e_\beta}{\mathrm{dt}} = \omega_e e_\alpha \end{cases} \quad (3)$$

For the sensorless vector control of PMSM, the sliding mode controller is adopted for use in the observer design and so is named the SMO. However, there are the shortcomings of chattering and phase lag in the traditional SMO. Figure 1 shows the traditional SMO, where the signum function is used as the switching function and the low-pass filter (LPF) is used to eliminate the chattering. The rotor position compensation is used for compensating the phase lag caused by the filter.

3 Proposed SMO

To eliminate the undesirable chattering, a new SMO which uses a sigmoid function as the switching function is proposed, as shown in Fig. 2.

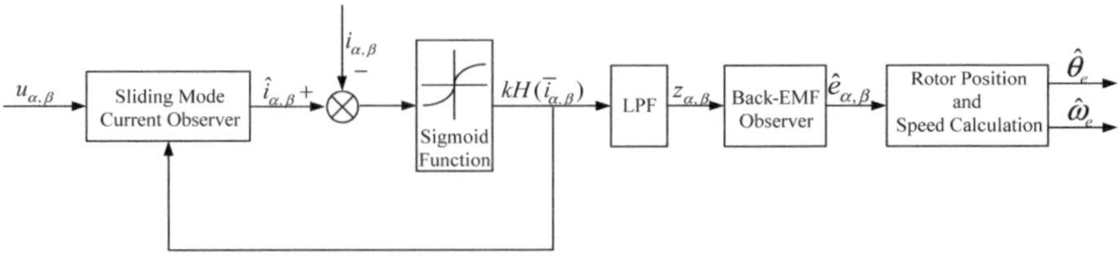

Fig. 2 Block diagram of proposed SMO

The sliding mode surface is defined as

$$S = [S_\alpha \quad S_\beta]^T = [\bar{\imath}_\alpha \quad \bar{\imath}_\beta]^T \tag{4}$$

where $\bar{\imath}_\alpha = \hat{\imath}_\alpha - i_\alpha$ and $\bar{\imath}_\beta = \hat{\imath}_\beta - i_\beta$ are the stator current errors. The proposed current sliding mode observer is defined as follows:

$$\begin{cases} \frac{d\hat{\imath}_\alpha}{dt} = -\frac{R_S}{L_S}\hat{\imath}_\alpha + \frac{u_\alpha}{L_S} - \frac{k}{L_S}H(\hat{\imath}_\alpha - i_\alpha) \\ \frac{d\hat{\imath}_\beta}{dt} = -\frac{R_S}{L_S}\hat{\imath}_\beta + \frac{u_\beta}{L_S} - \frac{k}{L_S}H(\hat{\imath}_\beta - i_\beta) \end{cases} \tag{5}$$

where $\hat{\imath}_\alpha$ and $\hat{\imath}_\beta$ are the stator current estimates, k is the switching gain, H is the sigmoid function, which is represented as

$$H(x) = \frac{2}{(1 + e^{-ax})} - 1 \tag{6}$$

where a is a positive constant used to regulate the slope of the sigmoid function.

Subtracting (1) from (5) yields the stator current errors equation:

$$\begin{cases} \frac{d\bar{\imath}_\alpha}{dt} = -\frac{R_S}{L_S}\bar{\imath}_\alpha + \frac{1}{L_S}(e_\alpha - kH(\hat{\imath}_\alpha - i_\alpha)) \\ \frac{d\bar{\imath}_\beta}{dt} = -\frac{R_S}{L_S}\bar{\imath}_\beta + \frac{1}{L_S}(e_\beta - kH(\hat{\imath}_\beta - i_\beta)) \end{cases} \tag{7}$$

In order to verify the stability of the SMO, the Lyapunov function V is chosen as

$$V = \frac{1}{2}S^T S \tag{8}$$

Its time derivative can be written as

$$\dot{V} = \dot{S}^T S = \dot{\bar{\imath}}_\alpha \bar{\imath}_\alpha + \dot{\bar{\imath}}_\beta \bar{\imath}_\beta \tag{9}$$

According to the Lyapunov's stability theory, the stabled observer must satisfy $V > 0$ and $\dot{V} < 0$. To satisfy the condition $\dot{V} < 0$, (9) is expressed as (10).

$$\dot{V} = -\frac{R_s}{L_s}\left(\bar{\iota}_\alpha^2 + \bar{\iota}_\beta^2\right) + \frac{1}{L_s}(e_\alpha \bar{\iota}_\alpha - k\bar{\iota}_\alpha H(\bar{\iota}_\alpha)) + \frac{1}{L_s}(e_\beta \bar{\iota}_\beta - k\bar{\iota}_\beta H(\bar{\iota}_\beta)) < 0 \tag{10}$$

The switching gain k can be derived from (10) as

$$k \geq \max(|e_\alpha|, |e_\beta|) = |\omega_e|\psi_f \tag{11}$$

k must be large enough to satisfy (11), however, if it is too large the chattering may lead to estimation errors. The following adaptive law is proposed:

$$k = l|\hat{\omega}_e|\psi_f \tag{12}$$

where $l = 1.5$ is a safety factor and $\hat{\omega}_e$ is the estimated speed.

Once sliding mode occurs, $S_{\alpha,\beta} = 0, \dot{S}_{\alpha,\beta} = 0$.Using the equivalent control method, the equivalent control values are

$$\begin{cases} u_{\alpha eq} = e_\alpha = kH(\bar{\iota}_\alpha)_{eq} \\ u_{\beta eq} = e_\beta = kH(\bar{\iota}_\beta)_{eq} \end{cases} \tag{13}$$

For the traditional SMO, a low-pass filter is usually adopted to extract e_α and e_β from the corresponding equivalent control values in (13). Suppose that the low-pass filter outputs are z_α, z_β, however, for high-performance applications, z_α, z_β cannot be used directly due to the presence of high frequency components. In order to improve the filter properties to obtain more accurate back-EMF, a back-EMF observer is designed to re-estimate the back-EMF. The back-EMF observer is as follows:

$$\begin{cases} \dot{\hat{e}}_\alpha = -\hat{\omega}_e \hat{e}_\beta - l_1(\hat{e}_\alpha - z_\alpha) \\ \dot{\hat{e}}_\beta = \hat{\omega}_e \hat{e}_\alpha - l_1(\hat{e}_\beta - z_\beta) \\ \dot{\hat{\omega}}_e = (\hat{e}_\alpha - z_\alpha)\hat{e}_\beta - (\hat{e}_\beta - z_\beta)\hat{e}_\alpha \end{cases} \tag{14}$$

where l_1 is the observer gain, and $\hat{e}_\alpha, \hat{e}_\beta$ are the estimated back-EMF. Assume $z_{\alpha,\beta} = e_{\alpha,\beta}$ for the rest of analysis, by subtracting (3) from (14), the mismatches in the back-EMF equations are

$$\begin{cases} \dot{\bar{e}}_\alpha = -\hat{\omega}_e \hat{e}_\beta + \omega_e e_\beta - l_1(\hat{e}_\alpha - e_\alpha) \\ \dot{\bar{e}}_\beta = \hat{\omega}_e \hat{e}_\alpha - \omega_e e_\alpha - l_1(\hat{e}_\beta - e_\beta) \\ \dot{\bar{\omega}}_e = (\hat{e}_\alpha - e_\alpha)\hat{e}_\beta - (\hat{e}_\beta - e_\beta)\hat{e}_\alpha \end{cases} \tag{15}$$

where $\bar{e}_\alpha = \hat{e}_\alpha - e_\alpha, \bar{e}_\beta = \hat{e}_\beta - e_\beta, \bar{\omega}_e = \hat{\omega}_e - \omega_e$ are the observer errors.

To prove the convergence of the observer, a Lyapunov function is defined as

$$V_1 = \frac{1}{2}(\bar{e}_\alpha^2 + \bar{e}_\beta^2 + \bar{\omega}_e^2) \tag{16}$$

The time derivative of (16) can be calculated as

$$\begin{aligned}\dot{V}_1 &= \dot{\bar{e}}_\alpha \bar{e}_\alpha + \dot{\bar{e}}_\beta \bar{e}_\beta + \dot{\bar{\omega}}_e \bar{\omega}_e \\ &= \omega_e \bar{e}_\beta (\hat{e}_\alpha - e_\alpha) - \omega_e \bar{e}_\alpha (\hat{e}_\beta - e_\beta) - l_1 \left(\bar{e}_\alpha^2 + \bar{e}_\beta^2\right) \\ &= -l_1 \left(\bar{e}_\alpha^2 + \bar{e}_\beta^2\right) \le 0\end{aligned} \quad (17)$$

It can be shown using Lyapunov function that the estimates $\hat{e}_\alpha$ and $\hat{e}_\beta$ tend to e_α and e_β asymptotically.

4 PLL for Position and Speed Extraction

An approach to rotor position estimation can be the use of an inverse trigonometric function directly from the back-EMF space vector. Motor speed then can be obtained as a derivative, but this commonly leads to noise issues. The technique adopted in this case is the phase and frequency tracking by means of a PLL (Fig. 3) operating on the estimated back-EMF,
where θ_e is the real rotor position, θ_e is the estimated rotor position, and K_P and K_I are the proportional and integral (real) gains for the PLL, respectively.

According to Fig. 3, the error signal Δe can be obtained:

$$\Delta e = -\hat{e}_\alpha \cos\theta_e - \hat{e}_\beta \sin\theta_e = \omega_e \psi_f \sin(\theta_e - \theta_e) \quad (18)$$

when $|\theta_e - \theta_e| < \pi/6$, (18) can be simplified as

$$\Delta e \approx \omega_e \psi_f (\theta_e - \theta_e) \quad (19)$$

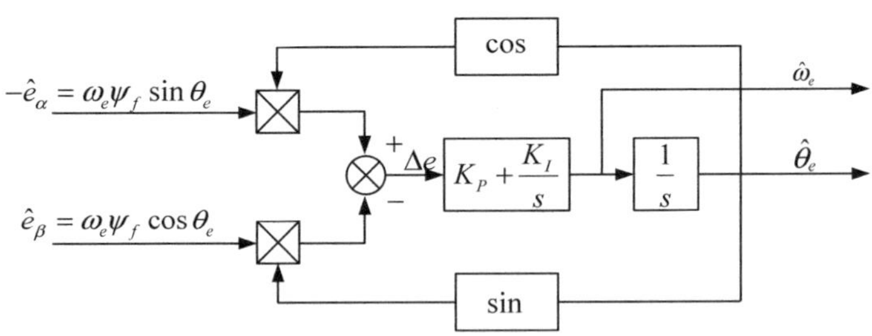

Fig. 3 Block diagram of PLL

The equivalent block diagram of the PLL is shown in Fig. 4; the PLL open-loop transfer function can be expressed as

$$G_{PLL}(s) = \omega_e \psi_f (K_P + \frac{K_I}{s}) \frac{1}{s} \tag{20}$$

Since the loop gain is dependent on speed, different dynamical responses would be obtained at the various speeds. To achieve a constant dynamic behavior of the PLL in all the operating range, input normalization can be done by scaling gain using the back-EMF vector magnitude. To reduce noise propagation, a modified vector magnitude calculation is proposed as shown in Fig. 5.

The equivalent block diagram of the input-normalized PLL is shown in Fig. 6. With normalized input, the error signal (19) becomes

$$\Delta e \approx \theta_e - \theta_e = \theta_e \tag{21}$$

The PLL open-loop transfer function (20) becomes

$$G_{PLL}(s) = (K_P + \frac{K_I}{s}) \frac{1}{s} \tag{22}$$

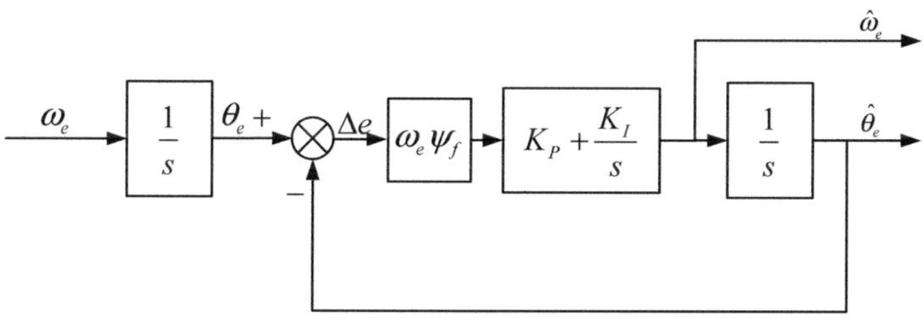

Fig. 4 Equivalent block diagram of PLL

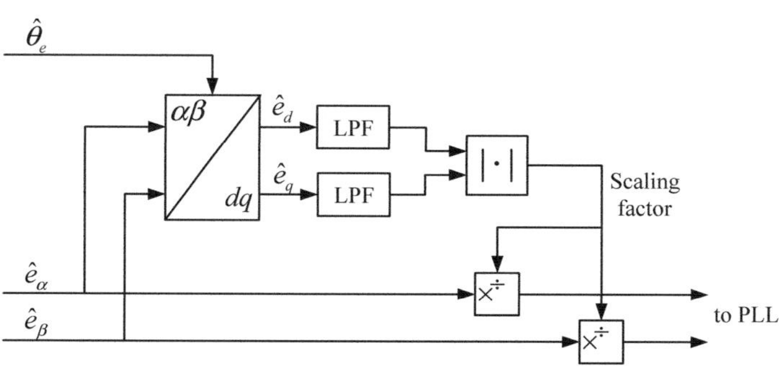

Fig. 5 Back-EMF estimates magnitude normalization for PLL

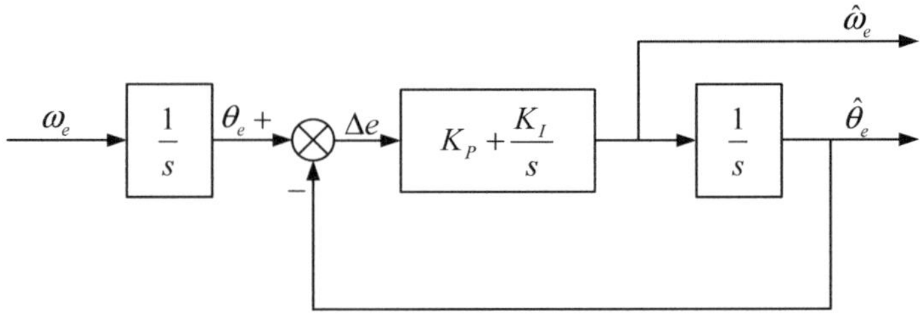

Fig. 6 Equivalent block diagram of input-normalized PLL

Thereby, the speed and position estimation method based on PLL can be expressed as

$$\begin{cases} \dot{\theta}_e = \hat{\omega}_e + K_P \theta_e \\ \dot{\hat{\omega}}_e = K_I \theta_e \end{cases} \tag{23}$$

Considering the speed estimator converges much faster than the speed varies, the error dynamic equations are obtained as (24), and in matrix form (25):

$$\begin{cases} \dot{\theta}_e = \bar{\omega}_e - K_P \theta_e \\ \dot{\bar{\omega}}_e = -K_I \theta_e \end{cases} \tag{24}$$

$$\begin{bmatrix} \dot{\theta}_e \\ \dot{\bar{\omega}}_e \end{bmatrix} = \begin{bmatrix} -K_P & 1 \\ -K_I & 0 \end{bmatrix} \begin{bmatrix} \theta_e \\ \bar{\omega}_e \end{bmatrix} \tag{25}$$

The characteristic polynomial of (25) is

$$C(s) = s^2 + K_P s + K_I \tag{26}$$

By properly selecting the gain K_P and K_I, a stable speed and position estimator is acquired.

5 Simulation Results

In order to verify the validity of proposed SMO, the simulation mode of PMSM sensorless vector control system is given in Fig. 7. The simulation parameters of PMSM are as follows: $R_S = 2.875\ \Omega, L_S = 8.5$ mH, $\psi_f = 0.175$ Wb, $J = 0.01$ kg.m^2, $p = 4$.

Figure 8 shows the comparison of simulation results of the estimated back-EMF between the traditional SMO and the proposed SMO at 500 r/min. It is very clear that in the proposed SMO there are no large ripples at the estimated back-EMF; the

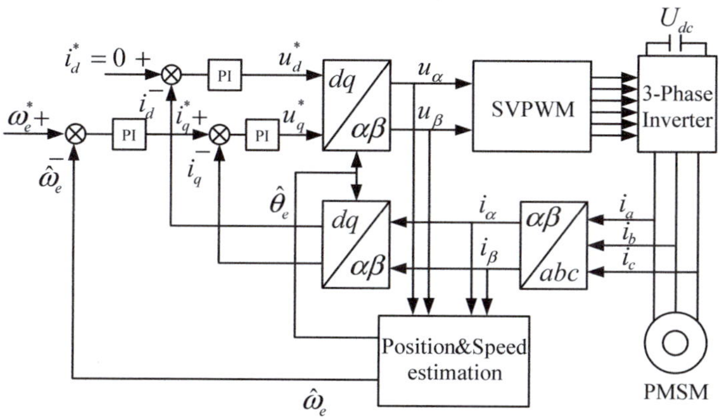

Fig. 7 Block diagram of sensorless vector control system

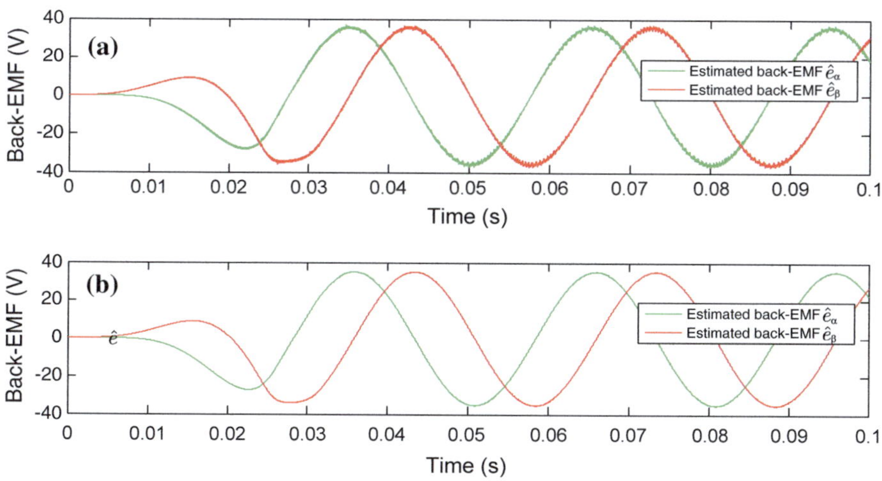

Fig. 8 Simulative comparison of back-EMF between traditional and proposed SMO

high frequency components are filtered apparently. The waveforms of the estimated back-EMF in the proposed SMO are smoother.

Figure 9 shows the comparison of simulation results of rotor position between the traditional SMO and the proposed SMO at 500 r/min. It is noticed that with the presence of the low-pass filter in the traditional SMO, even after the rotor position compensation, the estimated rotor position still exist significant phase lag compared with the real rotor position. After the adoption of the proposed SMO, the phase lag can be eliminated and the estimation error becomes smaller.

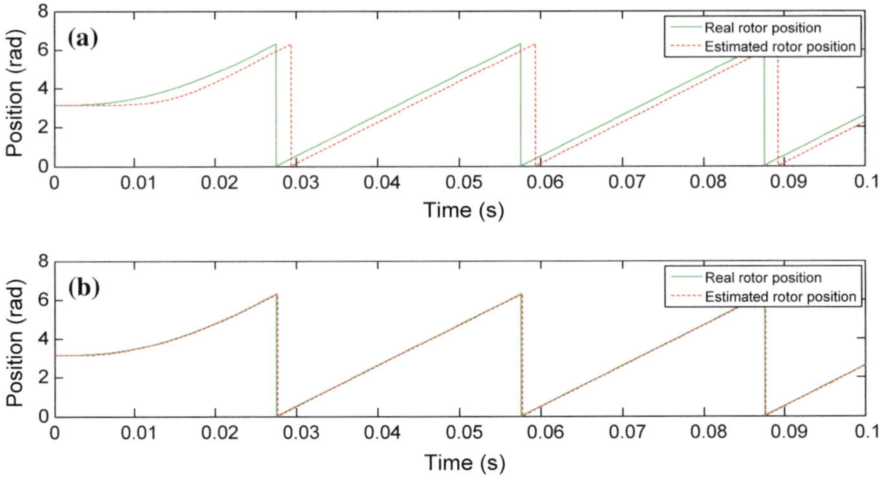

Fig. 9 Simulative comparison of rotor position between traditional and proposed SMO

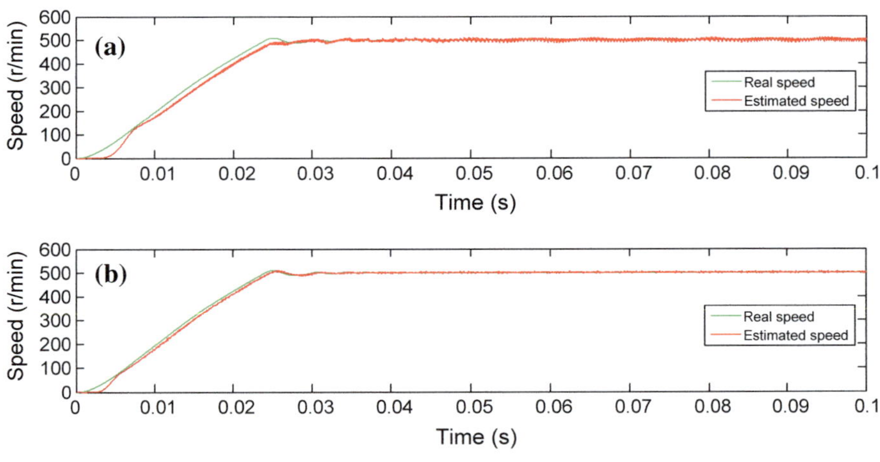

Fig. 10 Simulative comparison of speed between traditional and proposed SMO

Figure 10 shows the comparison of simulation results of speed between the traditional SMO and the proposed SMO at 500 r/min. Simulation results show that in the proposed SMO, the estimated speed can preferably track the change of the real speed, and compared with the traditional SMO, the fluctuation is smaller and the steady-state performance is better.

6 Conclusions

This paper proposes a new sensorless control algorithm for PMSM based on a new SMO. Using Lyapunov function theoretical analyzes the stability conditions of the proposed SMO. The chattering problem is resolved using a sigmoid function as the switching function and the use of the back-EMF observer. The PLL method improves the estimation precision and reduces noise. By simulation analysis proves that the proposed algorithms improve the observation accuracy and the steady-state performance of the system effectively.

Acknowledgments This work is supported by National Natural Science Foundation of China (grant number 61573170), Natural Science Foundation of Jiangsu Province (grant number BK20150530), Jiangsu Province Colleges and Universities Natural Science Research Project (grant number 12KJB470002), A Project Funded by the Priority Academic Program Development of Jiangsu Higher Education Institutions (PAPD), and the Professional Research Foundation for Advanced Talents of Jiangsu University (grant number 14JDG077).

References

1. Naidu M, Bose BK (1992) Rotor position estimation scheme of a permanent magnet synchronous machine for high performance variable speed drive. IEEE IAS 48–53
2. Lu KY, Lei X, Blaabjerg F (2013) Artificial inductance concept to compensate nonlinear inductance effects in the back EMF-based sensorless control method for PMSM. IEEE Trans Energy Convers 28(3):593–600
3. Bado A, Bolognani S, Zigliitio M (1992) Effective estimation of speed and rotor position of a pm synchronous motor drive by a Kalman filtering technique. In: Power electronics specialists conferences. IEEE, 951–957
4. Dhaouadir R, Mohan N (1991) Design and implementation of an extended Kalman filter for the state estimation of a permanent magnet synchronous moto. IEEE Trans Power Electron 6 (3):491–497
5. Haizhu X, Shunyi X, Linsen Z et al (2012) Investigation of hybrid sensorless control approach for dual-rotor PMSM. Electric Mach Control 16(7):12–16
6. Feng Q, Yikang H, Hongping J (2007) Investigation of the sensorless control for PMSM based on a hybrid rotor position self-sensing approach. Proc CSEE 27(3):13–17
7. Zhe S, Rongxiang Z, Ruzhen D (2007) Research on sensorless control method of PMSM based on a adaptive sliding mode observer. Proc CSEE 27(1):23–27
8. Lei H, Guangzhou Z, Heng N (2007) Sensorless control of interior permanent magnet synchronous motor by estimation of an extended electromotive force. Proc CSEE 27(9):59–63
9. Ran L, Guangzhou Z, Shaojuan X (2012) Sensor-less control of permanent magnet synchronous motor based on extended sliding mode observer. Trans China Electrotechnical Soc 27(3):79–85
10. Zhao Y, Qiao W, Wu L (2013) An adaptive quasi sliding mode position observer-based sensorless control for interior permanent magnet synchronous motors. IEEE Trans Power Electron 28(12):5618–5629
11. Jinkun L (2012) Sliding mode control design and matlab simulation. Tsing Hua University Press, Beijing, pp 4–6

12. Jie S, Wei C, Hongwei F, Yuejin Z (2011) Vector control of permanent magnet synchronous motor based on sliding mode observer. Electric Mach Control Appl 38(1):38–42
13. Wen D, Deliang L, Zhanqiang L (2012) Position sensorless control of PMSM using sliding mode observer with two-stage filter. Electric Mach Control 16(11):1–10
14. Kim H, Son J, Lee J (2011) A high-speed sliding-mode observer for the sensorless speed control of a PMSM. IEEE Trans Industr Electron 38(9):4069–4077

Tracking Control of a Nonminimum Phase Inverted Pendulum

Linqi Ye, Qun Zong, Xiuyun Zhang, Dandan Wang and Qi Dong

Abstract Three methods are investigated for the tracking problem of the famous cart–pole system (a kind of planar inverted pendulum). The output is required to track a sinusoid signal. Control design is based on the linearized model. First, we show that using output error and states feedback, approximate tracking can be achieved with bounded tracking error. Then exact tracking via output regulation is investigated. By constructing a regulator equation, the equivalent input and equivalent states which are needed to maintain output at the reference trajectory can be calculated. We show that the tracking problem is equivalent to the stabilizing problem in the states error coordinate. Finally, we study exact tracking via stable system center method. Because of the nonminimum phase property, a bounded solution for the internal dynamics is required and is estimated by stable system center method. Then the tracking problem can also be transformed into a stabilizing problem. Simulations are made for each method.

Keywords Tracking control · Nonminimum phase · Output regulation · Stable system center · Inverted pendulum

1 Introduction

Tracking control is much more complicated than stabilizing control, especially when the system is nonminimum phase. Stabilizing control only requires maintaining the output at a set-point, i.e., the equilibrium. For linear system, the stabilizing control theory has already been very perfect. Through proper coordinate change, the equilibrium can be moved to zero, and then various linear control

L. Ye · Q. Zong (✉) · X. Zhang · D. Wang · Q. Dong
School of Electrical Engineering and Automation, Tianjin University,
Tianjin 300072, China
e-mail: zongqun@tju.edu.cn

Y. Jia et al. (eds.), *Proceedings of 2016 Chinese Intelligent Systems Conference*, Lecture Notes in Electrical Engineering 405,
DOI 10.1007/978-981-10-2335-4_32

methods such as pole assignment and LQR can be used. So to say, it is not difficult for an engineer with basic control knowledge to accomplish the task of stabilizing a linear system. However, when it turns to the tracking problem, which aims at keeping the output moving along a varying trajectory (such as a sinusoid signal), things become much more challenging even for linear system.

The most widely applied methods, PID control and the basic feedback control method, and pole assignment, are not tracking control methods. They can be used to set-point control and achieve zero steady-state error, but when used in tracking control, the error increases as the tracking signal frequency goes up. To eliminate the tracking error, researchers have done amount of work in the past years. Summarizing the existed literatures, we may find that inversion seems necessary for exact tracking. Recall stabilizing control, the equilibrium to be achieved is known, which means we know the value every state should reach. Finally, the output will reach the set-point and other states also reach their steady values. What about tracking control? When the output exactly moves along the reference trajectory, we can imagine that other states should also move along certain trajectories. So this is the case for exact tracking: every state does not converge to fixed point but converges to fixed trajectory. In other words, all states as well as the input have specific trajectories to maintain the output at the reference trajectory. If we can calculate these trajectories, then tracking problems can be converted into stabilizing problem. This idea is the key for exact tracking.

In a manner of speaking, exact tracking is indeed an inversion problem, that is, giving a plant and a reference output, to calculate the equivalent input and equivalent states. A similar concept, named as stable inversion, has been widely studied [1, 2]. However, what we focused on is another method, output regulation theory [3, 4], which aims to achieve asymptotic tracking and disturbance rejection for a class of reference trajectories and disturbances. Suppose the reference output is generated by a known exosystem; then regulator equation can be constructed based on the plant and exosystem. The solution of the regulator equation directly leads to equivalent input and equivalent states. It should be noted that, whether the system is minimum phase or nonminimum phase, it has no influence on using output regulation theory.

The application of differential geometry theory to nonlinear control [5] brings us a new perspective to look at the problem of tracking control. Now we know that a system is composed of external and internal states. The dynamics of the internal states, i.e., the internal dynamics, is only driven by the external states. When it comes to minimum phase system, the internal dynamics is naturally stable, thus we only need to design tracking controller for the external states. This is not difficult because the external states are derivatives of the output which implies the equivalent external states are simply the derivatives of the reference output. But for nonminimum phase system, the internal dynamics is not stable, which means the internal states will not converge to their equivalent trajectories naturally. We need to force it. The equivalent internal states value is a bounded solution of the internal dynamics, and is also called ideal internal dynamics (IID). Stable system center [6, 7] is a method to estimate the IID asymptotically. When IID is obtained, we then know the equivalent values of all states and the tracking problem can be transformed into stabilizing problem.

Inverted pendulum is a classical nonlinear, unstable, nonminimum phase, underactuated system which has been widely used in control education and research to demonstrate the effectiveness of various control methods. References [8–13] have studied the stabilizing problem of inverted pendulum while the tracking problem has been studied in [14–16]. Here we will use it as an example to study some tracking control methods. Our methods are based on the linearized model.

2 Model Description

The cart–pole system consists of a cart and an inverted pendulum as shown in Fig. 1.

Here u is an external force that moves the cart in the horizontal plane, s is the cart position, θ is the pole angle, M is the mass of cart, m is the mass of pole, and l is the half length of pole. According to [8], the motion equations of the inverted pendulum are

$$\begin{aligned} &(M+m)\ddot{s} - ml\cos\theta\ddot{\theta} + ml\sin\theta\ddot{\theta}^2 = u \\ &I\ddot{\theta} - ml\cos\theta\ddot{s} - mgl\sin\theta = 0 \end{aligned} \tag{1}$$

where $I = 4ml^2/3$ is the moment of inertia of the pendulum with respect to the pivot. Define $v = \dot{s}, \omega = \theta$ as the cart velocity and angle acceleration, where (1) can be written in state space form

$$\begin{aligned} &\dot{s} = v, \quad \dot{v} = \frac{mg\sin\theta\cos\theta - 4ml\omega^2\sin\theta/3}{4(M+m)/3 - m\cos^2\theta} + \frac{4/3}{4(M+m)/3 - m\cos^2\theta}u \\ &\ddot{\theta} = \omega, \quad \dot{\omega} = \frac{(M+m)g\sin\theta - ml\omega^2\sin\theta\cos\theta}{4(M+m)l/3 - ml\cos^2\theta} + \frac{\cos\theta}{4(M+m)l/3 - ml\cos^2\theta}u \end{aligned} \tag{2}$$

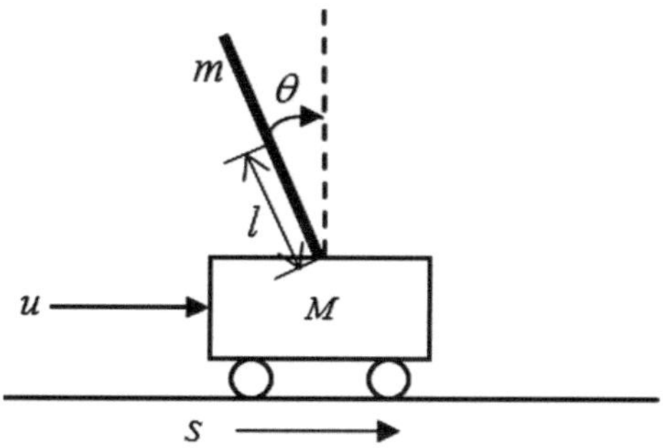

Fig. 1 Cart–pole system

where $g = 9.8\ \text{m/s}^2$ is the acceleration due to gravity, and other parameters are selected as $m = 1\,\text{kg}, M = 10\,\text{kg}, l = 1\,\text{m}$.

It can be verified that $(s^*, 0, 0, 0)$ is the equilibrium set (i.e., the system is in equilibrium when all states derivatives equal zero). The linearized model of the nonlinear model (2) in the equilibrium $(s^*, 0, 0, 0)$ is

$$\begin{aligned} \dot{s} &= v, \quad \dot{v} = 0.7171\theta + 0.09756u \\ \dot{\theta} &= \omega, \quad \dot{\omega} = 7.888\theta + 0.07317u \end{aligned} \tag{3}$$

This linearized model will be used later for tracking controller design, while the nonlinear model is used for simulation. The control objective is to track sinusoid trajectory for the cart position and pole angle, respectively. Two cases are considered here:

Case 1: Tracking control of the cart position with the desired trajectory $s_d = \sin(t)$.
Case 2: Tracking control of the pole angle with the desired trajectory $\theta_d = 0.2\sin(t)$.

3 Approximate Tracking Control

Consider the following nth-order SISO linear system:

$$\dot{x} = Ax + Bu, \quad y = x_1 \tag{4}$$

where $x = [x_1, x_2, \ldots, x_n]^T$ is the states vector, $u, y \in R^1$ are the input and the output, respectively. Suppose the desired tracking trajectory is y_d, where y_d and $\dot{y}_d$ are bounded, and denote the tracking error as $\tilde{x}_1 = y - y_d$.

Theorem 1 *If the feedback control $u = -Kx$ can stabilize system (4), then the control law*

$$u = -K\tilde{x} \tag{5}$$

with $\tilde{x} = [\tilde{x}_1, x_2, \ldots, x_n]^T$ can achieve approximate tracking with bounded tracking error for system (4).

Proof First, it indicates that $A - BK$ is Hurwitz since $u = -Kx$ can stabilize system (4). And if control law $u = -K\tilde{x}$ is applied, the closed-loop system becomes

$$\dot{\tilde{x}} = (A - BK)\tilde{x} + d \tag{6}$$

where $d = AY_d - \dot{Y}_d$ with $Y_d = [y_d, 0, \ldots, 0]^T$ can be seen as a perturbation. Since $A - BK$ is Hurwitz, according to the input-to-state stability theorem [17], $\tilde{x}$ will

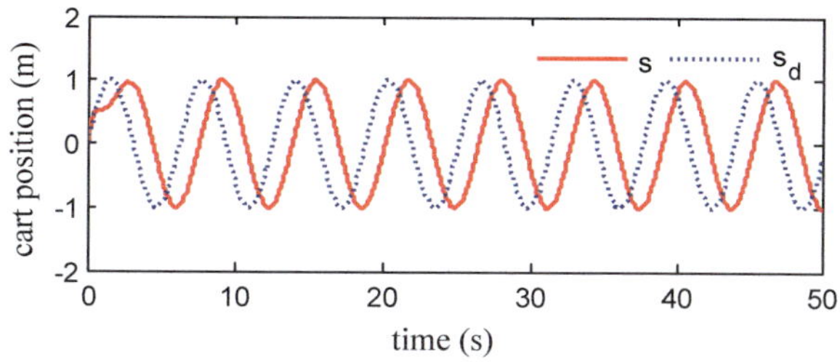

Fig. 2 Approximate tracking of cart position

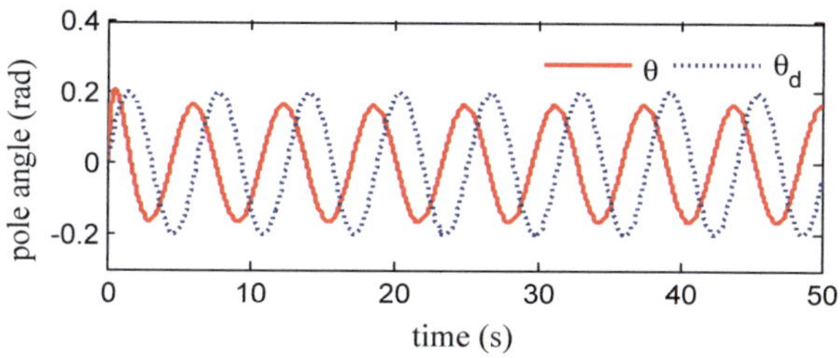

Fig. 3 Approximate tracking of pole angle

remain bounded for bounded d. Thus approximate tracking is achieved. This completes the proof.

Remark For the inverted pendulum model (4), the first column of A equals zero which indicates that $d = -\dot{Y}_d$. Thus when y_d is a constant, i.e., $d = 0$, control law $u = -K\tilde{x}$ can achieve asymptotic tracking. However, as the frequency of y_d increases, which means the upper bound of d increases, the tracking error will go up.

For case 1, we have $\tilde{x} = [s - s_d, v, \theta, \omega]^T$, and the control gain is selected as $K = [-200.8, \quad -248.2, \quad 1851.6, \quad 590.6]$ to place the closed-loop poles at $[-8, -9, -1 \pm i]$. The simulation result is shown in Fig. 2, from which we can see there exists a serious phase lag.

For case 2, we have $\tilde{x} = [s, v, \theta - \theta_d, \omega]^T$ and K is the same as that in case 1. The simulation result is presented in Fig. 3. Also a phase lag is observed.

4 Exact Tracking Control via Output Regulation

As mentioned in the former section, using output error and states feedback, approximate tracking with bounded tracking error can be achieved. In this section we will illustrate that using output regulation theory, output tracking problem can be transformed into a stabilizing problem.

First, let us consider how to transform a tracking problem into a stabilizing problem. Consider the nth-order SISO linear system as follows:

$$\dot{x} = Ax + Bu, \quad y = Cx \tag{7}$$

where $x = [x_1, x_2, \ldots, x_n]^T$ is the states vector, $u, y \in R^1$ are the input and output, respectively.

Suppose the control objective is to let y track y_d. To maintain y at y_d, we need certain quantity of control input and the states should also maintain in certain trajectories. Figuratively, we may call them equivalent input and equivalent states, and denote by x_d, u_d, respectively. Obviously, x_d, u_d should satisfy the following equation:

$$\dot{x}_d = Ax_d + Bu_d, \quad y_d = Cx_d \tag{8}$$

Define the states tracking error $\tilde{x} = x - x_d$, output tracking error $e = y - y_d$, and new control input $v = u - u_d$. Subtracting (8) from (7), we can obtain the stabilizing form of the original tracking problem:

$$\dot{\tilde{x}} = A\tilde{x} + Bv, \quad e = C\tilde{x} \tag{9}$$

Therefore, the original tracking problem to let y track y_d is equivalent to the stabilizing problem of system (9). And it is interesting to note that system (9) has identical form to the original system (7). So we may say that the tracking problem of a linear system is nothing different from its stabilizing problem; the only thing we need to do is to calculate the equivalent input and equivalent states, and output regulation theory can help us realize it.

To use output regulation theory, we need to know the exosystem, which generates the reference output. We may suppose the exosystem is

$$\dot{w} = Sw, \quad y_d = Qw \tag{10}$$

Then the tracking problem is equivalent to the following output regulation problem:

$$\dot{x} = Ax + Bu, \quad e = Cx - Qw \tag{11}$$

The equivalent input and equivalent states u_d, x_d should meet

$$\dot{x}_d = Ax_d + Bu_d, \quad e = Cx_d - Qw \tag{12}$$

Assume the equivalent input and equivalent states are $x_d = Xw, u_d = Uw$, where X, U are matrices needed to be solved. Substituting $x_d = Xw, u_d = Uw$ into (12) yields the regulator equation:

$$XS = AX + BU, \quad 0 = CX - Q \tag{13}$$

Fig. 4 Cart position via output regulation

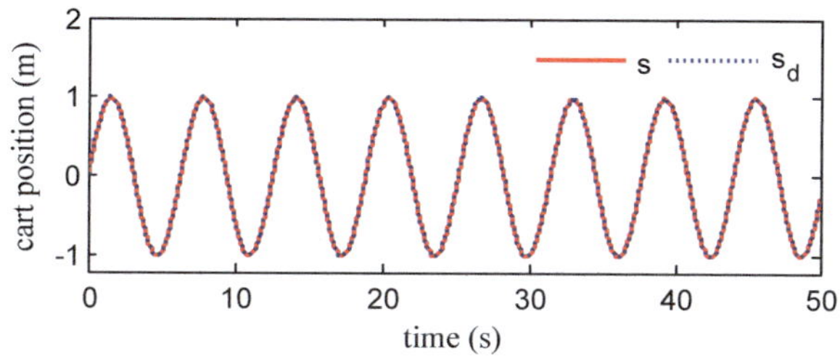

Fig. 5 Pole angle via output regulation

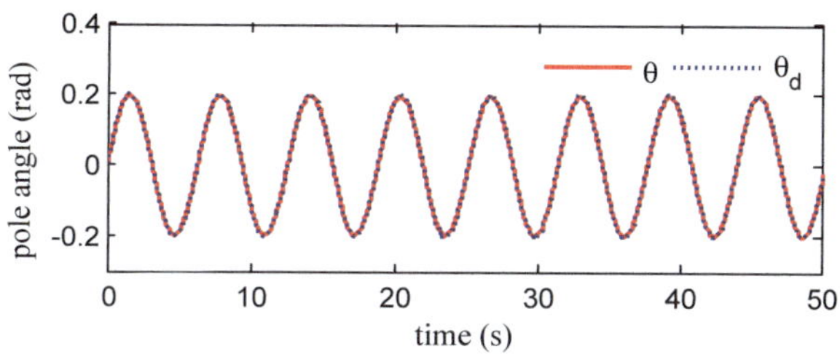

It is a matrix equation and can be easily solved, and then we can obtain the equivalent input and equivalent states.

As for the stabilizing of system (9), we still use pole assignment. Letting $v = -K\tilde{x}$ with $\tilde{x} = x - x_d$, and K ensures $A - BK$ be Hurwitz. Combine $v = u - u_d$ we can write the control law as

$$u = -K\tilde{x} + u_d \tag{14}$$

This yields the asymptotic stable closed-loop system $\dot{\tilde{x}} = (A - BK)\tilde{x}, \quad e = C\tilde{x}$. Comparing (14) with (5), we can find that the asymptotic control law adds an equivalent input and use error feedback of all states.

Now let us turn to the tracking problem of the inverted pendulum. The exosystems in case 1 and case 2 are the same: $\dot{w}_1 = w_2, w_2 = -w_1$.

For case 1, we have $s_d = w_1$, i.e., $Q = [1 \quad 0]$. The solution of the regulator Eq. (13) is $X = [1, 0;\ 0, 1;\ 0.0898,\ 0; 0,\ 0.0898], U = [-10.9102,\ 0]$. Then control law (14) is applied with the same K as before. The simulation result is shown in Fig. 4. We can see that almost perfect tracking is achieved.

For case 2, we have $\theta_d = 0.2w_1$, i.e., $Q = [0.2 \quad 0]$. The solution of the regulator Eq. (13) is $X = [2.2267\ , 0\ ;\ 0,\ 2.2267;\ 0.2,\ 0\ ;\ 0,\ 0.2], U = [-24.2933,\ 0]$. And also the control law (14) is applied with the same K as before. The simulation result is shown in Fig. 5. The tracking result is also very perfect.

5 Exact Tracking Control via Stable System Center

Unlike output regulation method, which directly based on system (7) and solve all the equivalent states and equivalent input by a regulator equation, we will consider how to solve the tracking problem based on the normal form in this section. First, let us introduce how to turn system (7) into a normal form.

For system (7), if $CA^{i-1}B=0(i=1,\ldots,r-1), CA^{r-1}B\neq 0$. Then we say system (7) has relative degree r, and $y,\dot{y},\ldots,y^{(r-1)}$ are called external states. By taking the following coordinate change,

$$\xi_i=y^{(i-1)}=CA^{i-1}x, i=1,2,...,r\ ; \quad \eta_i\begin{cases}\in\{x_1,x_2,...,x_n\}\\ \notin span\,\{\xi_1,\xi_2,...,\xi_r\}\end{cases}, \quad i=1,2,...,n-r \tag{15}$$

System (7) can be converted into the normal form as

$$\begin{aligned} y^{(r)}&=E\xi+F\eta+\alpha u\\ \dot{\eta}&=R\xi+S\eta+Tu \end{aligned} \tag{16}$$

where $\xi=[\xi_1,\xi_2,\ldots,\xi_r]^T$ is the external states vector and $\eta=[\eta_1,\eta_2,\ldots,\eta_{n-r}]^T$ is the internal states vector.

For simplicity, denote $z=[\xi,\eta]^T$; then the normal form (16) can be written as

$$\dot{z}=A_1z+B_1u, \quad y=z_1 \tag{17}$$

According to the first equation of (16), when y moves along y_d, the equivalent input is given by

$$u_d=\alpha^{-1}\left(y_d^{(r)}-E\xi_d-F\eta_d\right) \tag{18}$$

with $\xi_d=\left[y_d,\dot{y}_d,\ldots,y_d^{(r-1)}\right]$. Substituting (18) into the second equation of (16) yields

$$\dot{\eta}_d=R\xi_d+S\eta_d+\alpha^{-1}T\left(y_d^{(r)}-E\xi_d-F\eta_d\right) \tag{19}$$

Denoting

$$Q=S-\alpha^{-1}TF, \quad r(t)=R\xi_d+\alpha^{-1}T\left(y_d^{(r)}-E\xi_d\right) \tag{20}$$

then (19) becomes

$$\dot{\eta}_d = Q\eta_d + r(t) \tag{21}$$

This is known as the internal dynamics. It seems that by giving any initial value $\eta_d(0)$ we can obtain a solution $\eta_d(t)$ through (21). But it should be noted that for nonminimum phase system, system (21) is unstable and cannot be used to generate a bounded $\eta_d(t)$. So this is the case: we need to calculate a bounded solution $\eta_d(t)$ that satisfies (21). This bounded solution is also called ideal internal dynamics (IID). Stable system center [6, 7] provides a method to estimate the IID.

Suppose the eigenpolynomial of the exosystem which generates the reference output y_d is

$$s^k + p_{k-1}s^{k-1} + \ldots + p_1 s + p_0 \tag{22}$$

Then the IID can be estimated by constructing the following causal equation

$$\hat{\eta}_d^{(k)} + c_{k-1}\hat{\eta}_d^{(k-1)} + \ldots + c_1\dot{\hat{\eta}}_d + c_0\hat{\eta}_d = -\left(P_{k-1}r^{(k-1)} + \ldots + P_1\dot{r} + P_0 r\right) \tag{23}$$

where $c_{k-1}, c_{k-2}, \ldots, c_0$ is a set of Hurwitz polynomial coefficients, and P_i is defined by

$$\begin{aligned}
&P_{k-1} = \left(I + c_{k-1}Q^{-1} + \ldots + c_0 Q^{-k}\right)\left(I + p_{k-1}Q^{-1} + \ldots + p_0 Q^{-k}\right)^{-1} - I \\
&P_{k-2} = c_{k-2}Q^{-1} + \ldots + c_0 Q^{-(k-1)} - (P_{k-1} + I)\left(p_{k-2}Q^{-1} + \ldots + p_0 Q^{-(k-1)}\right) \\
&\ldots \\
&P_1 = c_1 Q^{-1} + c_0 Q^{-2} - (P_{k-1} + I)\left(p_1 Q^{-1} + p_0 Q^{-2}\right) \\
&P_0 = c_0 Q^{-1} - (P_{k-1} + I)p_0 Q^{-1}
\end{aligned} \tag{24}$$

Remark Since the inversion of Q is used, Q must be nonsingular, which indicates that this method can be only used to zero dynamics without zero eigenvalues.

To evaluate the estimate precision, we define

$$e = \dot{\hat{\eta}}_d - Q\hat{\eta}_d - r \tag{25}$$

as the estimate error.

Theorem 2 *The estimate error e satisfies $e^{(k)} + c_{k-1}e^{(k-1)} + \ldots + c_0 e = 0$. And since $c_{k-1}, c_{k-2}, \ldots, c_0$ is a set of Hurwitz polynomial coefficients, e will converge to zero asymptotically.*

Proof From (22) we have $y_d^{(k)} + p_{k-1} y_d^{(k-1)} + \ldots + p_1 \dot{y}_d + p_0 y_d = 0$. And from (20) we know that r is a combination of y_d and its derivatives, thus r also satisfies $r^{(k)} + p_{k-1} r^{(k-1)} + \ldots + p_1 \dot{r} + p_0 r = 0$. According to (23) we have

$$\begin{aligned} &\hat{\eta}_d^{(k)} + c_{k-1}\hat{\eta}_d^{(k-1)} + \ldots + c_1 \dot{\hat{\eta}}_d + c_0 \hat{\eta}_d = -\left(P_{k-1} r^{(k-1)} + \ldots + P_1 \dot{r} + P_0 r\right) \\ &\Rightarrow \hat{\eta}_d^{(k+1)} + c_{k-1}\hat{\eta}_d^{(k)} + \ldots + c_1 \hat{\eta}_d^{(2)} + c_0 \dot{\hat{\eta}}_d = -\left(P_{k-1} r^{(k)} + \ldots + P_1 r^{(2)} + P_0 \dot{r}\right) \end{aligned} \tag{26}$$

Combining (24), (25), and (26), it follows that

$$\begin{aligned} e^{(k)} + c_{k-1} e^{(k-1)} + \ldots + c_0 e &= \hat{\eta}_d^{(k+1)} + c_{k-1}\hat{\eta}_d^{(k)} + \ldots + c_0 \hat{\eta}_d^{(1)} - Q\left(\hat{\eta}_d^{(k)} + c_{k-1}\hat{\eta}_d^{(k-1)} + \ldots + c_0 \hat{\eta}_d\right) - \left(r^{(k)} + c_{k-1} r^{(k-1)} + \ldots + c_0 r\right) \\ &= -\left(P_{k-1} r^{(k)} + \ldots + P_1 r^{(2)} + P_0 r^{(1)}\right) + Q\left(P_{k-1} r^{(k-1)} + \ldots + P_1 r^{(1)} + P_0 r\right) - \left(r^{(k)} + c_{k-1} r^{(k-1)} + \ldots + c_0 r\right) \\ &= (-I - P_{k-1}) r^{(k)} + (QP_{k-1} - c_{k-1} - P_{k-2}) r^{(k-1)} + \ldots + (QP_1 - c_1 - P_0) r^{(1)} + (QP_0 - c_0) r \\ &= -\left(I + c_{k-1} Q^{-1} + \ldots + c_0 Q^{-k}\right)\left(I + p_{k-1} Q^{-1} + \ldots + p_0 Q^{-k}\right)^{-1}\left(r^{(k)} + p_{k-1} r^{(k-1)} + \ldots + p_0 r\right) = 0 \end{aligned} \tag{27}$$

This completes the proof.

According to (18), the estimate of the equivalent input is defined by

$$\hat{u}_d = \alpha^{-1}\left(y_d^{(r)} - E\xi_d - F\hat{\eta}_d\right) \tag{28}$$

Denote the states tracking error $\tilde{z} = [\xi, \tilde{\eta}]^T$ with $\xi = \xi - \xi_d$, $\tilde{\eta} = \eta - \hat{\eta}_d$, and the output tracking error $\tilde{y} = y - y_d$.

Theorem 3 *The control law*

$$u = -K\tilde{z} + \hat{u}_d \tag{29}$$

with $A_1 - B_1 K$ be Hurwitz can realize asymptotic tracking for system (17).

Proof Combining (20), (25), and (28), we will derive that

$$e = \dot{\hat{\eta}}_d - Q\hat{\eta}_d - r = \dot{\hat{\eta}}_d - R\xi_d - S\hat{\eta}_d - T\hat{u}_d \tag{30}$$

Define a new control input $v = u - u_d$, then from (16) we have

$$\begin{aligned} &\tilde{y}^{(r)} = E\xi + F\eta + \alpha u - y_d^{(r)} = E(\xi + \xi_d) + F(\tilde{\eta} + \hat{\eta}_d) + \alpha(v + \hat{u}_d) - y_d^{(r)} = E\xi + F\tilde{\eta} + \alpha v \\ &\dot{\tilde{\eta}} = R\xi + S\eta + Tu - \dot{\hat{\eta}}_d = R(\xi + \xi_d) + S(\tilde{\eta} + \hat{\eta}_d) + T(v + \hat{u}_d) - \dot{\hat{\eta}}_d \\ &\quad = R\xi + S\tilde{\eta} + Tv - (\dot{\hat{\eta}}_d - R\xi_d - S\hat{\eta}_d - T\hat{u}_d) = R\xi + S\tilde{\eta} + Tv - e \end{aligned} \tag{31}$$

which can be written as $\dot{\tilde{z}} = A_1\tilde{z} + B_1 v - [0_r; e]$ according to (17). Substituting $v = u - \hat{u}_d = -K\tilde{z}$ into it yields $\dot{\tilde{z}} = (A_1 - B_1 K)\tilde{z} - [0_r; e]$. Combine that $A_1 - B_1 K$ is

Hurwitz and $\lim_{t\to\infty} e \to 0$ from Theorem 2 we can obtain that $\lim_{t\to\infty} \tilde{z} \to 0$. This completes the proof.

Consider the linearized model (3), the equivalent input and equivalent states are defined by

$$\begin{aligned} \dot{s}_d &= v_d\,,\ \dot{v}_d = 0.71710_d + 0.09756u_d \\ \theta_d &= \omega_d,\ \dot{\omega}_d = 7.888\theta_d + 0.07317u_d \end{aligned} \tag{32}$$

In case 1, s and v are the external states. From the first two equations of (32) we have $u_d = (\ddot{s}_d - 0.71710_d)/0.09756$. Substituting it into the last two equations of (32) yields

$$\begin{bmatrix} \theta_d \\ \dot{\omega}_d \end{bmatrix} = \begin{bmatrix} 0 & 1 \\ 7.3502 & 0 \end{bmatrix} \begin{bmatrix} \theta_d \\ \omega_d \end{bmatrix} + \begin{bmatrix} 0 \\ 0.75\ddot{s}_d \end{bmatrix} \tag{33}$$

This is the internal dynamics with respect to output s. Since matrix $[0\ ,\ 1\ ;\ 7.3502\ ,\ 0]$ has eigenvalues ± 2.7111, the internal dynamics is unstable. We use (23) to estimate the IID, where $p_0 = 1, p_1 = 0, r(t) = [0;\ -0.75\sin t]$. Choose $c_0 = 1, c_1 = 2$, using (24) P_i is calculated as $P_0 = [-0.2395, 0; 0, -0.2395]$, $P_1 = [0, 0.2395; 1.7605, 0]$. So the IID estimator is designed as

$$\begin{bmatrix} \ddot{\theta}_d + 2\dot{\theta}_d + \theta_d \\ \ddot{\hat{\omega}}_d + 2\dot{\hat{\omega}}_d + \hat{\omega}_d \end{bmatrix} = \begin{bmatrix} 0.1796\ \cos\ t \\ -0.1796\ \sin\ t \end{bmatrix} \tag{34}$$

This together with (29) consists of the control law. Choose the same control gain as before and the initial value $\theta_d(0) = \dot{\theta}_d(0) = \hat{\omega}_d(0) = \dot{\hat{\omega}}_d(0)$. The simulation result is shown in Fig. 6, from which we can see that the IID estimator error converges to zero as well as the cart position tracking error within 10 s.

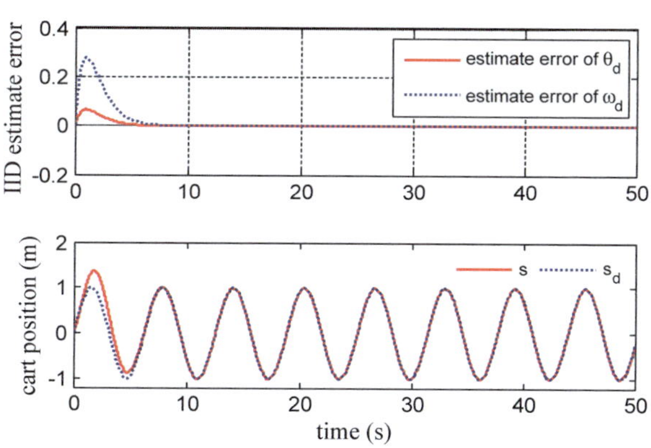

Fig. 6 ID estimator error and cart position via stable system center

In case 2, θ, ω are the external states. From the last two equations of (32) we have $u_d = (\dot{\theta}_d - 7.888\theta_d)/0.07317$. Substituting it into the first two equations of (32) yields

$$\begin{bmatrix} \dot{s}_d \\ \dot{v}_d \end{bmatrix} = \begin{bmatrix} 0 & 1 \\ 0 & 0 \end{bmatrix} \begin{bmatrix} s_d \\ v_d \end{bmatrix} + \begin{bmatrix} 0 \\ -9.8\theta_d + 1.333\dot{\theta}_d \end{bmatrix} \tag{35}$$

This is the internal dynamics with respect to output θ. Since matrix $[0, 1; 0, 0]$ has zero eigenvalue, it cannot be inverted. Thus stable system center method cannot be used in this case.

6 Conclusion

The approximate tracking method is simplest and is suitable for slow varying trajectory tracking, but tracking error increases as the tracking frequency rises. Output regulation and stable system center method can achieve asymptotic tracking. They both need to know the exosystem of the reference output, and both focus on transforming the tracking problem into stabilizing problem. Output regulation method constructs regulator equation and calculates the equivalent input and equivalent states. Stable system center method is based on the system normal form and tries to estimate the ideal internal dynamics, but it can be only used to system with zero dynamics that has no zero eigenvalues. Although simulation results exhibit good performance, we should notice that these three methods are all based on the linearized model. Exact tracking for wider range and higher frequency reference signals still remain a challenge to the nonlinear inverted pendulum.

Acknowledgments This work is supported by NSFC under Grant No. 61273092.

References

1. Hunt LR, Meyer G (1997) Stable inversion for nonlinear systems. Automatica 33 (8):1549–1554
2. Devasia S, Chen D, Paden B (1996) Nonlinear inversion-based output tracking. Autom Control IEEE Trans 41(7):930–942
3. Byrnes CI, Priscoli FD, Isidori A (1997) Output regulation of nonlinear systems. In: Output regulation of uncertain nonlinear systems. Birkhäuser Boston, pp 131–140
4. Jie H, Chen Z (2005) A general framework for tackling the output regulation problem. IEEE Trans Autom Control 49(12):2203–2218
5. Isidori A (1999) Nonlinear control systems II. Springer, London
6. Shkolnikov IA, Shtessel YB (2001) Tracking controller design for a class of nonminimum-phase systems via the method of system center. IEEE Trans Autom Control 46(10):1639–1643
7. Shkolnikov I, Shtessel A et al (2002) Tracking in a class of nonminimum-phase systems with nonlinear internal dynamics via sliding mode control using method of system center. Automatica 38(5):837–842

8. Landry M, Campbell SA, Morris K et al (2005) Dynamics of an inverted pendulum with delayed feedback control. SIAM J Appl Dyn Syst 4(2):333–351
9. Angeli D (1999) Almost global stabilization of the inverted pendulum via continuous state feedback. Automatica 37(01):1103–1108
10. Shiriaev A, Pogromsky A, Ludvigsen H et al (2000) On global properties of passivity-based control of an inverted pendulum. Int J Robust Nonlinear Control 3(4):2513–2518 vol.3
11. Lozano R, Fantoni I, Dan JB (2000) Stabilization of the inverted pendulum around its homoclinic orbit. Syst Control Lett 40(3):197–204
12. Olfati-Saber R (2010) Fixed point controllers and stabilization of the cart-pole system and the rotating pendulum. In: Proceedings of the, IEEE conference on decision and control, pp 1174–1181
13. Bedrossian NS (1992) Approximate feedback linearization: the cart-pole example. In: Proceedings of IEEE International Conference on Robotics and Automation 1987-1992, vol 3. IEEE
14. Gurumoorthy R, Sanders SR (1993) Controlling non-minimum phase nonlinear systems—the inverted pendulum on a cart example. In: American control conference, pp 680–685
15. Yan Q (2004) Output tracking of underactuated rotary inverted pendulum by nonlinear controller. In: IEEE conference on decision & control, vol. 3, pp 2395–2400
16. Huang J (2000) Asymptotic tracking of a nonminimum phase nonlinear system with nonhyperbolic zero dynamics. IEEE Trans Autom Control 45(3):542–546
17. Khalil H (2002) Nonlinear systems. Prentice Hall

Research on Sliding-Mode Control Technology of High-Performance LED Lighting Circuit

Kai-he Sun, Song-yin Cao, Yu Fang and Jin-yan Zheng

Abstract In order to adapt to the dimming characteristics of load led lamp, and to solve the shortcomings of slow dynamic response and poor dynamic performance on control parameter of the traditional PI control, a sliding-mode controller based on half-bridge LLC resonant converter with current negative feedback is studied in this paper. Two fixed frequencies are set according to the constant current characteristic of the LED lamp, and the switching of the working state under the two operating frequencies of the LLC converter is realized by sliding-mode control. The LED lamp dimming can be realized when the reference current is changed. In this paper, the large-signal model of LLC resonant converter is derived based on the extended description function method, and the sliding surface equations and control parameters of the system are deduced according to the model of the system. Then, the simulation circuit of 240 W LLC resonant converter with PI control and sliding-mode control is built. According to the simulation results, when the load current jumps, the response speed of sliding-mode control is faster than that of PI control, the dynamic performance of sliding-mode control is higher, and the system is more robust.

Keywords LED dimming · LLC resonant converter · Large-signal model · Sliding-mode control

---

K.-h. Sun (✉) · S.-y. Cao · Y. Fang · J.-y. Zheng
College of Information Engineering, Yangzhou University, Yangzhou 225127, China
e-mail: sunkaihe32@163.com

S.-y. Cao
e-mail: sycao@yzu.edu.cn

Y. Fang
e-mail: yzfangyu@126.com

J.-y. Zheng
e-mail: zhengjinyannt@163.com

Y. Jia et al. (eds.), *Proceedings of 2016 Chinese Intelligent Systems Conference*, Lecture Notes in Electrical Engineering 405,
DOI 10.1007/978-981-10-2335-4_33

1 Introduction

LED (light emitting diode), with small size, rich colors, and rapid response, is energy-saving, environmentally friendly, long-life, safe, and light intensity adjustable. With these advantages, it has attracted global attention and has been recognized as a new green light source in the world [1]. The traditional light sources, with high power loss and low efficiency, cannot be realized the light brightness adjustment, which not only causes the energy waste, but also affects people's work efficiency. Therefore, more and more attentions have been paid to the research on the control of high-performance LED lighting circuit [2, 3]. With the increasing requirement on the efficiency of the LED drive power supply, the LLC resonant converter has been widely studied and concerned. LLC resonant circuit can realize the soft-switching of the switch device on primary side and diodes on secondary side, along with a good load regulation capability and capacity to adapt the load. Therefore, LLC resonant circuit is in line with the requirements of high efficiency, high power density, and high dynamic response performance on dimming driving circuit [4, 5].

PID negative feedback control method is generally used in traditional LED driving power supply, which has a good performance in the steady state. However, the dynamic response performance of PID control is poor [6]. This paper designs a sliding-mode controller based on half-bridge LLC resonant converter with current negative feedback [7, 8]. Two fixed frequencies are set according to the constant current characteristic of the LED lamp, and the LLC converter works in turn under the two operating frequencies by sliding-mode control. In addition, LED light is adjusted by the changing of the reference current.

2 Overall Scheme of Lighting Circuit

The general design scheme of LED dimming circuit is shown in Fig. 1. The structure of the main circuit includes the AC/DC power factor correction (PFC) module, DC/DC driving module, output current sampling module, ambient light sampling module, and DSP control module. The PFC module realizes the power factor correction and DC boost using the boost circuit, DC/DC driving

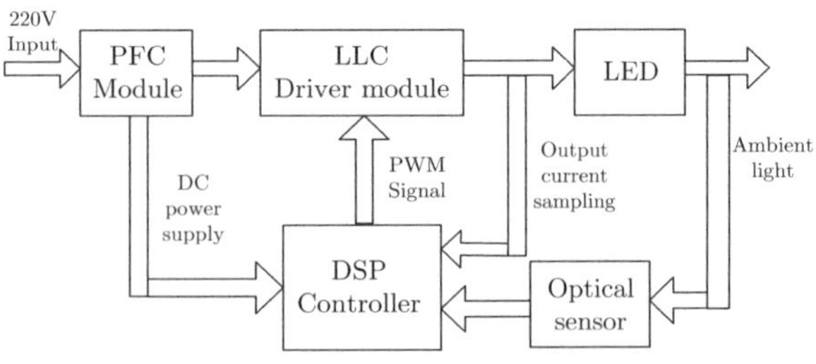

Fig. 1 Block diagram of the overall circuit design

circuit module adopts half-bridge LLC resonant converter topology, and the main control circuit realizes the digital control using DSP (digital signal processing).

3 The Large-Signal Model of Half-Bridge LLC Resonant Circuit

The structure of half-bridge LLC resonant circuit is shown in Fig. 2. Assuming that the circuit components are ideal, the circuit structure can be equivalent as the resonant network model shown in Fig. 3.

Select the resonant inductor current i_r, the transformer primary current i_p, the resonant capacitor voltage v_{cr}, the output filter capacitor voltage V_o as the state variables. Nonlinear differential equations can be listed according to the equivalent circuit model, and the characteristics of the system circuit are described in mathematical expressions (1):

$$\begin{cases} v_{in} = v_{cr} + L_r \frac{di_r}{dt} + L_m \frac{d(i_r - i_p)}{dt} \\ \mathrm{sgn}(i_p) \cdot nv_o = L_m \frac{d(i_r - i_p)}{dt} \\ i_r = C_r \frac{dv_{cr}}{dt} \\ n \cdot |i_p| = C_o \frac{dv_o}{dt} + \frac{v_o}{R_o} \end{cases} \tag{1}$$

where v_{in} is the equivalent input voltage of the LLC resonant converter, and n is the transformer turn ratio.

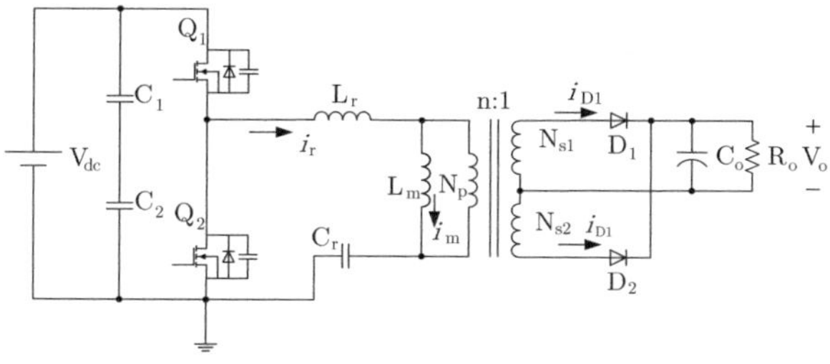

Fig. 2 Half-bridge LLC resonant converter topology

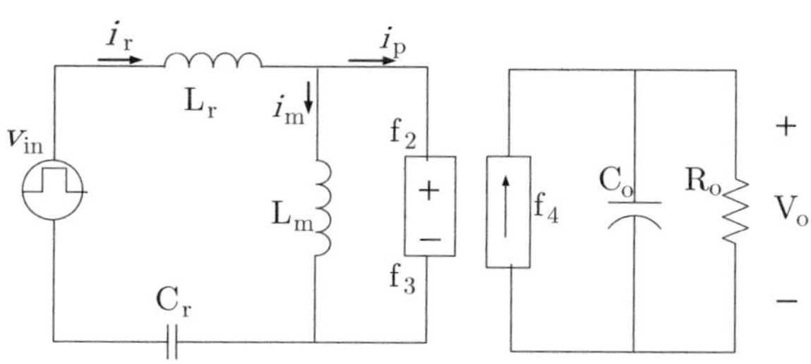

Fig. 3 Equivalent half-bridge LLC resonant network model

Based on the extended description function method, if we only take the fundamental component into consideration, the resonant inductor current i_r, transformer current i_p, and resonant capacitor voltage v_{cr} of the resonant tank circuit can be approximately expressed by sine fundamental component and cosine fundamental component [9, 10]. The resonant tank circuit variables can be approximated to sinusoidal fundamental component amplitude and phase relative to the input voltage v_{in}, as shown in formula (2):

$$\begin{cases} i_r = i_{r\ \sin} \sin \omega t + i_{r\ \cos} \cos \omega t = i_{rs} \sin(\omega t + \varphi_r) \\ i_p = i_{p\ \sin} \sin \omega t + i_{p\ \cos} \cos \omega t = i_{ps} \sin(\omega t + \varphi_p) \\ v_{cr} = v_{cr\ \sin} \sin \omega t + v_{cr\ \cos} \cos \omega t = v_{crs} \sin(\omega t + \varphi_{cr}) \end{cases} \tag{2}$$

The nonlinear variables v_{in}, $\mathrm{sgn}(i_p)$ and $|i_p|$ in formula (1) can be approximated to the direct current component and the fundamental component by Fourier transformation, and thus formula (3) is obtained:

$$\begin{cases} v_{in} \approx f_1(v_g) \cdot \sin \omega t \\ \mathrm{sgn}(i_p) \approx f_2(\varphi_p) \cdot \sin \omega t + f_3(\varphi_p) \cdot \cos \omega t \\ |i_p| \approx f_4(i_{ps}) \end{cases} \tag{3}$$

where

$$f_1(v_g) = \frac{2}{\pi} v_g,\ f_2(\varphi_p) = \frac{4}{\pi} \cos \varphi_p,\ f_3(\varphi_p) = \frac{4}{\pi} \sin \varphi_p,\ f_4(i_{ps}) = \frac{2}{\pi} i_{ps}$$

According to the principle of harmonic balance, the coefficients of the corresponding fundamental component on the left and right sides of the equation are equal. Then, the large-signal model of the system can be obtained:

$$\begin{cases} \frac{di_{rs}}{dt} = \frac{1}{L_r}\left(\frac{2}{\pi} v_g \cdot \cos \varphi_r - v_{crs} \cdot \cos(\varphi_r - \varphi_{cr}) - \frac{4}{\pi} n v_o \cdot \cos(\varphi_r - \varphi_p)\right) \\ \frac{di_{ps}}{dt} = \frac{L_r + L_m}{L_r \cdot L_m}\left(\frac{L_m}{L_r + L_m} \cdot \frac{2}{\pi} v_g \cdot \cos \varphi_p - \frac{L_m}{L_r + L_m} \cdot v_{crs} \cdot \cos(\varphi_p - \varphi_{cr}) - \frac{4}{\pi} n v_o\right) \\ \frac{dv_{crs}}{dt} = \frac{1}{C_r} \cdot i_{rs} \cdot \cos(\varphi_r - \varphi_{cr}) \\ \frac{dv_o}{dt} = \frac{1}{C_o}\left(\frac{2}{\pi} n \cdot i_{ps} - \frac{v_o}{R_o}\right) \\ \frac{d\varphi_r}{dt} = \frac{1}{i_{rs} \cdot L_r}\left(-\frac{2}{\pi} v_g \cdot \sin \varphi_r + v_{crs} \cdot \sin(\varphi_r - \varphi_{cr}) + \frac{4}{\pi} n v_o \cdot \sin(\varphi_r - \varphi_p)\right) - \omega \\ \frac{d\varphi_p}{dt} = \frac{L_r + L_m}{i_{ps} \cdot L_r \cdot L_m}\left(-\frac{L_m}{L_r + L_m} \cdot \frac{2}{\pi} v_g \cdot \sin \varphi_p + \frac{L_m}{L_r + L_m} \cdot v_{crs} \cdot \sin(\varphi_p - \varphi_{cr})\right) - \omega \\ \frac{d\varphi_{cr}}{dt} = \frac{1}{v_{crs} \cdot C_r} \cdot i_{rs} \cdot \sin(\varphi_r - \varphi_{cr}) - \omega \end{cases} \tag{4}$$

4 Sliding-Mode Controller of the LLC Resonant Converter

The sliding-mode control circuit is shown in Fig. 4. In order to enable sliding-mode control to be applied in the LLC resonant converter, in this paper, LLC converter only works on two specific discrete switching frequencies, and the output signal of the sliding-mode controller is used to control the switching under these two switching frequencies of the LLC converter. With the output current and output filter capacitor current as feedback variables, the sliding-mode control signal u generated by the hysteresis comparator is used to control the switching of switching frequency [11, 12]. Thus, the following definition can be obtained as in formula (5):

$$f = f_{\max} + u \cdot (f_{\min} - f_{\max})$$
$$u = \begin{cases} 1 \\ 0 \end{cases} \tag{5}$$

4.1 Theory of Selecting Fixed Frequency Points

LLC constant pressure design is usually adopted as formula (5) as shown by the normalized AC gain equation, such as formula (6):

$$M_{ac}(f_n, k, Q) = \frac{2nV_o}{V_{in}} = \frac{1}{\sqrt{(1 + k - \frac{k}{f_n^2})^2 + Q^2(f_n - \frac{1}{f_n})^2}} \tag{6}$$

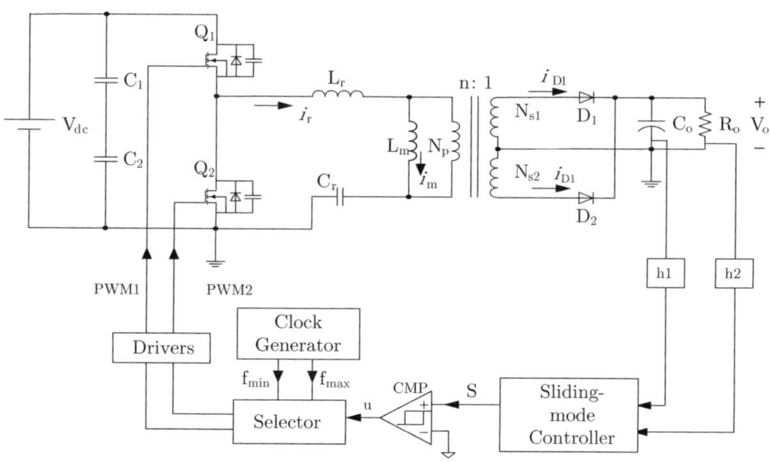

Fig. 4 Sliding-mode control scheme for LLC resonant converter

where f_n is the normalized frequency and k is the inductance coefficient. Quality factor Q in formula (6) is shown in formula (7):

$$Q = \frac{\sqrt{L_r/C_r}}{8n^2R/\pi^2} \tag{7}$$

However, LED driver generally requires constant current output. In order to get the constant current characteristics of the output of the LED driver, by substituting $V_o = R \cdot I_o$ and formula (7) into voltage gain Eq. (6), the output constant current characteristics of LED driver can be obtained as formula (8). And the characteristic curve of the output voltage and frequency is shown in Fig. 5, taking inductance coefficient k as 0.25:

$$V_o = \sqrt{\frac{1 - \left(\frac{2n}{V_{in}}\right)^2 \left(f_n - \frac{1}{f_n}\right)^2 \frac{\pi^4 L_r I_o^2}{64 C_r n^4}}{\left(\frac{2n}{V_{in}}\right)^2 \left(1 + k - \frac{k}{f_n^2}\right)^2}} \tag{8}$$

In Fig. 5, each curve corresponds to a different output current. The horizontal coordinate indicates the ratio of the working frequency to the resonant frequency. The vertical coordinate indicates the value of the output voltage. The $f_{\min}$ and $f_{\max}$ are the two fixed frequency points of the LLC converter. Known by the characteristics of LLC, when the converter is at full load, it works at the resonant frequency f_r, with the minimum loss and highest efficiency. And the LLC circuit can obtain the constant current output characteristics when working on the resonant points. Therefore, the minimum operating frequency $f_{\min}$ is located at the resonant frequency. For white LED load, data test and analysis conducted on the 1 W high power LED is shown in Table 1. When the load voltage is below 2.6 V, the current

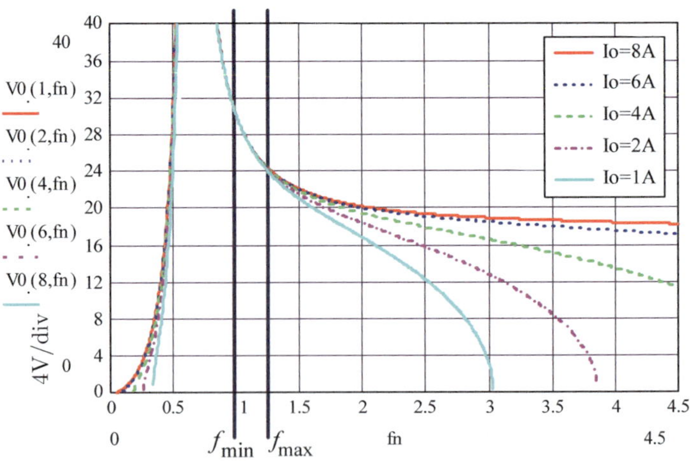

Fig. 5 Normalized dc current curves

Table 1 LED load characteristics

Voltage	3.3 V	3.2 V	3.0 V	2.9 V	2.8 V	2.6 V	2.5 V
Current	300 mA	247 mA	135 mA	95 mA	55 mA	5 mA	0 mA

flowing through the LED is already very small. As the rated output voltage of the LED driver power supply is 30 V, the load in this paper uses nine LEDs of 1 W in series. When voltage output on the load is 23 V, the current flowing through each LED is substantially zero. So the maximum operating frequency point can be determined according to the minimum voltage output of the LED load.

4.2 *The Design of Sliding-Mode Surface*

Usually, the sliding-mode surface uses the linear combination of the system variables, and the output voltage V_o is taken as the output control variable. Due to the sensitive nature of the LED load to the current, the output current I_o is selected as the control variable. According to the large-signal model of the system, the relative order of the system is 3, so it can be assumed that the response characteristic of the output current is as formula (9):

$$a_3\frac{d^3I_o}{dt^3}+a_2\frac{d^3I_o}{dt^2}+a_1\frac{dI_o}{dt}+a_0(I_o-I_{ref})=0 \tag{9}$$

According to the large-signal model of the system, the following equation related to *Io* can be deduced:

$$a_3'\frac{d^3I_o}{dt^3}+a_2'\frac{d^3I_o}{dt^2}+a_1'\frac{dI_o}{dt}+a_0'\frac{d^2i_{pm}}{dt^2}=0 \tag{10}$$

Ideally, the invariance conditions of sliding-mode region are $S=0$ and $\dot{S}=0$. The sliding surface equation can be written as follows:

$$S=k_p(\mathrm{I_o}-\mathrm{I_{ref}})+k_i\int(\mathrm{I_o}-\mathrm{I_{ref}})\mathrm{dt}+k_d\frac{dI_o}{dt}+k_c\frac{di_{pm}}{dt} \tag{11}$$

According to formula (11), the corresponding block diagram of sliding-mode controller can be obtained, as shown in Fig. 6 [13, 14]. Sampling output current Io and output filter capacitor current i_c are feedback variables, and the output filter capacitor current i_c reflects the feedback variable of the transformer primary side current i_p.

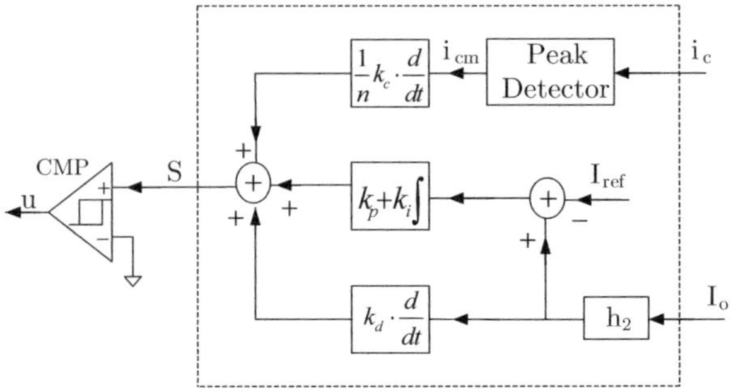

Fig. 6 Sliding-mode control block diagram

The sliding-mode control law of the system can be obtained by the existence condition $S \cdot \dot{S} < 0$ of sliding-mode control:

$$u = \begin{cases} 0 & for \quad S > 0 \\ 1 & for \quad S < 0 \end{cases} \tag{12}$$

But in the practical system, a hysteresis comparator is needed to ensure the feasibility of the sliding-mode control. Therefore, the control method can be optimized as

$$u = \begin{cases} 0 & for \quad S > \sigma \\ 1 & for \quad S < -\sigma \end{cases} \tag{13}$$

The σ in the above formula is the hysteresis band width.

5 Simulation Results

In order to verify the feasibility and the fast transient response characteristics of the sliding-mode control scheme, a 240 W LLC resonant converter prototype is built in this paper. The parameters are as follows in Tables 2 and 3.

Table 2 Parameters of power stage

Symbol	Description	Value	Unit
V_{in}	Input voltage	390–430	V
V_o	Output voltage	30	V
P_o	Output power	240	W
n	Transformer ratio	33:5	–
C_r	Resonant capacitor	47	nF
L_r	Resonant inductance	158	μH
L_m	Magnetizing inductance	700	μH
C_o	Output capacitor	2000	μF

Table 3 Parameters of pi controller and sliding-mode controller

Control mode	Symbol	Description	Value	Unit
PI Controller	K_p	Proportional gain	60	–
	K_i	Integral gain	600000	–
Sliding-mode controller	K_p	Proportional gain	50	–
	K_i	Integral gain	5000	–
	K_d	Differential gain	0.0001	–
	K_c	Current peak gain	0.006	–
	σ	Hysteresis band width	0.005	–
	f_{min}	Low fixed frequency	58	kHz
	f_{max}	High fixed frequency	200	kHz

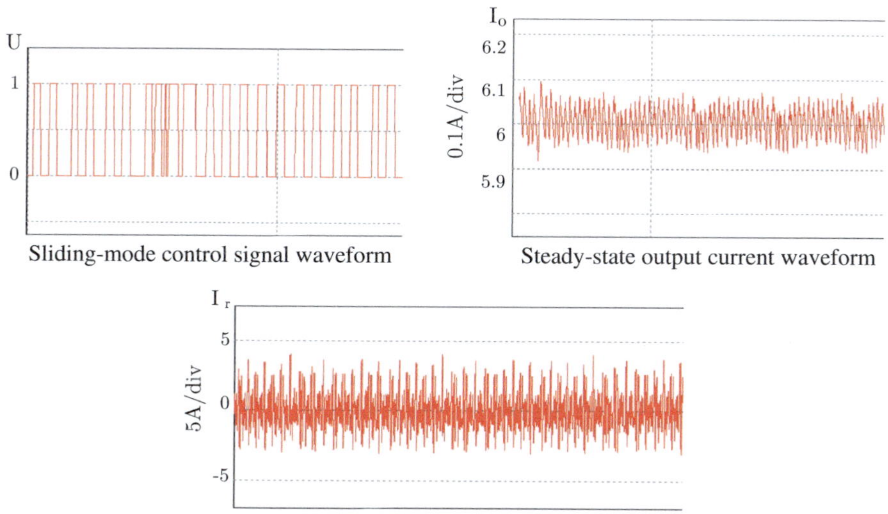

Fig. 7 Experimental waveforms of steady-state output current, resonant cavity current, and sliding-mode control signal (input voltage: 430 V)

As the load of the simulation circuit is a pure resistive load, the selection of high fixed frequency will be larger.

The PI control and sliding-mode control are simulated on the prototype. The sliding-mode controller is shown in Fig. 6. Figure 7 shows the waveforms of sliding-mode control signal, the steady-state output current, and the resonant cavity current. The feasibility of the proposed sliding-mode control scheme is verified by the simulation waveforms.

Figure 8 shows the output current transient response simulation waveforms using the traditional PI control, respectively, as shown in Fig. 8a, b. Figure 9 shows the simulation waveforms of output current transient response using sliding-mode

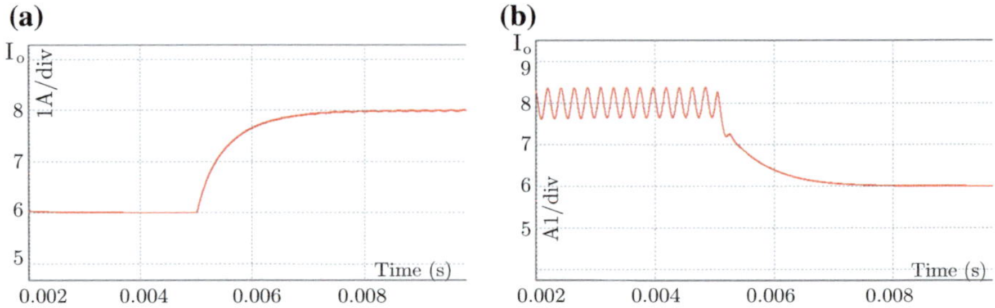

Fig. 8 Output current transient response waveforms with conventional PI control

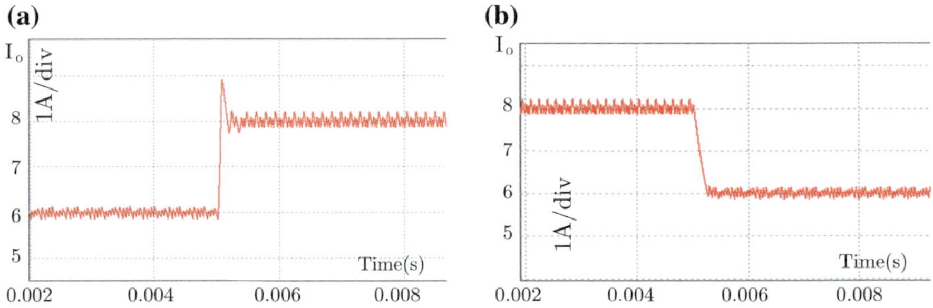

Fig. 9 Output current transient response waveforms with sliding-mode control

Table 4 Comparison of transient response between pi control and sliding-mode control

Load current variation	Transient response	PI controller	Sliding-mode controller	Unit
From 6 A to 8 A(brighten)	Convergence time	2996	187	μS
	Overshoot	0	0.89	A
From 8 A to 6 A(darken)	Convergence time	3205	261	μS
	Overshoot	0	0.03	A

control, respectively, as shown in Fig. 8a, b. The specific dynamic performance has been compared in Table 3.

It can be seen from Figs. 8 and 9, and Table 4 that the response speed of sliding-mode control is rapider than that of PI control when the output current is changed. However, due to the chattering of the sliding-mode controller, there is a relatively large ripple in the output steady-state waveform of the sliding-mode controller. In order to reduce the output ripple, the capacitance value of the output capacitor should be increased as much as possible. In addition, the selection of span between the two fixed frequencies should be reasonable, and try to minimize the span between the two switching frequencies at the request of the requirements of the application.

6 Conclusion

Aimed at LED dimming, a sliding-mode controller based on half-bridge LLC resonant converter with current negative feedback is studied in this paper. In order to realize the sliding-mode control of the LLC frequency-changing circuit, two fixed operating frequencies are set according to the constant current characteristic of the LED, thus realizing the switching between the two operating frequencies through the output of the sliding-mode controller. Therefore, the large-signal model of the system is given, and the sliding-mode surface equation of the system is derived based on the large-signal model. The experimental results show that when the reference current is changed, which is realized by adjusting the brightness of LED lamp, the sliding-mode control strategy can achieve fast dynamic response and can be more robust, thus obtaining a high-performance dimming performance.

References

1. Magesh Kannan P, Nagarajan G (2015) VLSI implementation of adaptive current controller for high bright LED lighting. In: 2015 3rd International Conference on Signal Processing, Communication and Networking (ICSCN), Chennai, pp 1–6
2. Farahat A, Florea A, Lastra JLM, Brañas C, Sánchez Francisco Javier Azcondo (2015) Energy efficiency considerations for led-based lighting of multipurpose outdoor environments. IEEE J Emerg Sel Top Power Electron 3:599–608
3. Li H, Liu Z, Sun Y, Lin Y (2015) Design of LED dimming system with the specific target parameter control function. In: 2015 12th China international forum on Solid State Lighting (SSLCHINA), Shenzhen, pp 106–110
4. Zahid ZU, Lai JJ, Huang XK, Madiwale S, Hou J (2014) Damping impact on dynamic analysis of LLC resonant converter. In: 2014 IEEE Applied Power Electronics Conference and Exposition—APEC 2014, Fort Worth, TX, pp 2834–2841
5. Dil Kumar TR, Mija SJ (2015) Dynamic SMC control scheme with adaptively tuned PID controller for speed control of DC motor. In: 2015 IEEE International Conference on Industrial Technology (ICIT), Seville, pp 187–191
6. Feng Weiyi, Lee Fred C, Mattavelli Paolo (2014) Optimal trajectory control of LLC resonant converters for LED PWM dimming. IEEE Trans Power Electron 29(2):979–987
7. Lijing D, Senchun C, Baihai Z, Kiong NS (2016) Sliding mode control for multi-agent systems under a time-varying topology. Int J Syst Sci 47, 2193–2200
8. Ma H, Liu Q, Guo J (2012) A sliding-mode control scheme for llc resonant DC/DC converter with fast transient response. In: IECON 2012-38th annual conference on IEEE industrial electronics society, Montreal, QC, 162–167
9. Buccella Concettina, Cecati Carlo, Latafat Hamed, Pepe Pierdomenico, Razi Kaveh (2015) Observer-based control of LLC DC/DC resonant converter using extended describing functions. IEEE Trans Power Electron 30(10):5881–5891
10. Cao Y, Chen XB (2014) An output-tracking-based discrete PID-sliding mode control for MIMO systems. IEEE/ASME Trans Mechatron 19(4):1183–1194
11. Yang Y, Wang W, Zhou T (2014) Sliding mode control based on three-sliding-surface for an inverted pendulum system. In: 2014 33rd Chinese Control Conference (CCC), Nanjing, pp 65–70

12. Minjae J, Hosung K, Joowon B (2012) Dynamic analysis and optimal design of high efficiency full bridge LLC resonant converter for server power system. In: Applied Power Electronics Conference and Exposition (APEC), Orlando, FL, pp 1292–1297
13. Xu SS, Chen S, Wu Z (2015) Study of nonsingular fast terminal sliding-mode fault-tolerant control. IEEE Trans Ind Electron 62(6):3906–3913
14. Lakhekar G, Deshpande R (2014) Diving control of autonomous underwater vehicles via fuzzy sliding mode technique. In: 2014 International Conference on Circuit, Power and Computing Technologies (ICCPCT), Nagercoil, pp 1027–1031

Globally Exponentially Stable Triangle Formation Control of Multi-robot Systems

Qin Wang, Qingguang Hua and Zuwen Chen

Abstract In this paper, the problem of formation control for nonholonomic robots is investigated. Based on the negative gradient method and the Lyapunov direct method, a globally and exponentially stable control scheme for multi-robot formation control system is designed. The proposed control law using the adaptive perturbation method can guarantee the globally exponential stability of the desired triangle and line formation, and the equilibrium set of the overall system is unique, which is exactly the desired formation set. Finally, some simulations illustrate the effectiveness and correctness of the proposed controllers.

Keywords Multi-robot systems · Lyapunov direct method · Globally exponential stability

1 Introduction

With the development of control theory and artificial intelligence, the multi-agent formation control has become a hot research field. The goal of formation control is to drive multiple robots to the desired shape. It is well known that robot groups are more successful than a single robot in many critical applications such as exploration, search and rescue. In recent years, control of multi-agent system has received increasing attention [1–3], and multi-robot with nonholonomic constraints also gained increasing attention.

The formation control problems are categorized according to variables, there are two research directions: position-based [4, 5] and distance-based [6, 7]. In distance-based formation control, control law is nonlinear even if agent models are linear, so it is a challenge to analyze the stability of the multi-agent system under a distance-based control law. What is more, most of the mobile models have to satisfy

Q. Wang (✉) · Q. Hua · Z. Chen
Department of Automation, College of Information Engineering, Yangzhou University, Yangzhou 225127, China
e-mail: qinwang@yzu.edu.cn

Y. Jia et al. (eds.), *Proceedings of 2016 Chinese Intelligent Systems Conference*, Lecture Notes in Electrical Engineering 405,
DOI 10.1007/978-981-10-2335-4_34

nonholonomic constraints, the design of control laws for distance-based formation is more difficult. Otherwise, there is a more challenging problem: how to design the controller to keep formation and avoids the collision between neighboring robots. Authors of [8] discussed triangle formation control problem with a common leader, the result showed that three agents converge exponentially fast to a desired triangular formation from any initially non-collinear position. It was also shown that there is a thin set of initially collinear formations which remain collinear and may drift off to infinity as $t \to \infty$. The results of [9] stated that the equilibrium set which was the gradient of the formation control systems converged to zero, and the equilibrium set was the union set combined by desired set of equilibrium points and the collinear set. In [10], it was shown that the tree structure was necessary and sufficient condition for distance-based formation stabilization with negative gradient control laws. The distributed formation control method for rigid formation graph based on the gradient control law was proposed in [11], and it was shown that the closed-loop system was locally asymptotically stable using the center manifold theory. The paper [12] showed that the gradient control can stabilize an acyclic triangular formation if the three agents are not initially collinear. The control law proposed in [13] was based on the adaptive perturbation methods, and it can guarantee that the equilibrium set of the overall system is unique, which exactly the desired formation is set. However, the nonholonomic constraints of models are not considered in the above papers.

In this paper, we consider the formation control problem for nonholonomic unicycles. Based on potential function and adaptive perturbation method, the globally stable formation control law is designed to guarantee the exponential convergence of the desired triangle or line formation shape. It has higher convergence rate compared with [11, 12] and there is no collision between neighboring robots.

2 Problem Statement

Consider a system of three nonholonomic robots operating in the same work space R^2. Agent's motion is described by the following nonholonomic kinematics:

$$\begin{aligned} \dot{x}_i &= u_i \cos\theta_i \\ \dot{y}_i &= u_i \sin\theta_i \\ \theta_i &= \omega_i. \end{aligned} \tag{1}$$

where $r_i = [r_{xi},\ r_{yi}]^T = [x_i, y_i]^T$ and $\theta_i \in (-\pi, \pi]$ are the position and steering angle of agent i (see Fig. 1), $u_i \in R$ and $\omega_i \in R$ are the translational and rotational velocity of agent i.

The information architecture is modeled as an undirected graph $G = (V, E)$, where $V = \{1, 2, 3\}$ is a set of vertices and $E \subseteq V \times V$ is a set of edges. Let N_i denote

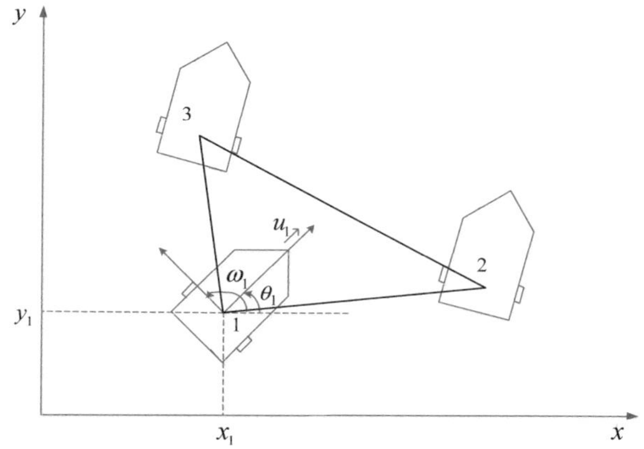

Fig. 1 Mathematical model of mobile robots

the set of neighboring vertices of vertex i. Let each edge $e_{ij} \in E$ in graph G be assigned to a scalar parameter $d_{ij} = d_{ji}$, representing the desired distance which robots i, j should preserve. Where d_{ij}, $(i, j) \in E$ are positive constants. Denote by $\alpha_{ij} = ||r_{ij}||^2$, the Euclidean distance between robots i and j with $j \in N_i$.

From the above, the problem of formation control is to design control law which can make every trajectory of overall system starting from any position converge to the desired formation, where $\lim_{t\to\infty}(||r_{ij}|| - d_{ij}) = 0$. And there is no collision between neighboring robots, which means there is no $t = t_1 \geq 0$, such as $||r_{ij}|| = 0$, $i = 1, 2, 3$.

3 Controller Design

We give the potential function between robots i and its neighboring robots j, whose definition had been given in [7] as follows:

$$V_{ij} = \frac{(||r_{ij}||^2 - d_{ij}^2)^2}{||r_{ij}||^2}. \tag{2}$$

We also define

$$\rho_{ij} = \frac{\partial V_{ij}}{\partial ||r_{ij}||^2} = \frac{||r_{ij}||^4 - d_{ij}^4}{||r_{ij}||^4}. \quad \rho = [\rho_{12} \quad \rho_{23} \quad \rho_{31}]^T.$$

Motivated by [13], we proposed a globally stable formation control strategy for nonholonomic robots system by adding an adaptive perturbation to the translational velocity control of any robot.

Then we give the control law with adaptive perturbation as follows:

$$u_1 = -sign(R_1^T \nabla_{r1} V_1)||\nabla_{r1} V_1||(\sum_{j \in N_1} ||r_{1j}||^2) - l_1|\rho_{12}|(a_1 + sign(R_1^T \nabla_{r1} V_1)). \quad (3)$$

$$u_2 = -sign(R_2^T \nabla_{r2} V_2)||\nabla_{r2} V_2||(\sum_{j \in N_2} ||r_{2j}||^2). \quad (4)$$

$$u_3 = -sign(R_3^T \nabla_{r3} V_3)||\nabla_{r3} V_3||(\sum_{j \in N_3} ||r_{3j}||^2). \quad (5)$$

$$\theta_i = -k_i(\theta_i - \theta_{id}), i = 1, 2, 3. j \in N_i \quad (6)$$

where $\nabla_{ri} V_i = 2 \sum_{j \in N_i} \rho_{ij} r_{ij}$, $\theta_{id} = \arctan2(\nabla_{ri} V_{iy}, \nabla_{ri} V_{ix})$, $R_i = [\cos\theta_i \quad \sin\theta_i]^T$, $\theta_{id} \in (-\pi, \pi]$. l_1, k_i are any positive constants, $i = 1, 2, 3$. a_1 is equal to $\sin(w_1 t)$ or $\cos(w_1 t)$ and w_1 is any positive constant. The arctan2 function can be expressed as in the paper [13].

4 Stability Analysis

Theorem 1 *Assume that system (1) is driven by the control law (3)–(6) with the potential function V_{ij} as in (2). Then, the desired formation of multi-robot systems is globally exponentially stable, the translational and rotational velocities of three nonholonomic robots converge to zero, and no collision between any pair of robots occurs during the motion.*

Proof Let the Lyapunov function candidate be chosen as following function:

$$V(r_{ij}(t)) = \sum_{i=1}^{3} \sum_{j \in N_i} V_{ij}. \quad (7)$$

Note that function V is smooth and hence regular, its generalized gradient is a singleton which is equal to its usual gradient everywhere in the state space. Then, the derivation of V is as follows,

$$\begin{aligned}\dot{V} \subset & -2\sum_{i}^{3} R_i^T \nabla_{r_i} V_i K[sign](R_i^T \nabla_{r_i} V_i)||\nabla_{r_i} V_i||(\sum_{j \in N_i} ||r_{ij}||^2) - 2l_1 a_1 R_1^T \nabla_{r_1} V_1 |\rho_{12}| \\ & - 2l_1 R_1^T \nabla_{r_1} V_1 K[sign](R_1^T \nabla_{r_1} V_1)|\rho_{12}|.\end{aligned} \quad (8)$$

where $K[f](x)$ is called Filippov set-valued mapping [14]. By using Theorem 1 in [15], the inclusions of the Filippov set can be calculated, then we have $R_i^T \nabla_{ri} V_i K[sign](R_i^T \nabla_{ri} V_i) = |R_i^T \nabla_{ri} V_i|$, thus it follows that

$$\dot{V} \leq -2 \sum_{i}^{3} (R_i^T \nabla_{ri} V_i)^2 (\sum_{j \in N_i} ||r_{ij}||^2) \leq 0. \tag{9}$$

Therefore, we have the following lemmas.

Lemma 1 *Consider that system (1) is driven by the control law (3)–(6) with potential function given by (2), and starting from any initial state in* $S = \{r_{ij} | V \leq V_0 < +\infty\}$. *Then it holds that* $0 < \zeta_1 \leq ||r_{ij}(t)||^2 \leq \zeta_2$.
where

$$\zeta_1 = \frac{1}{2}\left(2d_{ij}^2 + V_0 - \sqrt{V_0^2 + 4d_{ij}^2 V_0^2}\right), \quad \zeta_2 = \frac{1}{2}\left(2d_{ij}^2 + V_0 + \sqrt{V_0^2 + 4d_{ij}^2 V_0^2}\right).$$

Proof For any $r_{ij}(0) \in S$ we have $V(r_{ij}(t)) \leq V(r_{ij}(0)) \leq V_0 < +\infty$ for all $t > 0$, since $V(r_{ij}(t)) = \sum_{i=1}^{3} V_{ij}$, hence $V_{ij} \leq V_0$. So that $\zeta_1 \leq ||r_{ij}||^2 \leq \zeta_2$, where $\zeta_1 = \frac{1}{2}\left(2d_{ij}^2 + V_0 - \sqrt{V_0^2 + 4d_{ij}^2 V_0^2}\right)$, It is easily seen that ζ_1 is strictly positive. Therefore, no collision occurs between any two adjacent robots.

Lemma 2 *Consider system (1) is driven by the control law (3)–(6) with potential function given by (2), and starting from any initial state in S. Then, the overall system converges exponentially fast to the set* $E = \{r_{ij} | \; ||r_{ij}|| = d_{ij}, i = 1, 2, 3\}$.

Proof Since (7), we have

$$\dot{V} \leq -4\rho' Q' Q\rho(||r_{12}||^2 + ||r_{23}||^2 + ||r_{31}||^2). \tag{10}$$

where

$$Q = \begin{bmatrix} -r_{12}^T R_1 & r_{12}^T R_2 & 0 \\ 0 & -r_{23}^T R_2 & r_{23}^T R_3 \\ r_{31}^T R_1 & 0 & -r_{31}^T R_3 \end{bmatrix}$$

The agents will not be collinear because of adaptive perturbation [16]. So that:

$$r_{12}^T R_1 \neq \frac{\rho_{31}}{\rho_{12}} r_{31}^T R_1, \quad r_{23}^T R_2 \neq \frac{\rho_{12}}{\rho_{23}} r_{12}^T R_2, \quad r_{31}^T R_3 \neq \frac{\rho_{23}}{\rho_{31}} r_{23}^T R_3,$$

Thus $r_{12}^T R_1 r_{23}^T R_2 r_{31}^T R_3 \neq r_{12}^T R_2 r_{23}^T R_3 r_{31}^T R_1$. Since Q is invertible matrix, thus $Q^T Q$ is positive definite, then we have

$$\begin{aligned}\dot{V} &\leq -4\rho^T Q^T Q\rho(||r_{12}||^2 + ||r_{23}||^2 + ||r_{31}||^2) \\ &\leq -\lambda(V_{12} + V_{23} + V_{31}) \\ &\leq -(\lambda/2)V.\end{aligned} \tag{11}$$

where λ is the smallest eigenvalue of $Q^T Q$.

In steady state $\dot{V} = -(\lambda/2)V = 0$, hence $||r_{ij}|| = d_{ij}$, which means the multi-robot formation system converge exponentially fast to the unique desired set $E = \{r_{ij}|\ ||r_{ij}|| = d_{ij}, i = 1, 2, 3\}$. In view of the definition of $u_i, i = 1, 2, 3$, the translational velocities of three robots converge to zero. Within E, the closed-loop equations for the orientation velocities have the form $\dot{\theta}_i = -k_i\theta_i$ for all i, and hence the orientation velocities of all robots also tend to zero.

5 Simulations

Case 1: When the initial position of three robots are $r_1 = [0.2, 0.3]^T$, $r_2 = [0.4, 0.5]^T$, $r_3 = [0.6, 0.7]^T$, and three robots are very close. The desired distance are $d_{12} = 3$, $d_{23} = 4$, $d_{31} = 5$. We chose $k_1 = k_2 = k_3 = 1$, $l_1 = 10$, $w_1 = 1.2$ and $a_1 = \cos(w_1 t)$. The movement trajectories of the robots and the distance $||r_{ij}||$ shown in Figs. 2 and 3, respectively, demonstrate that three robots achieve the desired formation and never collide between any pair of adjacent robots. In Figs. 4 and 5, all the robots' velocities converge to zero.

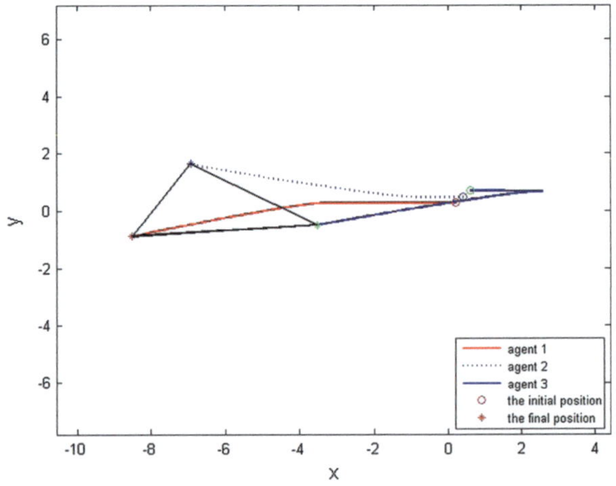

Fig. 2 The movement trajectories of the robots

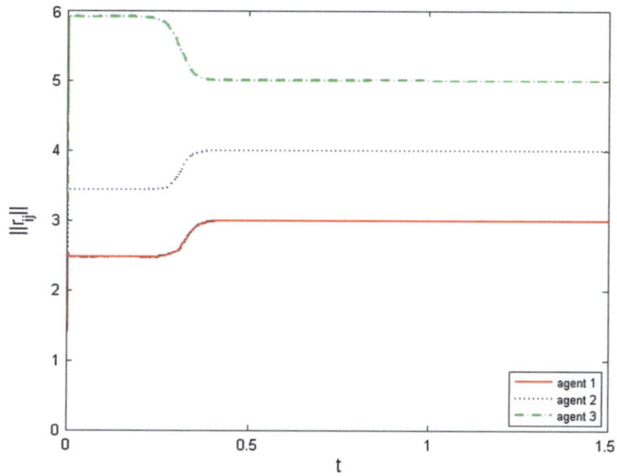

Fig. 3 The distance of three robots

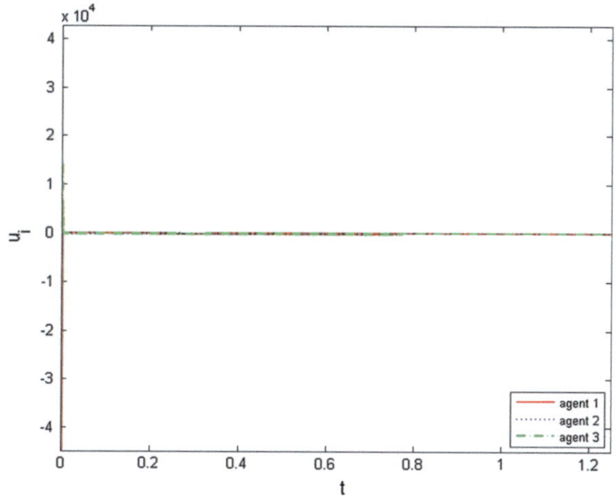

Fig. 4 The translational velocity of three robots

Case 2: When the initial position of three robots are $r_1 = [1, 0]^T$, $r_2 = [2, 0]^T$, $r_3 = [0, 3]^T$, the desired distance are $d_{12} = 2$, $d_{23} = 3$, $d_{31} = 5$, which means the desired formation are collinear. We chose $k_1 = 1.4$, $k_2 = 1.4$, $k_3 = 1.4$, $l_1 = 40$, and $a_1 = \cos(w_1 t)$, $w_1 = 1.2$. The movement trajectories of the robots and the distance $||r_{ij}||$ shown in Figs. 6 and 7, respectively, demonstrate that three robots achieve the desired formation and never collide between any pair robots. In Figs. 8 and 9, all the robots' velocities converge to zero.

Fig. 5 The rotational velocity of three robots

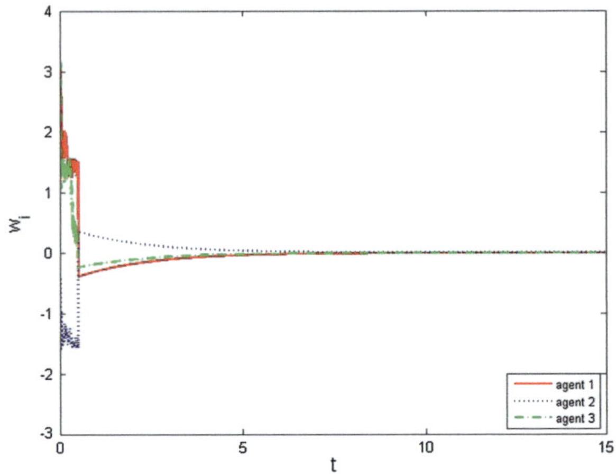

Fig. 6 The movement trajectories of the robots

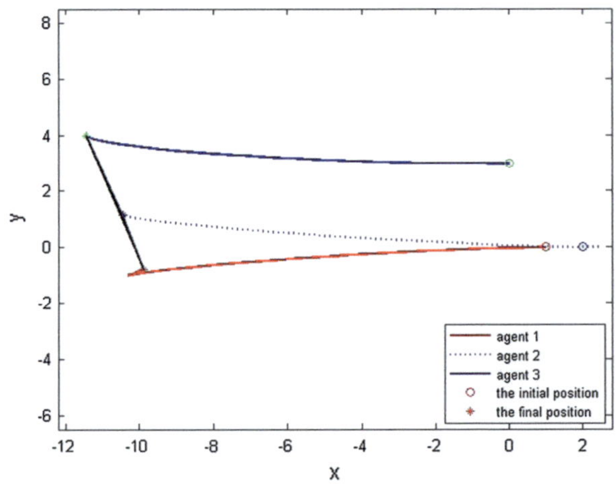

Fig. 7 The distance of three robots

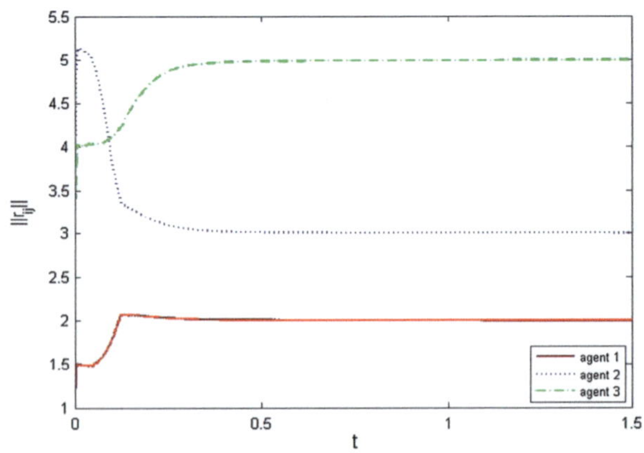

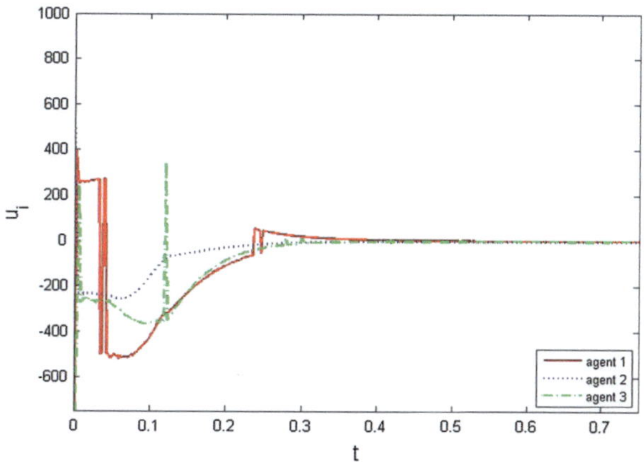

Fig. 8 The translational velocity of three robots

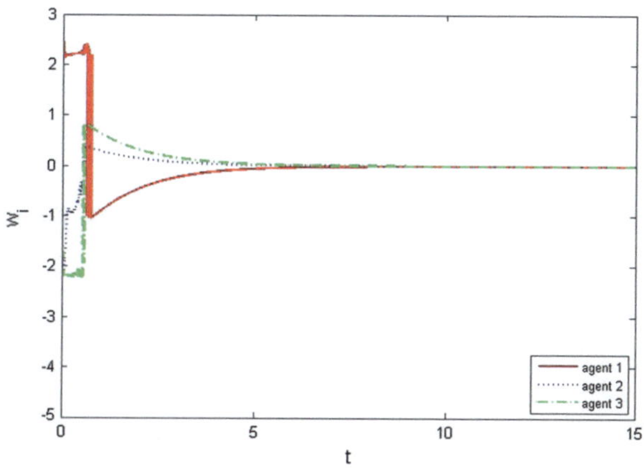

Fig. 9 The rotational velocity of three robots

Acknowledgments This work is supported by National Nature Science Foundation under grants 61503329 and 61473249, the Natural Science Foundation of Jiangsu Province BK20140490, the Natural Science Foundation of the Jiangsu Higher Education Institutions of China under grant 14KJD120003.

References

1. Millan P, Orihuela L, Jurado I, Rodriguez F (2014) Rubio. Formation control of autonomous underwater vehicles subject to communication delays. IEEE Trans Control Syst Technol 22 (2):770–777
2. Bayezit I, Fidan B (2013) Distributed cohesive motion control of flight vehicle formations. IEEE Trans Ind Electron 60(12):5763–5772
3. Fidan B, Gazi V, Zhai S, Cen N, Karatas E (2013) Single-view distance-estimation-based formation control of robotic swarms. IEEE Trans Ind Electron 60(12):5781–5791

4. Wen W, Duan Z, Ren W, Chen G (2014) Distributed consensus of mufti-agent systems with general linear node dynamics and intermittent communications. Int J Robust Nonlinear Control 24(16):2438–2457
5. Coogan S, Arcak M (2012) Scaling the size of a formation using relative position feedback. Automatica 48(10):2677–2685
6. Oh KK, Ahn HS (2014) Distance-based undirected formations of single- and double-integrator modeled robots in n-dimensional space. Int J Robust Nonlinear Control 24(12):1809–1820
7. Belabbas MA, Mou S, Morse AS, Anderson BDO (2012) Robustness issues with undirected formations. In: Proceedings of the 51st IEEE conference on decision and control, pp 1445 –1450
8. Anderson BDO, Yu C, Dasgupta S, Morse AS (2007) Control of a three-coleader formation in the plane. Syst Control Lett 56:573–578
9. Cao M, Yu C, Morse AS, Anderson BDO (2007) Controlling a triangular formation of mobile autonomous robots. In: Proceedings of the 46th IEEE conference on decision and control
10. Dimarogonas DV, Johansson KH (2008) On the stability of distance based formation control. In: 47th IEEE conference on decision and control, pp 1200–1205
11. Krick L, Broucke M, Francis B (2009) Stabilization of infinitesimally rigid formations of multi-robot networks. Int J Control 82(3):423–439
12. Cao M, Yu C, Anderson BDO (2011) Formation control using range-only measurements. Automatica 4:776–781
13. Tian YP, Wang Q (2013) Global stabilization of rigid formations in the plane. Automatica 49 (5):1436–1441
14. Filippov A (1988) Differential equations with discontinuous right-hand sides. Kluwer, Norwell, MA
15. Paden B, Sastry SS (1987) A calculus for computing Filippov's differential inclusion with application to the variable structure control of robot manipulators. IEEE Trans Circuits Syst 34 (1):73–82
16. Krstic M, Kanellakopoulos I, Kokotovic PV (1995) Nonlinear and adaptive control design. Wiley, New York

An Improved Algorithm for Siphons and Minimal Siphons in Petri Nets Based on Semi-tensor Product of Matrices

Jingjing Wang, Xiaoguang Han, Zengqiang Chen and Qing Zhang

Abstract This paper proposes an improved algorithm for enumerating siphons and minimal siphons of Petri nets (PNs) in the framework of [18]. First, the logical equation of each transition is converted into a matrix equation by using the semi-tensor product (STP) of matrices, and then the matrix equation group of the PNs is obtained. Second, an improved algorithm is proposed to calculate the siphons and minimal siphons, respectively. Finally, two illustrative examples are presented to show the efficiency and application of the improved algorithm.

Keywords Petri nets (PNs) · Siphons · Minimal siphons · Semi-tensor product (STP) of matrices

1 Introduction

Petri nets (PNs) were first introduced by Petri in 1962 and then developed by some scholars [1–3]. As graphical and mathematical tools, PNs provide a uniform environment for modeling, analysis, and design of discrete event systems (DESs) [4]. PNs have been successfully applied to many fields such as flexible manufacturing systems [5–7], communication systems [8], concurrent systems [9, 10], and so on.

Briefly, a siphon is a subset of places whose input transitions are also output transitions, which implies that a siphon will remain permanently unmarked once it loses all tokens. As structural objects of PNs, siphons intimately related to liveness [11] and deadlock-freedom [12], which are very important properties when using

J. Wang · X. Han · Z. Chen (✉)
College of Computer and Control Engineering, Nankai University, Tianjin 300350, China
e-mail: chenzq@nankai.edu.cn

J. Wang · X. Han · Z. Chen
Tianjin Key Laboratory of Intelligent Robotics, Tianjin 300350, China

Z. Chen · Q. Zhang
College of Science, Civil Aviation University of China, Tianjin 300300, China

Y. Jia et al. (eds.), *Proceedings of 2016 Chinese Intelligent Systems Conference*, Lecture Notes in Electrical Engineering 405,
DOI 10.1007/978-981-10-2335-4_35

PNs to analysis real systems. For this reason, developing effective and efficient algorithms to compute siphons is very instructive. A minimal siphon is a siphon such that any proper subset of it is not a siphon, and the development of deadlock control policies is closely related to the computation of minimal siphons. It is well-known that enumerating minimal siphons for PNs is an NP-complete problem [12, 13].

A large amount of algorithms have been proposed to calculate siphons and minimal siphons. Siphons and traps are corresponded to the roots of some logical equations in [14]. By resorting to binary decision diagrams (BDD), an efficient approach to compute minimal siphons is provided in [15]. A sign incidence matrix is introduced to compute the basis siphons in [16].

Recently, a novel approach is proposed based on semi-tensor product (STP) of matrices in [17]. Cheng first proposed STP in [18]. STP is a generalization of the conventional matrix product and has been successfully applied to many fields [19–21]. By resorting to STP, the authors establish siphon equation (SE) of PNs in [17], which can be used to calculate the siphons and minimal siphons for any PNs. The matrix framework of calculating siphons and minimal siphons is established in [17]. However, the dimension of SM is 2×2^n, where n denotes the number of places in PNs, which means it suffers from exponential explosion problem. For the dimension extension property of STP, the process of establishing SE consumes a lot of computer memory. So, it is difficult or even impossible to establish the SE when the size of the PN is large.

An efficiency algorithm is significant when using PNs to model the real-word systems. In a PN, the number of output places and input places of a transition is very small, so it is easy to establish the equation which is only related to one transition. In the framework of [18], the matrix equation group is established to calculate the siphons and minimal siphons. The solutions of the matrix equation group can be obtained by combining the solutions of every equation in it. On the basis of the above discussion, an improved algorithm is proposed in this paper.

The main contributions of this paper are as follows:

(1) Convert the SE in [17] into the matrix equation group, each equation in the matrix equation group is easy to established and solved.
(2) The solution set of the equation in the matrix equation group is represented in a novel way, which is facilitated to get the solutions of the matrix equation group.
(3) An improved algorithm is proposed in this paper to calculate all siphons and minimal siphons for any ordinary PNs.

This paper is organized as follows. In Sect. 2, the definitions and related concepts of STP and PNs are given, respectively. Section 3 first introduces the theoretical results proposed in [17], and then proposes some new theoretical results; an improved algorithm is presented at last. In Sect. 4, two illustrative examples are presented to show the efficiency and application of the improved algorithm, and conclusion is obtained in Sect. 5.

2 Preliminaries

2.1 Notations

In this subsection, we introduce some notations that will be used in this paper.

(1) $\mathcal{M}_{m \times n}$ is the set of $m \times n$ real matrices.
(2) $\mathcal{W}_{m \times n}$ is the set of $m \times n$ complex matrices.
(3) $Col_i(M)$ is the i-th column of matrix M.
(4) $Row_i(M)$ is the i-th row of matrix M.
(5) $\mathcal{D}: = \{0, 1\}$.
(6) $|x|$ is the cardinality of set X.
(7) $\mathcal{R}^n$ is the set of all real vectors of dimension n.
(8) $\mathcal{C}^n$ is the set of all complex vectors of dimension n.
(9) $\Delta_n \triangleq \{\delta_n^1, \delta_n^2, \ldots, \delta_n^n\}$, where δ_n^k denotes the k-th column of the identity matrix I_n, $k = 1, 2, \ldots, n$. When $n = 2$, we have $\Delta_2 = \{\delta_2^1, \delta_2^2\}$. For compactness, let $\Delta_2 = \Delta$.
(10) $L \in \mathcal{M}_{m \times n}$ is a logical matrix, and its columns are the form of δ_n^k. Denote its columns by $Col(L)$, and then we have $Col(L) \subset \Delta_m$. We denote the set of $m \times n$ logical matrices by $\mathcal{L}_{m \times n}$.

If $L \in \mathcal{L}_{m \times n}$, then it can be expressed as $L = [\delta_m^{i_1}, \delta_m^{i_2}, \ldots, \delta_m^{i_n}]$ and it is briefly denoted as $L = \delta_m[i_1, i_2, \ldots, i_n]$, $i_k \in \{1, 2, \ldots, m\}$.

2.2 Semi-tensor Product of Matrices

In this subsection, we introduce some preliminaries about STP, which will be used in this paper. For more details on STP, we refer to [18].

Definition 1 ([18]) Let $A \in \mathcal{M}_{m \times n}$, $B \in \mathcal{M}_{p \times q}$. Then the semi-tensor product (STP) of A and B is defined as

$$A \ltimes B = (A \otimes I_{t/n})(B \otimes I_{t/p}) \tag{1}$$

where $t = lcp(n, p)$ is the least common multiple of n and p, $\otimes$ is the Kronecker product.

Remark 1 The STP is a generalization of the conventional matrix product, and when $n = p$, we have $A \ltimes B = AB$. Throughout this paper the matrix product is assumed to be the STP. We usually omit the symbol "$\ltimes$" hereinafter.

Definition 2 ([18]) A swap matrix $W_{[m,n]}$ is a $mn \times mn$ matrix, which is defined as

$$\begin{aligned} W_{[m,n]} = \delta_{mn}[&1, m+1, 2m+1, \ldots, (n-1)m+1, \\ &2, m+2, 2m+2, \ldots (n-1)m+2, \ldots \\ &m, 2m, 3m, \ldots, nm] \end{aligned} \quad (2)$$

Remark 2 When $m=n$, we denote $W_{[m,n]} = W_{[n]}$.

Lemma 1 ([18]) *Let* $X \in \mathcal{R}^n$ and $Y \in \mathcal{R}^m$ *be two column vectors. Then*

$$W_{[m,n]}XY = YX, W_{[n,m]}YX = XY. \quad (3)$$

Lemma 2 ([18]) *Assume* $A \in \mathcal{M}_{m \times n}$ *is given.*

(1) Let $X \in \mathcal{R}^t$ be a row vector. Then

$$AX = X(I_t \otimes A). \quad (4)$$

(2) Let $X \in \mathcal{R}^t$ be a column vector. Then

$$XA = (I_t \otimes A)X. \quad (5)$$

Lemma 3 ([18]) *Let* $M_r = \delta_4[1,4]$ *be a power-reducing matrix. Then*

$$X^2 = M_r X \quad (6)$$

Lemma 4 ([18]) *Assume that the logical variables* $x, y \in \Delta$, *the disjunction of them is* $x \vee y$; *conjunction is* $x \wedge y$; *implication is* $x \rightarrow y$. *Then*

$$x \vee y = M_d xy; x \wedge y = M_c xy; x \rightarrow y = M_i xy \quad (7)$$

where $M_d = \delta_2[1,1,1,2]$, $M_c = \delta_2[1,2,2,2]$ *and* $M_d = \delta_2[1,2,1,1]$.

Lemma 5 ([18]) *Assume that* $X = x_1 x_2 \ldots x_k = \delta_{2^k}^i$ *and define* $B_0 :=2^k - i$. *Then* x_j *can be calculated inductively (in scalar form) as*

$$\begin{cases} x_j &= \left[\frac{B_{j-1}}{2^{k-j}}\right] \\ B_j = B_{j-1} - x_j * 2^{k-j}, j = 1, 2, \ldots, k \end{cases} \quad (8)$$

where $[a]$ *is the largest integer less than or equal to* a.

2.3 Petri Nets

This subsection introduces some definitions about PNs that will be used in this paper. For more details of PNs can be found in [1, 2].

A generalized Petri net is a 4-tuple expressed as $PN = (P, T, F, W)$, where $P = \{p_1, p_2, \ldots, p_n\}$ is a finite set of places. $T = \{t_1, t_2, \ldots, t_n\}$ is a finite set of transitions. $F \subseteq (P \times T) \cup (T \times P)$ is a flow relation of the net, represented by arcs with arrows from places to transitions or from transitions to places. $W: F \to \{1, 2, 3, \ldots\}$ is a weight function that assigns a weight to an arc. Moreover, $P \cap T = \emptyset$, $P \cup T \neq \emptyset$. If the weight functions of a PN are all equal to one, we say it is an ordinary PN and use $N = (P, T, F)$ to denote it. In this paper, we investigate only ordinary PNs and its abbreviation is PNs if no confusion is raised. Assume that $x \in P \cup T$, which means that x is a place or a transition in PN. Then, the preset and the postset of x can be defined as ${}^{\bullet}x = \{y \in P \cup T | (y, x) \in F\}$ and $x^{\bullet} = \{y \in P \cup T | (x, y) \in F\}$, respectively. A siphon is a nonempty subset of places $S \subseteq P$ such that ${}^{\bullet}S \subseteq S^{\bullet}$, and a trap is a subset of places $S \subseteq P$ such that $S^{\bullet} \subseteq {}^{\bullet}S$. The number of siphons in a PN is grows exponentially with the number of the places and the number of the transitions.

A siphon is a basis siphon if it cannot be represented as a union of other siphons. All siphons in a PN can be generated by the union of some basis siphons. A siphon is said to be a minimal siphon if it does not contain any other siphon.

It is obviously that siphons and traps are opposite concepts and all the above definitions and properties hold also for traps. This paper only considers the siphons and minimal siphons.

3 Approach of Calculating Siphons and Minimal Siphons of Petri Nets

3.1 Some Theoretical Results Proposed in [18]

In this subsection, some theoretical results are given which have been proposed in [17].

Let logical variables set be $X = \{x_1, x_2, \ldots x_n\}$, where $x_i \in \mathcal{D}$, $i = 1, 2, \ldots, n$, and $n = |P|$. Assume that $x_i = 1$, if and only if $p_i \in S$, where S is the subset of places. There is a bijection between X and S

$$f: X \to S, \text{ when } x_i = 1, f(x_i) = p_i \in S \tag{9}$$

Based on the above bijection, the following results can be obtained.

Lemma 6 ([17]) *All siphons in* $N = (P, T, F)$ *are obtained by the roots of the following logical equation:*

$$\wedge_{j=1}^{m}(\vee\left[X_j^{\bullet}\right] \rightarrow \vee\left[X_j\right]) = 1 \tag{10}$$

where $m = |T|$, $X_j^{\bullet}$ *and* X_j *denote the subsets of* X *corresponding to* $t_j^{\bullet}$ *and* ${}^{\bullet}t_j$.

Lemma 7 ([17]) *The logical Eq. (10) can be converted equivalently into the following matrix equation, i.e.,*

$$Lx = b \tag{11}$$

where the logical matrix $L \in \mathcal{L}_{2\times 2^n}$ *is unique,* $x = \ltimes_{k=1}^{n} x_k$, $b = \delta_2^1$. *Equation (11) is the siphon equation (SE) of PN, and L is the siphon matrix (SM) of PN.*

Lemma 8 ([17]) *Given a Petri net* $N = (P, T, F)$ *in which the SE is represented by (11). Then all siphons of* $N = (P, T, F)$ *are obtained by all nonzero solutions of the SE (11).*

Lemma 9 ([17]) *Let* $L = \delta_2[i_1, i_2, \ldots, i_{2^n}]$, *and then* $X = \ltimes_{t=1}^{n} x_t = \delta_{2^n}^{k}$ *is a solution of SE (11) if and only if* $i_k = 1$.

3.2 Main Results for the Improved Algorithm

This subsection introduces some results which are used to propose the improved algorithm.

It is obvious that (10) is equal to the following logical equation group:

$$\begin{cases} \vee\left[X_1^{\bullet}\right] \rightarrow \vee\left[X_1\right] = 1 \\ \vee\left[X_2^{\bullet}\right] \rightarrow \vee\left[X_2\right] = 1 \\ \vdots \\ \vee\left[X_m^{\bullet}\right] \rightarrow \vee\left[X_m\right] = 1 \end{cases}. \tag{12}$$

Every equation in the logical equation group (12) can be converted equivalently into a matrix equation like (11), so from Lemmas 7 and 8, we can get

Theorem 1 *All siphon in* $N = (P, T, F)$ *are obtained by nonzero solutions of the following matrix equation group:*

$$\begin{cases} L_1 x_1' = b \\ L_2 x_2' = b \\ \vdots \\ L_m x_m' = b \end{cases} \tag{13}$$

where $L_i x_i' = b(i = 1, 2, \ldots, m)$ *is converted equivalently by the logical equation* $\vee[X_i^\bullet] \to \vee[X_i] = 1$, $x_i' = x_{i_1} \ltimes x_{i_2} \ltimes \cdots \ltimes x_{i_{q_i}}$. *And* $x_{i_1}, x_{i_2}, \ldots x_{i_{q_i}}$ *are the logical variables that are contained in the union of set* $X_j^\bullet$ *and* X_j, *and* q_i *is the number of the logical variables contained in the union of set* $X_j^\bullet$ *and* X_j.

By utilizing Lemma 9, the solutions of $L_i x_i' = b(i = 1, 2, \ldots, m)$ in (13) can be obtained. Assume that the solutions of $L_i x_i' = b$ are $x_i' = \delta_{2^{q_i}}^{i_1}, \delta_{2^{q_i}}^{i_2}, \ldots, \delta_{2^{q_i}}^{i_k}$. The vector forms of them are $\alpha_{i_1}, \alpha_{i_2}, \ldots, \alpha_{i_k}$ by applying Lemma 5.

The logical variables $x_{i_1}, x_{i_2}, \ldots x_{i_{q_i}}$ correspond to the places $p_{i_1}, p_{i_2}, \ldots, p_{i_{q_i}}$ in PN. There is a bijection between α_{i_j} and S_{i_j}, which is defined by the relationship as follows:

$$S_{i_j} = f(\alpha_{i_j}) = \left\{ p_{i_r} \in P \middle| x_{i_r}^{i_j} = 1 \right\} \tag{14}$$

where $S_{i_j} \subseteq P$, $\alpha_{i_j} = (x_{i_1}^{i_j}, x_{i_2}^{i_j} \ldots, x_{i_{q_i}}^{i_j})^T$, $x_{i_r}^{i_j} \in \mathcal{D} = \{0, 1\}$, $r = 1, 2, \ldots, q_i$.

For solution's vector form $\alpha_{i_j} = (x_{i_1}^{i_j}, x_{i_2}^{i_j} \ldots, x_{i_{q_i}}^{i_j})^T$, we know $x_{i_k}^{i_j} = 1$ indicates that p_{i_k} must be contained in S_{i_j}, $x_{i_k}^{i_j} = 0$ indicates that p_{i_k} cannot be contained in S_{i_j}, and $x_{i_k'}(i_k' \neq i_1, i_2, \ldots, i_{q_i})$ respond the place that is not contained in the union of set $X_j^\bullet$ and X_j, the values of them do not influence the validity of $L_i x_i' = b$. Therefore, no matter the responding places are contained in S_{i_j} or not, $\vee[X_i^\bullet] \to \vee[X_i] = 1$ holds always. It is rational to use the combination idea to get the solutions of (12) and calculate siphons of PNs. In order to introduce the process of the combination, the following definitions are given.

Definition 3 A matrix $A_{c_i \times n}^i \in \mathcal{W}_{m \times n}$ is said to be the solution matrix of $L_i x_i' = b$, where c_i is the number of the solutions of $L_i x_i' = b$ and $A_{c_i \times n}^i$ is constructed as follows:

(1) Consider the solutions of $L_i x_i' = b$ and the vector forms of them are:

$$\alpha_{i_j} = (x_{i_1}^{i_j}, x_{i_2}^{i_j} \ldots, x_{i_{q_i}}^{i_j})^T, j = 1, 2, \ldots, c_i.$$

(2) Construct a row vector $\bar{\alpha}_{i_j} = (\bar{x}_1^{i_j}, \bar{x}_2^{i_j}, \ldots, \bar{x}_n^{i_j})$ as follows:

$$\begin{cases} \bar{x}_{i_k}^{i_j} = 1, & if \quad x_{i_k}^{i_j} = 1 \\ \bar{x}_{i_k}^{i_j} = -1, & if \quad x_{i_k}^{i_j} = 0 \\ \bar{x}_{i_k}^{i_j} = 1i, & if \quad i_k \textit{ are not contained in the union of set } X_i^\bullet \textit{ and } X_i \end{cases}$$

where $1i$ is the complex variable and i is the imaginary unit.

(3) Let the row vectors formed in (2) formed a matrix, and say that matrix is the solution matrix of $L_i x_i' = b$.

From Definition 3 we know that, for $a_{i_j,i_k} \in A^i_{c_i \times n}$, $a_{i_j,i_k} = 1$ means that p_{i_k} must be contained in S_{i_j}; $a_{i_j,i_k} = -1$ means that p_{i_k} cannot be contained in S_{i_j}; and $a_{i_j,i_k} = 1i$ means that p_{i_k} is not contained in the union of set $t_j^{\bullet}$ and ${}^{\bullet}t_j$, so p_{i_k} can be contained in S_{i_j} or not. We give the combination idea as follows to get the solutions of (12).

Definition 4 For solution matrixes $A^1_{c_1 \times n}, A^2_{c_2 \times n}, \ldots, A^m_{c_m \times n}$ the combination matrix of them is C. And the combination matrix C is construed as follows:

(1) $\mathrm{Row_i}(C) = \mathrm{Row}_{s_1}(A^1) + \mathrm{Row}_{s_2}(A^2) + \cdots + \mathrm{Row}_{s_m}(A^m)$, where $s_1 = 1, 2, \ldots q_1, \ldots, s_m = 1, 2, \ldots q_m$.

(2) Consider $c_{ij} \in \mathrm{Row_i}(\mathrm{C})$, its real part is expressed as $Re(c_{ij})$ and its imaginary part is expressed as $Im(c_{ij})$. Then using $abs(Re(c_{ij}))$ indicates the absolute of $Re(c_{ij})$. If there exist $c_{ij} \in \mathrm{Row_i}(C)$, which is not satisfies the equation as $abs(Re(c_{ij}) + Im(c_{ij})) \neq m$, then the $\mathrm{Row_i}(C)$ will be deleted from C.

Equation (1) in Definition 4 indicates that each solution of the equation in (13) is combined with other equations' solutions, which means that all combinations are considered. Equation (2) in Definition 4 indicates the combination matrix C only includes the combinations of (1) which are no conflict exist. For example,

$$\mathrm{Row_i}(C) = \mathrm{Row}_{s_1}(A^1) + \mathrm{Row}_{s_2}(A^2),$$

where $a^1_{s_1 j} \in \mathrm{Row}_{s_1}(A^1) = 1$, and $a^2_{s_2 j} \in \mathrm{Row}_{s_2}(\mathrm{A}^2) = -1$, which means that the place p_j must be contained in $S^1_{s_1}$ but cannot be contained in $S^2_{s_2}$, so there exists a conflict between $S^1_{s_1}$ and $S^2_{s_2}$, which means the two solution sets cannot be combined. On the other hand, $a^1_{s_1 j} + a^2_{s_2 j} = 0$, $abs(Re(c_{ij})) + Im(c_{ij}) < m$, where $m = 2$,$\mathrm{Row_i}(C)$ will be deleted from C.

Then we convert the combination matrix C into matrix D^0 as following:

For $Re(c_{ij}) > 0$ means that p_j contained in siphon S_i, we define $d^0_{ij} = 1$, otherwise, we define $d^0_{ij} = 0$.

Theorem 2 *The row vectors of matrix D^0 are the solutions of (13).*

Proof There is no isolated point exists in net, from the definition of PNs, every place of a PN must belong to a transition's preset or postset. So from Definition 4, we have $Re(c_{ij}) \neq 0$. For $Re(c_{ij}) > 0$ means that p_j contained in siphon S_i, $Re(c_{ij}) < 0$ means that p_j not contained in siphon S_i, so from Lemmas 8 and 9, we know that the row vectors of matrix D^0 are the solutions of (13), and we can get the solutions of (12).

Remark 4 When the transition number $m = |T|$ is large, we can first divide solution matrixes of $L_i x_i' = b$, $(i = 1, 2, \ldots, m)$ into several groups, and then we do combinations in every group to get the combination matrixes. By repeating the process, we will finally get the combination matrix C.

Using the MATLAB function *Unique* on D^0, we can get the solution matrix D with unique rows and the rows of D are in sorted order, and the number of "1" in the first row is the least, which means the places contained in S_1 is the least. So, siphon S_1 must be the minimal siphon that contains the fewest places, and we can use this property to calculate the minimal siphons.

3.3 The Design of the Improved Algorithm

In this subsection, the improved algorithm to calculate siphons and minimal siphons in PNs is presented. The detailed steps are as follows:

Algorithm Calculating all siphons and all minimal siphons in PNs.

Step 1: Calculate the set of input and output places of transition t_j, then we have $X_j^{\bullet}$ and X_j, respectively, and construct the logical Eq. (12).

Step 2: By applying Theorem 1 convert the logical equation group (12) into the matrix equation group (13).

Step 3: By using Definition 3, we get the solution matrix $A^i_{c_i \times n}$ of equation $L_i x_i^{'} = b$.

Step 4: Combine the solution matrixes $A^i_{c_i \times n}$, $(i = 1, 2, \ldots, m)$ to get the combination matrix C by applying Definition 4. Convert the combination matrix C into solution matrix D^0, and every nonzero row vectors of D^0 is a solution of (13).

Step 5: Using MATLAB function *Unique*, we get the sorted solution matrix D, and all siphons of PN can be got by nonzero row vectors of D.

Step 6: Let B_k is the set of all siphons. That is, $B_k = \{S_1, S_2, \ldots, S_k\}$, where the subset of places S_i is a siphon of PN which is obtained by nonzero row vector $\mathrm{Row_i}(D)$. We let $B_0 \triangleq \{S_1\}$.

Step 7: For each siphon $S_i \in B_k$, $i = (2, 3, \ldots, k)$, if there exist $S_t \in B_{i-2}$, such that S_t is a subset of S_i, i.e., $S_t \subset S_i$, then $B_{i-1} \triangleq B_{i-2}$, otherwise $B_{i-1} \triangleq B_{i-2} \cup \{S_i\}, i = 2, 3, \ldots, k$.

Step 8: By repeating Step 7, we can ultimately obtain the set B_{k-1}, which consists of all minimal siphons of PN.

4 The Illustrative Examples

In this section, two illustrative examples are presented. Example 1 shows the process of the improved algorithm, and Example 2 displays the application and the efficiency of the improved algorithm.

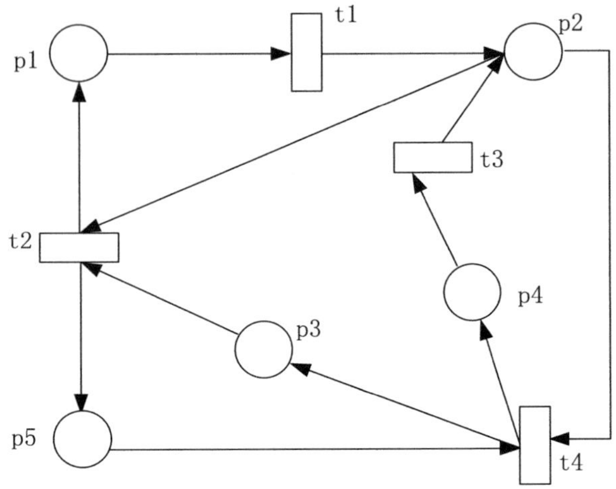

Fig. 1 The PN of Example 1

Example 1 Consider the PN in Fig. 1, which from [22].

Step 1: By using Lemma 6 we can obtain the logical equations:

$$\begin{cases} x_2 & \rightarrow & x_1 & = 1 \\ (x_1 \vee x_5) & \rightarrow & (x_2 \vee x_3) & = 1 \\ x_2 & \rightarrow & x_4 & = 1 \\ (x_3 \vee x_4) & \rightarrow & (x_2 \vee x_5) & = 1 \end{cases}$$

Step 2: Convert the logical equations into matrix equation group:

$$\begin{cases} L_1 \ltimes x_1 \ltimes x_2 = b \\ L_2 \ltimes x_1 \ltimes x_2 \ltimes x_3 \ltimes x_5 = b \\ L_3 \ltimes x_2 \ltimes x_4 = b \\ L_4 \ltimes x_2 \ltimes x_3 \ltimes x_4 \ltimes x_5 = b \end{cases}$$

where $L_1 = \delta_2[1, 1, 2, 1]$, $L_2 = \delta_2[1, 1, 1, 1, 1, 1, 2, 2, 1, 1, 1, 1, 1, 1, 2, 1]$,

$$L_3 = \delta_2[1, 2, 1, 1], L_4 = \delta_2[1, 1, 1, 1, 1, 1, 1, 1, 1, 2, 1, 2, 1, 2, 1, 1].$$

The solutions of $L_1 \ltimes x_1 \ltimes x_2 = b$ as $(x_1, x_2) = (1, 1), (1, 0), (0, 0)$.
The solutions of $L_2 \ltimes x_1 \ltimes x_2 \ltimes x_3 \ltimes x_5 = b$ as

$$\begin{aligned} (x_1, x_2, x_3, x_5) = {} & (1, 1, 1, 1), (1, 1, 1, 0), (1, 1, 0, 1), (1, 1, 0, 0), (1, 0, 1, 1), (1, 0, 1, 0), (0, 1, 1, 1) \\ & (0, 1, 1, 0), (0, 1, 0, 1), (0, 1, 0, 0), (0, 0, 1, 1), (0, 0, 1, 0), (0, 0, 0, 0). \end{aligned}$$

The solutions of $L_3 \ltimes x_2 \ltimes x_4 = b$ as $(x_2, x_4) = (1, 1), (0, 1), (0, 0)$.

The solutions of $L_4 \ltimes x_2 \ltimes x_3 \ltimes x_4 \ltimes x_5 = b$ as

$$(x_2,x_3,x_4,x_5) = (1,1,1,1),(1,1,1,0),(1,1,0,1),(1,1,0,0),(1,0,1,1),(1,0,1,0),\\(1,0,0,1),(1,0,0,0),(0,1,1,1),(0,1,0,1),(0,0,1,1),(0,0,0,1),(0,0,0,0).$$

Step 3: Calculate the solution matrix $A^i_{c_i \times n}$ of equation $L_i x_i^{'} = b,(i = 1,2,3,4)$.

For the solution of $L_1 \ltimes x_1 \ltimes x_2 = b$ as $(x_1,x_2) = (1,0)$, from Definition 3, we have $\bar{x}_1 = 1$ and $\bar{x}_2 = -1$, and the logical variables x_3,x_4,x_5 are not contained in $\{x_1, x_2\}$, so $\bar{x}_3 = \bar{x}_4 = \bar{x}_5 = 1i$. So we have the row vector:

$$\bar{\alpha} = (\bar{x}_1,\bar{x}_2,\bar{x}_3,\bar{x}_4,\bar{x}_5) = (1,-1,1i,1i,1i).$$

Do the same operation for other solutions, we can get the solution matrix formed by the row vectors is A^1.

And other solution matrixes can be obtained at the same way.

$$A^1 = \begin{bmatrix} 1 & 1 & 1i & 1i & 1i \\ 1 & -1 & 1i & 1i & 1i \\ -1 & -1 & 1i & 1i & 1i \end{bmatrix}, \quad A^3 = \begin{bmatrix} 1i & 1 & 1i & 1 & 1i \\ 1i & -1 & 1i & 1 & 1i \\ 1i & -1 & 1i & -1 & 1i \end{bmatrix}$$

$$A^2 = \begin{bmatrix} 1 & 1 & 1 & 1i & 1 \\ 1 & 1 & 1 & 1i & -1 \\ 1 & 1 & -1 & 1i & 1 \\ 1 & 1 & -1 & 1i & -1 \\ 1 & -1 & 1 & 1i & 1 \\ 1 & -1 & 1 & 1i & -1 \\ -1 & 1 & 1 & 1i & 1 \\ -1 & 1 & 1 & 1i & -1 \\ -1 & 1 & -1 & 1i & 1 \\ -1 & 1 & -1 & 1i & -1 \\ -1 & -1 & 1 & 1i & 1 \\ -1 & -1 & 1 & 1i & -1 \\ -1 & -1 & -1 & 1i & -1 \end{bmatrix}. \quad A^4 = \begin{bmatrix} 1i & 1 & 1 & 1 & 1 \\ 1i & 1 & 1 & 1 & -1 \\ 1i & 1 & 1 & -1 & 1 \\ 1i & 1 & 1 & -1 & -1 \\ 1i & 1 & -1 & 1 & 1 \\ 1i & 1 & -1 & 1 & -1 \\ 1i & 1 & -1 & -1 & 1 \\ 1i & 1 & -1 & -1 & -1 \\ 1i & -1 & 1 & 1 & 1 \\ 1i & -1 & 1 & -1 & 1 \\ 1i & -1 & -1 & 1 & 1 \\ 1i & -1 & -1 & -1 & 1 \\ 1i & -1 & -1 & -1 & -1 \end{bmatrix}$$

Step 4: Combine the solution matrices $A^i_{c_i \times n}$, $(i = 1,2,3,4)$ and then use the function *Unique* to get the solution matrix D of the matrix equation group in Step 2:

Consider the following combinations,

(1) $\text{Row}_1(A^1) + \text{Row}_1(A^2) + \text{Row}_1(A^3) + \text{Row}_1(A^4)$
$= [2+2i \quad 4 \quad 2+2i \quad 2+2i \quad 2+2i]$.
For $Re(2+2\text{i}) + Im(2+2\text{i}) = 4, Re(4) + Im(4) = 4$, there is

$$\begin{aligned}\text{Row}_1(\text{C}) &= \text{Row}_1(A^1) + \text{Row}_1(A^2) + \text{Row}_1(A^3) + \text{Row}_1(A^4)\\ &= [2+2i \quad 4 \quad 2+2i \quad 2+2i \quad 2+2i].\end{aligned}$$

(2) $\text{Row}_1(A^1) + \text{Row}_1(A^2) + \text{Row}_1(A^3) + \text{Row}_2(A^4)$
$= [2+2i \quad 4 \quad 2+2i \quad 2+2i \quad 2+2i]$

For $Re(1+2\text{i}) + Im(1+2\text{i}) = 3 \neq 4$, so this row vector is not contained in solution matrix C.

Repeat the above process and we will get the solution matrix C as follows:

$$C = \begin{bmatrix} 2+2i & 4 & 2+2i & 2+2i & 2+2i \\ 2+2i & 4 & 2+2i & 2+2i & -2+2i \\ 2+2i & 4 & -2+2i & 2+2i & 2+2i \\ 2+2i & 4 & -2+2i & 2+2i & -2+2i \\ 2+2i & -3+i & 2+2i & 2+2i & 2+2i \\ 2+2i & -4 & 2+2i & -2+2i & 2+2i \\ -2+2i & -4 & 2+2i & 2+2i & 2+2i \\ -2+2i & -4 & 2+2i & -2+2i & 2+2i \\ -2+2i & -4 & -2+2i & -2+2i & -2+2i \end{bmatrix}$$

Then convert combination matrix C to solution matrix D^0, and use function *Unique* to get the sorted matrix D.

$$D^0 = \begin{bmatrix} 1 & 1 & 1 & 1 & 1 \\ 1 & 1 & 1 & 1 & 0 \\ 1 & 1 & 0 & 1 & 1 \\ 1 & 1 & 0 & 1 & 0 \\ 1 & 0 & 1 & 1 & 1 \\ 1 & 0 & 1 & 0 & 1 \\ 0 & 0 & 1 & 1 & 1 \\ 0 & 0 & 1 & 0 & 1 \\ 0 & 0 & 0 & 0 & 0 \end{bmatrix}, D = \begin{bmatrix} 0 & 0 & 0 & 0 & 0 \\ 0 & 0 & 1 & 0 & 1 \\ 0 & 0 & 1 & 1 & 1 \\ 1 & 0 & 1 & 0 & 1 \\ 1 & 0 & 1 & 1 & 1 \\ 1 & 1 & 0 & 1 & 0 \\ 1 & 1 & 0 & 1 & 1 \\ 1 & 1 & 1 & 1 & 0 \\ 1 & 1 & 1 & 1 & 1 \end{bmatrix}.$$

Every nonzero row vector in D is a nonzero solution of the equations, so all siphons in the PN are

$$S_1 = \{p_3, p_5\}, S_2 = \{p_3, p_4, p_5\}, S_3 = \{p_1, p_3, p_5\}, S_4 = \{p_1, p_3, p_4, p_5\}$$
$$S_5 = \{p_1, p_2, p_4\}, S_6 = \{p_1, p_2, p_4, p_5\}, S_7 = \{p_1, p_2, p_3, p_4\},$$
$$S_8 = \{p_1, p_2, p_3, p_4, p_5\}.$$

Step 5: Get all the minimal siphons by using Step 6 and Step 7 in the Algorithm. They are

$$S_1 = \{p_3, p_5\}, S_5 = \{p_1, p_2, p_4\}.$$

Example 2 Consider the PN in Fig. 2, which is used to model the concurrent execution of working process in flexible manufacturing systems (FMS). Figure 2 is

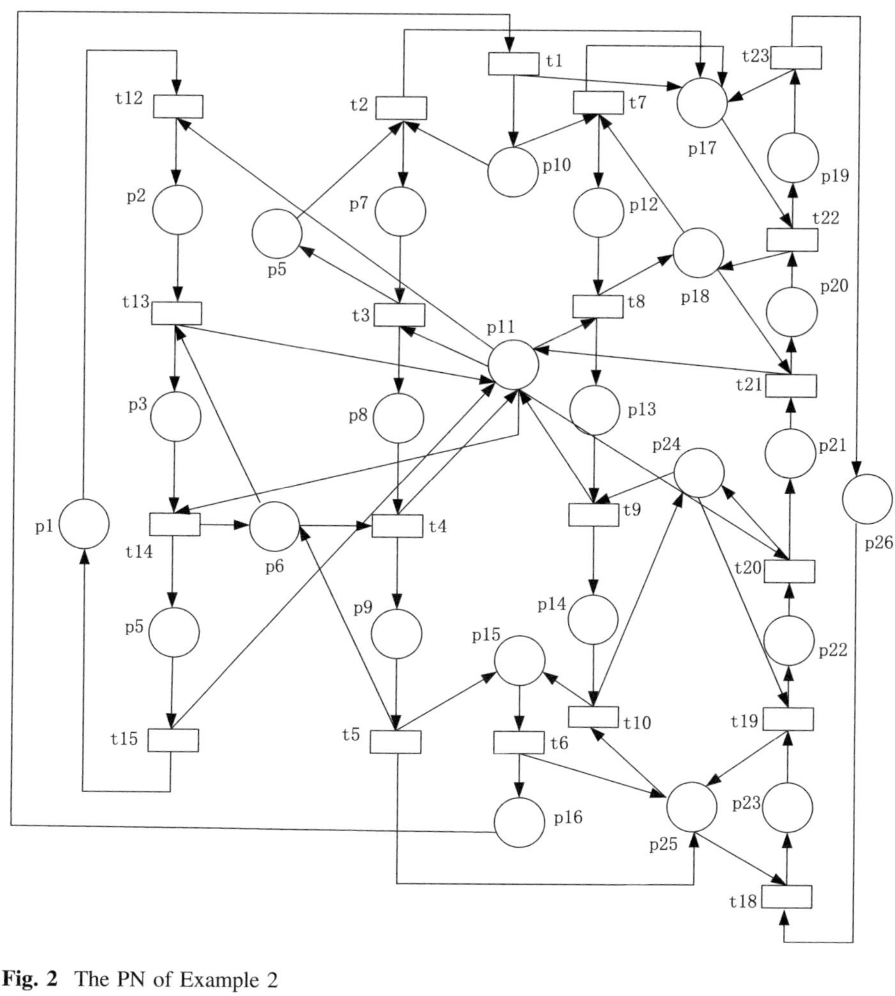

Fig. 2 The PN of Example 2

Table 1 Set of minimal siphons in Fig. 2

i	S_i	i	S_i
1	$\{p_{15},p_{22},p_{24},p_{25}\}$	15	$\{p_2,p_4,p_5,p_8,p_{11},p_{15},p_{17},p_{18},p_{19},p_{24},p_{25}\}$
2	$\{p_2,p_4,p_8,p_{11},p_{13},p_{18},p_{20}\}$	16	$\{p_4,p_6,p_{11},p_{15},p_{21},p_{24},p_{25}\}$
3	$\{p_2,p_4,p_5,p_8,p_{11},p_{13},p_{17},p_{18},p_{19}\}$	17	$\{p_4,p_5,p_6,p_{11},p_{15},p_{18},p_{20},p_{24},p_{25}\}$
4	$\{p_4,p_6,p_9,p_{11},p_{13},p_{21}\}$	18	$\{p_4,p_5,p_6,p_{11},p_{15},p_{17},p_{18},p_{19},p_{24},p_{25}\}$
5	$\{p_4,p_6,p_9,p_{11},p_{13},p_{18},p_{20}\}$	19	$\{p_1,p_2,p_3,p_4\}$
6	$\{p_4,p_5,p_6,p_9,p_{11},p_{13},p_{17},p_{18},p_{19}\}$	20	$\{p_2,p_4,p_8,p_{11},p_{13},p_{21}\}$
7	$\{p_2,p_4,p_8,p_{11},p_{14},p_{21},p_{24}\}$	21	$\{p_3,p_6,p_9\}$
8	$\{p_2,p_4,p_8,p_{11},p_{14},p_{18},p_{20},p_{24}\}$	22	$\{p_5,p_7\}$
9	$\{p_2,p_4,p_5,p_8,p_{11},p_{14},p_{17},p_{18},p_{19},p_{24}\}$	23	$\{p_7,p_8,p_9,p_{10},p_{12},p_{13},p_{14},p_{15},p_{16}\}$
10	$\{p_4,p_6,p_9,p_{11},p_{14},p_{21},p_{24}\}$	24	$\{p_{10},p_{17},p_{19}\}$
11	$\{p_4,p_6,p_9,p_{11},p_{14},p_{18},p_{20},p_{24}\}$	25	$\{p_{12},p_{18},p_{20}\}$
12	$\{p_4,p_5,p_6,p_9,p_{11},p_{14},p_{17},p_{18},p_{19},p_{24}\}$	26	$\{p_{14},p_{22},p_{24}\}$
13	$\{p_2,p_4,p_8,p_{11},p_{15},p_{18},p_{21},p_{24},p_{25}\}$	27	$\{p_{15},p_{23},p_{25}\}$
14	$\{p_2,p_4,p_8,p_{11},p_{15},p_{18},p_{20},p_{24},p_{25}\}$	28	$\{p_{19},p_{20},p_{21},p_{22},p_{23},p_{26}\}$

derived from Fig. 8 in [23], and the places and transitions are renamed. The PN in Fig. 2 belongs to a particular class of nets that we call Systems of Simple Sequential Processes with Resources (S^3PR) and it is a classical instance for FMS.

The algorithm proposed in [17] cannot calculate the siphons and minimal siphons of the PN in Fig. 2, for the number of the places is too large, the computer is out of memory. However, the improved algorithm can calculate the siphons and minimal siphons in 6 min, which is not bad for the size of the PN and the results are as follows:

The number of siphons of this PN is 503488, which is very large. The number of minimal siphons is 28 and all minimal siphons shown in Table 1.

The first 18 minimal siphons are strict minimal siphons, and the last 10 minimal siphons also are traps. The calculation results of the strict minimal siphons are consistent with the result of [23].

Remark 5 A MATLAB toolbox has been created by Daizhan Cheng and Hongsheng Qi for the computation of STP at http://lsc.amss.ac.cn/dcheng/stp/STP.zip. The numerical computation in this paper is based on this MATLAB toolbox.

5 Conclusion

In this paper, we have investigated the problem of calculating siphons and minimal siphons in PNs. The logical equation group was converted into matrix equation group; the solutions of the matrix equation group were obtained by combing the solutions of each equation in the matrix equation group. An improved algorithm was proposed to calculate all siphons and minimal siphons. Two illustrative examples are presented to show the efficiency and application of the improved algorithm.

Acknowledgments This work was supported in part by the Natural Science Foundation of China Under Grants of 61573199 and 61573200, the Tianjin Natural Science Foundation Grant No. 14JCYBJC18700.

References

1. Zurawsk Richard, Mengchu Zhou (1994) Petri nets and industrial applications: a tutorial. IEEE Trans Industr Electron 41(6):567–583
2. Radford P (1981) Petri net theory and the modeling of systems. Prentice Hall PTR
3. Murata Tadao (1989) Petri nets: properties, analysis and applications. Proc IEEE 77(4):541–580
4. Cassandras Christos G, Lafortune Stephane (2001) Introduction to discrete event systems. Springer, US
5. Ezpeleta J, Colom JM, Martinez J (1992) A Petri net based deadlock prevention policy for flexible manufacturing systems. IEEE Trans Robot Autom 173−184
6. Zhiwu Li, Mengchu Zhou (2004) Elementary siphons of petri nets and their application to deadlock prevention in flexible manufacturing systems. IEEE Trans Syst Man Cybern Part A Syst Hum 34(1):38–51
7. Huixia Liu, Keyi Xing, Weimin Wu et al (2015) Deadlock prevention for flexible manufacturing systems via controllable siphon basis of Petri nets. IEEE Trans Syst Man Cybern Syst 45(3):519–529
8. Tianhua Xu, Tao Tang (2007) The modeling and analysis of Data Communication System (DCS) in Communication Based Train Control (CBTC) with colored Petri nets. In: Eighth international symposium on autonomous decentralized systems
9. Yufeng Chen, Zhizu Li, Barkaoui Kamel et al (2015) On the enforcement of a class of nonlinear constraints on Petri nets. Automatica 55:116–124
10. Hesuan Hu, Zhiwu Li (2008) Efficient deadlock prevention policy in automated manufacturing systems using exhausted resources. Int J Adv Manuf Technol 40(5):566–571
11. Barkaoui K, Minoux M (1992) A polynomial-time graph algorithm to decide liveness of some basic classes of bounded Petri nets. In: International conference on application and theory of Petri nets. Springer-Verlag, pp 62−75
12. Zhiwu Li, Mengchu Zhou (2008) On siphon computation for deadlock control in a class of Petri nets. Syst Man Cybern Part A Syst Hum IEEE Trans 38(3):667–679
13. Yamaguchi Masahiro, Watanabe Toshimasa (1999) Time complexity analysis of the minimal siphon extraction problem of Petri nets. IEICE Trans Fundam Electron Commun Comput 82 (11):2558–2565
14. Wegrzyn Agnieszka, Karatkevich Andrei, Bieganowski Jacek (2004) Detection of deadlocks and traps in Petri nets by means of thelen's prime implicant method. Int J Appl Math Comput Sci 14(1):113–121
15. Yufeng Chen, Gaiyun Liu (2013) Computation of minimal siphons in petri nets by using binary decision diagrams. Acm Trans Embed Comput Syst 12(1):1–15
16. Erwin boer R, Murata Tadao (1994) Generating basis siphons and traps of Petri nets using sign incidence matrix. IEEE Trans Circuits Syst I Fundam Theory Appl 41(4):266–271

17. Xiaoguang Han, Zengqiang Chen, Zhongxin Liu, et al (2015) Calculation of siphons and minimal siphons in Petri nets based on semi-tensor product of matrices. IEEE Trans Syst Man, Cybern: Syst in press, doi:10.1109/TSMC.2015.2507162
18. Daizhan Cheng, Hong Qi, Yin Zhao (2012) An introduction to semi-tensor product of matrices and its applications. In: An introduction to semi-tensor product of matrices and its applications, p 612
19. Daizhan Cheng, Hongsheng Qi, Zhiqiang Li (2011) Analysis and control of Boolean networks. In: Communications & Control Engineering
20. Daizhan Cheng, Fenghua He, Hongsheng Qi et al (2015) Modeling, analysis and control of networked evolutionary games. IEEE Trans Autom Control 60(9):2402–2415
21. Xiangru Xu, Yiguang Hong (2012) Matrix expression and reachability analysis of finite automata. J Control Theory Appl 10(2):210–215
22. Xiaoguang Han, Zengqiang Chen, Zhongxin Liu, et al (2016) Calculation Basis siphons of Petri nets based on semi-tensor product of matrices. In: Proceeding of 35th chinese control conference
23. Ezpeleta J, Colom JM, Martinez J (1992) A Petri net based deadlock prevention policy for flexible manufacturing systems. IEEE Trans Robot Autom173−184

A Novel MPPT Control Algorithm Based on Peak Current

Mingming Ma, Yu Fang, Jinyan Zheng, Songyin Cao and Yong Xie

Abstract Aiming at the MPPT (maximum power point tracking) control of the fly-back type grid-connected microinverter, a novel maximum power point tracking control algorithm based on peak current sampling is studied, and a current sampling circuit composed of blocking diode and bias diode is proposed. Compared with maximum power point tracking methods of the conventional MPPT, the method in this paper does not need to calculate the actual power by multiplying the input voltage and the corresponding input current of the photovoltaic components. And the maximum power point of PV modules can be obtained according to comparing the detected sampling peak current of the switch tube. Therefore, the input voltage sampling circuit, the cost and the resources of the control chip are saved, and the complexity of the computation on maximum power point is simplified, thus ultimately achieving the MPPT fast control algorithm. The method in this paper has no need of sampling resistance or current Holzer, which improves the efficiency of grid-connected microinverter. The experimental results verify the correctness of the method studied in this paper.

Keywords Grid-connected power generation microinverter · Maximum power point tracking · Peak current sampling

M. Ma (✉) · Y. Fang · J. Zheng · S. Cao · Y. Xie
College of Information Engineering, Yangzhou University, 225127 Yangzhou, China
e-mail: ma_ming_ming@126.com

Y. Fang
e-mail: yzfangyu@126.com

J. Zheng
e-mail: zhengjinyannt@163.com

S. Cao
e-mail: caosongyin@163.com

Y. Xie
e-mail: yzxieyong@126.com

Y. Jia et al. (eds.), *Proceedings of 2016 Chinese Intelligent Systems Conference*, Lecture Notes in Electrical Engineering 405,
DOI 10.1007/978-981-10-2335-4_36

1 Introduction

In recent years, the traditional fossil energy is gradually exhausted, and the development of photovoltaic power generation industry is showing a trend of increasing year by year. With the continuous progress of technology, photovoltaic inverter is also moving towards the direction of miniaturization and modularization. The power density and conversion efficiency of the photovoltaic inverter are becoming higher and higher, the circuit structure is becoming more simplified and the cost of hardware is increasingly low [1, 2].

In traditional photovoltaic grid-connected microinverter module, the calculation of the maximum power point often used is to sample the output voltage and output current of PV module to calculate the actual output power of photovoltaic panels. Then, through three-point comparison method, mountain climbing method, perturbation method and other means, the maximum power point tracking is achieved. It is necessary to sample the output voltage and the output current to realize the control. In the photovoltaic grid-connected microinverter, sampling of the output current of PV module is commonly converting the current signal into voltage signal through sampling resistance. When the current flows through the sampling resistance, it will generate heat loss on the resistance, thus reducing the energy conversion efficiency of the whole circuit. In order to reduce loss as far as possible, high power sampling resistance with small values is needed, but the disadvantage is that pressure drop on the sampling resistance will be smaller. Therefore, an amplifying circuit that amplifies the small voltage signals on sampling resistance into analog signals for digital signal processing (DSP) is needed, which increases the complexity of the circuit. Another common current sampling device is the current transformer, which converses the high current signal into a small current signal to further conditioning the small signal with the characteristics that primary side and secondary side current of the transformer is inversely proportional to the number of coil. The advantage of this sampling method is that isolate sampling can be realized with high precision. However, the cost is high, the circuit is complex, and it is only suitable for the sampling of high frequency pulse current. In this paper, a novel current sampling method is proposed, on the basis of which a fast MPPT maximum power point algorithm is presented [3, 4].

2 Review of Traditional Technology

The traditional maximum power point tracking obtains the output power of the PV module by sampling the average input voltage and average input current of photovoltaic module, and then multiplying the average input voltage and average input current. This method requires the voltage sampling circuit and the current sampling circuit to sample the average input voltage and the average input current, whose hardware cost is relatively high [5, 6].

Another method of calculating the maximum power point is: get the peak current in main switches of the fly-back type microinverter by the current transformer, and obtain a transient output power of the photovoltaic module by multiplying the peak current and the sampling input voltage of the photovoltaic battery component. By comparing this instantaneous power, the maximum power point is obtained. However, this method has a serious flaw: on the condition that the switching frequency is certain, the instantaneous input power is of increasing function relation with the peak current and has nothing to do with the input voltage. But in the maximum power tracking control, the output voltage of the PV module changes in real-time, so the calculated maximum power point using this method is not correct [7].

Both the above two traditional methods of calculating the maximum power point need to collect the output voltage of PV module.

3 Principles of a Novel Current Sampling Circuit

A novel current sampling circuit is shown in Figs. 1 and 2. The output of the PWM generator is connected to the input of the driving circuit. The output of the driving circuit is divided into two parts, one is connected to the door limit current resistance of switch tubes in the main circuit of the switching power supply to control the on-off of MOSFET switches, the other connected to the current limiting resistor is divided into two parts, respectively, connected to the anode of blocking diode D2 and the anode of biased diodes D3. The cathode of blocking diode D2 is connected to the drain of MOSFET switch tube in the main circuit of the switching power supply. The cathode of biased diodes D3 is connected to the positive phase terminal of the amplifier, sending signal to a microcontroller or DSP by amplifying and conditioning circuit. When the PWM generator generates a pulse control signal at a high level, the switch tube in the main circuit of the switching power supply is turned on, current in the main circuit flows through the switch tube, and a voltage drop lands on the on-resistance of the switch. Here, the blocking diode conducts

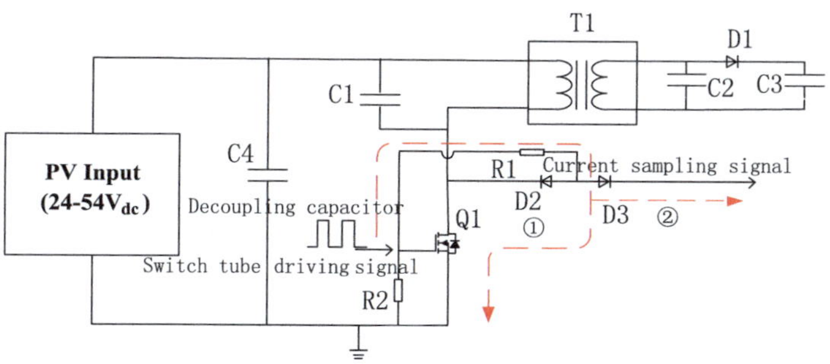

Fig. 1 Circuit schematic with switching tube voltage drop

Fig. 2 Principle diagram of signal amplification circuit

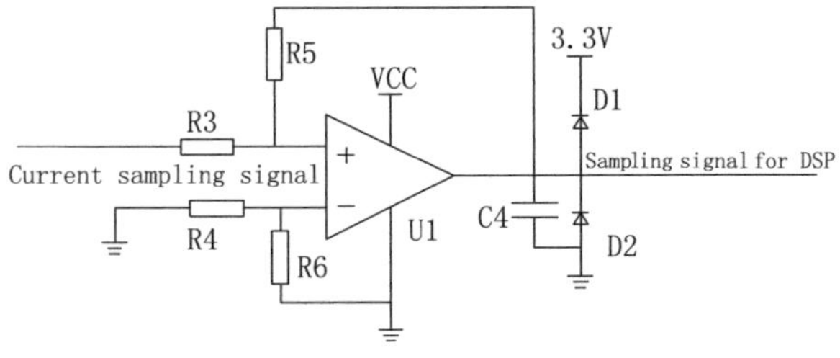

under positive voltage, and a voltage value is got on the anode, which is equal to the sum of the voltage drop on the on-resistance and the on-state voltage drop on the diode, and then it is sent to the positive input of amplifier after voltage drop compensation by bias diode. As long as the bias diode and blocking diode select the same type, the input voltage of the amplifier can be regarded as equal to the voltage drop on the on-resistance of MOSFET, thus accurately achieving the current through the switch tube.

4 A Novel Method for Calculating the Maximum Power Point

4.1 Theoretical Analysis of a New Method for Calculating the Maximum Power Point

The novel maximum power point tracking algorithm in this paper is to achieve the maximum power point tracking by sampling the peak current of the fly-back circuit. Specific analyses are as follows: the output waveform from inverter to the grid is a sine wave synchronized with the grid voltage. The output current is the largest at the crests and troughs of sine wave, and peak current flowing at the fly-back circuit at this moment is also the largest. Assuming that the peak current in the fly-back circuit is I_{pk}, average current in a switching period is I_p, output voltage of the PV module is U_i, output current of the PV module is I_i, formula 1 can be obtained

$$P_i = U_i \cdot I_i = \frac{2}{\pi} \cdot U_i \cdot I_p \tag{1}$$

Within a single switching cycle, the turn-on time of the switch is t_{on} and the single cycle time is T_s, then the relationship between the peak current I and the average current I_p in a single cycle can be obtained, as shown in formula 2

$$\frac{1}{2} \cdot t_{on} \cdot I_{pk} = I_p \cdot T_s \tag{2}$$

The primary inductance of the fly-back transformer is L_m. According to the relationship between the input voltage U_i and the peak current I_{pk}, the formula 3 can be got

$$\frac{U_i}{L_m} \cdot t_{on} = I_{pk} \tag{3}$$

Deform the formula 3, thus the formula 4 is obtained

$$\mathrm{t}_{on} = I_{pk} \cdot \frac{L_m}{U_i} \tag{4}$$

Deform the formula 1, thus the formula 5 is obtained

$$I_p = \frac{\pi \cdot I_i}{2} \tag{5}$$

Substitute formulas 5 and 4 in formula 2, thus the formula 6 is obtained

$$P_i = I_i \cdot U_i = \frac{L_m \cdot I_{pk}^2}{\pi \cdot T_s} \tag{6}$$

As can be seen from formula 6, the input power P_i of the inverter is proportional to I_{pk}^2, and I_{pk}^2 is an increasing function of I_{pk} ($I_{pk} > 0$). Based on the above analyses, this paper proposes: in maximum power tracking, maximum power point tracking can be achieved by tracking the maximum I_{pk}, instead of multiplying input voltage and input current to obtain the specific power, thus saving the resources of the program and improve the speed of algorithm. At the same time, with no need to sample the input voltage, the MPPT control algorithm can reduce the cost.

4.2 *Program Design of the Novel MPPT Algorithm*

Figure 3 is the flow chart of MPPT control, power-on detection is conducted after the completion of the software startup program. If the test is passed, the software gives a very small initial value for the output given current. After a period of time, the output given current increases $\triangle I$. Compare the former and after the latter peak current, if the latter peak current is larger than the former, then the output given current I_f increases $\triangle I$; if the latter peak current is smaller than the former, then the output given current I_f reduces $\triangle I$. Finally, the output given current I_f will swing within a very small range, thus ultimately achieving the maximum power tracking.

According to the above theoretical analyses, it can be found that there is a certain error in the actual test, when only maximum peak current I_{pk} is calculated in the entire power grid cycle. So, multiple averaging is needed to reduce the interference.

Fig. 3 MPPT program block diagram

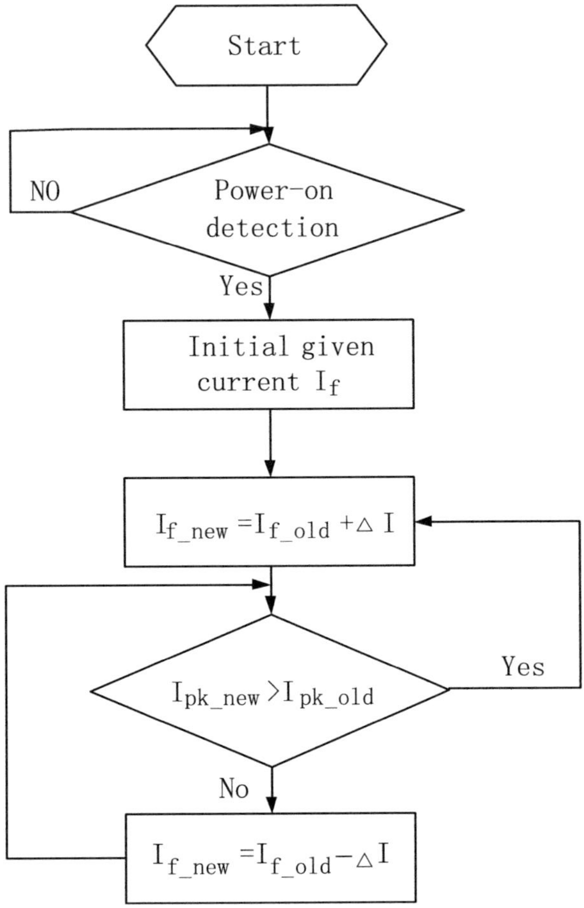

However, this will reduce the maximum power tracking speed. On this basis, further research is studied to get the formula 7

$$I_{\text{total}} = \sum_{i=1}^{N} I_{pki} = \text{k} \cdot I_{pk} \tag{7}$$

I_{pki} represents the peak current value of the transformer primary side in each switching cycle. N represents the sampling times of peak current in a power grid period, and I_{pk} is the maximum value of peak current in a single grid period, and k is constant. It can be seen from the formula 7 that the sum of peak current of the inverter in a single grid period, the single is the monotonic increasing function of the maximum peak current I_{pk} in a single cycle ($I_{pk} > 0$). The inverse function of monotone increasing function is also monotone increasing function, so I_{pk} is also a monotonic increasing function of I_{total}. Therefore, we can know that the I_{total} is a

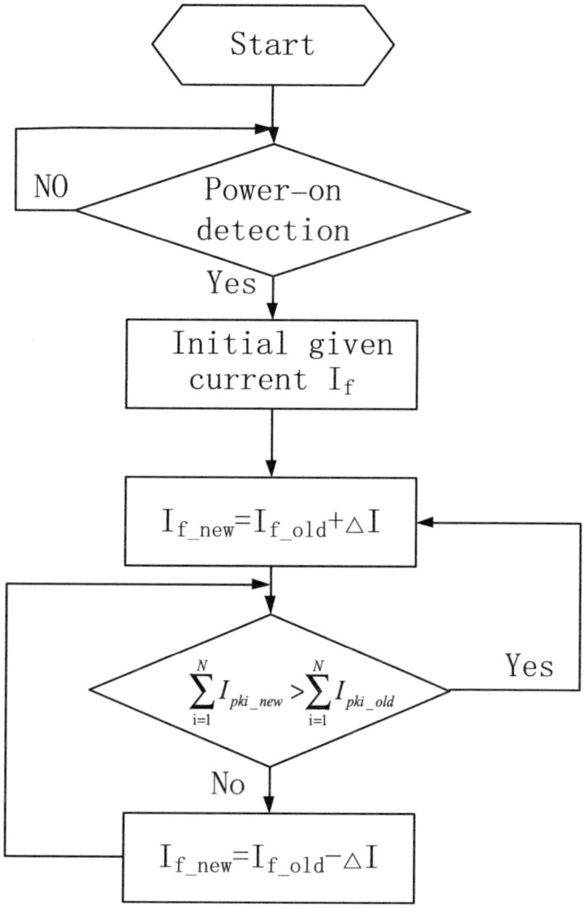

Fig. 4 Improved block diagram of MPPT program

monotonic increasing function of photovoltaic inverter P_i ($I_{pki} > 0$). The method can effectively improve the anti-interference ability and enhance the stability of the system by summing the sampled peak current of each switching period. MPPT algorithm procedure is shown as follows (Fig. 4).

5 Experimental Results

5.1 Test Waveform of the Novel Current Sampling Circuit

As can be seen from Fig. 5, the photovoltaic grid-connected microinverter begins to output power at the point A, and it reaches the maximum power point at the point B. After point B, the PV voltage is changed within a small range.

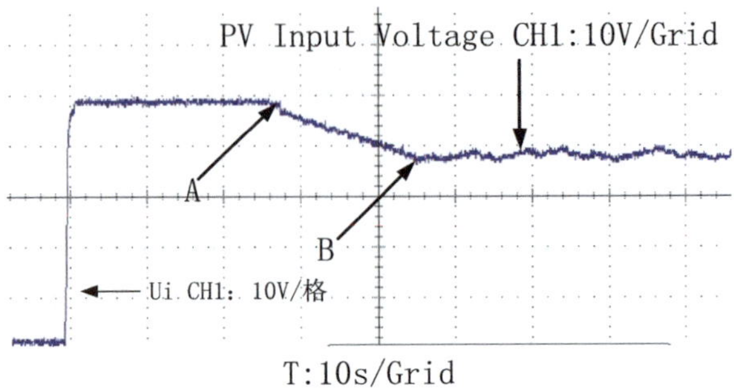

Fig. 5 Output voltage waveform of PV module

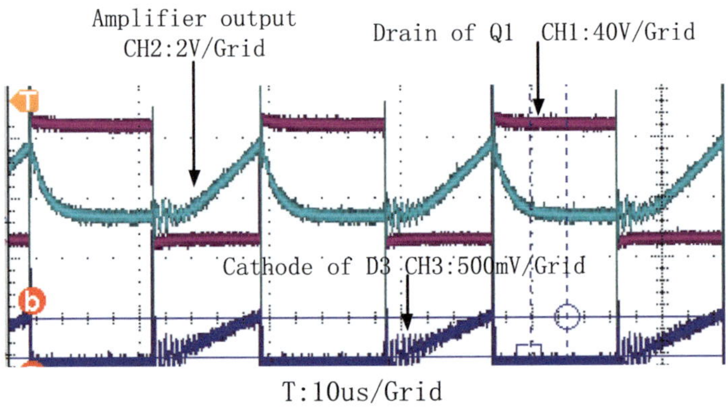

Fig. 6 Peak current sampling waveform

As shown in Fig. 6, when the switch tube gate is at a high level, the drain of Q1, which is shown in the CH1 is grounded. At this time, cathode voltage of D3, which is shown in the CH3 begins to rise, and output waveform CH2 of the differential amplifier circuit synchronously rise. When the voltage reaches its peak value, the ADC pin of the DSP begins to sample.

5.2 Test Waveform of the Novel MPPT Algorithm

As shown in Fig. 7, when the grid-connected power generation function is started, the grid-connected current increases from zero. After about 20 s, the output power tracks the maximum power point and output stably.

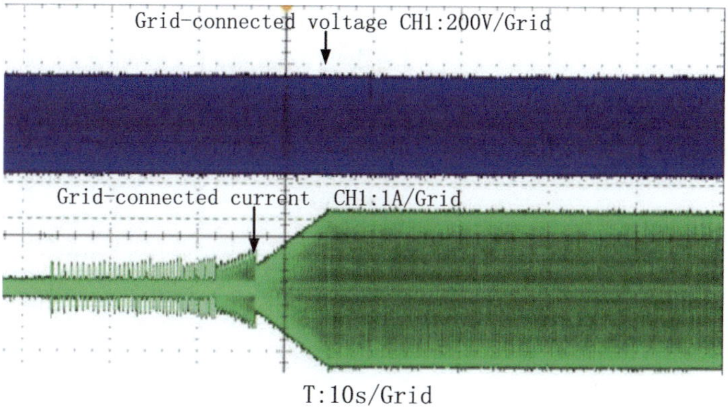

Fig. 7 Maximum power tracking waveform

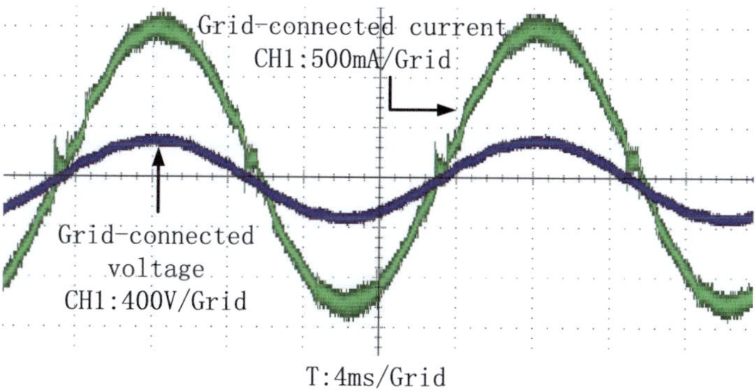

Fig. 8 The point waveform of maximum power

Figure 8 is the grid-connected current and the grid-connected voltage waveform of PV inverter under the maximum output power. It can be seen from the figure that the grid-connected current and the grid-connected voltage are synchronous.

Figure 9 is the method which multiplies the peak current of the transformer primary side and sampled input voltage of the photovoltaic components to obtain the instantaneous output power of the PV module, thus realizing the maximum power point tracking, as mentioned in above traditional technology review. Figure 10 is the method proposed in this paper, which compares peak currents to achieve the maximum power tracking. By comparing these two figures, it is proved that the method that compares peak currents to achieve the maximum power tracking is more accurate.

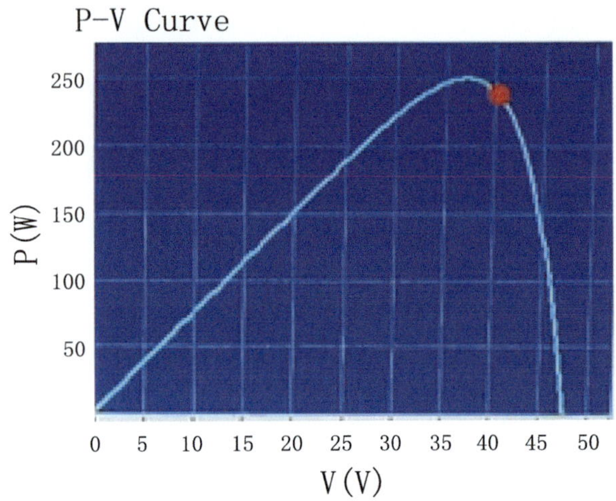

Fig. 9 MPPT by instantaneous output power of the PV module

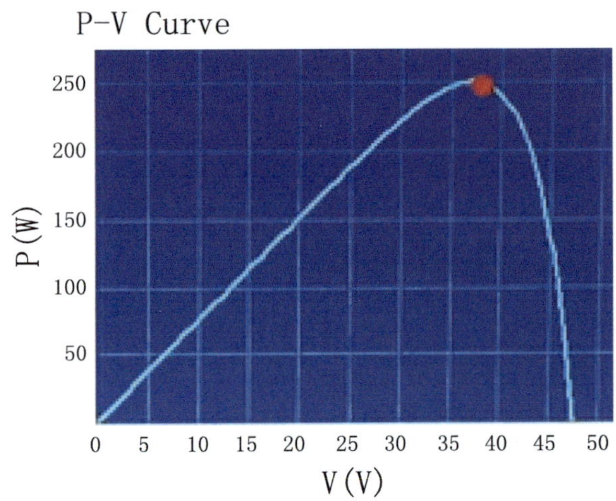

Fig. 10 MPPT by peak currents of the PV module

6 Conclusion

Aiming at the maximum power point calculation method in the fly-back grid-connected power generation microinverter, a current detection circuit composed of the blocking diode and bias diode is proposed in this paper. And a novel fast algorithm for maximum power point is also presented. The proposed current detecting circuit, which is composed of a blocking diode and a bias diode, has high precision in current sampling with low cost. The studied maximum power point calculation method has the advantages of little program code and high computational efficiency. Maximum power point tracking (MPPT) control of maximum power point fast algorithm based on peak current can realize the efficient energy transfer, with no need to introduce the input voltage of the PV module for computing, which saves the cost, thus improving the cost performance of grid-connected photovoltaic microinverter.

References

1. Kwon J-M, Kwon B-H, Nam K-H (2008) Three-phase photovoltaic system with three-level boosting MPPT control. IEEE Trans Power Electron 23(05):2319–2325
2. Soomro AM, Syed AA, Khizer AN (2014) Input series output parallel based multi-modular boost half bridge DC-DC converter with a dynamic control scheme. Inform Electron Eng (04)
3. Zhang F, Zhu XD, Wang MJ, Gao C (2014) Approach to advanced photocatalysts. In: Special topic progress of the materials genome project (1)
4. Huang Y, Zhang X-L, Liu WN, Wang X-M, Han, J-K (2016) Microstructure and mechanical properties of NANOBAIN steel (03)
5. Dongol D, Bollin E, Feldmann T (2015) An overview to the concept of smart coupling and battery management for grid connected photovoltaic battery system. Energy Efficient Technol (04)
6. Wang ZB, Zhao L, Wang SG, Zhang JW, Wang B, Wang LJ (2014) COMPASS time synchronization and dissemination—Toward centimetre positioning accuracy. Article Astrometry (09)
7. Ye CJ, Huang MX (2014) Multi-objective optimal configuration of current limiting strategies (09)

Distributed Finite-Time Formation Control for Multiple Nonholonomic Mobile Robots

Miao Li, Zhongxin Liu and Zengqiang Chen

Abstract In this paper, the finite-time formation control problem for a group of nonholonomic mobile robots is considered. A distributed finite-time estimator is proposed to estimate leader's state in finite time. Then, based on the estimated values of estimator, a distributed finite-time formation control law is designed. With the help of finite-time Lyapunov theory and graph theory, rigorous proof shows that the group of mobile robots can converge to desired formation pattern and its centroid can converge to the desired trajectory in finite time. Simulations are given to verify the effectiveness of the method.

Keywords Nonholonomic mobile robot · Formation control · Distributed control · Finite-time stability

1 Introduction

Recently, the cooperative control of multiple mobile robots has received a great deal of attention due to its wide application prospects, including cooperative surveillance, large object move, troop hunting, etc. Compared with single robot, multiple mobile robots can finish certain works more accurately, efficiently and robustly. As a typical cooperative control problem, formation control of multiple mobile robots is the foundation of other coordination problems. The control strategies are mainly behavior-based control [1], leader–follower-based control [2], artificial potential-based control [3, 4], virtual structure-based control [5], etc.

M. Li · Z. Liu (✉) · Z. Chen
College of Computer and Control Engineering, Nankai University, Tianjin 300353, China
e-mail: lzhx@nankai.edu.cn

M. Li · Z. Liu · Z. Chen
Tianjin Key Laboratory of Intelligent Robotics, Tianjin 300353, China

Z. Chen
College of Science, Civil Aviation University of China, Tianjin 300300, China

Y. Jia et al. (eds.), *Proceedings of 2016 Chinese Intelligent Systems Conference*, Lecture Notes in Electrical Engineering 405,
DOI 10.1007/978-981-10-2335-4_37

Due to the existence of nonholonomic constraint and nonlinear characteristics, the control of mobile robots is more difficult. With the development of cooperative control theory in networked system, these strategies have been used to realize the formation control of multiple mobile robots. In [6, 7], the theories of cascaded systems and communication graph are applied to design control law for nonholonomic kinematic systems, which can guarantee the system globally exponential stability. In [8], based on a new transformation, a distributed control scheme is developed to achieve the desired control objectives.

Most of control algorithms proposed in previous works can guarantee that multi-agent systems asymptotically converge to desired motion, that is, convergence time is infinite. Obviously, finite-time control algorithms with a faster convergence rate are more optimal. Previous works on finite-time control are mainly about first-order and second-order multi-agent system. In [9], finite-time state consensus problem for first-order multi-agent systems is researched and distributed protocols are presented. In [10–12], a distributed protocol is proposed to solve the finite-time consensus problem for second-order agent dynamics. For multiple nonholonomic mobile robots, some finite-time control algorithms were prposed. In [13], the problem of finite-time leader-following consensus is discussed, but one of control inputs is forced to be the reference signal artificially for a given period of time. In [14, 15], the finite-time tracking control problem is solved when the leader's state is available to each follower. In [16], a kind of finite-time formation control algorithm is proposed. However, it excludes the case where the observer's value is zero.

Inspired by aforementioned articles, we mainly investigate the distributed finite-time formation control problem for multiple mobile robots. The contributions of this paper are as follows. The control algorithm is distributed. Each robot can only obtain its neighbors' information. Besides the desired trajectory is considered as a virtual leader whose information is available to a subset of mobile robots. We design an estimator to estimate the leader's state in finite time. And based on the estimated values of estimator, a finite-time formation controller is designed for each mobile robot, the controller can guarantee that the group of mobile robots converges to desired formation, and the centroid of formation converges to desired trajectory in finite time.

The rest of this paper is organized as follows. In Sect. 2, the model of nonholonomic mobile robots and preliminaries used in this paper are given. In Sect. 3, some main results on distributed formation control problem for multiple mobile robots are established. The simulation results are presented in Sect. 4. Finally, conclusion is drawn in Sect. 5.

2 Problem Statement

Notation Let $\text{diag}(w) = \text{diag}(w_1, \ldots, w_n)$, $1_n = [1, 1 \ldots 1]^T \in R^n$. $\lambda_{\min}(\cdot)$ and $\lambda_{\max}(\cdot)$ are the smallest and largest eigenvalues of the matrix, respectively. For any vectors $a = [a_1, \ldots, a_n]^T$ and $b = [b_1, \ldots, b_n]^T$, denote $a \odot b = [a_1 b_1, \ldots, a_n b_n]^T$, Let $\text{sig}(x)^\alpha = \text{sign}(x)|x|^\alpha$, where $\alpha > 0$, $x \in R$, and $\text{sign}(\cdot)$ is the sign function, If $x_{i*} = [x_{i1}, \ldots, x_{in}]^T \in R^n$, then $\text{sig}(x_{i*})^\alpha = \text{sign}(x_{i*}) \odot |x_{i*}|^\alpha$, where $\text{sign}(x_{i*}) = [\text{sign}(x_{i1}), \ldots, \text{sign}(x_{in})]^T$ and $|x_{i*}|^\alpha = [|x_{i1}|^\alpha, \ldots, |x_{in}|^\alpha]^T$.

2.1 Algebraic Graph Theory

As done in [17], for a group of n nonholonomic mobile robots, if each mobile robot is a node, we can represent the communication between the robots by a weighted directed graph $G = (\mathcal{V}, E, A)$, where $\mathcal{V} = \{v_1, v_2, \ldots, v_n\}$ is a node set representing the mobile robots, and $E \subseteq \mathcal{V} \times \mathcal{V}$ is an edge set. The weighted adjacency matrix is defined as $A = [a_{ij}] \in R^{n \times n}$, where a_{ij} denotes the edge weight from v_i to v_j. if $(v_i, v_j) \in E$, then $a_{ij} > 0$, Otherwise $a_{ij} = 0$. The set of neighbors of node v_i is denoted by $N_i = \{v_j \in \mathcal{V}: (i, j) \in E, i \neq j\}$. Furthermore, The matrix $L = D - A$ is called Laplacian matrix of G, where $D = \text{diag}\{d_1, d_2, \ldots, d_n\}$ with $d_i = \sum_{j=1}^n a_{ij}$. Weighted directed graph G is said to satisfy the detail-balanced condition [17] in weights if there exist some scalars $w_i > 0$ with $i = 1, 2 \ldots, n$ satisfying $w_i a_{ij} = w_j a_{ji}$ for all $i, j = 1, 2 \ldots, n$.

Assume that a desired trajectory is regard as a virtual leader. Let B denote the connection weight between robot i and the virtual leader, where $B = \text{diag}(b_1, b_2, \ldots, b_n)$. If robot i can obtain information of the virtual leader, then $b_i = 1$, otherwise $b_i = 0$.

2.2 Problem Formulation

Consider a group of n nonholonomic mobile robots labeled as $i = 1, 2, \ldots, n$. The kinematics of robot i can be described by:

$$\begin{cases} \dot{x}_i = v_i \cos\theta_i \\ \dot{y}_i = v_i \sin\theta_i \\ \dot{\theta}_i = \omega_i \end{cases} \tag{1}$$

where (x_i, y_i) is the position coordinates of robot i, θ_i is the heading angle, and (v_i, ω_i) represent the linear velocity and the angular velocity of robot i. It is well-known that the system (1) satisfies the pure roll without slipping constraint, i.e., $\dot{y}_i \cos\theta_i - \dot{x}_i \sin\theta_i = 0$.

Suppose that the n mobile robots converge to the desired formation pattern F, (p_{0x}, p_{0y}) is the centroid of the formation pattern F, (p_{ix}, p_{iy}) is the position vectors of robots i relative to the centroid of formation pattern F, which satisfies

$$\sum_{i=1}^{n} \frac{p_{ix}}{n} = p_{0x}, \sum_{i=1}^{n} \frac{p_{iy}}{n} = p_{0y}$$

Without loss of generality, we assume that $p_{0x} = 0, p_{0y} = 0$. The desired trajectory is considered as a virtual leader 0, which can be described by:

$$\begin{cases} \dot{x}_0 = v_0(t)\cos\theta_0 \\ \dot{y}_0 = v_0(t)\sin\theta_0 \\ \theta_0 = \omega_0(t) \end{cases} \tag{2}$$

where $(v_0(t), \omega_0(t))$ are known time-varying functions.

Now, our control problem is defined as designing a controller for mobile robot i based on its state and its neighbors' state, such that the group of robots converges to a desired formation pattern F and the centroid of formation converges to the desired trajectory in finite-time T, that is to say,

$$\begin{aligned} &\lim_{t \to T} \begin{bmatrix} x_i(t) - x_j(t) \\ y_i(t) - y_j(t) \end{bmatrix} = \begin{bmatrix} p_{ix} - p_{jx} \\ p_{iy} - p_{jy} \end{bmatrix} \\ &\lim_{t \to T} (\theta_i - \theta_0) = 0 \\ &\lim_{t \to T} \left(\sum_{i=1}^{n} \frac{x_i(t)}{n} - x_0(t) \right) = 0 \\ &\lim_{t \to T} \left(\sum_{i=1}^{n} \frac{y_i(t)}{n} - y_0(t) \right) = 0 \end{aligned}$$

Before the introduction of the distributed control algorithm, some necessary assumptions which mobile robot system should satisfy, and lemmas are listed.

Assumption 1 $|u_{10}| > 0$, $|\dot{u}_{10}| \le \kappa$, $|\dot{u}_{20}| \le \kappa$.

Assumption 2 Weighted directed graph G is strongly connected and detail-balanced, and there is at least one robot which can obtain the leader's information.

Lemma 1 ([18]) *For any* $x_i \in R$, $i = 1, 2, \ldots n$, $0 < p \le 1$, *the following inequalities hold*

$$\left(\sum_{i=1}^{n} |x_i| \right)^p \le \sum_{i=1}^{n} |x_i|^p \le n^{1-p} \left(\sum_{i=1}^{n} |x_i| \right)^p$$

Lemma 2 ([19]) *Consider the system*

$$\dot{x}=f(x,t),\ f(0,t)=0,\ x\in R^n$$

Suppose there are C^1 *function* $V(x)$ *defined on a neighborhood* $\hat{U}\subset R^n$ *of the origin. If the function* $V(x)$ *is positive definite and* $\dot{V}(x)+cV^{\alpha}(x)\le 0$ *on* $\hat{U}$ *where* $c>0$ *and* $0<\alpha<1$ *are real numbers, then the origin of system is locally finite time stable. The setting time depending on the initial state* $x(0)=x_0$, *satisfies* $T(x_0)\le \frac{V(x_0)^{1-\alpha}}{c(1-\alpha)}$. *If* $V(x)$ *is also radially unbounded and* $\hat{U}=R^n$, *the system is globally finite time stable.*

Lemma 3 ([20]) *If Assumption 2 holds, then the matrix* $diag(w)(L+B)$ *is positive definite.*

3 Main Results

In this section, the finite-time formation control problem for system (1) is solved. First, a finite-time distributed estimator is proposed to estimate the virtual leader's state. Second, a distributed formation control law is designed to make the mobile robots converge to the desired state in finite time.

For the controller design, we define the following transformation:

$$\begin{cases} x_{1i}=\theta_i \\ x_{2i}=(x_i-p_{ix})\sin\theta_i-(y_i-p_{iy})\cos\theta_i \\ x_{3i}=(x_i-p_{ix})\cos\theta_i+(y_i-p_{iy})\sin\theta_i \\ u_{1i}=\omega_i \\ u_{2i}=v_i-x_{2i}\omega_i \end{cases} \tag{3}$$

Using the above-coordinate transformation, the system (1) is transferred into the following form:

$$\begin{cases} \dot{x}_{1i}=u_{1i} \\ \dot{x}_{2i}=u_{1i}x_{3i} \\ \dot{x}_{3i}=u_{2i} \end{cases} \tag{4}$$

where $0\le i\le n$, $u_i=[u_{1i},u_{2i}]^T$ and $x_i=[x_{1i},x_{2i},x_{3i}]^T$ are the control input and state of mobile robot i after the coordinate transformation, respectively.

Defining the formation error vector as $\tilde{x}_i=[\tilde{x}_{1i},\tilde{x}_{2i},\tilde{x}_{3i}]^T$, the formation error dynamics can be derived as following:

$$\begin{cases} \dot{\tilde{x}}_{1i} = u_{1i} - u_{10} \\ \dot{\tilde{x}}_{2i} = u_{10}\tilde{x}_{3i} + x_{3i}(u_{1i} - u_{10}) \\ \dot{\tilde{x}}_{3i} = u_{2i} - u_{20} \end{cases} \tag{5}$$

We divide formation error dynamics (5) into a first-order subsystem (6),

$$\dot{\tilde{x}}_{1i} = u_{1i} - u_{10} \tag{6}$$

and a second-order subsystem (7)

$$\begin{cases} \dot{\tilde{x}}_{2i} = u_{10}\tilde{x}_{3i} + x_{3i}(u_{1i} - u_{10}) \\ \dot{\tilde{x}}_{3i} = u_{2i} - u_{20} \end{cases} \tag{7}$$

Based on the transformation (3), the objective is converted into designing a distributed control laws $u_i = [u_{1i}, u_{2i}]^T$ to stabilize the closed-loop system (6) and (7) in a finite time.

3.1 Distributed Finite-Time Estimator

In the following, based on sliding mode control method, we construct distributed finite-time estimator for each mobile robot to estimate the virtual leader's information.

Theorem 1 *For mobile robot system* (6)–(7)*, if Assumption* 1 *and* 2 *hold, we design the distributed finite-time estimator as follows:*

$$\begin{cases} \dot{\hat{u}}_{1i} = -k_1\left(\sum\limits_{j\in N_i} a_{ij}(\hat{u}_{1i} - \hat{u}_{1j}) + b_i(\hat{u}_{1i} - u_{10})\right) - k_2\mathrm{sign}\left(\sum\limits_{j\in N_i} a_{ij}(\hat{u}_{1i} - \hat{u}_{1j}) + b_i(\hat{u}_{1i} - u_{10})\right) \\ \dot{\hat{u}}_{2i} = -k_1\left(\sum\limits_{j\in N_i} a_{ij}(\hat{u}_{2i} - \hat{u}_{2j}) + b_i(\hat{u}_{2i} - u_{20})\right) - k_2\mathrm{sign}\left(\sum\limits_{j\in N_i} a_{ij}(\hat{u}_{2i} - \hat{u}_{2j}) + b_i(\hat{u}_{2i} - u_{20})\right) \end{cases} \tag{8}$$

where $i = 1, 2\ldots, n$, $k_1, k_2 > 0$, $\hat{u}_{1i}, \hat{u}_{2i}$ *are the estimated information of the leader for the robot i, and then* $\hat{u}_{1i} = u_{10}, \hat{u}_{2i} = u_{20}$ *for* $\forall t > T_1$.

Proof Let $k_2 > \kappa$, $\hat{u}_{1*} = [\hat{u}_{11}, \ldots, \hat{u}_{1n}]^T$, $\hat{u}_{2*} = [\hat{u}_{21}, \ldots, \hat{u}_{2n}]^T$, $\bar{u}_{1*} = \hat{u}_{1*} - u_{10}1_n$, $\bar{u}_{2*} = \hat{u}_{2*} - u_{20}1_n$, $N = \mathrm{diag}(L + B, L + B)$, $\bar{u} = [\bar{u}_{1*}, \bar{u}_{2*}]^T$, $f_0 = [u_{10}1_n^T, u_{20}1_n^T]^T$

From (8), the estimator error can be rewritten in a compact form as

$$\dot{\bar{u}} = -k_1 N\bar{u} - k_2\mathrm{sign}(N\bar{u}) - \dot{f}_0 \tag{9}$$

Choose the Lyapunov candidate function as

$$V_1 = \frac{1}{2}\bar{u}^T M^T \bar{u}$$

The derivative of V_1 along system (9) is

$$\begin{aligned}\dot{V}_1 =& -k_1\bar{u}^T N^T WN\bar{u} - k_1\bar{u}^T N^T W\mathrm{sign}(N\bar{u}) - \bar{u}^T N^T W\dot{f}_0 \\ \leq& -\lambda_{\min}(W)\left(k_1\bar{u}^T N^2\bar{u} + (k_2-\kappa)\left\|\bar{u}^T N\right\|_1\right) \\ \leq& -\lambda_{\min}(W)\left(k_1\bar{u}^T N^2\bar{u} + (k_2-\kappa)\left\|\bar{u}^T N\right\|_2\right) \\ \leq& -\lambda_{\min}(W)\left(\frac{2\lambda_{\min}(M)k_1V_1}{\lambda_{\max}^2(W)} + \frac{\sqrt{2}(k_2-\kappa)\sqrt{\lambda_{\min}(M)}}{\lambda_{\max}(W)}\sqrt{V_1}\right)\end{aligned}$$

By calculating the differential inequality, we can get

$$\sqrt{V_1} \leq \left(\sqrt{V_1(0)} + \frac{(k_2-\kappa)\lambda_{\max}(W)}{\sqrt{2k_1\lambda_{\min}(M)}}\right)e^{-\lambda_{\min}(W)\frac{\lambda_{\min}(M)}{\lambda_{\max}^2(W)}t} - \frac{(k_2-\kappa)\lambda_{\max}(W)}{\sqrt{2k_1\lambda_{\min}(M)}}$$

Since $k_2 > \kappa$, we can conclude that V_1 reaches zero in finite time, which implies that $V_1 = 0$, $\forall t \geq T_1$, where $T_1 = \frac{\lambda_{\max}^2(W)}{\lambda_{\min}(W)\lambda_{\min}(M)}\ln(1 + \frac{\sqrt{2k_1\lambda_{\min}(M)V_1(0)}}{(k_2-\kappa)\lambda_{\max}(W)})$. That is to say, $\hat{u}_{1i} = u_{10}$, $\hat{u}_{2i} = u_{20}$, $\forall t \geq T_1$, $i = 1, 2 \ldots, n$. This completes the proof.

Remark 1 From the expression of convergence time T_1, the parameter k_2 can affect the convergence speed of the system (9). With a larger k_2, the convergence speed is faster, but chattering phenomenon of system (9) will be more obvious caused by the discontinuity of $\mathrm{sign}(\cdot)$ function. So, we need to consider the control precision and convergence speed comprehensively when choosing parameter k_2.

3.2 Distributed Finite-Time Formation Control Law

Based on the estimated values for desired state of the virtual leader 0, a finite-time formation controller is designed for each mobile robot.

Theorem 2 *Consider mobile robot system* (6)–(7), *if Assumption* 1 *and* 2 *hold, we design finite-time control law as follows:*

$$u_{1i} = \hat{u}_{1i} - \mathrm{sig}(\varepsilon_{1i})^{\alpha} \tag{10}$$

$$u_{2i} = \hat{u}_{2i} - \hat{u}_{1i}\mathrm{sig}(\varepsilon_{2i})^{\frac{\alpha}{2-\alpha}} - \mathrm{sig}(\varepsilon_{3i})^{\alpha} \tag{11}$$

where $i=1,2\ldots,n$, $0<\alpha<1$, $\varepsilon_{1i}=\sum_{j\in N_i}a_{ij}(x_{1i}-x_{1j})+b_i(x_{1i}-x_{10})$, $\varepsilon_{2i}=\sum_{j\in N_i}a_{ij}(x_{2i}-x_{2j})+b_i(x_{2i}-x_{20})$, $\varepsilon_{3i}=\sum_{j\in N_i}a_{ij}(x_{3i}-x_{3j})+b_i(x_{3i}-x_{30})$, *then* $x_{1i}=x_{10}$, $x_{2i}=x_{20}$ *and* $x_{3i}=x_{30}$ *for* $\forall t>T_1+T_2+T_3$.

Proof In the following part, we will prove Theorem 2 in two steps. Firstly, we will prove subsystem (6) is not divergent when $t<T_1$, and subsystem (6) is finite-time stability when $t>T_1$. Secondly, we will prove subsystem (7) is not divergent when $t<T_1+T_2$, and subsystem (7) is finite-time stability when $t>T_1+T_2$.

First, we prove subsystem (6) is not divergent when $t<T_1$, and subsystem (6) is finite-time stability when $t>T_1$.

Let $x_{1*}=[x_{11},\cdots,x_{1n}]^T$, $\tilde{x}_{1*}=x_{1*}-1_n x_{10}$, $\varepsilon_{1*}=[\varepsilon_{11},\ldots,\varepsilon_{1n}]^T$. Consider the Lyapunov function

$$V_2=\frac{1}{2}\tilde{x}_{1*}^T\tilde{x}_{1*}$$

The derivative of V_2 along system (6) is

$$\dot{V}_2=\tilde{x}_{1*}^T\dot{\tilde{x}}_{1*}=\tilde{x}_{1*}^T(\hat{u}_{1*}-u_{10}1_n)-\tilde{x}_{1*}^T\mathrm{sig}(\varepsilon_{1*})^\alpha$$

Since $\hat{u}_{1i}$ converges to zero in finite time T_1, $\hat{u}_{1i}$ will be bounded in any time, i.e., there is a positive constant C_1, such that $\|\hat{u}_{1*}-u_{10}1_n\|_2\le C_1$, we have

$$\dot{V}_2\le C_1\left\|\tilde{x}_{1*}^T\right\|_2+n^{1-\alpha}\|(L+B)\|_F^\alpha\|\tilde{x}_{1*}\|_2^{1+\alpha}$$

When $\left\|\tilde{x}_{1*}^T\right\|_2\le 1$, we have

$$\dot{V}_2\le L_1$$

where $L_1=n^{1-\alpha}\|(L+B)\|_F^\alpha+C_1$. However, when $\left\|\tilde{x}_{1*}^T\right\|_2=\eta_2\ge 1$, we have

$$\begin{aligned}\dot{V}_2&\le C_1\eta_2+n^{1-\alpha}\|(L+B)\|_F^\alpha\eta_2^{1+\alpha}\\&\le 2C_1V_2+2n^{1-\alpha}\|(L+B)\|_F^\alpha V_2\\&\le L_2V_2\end{aligned}$$

where $L_2=2C_2+2n^{1-\alpha}\|(L+B)\|_F^\alpha$. Thus, for any $\tilde{x}_{1*}$, we have $\dot{V}_2\le L_2V_2+L_1$. After some manipulation, we can further get $V_2\le\left(V_2(0)+\frac{L_1}{L_2}\right)e^{L_2t}-\frac{L_1}{L_2}$. Hence, $\tilde{x}_{1*}$ is bounded when $t<T_1$.

From Theorem 1, when $t \geq T_1$, $\hat{u}_{1i} = u_{10}$, then control law (10) is equivalent to the following:

$$u_{1i} = u_{10} - \text{sig}(\varepsilon_{1i})^{\alpha}$$

Closed-loop system (6) can be rewritten in a compact form as:

$$\dot{\tilde{x}}_{1*} = -\text{sig}(\varepsilon_{1*})^{\alpha} \tag{12}$$

Choose the Lyapunov candidate function as

$$V_3 = \frac{1}{2} x_{1*}^T (L+B)^T \text{diag}(w) x_{1*}$$

The derivative of V_3 along system (11) satisfies

$$\begin{aligned}\dot{V}_3 &= -x_{1*}^T (L+B)^T \text{diag}(w) \text{sig}(\varepsilon_{1*})^{\alpha} \\ &\leq -\lambda_{\min}(\text{diag}(w)) \left(\frac{2\lambda_{\min}(\text{diag}(w)(L+B))}{\lambda_{\max}^2(\text{diag}(w))} \right)^{\frac{1+\alpha}{2}} V_3^{\frac{1+\alpha}{2}}\end{aligned}$$

where $\rho_1 > 0, 0 < \frac{1+\alpha}{2} < 1$. By Lemma 2, we have V_3 will converge to zero in finite time, it can be concluded that $x_{1i} = x_{10}$, $\forall t > T_1 + T_2$, $i = 1, 2 \ldots, n$. where $T_2 \leq \frac{2^{\frac{1-\alpha}{2}} \lambda_{\max}^{1+\alpha}(\text{diag}(w)) V_3(0)^{\frac{1-\alpha}{2}}}{(1-\alpha)\lambda_{\min}(\text{diag}(w)) \left(\lambda_{\min}(\text{diag}(w)(L+B))\right)^{\frac{1+\alpha}{2}}}$.

Next, we will prove subsystem (7) is not divergent when $t < T_1 + T_2$, and subsystem (7) is finite-time stability when $t > T_1 + T_2$.

Let $x_{2*} = [x_{21}, \ldots, x_{2n}]^T$, $\tilde{x}_{2*} = x_{2*} - 1_n x_{20}$, $x_{3*} = [x_{31}, \ldots, x_{3n}]^T$, $\tilde{x}_{3*} = x_{3*} - 1_n x_{30}$. Consider the Lyapunov function

$$V_4 = \frac{1}{2} \tilde{x}_{2*}^T \tilde{x}_{2*} + \frac{1}{2} \tilde{x}_{3*}^T \tilde{x}_{3*}$$

The derivative of V_4 along system (7) is

$$\begin{aligned}\dot{V}_4 &= \tilde{x}_{2*}^T \dot{\tilde{x}}_{2*} + \tilde{x}_{3*}^T \dot{\tilde{x}}_{3*} \\ &= u_{10} \tilde{x}_{2*}^T \tilde{x}_{3*} + \tilde{x}_{2*}^T \tilde{u}_{1*} \tilde{x}_{3*} + \tilde{x}_{2*}^T \tilde{u}_{1*} x_{30} + \tilde{x}_{3*}^T \left(\hat{u}_{2*} - u_{20} 1_n - \hat{u}_{1*} \text{sig}(\varepsilon_{2*})^{\frac{\alpha}{2-\alpha}} - \text{sig}(\varepsilon_{3*})^{\alpha} \right)\end{aligned}$$

Since $u_{1i}, \hat{u}_{1i}$ converge to u_{10} and $\hat{u}_{2i}$ converges to u_{20} in finite time $T_1 + T_2$, $u_{1i}, \hat{u}_{1i}$ and $\hat{u}_{2i}$ will be bounded in any time, i.e., there is a positive constant C_2, such that $\|\hat{u}_{1*} - u_{10} 1_n\|_2 \leq C_2$, $\|\tilde{u}_{1*}\|_F \leq C_2$, $\|\hat{u}_{1*}\|_F \leq C_2$, $\|\hat{u}_{2*} - u_{20} 1_n\|_2 \leq C_2$.

$$\dot{V}_4 \le C_2\left\|\tilde{x}_{2*}^T\right\|_2\|\tilde{x}_{3*}\|_2 + C_2\left\|\tilde{x}_{2*}^T\right\|_2\|\tilde{x}_{3*}\|_2 + C_2^2\left\|\tilde{x}_{2*}^T\right\|_2 + C_2\left\|\tilde{x}_{3*}^T\right\|_2$$
$$+ C_2 n^{\frac{2-2\alpha}{2-\alpha}}\|(L+B)\|_F^{\frac{\alpha}{2-\alpha}}\|\tilde{x}_{3*}\|_2\|x_{2*}\|_2^{\frac{\alpha}{2-\alpha}} + n^{1-\alpha}\|(L+B)\|_F^{\alpha}\|\tilde{x}_{3*}\|_2^{1+\alpha}$$

When $\left\|\left(\tilde{x}_{2*}^T, \tilde{x}_{3*}^T\right)^T\right\|_2 \le 1$, we have

$$\dot{V}_4 \le L_3$$

where $L_3 = n^{1-\alpha}\|(L+B)\|_F^{\alpha} + 3C_2 + C_2^2 + C_2 n^{\frac{2-2\alpha}{2-\alpha}}\|(L+B)\|_F^{\frac{\alpha}{2-\alpha}}$. However, when $\left\|\left(\tilde{x}_{2*}^T, \tilde{x}_{3*}^T\right)^T\right\|_2 = \eta_2 \ge 1$, we have

$$\dot{V}_4 \le C_2\eta_2^2 + C_2\eta_2^2 + C_2^2\eta_2 + C_2\eta_2 + C_2 n^{\frac{2-2\alpha}{2-\alpha}}\|(L+B)\|_F^{\frac{\alpha}{2-\alpha}}\eta_2^{\frac{2}{2-\alpha}} + n^{1-\alpha}\|(L+B)\|_F^{\alpha}\eta_2^{1+\alpha}$$
$$\le 6C_2V_4 + 2n^{1-\alpha}\|(L+B)\|_F^{\alpha}V_4 + 2C_2^2V_4 + 2C_2 n^{\frac{2-2\alpha}{2-\alpha}}\|(L+B)\|_F^{\frac{\alpha}{2-\alpha}}V_4$$
$$\le L_4V_4$$

where $L_4 = 6\mathrm{C}_2 + 2n^{1-\alpha}\|(L+B)\|_F^{\alpha} + 2\mathrm{C}_2^2 + 2\mathrm{C}_2 n^{\frac{2-2\alpha}{2-\alpha}}\|(L+B)\|_F^{\frac{\alpha}{2-\alpha}}$. Thus, for any $\tilde{x}_{2*}$ and $\tilde{x}_{3*}$, we have $\dot{V}_4 \le L_4V_4 + L_3$. After some manipulation, we can further get $V_4 \le (V_4(0) + \frac{L_4}{L_3})e^{L_4t} - \frac{L_4}{L_3}$. Hence, $\tilde{x}_{2*}, \tilde{x}_{3*}$ is bounded when $t < T_1 + T_2$.

When $t \ge T_1 + T_2$, $\hat{u}_{1i} = u_{10}$, $\hat{u}_{2i} = u_{20}$, $u_{1i} = u_{10}$, then control law (11) becomes

$$u_{2i} = u_{20} - u_{10}\mathrm{sig}(\varepsilon_{2i})^{\frac{\alpha}{2-\alpha}} - \mathrm{sig}(\varepsilon_{3i})^{\alpha}$$

Closed-loop system (7) can be obtained as

$$\begin{cases} \dot{\varepsilon}_{2*} = u_{10}\varepsilon_{3*} \\ \dot{\varepsilon}_{3*} = -(L+B)\mathrm{sig}(\varepsilon_{3*})^{\alpha} - u_{10}(L+B)\mathrm{sig}(\varepsilon_{2*})^{\frac{\alpha}{2-\alpha}} \end{cases} \tag{13}$$

Choose the Lyapunov candidate function as

$$V_5 = \frac{1}{2}\varepsilon_{3*}^T((L+B)\mathrm{diag}(w))^{-1}\varepsilon_{3*} + \frac{2-\alpha}{2}\left(\left|\varepsilon_{2*}^{\frac{1}{2-\alpha}}\right|\right)^T(\mathrm{diag}(w))^{-1}\left(\left|\varepsilon_{2*}^{\frac{1}{2-\alpha}}\right|\right)$$

Let $f(\varepsilon_{3*}) = -\varepsilon_{3*}^T(\mathrm{diag}(w))^{-1}\mathrm{sig}(\varepsilon_{3*})^{\alpha}$. For any $k > 0$, we obtain

$$V_5\left(k^{2-\alpha}\varepsilon_{2*}, k\varepsilon_{3*}\right) = k^2V_5(\varepsilon_{2*}, \varepsilon_{3*}) \tag{14}$$

and

$$f(k\varepsilon_{3*}) = k^{1+\alpha}f(\varepsilon_{3*}) \tag{15}$$

When $V_5(\varepsilon_{2*},\varepsilon_{3*})>0$, let $k=(V_5(\varepsilon_{2*},\varepsilon_{3*}))^{-\frac{1}{2}}$ in (15), we can get

$$\begin{aligned}\frac{f(\varepsilon_{3*})}{(V_5(\varepsilon_{2*},\varepsilon_{3*}))^{\frac{1+\alpha}{2}}}&=f\left((V_5(\varepsilon_{2*},\varepsilon_{3*}))^{-\frac{1}{2}}\varepsilon_{3*}\right)\\ &\leq \max f\left((V_5(\varepsilon_{2*},\varepsilon_{3*}))^{-\frac{1}{2}}\varepsilon_{3*}\right)\\ &=-c\end{aligned}$$

where $c\geq 0$. Since $\{(\varepsilon_{2*}^T,\varepsilon_{3*}^T)^T: c=0\}=\{(0^T,0^T)^T\}$, we have $V_5(\varepsilon_{2*},\varepsilon_{3*})=0$. Conflict with prerequisite condition $V_5(\varepsilon_{2*},\varepsilon_{3*})>0$. So, it is easy to obtain that $c>0$. Then, we obtain $f(\varepsilon_{3*})\leq -c(V_5(\varepsilon_{2*},\varepsilon_{3*}))^{\frac{1+\alpha}{2}}$ with $\alpha\in(0,1)$. Second, we will prove that when $V_5(\varepsilon_{2*},\varepsilon_{3*})=0$, we have $(\varepsilon_{3*}^T,\varepsilon_{2*}^T)^T=(0^T,0^T)^T$. Then it can be obtained that $f(\varepsilon_{3*})\leq -c(V_5(\varepsilon_{2*},\varepsilon_{3*}))^{\frac{1+\alpha}{2}}$.

Moreover, the derivative of V_5 along system (13) is

$$\begin{aligned}\dot{V}_5&=\varepsilon_{3*}^T((L+B)\mathrm{diag}(w))^{-1}\dot{\varepsilon}_{3*}+\left(\left|\varepsilon_{2*}^{\frac{1}{2-\alpha}}\right|\right)^T(\mathrm{diag}(w))^{-1}(\mathrm{sig}(\varepsilon_{2*})^{\frac{\alpha-1}{2-\alpha}}\odot\dot{\varepsilon}_{2*})\\ &=-\varepsilon_{3*}^T(\mathrm{diag}(w))^{-1}\mathrm{sig}(\varepsilon_{3*})^{\alpha}-u_{10}\varepsilon_{3*}^T(\mathrm{diag}(w))^{-1}\mathrm{sig}(\varepsilon_{2*})^{\frac{\alpha}{2-\alpha}}\\ &\quad+u_{10}\left(\mathrm{sig}(\varepsilon_{2*})^{\frac{\alpha}{2-\alpha}}\right)^T(\mathrm{diag}(w))^{-1}\varepsilon_{3*}\\ &=-\varepsilon_{3*}^T(\mathrm{diag}(w))^{-1}\mathrm{sig}(\varepsilon_{3*})^{\alpha}\leq 0\end{aligned}$$

Hence, $\dot{V}_5(\varepsilon_{2*},\varepsilon_{3*})=f(\varepsilon_{3*})\leq -c(V_5(\varepsilon_{2*},\varepsilon_{3*}))^{\frac{1+\alpha}{2}}$. By lemma 2, we have V_5 will converge to 0 in finite time T_3, it can be concluded that $x_{2i}=x_{20}$, $x_{3i}=x_{30}$, $\forall t>T_1+T_2+T_3$, $i=1,2\ldots,n$, where $T_3=\frac{2}{c(1-\alpha)}(V_5(\varepsilon_{2*}(0),\varepsilon_{3*}(0)))^{\frac{1-\alpha}{2}}$. This completes the proof.

Hence, by Theorems 1 and 2 we conclude that nonholonomic mobile robot system (1) is finite-time stability under distributed control laws (10) and (11). Thus, the formation control problem is solvable in finite time $T=T_1+T_2+T_3$.

Remark 2 $|\dot{u}_{10}|\leq\kappa$, $|\dot{u}_{20}|\leq\kappa$, it means the line velocity and angular velocity of virtual leader not only can be constants, but also can be time-varying functions. So, the control algorithm proposed in this paper can guarantee the group of mobile robots moves along with the desired trajectory generated by a virtual leader whose line velocity and angular velocity are varying.

4 Simulation Results

In this section, some simulation results are presented to illustrate the effectiveness of the proposed theoretical results.

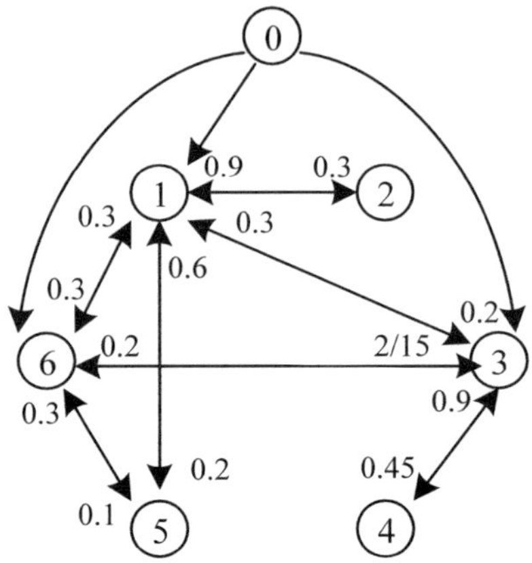

Fig. 1 The desired formation pattern and the information exchange among mobile robots

We consider mobile robots system (1) with $n = 6$, the information exchange topology among mobile robots is shown in Fig. 1, define $\mathrm{diag}(w) = \mathrm{diag}(3, 1, 2, 1, 1, 3)$. Obviously, $\mathrm{diag}(w)(L + B)$ is positive definite. Hence Assumption 2 holds.

Example 1 The desired trajectory is generated by system (1) with $v_0 = 5$ m/s, $\omega_0 = 0.5$ rad/s. The initial state of the virtual leader 0 is $[x_0(0), y_0(0), \theta_0(0)]^T = [0, -10, 0]^T$. The desired formation pattern F is $(p_{1x}, p_{1y}) = (-1, \sqrt{3})$, $(p_{2x}, p_{2y}) = (1, \sqrt{3})$, $(p_{3x}, p_{3y}) = (2, 0)$, $(p_{4x}, p_{4y}) = (1, -\sqrt{3})$, $(p_{5x}, p_{5y}) = (-1, -\sqrt{3})$ and $(p_{6x}, p_{6y}) = (-2, 0)$. According to the procedure detailed in the proof of Theorem 1, estimator gain can be calculated. So, the gain parameters for each mobile robot are chosen as $k_1 = 3$, $k_2 = 0.5$, $\alpha = 0.8$. The response curves of each mobile robot system (1) under control laws (10) and (11) are given in Fig. 2, 3 and 4. It is shown from Fig. 2 that the formation pattern of six robots is converge to the desired formation and the centroid of formation converges to the desired trajectory. Figure 3 shows that the angular velocity ω_i and velocity v_i of six mobile robots converge to angular velocity ω_0 and line velocity v_0 of the virtual leader. Position errors and heading errors are presented in Fig. 4, from which it is easy to see that after a few seconds the position error and heading error converge to (p_{ix}, p_{iy}) and 0, respectively.

Example 2 The desired trajectory is generated by system (1) with $\omega_0 = 1$ rad/s, $v_0 = 3 \sin\, t \cos\, t$ m/s. The initial state of the leader 0 is $[x_0(0), y_0(0), \theta_0(0)]^T = [1, 0, 0]^T$. The desired formation pattern F is $(p_{1x}, p_{1y}) = \frac{1}{5\sqrt{3}}(-1, \sqrt{3})$, $(p_{2x}, p_{2y}) = \frac{1}{5\sqrt{3}}(1, \sqrt{3})$, $(p_{3x}, p_{3y}) = \frac{1}{5\sqrt{3}}(2, 0)$, $(p_{4x}, p_{4y}) = \frac{1}{5\sqrt{3}}(1, -\sqrt{3})$, $(p_{5x}, p_{5y}) = \frac{1}{5\sqrt{3}}(-1, -\sqrt{3})$ and $(p_{6x}, p_{6y}) = \frac{1}{5\sqrt{3}}(-2, 0)$. Following the procedure detailed in

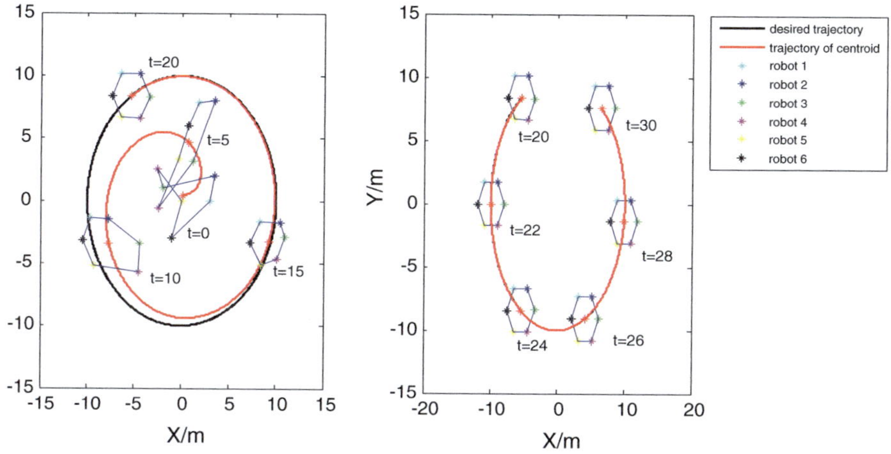

Fig. 2 Formation pattern of six robots at some time, the desired trajectory and the trajectory of centroid formation

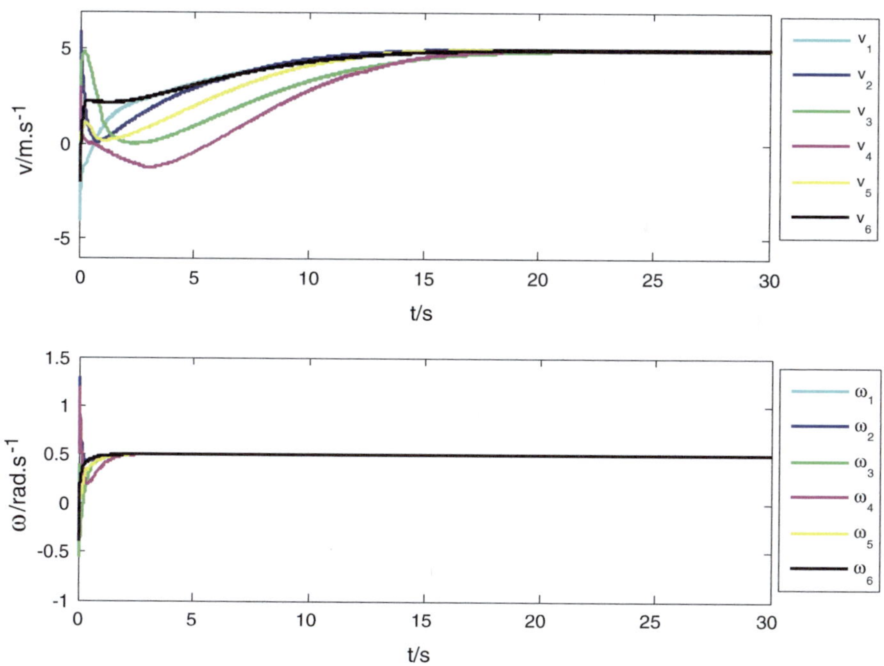

Fig. 3 Line velocity and angular velocity of six robot

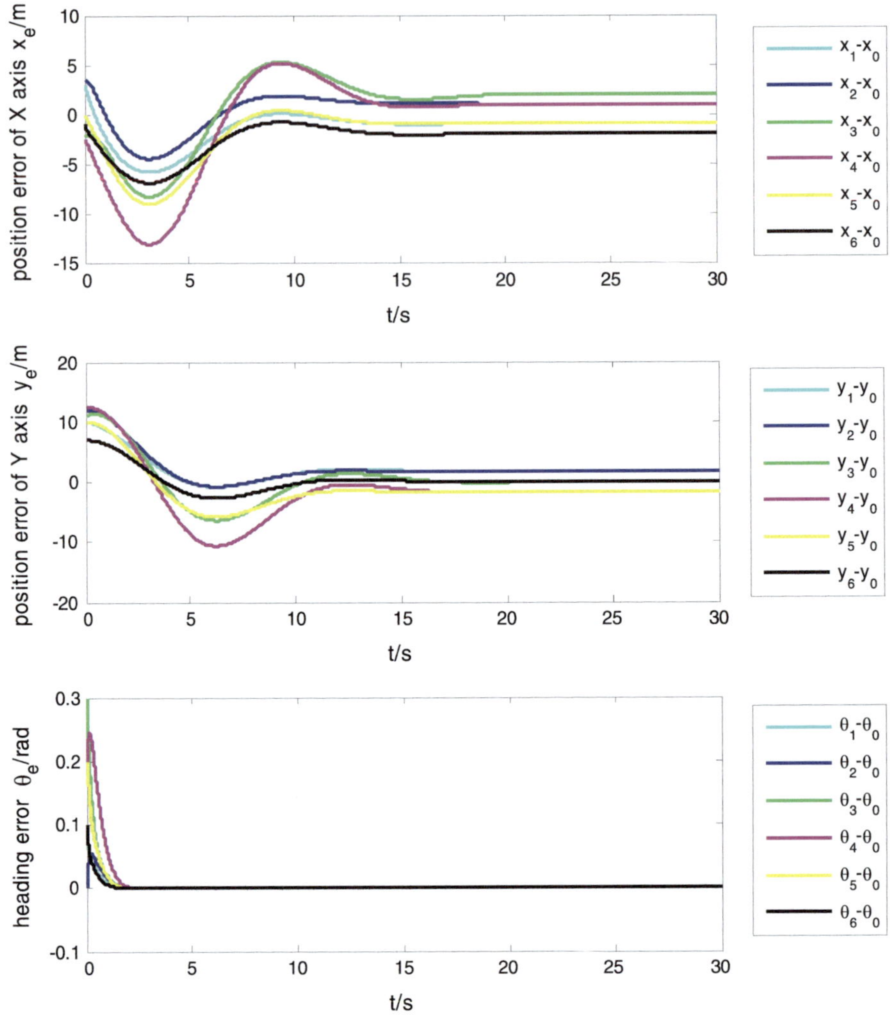

Fig. 4 Position errors and heading errors of six robots

the proof of Theorem 1, the estimator gain can be calculated. So, the gain parameter for each robot are chosen as $k_1 = 1, k_2 = 9, \alpha = 0.6$. The response curves of each mobile robot system (1) under control laws (10) and (11) are given in Figs. 5, 6 and 7, form which it is easy to see that six mobile robots achieve desired formation pattern, the trajectory of centroid formation pattern converges to desired trajectory, and the angular velocity ω_i and velocity v_i of six mobile robots converge to ω_0, v_0.

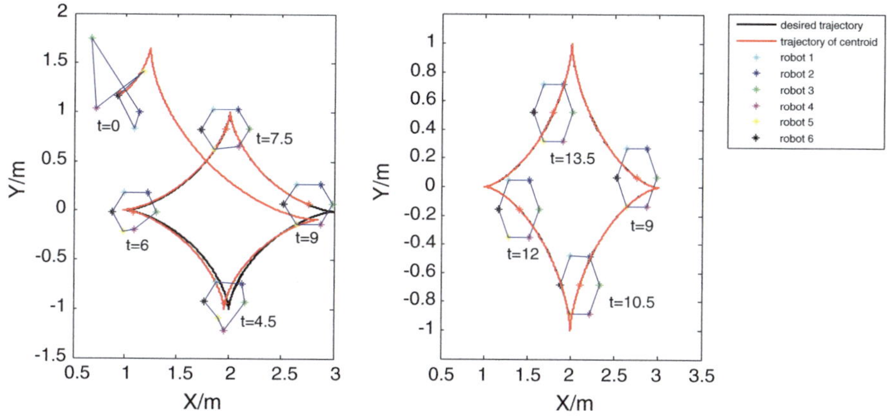

Fig. 5 Formation pattern of six robots at some time, the desired trajectory and the trajectory of centroid of formation

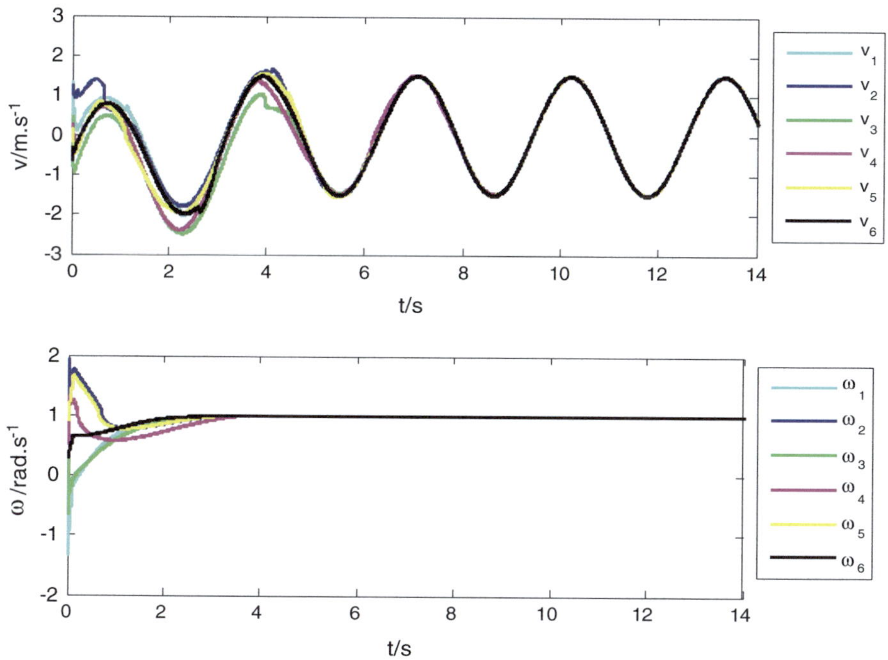

Fig. 6 Line velocity and angular velocity of six robots

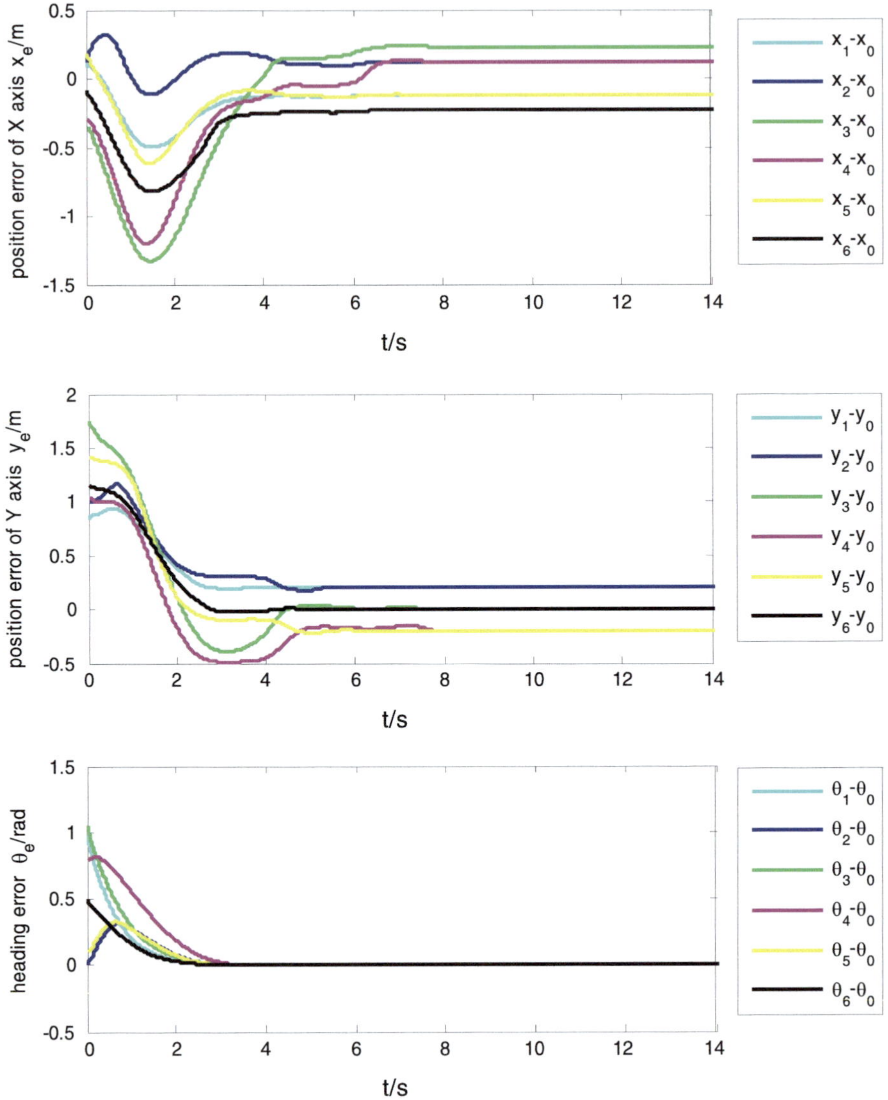

Fig. 7 Position error and heading error of six robots

5 Conclusion

In this paper, the distributed finite-time formation control problem for multiple nonholonomic mobile robots has been studied. And a distributed finite-time estimator and a distributed finite-time formation controller are proposed to solve it. The simulation results show the feasibility of the proposed algorithm. However, the communication topology is connected and balanced-detail and the communication delay is not considered. So, the future work will focus on studying the unified controller to overcome these constraints.

Acknowledgments This research is supported by the National Natural Science Foundation of China (Grant No.61573200,61273138), and the Tianjin Natural Science Foundation of China (Grant No.14JCYBJC18700, 14JCZDJC39300).

References

1. Balch Turker, Arkin Ronald C (1998) Behavior-based formation control for multi-robot teams [J]. IEEE Trans Robot Autom 14(6):926–939
2. Desai Jaydev P, Ostrowski James P, Kumar Vijay (2001) Modeling and control of formations of nonholonomic mobile robots [J]. IEEE Trans Robot Autom 17(6):905–908
3. Wei Ren, Beard Randal W (2004) Formation feedback control for multiple spacecraft via virtual structures. IEE Proc Control Theory Appl 151(3):357–368
4. Kar-Han Tan, Anthony Lewis M (1997) Virtual structures for high-precision cooperative mobile robotic control [J]. Auton Robots 4(4):387–403
5. Khatib Oussama (1986) Real-time obstacle avoidance for manipulators and mobile robots [J]. Int J Robot Res 5(1):290–298
6. Kecai Cao, Bin Jiang, Yangquan Chen (2015) Cooperative control design for non-holonomic chained-form systems [J]. Int J Syst Sci 46(9):1525–1539
7. Chen C, Xing Y, Djapic V (2014) Distributed formation tracking control of multiple mobile robotic systems [C]. In: 53rd IEEE Conference on Decision and Control. pp 3695–3700
8. Zhaoxia Peng, Guoguan Wen, Ahmed Rahmani (2013) Distributed consensus-based formation control for multiple nonholonomic mobile robots with a specified reference trajectory [J]. Int J Syst Sci 46(8):1447–1457
9. Long Wang, Feng Xiao (2010) Finite-time consensus problems for networks of dynamic agents [J]. IEEE Trans Autom Control 55(4):950–955
10. Wang X, Hong Y (2008) Finite-time consensus for multi-agent networks with second-order agent dynamics [C]. In: 17th IFAC World Congress. 15185–15190
11. Li Shihua Du, Haibo Lin Xiangze (2011) Finite-time consensus algorithm for multi-agent systems with double-integrator dynamics[J]. Automatica 47(8):1706–1712
12. Xiaoqing Lu, Shihua Chen, Jinhu Lü (2013) Finite-time tracking for double-integrator multi-agent systems with bounded control input [J]. IET Control Theory Appl 7(11):1562–1573
13. Wang J, Gao T, Zhang G (2014) Finite-time leader-following consensus for multiple non-holonomic agents [C]. In: Proceedings of the 33rd Chinese Control Conference. 1580–1585
14. Ou Meiying Du, Haibo Li Shihua (2012) Finite-time tracking control of multiple nonholonomic mobile robots [J]. J Franklin Inst 349(9):2834–2860
15. Bayat F, Mobayen S, Javadi S (2016) Finite-time tracking control of nth-order chained-form non-holonomic systems in the presence of disturbances [J]. ISA Trans
16. Ou Meiying Du, Haibo Li Shihua (2014) Finite-time formation control of multiple nonholonomic mobile robots [J]. Int J Robust Nonlinear Control 24(1):140–165
17. Wenjie Dong, Yi Guo, Jay Farrell (2006) Formation control of nonholonomic mobile robots [C]. Am Control Conf 2:63–69
18. Hardy GH, Littlewood JE, Pólya G (1952) Inequalities[M]. Cambridge University Press, Cambridge
19. Yiguang Hong, Daizhan Chen (2005) Analysis and control of nonlinear systems[M]. Science Press, Beijing
20. Yanjiao Zhang, Ying Yang, Zhao Yu et al (2013) Distributed finite-time tracking control for nonlinear multi-agent systems subject to external disturbances [J]. Int J Control 86(1):29–40

Adaptive Tracking Control for Differential-Drive Mobile Robots with Multi Constraint Conditions

Liang Yang and Yingmin Jia

Abstract To solve the trajectory tracking problem, we propose an adaptive controller with input constraint for the differential-drive mobile robots containing uncertain parameters. Two inequalities guarantee the satisfaction of input constraints, and the adaptive control method can adjust the uncertain parameters of the kinematic model on line. The system stability is proved by the Lyapunov stability theory. Simulation results verified the effectiveness of the former proposed method.

Keywords Adaptive tracking control · Uncertain parameters · Input constraints

1 Introduction

In recent years, the control problem of nonholonomic systems has been studied extensively, and the differential-drive mobile robots have been paid more attention by researchers as a typical nonholonomic system. According to the Brockett condition, any control law of smooth and non time-varying cannot solve the stabilization problem of nonholonomic systems. The control problem is more complicated.

At present, there have been many researches on trajectory tracking of nonholonomic mobile robots, such as backstepping method [1], sliding mode control [2], feedback linearization [3], fuzzy logic control [4], neural network-based control [5], and their combinations [6]. In practice, there are uncertain parameters in the robot system, and the adaptive method is an effective way to solve the control problem of the robot system. In [7], an adaptive tracking controller for the kinematic model with uncertain parameters is designed. In [8], the backstepping method is used to design an adaptive tracking controller for the differential-drive mobile robots with uncertain parameters. In addition to uncertain parameters, the robot

L. Yang (✉) · Y. Jia
The Seventh Research Division and the Center for Information and Control, Beihang University (BUAA), Beijing 100191, China
e-mail: yangliang7@126.com

Y. Jia et al. (eds.), *Proceedings of 2016 Chinese Intelligent Systems Conference*, Lecture Notes in Electrical Engineering 405,
DOI 10.1007/978-981-10-2335-4_38

system also has the problem of input constraints. For example, the linear velocity and angular velocity of the mobile robot is bounded. In [9], the backstepping technique is used to solve the tracking control problem for the nonholonomic mobile robots with input constraint. In [10], a trajectory tracking controller is designed based on the control Lyapunov function. In [11], the tracking controller is designed using two first-order filters with bounded velocities and accelerations. But no article has discussed the trajectory tracking control problem of mobile robots under the above two kinds of constraints. In this paper, we design an adaptive controller for the kinematic model of the differential-drive mobile robots with uncertain parameters and input constraints to solve the tracking control problem.

The arrangement of this paper is as follows: in Sect. 2, the trajectory tracking error equation under the condition of input constraints and uncertain parameters are stated. In Sect. 3, the control law is given based on the adaptive control method. Simulation results are provided in Sect. 4, and the conclusions are stated in Sect. 5.

2 Problem Statement

The differential-drive mobile robot is considered as shown in Fig. 1. The kinematic model is obtained:

$$\begin{aligned} \dot{x} &= v \cos \theta \\ \dot{y} &= v \sin \theta \\ \dot{\theta} &= \omega \end{aligned} \tag{1}$$

where (x, y) are the position of the mobile robot, θ is the orientation and v, ω are the linear and angular velocities, respectively.

The kinematic model of the reference robot is taken as

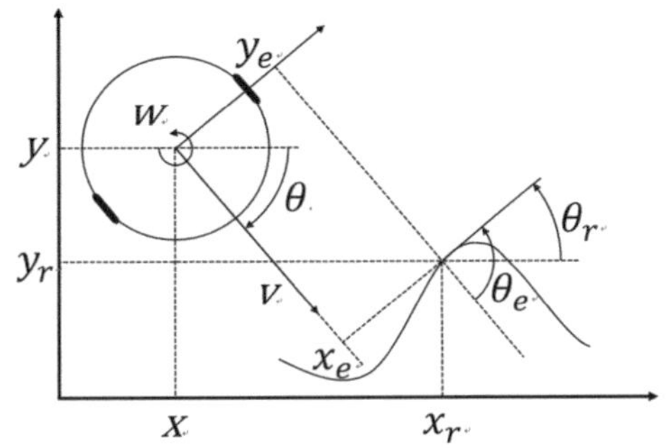

Fig. 1 Trajectory tracking configuration of the differential-drive mobile robot

$$\begin{aligned} \dot{x}_r &= v_r \cos\theta_r \\ \dot{y}_r &= v_r \sin\theta_r \\ \dot{\theta}_r &= \omega_r \end{aligned} \tag{2}$$

where $(x_r, y_r, \theta_r, v_r, \omega_r)$ are the desired values for $(x, y, \theta, v, \omega)$.

We define the tracking errors as follows

$$\begin{bmatrix} x_e \\ y_e \\ \theta_e \end{bmatrix} = \begin{bmatrix} \cos\theta & \sin\theta & 0 \\ -\sin\theta & \cos\theta & 0 \\ 0 & 0 & 1 \end{bmatrix} \begin{bmatrix} x_r - x \\ y_r - y \\ \theta_r - \theta \end{bmatrix} \tag{3}$$

differentiating both sides of (3), we can easily get the differential equation of tracking error

$$\begin{aligned} \dot{x}_e &= v_r \cos\theta_e - v + \omega y_e \\ \dot{y}_e &= v_r \sin\theta_e - \omega x_e \\ \theta_e &= \omega_r - \omega \end{aligned} \tag{4}$$

In practice, the trajectory tracking control problem of differential-drive mobile robots will be affected by some constraint conditions, we discuss two cases here. We set up ω_L and ω_R, respectively for the left and right angular velocities of the driving wheels. In general, the linear velocity v and angular velocity ω of the mobile robot are related to the angular velocities of the driving wheel

$$\begin{bmatrix} \omega_R \\ \omega_L \end{bmatrix} = \begin{bmatrix} \frac{1}{r} & \frac{b}{r} \\ \frac{1}{r} & -\frac{b}{r} \end{bmatrix} \begin{bmatrix} v \\ \omega \end{bmatrix} \tag{5}$$

where b is half of the distance between the two driving wheels and r is the radius of driving wheel.

Assumption 1 Uncertain parameters: when the parameters of r and b are uncertain we set

$$\begin{aligned} \alpha_1 &= \frac{1}{r} \\ \alpha_2 &= \frac{b}{r} \end{aligned} \tag{6}$$

Then, w_R and w_L are chosen as follows

$$\begin{bmatrix} \omega_R \\ \omega_L \end{bmatrix} = \begin{bmatrix} \hat{\alpha}_1 & \hat{\alpha}_2 \\ \hat{\alpha}_1 & -\hat{\alpha}_2 \end{bmatrix} \begin{bmatrix} v \\ \omega \end{bmatrix} = \begin{bmatrix} \alpha_1 + \tilde{\alpha}_1 & \alpha_2 + \tilde{\alpha}_2 \\ \alpha_1 + \tilde{\alpha}_1 & -\alpha_2 - \tilde{\alpha}_2 \end{bmatrix} \begin{bmatrix} v \\ \omega \end{bmatrix} \tag{7}$$

where $\hat{\alpha}_1, \hat{\alpha}_2$ are the estimated values of α_1, α_2, respectively. And we assume α_1, α_2 are bounded.

Substituting (5) into (4), we get

$$\begin{bmatrix} \dot{x}_e \\ \dot{y}_e \\ \theta_e \end{bmatrix} = \omega_R \begin{bmatrix} \frac{r}{2b}(y_e - b) \\ -\frac{r}{2b}x_e \\ -\frac{r}{2b} \end{bmatrix} + \omega_L \begin{bmatrix} -\frac{r}{2b}(y_e + b) \\ \frac{r}{2b}x_e \\ \frac{r}{2b} \end{bmatrix} + \begin{bmatrix} v_r \cos\theta_e \\ v_r \sin\theta_e \\ \omega_r \end{bmatrix} \tag{8}$$

Substituting (7) into (8), we get

$$\begin{bmatrix} \dot{x}_e \\ \dot{y}_e \\ \theta_e \end{bmatrix} = (1 + \frac{\tilde{\alpha}_1}{\alpha_1})v \begin{bmatrix} -1 \\ 0 \\ 0 \end{bmatrix} + (1 + \frac{\tilde{\alpha}_2}{\alpha_2})\omega \begin{bmatrix} y_e \\ -x_e \\ -1 \end{bmatrix} + \begin{bmatrix} v_r \cos\theta_e \\ v_r \sin\theta_e \\ \omega_r \end{bmatrix} \tag{9}$$

Assumption 2 The robots is subject to the input constraints

$$|v| \le \bar{v}, |\omega| \le \bar{\omega} \tag{10}$$

where $\bar{v}$ and $\bar{\omega}$ are positive constants.

The control objective is to find v and ω for the differential-drive mobile robot with the unknown parameters r, b and the input constraint (10), such that

$$\lim_{t \to \infty} (x_e^2 + y_e^2 + \theta_e^2) = 0 \tag{11}$$

3 Controller Design Under Multi Constraint Conditions

We designed a controller as follows

$$\begin{aligned} v &= v_r \cos\theta_e + k_1 \frac{2}{\pi} \arctan(x_e) + \phi_1(\hat{\alpha}_1) \subsetneq \\ \omega &= \omega_r + \frac{2k_2 v_r y_e \cos\frac{\theta_e}{2}}{\sqrt{1 + x_e^2 + y_e^2}} + \frac{k_3 \sin\frac{\theta_e}{2}}{\sqrt{1 + x_e^2 + y_e^2}} + \phi_2(\hat{\alpha}_2) \end{aligned} \tag{12}$$

where k_1, k_2, k_3 are positive feedback parameters, and ϕ_1, ϕ_2 are the function of $\hat{\alpha}_1$ and $\hat{\alpha}_2$, respectively.

Remark The design of controller is referred to [12]. Two functions containing the estimate of $\hat{\alpha}_1, \hat{\alpha}_2$ are given in v, ω. In the following, we can see this new design method easily solving the control problem of the differential-drive mobile robots with input constraints and uncertain parameters.

Theorem *The tracking control problem under the Assumption* 1 and 2 *can be solved by applying the controller* (12), *if the parameters satisfied*

$$\begin{aligned} &\bar{v}_r + 2k_1 \leq \bar{v} \\ &\bar{\omega}_r + 2k_2\bar{v}_r + 2k_3 \leq \bar{\omega} \end{aligned} \tag{13}$$

Proof We define V as

$$V = \sqrt{1 + x_e^2 + y_e^2} - 1 + \frac{2(1 - \cos\frac{\theta_e}{2})}{k_2} + \frac{\tilde{\alpha}_1^2}{2\gamma_1\alpha_1} + \frac{\tilde{\alpha}_2^2}{2\gamma_2\alpha_2} \tag{14}$$

With positive constants γ_1, γ_2.

The derivative of V is

$$\begin{aligned} \dot{V} &= \frac{x_e\dot{x}_e + y_e\dot{y}_e}{\sqrt{1 + x_e^2 + y_e^2}} + \frac{\sin\frac{\theta_e}{2}}{k_2}\theta_e + \frac{\tilde{\alpha}_1}{\gamma_1\alpha_1}\dot{\hat{\alpha}}_1 + \frac{\tilde{\alpha}_2}{\gamma_2\alpha_2}\dot{\hat{\alpha}}_2 \\ &= \frac{1}{\sqrt{1 + x_e^2 + y_e^2}}\Big(- k_1\frac{2}{\pi}\arctan(x_e)x_e - \frac{k_3}{k_2}\sin^2\frac{\theta_e}{2} - \phi_1(\hat{\alpha}_1)x_e \\ &\quad - \frac{\sqrt{1 + x_e^2 + y_e^2}\sin\frac{\theta_e}{2}\phi_2(\hat{\alpha}_2)}{k_2}\Big) + \frac{\tilde{\alpha}_1}{\alpha_1}\Big(\frac{1}{\gamma_1}\dot{\hat{\alpha}}_1 - \frac{x_e v}{\sqrt{1 + x_e^2 + y_e^2}}\Big) \\ &\quad + \frac{\tilde{\alpha}_2}{\alpha_2}\Big(\frac{1}{\gamma_2}\dot{\hat{\alpha}}_2 - \frac{\sin\frac{\theta_e}{2}\omega}{k_2}\Big) \end{aligned} \tag{15}$$

We set

$$\begin{aligned} \phi_1(\hat{\alpha}_1) &= - \hat{\alpha}_1\frac{2}{\pi}\arctan(x_e) \\ \phi_2(\hat{\alpha}_2) &= - \frac{\hat{\alpha}_2\sin\frac{\theta_e}{2}}{\sqrt{1 + x_e^2 + y_e^2}} \end{aligned} \tag{16}$$

Now, the parameter update rules are chosen as

$$\begin{aligned} \frac{1}{\gamma_1}\dot{\hat{\alpha}}_1 &= \frac{x_e v}{\sqrt{1 + x_e^2 + y_e^2}} \\ \frac{1}{\gamma_2}\dot{\hat{\alpha}}_2 &= \frac{\sin\frac{\theta_e}{2}\omega}{k_2} \end{aligned} \tag{17}$$

Then

$$\dot{V} = \frac{1}{\sqrt{1+x_e^2+y_e^2}}(-(k_1-\hat{\alpha}_1)\frac{2}{\pi}\arctan(x_e)x_e - \frac{k_3-\hat{\alpha}_2}{k_2}\sin^2\frac{\theta_e}{2}) \le 0 \tag{18}$$

when

$$\hat{\alpha}_1 \le k_1, \hat{\alpha}_2 \le k_3 \tag{19}$$

As $V \ge 0$ and $\dot{V} \le 0$, $\lim_{t\to\infty}\int_0^t \dot{V}dt$ has a finite limit. Also as $V(t) \le V(0)$, x_e and y_e are bounded. Since $\alpha_1, \hat{\alpha}_1, \alpha_2$ and $\hat{\alpha}_2$ are bounded, then in view of (9), (12), $\dot{x}_e, \dot{y}_e$ and θ_e are bounded. The derivative of (18) is

$$\begin{aligned}\ddot{V} = &\frac{1}{\sqrt{1+x_e^2+y_e^2}}(-(k_1-\hat{\alpha}_1)\frac{2}{\pi}\frac{x_e^2\dot{x}_e}{1+x_e^2} - (k_1-\hat{\alpha}_1)\frac{2}{\pi}\arctan(x_e)\dot{x}_e + \dot{\hat{\alpha}}_1\frac{2}{\pi} \\ &\arctan(x_e)x_e - \frac{k_3-\hat{\alpha}_2}{k_2}\sin\frac{\theta_e}{2}\cos\frac{\theta_e}{2}\theta_e + \frac{\dot{\hat{\alpha}}_2}{k_2}\sin^2\frac{\theta_e}{2}) - \frac{x_e\dot{x}_e + y_e\dot{y}_e}{(1+x_e^2+y_e^2)^{3/2}}\end{aligned} \tag{20}$$

We get $\ddot{V}$ is bounded. Subsequently, $\dot{V}$ is uniformly continuous. Applying the Barbalat's lemma, we get

$$\lim_{t\to\infty}\dot{V} = \lim_{t\to\infty}\frac{1}{\sqrt{1+x_e^2+y_e^2}}(-(k_1-\hat{\alpha}_1)\frac{2}{\pi}\arctan(x_e)x_e - \frac{k_3-\hat{\alpha}_2}{k_2}\sin^2\frac{\theta_e}{2}) = 0 \tag{21}$$

We set $\theta_e \in [-\pi, \pi)$, then $\lim_{t\to\infty}|x_e| = 0$, $\lim_{t\to\infty}|\theta_e| = 0$. Further, we need to prove $\lim_{t\to\infty}|y_e| = 0$.

From (9) and (12), we get

$$\begin{aligned}\theta_e &= -(1+\frac{\tilde{\alpha}_2}{\alpha_2})\omega + \omega_r \\ &= -\frac{1}{\sqrt{1+x_e^2+y_e^2}}(2k_2v_ry_e\cos\frac{\theta_e}{2} + k_3\sin\frac{\theta_e}{2} - \hat{\alpha}_2\sin\frac{\theta_e}{2}) - \frac{\tilde{\alpha}_2}{\alpha_2}\omega\end{aligned} \tag{22}$$

As $\lim_{t\to\infty}\tilde{\alpha}_2 = 0$, $\lim_{t\to\infty}|\theta_e| = 0$

When $t \to \infty$, (22) into

$$0 = -\frac{2k_2v_ry_e}{\sqrt{1+x_e^2+y_e^2}} \tag{23}$$

If $\lim_{t\to\infty}v_r \ne 0$, then $\lim_{t\to\infty}|y_e| = 0$.

$$
\begin{aligned}
&|v| \le |v_r \cos\theta_e| + \left|k_1 \frac{2}{\pi}\arctan(x_e)\right| + \left|\hat{\alpha}_1 \frac{2}{\pi}\arctan(x_e)\right| \le \bar{v}_r + 2k_1 \le \bar{v} \\
&|\omega| \le |\omega_r| + \left|\frac{2k_2 v_r y_e \cos\frac{\theta_e}{2}}{\sqrt{1+x_e^2+y_e^2}}\right| + \left|\frac{k_3 \sin\frac{\theta_e}{2}}{\sqrt{1+x_e^2+y_e^2}}\right| + \left|\frac{\hat{\alpha}_2 \sin\frac{\theta_e}{2}}{\sqrt{1+x_e^2+y_e^2}}\right| \\
&\le \bar{\omega}_r + 2k_2\bar{v}_r + 2k_3 \le \bar{\omega}
\end{aligned} \tag{24}
$$

If we choose the control inputs as (12) and the parameter update rules as (16), at the same time, the input constraints satisfying (13), the adaptive tracking control problem can be solved.

4 Simulation Results

In the simulation, we choose the parameters as $r = 0.25$, $b = 1$, $k_1 = 10$, $k_2 = 2$, $k_3 = 10$, $\gamma_1 = 0.75$, $\gamma_2 = 3$.

The reference trajectory is defined as follows:

$$
v_r = \begin{cases}
0.25(1 - \cos\frac{\pi t}{5}) & 0 \le t < 5 \\
0.5 & 5 \le t < 20 \\
0.25(1 + \cos\frac{\pi t}{5}) & 20 \le t < 25 \\
0.15(1 - \cos\frac{2\pi t}{5}) & 25 \le t < 35 \\
0.25(1 + \cos\frac{\pi t}{5}) & 35 \le t < 40 \\
0.15(1 - \cos\frac{2\pi t}{5}) & 40 \le t < 45
\end{cases}
$$

$$
\omega_r = \begin{cases}
0 & 0 \le t < 25 \\
-\frac{1}{10}(1 - \cos\frac{2\pi t}{5}) & 25 \le t < 30 \\
\frac{1}{10}(1 - \cos\frac{2\pi t}{5}) & 30 \le t < 35 \\
0 & 35 \le t < 45
\end{cases}
$$

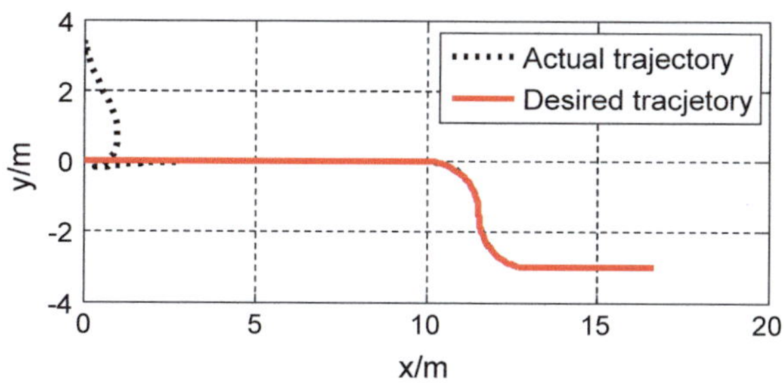

Fig. 2 Actual trajectory and desired trajectory

Fig. 3 Tracking errors x_e y_e θ_e

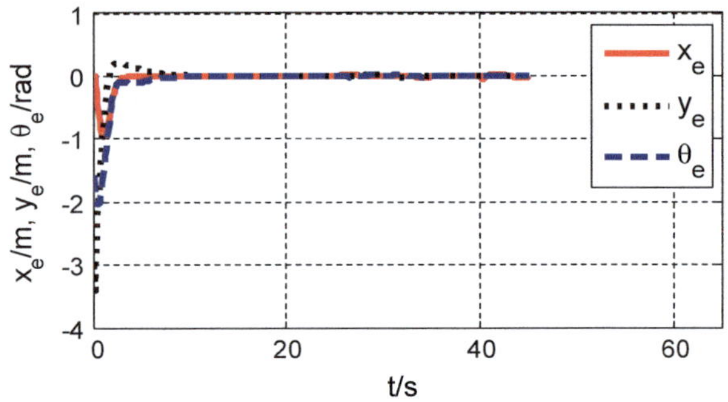

Fig. 4 Estimated parameters $\hat{\alpha}_1$ $\hat{\alpha}_2$

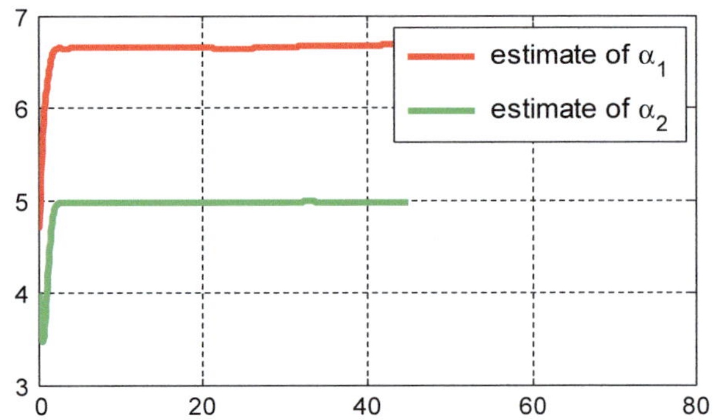

The simulation results are shown in Figs. 2, 3 and 4. From the results, the trajectory of the controlled robot is coincident with the reference trajectory. In the process of tracking, the tracking error converges to 0. In addition, the control method can adjust the unknown parameters of the model on line.

5 Conclusion

An adaptive tracking controller for differential-drive mobile robots with uncertain parameters and input constraints is proposed in this paper. Simulation results verified the effectiveness of our proposed method. In the future work, the tracking controller for the dynamic model with uncertain parameters and input constraints should be discussed.

References

1. Jiangdagger ZP, Nijmeijer H (1997) Tracking control of mobile robots: a case study in backstepping *[J]. Automatica 33(7):1393–1399
2. Yang JM, Choi IH, Kim JH (1998) Sliding mode control of a nonholonomic wheeled mobile robot for trajectory tracking[C]. IEEE Int Conf Robot Autom 4:2983–2988
3. Ying H (1999) Analytical analysis and feedback linearization tracking control of the general Takagi-Sugeno fuzzy dynamic systems[J]. IEEE Trans Syst Man Cybern Part C Appl Rev 29 (2):290–298
4. MartíNez-Alfaro H (1998) Mobile robot path planning and tracking using simulated annealing and fuzzy logic control[J]. Expert Syst Appl 15(3):421–429
5. Yoo SJ, Choi YH, Jin BP (2006) Generalized predictive control based on self-recurrent wavelet neural network for stable path tracking of mobile robots: adaptive learning rates approach[J]. IEEE Trans Circuits Syst I Regul Pap 53(6):1381–1394
6. Ryu SH, Park JH (2001) Fuzzy logic based tuning of sliding mode controller for robot trajectory control[C]. In: IEEE International Conference on Robotics & Automation, vol 3. pp 2974–2979
7. Fukao T, Nakagawa H, Adachi N (2000) Adaptive tracking control of a nonholonomic mobile robot[J]. IEEE Trans Robot Autom 16(5):609–615
8. Wang H, Fukao T, Adachi N (2001) Adaptive tracking control of nonholonomic mobile robots: a backstepping approach. Jrsj 19(2):271–276
9. Lee TC, Song KT, Lee CH et al (1999) Tracking control of mobile robots using saturation feedback controller[C]. In: Proceedings of 1999 IEEE international conference on robotics and automation, 1999. IEEE. pp 2639–2644 vol.4
10. Ren W, Beard RW (2003) CLF-based tracking control for UAV kinematic models with saturation constraints[C]. IEEE Conf Decis Control 3924–3929
11. Chen X, Jia Y, Matsuno F (2014) Tracking control of nonholonomic mobile robots with velocity and acceleration constraints[J]. In: Proceedings of the American Control Conference. pp 880–884
12. Hai-Tao MA, Yong W (2009) Trajectory tracking control of nonholonomic mobile robot with input saturation constraints[J]. J Univ Sci Technol China 5(5):499–503

Point Cloud Segmentation Based on FPFH Features

Tianyu Zhao, Haisheng Li, Qiang Cai and Qian Cao

Abstract Point cloud segmentation is a key part of geometric processing. In this paper, a point cloud segmentation method based on the Fast Point Feature Histograms (FPFH) is proposed. Fast Point Feature Histograms (FPFH) is used to extract features of point clouds, and then Gaussian Mixture Model (GMM) is employed to cluster the point clouds. Experimental results on SHREC2014 dataset show the effectiveness of the proposed method. It can avoid both under- segmentation and over-segmentation.

Keywords Point cloud · GMM model · FPFH · Segmentation

1 Introduction

The point cloud is a large set of data to express the spatial distribution of the target and characterized the target surface in the same spatial reference system. The point cloud can be created by 3D scanners such as 3D laser scanner and Microsoft Kinect, which measure a large number of points on the surface of the object, and these points are often output as a data file.

The point cloud model is a CAD model which uses the discrete points to describe the surface information of the object. This model does not need to save topology data and reduce the consumption of storage space and the workload of the algorithm. At present, with the development of measurement technology, the amount of data becomes larger, and the details are described more and more abundant. The geometric processing of point cloud model has become a hot spot of

T. Zhao · H. Li (✉) · Q. Cai · Q. Cao
School of Computer and Information Engineering, Beijing Technology and Business University, Beijing 100048, China
e-mail: lihsh@th.btbu.edu.cn

T. Zhao · H. Li · Q. Cai · Q. Cao
Beijing Key Laboratory of Big Data Technology for Food Safety, Beijing 100048, China

Y. Jia et al. (eds.), *Proceedings of 2016 Chinese Intelligent Systems Conference*, Lecture Notes in Electrical Engineering 405,
DOI 10.1007/978-981-10-2335-4_39

research. The segmentation of point clouds is a significant part of geometric processing.

The segmentation is a process to divide the point of the point clouds into the small, coherent and connected subset. After this process, the points with similar attributes are classified as a class. The subsets of these points should have different features.

The rest of the paper is organized as follows. Section 2 introduces the method of the segmentation of the point clouds. The processing of the segmentation based on the Fast Point Feature Histograms (FPFH) [1] features and the clustering method-Gaussian Mixture Model (GMM) are described in Sect. 3. Experimental results are shown in Sect. 4.

2 Related Work

Commonly, the segmentation of point clouds algorithm mainly includes edge-based segmentation algorithm, region-based segmentation algorithm, clustering-based segmentation algorithm, and graph-based segmentation algorithm.

Edge-based methods [2–5] are starting from the perspective of pure mathematics, that normal vector of measuring point or sudden change of curvature is a boundary with another region, and the region surrounded by closed borders as the final segmentation result. The key is how to identify the boundary part. Gong et al. [2] introduced the space grid division of point clouds, using octree to organize grid structure, and the normal vector deviation has become as the basis of grid subdivision and feature extraction. Dong et al. [3] found the points which have a large curvature variation are extracted as the boundary points, thus the point cloud data is divided into multiple regions. Yinglin and Dongri [4] firstly used the grid division, and then calculated the difference between the current grid curvature and the adjacent grid curvature. Utilizing the difference to extract the characteristic grid. Finally, according to the characteristic grid to get the boundary of point clouds, and achieved the region segmentation of spatial scattered point clouds. Mo and Yin [5] adopt 3D active contour model to make the segmentation of point clouds. In order to remove the influence of noise, the algorithm constructs the signed distance function to estimate the mean curvature of the point clouds.

Region-based methods include bottom-up and top-down two ways. Bottom-up algorithm is commonly known as the region growth method. Rabbani et al. [6] used the normal vectors of the points and the redundancy as the region growing proof to realize the scattered point cloud segmentation. Shixiong and Wang [7] proposed that the boundary is determined by the curvature change of the main direction in the neighborhood region of the sampling points. Top-down algorithm is also known as hierarchical decomposition methods. Firstly, it is assumed that all points belonging to the same patch. Then the octree [8], KDtree [9] and the other spatial hierarchy tree can be used to solve hierarchical subdivision.

The regional segmentation of point clouds are regarded as the classification process of data points with certain geometric parameters in clustering-based methods. There are some algorithms in the segmentation process such as surface element categories [10], mean shift clustering [11], self-organizing feature mapping [12], spectral clustering [13], K-means clustering [14] and fuzzy clustering [15].

Graph-based methods are inspired by the structure of the neighbor graph. The points in the same segmentation region are more closely connected than the points in the different segmentation regions. Therefore, the boundary of the two segmentation regions must be the weakest link. A neighborhood graph is a property map of the point clouds. Each point of the point clouds is a node in this graph, and each edge has a weight, which represents the similarity of a pair of points in the point clouds. The segmentation processing should ensure that the similarity of the points to reach maximum in the same segmentation patch, and the similarity of the points to reach minimum between the different segmentation patches. In some methods, the segmentation of point clouds can be solved by using probabilistic stochastic model [16], such as Conditional Random Fields (CRF) [17], Markov Random Fields (MRF) [18].

The segmentation methods mentioned above have some disadvantages, such as the detection based on the region edge general need to find the boundary points using the normal vectors or other features. It is difficult to implement and maybe generate some gap. If the selected features are relatively few, the interpolation problem will become difficult. Region growing-based methods are easily influenced by noise, although the edge of these methods are closed, sometimes the segmentation regions will be deformed; clustering-based methods have certain advantages for the more obvious surface. If the surfaces to be processed are more complicated, it is hard to directly determine the type of the surface and the classification of the surface. Graph-based methods can solve the problem about the complex surface's segmentation, but these methods usually need a special sensor system to finish the process.

3 The Proposed Method

3.1 Feature Description and Extraction

Feature description and extraction of point cloud plays a key role in the point cloud information processing. For example, point cloud recognition, point cloud segmentation and point cloud surface reconstruction algorithms also use the results from the features of the description.

3.1.1 Normal Estimation

The relative normal of the point cloud data is an important geometric property in the surface. Given a geometric surface, the normal of a point is a vector that is perpendicular to the point, and it is easy to calculate. In the point cloud data, there is not a real surface. The normal of the point can be get through the following two ways:

(1) The surface of the point is obtained by the surface reconstruction technique, and the vector of the surface can be calculated.
(2) Directly using point cloud data to approximate estimation of surface normal.

There are many approaches to estimate the normal. One of the relatively simple expressions is as follows: if we want to obtain the surface normal of a point, we should seek for the point where the approximation surface tangent plane normal. So we can convert this problem into the plane fitting of the least square method.

(1) Search for neighbor elements of the sample point, determining the K neighbor of the sample point;
(2) Estimate the 3D centroid coordinates of the sample point's neighbor elements;
(3) Use the formula to calculate the sample point and the neighboring points corresponding covariance matrix, and the eigenvalues and eigenvectors of the covariance matrix;

$$C = \frac{1}{k}\sum_{i=1}^{k}(P_i - \overline{P})^T(P_i - \overline{P}), C \cdot \overrightarrow{v_j} = \lambda_i \cdot \overrightarrow{v_i}, j \in \{0, 1, 2\} \tag{1}$$

K is the number of the point P_i's neighbor points, $\overline{P}$ is the 3D centroid of neighbor elements, λ_i is the i-th eigenvalue of covariance matrix, $\overrightarrow{v_i}$ is the i-th eigenvector of covariance matrix.

(4) Analysis eigenvalues and eigenvectors of covariance matrix, the eigenvector corresponding to the largest eigenvalue is as the estimated normal of the sample point.

3.1.2 Fast Point Feature Histograms (FPFH)

FPFH is based on the Point Feature Histograms (PFH) [1]; the main idea of PFH is using a histogram to display the geometric feature information of a sample point's adjacent local region. To compare the PFH between two points, and then the correspondence relation between two points is determined by the PFH's similarity.

The idea of Fast Point Feature Histograms (FPFH) is to calculate the simplified point feature histogram (SPFH) of each point and its K neighbors, and the final fast point feature histogram (FPFH) of all the SPFH sets is synthesized by Eq. (2). The calculation steps of FPFH are as follows:

(1) Calculate the three eigenvalues between each point p_q and its K neighbors, and compute the SPFH of the point p_q.
(2) Determine the K neighbor region of every k neighbors, and use the first step to get the SPFH
(3) The final FPFH is calculated by the following equation

$$FPFH(p_q) = SPFH(p_q) + \frac{1}{k}\sum_{i=1}^{k}\frac{1}{\omega_i}\cdot SPFH(p_i) \tag{2}$$

In a given metric space, the weight ω_i expresses the distance between the sample point p_q and the neighbor point p_i[1]. An influence region diagram illustrating the FPFH computation is presented in Fig. 1. Each query point (red) is connected only to its direct k-neighbors (enclosed by the gray circle). Each direct neighbor is connected to its own neighbors and the resulted histograms are weighted together with the histogram of the query point to form the FPFH. The connections marked with 2 will contribute to the FPFH twice.

The value interval of three feature values will be divided into 11 intervals to get 33 ranges, and the three feature values of each point can fall into three different ranges.

3.2 Gaussian Mixture Model (GMM)

Every Gaussian Mixture Model (GMM) is made up of Gaussian Models. Each Gaussian is called a "Component". These Components constitute the probability density function of GMM by Eq. (3).

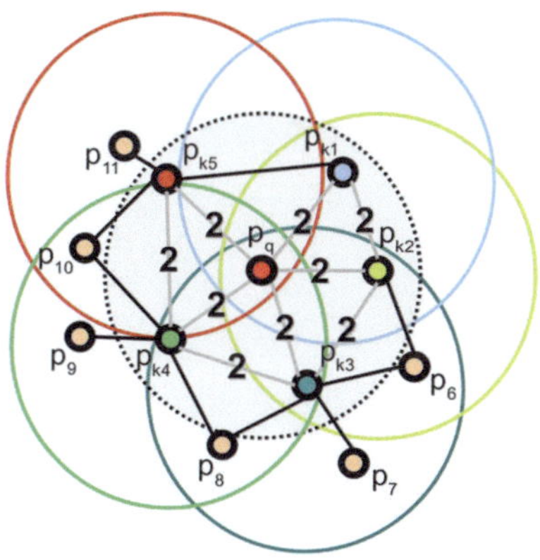

Fig. 1 Effected region of FPFH calculation of query point p_q

$$p(x) = \sum_{k=1}^{K} p(k)p(x|k) = \sum_{k=1}^{K} \pi_k N(x|\mu_k, \textstyle\sum_k) \tag{3}$$

The estimated quantity of the model parameters: Impact factors for each class (π_k); the mean of each class (μ_k); Covariance matrix ($\sum_k$).

If we want to take a random point from the distribution of GMM, in fact, it can be divided into two steps. First, randomly choose a point from the k Gaussian Component, the probability that each component is selected is π_k. After checked the component, we only need to select a point from the distribution of this Component. Then, this problem has turned to a normal Gaussian distribution problem.

GMM log-likelihood function

$$\sum_{i=1}^{N} \log\left\{ \sum_{k=1}^{K} \pi_k N(x_i|\mu_k, \textstyle\sum_k) \right\} \tag{4}$$

There is an additive in the logarithmic function, so we cannot directly obtain maximum value with derivative solution equation approach. In order to solve this problem, we choose the method of the randomly selecting of Gaussian mixture model (GMM). The main steps are as follows:

(1) Estimate the probability that the data is generated by each Component; for each data x_i, it is generated by the k-th component:

$$\gamma(i,k) = \frac{\pi_k N(x_i|\mu_k, \sum_k)}{\sum_{j=1}^{k} \pi_j N(x_i|\mu_j, \sum_j)} \tag{5}$$

In this equation, we also need to use iterative method to estimate the values of μ_k, $\sum_k$. We use the last iteration value of μ_k,$\sum_k$ to calculate the $\gamma(i,k)$.

(2) Component generates $\gamma(i,k)x_i \ldots \gamma(N,k)x_N$, and each Component is a standard Gaussian distribution. It is easy to compute the Maximum likelihood parameter.

$$\mu_k = \frac{1}{N_k} \sum_{i=1}^{N} \gamma(i,k)x_i \tag{6}$$

$$\textstyle\sum_k = \frac{1}{N_k} \sum_{i=1}^{N} \gamma(i,k)(x_i - \mu_k)(x_i - \mu_k)^T \tag{7}$$

$$N_k = \sum_{i=1}^{N} \gamma(i,k) \tag{8}$$

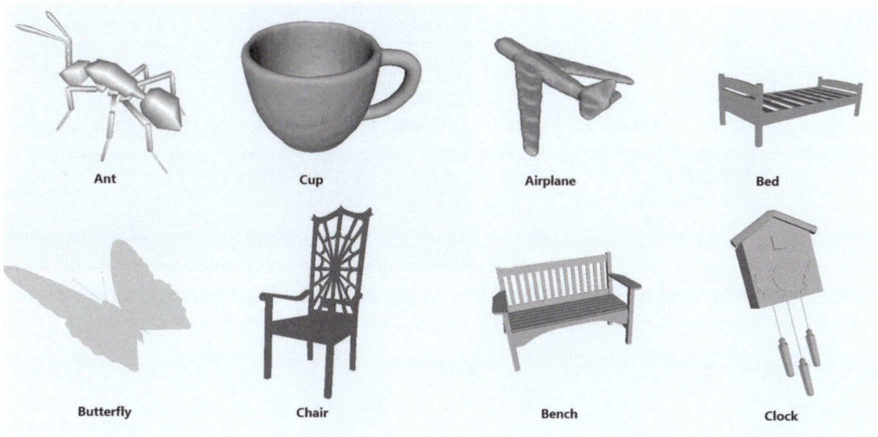

Fig. 2 Example models in the dataset

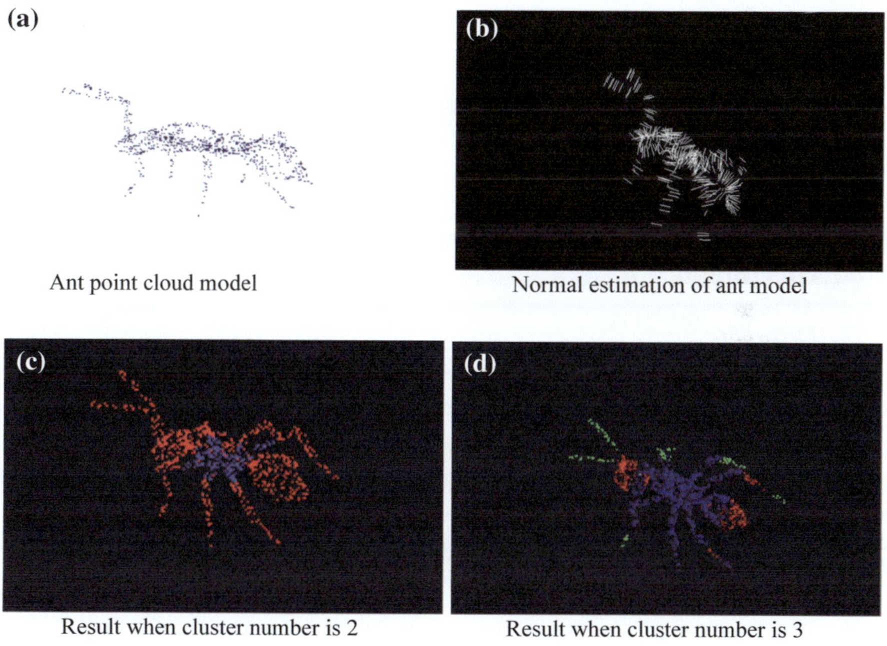

Fig. 3 Segmentation results of ant model

And π_k can be estimated as N_k/N.

(3) Repeat the previous two steps, until the value of the likelihood function converges.

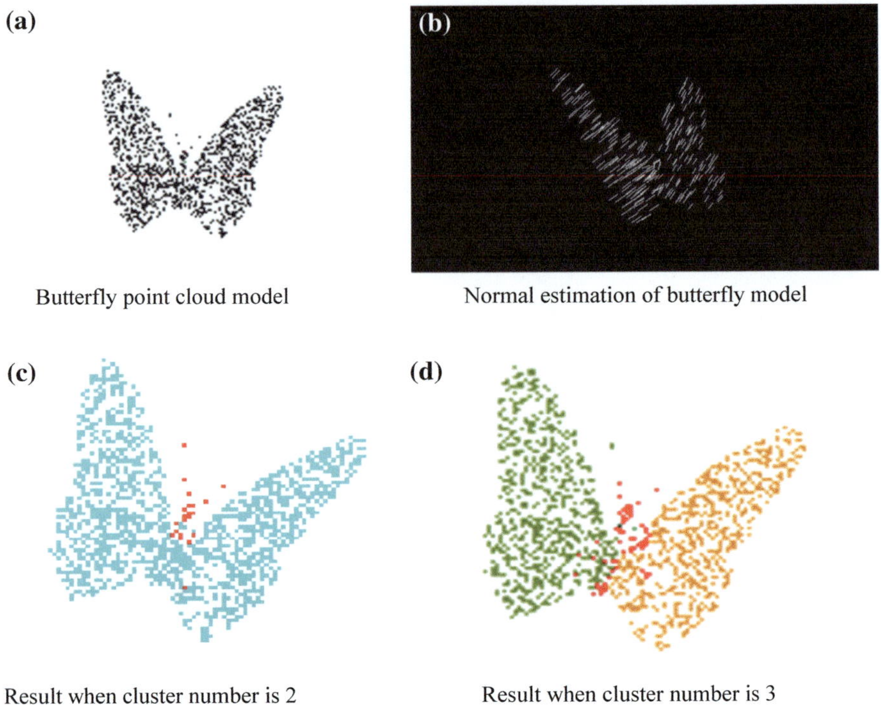

Fig. 4 Segmentation results of butterfly model

4 Experiments

4.1 Datasets

SHREC 2014 LSSTB_TARGET_MODELS [19] are chosen as the experimental dataset. This dataset contains 171 classes and 8987 models. Each class has 53 models; the number of vertices of each model is 5233. The model files were stored in. OFF format ASCII text files. Figure 2 shows eight types of models in the dataset.

4.2 Experimental Results

The ant point cloud model is shown in Fig. 3a. Normal estimation result of the given ant model is illustrated in Fig. 3b. Segmentation results when cluster number is 2 and 3 are shown in Fig. 3c, d, respectively.

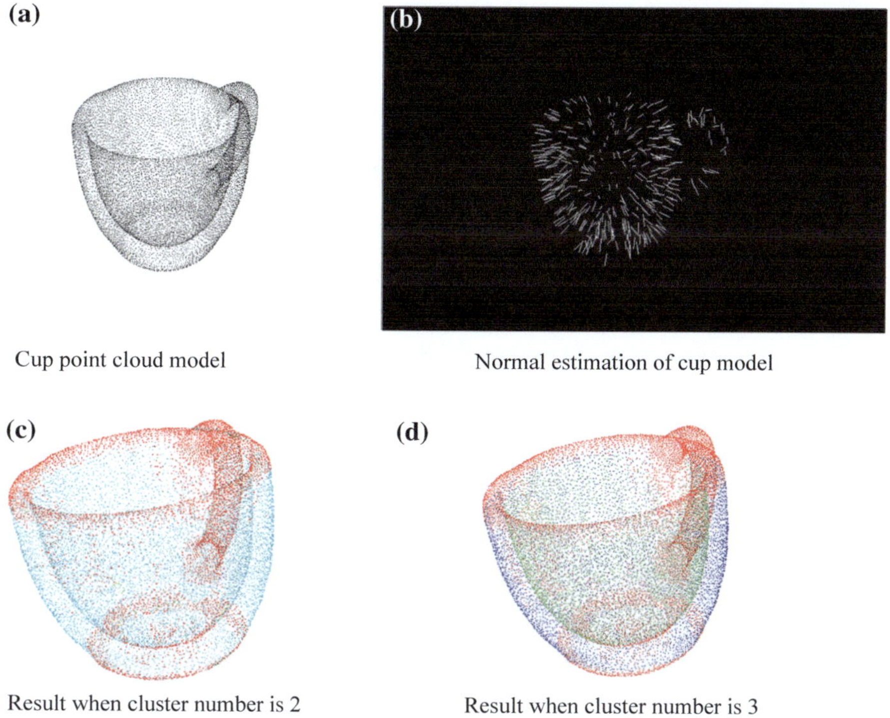

Fig. 5 Segmentation results of cup model

The segmentation result of butterfly model and cup model are shown in Figs. 4 and 5. The results show the effective of the proposed method.

5 Conclusions

A point cloud segmentation method based on the Fast Point Feature Histograms (FPFH) is proposed. Experimental results on SHREC2014 dataset show effective of the given method. The future work will focus on the improving the distance calculated method in FPFH feature extraction and to get more precise segmentation results.

Acknowledgments This work was partially supported by Beijing Natural Science Foundation (4162019) and Scientific Research Common Program of Beijing Municipal Commission of Education (No. KM201410011005).

References

1. Rusu RB, Blodow N, Marton ZC, Beetz M (2009) Fast point feature histograms (FPFH) for 3D registration. In: Proceedings of the IEEE international conference on robotics and automation(ICRA). Kobe, Japan, 12–17 May 2009
2. Gong J, Ke S, Zhu Q et al (2012) An efficient management method for point cloud data based on octree and 3DR-tree[J]. Acta Geod Cartogr Sin 41(4):597–604
3. Dong M, Zheng K, Yao B (2005) Study on the region segmentation of the point cloud data from the surface reconstruction. Chin J Image Graph 10(5):575–578
4. Yinglin K, Dongri S (2005) Segmentation point cloud data based edge feature. J Zhejiang Univ (Engineering Science Edition) 39(3):377–396
5. Mo K, Yin Z (2011) Defect segmentation method 3D active contour model based on point cloud. J Huazhong Univ Sci Technol (NATURAL SCIENCE EDITION) 39(1):82–85
6. Rabbani T, Van den Heuvel FA, Vosselman MG (2006) Segmentation of point clouds using smoothness constraint international archives of photogrammetry. Remote Sens Spat Inf Sci 36(5):248–253
7. Shixiong W, Wang C (2007) Data segmentation of scattered noise points. Chin J Mech Eng 43(2):230–233
8. Huan F, Liang L, Wang F (2012) the local convexity and octree point cloud segmentation algorithm. J Xi'an Jiaotong Univ 46(10):60–65
9. Junfeng L, Yueling Z, Jinsheng X et al (2015) Improved of SIFT matching algorithm based on main gradient of direction. Comput Eng Appl 51(13):149–152
10. Besl P, Jain R (1987) Segmentation and classification of range images. IEEE Trans Pattern Anal Mach Intell 9(5):608–620
11. Dai HT, Tang Z, Zhang Z (2011) Mean shift clustering based image segmentation. Commun Technol 44(24012):117–120
12. Wu H (2006) Research on clustering algorithm based on Self-organizing feature mapping [D]. Chongqing Univ
13. Ma T, Long X, Feng L etc (2012) Spectral clustering segmentation of point cloud model. J Comput Aided Des and Comput Graph 24(12):1549–1558
14. Sun X, Sun H, Li H (2006) Segmentation of point. Comput Eng Appl 10:42–45
15. Ghaffarian S (2014) Automatic histogram-based fuzzy C-means clustering for remote sensing imagery[J]. ISPRS J Photogramm Remote Sens 97:46–57
16. Nguyen A, Le B (2013) 3D Point cloud segmentation: a survey. In: 2013 6th IEEE conference on robotics, automation and mechatronics (RAM)
17. Rusu RB, Holzbach A, Blodow N, Beetz M (2009) Fast geometric point labeling using conditional random fields. In: Proceedings of the 22nd IEEE/RSJ international conference on intelligent robots and systems (IROS). St. Louis, MO, USA
18. Schoenberg J, Nathan A, Campbell M (2010) Segmentation of dense range information in complex urban scenes. In: Proceedings of the IEEE/RSJ international conference on intelligent robots and systems (IROS). Taipei, pp 2033–2038
19. Li B, Lu Y, Li C, Godil A, Schreck T, Aono M, Chen Q, Chowdhury NK, Fang B, Furuya T, Johan H, Kosaka R, Koyanagi H, Ohbuchi R, Tatsuma A (2014) SHREC' 14 track: large scale comprehensive 3D shape retrieval. Eurographics Workshop on 3D Object Retrieval 2014 (3DOR 2014), pp 131–140

Salient Object Detection Based on RGBD Images

Qiang Cai, Liwei Wei, Haisheng Li and Jian Cao

Abstract Salient object detection is very important in many image and vision-related applications. We add the depth clue into the detection method to extract salient objects. In low-level feature extract part, we extract the depth edge and corner clue, combining with color image features to form a 55 dimensions' vector. In the high-level prior part, the depth prior is used to predict the probability together with the other three priors. The experiment result showed that with the depth clue, the salient detection result is improved.

Keywords Salient object detection · Depth image · High-level prior · Low-level feature

1 Introduction

Saliency detection is a mechanism to extract pertinent, attention grabbing regions of the scene fast without conscious awareness [1]. It has a large number of applications in computer vision and image processing tasks, such as visual content compression and summarization [2], information retrieval [3] and content-aware image resizing [4], etc. There are two research branches about saliency detection: fixation prediction and salient object segmentation. The goal of fixation prediction is to predict the actual human eye gaze patterns of an image, whereas the goal of salient object segmentation is to segment the salient object out of the background. Here we focus on the latter one saliency detection.

There are lots of saliency detection methods making use of the color information on 2D images. Generally, saliency detection models can be categorized into bottom-up model using low-level features and top-down model using high-level

Q. Cai · L. Wei (✉) · H. Li · J. Cao
Beijing Key Laboratory of Big Data Technology for Food Safety,
School of Computer and Information Engineering, Beijing Technology
and Business University, Beijing 100048, China
e-mail: weiliwei_wlw@hotmail.com

Y. Jia et al. (eds.), *Proceedings of 2016 Chinese Intelligent Systems Conference*, Lecture Notes in Electrical Engineering 405,
DOI 10.1007/978-981-10-2335-4_40

prior. The bottom-up method computes the object salience from different feature maps, such as such as color, intensity, orientation, and so on. The top-down method uses accumulated statistical knowledge of the visual features of the desired target and background clutter accumulated for a long period to tune the bottom-up maps. Inspired by premier work [5, 6] and thus to enhance the ability of salient object detection, we combine these two factors together in these paper.

Depth information are import clues to perceive object in human visual system, which did not get enough attentions before. They provide additional important information about contents in the visual field and can be regarded as relevant features for saliency detection. With the advent of depth image acquisition device, there arise some RGBD (color and depth) image datasets for salient object detection [7, 8] as well as come depth-related work [9, 10].

2 High-Level Prior and Low-Level Feature Diffusion Based on RGBD Map

We first use the SLIC algorithm [11] to partition the input depth image into $N = 200$ nonoverlapping patches. For each patch, a D-dimension feature vector is extracted and denoted as $f_i \in R^D$. The ensemble of feature vectors forms a matrix representation of I, denoted as $F = [f_1, f_2, \ldots, f_N] \in \Re^D$. Here we use the low-rank matrix salient map generation method which decomposes the matrix F into two parts. For more details, we recommend readers to [12]. What we mainly focus in the work here is how to generate these features.

2.1 Low-Level Features

Depth edge and corner feature. To understand the image in human ways, we choose to extract more abstract features to compensate the concrete features. We use the Canny edge detection algorithm [13] to extract the edges and corners from the depth image, forming 2D features. Figure 1 shows the original image, depth image, depth edge image and corner image. Edge and corner information from depth image is showed in Fig. 1c, d;

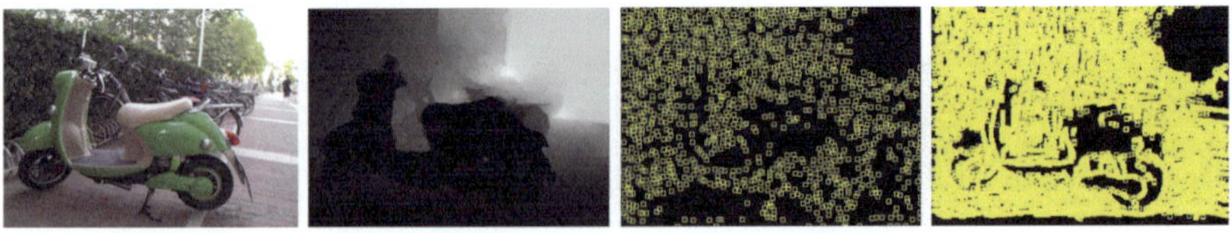

(a) RGB (b) depth (c) corner (d) edge

Fig. 1 Color image, depth image, depth corner image, and edge image

Color feature. From color images, three RGB color values as well as the hue and the saturation attributes are extracted for each pixel, producing 5D features. Each feature is normalized by subtracting its median value over the entire image;

Steerable pyramids [14]. We set the steerable pyramid filters with four directions for three different scales on the color image, getting 12 filter responses at each location;

Gabor filters [15]. Gabor filter were extracted responses with 12 orientations and 3 scales are extracted. The scaling factor is 2 and the bandwidth of the smallest filter is chosen to be 8. All those 55 features are then formed to a whole feature vector, which captures color, edge, texture, and depth that are the most common low-level visual features.

2.2 High-Level Prior

Depth prior. Normally we pay more attention to the objects which are close to us. The distance decides whether the object is salient or not. As we have divided the images into superpixels, thus these superpixels have their own depth, the formulation below is to explain the depth difference between area s_k and s_i

$$F_{dc}(s_k, s_i) = d_i - d_k \tag{1}$$

Here d represents the center of area r. The probability of near the camera axis is

$$p_d(x) = \exp\left(F/\sigma^2\right). \tag{2}$$

Background prior [16]. It is defined to find the background information in an image. It calculates the probabilities of image regions connected to image boundaries and the probabilities are called background weighted contrast and which is calculates as follows

$$P_i^{bg} = 1 - \exp\left[-\frac{\text{Bnd Con}^2}{2\sigma^2 \text{bndCon}}\right] \tag{3}$$

Here p stands for the probability of each superpixel belonging to the background. Pi is the superpixel. And $\text{BndCon}^2(p_i)$ stands for how likely the superpixel Pi is connected to the borders.

Color prior. It has been found that warm color is the most noticed color in human vision [17]. Thus we set the color prior for saliency to be

$$p_c(x) = \exp\left((h_S(c_x) - h_B(c_x))/\sigma_3^2\right) \tag{4}$$

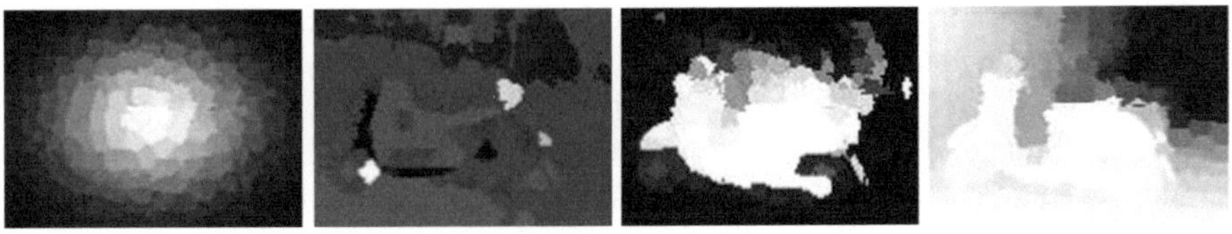

(a) center prior (b) color prior (c) background prior (d) depth prior

Fig. 2 Center prior, color prior, background prior and depth prior

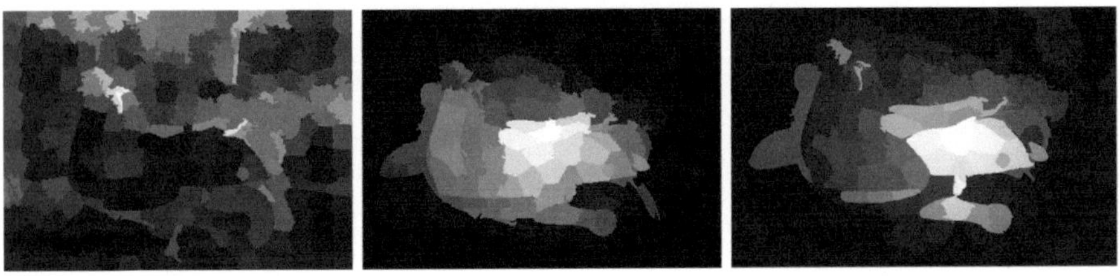

(a) final feature map (b) final prior map (c) final salient map

Fig. 3 Final maps

where $c(x)$ is the color at location x. h_s and h_B indicate the histogram value calculated from the certain datasets and they are stands for the salient object color histogram and background histogram.

Location prior. Objects near the image center are more attractive to people [18]. Therefore, we generate a prior map using a Gaussian distribution based on the distances of the pixels to the image center, in which

$$p_l(x) = \exp\left(-d(x, c)/\sigma_1^2\right) \tag{5}$$

Figure 2 shows the center prior, color prior, background prior, and depth prior. And the final prior value is formed by these four prior values, the formulation is

$$P(s_k) = P_d(s_k) \bullet P_{bg}(s_k) \bullet P_c(s_k) \bullet P_l(s_k) \tag{6}$$

And the final prior map we got is shown in Fig. 3b. After combining final feature map and final prior map, we finally get the salient map as shown in Fig. 3c.

3 Experiment and Evaluation

Datasets. We choose the dataset introduce in [7] which contains 1000 pair color and depth images captured by Microsoft Kinect SDK. It includes more than 400 kinds of common objects captured in 11 types of scenes under different illumination conditions.

Evaluation. In the salient object detection literature such as [21], different measures are proposed to evaluate saliency maps. Our method is evaluated by three metrics: precision–recall curve, F-measure, and receiver operating characteristic curve (ROC); the F-measure is formulated as

$$F_{\beta} = \frac{\left(1+\beta^{2}\right) \bullet \text{precision} \bullet \text{recall}}{\beta^{2} \quad \text{precision} + \text{recall}} \tag{7}$$

the range of β^2 is set to (0, 1.0), we set $\beta^2 = 0.3$ in our experiments to stress precision more than recall. And by thresholding the saliency maps and plotting true positive rate versus false positive rate, we get the ROC curve.

We compare our method with some other methods. We do not intend to emphasize that our method outperforms the others as we use additional depth information. The goal is to show that additional depth information can improve the result of salient object detection. To reduce errors, we use the authors' original codes to implement corresponding methods on the same dataset.

The quantitative evaluation results are shown in Fig. 4. The precision–recall curve, ROC curve, and F-measure all demonstrate that our method can achieve high precision and recall rate at the same time, which indicates that our color plus

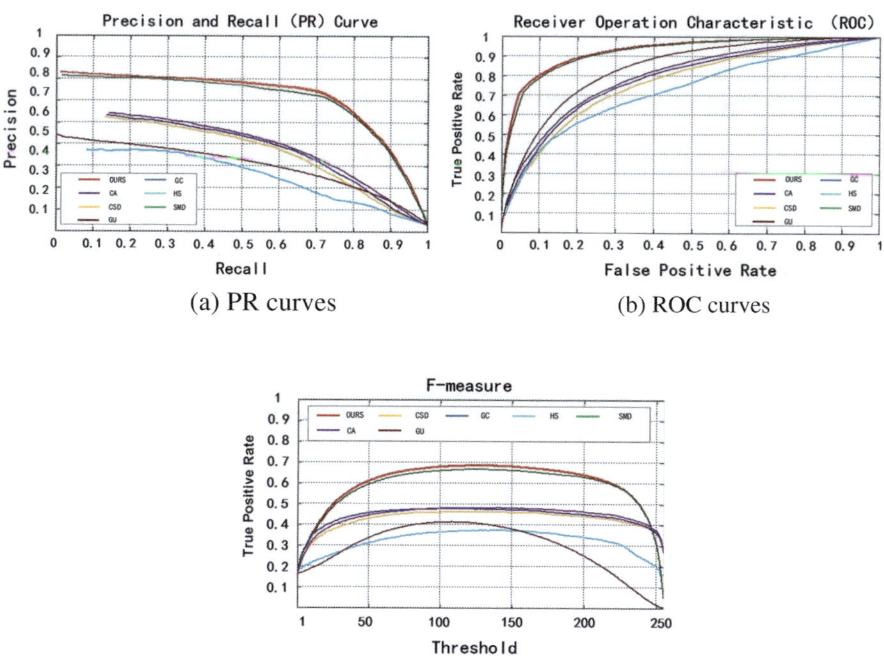

(a) PR curves

(b) ROC curves

(c) F-measure curves

Fig. 4 The final curves PR curves

Fig. 5 The detection result compared with others. RGB GT OURS SMD [12] CSD [18] GC [19] GU [19] CA [20] HS [21]

depth-based method greatly outperforms the other methods without introducing depth information.

From Fig. 5, we can see with the depth information the result is improved.

4 Conclusion

We add the depth clue into the detection method to extract salient objects in challenging images. In the feature extract part, we use the depth edge and corner clue, we combine color and depth features and formed a 55 dimensions' vector. In the high-level prior part, the depth prior is combined with other prior to predict the probability thus to form the saliency map. Next step, we will further explore how to improve saliency detection by combining depth and other clues.

Acknowledgments This work was supported by Beijing Natural Science Foundation (No. 4162019) and General Project of Beijing Municipal Education Commission Science and Technology Development Plans (No. SQKM201610011010).

References

1. Treisman AM, Gelade G (1980) A feature-integration theory of attention. Cogn Psychol 12 (1):97–136
2. Ma Y-F, Lu L, Zhang H-J, Li M (2002) A user attention model for video summarization. In: ACM Multi-media
3. Li L, Jiang S, Zha Z-J, Wu Z, Huang Q (2013) Partial-duplicate image retrieval via saliency-guided visual matching. IEEE MM 20(3):13–23
4. Avidan S, Shamir A (2007) Seam carving for content-aware image resizing. In: ACM TOG, vol 26. p 10
5. Navalpakkam V, Itti L (2006) An integrated model of top-down and bottom-up attention for optimizing detection speed. In: CVPR. pp 2049–2056
6. Borji A (2012) Boosting bottom-up and top-down visual features for saliency estimation[C]. In: 2012 IEEE conference on computer vision and pattern recognition (CVPR). IEEE, pp 438–445
7. Peng H, Li B, Xiong W, Hu W, Ji R (2014) Rgbd salient object detection: a benchmark and algorithms. In: European conference on computer vision (ECCV). pp 92–109
8. Ju R, Liu Y, Ren T et al (2015) Depth-aware salient object detection using anisotropic center-surround difference[J]. Sig Process Image Commun 38:115–126
9. Desingh K, Madhava Krishna K, Rajan D et al (2013) Depth really matters: improving visual salient region detection with depth[C]. BMVC
10. Lang C, Nguyen T V, Katti H et al (2012) Depth matters: Influence of depth cues on visual saliency. Computer Vision–ECCV 2012. Springer, Berlin, pp 101–115
11. Achanta R, Shaji A, Smith K, Lucchi A, Fua P, Su¨sstrunk S (2012) Slic superpixels compared to state-of-the-art superpixel methods. IEEE TPAMI 34(11):2274–2282
12. Peng H, Li B, Ling H, Hu W, Xiong W, Maybank SJ (1986) Salient object detection via structured matrix decomposition. Canny J. A computational approach to edge detection[J]. In: IEEE Transactions on Pattern Analysis and Machine Intelligence, no 6. pp 679–698
13. Harris C, Stephens MA (1988) Combined corner and edge detector[C]. Alvey Vis Conf 15:50
14. Simoncelli EP, Freeman WT (1995) The steerable pyramid: a flexible architecture for multi-scale derivative computation. In: ICIP
15. Feichtinger HG, Strohmer T (1998) Gabor analysis and al-gorithms: theory and applications
16. Zhu W, Liang S, Wei Y et al (2014) Saliency optimization from robust background detection [C]. In: Proceedings of the IEEE conference on computer vision and pattern recognition. pp 2814–2821

17. Shen X Wu Y (2012) A unified approach to salient object detection via low rank matrix recovery. In: CVPR. pp 2296–2303
18. Judd T, Ehinger K, Durand F, Torralba A (2009) Learning to predict where humans look. In: ICCV
19. Cheng MM, Warrell J, Lin WY, Zheng S, Vineet V, Crook N: Efficient salient region detection with soft image abstraction. In: ICCV
20. Goferman S, Manor LZ, Tal A (2010) Context-aware saliency detection. In: CVPR
21. Yan Q, Xu L, Shi J, Jia J (2013) Hierarchical saliency detection. In: CVPR

FPGA Design of MB-OFDM UWB Baseband System Based on Parallel Structure

Shi-jie Ren, Xin Su, Zhan Xu and Xiang-yuan Bu

Abstract A design method of multiband orthogonal frequency division multiplexing ultra wideband (MB-OFDM UWB) baseband system using parallel structure is proposed. FPGA is used to design the transmitter and the receiver. The input of digital to analog conversion (DAC) module, the output of the analog to digital conversion (ADC) module, the synchronization module, the carrier frequency offset (CFO) estimation, and compensation module are all made up of four-channel parallel structures. The simulation results prove that, when the CFO and the sampling frequency offset (SFO) are up to $\pm$ 20 ppm, in additive white Gaussian noise (AWGN) channel, CM1 or CM2 channel, the scheme ensures the low bit error rate (BER). It is suitable for high-speed MB-OFDM UWB system.

Keywords MB-OFDM UWB · ECMA-368 · FPGA · Carrier frequency offset · Sampling frequency offset · Residual phase · Signal-to-noise ratio (SNR)

S. Ren · X. Bu
School of Information and Electronics, Beijing Institute of Technology, Beijing 100081, China

S. Ren
Institute of Physics Science and Information Engineering, Liaocheng University, Liaocheng 252000, China

X. Su
Media Institute of Technology, Liaocheng University, Liaocheng 252000, China

Z. Xu (✉)
School of Information and Communication Engineering, Beijing Information Science and Technology University, Beijing 100192, China
e-mail: xuzhan_work@126.com

Y. Jia et al. (eds.), *Proceedings of 2016 Chinese Intelligent Systems Conference*, Lecture Notes in Electrical Engineering 405,
DOI 10.1007/978-981-10-2335-4_41

1 Introduction

Because of the high data rate, high spectrum efficiency, strong anti-multipath ability, MB-OFDM UWB system has become an important technology for wireless communication. The available frequency band is 3.1–10.6 GHz. Its working bandwidth is more than 500 MHz, so it can provide high channel capacity. The lowest transmission rate of is 53.3 Mbps, and the highest rate is up to 480Mbps. In order to avoid interference with other wireless communication systems, the maximum average power is −41.3 dBm. The high speed and low power consumption is a challenge for hardware design.

Cheol-Ho SHIN designed a MB-OFDM UWB receiver baseband system according to the IEEE802.15.3a standard [1]. They realized the four-channel parallel structure synchronous and four-channel parallel FFT. Under AWGN channel with CFO and SFO, compared to the single-channel structure, four-channel parallel structures there are SNR losses less than 0.25 dB. Wen-Hua Wu designed MB-OFDM UWB receiver baseband system at 480 Mbps transmitter rate according to ECMA-368 standard. The frame synchronization detection of the receiver uses four-channel parallel structure [2]. The above documents did not introduce the parallel design of transmitter.

Based on the ECMA-368 standard, MB-OFDM UWB baseband system is designed in this paper. Its information rate is up to 106.7 Mbps. Many modules are designed using parallel structure, for example the DAC module of transmitter, the ADC and synchronization module of the receiver, the CFO estimation module and compensation module. The clock of DAC and ADC is 528 MHz in our work. Because of the parallel structure, 528 MHz clock frequency is enough for every module. The simulation results demonstrate that the BER of the MB-OFDM UWB baseband system is below 10^{-6} in AWGN, CM1, and CM2 channel when SNR is great than 16 dB.

2 MB-OFDM UWB Baseband System

MB-OFDM UWB baseband system completes signal processing key technology including coding, modulation, demodulation, frame synchronization and so on. The hardware platform of the system is comprised of the following components:

(1) the main chip. The main processing chip FPGA is Virtex-6 series XC6VLX240T module produced by Xilinx Company. The DA module is AD9739 which has high performance, high frequency, dual channels, 14 bit DAC, and the maximum sampling rate is up to 2.5 Gsps. The AD module is EV10AQ190, which has high performance, high frequency, four channels, 10 bit ADC, and the maximum sampling rate is up to 5 Gsps. The sampling rate used in DA module and AD module is 528 Msps in our work.

(2) clock. In this system, the external clock frequency of AD9739 is 2.112 GHz, and the EV10AQ190 is 1.056 GHz. The main clock of 528 MHz is provided by AD9739. The other clocks are generated from the multilevel MMCM IP core.
(3) the communications port. Communication port includes HR911130C Gigabit Ethernet port and a serial port. The former is used to transmit the PSDU data, and the latter is used to pass the BER data which are counted by FPGA to the host computer.

3 Design Scheme of MB-OFDM UWB Transmitter

At the transmitter end, data is made up of PLCP preamble, PLCP header, and PSDU. Preamble is stored in ROM, which does not need modulation, and will make up frame directly in the time domain. While the PLCP header and PSDU are the original source of information, which will be coded and modulated before to make up frame.

At the transmitter, the baseband process is as follows: FPGA receives the source bit by Ethernet port from the host computer. Every 376 bits data consists of a basic processing block. Each block is coded by CRC module and then is scrambled to 400bits. When accept 1 bit data, (3, 1, 9) convolution code module output 3 bits. Interleaving module has three parallel outputs. The asynchronous clock conversion module gives two parallel data as outputs. The QPSK mapping module completes the constellation mapping. In the pilot insertion module, each of the 100 data subcarriers is inserted by 10 pilots and 12 guard subcarriers, so they are expanded into 128 subcarriers. The ping-pong structure of dual port RAM completes the rate conversion. In time domain, the IFFT module output signal is made up of 128 bits per symbol. These symbols add 32 zero postfix (ZP) and 4 frequency hopping interval to form 164 bits OFDM symbol. Time domain expansion is used. According to the time sequence, PLCP preamble, PLCP header, and PSDU form the frame structure. Under 132 MHz clock, four-channel parallel data is given as input to the Oserdes module. After parallel to serial changing, data are sent to DAC with 528Msps rate. Baseband processing design principle is as shown in Fig. 1.

4 The Design Scheme of MB-OFDM UWB Receiver

At the receiver end, ADC sampling rate is 528 Msps. I channel data and Q channel data are sampled. Sampling data is changed from serial to parallel through Iserdes module. The frame synchronization module carries out frame detection. CFO synchronization module estimates and compensates the carrier frequency offset. In time domain, Frame dismantling module removes the ZP from frame. FFT module

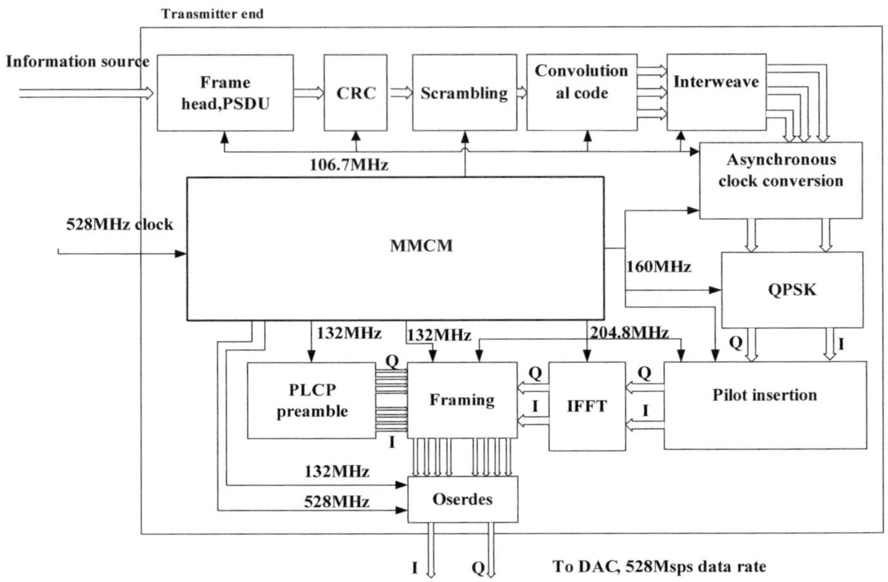

Fig. 1 Baseband processing design principle of transmitter

gives I channel data and Q channel data as outputs. Based on each OFDM symbol, it carries out channel equalization, sampling frequency synchronization, and the residual phase tracking. The carrier recovery module extracts 100 data subcarriers from each OFDM symbol. Soft demodulation is carried out in QPSK inverse mapping module. Viterbi decoding is the inverse of convolution. Inverse scrambling and inverse CRC is also needed according to the transmitter end. Demodulation software design principles are shown in Fig. 2.

4.1 Frame Synchronization Design Based on Four-Channel Parallel Structure

The input of frame synchronization module is four-channel parallel data.

Frame synchronization includes four steps.

The first step is one bit quantity. I channel data $y_I(n)$ and Q channel data $y_Q(n)$ are all quantified to one bit.

$$Q(n) = \text{sign}[y_I(n)] + j^* \text{sign}[y_Q(n)] \quad (1)$$

set $y_Q(\text{n}) = 0$, then

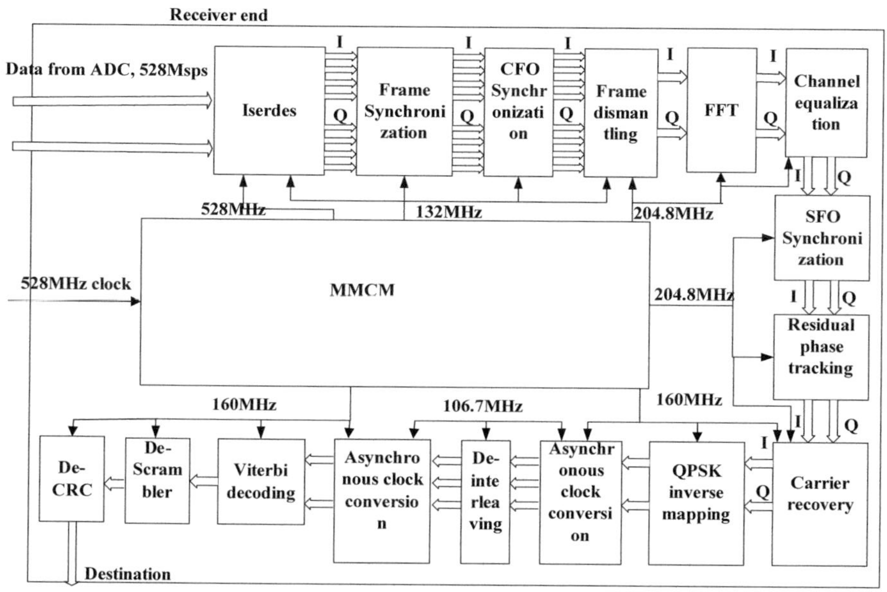

Fig. 2 Baseband processing design principle of receiver

$$Q(\mathrm{n}) = \mathrm{sign}[y_I(\mathrm{n})] \tag{2}$$

At low SNR, 1 bit quantification brings about 2 dB loss of SNR [3].

The second step is to slide window and calculate correlation between $Q(n)$ and preamble symbol $C(n)$.

$$\Lambda_{cc}(d) = \sum_{n=0}^{N-1} Q(n+d)\mathrm{C}^*(n), d = 0, 1, 2, \ldots, M \tag{3}$$

N is IFFT points, namely the number of subcarriers. N is 128. M is the total sampling point of OFDM symbol. Parallel computing structure is used to calculate the correlation Λ_{cc} shown as Eqs. (4)–(6).

The serial data is changed into four-channel parallel data.

$$\begin{aligned} A_1(k) &= Q(4k) \\ A_2(k) &= Q(4k+1) \\ A_3(k) &= Q(4k+2) \\ A_4(k) &= Q(4k+3) \\ k &= 0, 1, 2, 3, \ldots \end{aligned} \tag{4}$$

Change the preamble symbol to four-channel parallel data.

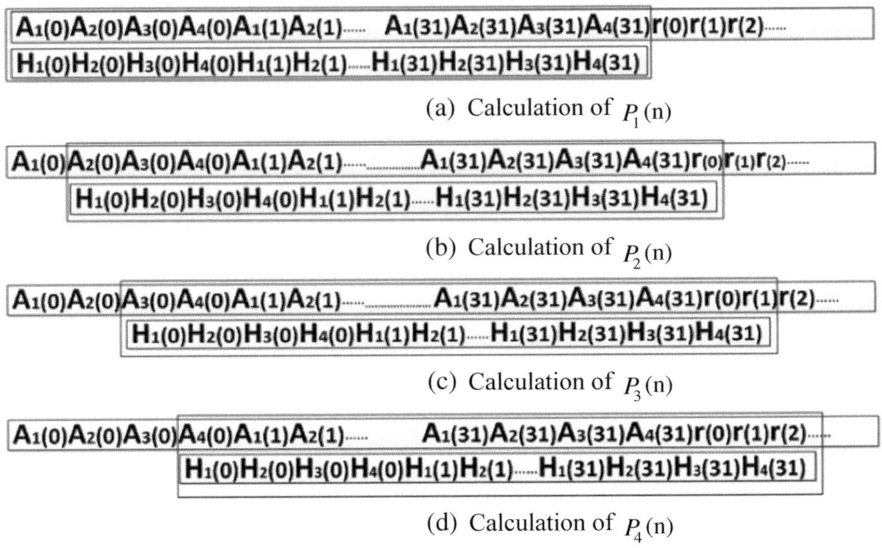

Fig. 3 Equivalent of sliding window correlation calculation

$$\begin{aligned} H_1(k) &= C(4k) \\ H_2(k) &= C(4k+1) \\ H_3(k) &= C(4k+2) \\ H_4(k) &= C(4k+3) \\ k &= 0, 1, 2, 3, \ldots \end{aligned} \tag{5}$$

It needs to complete four correlation calculations every clock cycle. Equivalent of sliding window correlation calculation is shown in Fig. 3.

where $P_1(n), P_2(n), P_3(n), P_4(n) (n = 0, 1, 2, 3, \ldots)$ are output data in every clock cycle.$r(0), r(1), r(2)$ are input data into the sliding window in the next clock cycle. $P_1(n), P_2(n), P_3(n), P_4$ are then changed to serial data $P(m) (m = 0, 1, 2, 3, \ldots)$.

$$\Lambda_{cc}(d) = P(m) \tag{6}$$

The third step is multipath energy accumulation.

Ultra wideband channel is a typical dense multipath channel. The strongest multipath components are likely to occur after the first path. In order to prevent the receiver lock to the strongest multipath without locking the first path, multipath energy accumulation method is used to reduce the influence of multipath. Multipath energy accumulation formula is shown as follows:

$$\Lambda(d) = \sum_{l=0}^{N_p-1} \Lambda_{cc}(d+l) \tag{7}$$

where $\Lambda(d)$ is accumulation result. N_p is the length of window. N_p must be large enough to contain the maximum number of multipath components. N_{zp} is the length of ZP. The length of the channel multipath delay is less than the length of ZP, namely $N_p \le N_{zp}$.

The fourth step is to determine the position of the frame head.

When multipath energy is accumulated more than the threshold value, determine the frame reaching. The point of the maximum value is the frame synchronization time named as $\widehat{d}$.

$$\widehat{d} = \arg\max_{d}\{\Lambda(\mathrm{d})\} \tag{8}$$

According to $\widehat{d}$, we can find the position of the frame head.

4.2 *CFO Estimation and Compensation Based on Four-Channel Parallel Structure*

The phase offset of received OFDM symbols is mainly caused by carrier frequency offset and sampling frequency offset.

Estimation and compensation of carrier frequency offset can be done in the time domain or in the frequency domain. We do it in time domain. In the same frequency band, two repeat OFDM symbols in preamble whose spacing is D OFDM symbols are used to complete the CFO estimation.

CFO estimation value is shown as [4]

$$\widehat{\Delta f}_{c,c_i} = -\frac{1}{2\pi DMT_s'}\angle R = -\frac{1}{2\pi DMT_s'}\arctan\left(\sum_{n=0}^{M-1} r_{i,n} r_{i+D,n}^{*}\right), \quad \mathrm{i} = 1,2,3,\ldots n = 0,1,2,\ldots M. \tag{9}$$

where i is the serial number of OFDM symbol. $r_{i,n}(\mathrm{i} = 1,2,3,\ldots, \mathrm{n} = 0,1,\ldots,\mathrm{M}-1)$ is the ith OFDM symbol. M is sampling point of OFDM symbol. T_s' is the clock of the receiver end. C_i is the number of frequency band.

$$c_i = \mathrm{mod}\,(\mathrm{i}-1,3)+1 \tag{10}$$

Using four-channel parallel structure compute the correlation of $r_{i,n}$ and $r_{i+D,n}$. Compensation of carrier frequency offset to OFDM symbol $r_{i,n}$ is shown as [5]:

$$\widehat{r}_{i,n} = r_{i,n} \times e^{-j2\pi\widehat{\Delta f}_{c,c_i}(\mathrm{iM}+\mathrm{n})T_s'} \tag{11}$$

The compensation process is also finished under four-channel parallel structure.

5 Simulation

In AWGN,CM1,CM2 channel [6], the CFO and SFO are up to ± 20 ppm. Receiver's SNR range is from −8.4 dB to 24 dB. The SNR range of simulation is from 0 dB to 25 dB. The simulation parameters are shown in Table 1.

CM1and CM2 Channel model parameters are shown in Table 2.

where Λ is the average arrival rate of cluster. λ is the average arrival rate of pulse. Γ is the power attenuation factor of cluster. γ is the power attenuation factor of pulse in the cluster. σ_1 is the channel coefficient standard deviation of cluster. σ_2 is the channel coefficient standard deviation within a cluster pulse. σ_x is the standard deviation of the channel amplitude gain.

The optimum suitable threshold value range is 0.75–0.78, and it is set to 0.78 in our work. As shown in Fig. 4, the detection probability simulation results are obtained in AWGN channel and 10 paths Rayleigh fading channels under different SNRs. As it can be seen, when the Eb/N0 is greater than 2 dB, the detection probability tends to 1.

The packet error rate is shown in Fig. 5.

In CM1,CM2 channel, when SNR is great than 8 dB, the packet error rate is below 8 %. This can ensure the accuracy of data transmission.

The BER of the system is shown in Fig. 6.

Table 1 simulation parameters

Parameters	Symbol and value
Data rate	106.7 Mbps
Modulation type	QPSK
Convolutional code	(3,1,9)
Time expansion factor	2
Load(Byte)	1024

Table 2 IEEE UWB channel model parameters

Parameters	CM1	CM2
$\Lambda(1/ns)$	0.0233	0.4
$\lambda(1/ns)$	2.5	0.5
Γ	7.1	5.5
γ	4.3	6.7
$\sigma_1(dB)$	3.4	3.4
$\sigma_2(dB)$	3.4	3.4
$\sigma_x(dB)$	3	3

Fig. 4 Frame detection probability under different SNR

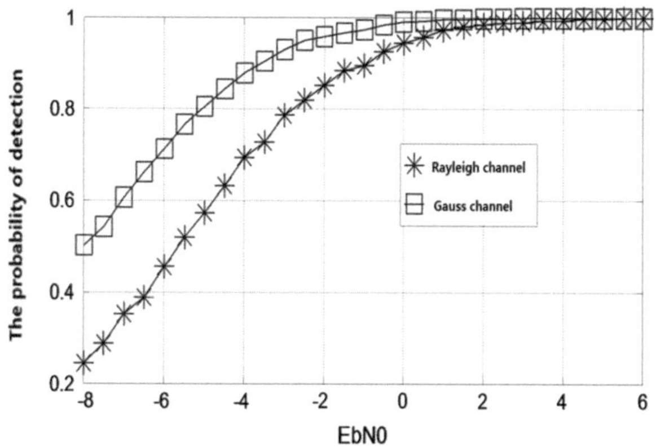

Fig. 5 The packet error rate

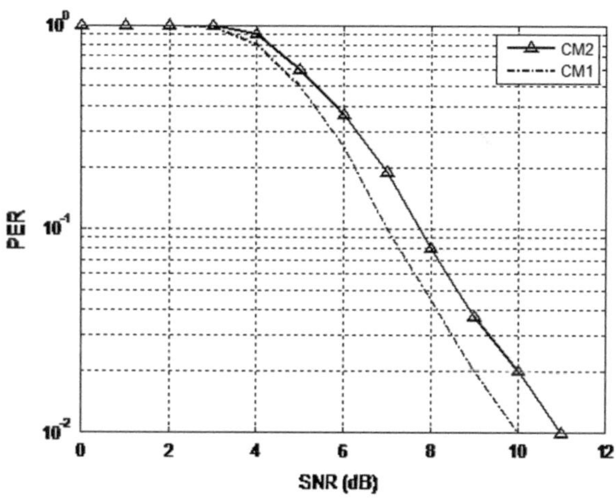

Fig. 6 BER of system

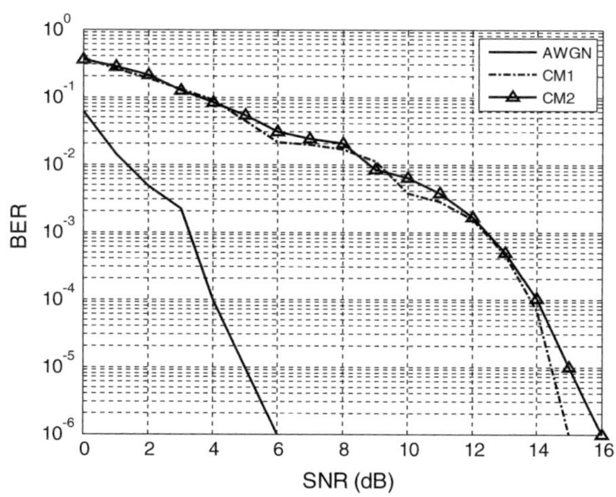

In AWGN, CM1, and CM2 channel, the scheme ensures the low BER. When SNR is great than 16 dB, the BER is less than 10^{-6}. The performance of the algorithm can meet the requirements of 106.7 Mbps communication.

6 Conclusion

This paper presents a parallel structure baseband design of MB-OFDM-UWB system using FPGA. Its supports data rate is up to 106.7 Mbps. The total power is less than 500mW. With CFO ($\pm$ 20 ppm) and SFO ($\pm$ 20 ppm) and in AWGN, CM1, CM2 channel, the performance of the system is perfect.

Acknowledgment This work was supported by the national natural science fund of Chinese (NO. 61402044), the Beijing city board of education scientific research plan (NO. KM201511232011).

References

1. Cheol-Ho S, Choi S, Leee H, Pack J-K (2007) A design and performance of 4-paralled MB-OFDM UWB receiver [J]. IEICE Trans Commun E90-B(3):672–675
2. WU W-H, WU Y-W, Ma H-P (2008) A 480 Mbps MB-OFDM-based UWB baseband inner transceiver [C]. In: IEEE asia pacific conference on circuits and systems. pp 164–167
3. Jin Z, Limin X, Zhan X (2012) Performance analysis of synchronization based on one-bit quantization in low SNR multipath regime [C]. Int Conf Wirel Commun Sig Proces 2012:25–27
4. Larenti N, Renna F (2008) Estimation of carrier and sampling frequency offset for ultra wide band multiband OFDM systems [C]. In: IEEE international conference on ultra-wideband. pp 49–54
5. Lin ZW, Xiaoming P, Chin K-BF (2012) Iterative sampling frequency offset estimation for MB-OFDM UWB systems with long transmission packet [J]. IEEE Trans Veh Technol 61 (4):1685–1697
6. Andreas F (2003) Molisch, channel models for ultra wideband personal area networks [J]. IEEE Wirel Commun 10(6):14–21

ST Segment Deviation Parameter Statistic Based on Spectrogram

Shi-jie Ren, Xin Su, Zhan Xu and Xiang-yuan Bu

Abstract The aim of this work is to detect the existence of ST segment deviation episodes in electrocardiogram (ECG) signals using spectrogram. Spectrogram is one kind of time–frequency distribution (TFD) which provides good aggregation property. Downloaded from MIT-BIH database, the experimental samples of ECG signals include 60 records without ST segment deviation and 60 records with ST segment deviation. We compare smoothed pseudo-Wigner–Ville distribution (SPWVD) with spectrogram of ECG signals. Spectrogram is used to statistic ST segment deviation in order to find out sensitive parameters. Fisher linear discriminate analysis is used to identify ST segment deviation episodes. The recognition rate of this method is up to 91.4 %. The investigation lays a basis for promoting the accuracy of ST segment deviation recognition.

Keywords Electrocardiogram (ECG) · ST segment deviation · Time–frequency distribution (TFD) · Spectrogram

S. Ren · X. Bu
School of Information and Electronics, Beijing Institute of Technology, Beijing 100081, China

S. Ren
Institute of Physics Science and Information Engineering, Liaocheng University, Liaocheng 252000, China

X. Su
Media Institute of Technology, Liaocheng University, Liaocheng 252000, China

Z. Xu (✉)
School of Information and Communication Engineering, Beijing Information Science and Technology University, Beijing 100192, China
e-mail: xuzhan_work@126.com

Y. Jia et al. (eds.), *Proceedings of 2016 Chinese Intelligent Systems Conference*, Lecture Notes in Electrical Engineering 405,
DOI 10.1007/978-981-10-2335-4_42

1 Introduction

ECG signal is a nonstationary process. Its frequency continuously varies with time. There is currently great interest in using linear and nonlinear signal processing techniques to characterize ST segment deviation detection. Time–frequency analysis displays time, frequency, and amplitude to characterize such processes. To analysis ECG signal using time–frequency distribution (TFD) systematically is prospective [1–3].

R.H. Clayton et al. have researched ECG signal using TFD [4]. TFD of recordings were estimated with the short time Fourier transform, Wigner–Ville, smoothed Wigner–Ville and Choi–Williams algorithms. The smoothed Wigner–-Ville distribution was found there are less cross terms.

A variety of analyzing wavelets has been proposed also in recent years for analysis of the ECG [5–8]. Time–frequency distributions of the ECG during the QRS may become another electrocardiographic indicator. Their work demonstrates the ability of the continuous wavelet transform to detect short lasting events of low amplitude superimposed on large signal deflections.

XU Liang et al. use smoothed pseudo-Wigner–Ville distribution (SPWVD) to study heart rate variability during ECG ST segment deviation episodes [9]. The recognition rate of the method is up to 89.7 % with Fisher discriminate.

In this paper, the first contribution is the analysis of spectrogram. The aggregation property between SPWD and spectrogram is compared. The latter has good aggregation property than the former. ST segment deviation is analyzed with spectrogram. The second contribution of this work is the ST segment deviation parameter statistic method based on spectrogram. We define and statistic ST segment deviation parameters based on spectrogram. The experimental results show that this statistic method works perfectly.

In Sect. 2, Normal sinus rhythm ECG signal in time domain is researched.

In Sect. 3, SPWVD, spectrogram, and Wigner–Hough transformation are compared.

In Sect. 4, feature parameters of ST segment deviation based on spectrogram are defined and calculated.

In Sect. 5, experiment on the recognition process of ST segment deviation.

In Sect. 6, we conclude the paper and outline areas for future research.

2 Normal Sinus Rhythm ECG Signal

2.1 Time Domain Normal Sinus Rhythm ECG Signal

In this work, normal sinus rhythm ECG signal is taken from MIT-BIH normal sinus rhythm database, which are made of 16265.dat, 16265.atr, 16265.hea. The individual recordings contain ECG signals sampled at 128 samples per second with 12-bit resolution over a range of ± 10 mV.

Fig. 1 Normal sinus rhythm ECG signal

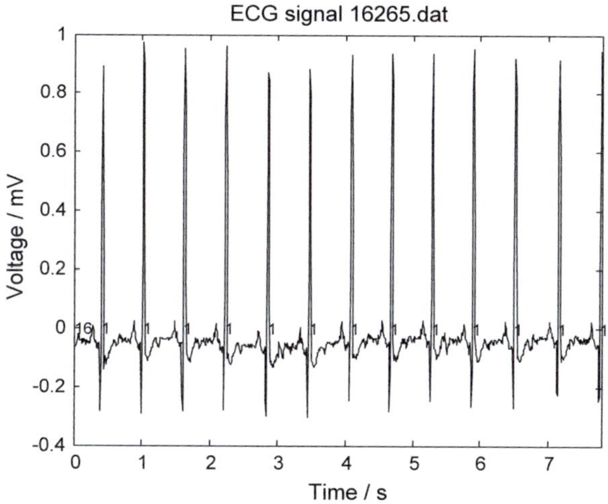

Part of the normal sinus rhythm ECG signal in time domain, named x(t), is shown in Fig. 1.

2.2 *Waves, Segments of ECG Signal*

Waves, segments, and intervals of ECG signal are shown in Fig. 2.

Waves of ECG signal include P wave, Q wave, R wave, S wave, and T wave. The P wave corresponds to atrium muscle depolarization voltage variation. The Q wave, R wave, and S wave (together treated as QRS complex wave) correspond to

Fig. 2 Waves, segments, and intervals of ECG signal

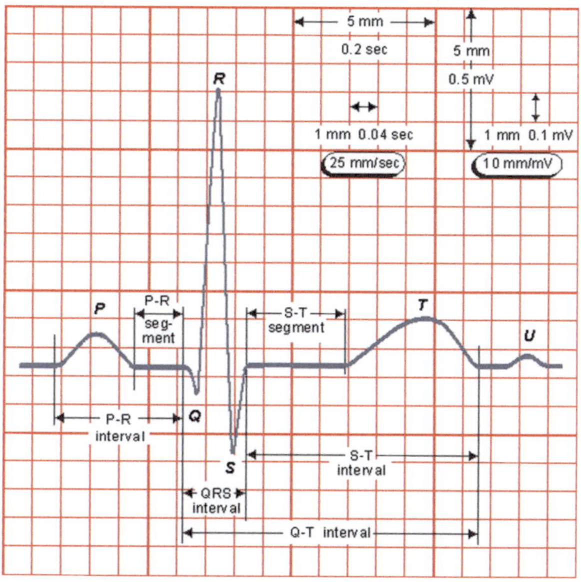

ventricle muscle depolarization voltage variation. The T wave corresponds to quick variation of ventricle muscle repolarization voltage.

Segments of ECG signal include PR segment and ST segment. The PR segment is usually flat segment between the end of the P wave and the start of the QRS complex. ST segment corresponds to slow variation of ventricle muscle repolarization voltage.

RR interval is the time between one R wave and its next R wave.

Several techniques are widely used to find the abnormal of ECG signal in time domain, and people have accumulated rich experience in this field. But sometimes, it is hard to find some kinds of abnormalities of ECG signal in time domain: for example ST segment deviation and so on. So, the time–frequency distributions are explored in this work.

The Time–Frequency Toolbox is a collection of about 100 scripts for MATLAB developed for the analysis of nonstationary signals using time–frequency distributions.

In Sect. 3, three kinds of time–frequency distributions of the normal sinus rhythm ECG signal are researched.

3 Time–Frequency Distribution of Normal Sinus Rhythm ECG Signal

We compared SPWVD and spectrogram with other distributions such as Wigner–Ville distribution, Choi–Williams distribution, etc. We found that the SPWVD and the spectrogram results in less cross terms, hence the current work focuses on SPWVD and spectrogram.

3.1 Smoothed Pseudo-Wigner–Ville Distribution

SPWVD of signal x(t) is defined as [10]

$$SPWVD_z(t,\omega) = \int_{-\infty}^{+\infty}\int_{-\infty}^{+\infty} g(u)h(\tau)z(t-u+\frac{\tau}{2})z^*(t-u-\frac{\tau}{2})e^{-j2\omega\tau}du d\tau \quad (1)$$

where, τ is the lag time, $g(u)$ and $h(\tau)$ are window functions, and $z(t)$ is the analytic signal of the time domain signal x(t).

$$z(t) = x(t) + j * H[x(t)] \quad (2)$$

where, $H[x(t)]$ is Hilbert transformation of $x(t)$.

SPWVD of the normal sinus rhythm ECG signal is shown in Figs. 3 and 4.

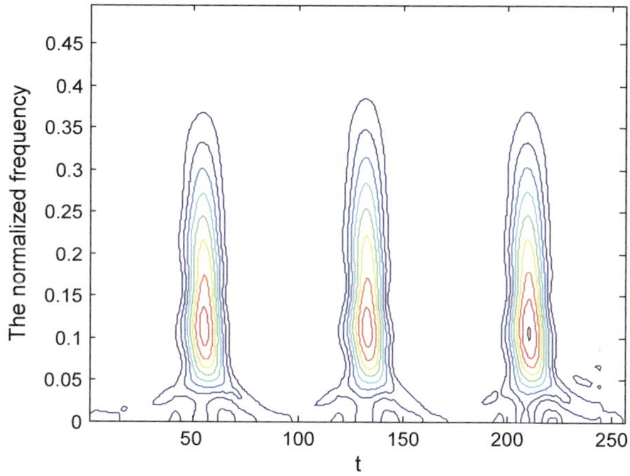

Fig. 3 Smoothed pseudo-Wigner–Ville distribution of ECG signal (contour map)

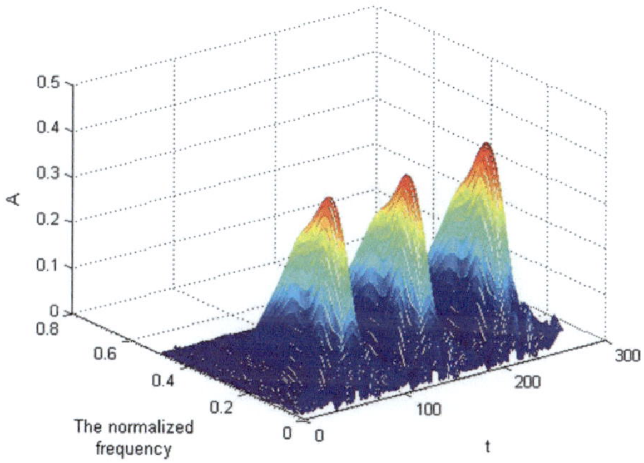

Fig. 4 Smoothed pseudo-Wigner–Ville distribution of ECG signal (3D Plot)

In Figs. 3 and 4, the X axis is time, the Y axis is the normalized frequency. In Fig. 4, the Z axis is amplitude.

According to Fig. 4, ECG signal is made up of many frequency components. SPWVD provides good aggregation, high resolution, and there are almost no cross terms.

3.2 *Spectrogram*

Spectrogram is defined as [11]

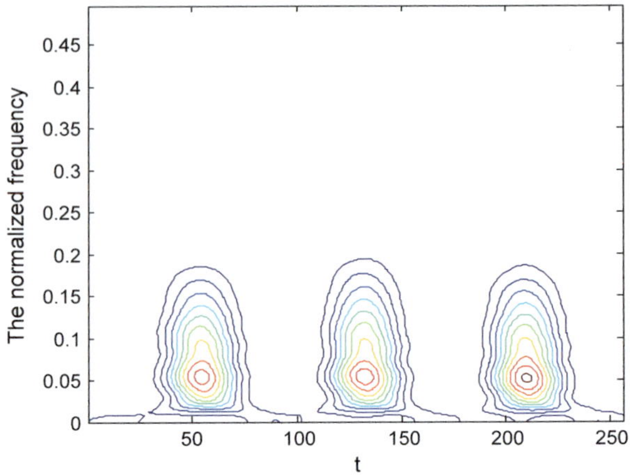

Fig. 5 Spectrogram of ECG signal (contour map)

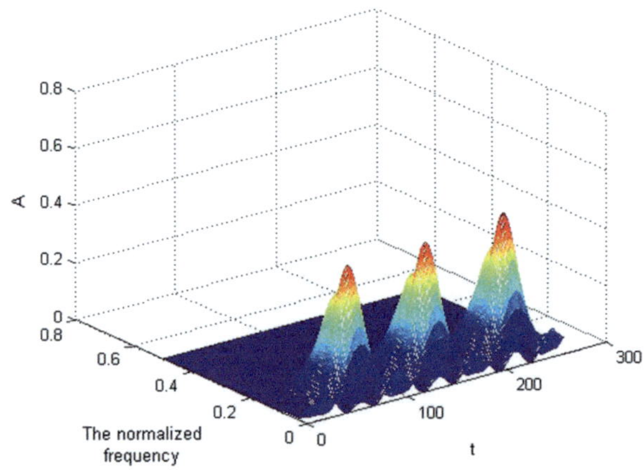

Fig. 6 Spectrogram of ECG signal (3D Plot)

$$S_z(t,f) = \int_{-\infty}^{+\infty} \int_{-\infty}^{+\infty} W_z(s,\ \xi) W_h(t-s, f-\xi) ds d\xi \qquad (3)$$

where, W_h is the Wigner–Ville distribution of window function h(t), W_z is the Wigner–Ville distribution of analytic signal z(t).

Spectrogram of the normal sinus rhythm ECG signal is shown in Figs. 5 and 6.

Spectrogram provides good aggregation, high resolution, and there are almost no cross terms. It is as excellent as smoothed pseudo-Wigner–Ville distribution [12, 13].

3.3 Wigner–Hough Transformation

Wigner–Hough transformation of complex signal z(t) is defined as [11]

$$WH_z(v_0, \beta) = \int_{-\infty}^{-\infty} \int_{T} z(t + \frac{\tau}{2}) z^*(t + \frac{\tau}{2}) e^{-j2\pi(v_0 + \beta t)} dt d\tau \quad (4)$$

where, τ is lag time, and z(t) is the analytic signal of the time domain signal x(t).

Wigner–Hough transformation to Wigner–Ville distribution of ECG signal is shown in Figs. 7 and 8.

In Fig. 8, the horizontal axis is rho, the vertical axis is theta, and the Z axis is the amplitude labeled A. rho and theta are all polar coordinate parameters.

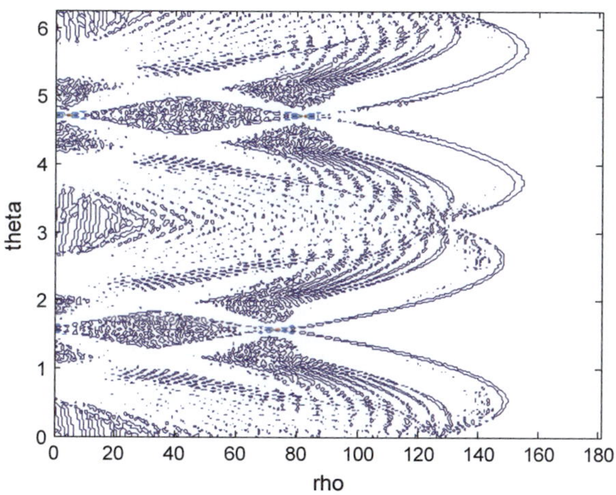

Fig. 7 Wigner–Hough transformation of ECG signal (contour map)

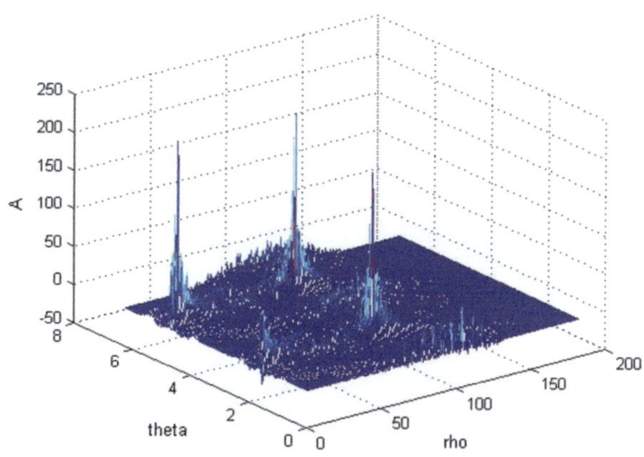

Fig. 8 Wigner–Hough transformation of ECG signal (3D Plot)

We can see many peaks obviously from Fig. 8. It means that ECG signal is made up of many frequency components, as shown in SPWVD and spectrogram.

4 ST Segment Deviation Parameter Calculation Based on Spectrogram

Heart rate variability (HRV) is the physiological phenomenon of variation in the time interval between heart beats. It is measured by the variation in the beat-to-beat interval. One kind of HRV is ST segment deviation. ST segment deviation refers to ST amplitude upward or downward migration.

ECG records are offered by MIT-BIH arrhythmia database. Among them there are 60 records without ST segment deviation, 30 records with ST segment deviation caused by rhythm change only, and 30 records with ST segment deviation caused by ischemia only.

Spectrogram of ECG is used in the study of ST segment deviation. The research includes feature parameter definition and feature parameters calculation.

4.1 Feature Parameter Definition of ST Segment Deviation

The frequency domain of ECG is divided into ultra-low-frequency band (ULF, 0–0.04 Hz), low-frequency band (LF, 0.04–0.15 Hz), high-frequency band (HF, 0.15–0.40 Hz), and all-frequency band (AF, 0–0.40 Hz) [14].

We define $P_{ULF}, P_{LF}, P_{HF}, P_{AF}$ to represent the power of ULF, LF, HF, and AF band. P_{LF} reflects the sympathetic strength. P_{HF} reflects the strength of vagus nerve [14].

Define equilibrium ratio

$$\frac{LF}{HF} = \frac{P_{LF}}{P_{HF}} \tag{5}$$

Equilibrium ratio reflects balance adjustment between sympathetic nerve and vagus [14].

In order to comparison, normalized power of the low-frequency band is defined as

$$LF_n = \frac{P_{LF}}{P_{LF} + P_{HF}} \tag{6}$$

Normalized power of the high-frequency band is defined as

$$HF_n = \frac{P_{HF}}{P_{LF} + P_{HF}} \tag{7}$$

PM_{ULF}, PM_{LF}, PM_{HF}, PM_{AF} are defined to represent the maximum power value of ULF, LF, HF, and AF band.

Instantaneous center frequency of the time–frequency distribution is defined as [9]

$$ICF(t) = \frac{\int_{f_1}^{f_2} f \cdot S_z(t, f)df}{\int_{f_1}^{f_2} S_z(t, f)df} \tag{8}$$

where S_z is spectrogram. ICF is a weighted average of the spectrum obtained. It is considered to reflect the relationship between sympathetic nerve and vague nerve.

Define ICF_{ULF}, ICF_{LF}, ICF_{HF}, ICF_{AF} to represent the ICF of ULF, LF, HF, and AF band.

Group delay of the ECG is defined as [9]

$$GD(f) = \frac{\int_{t_1}^{t_2} t \cdot S_z(t, f)dt}{\int_{t_1}^{t_2} S_z(t, f)dt} \tag{9}$$

GD(f) reflects the energy distribution of ECG signal.

In order to compare, normalized group delay of the ECG is defined as

$$GDP = \frac{GD(f)}{t_2 - t_1} \tag{10}$$

Define GDP_{ULF}, GDP_{LF}, GDP_{HF}, GDP_{AF} to represent the GDP of ULF, LF, HF, and AF band.

We define RR to represent the RR interval.

4.2 Feature Parameter Calculation

We calculate feature parameters in different time periods and different frequency bands to reveal ST segment deviation. The parameters are listed in Table 1.

Table 1 Feature parameters

Serial no.	Feature parameters	Serial no.	Feature parameters
1	P_{ULF}	11	HF_n
2	P_{LF}	12	ICF_{ULF}
3	P_{HF}	13	ICF_{LF}
4	P_{AF}	14	ICF_{HF}
5	PM_{ULF}	15	ICF_{AF}
6	PM_{LF}	16	GDP_{ULF}
7	PM_{HF}	17	GDP_{LF}
8	PM_{AF}	18	GDP_{HF}
9	LF/HF	19	GDP_{AF}

5 Experimental Results

Statistical tool SPSS 19.0 is used to calculate these feature parameters.

5.1 *Significant Differences of Feature Parameters*

Statistic feature parameters based on spectrogram at different temporal divisions and different frequency bands. There are significant differences between the parameters of test if $P < 0.05$ according to the calculated results. Statistical analysis revealed that eight parameters are sensitive for the detection of ST segment deviation as shown in Table 2.

Table 2 Discriminate of ST segment deviation

Parameters with significant difference and constant	Discriminate coefficient of ST segment without deviation	Discriminate coefficient of ST segment with deviation
RR	46.352	41.687
P_{LF}	2.504	−2.842
P_{AF}	−2.913	4.467
PM_{ULF}	−8.165	−12.459
LF/HF	−1.632	−2.486
GDP_{HF}	−54.122	−37.354
GDP_{AF}	156.321	146.635
ICF_{AF}	−34.562	−33.568
Constant	−65.765	−74.349

5.2 Fisher Linear Discriminate Analysis

The task of discriminate analysis is to classify object. It is our purpose to recognize the deviation of ST segment. One linear discriminate analysis is Fisher discriminate.

Based on the content of Table 2 define Fisher discriminate function N, to represent normal group without ST segment deviation.

$$\begin{aligned} N = & 46.352 \cdot RR + 2.504 \cdot P_{LF} - 2.913 \cdot P_{AF} - 8.165 \cdot PM_{ULF} - 1.632 \cdot LF/HF \\ & - 54.122 \cdot GDP_{HF} + 156.321 \cdot GDP_{AF} - 34.562 \cdot ICF_{AF} - 65.765 \end{aligned} \quad (11)$$

Define Fisher discriminate function D to represent abnormal group with ST segment deviation.

$$\begin{aligned} D = & 41.687 \cdot RR - 2.842 \cdot P_{LF} + 4.467 \cdot P_{AF} - 12.459 \cdot PM_{ULF} - 2.486 \cdot LF/HF \\ & - 37.354 \cdot GDP_{HF} + 146.635 \cdot GDP_{AF} - 33.568 \cdot ICF_{AF} - 74.349 \end{aligned} \quad (12)$$

If $N > D$, we judge the record belong to normal ST segment, else we judge the record belong to ST segment with deviation. The recognition rate of this discriminate is up to 91.4 %. The discriminate based on spectrogram provides good performance as compared to SPWVD.

6 Conclusions

We give the statistics about time–frequency analysis for the various data files used for the experimentation. The results show that during the process of ST segment deviation the majority of the feature parameters changed. These parameters show significant differences. GDP, RR, and ICF play an important role in Fisher discriminate function.

As a non-morphological method, feature parameters analysis based on time–-frequency method is useful in identifying the presence of ST segment deviation episodes. The recognition rate of Fisher linear discriminate is high. The investigation lays a basis for promoting the accuracy of ST segment deviation detection.

To analyze the incentives of ST segment deviation and to classify them may be looked into as part of future work.

Acknowledgments The ECG signal is offered by the MIT-BIH normal sinus rhythm database and arrhythmia database. The authors wish to thank the staff of the MIT-BIH database laboratory.

This work was supported by the national natural science fund of China (NO. 61402044), the China 863 plan program (NO. 2015AA01A706), the Beijing municipal education commission scientific research plan (NO. KM201511232011), and Liaocheng University scientific research fund.

References

1. Mahmoud Seedahmed S, Hussain Zahir M, Cosic Irena, Fang Q (2006) Time-frequency analysis of normal and abnormal biological signals. Biomed Signal Process Control 1 (1):33–43
2. Matz Gerald, Hlawatsch Franz (2003) Wigner distributions (nearly) everywhere: time–frequency analysis of signals, systems, random processes, signal spaces, and frames. Signal Process 83(7):1355–1378
3. Colak Omer H (2009) Preprocessing effects in time–frequency distributions and spectral analysis of heart rate variability. Digit Signal Process 19(4):731–739
4. Clayton RH, Murray A (1998) Comparison of techniques for time-frequency analysis of the ECG during fibrillation. IEE Proc Sci Meas Technol 145(6):301–306
5. Gramatikov B, Brinker J, Yi-chun S, Thakor NV (2000) Wavelet analysis and time–frequency distributions of the body surface ECG before and after angioplasty. Comput Methods Program Biomed 62:87–98, (2000)
6. Tirtom Huseyin, Engin Mehmet, Engin EZ (2008) Enhancement of time-frequency properties of ECG for detecting micropotentials by wavelet transform based method". Expert Syst Appl 34:746–753
7. Gramatikov Boris, Brinker Jeffrey, Yi-chun Sun, Thakor Nitish V (2000) Wavelet analysis and time-frequency distributions of the body surface ECG before and after angioplasty. Comput Methods Program Biomed 62:87–98
8. Gramatikov B, Georgiev I (1995) Wavelets as an alternative to STFT in signal-averaged electrocardiography. Med Biol Eng Comput 33:482–487
9. Liang XU, Xing WANG, Zhong-wei SUN, Peng Y (2008) Time-frequency method based analysis of heart rate variability during ECG ST-segment deviation episodes. Chin J Biomed Eng 27(1):23–28
10. Zhang X (2002) Modern signal processing, 2nd edn. Tsinghua Press
11. Zhe-xue GE, Zhong-sheng C (2006) Matlab time-frequency technology and its applications. Posts and Telecom Press
12. Marchant BP (2003) Time–frequency analysis for biosystems engineering. Biosyst Eng 85 (3):261–281
13. Debbal SM, Bereksi-Reguig F (2007) Time-frequency analysis of the first and the second heartbeat sounds. Appl Math Comput 184(2):1041–1052
14. Jing-xia SUN, Yan-qiang BAI (2001) Research progress of methods of heart rate variability analysis. Space Med Med Eng 14(3):230–234

Event-Triggered Consensus Control of Nonlinear Multi-agent Systems with External Disturbance

Yang Liu, Jun Gao and Xiaohui Hou

Abstract This paper investigates the consensus problem for a leader-following nonlinear multi-agent system with external disturbance by using the event-triggered control strategy. First, in order to transform the consensus control problem of disturbed system into the H_∞ problem, a controlled output function is defined. Then a distributed event-triggered protocol is designed, and a sufficient condition is given to ensure that the nonlinear multi-agent system can reach consensus with the desired disturbance attenuation ability.

Keywords Consensus · Multi-agent system · Event-triggered control · Nonlinear dynamics · External disturbance

1 Introduction

A multi-agent system consists of multiple independent agents which can act consistently through the transmission among agents. There have been inspiring successes in the areas such as sensor networks, consensus and formation control of vehicles, and cooperative control of robots by using decentralized strategy.

In fact, real multi-agent systems are often equipped with microprocessors that gather information and actuate the agent controller updates with limited on-board

Y. Liu (✉)
The Seventh Research Division and the Center for Information and Control, Beihang University, Beijing 100191, People's Republic of China
e-mail: ylbuaa@163.com

Y. Liu · X. Hou
School of Automation Science and Electrical Engineering, Beihang University, Beijing 100191, People's Republic of China

J. Gao
School of Mathematics and Systems Science, Beihang University, Beijing 100191, People's Republic of China

Y. Jia et al. (eds.), *Proceedings of 2016 Chinese Intelligent Systems Conference*, Lecture Notes in Electrical Engineering 405,
DOI 10.1007/978-981-10-2335-4_43

energy and resources. This brings some problems such as the waste of communication channel and the request for high processing speed of the microprocessors. In view of this, the event-triggered control has been proposed in consensus control of multi-agent system during recent years [1–13]. The event-triggered control indicates that an event condition, which is usually related to the states of the system, triggers the control task execution. Centralized or decentralized consensus protocol based on event-triggered control method is proposed in [2]. In [3], Meng and Chen designed a new event detector to accomplish event-triggered consensus control of multi-agent systems. Meanwhile, event-triggered consensus problem of discrete multi-agent systems with time delay is studied in [4–6]. Obviously, the event-triggered strategy can reduce the resource occupation in multi-agent systems with minimal loss of system behavior. However, the nonlinear system is less investigated in the recent event-triggered consensus research, and the effect of external disturbances is not taken into account.

Motivated by the above work, we consider the event-triggered consensus problem of a disturbed nonlinear multi-agent system. In the following section, a new output $z(t)$ is defined to transform the consensus problem of states into the convergence analysis of $z(t)$ to zero, and then the consensus study of disturbed system is reformulated as the H_∞ control problem of a nonlinear system. To proceed, we design a distributed consensus controller with a new event-triggered condition. Then, sufficient condition is given to ensure that the closed-loop multi-agent system reaches the consensus result with a desired H_∞ performance level.

2 Problem Statements

Consider a leader-following multi-agent system consisting of n following agents and one leader. The ith following agent is modeled by the nonlinear dynamics with unknown external disturbance

$$\dot{x}_i(t) = f(x_i(t), t) + u_i(t) + \omega_i(t) \tag{1}$$

where $t \in [0, \infty)$ is the time variable, $x_i(t)$, $u_i(t)$, $\omega_i(t) \in R^m$ are, respectively, the state, the input, and the external disturbance of agent i, and $f: R^m \times R^+ \to R^m$ is a smooth nonlinear vector-valued function. It is assumed that $\omega_i(t) \in L_2[0, \infty)$, where $L_2[0, \infty)$ is the space of square-integrable vector functions over $[0, \infty)$. This indicates that the energy-limited external disturbance is considered in this paper.

The dynamics of leader is given by

$$\dot{x}_0(t) = f(x_0(t), t) \tag{2}$$

with $x_0(t)$ denoting the state of leader. It is worth pointing out that the nonlinear dynamics is allowed to be unknown, which would not appear in the consensus controller design.

Assumption 1 It is assumed that $f: R^m \times R^+ \to R^m$ satisfies the following Lipschitz condition

$$||f(x, t) - f(y, t)|| \le \beta ||x - y||, \quad \forall x, y \in R^m, t \ge 0, \tag{3}$$

where $\beta > 0$ is a constant scalar.

Definition 1 The multi-agent system (1) is said to asymptotically achieve consensus if all the following agents' states satisfy

$$\lim_{t \to \infty} x_i(t) = x_0(t), \quad i = 1, 2, \ldots, n \tag{4}$$

Directed (or undirected) graphs are used to model the interaction topologies of n following systems (1). Let $G = (V, E, A)$ be a weighted directed graph of order n, where the set of nodes is described by V, the set of directed edges is $E \subseteq V \times V$, and a weighted adjacency matrix $A = [a_{ij}]$ is defined with nonnegative adjacency weights a_{ij}. It is stipulated that the adjacency weights associated with edges are positive, i.e., $(v_i, v_j) \in E \Leftrightarrow a_{ij} > 0$. In particular, node v_i represents the ith agent, and edge (v_i, v_j) represents that information is transferred from the jth agent to the ith one. The Laplacian of a weighted graph G is defined as $L = D - A$, where $D = \text{diag}\{d_1, d_2, \ldots, d_n\}$ is the degree matrix of G, whose diagonal element is $d_i = \sum_{j=1}^{n} a_{ij}$. Further, if a graph has the property that $a_{ij} = a_{ji}$ always holds, then it is called undirected.

Considering the leading role of leader (2), another node v_0 is added to represent the given leader. Then the edge (v_i, v_0) represents that agent i can obtain the leader's state information, and the corresponding weight a_{i0} is positive. To summarize, the set of neighbors of agent i is denoted by $N_i = \{v_j \in \{V \cup v_0\}: (v_i, v_j) \in E\}$. Define $H = L + \Lambda$ as the interaction matrix of the whole leader-following system (1)–(2), where $\Lambda = \text{diag}\{a_{i0}\}_{i=1}^{n}$.

Lemma 1 *Consider a network with undirected communication among the followers. If at least one agent in each connected component of G is connected to the leader, then the symmetric interaction matrix H is positive definite. Otherwise, H has at least one zero eigenvalue.*

Lemma 2 *If the interaction graph of the leader–follower system with directed communication has a spanning tree with the leader as root, then the negative interaction matrix H is Hurwitz stable. Otherwise, H will have at least one zero eigenvalue.*

3 Protocol Design and Consensus Analysis

3.1 Problem Reformulation

In order to quantitatively measure the effect of external disturbance to the consensus performance, we define the following controlled output functions

$$z_i(t) = x_i(t) - \frac{1}{n}\sum_{j=1}^{n} x_j(t), \quad i = 1, 2, \ldots, n,$$

whose compact form is

$$z(t) = (L_c \otimes I_m)x(t), \tag{5}$$

where $z(t) = [z_1^T(t)\, z_2^T(t) \ldots z_n^T(t)]^T$, $x(t) = [x_1^T(t)\, x_2^T(t) \ldots x_n^T(t)]^T$, L_c is a symmetric matrix with diagonal elements (n − 1)/n, and all the other elements −1/n. Meanwhile, the norm of $z(t)$ can reflect the degree of state deviations. Therefore, combing system (1) with the controlled output (5), we transform the consensus control of disturbed system (1) into the H_∞ control problem of the following system:

$$\begin{cases} \dot{x}(t) = F(x(t)) + u(t) + \omega(t) \\ z(t) = (L_c \otimes I_m)x(t) \end{cases} \tag{6}$$

where

$$\begin{aligned} F(x(t)) &= [f^T(x_1(t)) \quad f^T(x_2(t)) \ldots f^T(x_n(t))]^T, \\ u(t) &= [u_1^T(t) \quad u_2^T(t) \ldots u_n^T(t)]^T, \\ \omega(t) &= [\omega_1^T(t) \quad \omega_2^T(t) \ldots \omega_n^T(t)]^T. \end{aligned}$$

Therefore, the present objective is to design an event-triggered protocol such that

$$||T_{z\omega}(s)||_\infty = \sup_{\mu \in R}\bar{\sigma}(T_{z\omega}(j\mu)) = \sup_{0 \neq \omega(t) \in L_2[0,\infty)} \frac{||z(t)||_2}{||\omega(t)||_2} < \gamma, \tag{7}$$

or equivalently, the closed-loop system satisfies

$$\int_0^\infty ||z(t)||^2 dt < \gamma^2 \int_0^\infty ||\omega(t)||^2 dt, \quad \forall \omega(t) \in L_2[0, \infty), \tag{8}$$

where $\gamma > 0$ is a given H_∞ performance index.

3.2 Event-Triggered Protocol Design

Denote $t_0^i,\ t_1^i,\ t_2^i,\ \ldots$ as the event-triggered time of agent i, where $t_0^i = 0$ for $\forall i$. Then we propose the following distributed event-triggered consensus protocol:

$$u_i(t) = \sum_{j \in N_i} a_{ij}(x_i(t_k^i) - x_j(t_{k_j(t)}^j)) + a_{i0}(x_i(t_k^i) - x_0(t)), \quad t \in [t_k^i, t_k^{i+1}), \tag{9}$$

where $a_{ij} > 0$ is the interaction strength between agents i and j; $a_{i0} > 0$ if and only if the ith agent can obtain the state information of the given leader, or else $a_{i0} = 0$; $k_j(t) = \arg \min_{l \in N: t \geq t_l^j} \{t - t_l^j\}$. It is obvious that $t_{k_j(t)}^j$ is the last event-triggered time of agent j.

Meanwhile, define

$$e_i(t) = x_i(t_k^i) - x_i(t), \quad t \in [t_k^i,\ t_{k+1}^i)$$

as the measurement error of the ith agent, arising from the event-triggered controller update. The following distributed event-triggered condition is designed to determine the discrete event-triggered time instants

$$||e_i(t)|| = ||\sum_{j \in N_i} a_{ij}(x_i(t) - x_j(t)) + a_{i0}(x_i(t) - x_0(t))||, \tag{10}$$

from which it is immediately derived that

$$||e_i(t)|| \leq ||\sum_{j \in N_i} a_{ij}(x_i(t) - x_j(t)) + a_{i0}(x_i(t) - x_0(t))||, \quad \forall t \in [0,\ \infty). \tag{11}$$

3.3 Consensus Conditions

To analyze the consensus performance of the closed-loop system, we first define

$$\bar{x}_i(t) = x_i(t) - x_0(t), \quad i = 1,\ 2,\ \ldots,\ n \tag{12}$$

as the disagreement state vector. Obviously, the strict consensus result (4) is realized if and only if

$$\lim_{t \to \infty} \bar{x}_i(t) = 0, \quad i = 1,\ 2,\ \ldots,\ n.$$

Accordingly, the nonzero consensus trajectory is transformed to the origin of $\bar{x}(t) = [\bar{x}_1^T(t) \quad \bar{x}_2^T(t) \cdots \bar{x}_n^T(t)]^T$. Note that the compact form of (12) is

$$\bar{x}(t) = x(t) - 1_n \otimes x_0(t). \tag{13}$$

To proceed, we focus on the derivation of the dynamic equation of $\bar{x}(t)$, by combining the controlled plant and the event-triggered consensus protocol.

First, we have

$$\begin{aligned} \dot{\bar{x}}_i(t) &= \dot{x}_i(t) - \dot{x}_0(t) \\ &= f(x_i(t)) + u_i(t) + \omega_i(t) - f(x_0(t)). \end{aligned} \tag{14}$$

Let $F(x_0(t)) = 1_n \otimes f(x_0(t))$ and $\bar{x}(t) = [\bar{x}_1^T(t) \quad \bar{x}_2^T(t) \cdots \bar{x}_n^T(t)]^T$. Then, from (14) and (9), the closed-loop system in terms of variable $\bar{x}(t)$ is calculated as

$$\begin{aligned} \dot{\bar{x}}(t) &= F(x(t)) - F(x_0(t)) + (L \otimes I_m)(e(t) + x(t)) + (\Lambda \otimes I_m)(e(t) + x(t) + 1_n \otimes x_0(t)) + \omega(t) \\ &= F(x(t)) - F(x_0(t)) + (L \otimes I_m)(e(t) + \bar{x}(t) - 1_n \otimes x_0(t)) + (\Lambda \otimes I_m)(e(t) + \bar{x}(t)) + \omega(t) \\ &= F(x(t)) - F(x_0(t)) + (L \otimes I_m)(e(t) + \bar{x}(t)) + (\Lambda \otimes I_m)(e(t) + \bar{x}(t)) + \omega(t) \\ &= F(x(t)) - F(x_0(t)) + [(L + \Lambda) \otimes I_m](e(t) + \bar{x}(t)) + \omega(t) \\ &= F(x(t)) - F(x_0(t)) + (H \otimes I_m)(e(t) + \bar{x}(t)) + \omega(t), \end{aligned}$$

where the property $L1_n = 0$ is used in the third step.

Furthermore, by (5), it is derived that

$$\begin{aligned} z(t) &= (L_c \otimes I_m)x(t) \\ &= (L_c \otimes I_m)(\bar{x}(t) + 1_n \otimes x_0(t)) \\ &= (L_c \otimes I_m)\bar{x}(t), \end{aligned} \tag{15}$$

where the fact $L_c 1_n = 0$ is also applied. To sum up, the original consensus study is reformulated as a standard H_∞ control problem

$$\begin{cases} \dot{\bar{x}}(t) = F(x(t)) - F(x_0(t)) + (H \otimes I_m)(e(t) + \bar{x}(t)) + \omega(t) \\ \qquad = F(x(t)) - F(x_0(t)) + B(e(t) + \bar{x}(t)) + \omega(t) \\ z(t) = (L_c \otimes I_m)\bar{x}(t) \\ \qquad = C\bar{x}(t) \end{cases}. \tag{16}$$

Theorem 1 *Under the event-triggered protocol (9)–(10), the disturbed multi-agent system (1) can achieve consensus performance with the desired H_∞ disturbance attenuation ability γ, if there exists a positive definite matrix $P \in R^{nm \times nm}$ and positive scalars α, δ satisfying the following linear matrix inequality*

$$\begin{bmatrix} PB + B^T P + C^T C + \alpha H^T H + 2\delta\beta I & PB & P & 0 \\ B^T P & -\alpha I & 0 & 0 \\ P & 0 & -\gamma^2 I & 0 \\ 0 & 0 & 0 & P - \delta I \end{bmatrix} < 0, \tag{17}$$

where positive constants β, γ are, respectively, the threshold in the event-triggered condition (10) and the given H_∞ index.

Proof First, we investigate the stability property of the closed-loop system, by analyzing system (16) without external disturbance. Define a Lyapunov function

$$V(t) = \bar{x}^T(t)P\bar{x}(t),$$

then its derivative along system (16) with $\omega(t) = 0$ is

$$\begin{aligned}\dot{V}(t) &= 2\bar{x}^T(t)P[F(x(t)) - F(x_0(t))] + \bar{x}^T(t)(PB + B^TP)\bar{x}(t) + 2\bar{x}^T(t)PBe(t)\\ &\le 2\bar{x}^T(t)P[F(x(t)) - F(x_0(t))] + \bar{x}^T(t)(PB + B^TP)\bar{x}(t) + \alpha^{-1}\bar{x}^T(t)PBB^TP\bar{x}(t)\\ &\quad + \alpha e^T(t)e(t)\end{aligned} \tag{18}$$

Furthermore, it is obtained that

$$\begin{aligned}&2\bar{x}^T(t)P[F(x(t)) - F(x_0(t))]\\ &\quad\le 2\delta\sum_{i=1}^n \bar{x}_i^T(t)[f(x_i(t)) - f(x_0(t))]\\ &\quad\le 2\delta\sum_{i=1}^n ||\bar{x}_i(t)|| \cdot ||f(x_i(t)) - f(x_0(t))||\\ &\quad\le 2\delta\sum_{i=1}^n ||\bar{x}_i(t)|| \cdot \beta||x_i(t) - x_0(t)||\\ &\quad= 2\delta\beta\sum_{i=1}^n ||\bar{x}_i(t)||^2\\ &\quad= 2\delta\beta\bar{x}^T(t)\bar{x}(t),\end{aligned} \tag{19}$$

where $\delta > 0$ is a scalar satisfying the matrix inequality

$$P < \delta I. \tag{20}$$

Note that (20) is included in the consensus condition (17). Meanwhile, denote matrix H as the following partitioned form

$$H = \begin{bmatrix} H_1 \\ H_2 \\ \vdots \\ H_n \end{bmatrix},$$

and from (11), we have

$$e_i^T(t)e_i(t) \le \bar{x}^T(t)H_i^TH_i\bar{x}(t).$$

Then it immediately yields that

$$\sum_{i=1}^{n} e_i^T(t)e_i(t) \le \sum_{i=1}^{n} \bar{x}^T(t)H_i^T H_i\bar{x}(t),$$

which results in

$$e^T(t)e(t) \le \bar{x}^T(t)H^T H\bar{x}(t). \tag{21}$$

Substituting (19) and (21) into (18) leads to

$$\begin{aligned}\dot{V}(t) &\le 2\delta\beta\bar{x}^T(t)\bar{x}(t) + \bar{x}^T(t)(PB + B^T P)\bar{x}(t) + \alpha^{-1}\bar{x}^T(t)PBB^T P\bar{x}(t) + \alpha\bar{x}^T(t)H^T H\bar{x}(t)\\ &= \bar{x}^T(t)(PB + B^T P + \alpha^{-1}PBB^T P + \alpha H^T H + 2\delta\beta I)\bar{x}(t)\\ &< 0\end{aligned}$$

by applying Schur complement lemma to the matrix inequality (17). This means that the closed-loop multi-agent system is asymptotically stable.

Subsequently, the disturbance attenuation performance of system (16) is given by considering the function

$$J_T = \int_0^T ||z(t)||^2 dt - \gamma^2 \int_0^T ||\omega(t)||^2 dt. \tag{22}$$

To be specific, we have the following result under the zero initial condition:

$$\begin{aligned}J_T &= \int_0^T (z^T(t)z(t) - \gamma^2\omega^T(t)\omega(t) + \dot{V}(t))dt - V(T)\\ &= \int_0^T \{\bar{x}^T(t)C^T C\bar{x}(t) - \gamma^2\omega^T(t)\omega(t) + 2\bar{x}^T(t)P[F(x(t)) - F(x_0(t))] + \bar{x}^T(t)(PB\\ &\quad + B^T P)\bar{x}(t) + 2\bar{x}^T(t)PBe(t) + 2\bar{x}^T(t)P\omega(t)\}dt - V(T)\\ &\le \int_0^T \{\bar{x}^T(t)C^T C\bar{x}(t) - \gamma^2\omega^T(t)\omega(t) + 2\delta\beta\bar{x}^T(t)\bar{x}(t) + \bar{x}^T(t)(PB + B^T P)\bar{x}(t)\\ &\quad + 2\bar{x}^T(t)PBe(t) + 2\bar{x}^T(t)P\omega(t) + \alpha e^T(t)e(t) - \alpha e^T(t)e(t)\}dt - V(T)\\ &\le \int_0^T \{\bar{x}^T(t)C^T C\bar{x}(t) - \gamma^2\omega^T(t)\omega(t) + 2\delta\beta\bar{x}^T(t)\bar{x}(t) + \bar{x}^T(t)(PB + B^T P)\bar{x}(t)\\ &\quad + 2\bar{x}^T(t)PBe(t) + 2\bar{x}^T(t)P\omega(t) + \alpha\bar{x}^T(t)H^T H\bar{x}(t) - \alpha e^T(t)e(t)\}dt - V(T)\\ &\le \bar{x}^T(t)\ e^T(t)\ \omega^T(t)] \begin{bmatrix} PB + B^T P + C^T C + \alpha H^T H + 2\delta\beta I & PB & P\\ B^T P & -\alpha I & 0\\ P & 0 & -\gamma^2 I\end{bmatrix}\\ &\quad \cdot \begin{bmatrix}\bar{x}(t)\\ e(t)\\ \omega(t)\end{bmatrix} - V(T)\\ &< 0\end{aligned}$$

where the last step is obtained by using Schur complement lemma to inequality (17). Let $t \to \infty$ in the above result, we have

$$||z(t)||_2^2 < \gamma^2 ||\omega(t)||_2^2,$$

and equivalently, $||T_{z\omega}(s)||_\infty < \gamma$ holds. To summarize, we have proved that if the linear matrix inequality (17) is feasible, then the closed-loop multi-agent system can achieve consensus performance with H_∞ disturbance attenuation level γ. □

4 Conclusions

This paper addresses the event-triggered consensus problem of a nonlinear leader-following multi-agent system with external disturbances. By using the event-triggered control strategy, we have reduced the occupation of system resource and the frequency of the information update. Future work will focus on the leaderless nonlinear multi-agent system with external disturbances.

Acknowledgments This work was supported by the National Basic Research Program of China (973 Prgram, 2012CB821200, 2012CB821201), the NSFC (61473015, 61134005, 61327807, 61520106010).

References

1. Dimarogonas DV, Johansson KH (2009) Event-triggered control for multi-agent systems. In: Proceedings of the 48th IEEE Conference on Decision and control, CDC/CCC, 2009 held jointly with the 2009 28th Chinese control conference, pp 7131–7136
2. Dimarogonas DV, Frazzoli E, Johansson KH (2012) Distributed event-triggered control for multi-agent systems. IEEE Trans Autom Control 57(5):1291–1297
3. Meng X, Chen T (2013) Event based agreement protocols for multi-agent networks. Automatica 49(7):2125–2132
4. Chen X, Hao F (2012) Event-triggered average consensus control for discrete-time multi-agent systems. IET Control Theor Appl 16(6):2493–2498
5. Liu Z, Chen Z, Yuan Z (2012) Event-triggered average-consensus of multi-agent systems with weighted and direct topology. J Syst Sci Complex 25(5):845–855
6. Seyboth GS, Dimarogonas DV, Johansson KH (2013) Event-based broadcasting for multi-agent average consensus. Automatica 49(1):245–252
7. Huang N, Duan Z, Zhao Y (2014) Event-triggered consensus for heterogeneous multi-agent systems. In: The 33rd Chinese control conference, pp 1259–1264
8. Liu Y, Jia Y (2010) Consensus problem of high-order multi-agent systems with external disturbances: an H∞ analysis approach. Int J Robust Nonlinear Control 20(14):1579–1593
9. Liu Y, Jia Y (2012) H_∞ consensus control for multi-agent systems with linear coupling dynamics and communication delays. Int J Syst Sci 43(1):50–62
10. Li L, Ho DWC, Huang C et al (2014) Event-triggered discrete-time multi-agent consensus with delayed quantized information. In: The 33rd Chinese control conference, pp 1722–1727

11. Jiang H, Zhou H, Jia Y, et al (2013) Event-triggered robust H_∞ control for linear systems with disturbance. In: The 32nd Chinese control conference, pp 2102–2107
12. Yan H, Shen Y, Zhang H et al (2014) Decentralized event-triggered consensus control for second-order multi-agent systems. Neurocomputing 133:18–24
13. Hu J (2012) Second-order event-triggered multi-agent consensus control. In: The 31st Chinese control conference, pp 6339–6344

Integrated Design of Fault Diagnosis and Reconfiguration for Satellite Control System

Xijun Liu and Chengrui Liu

Abstract An integrated design of fault diagnosis and reconfiguration method is studied for actuator failures of satellite control system. First, the effectiveness factor and the controller gain are taken as a whole to design the combined control law, which is obtained by placing the closed-loop system poles according to the theory of regional pole assignment. And this paper uses perturbation linearization method to solve the generating bilinear matrix inequalities(BMI). Then the effectiveness factor is obtained with a two-stage Kalman filter and thus the control law can be solved. Finally, the effectiveness of the provided method is illustrated by a simulation.

Keywords Regional pole assignment · Fault diagnosis · Reconfiguration · Satellite control system · Effectiveness factor

1 Introduction

Nowadays the satellite is becoming more complex and large scale, which results in higher requirements in reliability and capability of independent operation of satellite control system. Plenty of reasons, such as frequent operation of actuator and harsh space environment, make it inevitable to fail on actuator. Actuator failure has been the main reason of failure in satellite control system [1]. However, the main fault-tolerant measure in satellite is component replacement, which means more weight and cost. How to get rid of dependence on hardware redundancy and make full use of satellite resource to realize the reconfiguration of control law is the problem that many experts dedicate to solve.

Fault diagnosis is generally recognized to originate from Beard's doctoral dissertation in 1971 [2]. From then on, plenty of fault diagnosis methods have been proposed. P.M. Frank thinks that all the methods can be divided into three

X. Liu (✉) · C. Liu
Beijing Institute of Control Engineering, Beijing 100190, China
e-mail: liuxijun9289@sina.com

Y. Jia et al. (eds.), *Proceedings of 2016 Chinese Intelligent Systems Conference*, Lecture Notes in Electrical Engineering 405,
DOI 10.1007/978-981-10-2335-4_44

categories: methods based on analytical model, methods based on knowledge, and methods based on signal processing [3]. Methods based on analytical model include methods of estimating parameters and methods of estimating states. Many scholars have made some progress based on these two methods [4–7]. The advantage of these two methods is that the failure can be described as a parameter or state, which can be easily estimated, and the estimation result can be used to adjust the control law to ensure excellent system performance. This is also one of the advantages of integrated design. Integrated design of fault diagnosis and reconfiguration can also ensure the real-time performance and rapidity of fault-tolerant control. Therefore, integrated design has drawn increasing attention.

Wang et al. [8] proposes an idea of integrated design based on quaternion. Han [9] designs an unknown input observer aiming at LTI system with disturbance. He designs an adaptive fault-tolerant control law according to obtained fault parameter. Dejun [10] uses state observer and Kalman filter simultaneously to estimate the information about the failure, which is used on following fault-tolerant control. Xueqin [11] estimates the effectiveness factor and state variable simultaneously based on bias-separated principle and uses the result to design a LQR to be the new controller.

This paper focuses on the satellite control system and studies an integrated design of fault diagnosis and reconfiguration method. First, the effectiveness factor and the controller gain are taken as a whole to design the combined control law, which is obtained by placing the closed-loop system poles according to the theory of regional pole assignment. This paper uses perturbation linearization method to solve the generating bilinear matrix inequalities (BMI). Then the effectiveness factor is obtained with a two-stage Kalman filter and thus the control law can be solved. Since all the poles of closed-loop system lie in the sector region, the closed-loop system's performances can be ensured. At the same time, the control law is adjusted in real-time according to the real-time estimated effectiveness factor, which ensures the real-time performance and rapidity. Finally, the effectiveness of the provided method is illustrated by a simulation.

2 Fault Modeling for Satellite Control System

Considering the rigid model, this paper takes thruster as the actuator. Choose the state variable as

$$x = \begin{bmatrix} \varphi & \theta & \psi & \dot{\varphi} & \dot{\theta} & \dot{\psi} \end{bmatrix}^T \tag{1}$$

So the state space model of the satellite control system can be written as

$$
\begin{aligned}
\dot{x} &= Ax + Bu + w \\
y &= Cx + v
\end{aligned} \tag{2}
$$

where

$$
A = \begin{bmatrix}
0 & 0 & 0 & 1 & 0 & 0 \\
0 & 0 & 0 & 0 & 1 & 0 \\
0 & 0 & 0 & 0 & 0 & 1 \\
\frac{I_z - I_y}{I_x}\omega_0^2 & 0 & 0 & 0 & 0 & \frac{I_x + I_z - I_y}{I_x}\omega_0 \\
0 & 0 & 0 & 0 & 0 & 0 \\
0 & 0 & \frac{I_x - I_y}{I_z}\omega_0^2 & \frac{I_y - I_x - I_z}{I_z}\omega_0 & 0 & 0
\end{bmatrix}, B = \begin{bmatrix}
0 & 0 & 0 \\
0 & 0 & 0 \\
0 & 0 & 0 \\
\frac{1}{I_x} & 0 & 0 \\
0 & \frac{1}{I_y} & 0 \\
0 & 0 & \frac{1}{I_z}
\end{bmatrix}
$$

$C = I_6$. ω_0 is orbit angular velocity, while $I_b = \mathrm{diag}(I_x, I_y, I_z)$ is the satellite's moment of inertial.

When failure occurs in the actuator, the state space model can be described as

$$
\begin{aligned}
\dot{x} &= Ax + B\Gamma u + w \\
y &= Cx + v
\end{aligned} \tag{3}
$$

while w, v are the uncorrelated white noise with zero mean and covariance matrix $\mathbf{W}$,$\mathbf{V}$, respectively. $\boldsymbol{\Gamma} = \mathrm{diag}(\gamma_1, \gamma_2, \gamma_3)$ is a diagonal matrix composed of three effectiveness factors. $\gamma_i (i = 1, 2, 3)$ describes the fault severity of the actuator. The main fault mode of actuators includes blockage and leakage. Blockage can be divided to partial blockage and complete blockage. When complete blockage occurs, component replacement should be taken into account. This paper aims to fault with partial blockage, that is to say the thrust produced by the thruster is smaller than the expected one. This condition can be described by the effectiveness factor as follows:

$$
T_{out} = \gamma T_0 \tag{4}
$$

where T_0 is the expected output moment. T_{out} is the actual output moment. γ is the effectiveness factor, whose value is between -1 and 1.

Considering that computer control system is adopted in the practical satellite control system, formula (3) requires a discretization. Suppose that sampling period is T, then the discrete model is as follows:

$$
\begin{aligned}
x(k+1) &= Fx(k) + G\Gamma u(k) + w_d \\
y(k) &= Cx(k) + v_d
\end{aligned} \tag{5}
$$

where $F = e^{AT}$, $G = \int_0^T e^{At} dt B$. w_d, v_d are the discrete noise, whose covariance matrices are $\mathbf{Q}$ and $\mathbf{R}$, respectively. Q and R satisfy

$$Q = Ew_d w_d^T = \int_0^T e^{At} W e^{A^T t} dt \qquad (6)$$
$$R = V$$

3 Integrated Design of Diagnosis and Reconfiguration

For system (3), an output feedback controller

$$u = Ky, \ K \in \kappa \qquad (7)$$

is designed to place poles of the closed-loop system

$$\dot{x} = (A + B\Gamma KC)x + w + B\Gamma Kv \qquad (8)$$

in the sector region as shown in Fig. 1. κ in formula (7) is a given set of bounded matrix, which is determined by physical characteristics of the actuator.

When all the closed-loop system's poles lie in the region as shown above, the system will have excellent rapid attenuation and smooth transition properties. It is easy to verify that (A, B) is controllable and (A, C) is observable for satellite control system, that is to say, it satisfies the condition of pole assignment. The theorem as follow can be easily obtained according to reference [12].

Theorem 1 *All the closed-loop system's poles lie in the region $S(\alpha, \theta, r)$ if and only if there exists a positive definite symmetric matrix such that*

Fig. 1 Sector region poles $S(\alpha, \theta, r)$

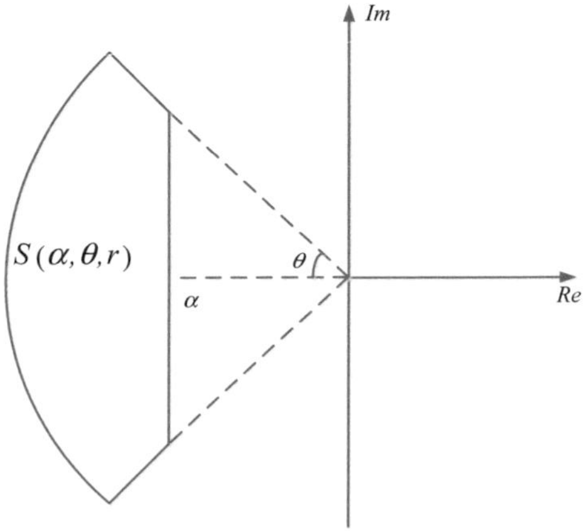

$$\begin{gathered}(A+B\Gamma KC)X+X(A+B\Gamma KC)^T+2\alpha X<0\\ \begin{bmatrix}-rX & (A+B\Gamma KC)X\\ X(A+B\Gamma KC)^T & -rX\end{bmatrix}<0\\ \begin{bmatrix}\sin\theta\left((A+B\Gamma KC)X+X(A+B\Gamma KC)^T\right) & \cos\theta\left((A+B\Gamma KC)X-X(A+B\Gamma KC)^T\right)\\ \cos\theta\left(X(A+B\Gamma KC)^T-(A+B\Gamma KC)X\right) & \sin\theta\left((A+B\Gamma KC)X+X(A+B\Gamma KC)^T\right)\end{bmatrix}<0\end{gathered} \tag{9}$$

It is easy to see from formula (9) that ΓK can be seen as a variable, which can be defined as $K^*=\Gamma K\cdot K^*$, called combined control law, can be obtained from solving formula (9), and the effectiveness factor can be obtained by some estimation methods. Thus, the controller gain K can be solved by $K=\Gamma^{-1}K^*$. The advantage is that the combined control law can be designed offline. In-orbit satellite just needs estimating the effectiveness factor Γ and adjusting the control law. This can reduce computing greatly while ensuring excellent system performance.

4 Solution to Combined Control Law Based on BMI

Since formula (9) is not standard LMIs, the LMI toolbox in MATLAB is not available. A new method is introduced to solve this problem as follows [13]:

Step 1: Let $j=1$, and take any initial combined control law $K_j^*=\Gamma K\in\kappa$

Step 2: Let $A_j=A+BK_j^*C$, solve the minimum problem with LMIs constraints as follows:

min t: t, X_1, X_2, X_3 subject to

$$\begin{gathered}A_jX_1^j+X_1^jA_j^T+2\alpha X_1^j<tI\\ \begin{bmatrix}-rX_2^j-tI & A_jX_2^j\\ X_2^jA_j^T & -rX_2^j-tI\end{bmatrix}<0\\ \begin{bmatrix}\sin\theta\left(A_jX_3^j+X_3^jA_j^T\right)-tI & \cos\theta\left(A_jX_3^j-X_3^jA_j^T\right)\\ \cos\theta\left(X_3^jA_j^T-A_jX_3^j\right) & \sin\theta\left({}_jX_3^j+X_3^jA_j^T\right)-tI\end{bmatrix}<0\\ t>-\varepsilon_1\\ X_i^j>I,\ i=1,2,3\end{gathered} \tag{10}$$

where ε_1 is a small positive number, such as 0.01.

The obtained minimum can be defined as $t_j=\min t$. If $t_j<0$, and all the poles of A_j lie in $S(\alpha,\theta,r)$, then K_j^* is the required value and the output controller gain $K=\Gamma^{-1}K_j^*$; otherwise return Step 1.

Step 3: If $t_j \geq 0$, let $X_i = X_i^j + \boldsymbol{\delta} X_i (i = 1, 2, 3)$, $K^* = K_j^* + \delta K^*$, and solve the perturbation linearization minimum problem of formula (9) concerning X_1, X_2, X_3, K^*

min t: δX_1, δX_2, δX_3, δK^* subject to

$$
\begin{aligned}
& \left[(A_j + \alpha I)\left(X_1^j + \delta X_1\right) + B\delta K^* C X_1^j \right] + [*]^T < tI \\
& \begin{bmatrix} -r\left(X_2^j + \delta X_2\right) - tI & A_j\left(X_2^j + \delta X_2\right) + B\delta K^* C X_1^j \\ * & -r\left(X_2^j + \delta X_2\right) - tI \end{bmatrix} < 0 \\
& \begin{bmatrix} \sin\theta\left(\left[A_j\left(X_3^j + \delta X_3\right) + B\delta K^* C X_3^j\right] + [*]^T\right) - tI & \psi_{12} \\ * & (1,1) \end{bmatrix} < 0 \\
& t > -\varepsilon_1 \\
& -0.2X_i^j < \delta X_i < 0.2X_i^j, i = 1, 2, 3 \\
& \left(\delta K^*\right)^T \delta K^* < 0.16\mathbf{I} \\
& \psi_{12} = \cos\theta\left(A_j\left(X_3^j + \delta X_3\right) - \left(X_3^j + \delta X_3\right)A_j^T + B\delta K^* C X_3^j - X_3^j\left(B\delta K^* C\right)^T\right)
\end{aligned}
\tag{11}
$$

where $[*]^T$ represents the same part as previous bracket. (1, 1) represents the same part as the element (1, 1) of matrix.

Step 4: If $\left|t_{j-1} - t_j\right| > \varepsilon_2$, and $j < N$(N is the maximum number of iteration), then let $K_{j+1}^* = K_j^* + \delta K^*$, $j = j + 1$, and return Step 2; otherwise return Step 1.

Suppose that t_0 is a large enough and positive number and ε_2 is a small positive number, such as 0.0001.

The algorithm above may converge to different local optimal point according to different initial value. Therefore, it is necessary to verify that whether all the poles of closed-loop system with combined control law K^* lie in the expected region. That whether K^* is the global optimal point is not that meaningful. So there exist many solutions for K^*, just take any one of them.

5 Estimation of Actuator Failure Effectiveness Factor

Since $\boldsymbol{\Gamma}$ is a diagonal matrix while $u(k)$ is a column vector, part of formula (5) can be transformed as follows:

$$
G\Gamma u(k) = G\begin{bmatrix} \gamma_1 & & \\ & \gamma_2 & \\ & & \gamma_3 \end{bmatrix}\begin{bmatrix} u_1 \\ u_2 \\ u_3 \end{bmatrix} = G\begin{bmatrix} u_1 & & \\ & u_2 & \\ & & u_3 \end{bmatrix}\begin{bmatrix} \gamma_1 \\ \gamma_2 \\ \gamma_3 \end{bmatrix} = GU\gamma \tag{12}
$$

So system (5) transforms into the following system:

$$\begin{aligned} x(k+1) &= Fx(k) + GU(k)\gamma + w_d \\ y(k) &= Cx(k) + v_d \end{aligned} \tag{13}$$

Based on bias-separated principle [6], a two-stage Kalman filter is used to estimate the effectiveness factor. The idea is that the effectiveness factor is taken as constant deviation to be estimated. The estimation result composes part of fault-tolerant control.

The iterative process to solve the effectiveness factor is shown as follows:

(1) Estimation of the effectiveness factor

$$\begin{aligned} \hat{\gamma}_{k+1|k} &= \hat{\gamma}_{k|k} \\ P^a_{k+1|k} &= P^a_{k|k} \\ \hat{\gamma}_{k+1|k+1} &= \hat{\gamma}_{k+1|k} + K^a_{k+1}\left(y_{k+1} - C\tilde{x}_{k+1|k} - H_{k+1|k}\hat{\gamma}_{k|k}\right) \\ K^a_{k+1} &= P^a_{k+1|k}H^T_{k+1|k}\left(H_{k+1|k}P^a_{k+1|k}H^T_{k+1|k} + CP^x_{k+1|k}C^T + R\right)^{-1} \\ P^a_{k+1|k+1} &= \left(I - K^a_{k+1}H_{k+1|k}\right)P^a_{k+1|k} \end{aligned} \tag{14}$$

(2) Unbiased estimation of state variable

$$\begin{aligned} \tilde{x}_{k+1|k} &= F\tilde{x}_{k|k} \\ P^x_{k+1|k} &= FP^x_{k|k}F^T + Q \\ \tilde{x}_{k+1|k+1} &= \tilde{x}_{k+1|k} + K^x_{k+1}\left(y_{k+1} - C\tilde{x}_{k+1|k}\right) \\ K^x_{k+1} &= P^x_{k+1|k}C^T\left(CP^x_{k+1|k}C^T + R\right)^{-1} \\ P^x_{k+1|k+1} &= \left(I - K^x_{k+1}C\right)P^x_{k+1|k} \end{aligned} \tag{15}$$

(3) Coupling relationship between state variable and the effectiveness factor

$$\begin{aligned} M_{k+1|k} &= FM_{k|k} + GU \\ H_{k+1|k} &= CM_{k+1|k} \\ M_{k+1|k+1} &= M_{k+1|k} - K^x_{k+1}H_{k+1|k} \end{aligned} \tag{16}$$

The effectiveness factor for multiplicative fault can be obtained by iterative process above, which can be used to solve the required output controller gain.

6 Simulation Verification

The satellite attitude dynamics model and a two-stage Kalman filter are built in MATLAB to verify the effectiveness of algorithm above. The effectiveness factor obtained from the two-stage Kalman filter and the combined control law are used to solve the controller gain in real-time. The simulation parameters

(1) Satellite's moment of inertial

$$I = \begin{bmatrix} 12.49 & 0.67 & 0.06 \\ 0.67 & 13.85 & 0.06 \\ 0.06 & 0.06 & 15.75 \end{bmatrix}$$

(2) Initial attitude angular velocity $\boldsymbol{\omega}_0 = [0.1 \quad 0.1 \quad 0.1]\text{deg/s}$
(3) Initial attitude angular $\alpha_0 = [3 \quad 4 \quad 5]\text{deg}$
(4) Orbit height H = 200 km
(5) Covariance matrices

$$W = diag([0.01^2 \quad 0.01^2 \quad 0.01^2 \quad 0.01^2 \quad 0.01^2 \quad 0.01^2])$$
$$V = 0.1^2 W$$

(6) Fault setting

$$\Gamma = \begin{bmatrix} 0.4 & & \\ & -0.8 & \\ & & 0.6 \end{bmatrix}$$

(7) Region pole assignment parameter setting

$$\alpha = 1.5,\ r = 5,\ \theta = \frac{\pi}{4}$$

After applying algorithm above, the result is as follows:

$$K^* = \begin{bmatrix} -31.8920 & 3.4354 & -12.3737 & -40.3101 & 2.5565 & -8.3189 \\ -2.6160 & -30.2464 & 10.9594 & -0.8283 & -41.5054 & 8.4730 \\ 10.4997 & -10.0078 & -28.7448 & 6.3435 & -6.4272 & -44.5035 \end{bmatrix}$$

If no measure is taken after fault occurs in the actuator, the satellite's attitude angle changes as Fig. 2. If fault-tolerant control above is adopted, the satellite's attitude angle changes as Fig. 3.

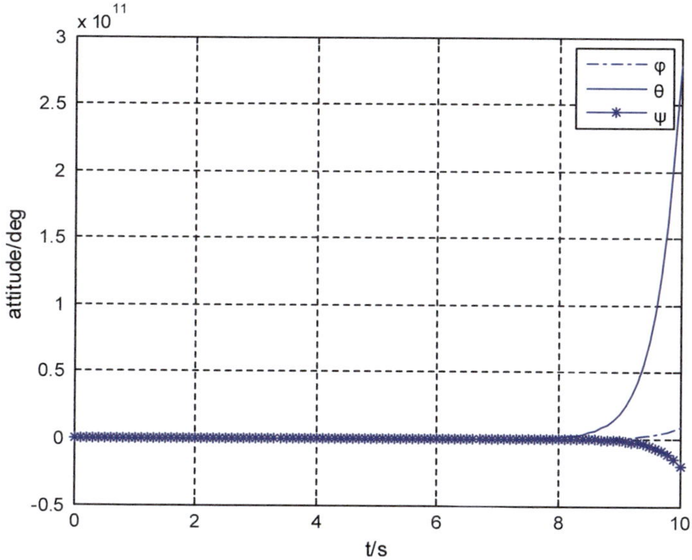

Fig. 2 The satellite's attitude angle if no measure is taken

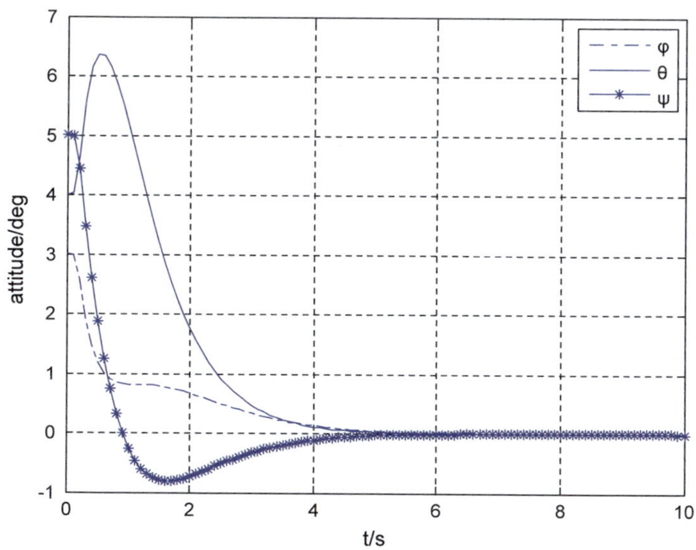

Fig. 3 The satellite's attitude angle if fault-tolerant control is taken

The estimation of effectiveness factor is as follows:

$$\Gamma = \begin{bmatrix} 0.3998 & & \\ & -0.7998 & \\ & & 0.6001 \end{bmatrix}$$

It is evident from Figs. 2 and 3 to draw the following conclusions. (1) Serious multiplicative fault may lead to the attitude's divergence. (2) After taking fault-tolerant measure above, the closed-loop system's performance is much better. (3) Given more simulations, slight fault $(0 < \gamma \leq 1)$ will lead to degradation of system performance, such as larger overshoot, longer setting time. But this will not lead to the attitude's divergence. The result of simulation above can effectively demonstrate the effectiveness of the provided method.

7 Conclusion

This paper researches an integrated design of fault diagnosis and reconfiguration method based on region pole assignment and two-stage Kalman filter. First, the effectiveness factor and the controller gain are taken as a whole to design the combined control law, and the BMIs produced by region pole assignment can be effectively solved by the perturbation linearization method. The obtained combined control law can ensure the closed-loop system with good performance. Then a two-stage Kalman filter is used to estimate the effectiveness factor and the output feedback control law is available now. This method combines the fault diagnosis with design of control law, which ensures the real-time and rapidity performance. At the same time, the combined control law can be designed offline, which greatly reduces computing on orbit. Finally, the simulation demonstrates the effectiveness of the provided method.

References

1. Ma G, Jiang Y, Qinglei H (2010) Time delay backstepping based fault tolerant attitude control of satellites. Acta Aeronautica et Astronautica Sinica 31(5):1066–1073
2. Beard RV (1971) Failure accommodation in linear system through self-reorganization. Report MVT-71-1, Man Vehicle Lab, MIT, Cambridge, Massachusetts, 1971
3. Frank PM () Analytical and qualitative model based fault diagnosis—a survey and some new result. Eur J Control
4. Zhang K, Jiang B (2010) Fault diagnosis observer-based output feedback fault tolerant control design. Acta Automatica Sinica 36(2):274–281
5. Xueqin C, Yingchun Z, Yunhai G, Huayi L (2007) Satellite attitude control system on-orbit reconfigurable faul-tolerant control based on the control effectiveness factor. J Astronautics 28 (1):94–98

6. Xueqin C, Feng W, Yingchun Z, Yunhai G (2009) Application of effectiveness factors to integrated FDD and FTC of satellite attitude control systems. Acta Aeronautica et Astronautica Sinica 30(3):476–483
7. Li D, Chen X, Li C (2015) Fault diagnosis of satellite actuator based on bias-separated theory. Syst Eng Electron 37(3):606–612
8. Wang F, Zong Q, Tian B (2014) Robust adaptive back-stepping flight control design for reentry RLV. Control Decis 29(1):12–18
9. Han J (1995) Extended state observer for a class of uncertain objects. Control Decis 10 (1):85–88
10. Dejun W (2004) Study on schemes of fault detection and diagnosis & fault tolerant control. Jilin University
11. Xueqin C (2008) The research of integrated fault detection and diagnosis and fault-tolerant control with its application in satellite attitude control. Harbin Institute of Technology
12. Yingmin J (2007) Robust H_{∞} control. Science Press, Beijing
13. Wang Y (2008) BMI-based output feedback control design with sector pole assignment. Acta Automatica Sinica 34(9):1192–1195

Distributed Optimization Over Weight-Balanced Digraphs with Event-Triggered Communication

Xiaowei Pan, Zhongxin Liu and Zengqiang Chen

Abstract This paper extends the event-triggered communication in consensus problems of multi-agent systems to the case of distributed continuous-time convex optimization over weight-balanced digraphs. We address problems whose global objective functions are a sum of local functions associated to each agent. We utilize the event-triggered communication technique to reduce the communication load and avoid Zeno behavior meanwhile. Based on Lyapunov approach, we prove that the Zero-Gradient-Sum (ZGS) algorithm combined with the event-triggered communication makes all agents' states converge to the optimal solution of the global objective function exponentially fast.

Keywords Distributed optimization · Zero-gradient-sum algorithm · Multi-agent system · Event-triggered · Consensus · Weight-balanced digraphs

1 Introduction

Distributed optimization based on multi-agent networks has attracted lots of attention in recent years due to its wide applications in the real world, such as in sensor networks [1], large-scale machine learning [2], and distributed parameter estimation [3]. Those problems share a common feature that they all can be cast as

---

X. Pan · Z. Liu (✉) · Z. Chen
College of Computer and Control Engineering, Nankai University,
Tianjin 300353, China
e-mail: lzhx@nankai.edu.cn

X. Pan · Z. Liu · Z. Chen
Tianjin Key Laboratory of Intelligent Robotics, Tianjin 300353, China

Z. Chen
College of Science, Civil Aviation University of China, Tianjin 300300, China

Y. Jia et al. (eds.), *Proceedings of 2016 Chinese Intelligent Systems Conference*, Lecture Notes in Electrical Engineering 405,
DOI 10.1007/978-981-10-2335-4_45

an optimization problem with a global objective function described as the sum of local functions, each of which is known only to one agent in the network. Agents need cooperate to solve the problem.

Many algorithms have been proposed to tackle this kind of problem. In [4], the distributed subgradient algorithm incorporating consensus-like protocol is proposed to solve unconstrained problems (see also [5, 6]) for the first time. Projection operator is applied to deal with constrained problems (see [7–9]) based on the work in [4]. They are all implemented in discrete time. Recently, a few works on continuous-time dynamics for distributed optimization are presented in [10, 11, 12, 13]. Those continuous-time algorithms can be analyzed via classical stability theory such as Lyapunov approach. In [11], the authors solve an equivalent form of the primal problem using the saddle points theory. The authors of [12] use sign functions to design the control law. All agents reach consensus in finite time and converge to the optimal solution together. The paper [13] designs an algorithm for a high-order continuous-time system. However, continuous communication is needed between neighbors in the realization process of the continuous-time algorithms. It may result in great communication burden, occupy lots of computation capacity, and also lead to masses of energy consumption. This is not desirable because each node is usually equipped with limited energy resource.

Event-triggered communication has been widely studied in recent years, especially in multi-agent systems (see [14–17]), in which the communication process is determined by the trigger condition with a prescribed threshold. However, only few works have been done for incorporating event-triggered schemes into distributed continuous-time optimization (see [18–21]). In [18], the authors add event-driven strategy to the algorithm proposed in [4]. The paper [19] studies the discrete communication in continuous-time algorithms based on the work of [11]. Both [20, 21] are built on the work in [10], while [20] takes a sample-based scheme with a constant sample period, and [21] designs an event-triggered communication law based on the idea from [16].

In this paper, in the first step, we combine the event-triggered scheme in [17] with the Zero-Gradient-Sum algorithm (ZGS) in [10]. We prove that these communication principles are well suitable for ZGS, with all states of agents converging to the optimal solution exponentially fast. Furthermore, we observe that a variation of the former event-triggered scheme can also be used in ZGS. The work [21] is most related to ours here. What differs is that, in [21], it needs to continuously monitor agents' states in order to check the trigger condition constantly, while we predict the next trigger time by solving optimization problems and only check the trigger condition when needed. The trigger conditions are different, too. Our proof is formulated almost the same as those in the literature, but in a different perspective, where we proceed without the assumption of receiving information from out-neighbors and sending information to in-neighbors. By doing this, we see the possibility to include event-triggered schemes in distributed continuous-time optimization algorithms.

The rest of the paper is organized as follows: Sect. 2 presents preliminaries and problem statement. Section 3 gives the algorithm with event-triggered communication and provides the proof of exponential convergence to the optimal solution of the problem. Section 4 illustrates the algorithm's behavior through simulations. Finally, Sect. 5 concludes the paper.

2 Preliminaries and Problem Statement

This section presents some preliminaries and notational conventions. Let $\mathbb{R}$, $\mathbb{R}_{>0}$, $\mathbb{R}_{\geq 0}$, and $\mathbb{Z}_{>0}$ denote the set of real, positive real, nonnegative real, and positive integer numbers, respectively. The transpose of a matrix A is A^{T}. We let $\mathbb{R}^m$ denote the set of $m \times 1$ real vectors, $\mathbb{R}^{n \times n}$ denote $n \times n$ real matrices, and I_n denote the identity matrix of dimension n. For two matrices A, B, we let $A \otimes B$ denote their Kronecker product. For $z \in \mathbb{R}^m$, $\|z\| = \sqrt{z^{\mathrm{T}} z}$ denotes the standard Euclidean norm. For vectors $z_1, \ldots, z_n$, $z = \left[z_1^{\mathrm{T}}, \ldots, z_n^{\mathrm{T}}\right]^{\mathrm{T}}$ is the aggregated vector. Let ∇f and $\nabla^2 f$ represent the gradient and the Hession of f, respectively.

2.1 Preliminaries

A digraph of n nodes is used to model a communication network. A weighted digraph is defined as $\mathcal{G} \triangleq (\mathcal{V}, \mathcal{E}, A)$, where $\mathcal{V} = \{\upsilon_1, \ldots, \upsilon_n\}$ is the set of nodes with the node indices belonging to a finite index set $\mathcal{I} = \{1, 2, \ldots, n\}$, $\mathcal{E} \subseteq \mathcal{V} \times \mathcal{V}$ is the set of edges, and $A = \left[a_{ij}\right]_{n \times n}$ is the weighted adjacency matrix. Given an edge $(\upsilon_j, \upsilon_i) \in \mathcal{E}$, we refer to υ_j as an in-neighbor of υ_i and υ_i as an out-neighbor of υ_j. A path from υ_j to υ_i is an ordered sequence of edges with the form $(\upsilon_j, \upsilon_{j_1}), (\upsilon_{j_1}, \upsilon_{j_2}), \cdots, (\upsilon_{j_m}, \upsilon_i)$, where $j_1, \ldots, j_m \in \mathcal{I}$, $m \in \mathbb{Z}_{>0}$. A digraph is strongly connected if there is a path between any pair of distinct nodes. The weighted adjacency matrix $A \in \mathbb{R}^{n \times n}$ satisfies $a_{ij} > 0$ if $(\upsilon_j, \upsilon_i) \in \mathcal{E}$ and $a_{ij} = 0$ otherwise.

The sets of in- and out-neighbors of a given node υ_i are defined as $\mathcal{N}_i^{in} = \{\upsilon_j \in \mathcal{V} \,|\, a_{ij} > 0\}$ and $\mathcal{N}_i^{out} = \{\upsilon_j \in \mathcal{V} \,|\, a_{ji} > 0\}$, respectively. The in-degree matrix is $D^{in} = diag\{d_1^{in}, \cdots, d_n^{in}\}$ and the out-degree matrix is $D^{out} = \mathrm{diag}\{d_1^{out}, \ldots, d_n^{out}\}$, where $d_i^{in} = \sum_{j=1}^{n} a_{ij}$ and $d_i^{out} = \sum_{j=1}^{n} a_{ji}$ for $i, j \in \mathcal{I}$. The weighted Laplacian matrix L is defined as $L = A - D^{in}$. For a strongly connected and weight-balanced digraph, zero is a simple eigenvalue of L and $L + L^{\mathrm{T}}$, and $L_s = -(L + L^{\mathrm{T}})$ is positive semidefinite. We order the eigenvalues of L_s as $0 = \lambda_1 < \lambda_2 \leq \cdots \leq \lambda_n$.

Note: Here the definition of L is different, but it does not change the property of L for weight-balanced digraphs except that there exists a minus. The reason why we use D^{in} to define L is that we disregard the assumption of receiving information from out-neighbors and sending information to in-neighbors for each agent.

Lemma 1 ([22]) *A twice continuously differentiable function* $f:\mathbb{R}^m \to \mathbb{R}$ *is strongly convex on* $\mathbb{R}^m$, *if there exists a constant* $\theta > 0$ *such that, for* $\forall x, y \in \mathbb{R}^m$, *the following equivalent conditions hold*

$$f(y) - f(x) - \nabla f(x)^{\mathrm{T}}(y - x) \geq \frac{\theta}{2}\|y - x\|^2,$$
$$(\nabla f(y) - \nabla f(x))^{\mathrm{T}} \geq \theta\|y - x\|^2,$$
$$\nabla^2 f(x) \geq \theta I_m.$$

Lemma 2 ([23]) *Assume* $f:\mathbb{R}^m \to \mathbb{R}$ *is a twice continuously differentiable strongly convex function on* $\mathbb{R}^n$ *and a sublevel set* $\mathcal{S} = \{x \in \mathbb{R}^m \,|\, f(x) \leq f(x_0)\}$, *where* x_0 *is a suitable starting point such that* $\mathcal{S}$ *must be closed. From (1),* $\mathcal{S}$ *is bounded, i.e.,* $\mathcal{S}$ *is a compact set, then there exists a constant* $\xi > 0$ *such that, for* $\forall x, y \in \mathcal{S}$, *the following equivalent conditions hold*

$$f(y) - f(x) - \nabla f(x)^{\mathrm{T}}(y - x) \leq \frac{\xi}{2}\|y - x\|^2,$$
$$\nabla^2 f(x) \leq \xi I_n.$$

2.2 Problem Statement

Consider a network of n agents interacting over a strongly connected and weight-balanced digraph $\mathcal{G}$. Each agent $i \in \mathcal{I}$ has a local cost function $f_i:\mathbb{R}^m \to \mathbb{R}$. The goal is to minimize the global objective function which is $F(x) = \sum_{i=1}^{n} f_i(x)$, where $x \in \mathbb{R}^m$ is the decision vector. To solve this problem in a distributed manner, let each agent keep an local estimation of x denoted as x_i for υ_i. So the problem is equivalent to that all agents reach consensus on the optimal solution of $F(x)$ finally.

Assumption 1 ([10]) For each $i \in \mathcal{I}$, the individual function f_i is twice continuously differentiable, strongly convex with convexity parameter $\mu_i > 0$ and has a locally Lipschitz Hessian $\nabla^2 f_i$.

Proposition 1 ([10]) *With Assumption* 1 , *there exists a unique* $x^* \in \mathbb{R}^m$ *such that for any* $x \in \mathbb{R}^m$, $F(x^*) \leq F(x)$ *and* $\nabla F(x^*) = 0$.

Lemma 3 ([10]) *Suppose A and B are two* $n \times n$ *positive semidefinite matrices, and have the same null space* $\Omega = \{x \,|\, Ax = 0\} = \{x \,|\, Bx = 0\}$. *Then, there exists a positive constant* μ *such that* $\mu A \leq B$.

Lemma 4 ([24]) *For any* $x, y \in \mathbb{R}$, *Young's inequality states that, for any* $\eta \in \mathbb{R}_{>0}$, $xy \leq \frac{1}{2\eta}x^2 + \frac{\eta}{2}y^2$.

A standard assumption in the classical control theory is that the data transmission required by the control or state estimation algorithm can be performed with infinite precision. However, due to the growth in communication technology, it is becoming more common to employ digital limited capacity communication networks for exchange of information between system components.

3 Zero-Gradient-Sum Algorithm with Event-Triggered Communication

In this section, we present our algorithm to solve the problem over strongly connected and weight-balanced digraph. First, the continuous-time ZGS algorithm for distributed optimization proposed by [10] is introduced as follows:

$$\begin{aligned} \dot{x}_i(t) &= (\nabla^2 f_i(x_i(t)))^{-1} u_i(t), \\ u_i(t) &= \sum_{\upsilon_j \in \mathcal{N}_i} a_{ij}(x_j(t) - x_i(t)), \\ x_i(0) &= x_i^*, \ i \in \mathcal{I} \end{aligned} \tag{1}$$

where $x_i(0)$ is the initial state of agent i and x_i^* is the minimizer of $f_i(x)$. This algorithm is designed for undirected graphs, but we can extend it to the weight-balanced digraphs (see [20]).

The execution of the aforementioned control strategy requires each agent communicates with their neighbors continuously. This may result in great communication burden, occupy lots of computation capacity, and also lead to masses of energy consumption. Therefore, it is crucial to bring down the frequency of unnecessary communication among agents.

3.1 Event-Triggered Communication Principles

The communication will be triggered at discrete time, if some predefined conditions are satisfied. The sequence of event-triggered executions is denoted as $\{t_k^j\}_{k=1}^{\infty}$, $j \in \mathcal{I}$. Meanwhile, the agents are triggered to update their controls. The event-triggered control law is presented as $u_i(t) = \sum_{j=1}^n L_{ij} x_j(t^j_{k_j(t)})$, $i \in \mathcal{I}$, where $k_j(t) = \arg\max_k \{t_k^j \le t\}$.

We say that agent υ_i is triggered at $t = t_k^i$ if υ_i renews its state and broadcasts the new state $x_i(t_k^i)$ to all its out-neighbors at the same time. Let $x(t) = [x_1^{\mathrm{T}}(t), \ldots, x_n^{\mathrm{T}}(t)]^{\mathrm{T}}$, $\hat{x}_i(t) = x_i(t^i_{k_i(t)})$ and $\hat{x}(t) = [\hat{x}_1^{\mathrm{T}}(t), \ldots, \hat{x}_n^{\mathrm{T}}(t)]^{\mathrm{T}}$, $e_i(t) = \hat{x}_i(t) - x_i(t)$ and $e(t) = [e_1^{\mathrm{T}}(t), \ldots, e_n^{\mathrm{T}}(t)]^{\mathrm{T}}$, $q_i(t) = \sum_{j=1}^n L_{ij} \|x_j(t) - x_i(t)\|^2$ and $\hat{q}_i(t) = \sum_{j=1}^n L_{ij} \|\hat{x}_j(t) - \hat{x}_i(t)\|^2$.

The designed principles should be free from Zeno behavior. That is, no infinite amount of communication events can occur during any finite period of time. Inspired by the work [17], we combine the proposed scheme in the paper with the distributed continuous-time optimization algorithm called ZGS.

The triggering condition is stated as follows: For agent υ_i, choose constants $\gamma_i \in (0, 1)$ and $0 < \varepsilon_i < \theta_i \sqrt{\gamma_i}/(2\,|L_{ii}|)$. With $t_1^i, \ldots, t_k^i$, the following triggering strategy is used to find t_{k+1}^i.

(a) Let $\tau_{k+1}^i = \max\{\tau \geq t_k^i : g_i(e_i(t)) \leq 0,\ \forall t \in [t_k^i, \tau]\}$, where $g_i(e_i(t)) = |e_i(t)| - \sqrt{\frac{\gamma_i}{4|L_{ii}|}\hat{q}_i(t)}$.

(b) If agent υ_i receives renewed information from its in-neighbors at some time $t \in (t_k^i, \tau_{k+1}^i]$ and either $t - t_k^i < \varepsilon_i$ or $g_i(e_i(t)) > 0$ is satisfied, then $t_{k+1}^i = t$. Otherwise $t_{k+1}^i = \tau_{k+1}^i$.

So, the control protocol with event-triggered communication is obtained as follows:

$$u(t) = L\hat{x}(t). \tag{2}$$

where $u(t) = [u_1^{\mathrm{T}}(t), \ldots, u_n^{\mathrm{T}}(t)]^{\mathrm{T}}$. The algorithm can be expressed in the following matrix form by replacing the $u_i(t)$ in (1) with (2):

$$\begin{aligned} \dot{x}(t) &= (\Gamma(x(t)))^{-1}(L \otimes I_n)\hat{x}(t), \\ x(0) &= x_0^*, \end{aligned} \tag{3}$$

where $\Gamma(x(t)) = \mathrm{diag}\{\nabla^2 f_1(x_1(t)), \ldots, \nabla^2 f_n(x_n(t))\}$.

3.2 Main Results

Theorem 1 *Suppose that $\mathcal{G}$ is strongly connected and weight-balanced. Under Assumption* 1, *with the algorithm above mentioned, we can get*

(a) *The states of all agents exponentially converge to the optimal solution x^* of the problem.*

(b) *In finite period of time, no agent will be triggered infinitely, i.e., the network is guaranteed to avoid Zeno behavior.*

Proof (a) The variable t is omitted for convenience and the proof is always the same. Choose the Lyapunov function as

$$V(x) = \sum_{i=1}^{n} (f_i(x^*) - f_i(x_i) - \nabla f_i(x_i)^{\mathrm{T}}(x^* - x_i)). \tag{4}$$

The time derivative of V is

$$\begin{aligned}\dot{V} &= \sum_{i=1}^{n} \left(-\nabla f_i(x_i)^{\mathrm{T}} \dot{x}_i - (\nabla^2 f_i(x_i)\dot{x}_i)^{\mathrm{T}}(x^* - x_i) + \nabla f_i(x_i)^{\mathrm{T}} \dot{x}_i \right) \\ &= \sum_{i=1}^{n} (\nabla^2 f_i(x_i)\dot{x}_i)^{\mathrm{T}}(x^* - x_i) \\ &= (x^*)^{\mathrm{T}} \sum_{i=1}^{n} \nabla^2 f_i(x_i)\dot{x}_i - \sum_{i=1}^{n} (\nabla^2 f_i(x_i)\dot{x}_i)^{\mathrm{T}} x_i.\end{aligned} \tag{5}$$

From the designed control law, it is obvious that $\nabla^2 f_i(x_i)\dot{x}_i = \sum_{j=1}^{n} L_{ij}\hat{x}_j$. So, the first item of (5) equals that $\sum_{i=1}^{n} \nabla^2 f_i(x_i)\dot{x}_i = \sum_{i=1}^{n}\sum_{j=1}^{n} L_{ij}\hat{x}_j = 0$, according to the property of the Laplacian L of weight-balanced digraphs. The second item of (5) is $\sum_{i=1}^{n} (\nabla^2 f_i(x_i)\dot{x}_i)^{\mathrm{T}} x_i = \sum_{i=1}^{n} (\sum_{j=1}^{n} L_{ij}\hat{x}_j)^{\mathrm{T}} x_i = \sum_{i=1}^{n}\sum_{j=1}^{n} L_{ij} x_i^{\mathrm{T}} \hat{x}_j = x^{\mathrm{T}}(L \otimes I_n)\hat{x}$.

So we get that

$$\begin{aligned}\dot{V} &= x^{\mathrm{T}}(L \otimes I_n)\hat{x} = (\hat{x} - e)^{\mathrm{T}}(L \otimes I_n)\hat{x} \\ &= \hat{x}^{\mathrm{T}}(L \otimes I_n)\hat{x} - e^{\mathrm{T}}(L \otimes I_n)\hat{x} \\ &= \frac{1}{2}\hat{x}^{\mathrm{T}}[(L + L^{\mathrm{T}}) \otimes I_n]\hat{x} - e^{\mathrm{T}}(L \otimes I_n)\hat{x}.\end{aligned} \tag{6}$$

Denoting $0 < \sigma = \sup\{\zeta : \zeta L^{\mathrm{T}} L \leq -(L^{\mathrm{T}} + L)\}$ and employing Young's inequality in the second item of (6), we have

$$-e^{\mathrm{T}}(L \otimes I_n)\hat{x} \leq \frac{1}{4}\hat{x}^{\mathrm{T}}[-(L + L^{\mathrm{T}}) \otimes I_n]\hat{x} + \frac{1}{\sigma} e^{\mathrm{T}} e. \tag{7}$$

Invoking the triggering condition, it is straightforward that $|e_i| \leq \sqrt{\frac{\gamma_i}{4|L_{ii}|}\hat{q}_i}$ or $|e_i| = 0$.

For the first case, we have

$$\begin{aligned}\frac{1}{\sigma} e^{\mathrm{T}} e &= \frac{1}{\sigma}\sum_{i=1}^{n} \|e_i\|^2 \leq \sum_{i=1}^{n} \frac{\gamma_i}{4\sigma|L_{ii}|}\hat{q}_i \\ &\leq \frac{\gamma_{\max}}{4\sigma\ell_{\min}} \sum_{i=1}^{n}\sum_{j=1}^{n} L_{ij}\|\hat{x}_j - \hat{x}_i\|^2 \\ &= \frac{\gamma_{\max}}{4\sigma\ell_{\min}} \sum_{i=1}^{n}\sum_{j=1}^{n} L_{ij}(\|\hat{x}_i\|^2 + \|\hat{x}_j\|^2 - 2\hat{x}_i^{\mathrm{T}}\hat{x}_j) \\ &= \frac{\gamma_{\max}}{4\sigma\ell_{\min}} \hat{x}^{\mathrm{T}}[-(L + L^{\mathrm{T}}) \otimes I_n]\hat{x},\end{aligned} \tag{8}$$

the last equation is obtained by using the property of the Laplacian L of weight-balanced digraph.

Substituting (7) together with (8) for the second item in (6) yields

$$\begin{aligned}\dot{V} &\leq \frac{1}{2}\hat{x}^{\mathrm{T}}[(L+L^{\mathrm{T}})\otimes I_n]\hat{x}+\frac{1}{4}\hat{x}^{\mathrm{T}}[-(L+L^{\mathrm{T}})\otimes I_n]\hat{x}+\frac{\gamma_{\max}}{4\sigma\ell_{\min}}\hat{x}^{\mathrm{T}}[-(L+L^{\mathrm{T}})\otimes I_n]\hat{x}\\ &= (\frac{1}{4}-\frac{\gamma_{\max}}{4\sigma\ell_{\min}})\hat{x}^{\mathrm{T}}[(L+L^{\mathrm{T}})\otimes I_n]\hat{x}.\end{aligned} \tag{9}$$

If $\frac{\gamma_{\max}}{\sigma\ell_{\min}}<1,\ \dot{V}\leq 0$.

For the second case, it is obvious that $\dot{V}\leq\frac{1}{4}\hat{x}^{\mathrm{T}}[(L+L^{\mathrm{T}})\otimes I_n]\hat{x}\leq 0$.

So what we get is that it means $\dot{V}\leq 0$ if $\frac{\gamma_{\max}}{\sigma\ell_{\min}}<1$. From LaSalle's invariance principle, we deduce that all $\hat{x}_i(t)$ reach consensus on a constant vector π, i.e., $\lim_{t\to\infty}\hat{x}_i(t)=\pi$. Then $\lim_{t\to\infty}\hat{q}_i(t)=\lim_{t\to\infty}\sum_{j=1}^{n}L_{ij}\|\hat{x}_j(t)-\hat{x}_i(t)\|^2=0$. So $\lim_{t\to\infty}e_i(t)=0$ and $\lim_{t\to\infty}x_i(t)=\lim_{t\to\infty}\hat{x}_i(t)-\lim_{t\to\infty}e_i(t)=\pi$. From the algorithm (3) and the initial values of x_i, we get that $\nabla F(\pi)=\sum_{i=1}^{n}\nabla f_i(\pi)=0$, which implies $\pi=x^*$. Thus $\lim_{t\to\infty}x_i(t)=x^*$.

Next, we will prove that all states reach consensus on the value x^* exponentially fast. For agent υ_i, we define the set

$$\mathcal{S}_i=\{x\in\mathbb{R}^m: f_i(x^*)-f_i(x)-\nabla f_i(x)^{\mathrm{T}}(x^*-x)\leq V(x(0))\},$$

and $\mathcal{S}=\operatorname{conv}(\cup_{i\in\mathcal{I}}\mathcal{S}_i)$, where conv denotes the convex hull. From Assumption 1, Lemma 1, $\dot{V}\leq 0$ and (4), we get that $\mathcal{S}$ is convex and compact, and $x_i(t),x^*\in\mathcal{S}_i\subset\mathcal{S}$, $\forall t\geq 0$, $i\in\mathcal{I}$. From Lemma 2, we get that there exists a positive constant ξ_i such that $\nabla^2 f_i(x_i)\leq\xi_i I_n$, $\forall x_i\in\mathcal{S}$.

Based on the same derivation as in [10], we get

$$V(x)\leq x^{\mathrm{T}}(P\otimes I_N)x, \tag{10}$$

where $P=[P_{ij}]\in\mathbb{R}^{n\times n}$ is a positive semidefinite matrix given by

$$P_{ij}=\begin{cases}(\frac{1}{2}-\frac{1}{n})\xi_i+\frac{1}{2n^2}\sum\limits_{r\in\mathcal{I}}\xi_r, & \text{if}\quad i=j\\ -\frac{\xi_i+\xi_j}{2n}+\frac{1}{2n^2}\sum\limits_{r\in\mathcal{I}}\xi_r, & \text{otherwise.}\end{cases} \tag{11}$$

Note that P and $-(L+L^{\mathrm{T}})$ are both positive semidefinite with the same null space: $\operatorname{span}\{1_n\}$. Then, based on Lemma 3, there exists a positive constant ρ defined as $\rho=\sup\{\mu:\mu P\leq-(L+L^{\mathrm{T}})\}$.

According to the former analysis, we get

$$\begin{aligned} V &\leq \frac{1}{\rho} x^{\mathrm{T}}[-(L+L^{\mathrm{T}}) \otimes I_n] x \\ &= \frac{1}{\rho}(\hat{x}-e)^{\mathrm{T}}[-(L+L^{\mathrm{T}}) \otimes I_n](\hat{x}-e) \\ &= \frac{1}{\rho}\hat{x}^{\mathrm{T}}[-(L+L^{\mathrm{T}}) \otimes I_n]\hat{x} + \frac{1}{\rho}e^{\mathrm{T}}[-(L+L^{\mathrm{T}}) \otimes I_n]e + \frac{2}{\rho}e^{\mathrm{T}}[(L+L^{\mathrm{T}}) \otimes I_n]\hat{x}. \end{aligned} \tag{12}$$

We know that the matrix $-(L+L^{\mathrm{T}})$ is positive semidefinite, so is the matrix $-(L+L^{\mathrm{T}}) \otimes I_n$. Obviously, the two matrices have the same largest eigenvalue, denoted as λ_n. The second item of (12) is changed into

$$\begin{aligned} \frac{1}{\rho}e^{\mathrm{T}}[-(L+L^{\mathrm{T}}) \otimes I_n]e &\leq \frac{\lambda_n}{\rho}e^{\mathrm{T}}e \\ &\leq \frac{\lambda_n \gamma_{\max}}{4\rho \ell_{\min}}\hat{x}^{\mathrm{T}}[-(L+L^{\mathrm{T}}) \otimes I_n]\hat{x}. \end{aligned} \tag{13}$$

For the last item of (12), we make some transformation, that is

$$\begin{aligned} \frac{2}{\rho}e^{\mathrm{T}}[(L+L^{\mathrm{T}}) \otimes I_n]\hat{x} &= \frac{2}{\rho}[e^{\mathrm{T}}(L \otimes I_n)\hat{x} + e^{\mathrm{T}}(L^{\mathrm{T}} \otimes I_n)\hat{x}] \\ &\leq \frac{2}{\rho}[\frac{1}{2}\hat{x}^{\mathrm{T}}[-(L+L^{\mathrm{T}}) \otimes I_n]\hat{x} + \frac{2}{\sigma}e^{\mathrm{T}}e] \\ &\leq (\frac{1}{\rho} + \frac{\gamma_{\max}}{\rho\sigma\ell_{\min}})\hat{x}^{\mathrm{T}}[-(L+L^{\mathrm{T}}) \otimes I_n]\hat{x}. \end{aligned} \tag{14}$$

Then the last inequality in (12) is reformulated as

$$V \leq (\frac{2}{\rho} + \frac{\gamma_{\max}}{\rho\sigma\ell_{\min}} + \frac{\lambda_n\gamma_{\max}}{4\rho\ell_{\min}})\hat{x}^{\mathrm{T}}[-(L+L^{\mathrm{T}}) \otimes I_n]\hat{x}. \tag{15}$$

Let $\beta = \frac{2}{\rho} + \frac{\gamma_{\max}}{\rho\sigma\ell_{\min}} + \frac{\lambda_n\gamma_{\max}}{4\rho\ell_{\min}}$, for simplicity's sake. Then

$$\hat{x}^{\mathrm{T}}[(L+L^{\mathrm{T}}) \otimes I_n]\hat{x} \leq -\frac{V}{\beta}, \tag{16}$$

Together with (9), we can obtain

$$\dot{V} \leq -\frac{1}{\beta}(\frac{1}{4} - \frac{\gamma_{\max}}{4\sigma\ell_{\min}})V. \tag{17}$$

It follows that

$$V(x(t)) \le e^{-\kappa t} V(x(0)), \tag{18}$$

where $\kappa = \frac{1}{\beta}\left(\frac{1}{4} - \frac{\gamma_{\max}}{4\sigma\ell_{\min}}\right) = \left(\frac{1}{4} - \frac{\gamma_{\max}}{4\sigma\ell_{\min}}\right) / \left(\frac{2}{\rho} + \frac{\gamma_{\max}}{\rho\sigma\ell_{\min}} + \frac{\lambda_n\gamma_{\max}}{4\rho\ell_{\min}}\right)$.

From (4) and Lemma 1, we have

$$V(x(t)) \ge \sum_{i=1}^{n} \frac{\theta_i}{2} \left\|x^* - x_i\right\|^2. \tag{19}$$

From (18), (19), and Lemma 2, we get

$$\begin{aligned} \sum_{i=1}^{n} \left\|x^* - x_i\right\|^2 &\le \frac{2}{\theta_{\min}} V(x(0)) e^{-\kappa t} \\ &\le \sum_{i=1}^{n} \frac{\xi_i}{\theta_{\min}} \left\|x^* - x_i(0)\right\|^2 e^{-\kappa t}, \end{aligned} \tag{20}$$

where $\theta_{\min} = \min\{\theta_1, \ldots, \theta_n\}$.

We can conclude that the states of all agents converge to the optimal solution x^* of the problem exponentially fast.

The proof of in this part is based on the paper [17]. To start with, given trigger times $t_1^i = 0, \cdots, t_k^i$, for agent υ_i, if it does not receive new information from its in-neighbors after t_k^i, we know that $t_{k+1}^i = \tau_{k+1}^i$. In this case, we have $\left\|e_i(\tau_{k+1}^i)\right\|^2 = \frac{\gamma_i}{4|L_{ii}|}\hat{q}_i(\tau_{k+1}^i)$, and, together with the designed control algorithm, the following result is obtained

$$\begin{aligned} \left\|e_i(\tau_{k+1}^i)\right\| &= \left\|\hat{x}_i(\tau_{k+1}^i) - x_i(\tau_{k+1}^i)\right\| = \left\|x_i(\tau_{k+1}^i) - x_i(t_k^i)\right\| \\ &= \left\|\int_{t_k^i}^{\tau_{k+1}^i} (\nabla^2 f_i(x_i(s)))^{-1} \textstyle\sum_{j=1}^{n} L_{ij}\hat{x}_j(s) ds\right\| \\ &\le \int_{t_k^i}^{\tau_{k+1}^i} (\nabla^2 f_i(x_i(s)))^{-1} \left\|\textstyle\sum_{j=1}^{n} a_{ij}(\hat{x}_j(s) - \hat{x}_i(s))\right\| ds \\ &\le \frac{1}{\theta_i} \int_{t_k^i}^{\tau_{k+1}^i} \left\|\sum_{j \ne i} L_{ij}(\hat{x}_j(s) - \hat{x}_i(s))\right\| ds \\ &\le \frac{\sqrt{n}}{\theta_i} \int_{t_k^i}^{\tau_{k+1}^i} \sqrt{L_{ii} \sum_{j \ne i} L_{ij} \left\|\hat{x}_j(s) - \hat{x}_i(s)\right\|^2} ds \\ &\le \frac{\sqrt{nL_{ii}}}{\theta_i} (\tau_{k+1}^i - t_k^i) \sqrt{\sum_{j \ne i} L_{ij} \left\|\hat{x}_j(t_k^i) - \hat{x}_i(t_k^i)\right\|^2} \\ &= \frac{\sqrt{nL_{ii}}}{\theta_i} (\tau_{k+1}^i - t_k^i) \sqrt{\hat{q}_i(\tau_{k+1}^i)}. \end{aligned} \tag{21}$$

Note that if $\sum_{j=1}^{n} L_{ij}\hat{x}_j(t_k^i) = 0$, no trigger will ever happen since $\dot{x}_i(t) = 0$ and $e_i(t) = 0$ for all $t \geq t_k^i$. Then we get

$$\tau_{k+1}^i - t_k^i \geq \theta_i \frac{\sqrt{\gamma_i}}{2|\sqrt{n}L_{ii}|}. \tag{22}$$

Let $\varepsilon_0 = \min\{\varepsilon_1, \ldots, \varepsilon_n\}$, and $\mathcal{U} = \left[T, T + \frac{1}{2}\varepsilon_0\right]$. Next, we will prove that, for agent υ_i, in any time interval of length $\frac{1}{2}\varepsilon_0$, there exist at most n triggers. Without loss of generality, let k_0 be a positive integer which satisfies $t_{k_0-1}^i < T$ and $t_{k_0}^i \geq T$. If there is no such k_0, then there is no triggers in $\mathcal{U}$.

If no information is received during the time period $(t_{k_0}^i, t_{k_0}^i + \varepsilon_i)$, then from (22), it holds that $t_{k_0+1}^i - t_{k_0}^i \geq \varepsilon_i \geq \varepsilon_0$. Hence, there is only one trigger in $\mathcal{U}$. Otherwise, suppose at least one in-neighbor of υ_i triggers at some time in $\mathcal{U}$, letting $t_1 \in (t_{k_0}^i, t_{k_0}^i + \varepsilon_i)$ be the first time when υ_i receives new information from its in-neighbors after $t_{k_0}^i$. Then, according to the second rule, we have $t_{k_0+1}^i = t_1$. There exists a nonempty set of all agents triggered at t_1, denoted by $\mathcal{Q}_1$.

If there are more triggers in $\mathcal{U}$, let t_2 be the next trigger time after t_1. There must be a nonempty set of agents triggered at t_2, denoted by $\mathcal{Q}_2$, such that $\mathcal{Q}_1 \cup \mathcal{Q}_2 - \mathcal{Q}_1 \neq \varnothing$. Actually, if not so, then $t_2 - t_1 \geq \varepsilon_0$ holds, which implies $t_2 \notin \mathcal{U}$. That is, the cardinality of $\mathcal{Q}_1 \cup \mathcal{Q}_2$ is not less than 2. Again in this way, let t_m be the $m+1$-th trigger time of υ_i and $\mathcal{Q}_m$ be the set of agents triggered at t_m. We have $\cup_{b=1}^{m} \mathcal{Q}_b \geq m$. Since there is at most n agents in the network, we can conclude that there are at most n triggers in $\mathcal{U}$. Thus the proof is completed.

3.3 A Variation of Principles

In this section, we get a variation of the principles given above. We show that the same results as Theorem 1 can be guaranteed and the proofs are similar as well. At any fixed time s, given $t_1^p, \ldots, t_h^p = t_{k_p(s)}^p$, for agent υ_p, pick $\gamma_p \in (0, 1)$ and $0 < \varepsilon_p < \theta_p \frac{\sqrt{\gamma_p}}{2|L_{pp}|}$, $t_{k_p(s)}^* = \max\limits_{\upsilon_j \in \mathcal{N}_p^{in}} t_{k_j(s)}^j$. Event-triggered principles are given as follows:

(1) At time $s = t_h^p$, check whether $\sum\limits_{j=1} L_{pj}\hat{x}_j(s) = 0$ or not;

(2) If the checking result of (1) is true, then agent υ_p does not trigger and its state remains constant until one of its in-neighbors triggers at time $t > t_{k_p(s)}^*$, then update time $s = t$ and $t_{k_p(s)}^*$, and go to step (1);

(3) If the checking result of (2) is false, then at time s, search τ_{h+1}^p, where $\tau_{h+1}^p = \max\left\{\tau \geq s\colon \left\|x_p(r) - x_p(t_h^p)\right\| \leq \sqrt{\frac{\gamma_p}{4|L_{pp}|}\hat{q}_p(t_{k_p(s)}^*)}, \forall r \in [t_{k_p(s)}^*, \tau]\right\}$, and go on to the next step;

(4) In case that none of the in-neighbors of υ_p is triggered during the time interval $(t^*_{k_p(s)}, \tau^p_{h+1})$, then υ_p triggers at time $t^p_{h+1} = \tau^p_{h+1}$;
(5) In case that there exists one in-neighbor of υ_p triggers at time $t \in (t^*_{k_p(s)}, \tau^p_{h+1})$: if $t - t^p_h \geq \varepsilon_p$ and $g_p(e_p(t)) < 0$, then update time $s = t$ and $t^*_{k_p(s)}$, and go to step (1); otherwise triggers at time $t^p_{h+1} = t$.

Theorem 2 *Under the proposed principles above and keeping all other conditions unchanged as before, the algorithm (3) guarantees that*

(a) *The states of all agents exponentially converge to the optimal solution x^* of the problem.*
(b) *In finite period of time, no agent will be triggered infinitely, i.e., the network is guaranteed to avoid Zeno behavior.*

Proof Theorem 2 can be seen as a variation of Theorem 1 due to the relation between the two kinds of principles. The analysis is quite similar and we can get the same convergence result as in Theorem 1, and the proofs progress alike. So we will not give the proof here.

Remark 2 The event-triggered principles here do not change the convergence results of the algorithm (3), because the requirements needed in the proof of Theorem 1 are not affected. The main difference between the former and the latter lies in (1) and (5). The latter includes the rule (1) to deal with the case when $\sum_{j=1} L_{pj}\hat{x}_j(s) = 0$, which says no trigger happens at time s for agent υ_p, and there is no need to calculate τ^p_{h+1}. The rule (5) aims to guarantee that $g_p(e_p(t)) \leq 0$, which plays an essential role in the proof of Theorem 1 as well as Theorem 2.

4 Simulations

In this section, we present numerical simulations to illustrate the main results proved in Sect. 3. Consider a multi-agent network with six agents and the communication topology is shown in Fig. 1. The corresponding weight matrix A and Laplacian matrix L are given by

$$A = \begin{bmatrix} 0 & 3 & 0 & 0 & 0 & 0 \\ 0 & 0 & 2 & 0 & 1 & 0 \\ 0 & 0 & 0 & 3 & 0 & 0 \\ 0 & 0 & 0 & 0 & 3 & 0 \\ 0 & 0 & 0 & 0 & 0 & 4 \\ 3 & 0 & 1 & 0 & 0 & 0 \end{bmatrix}, \quad L = \begin{bmatrix} -3 & 3 & 0 & 0 & 0 & 0 \\ 0 & -3 & 2 & 0 & 1 & 0 \\ 0 & 0 & -3 & 3 & 0 & 0 \\ 0 & 0 & 0 & -3 & 3 & 0 \\ 0 & 0 & 0 & 0 & -4 & 4 \\ 3 & 0 & 1 & 0 & 0 & -4 \end{bmatrix}.$$

Fig. 1 The communication topology

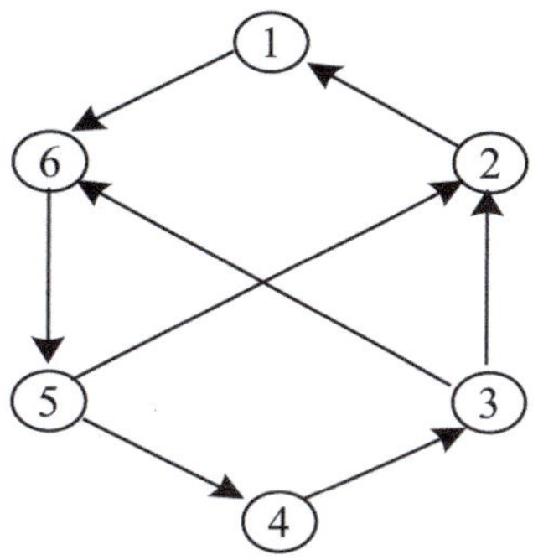

Fig. 2 The evolution of system states with event-triggered communication

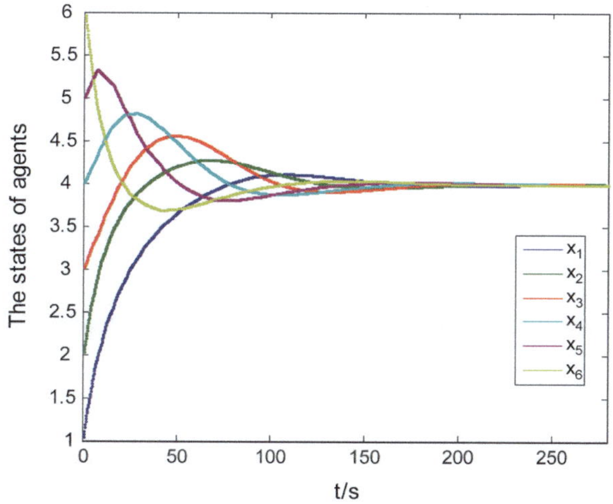

The optimization problem is to minimize the objective function $\sum_{i=1}^{6} f_i(x)$. Let each agent i have a local cost function defined as $f_i(x) = (x - i)^4 + 8i(x - i)^2$, and $f_i(x)$ satisfies Assumption 1, for all $i \in \{1, 2, 3, 4, 5, 6\}$.

As shown in Fig. 2, states of agents reach consensus finally. Figure 3 says that control inputs of agents go to zero eventually. Figure 4 indicates that the measurement error $\|e_i(t)\| = \|\hat{x}_i(t) - x_i(t)\|$ converges to zero for all agents. Figure 5 shows that no Zeno behavior happens with the proposed principles.

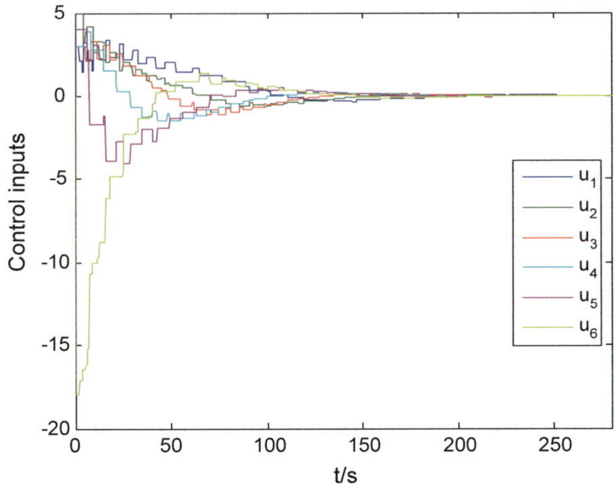

Fig. 3 Control inputs of all agents along time with event-triggered communication

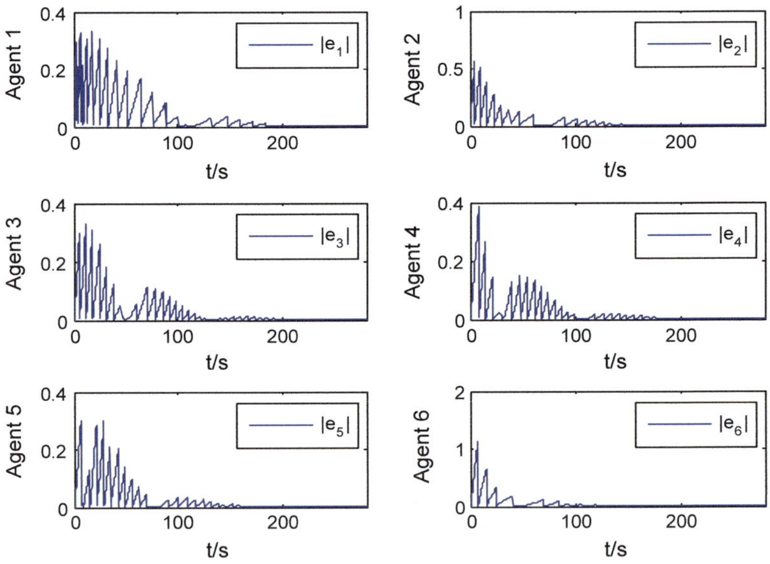

Fig. 4 The measurement error of agents' states caused by event-triggered communication

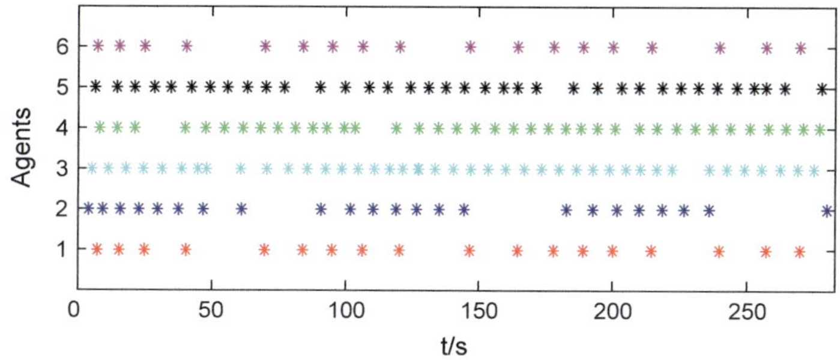

Fig. 5 The trigger time of each agent

5 Conclusions

This paper combines the event-triggered communication in consensus problem of multi-agent system with distributed continuous-time convex optimization algorithms over weight-balanced digraphs. We address the problem whose global objective function is a sum of local functions associated to each agent. We utilize the event-triggered communication technique to reduce the communication load and avoid Zeno behavior meanwhile. Based on Lyapunov analysis approach, we prove that the Zero-Gradient-Sum (ZGS) algorithm with the event-triggered communication makes all agents' states converge to the optimal solution of the global objective function exponentially fast.

Acknowledgments This research is supported by the National Natural Science Foundation of China (Grant No. 61573200, 61273138), and the Tianjin Natural Science Foundation of China (Grant No. 14JCYBJC18700, 14JCZDJC39300).

References

1. Rabbat M, Nowak R (2004) Distributed optimization in sensor networks. In: International symposium on information processing in sensor networks. IEEE, pp 20–27
2. Tsianos KI, Lawlor S, Rabbat M (2012) Consensus-based distributed optimization: practical issues and applications in large scale machine learning. Commun Control Comput:1543–1550
3. Ram S Sundhar, Veeravalli Venugopal V, Nedic Angelia (2010) Distributed and recursive parameter estimation in parametrized linear state-space models. IEEE Trans Autom Control 55(2):488–492
4. Nedic Angelia, Ozdaglar Asuman (2009) Distributed subgradient methods for multi-agent optimization. IEEE Trans Autom Control 54(1):48–61
5. Lobel Ilan, Ozdaglar Asuman (2011) Distributed subgradient methods for convex optimization over random networks. IEEE Trans Autom Control 56(6):1291–1306
6. Yanhui Yin, Zhongxin Liu, Zengqiang Chen (2015) Distributed subgradient algorithm for multi-agent optimization with directed communication topology. IFAC Papersonline 48(28):863–868

7. Nedic Angelia, Ozdaglar Asuman, Parrilo Pablo A (2010) Constrained consensus and optimization in multi-agent networks. IEEE Trans Autom Control 55(4):922–938
8. Duchi John C, Agarwal Alekh, Wainwright Martin J (2012) Dual averaging for distributed optimization: convergence analysis and network scaling. IEEE Trans Autom Control 57(3):592–606
9. Lee Soomin, Nedic Angelia (2013) Distributed random projection algorithm for convex optimization. IEEE J Sel Top Signal Process 7(2):221–229
10. Jie Lu, Tang Choon Yik (2012) Zero-gradient-sum algorithms for distributed convex optimization: the continuous-time case. IEEE Trans Autom Control 57(9):2348–2354
11. Gharesifard Bahman, Cortes Jorge (2014) Distributed continuous-time convex optimization on weight-balanced digraphs. IEEE Trans Autom Control 59(3):781–786
12. Lin P, Ren W, Song Y, Farrell JA (2014) Distributed optimization with the consideration of adaptivity and finite-time convergence. In: American Control Conference – ACC, pp 3177–3182
13. Yuheng Z, Yiguang H (2015) Distributed optimization design for high-order multi-agent systems. In: Proceedings of the 34th Chinese control conference, pp 7251–7256
14. Dimarogonas Dimos V, Frazzoli Emilio, Johansson Karl Henrik (2012) Distributed event-triggered control for multi-agent systems. IEEE Trans Autom Control 57(5):1291–1297
15. Liu Zhongxin, Chen Zengqiang, Yuan Zhuzhi (2012) Event-triggered average-consensus of multi-agent systems with weighted and direct topology. J Syst Sci Complex 25(5):845–855
16. Nowzari Cameron, Cortes Jorge (2016) Distributed event-triggered coordination for average consensus on weight-balanced digraphs. Automatica 68:237–244
17. Yi X, Lu W, Chen T Distributed event-triggered consensus for multi-agent systems with directed topologies. http://arxiv.org/abs/1407.3075
18. Zhong Minyi, Cassandras Christos G (2010) Asynchronous distributed optimization with event-driven communication. IEEE Trans Autom Control 55(12):2735–2750
19. Kia Solmaz S, Cortes Jorge, Martinez Sonia (2015) Distributed convex optimization via continuous-time coordination algorithms with discrete-time communication. Automatica 55:254–264
20. Chen Weisheng, Ren Wei (2016) Event-triggered zero-gradient-sum distributed consensus optimization over directed networks. Automatica 65:90–97
21. Liu J, Chen W (2016) Distributed convex optimization with event-triggered communication in networked systems. Int J Syst Sci:1–12
22. Nesterov Yurii (2004) Introductory lectures on convex optimization: a basic course. Kluwer, Norwell, MA
23. Boyd Stephen, Vandenberghe Lieven (2004) Convex optimization. Cambridge University Press, New York, NY
24. Hardy Godfrey H, Littlewood John E, Pólya George (1952) Inequalities. Cambridge University Press, Cambridge, UK

A Novel Safety Assessment Approach Based on Evolutionary Clustering Learning

Yi Chai, Zhaodong Liu, Hongpeng Yin and Yanxia Li

Abstract The safety risk assessment is a structured and systematic methodology aiming at enhancing the complex engineering system safety. It has been gradually and broadly used in the industrial process control system nowadays around the world. In this paper, a novel safety assessment approach based on evolutionary dictionary learning and fault tree analysis for the complex engineering system is proposed. First, historical signals are utilized to conduct the clustering learning dictionaries by norm similarity matching model and patch-based evolutionary dictionary learning algorithm. Second, the support vector machine method is employed to identify and reflect the normal and fault operating states. Third, an improved safety risks method is proposed to reflect the probable hazards of different faults on the basis of the fault tree analysis. Finally, this processing on online signals is to offer an effective safety assessment index and update the evolutionary dictionaries and safety routing metrics. The related experiments are constructed to demonstrate that our proposed approach can achieve high performance.

Keywords Safety risk assessment · Evolutionary dictionary learning · Support vector machine

1 Introduction

In the complex engineering systems, the ultimate principle of safety assessment mainly evaluates the failure risk, failure probability, and identifies the measurement of possible hazards. The unexpected hazardous accidents may cause huge economic losses, environmental contamination, and human injuries, due to component degradation, exogenous changes, and operational mistakes. The research of resolving safety problems draws many attentions of experts and scholars. In the last decades, many variations of technologies, including data-driven based and

---

Y. Chai · Z. Liu (✉) · H. Yin · Y. Li
School of Automation, Chongqing University, Chongqing 400044, China
e-mail: liuzhaodong@cqu.edu.cn

Y. Jia et al. (eds.), *Proceedings of 2016 Chinese Intelligent Systems Conference*, Lecture Notes in Electrical Engineering 405,
DOI 10.1007/978-981-10-2335-4_46

model-driven safety assessment methods, have been proposed. However, there are still many open-ended problems of safety assessment in the complex engineering systems.

In a real-world application, the growth of measured data has imposed an increasing need on data-driven based safety assessment methods for proactive maintenance scheme under dynamic operational and environmental changes. Probabilistic Risk Assessment (PRA) is a comprehensive, structured, and logical analysis method aimed at identifying and assessing risks. Many PRA approaches have been exploited, including failure modes and effects analysis (FMEA), information fusion based safety assessment method, fault tree analysis (FTA), and statistical probability method [1–3]. The conventional safety assessment approaches could effectively instruct the evaluation process of safety risk and offers an assistant decision-making for the complex engineering systems.

However, in real applications, data information from the complex engineering systems is not only collected over a period of time, but also the relationship between data points can change over that period. It is natural to expect that the measurement and evaluation of safety risks should reflect the change over a series of time steps. In this paper, a novel safety assessment approach based on evolutionary dictionary learning and fault tree analysis is proposed. First, historical signals are utilized to generate the representation dictionaries by evolutionary dictionary learning algorithm. Second, the support vector machine method is employed to separate the normal and fault operating conditions. Thirdly, according to the fault tree analysis of complex system, an improved safety risks method is proposed to reflect the probable hazards of different faults. Finally, this process on online signals is to offer an effective safety assessment index, update the representation dictionaries and safety routing metrics.

2 The Safety Assessment Method

Many of the safety assessment and prognosis methods used in the current complex engineering systems have been incorporated into safety pre-warning scheme. Their associated hidden variables indicate some important process behaviors of the complex system. Furthermore, the relationship between the variables and monitoring parameters may be changed and evolved over a series of time steps. Thus, how to measure the serious consequences of various failures, and compute the probability of occurrence remains an open-ended problem. This paper proposes a novel safety assessment approach to solve these problems. The framework of the proposed safety assessment method is shown in Fig. 1.

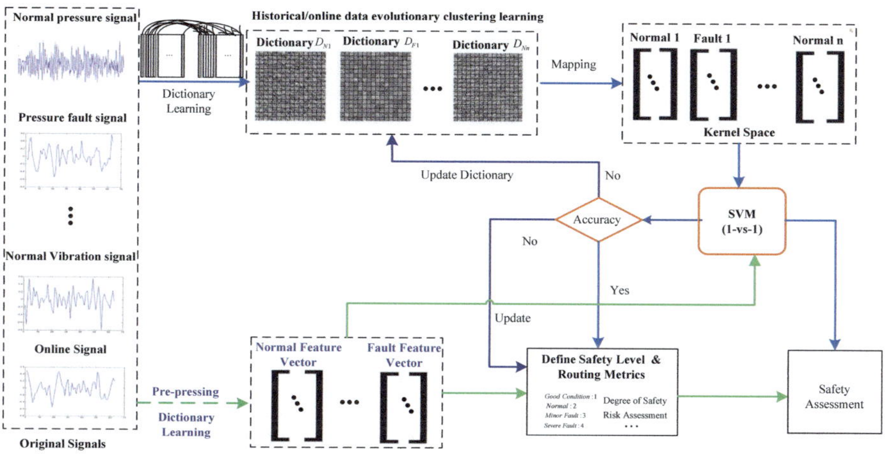

Fig. 1 Block diagram of the proposed method

2.1 The Dictionary Learning Algorithm and SVM

There are many monitoring signals in the complex engineering systems. How to represent these signals and reflect the operating state is a critical step. In this section, the representation and classifying algorithms, the evolutionary dictionary learning method and Support Vector Machines (SVM), are illustrated in details. Consider the monitoring signal acquisition model

$$X = DZ + \varepsilon \tag{1}$$

where, $X \in R^N$ is the discretized signals, $Z \in R^M$ is the acquired k-space features, $D \in R^{M \times N}$ is the dictionary operator, and ε is the noise vector. A typical algorithm attempts to solve the following problem:

$$\widehat{X} = \arg \min_{x} \{\frac{1}{2} \|X - DZ\|^2 + \lambda \|Z\|_1\}, \tag{2}$$

where λ is the regularization parameter, $\|\bullet\|$ is the standard Euclidean norm. In practice, a guide signal can be separated by patches. Given the signal patches $X = \{x_1, \cdots, x_i, \cdots\}$, one can formularize Eq. (2) as:

$$\hat{x}_i = \arg \min_{x_i} \{\frac{1}{2} \|x_i - d_i z_i\|^2 + \lambda \|z_i\|_1\} \tag{3}$$

It might be worth noting that when the feature dimension is large, solving the lasso problem is very time-consuming. In this work, the proposed algorithm assumes that a guide signal is available to give a good estimate of nonlocal similarity for signal patches. This nonlocal similarity information is integrated as prior

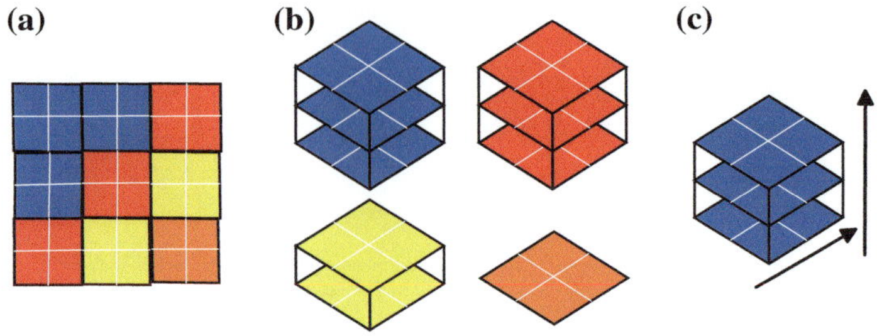

Fig. 2 Group signal patches. **a** A signal; **b** four groups; **c** group dimension

information into the proposed patch-based nonlocal operator. Patch grouping is shown in Fig. 2.

For a given signal $X \in R^N$, we decompose it into patches with fixed size $L \times 1$. The ν_j group of signal patches is denoted as F_{ν_j}, where ν_j stores the index of patches. Let $D_{2\mathrm{d}}$ be a 2-dimension transform, we define the nonlocal operator as

$$A_j = D_{2\mathrm{d}} F_{\nu_j} \tag{4}$$

An optimal grouping is that

$$z_j = A_j x_i \tag{5}$$

The adjoint operator A_j^T is $A_j^T = F_{\nu_j}^T D_{2\mathrm{d}}^T$ and it satisfies

$$\sum_j A_j^T A_j = O = [\vec{o}_1 \; \ldots \; \vec{o}_n \; \ldots \; \vec{o}_N]^T \tag{6}$$

The invertibility of O requires that each patch must be contained in at least one group. Similar patches are grouped to produce sparse coefficients since it shows great potentials to represent the original signal, as shown in Fig. 3. For a search

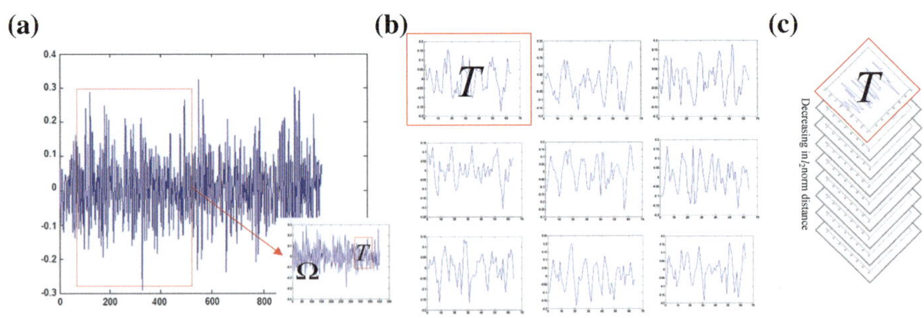

Fig. 3 Illustration of the similar patches found via block matching and the sparisity results. **a** A search region Ω and the reference patch T; **b** similar patches found by the l_2 norm distance. **c** 2-dimension array stacked from the similar patches

region Ω and the reference patch T, the similarity between the reference patch and a candidate patch is measured by l_2 norm distance

$$c(T,\alpha) = \left\| T(f,L) - \alpha(\hat{f},L) \right\|_2 \tag{7}$$

In the following section, we will illustrate the dictionary learning processing by KSVD in details [4]. For simplicity, we shall assume that all signals have the same co-rank $k - L$. Our goal is to find the dictionary giving rise to these signals. Assuming an initial estimation D_0 of the analysis operator, the optimization scheme is based on a two-phase block-coordinate relaxation approach. The detailed description is provided in Algorithm 1.

Algorithm 1 Analysis K-SVD

---

1. Input: Signals $X \in R^N$, initial dictionary D_0, and number of iterations n;
2. Output: Dictionary D and signal set X minimizing Eq. (3);
3. Find the similar patches via block matching via Eq. (7);
4. for $n = 1, 2, \cdots$ do
5. Analysis Pursuit: $\forall i : \{\hat{x}_i, \sup ports \hat{\Lambda}_i\} := Arg \min\limits_{x,\Lambda} \left\{ \left\| dz - x_i \right\|_2 + \lambda \left\| z \right\|_1 \right\}$
6. for $j = 1, \cdots, p$ do
7. Extract Relevant Examples:
   $J := indices\ of\ thecolumns\ of\ \hat{X}\ orthogonal\ to\ w_j$
8. Compute row: $\hat{w}_j := Arg \min\limits_{w} \left\| w^T X_J \right\|_2 \quad s.t. \ \left\| w \right\|_2 = 1$
9. Update row: $D\{j - th\ row\} := w^T{}_j$
10. end for
11. end for

Next, SVM is employed to identify the intact and contingent operating states [5]. The most common cases of nonlinear SVM are soft margin SVM (l_1-SVM) and l_2-SVM. In this paper, l_2-SVM is used to identify the safety case. The objective function of the corresponding optimal separating hyper plane of l_2-SVM is:

$$\min_{w,b,\xi} \frac{1}{2} A^T A + C \sum_{i=1}^{N} \xi_i \quad \text{s.t. } y_i(A^T \Phi(h_i) + d) \geq 1 - \xi_i \quad \text{for } i = 1, \ \ldots, \ N \tag{8}$$

where A is an m-dimensional vector, d is a constant. C is the penalty coefficient. $\xi_i(A, h_i, y_i) = (\max(0, 1 - y_i A^T A h_i))^2$ is the loss function. In brief, SVM is a contemporary machine learning tool, which has quickly evolved as an active area of research for its theoretical foundation.

2.2 The Measurement of Safety Assessment

In the safety assessment framework, fault tree analysis is the first and crucial step. The fundamental concept in fault tree analysis is the translation of a physical system into a structured logic diagram. TOP events are taken from a preliminary hazard analysis, however, informal it was; these events are usually strongly undesired system states. This logic diagram is constructed using the event symbols and logic symbols in Fig. 4.

Generally, a general fault tree analysis [1–3] is comprised of four steps: System definition, Fault tree construction, Qualitative Evaluation and Quantitative Evaluation. As shown in Fig. 4, the local model represents the combinations of events which lead to the top undesirable event. The top event can be expanded as:

$$\begin{aligned} T_{top-event} &= E_{\text{intermediate}-event} \cup (or\ \cap) X_{C_{event}} \cup (or\ \cap) X_{D_{event}} \\ &= X_{A_{event}} \cup (or\ \cap) X_{B_{event}} \cup (or\ \cap) X_{C_{event}} \cup (or\ \cap) X_{D_{event}} \end{aligned} \tag{9}$$

Given p_x, the failure probability of X_i, the failure probability of top event T is:

$$P_T(p_{x_1}, p_{x_2}, p_{x_3}, p_{x_4}) = 1 - (1 - p_{x_1} p_{x_2})(1 - p_{x_3} p_{x_4}) \tag{10}$$

Given the probability of evolution of fault classification q_{x_i} and the probability of basic event p_{x_i}, the safe degree of safety assessment can be defined as:

$$S = (1 - q_{x_1} p_{x_1}) \times (1 - q_{x_2} p_{x_2}) \times \cdots\cdots \times (1 - q_{x_n} p_{x_n}) \tag{11}$$

Meanwhile, the probability of evolution of fault classification is $q_{x_i} = \frac{1}{i-1} \sum_{i-1} (q_{x_1} + q_{x_2} + \cdots q_{x_{i-1}})$. The evaluation values D from expert are divided into five levels according to the condition attributes and frequency. The weights w are conducted via expert experience and hazard rating as shown in Table 4. Therefore, the index of safety risk can be defined as:

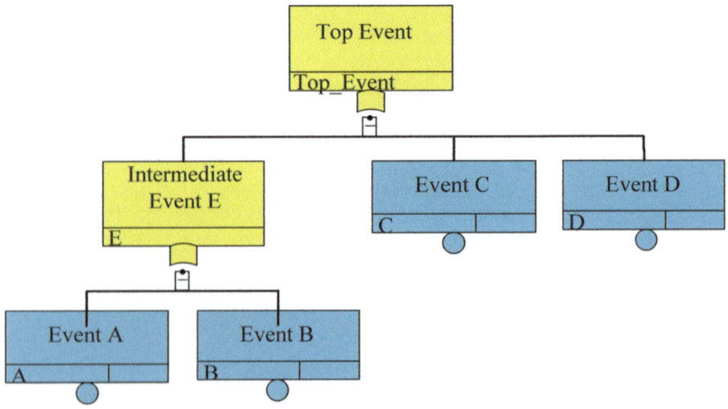

Fig. 4 The typical fault tree structure and symbols

$$T = \frac{1}{5}\sum_{i=1}^{N} w_i D_i \tag{12}$$

In this paper, we will use the hydraulic system as an example to demonstrate the effectiveness of our proposed safety assessment framework, which are intuitively displayed in the following section.

3 Experiments and Analysis

In this paper, a novel safety risk assessment algorithm based on evolutionary dictionary learning and fault tree analysis for the complex engineering systems is proposed. All the experiments are implemented in Matlab 2010a and on a Pentium (R) 2.5 GHz PC with 2.00 GB RAM. The source data are utilized to test and verify the performance, which are obtained from http://splab.cz/en/download/databaze/multispec.

In this work, the historical signals and online data are patched by a l_2 distance model, as shown in Fig. 3. The separated patches are used to generate the clustering dictionaries via KSVD. The combination of different signals can reflect the operational states of the hydraulic system. The learning dictionaries and features are shown in Fig. 5.

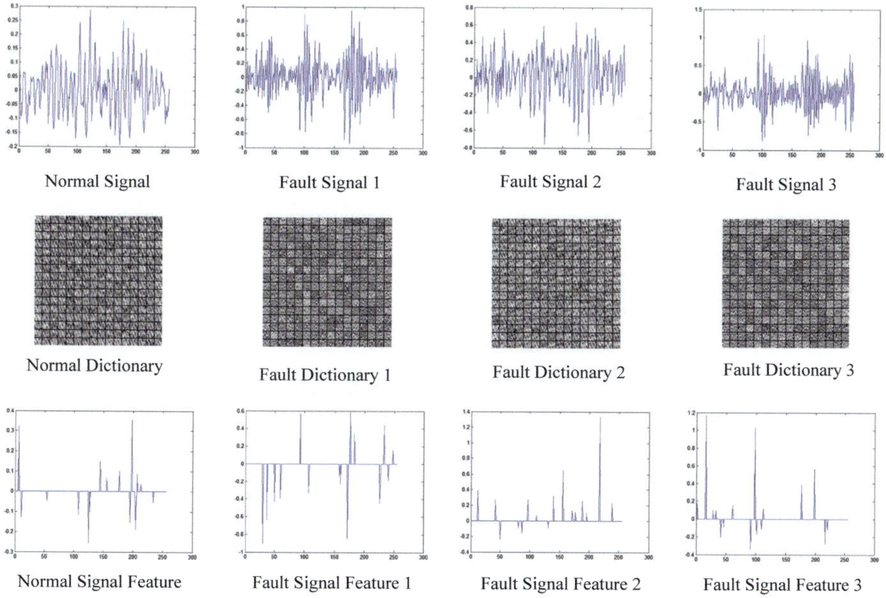

Fig. 5 The learning dictionaries and signal features

Table 1 The classification rates of different signals by SVM

Signals	Train	Test	Accuracy (%)	Time (s)
Normal	44800	4800	94.55	29.5893
Fault 1		4800	98.50	17.7595
Fault 2		4800	97.45	18.1528
Fault 3		4800	97.03	15.0764

As shown in Fig. 5, various signals are used to generate different dictionaries by patches-based KSVD learning algorithm. Furthermore, the representing features are clearly distinguishable under the different building dictionaries. Therefore, the classification rates of different signals for various operating states with high accuracy are obtained by SVM, which can be shown in Table 1.

As shown in Table 1, the operating states can be effectively identified by SVM method. It can also achieve a fast speed. Given these, in order to offer an intuitive display to the index of safety assessment, we turn to describe the main part of the safety risk assessment approach based on fault tree analysis. The minimal cut sets, minimal path sets, and the common cause failures can be acquired. The minimal cut sets include:$\{X_1, X_2, X_3\}$, $\{X_4, X_5, X_6\}$ and $\{X_7, X_8, X_9\}$. Therefore, the minimal cut sets and their representative sequential pattern are shown in Table 2.

As mentioned above, the probability of basic events is a key indicator to measure the safety risk assessment for the hydraulic system. In this work, we define our probability of the basic events to consult the reference [2] from NASA, which are acquired by statistics from amount of the complex industrial systems. Therefore, the probability of basic events can be shown in Table 3.

In addition, the logic gates are all OR gates according to the definition of symbol. When integrating the probability and the minimal cut sets, the safe-degree of the hydraulic system can be computed as:

$$S = \sum_{i=1}^{3} (1 - q_i P_i) = 0.99984 \tag{13}$$

It is nature that the hydraulic system has small fault probability with high performance. The evaluation values from expert are divided into five levels (1: weak; 2: weaker; 3: middle: 4: stronger; 5: strong). The weights are conducted via expert experience and hazard rating as shown in Table 4.

Table 2 The minimal cut sets and their representative sequential pattern

Pattern	Minimal cut sets	Events
1	$\{X_1, X_2, X_3\}$	Filter with dirt + the insecure between pump body and cover + the rollback of electric joint slack → pressure fluctuation
2	$\{X_4, X_5, X_6\}$	The looseness of pump gear + the leak of oil seal + the abnormal of pressure vingve → fuel shortage
3	$\{X_7, X_8, X_9\}$	The looseness of bearing cap + the fracture malfunction of driving shaft + oil overheat → pump vibration

Table 3 The probability of basic events

Number	Basic events	Probabilities
Event 1	The filter with dirt	$p_1 = 0.000005$
Event 2	The insecure between pump body and cover	$p_2 = 0.000005$
Event 3	The rollback of electric joint slack	$p_3 = 0.0000074$
Event 4	The looseness of pump gear	$p_4 = 0.0000036$
Event 5	The leak of oil seal	$p_5 = 0.0000085$
Event 6	The abnormal of pressure vingve	$p_6 = 0.0000091$
Event 7	The looseness of bearing cap	$p_7 = 0.0000028$
Event 8	The fracture malfunction of driving shaft	$p_8 = 0.000001$
Event 9	The oil overheat	$p_9 = 0.000008$

Table 4 The weights via expert experience and hazard rating

Subsystem	S_1	S_2	S_3
Weights	0.38	0.27	0.45
C_{X_1}	4	2	2
C_{X_2}	5	1	3
C_{X_3}	4	2	2
C_{X_4}	1	5	2
C_{X_5}	2	4	3
C_{X_6}	3	4	3
C_{X_7}	2	2	5
C_{X_8}	1	2	4
C_{X_9}	2	2	4

In this work, we assume that the three intermediate events (S_1, S_2, S_3) lead to the pump failure of the hydraulic system. Therefore, the safety risk index of the hydraulic system can be computed as:

$$T = \frac{1}{5}\sum_{i=1}^{3} w_i D_i = \frac{1}{5}(0.38 + 0.27 + 0.45) \times 4 = 0.8 \tag{14}$$

The index demonstrates that the hydraulic system has a high hazardous grade. In real applications, security and stability control measurements should be done for the hydraulic system.

4 Conclusions

This paper proposes a novel safety assessment approach for the complex engineering systems to measure the serious consequences of various failures, and compute the probability of occurrence. First, when running against the historical signals, the clustering learning dictionaries of normal and failure states can be obtained by evolutionary dictionary learning algorithm and KSVD. Meanwhile, the metrics of safety risk are constructed by analyzing fault tree and the risk probability. Second, the online signals are employed to update the representation dictionaries, and the metrics of safety assessment. This novel approach is not limited to the complex engineering systems. It also can extend to the other applications.

Acknowledgments This research is supported by National Natural Science Foundation of China (61374135), Chongqing Nature Science Foundation (cstc2015jcyjB0569), China Central Universities Foundation (106112015CDJXY170003), and Chongqing Graduate Student Research Innovation Project (CYB14023).

References

1. Pat-Cornell E, Dillon R (2001) Probabilistic risk analysis for the NASA space shuttle: a brief history and current work. Reliab Eng Syst Saf 74(3):345–352
2. Hu J et al (2010) An integrated method for safety pre-warning of complex system. Saf Sci 48 (5):580–597
3. Stamatelatos M et al (2011) Probabilistic risk assessment procedures guide for NASA managers and practitioners. *NASA*
4. Rubinstein R, Peleg T, Elad M (2013) Analysis K-SVD: a dictionary-learning algorithm for the analysis sparse model. IEEE Trans Signal Process 61(3):661–677
5. Hsu CW, Lin CJ (2002) A comparison of methods for multiclass support vector machines. IEEE Trans Neural Netw 13(2):415–425

Quantified Living Habits Using RTI Based Target Footprint Data

Weijia Zhang, Zhichao Tan, Guoli Wang and Xuemei Guo

Abstract Providing personalized healthcare for elders is more and more necessary in aging society. It is the premise to quantify their living habits properly. In this paper, a classification algorithm is used to transform footprints of elder into daily activities by combining point of interest. A concept of activity matrix and vector is proposed to quantify daily life, and then a clustering algorithm based on similarity is put forward to realize abnormal behaviors recognition. Finally, a experiment is given to illustrate the effectiveness of the proposed methods.

Keywords Indoor target localization · Quantified self · Abnormal behaviors analysis · Radio Tomographic Imaging

1 Introduction

With the aging society coming, the demands on personalized healthcare for daily living of elder becomes increasing [1, 2], which invokes for more intelligent technologies accordingly. Quantified daily living and abnormal behavior analysis are the fundamental yet important components of elder healthcare. Footprint data of elder are essential cues characterizing their living habits, which fully involves spatial–temporal information about daily living events, thus can offer the healthcare status of elder behind these events and play an indispensable role in elder healthcare applications. More importantly, the indoor footprints of elder are the natural consequences of their daily activities without any interference, therefore, they can be collected anytime with the guaranteed completeness as well as the stability. In addition, the

W. Zhang (✉) · Z. Tan
School of Electronics and Information Technology, Sun Yat-sen University, Guangzhou 510006, China
e-mail: xjrs_zhangwj@163.com

G. Wang · X. Guo
School of Data and Computer Science, Sun Yat-sen University, Guangzhou 510006, China

Y. Jia et al. (eds.), *Proceedings of 2016 Chinese Intelligent Systems Conference*, Lecture Notes in Electrical Engineering 405,
DOI 10.1007/978-981-10-2335-4_47

concise and simple structure of the footprint data benefits the storage, transmission, processing as well as analysis in an efficient fashion. Moreover, raw footprint data are less privacy concerns with the desirable robustness to indoor environment changes, thus their applicability is more ubiquitous.

This paper focuses on exploring the use of target-induced RF shadow fading image in quantified daily living and abnormal behavior analysis for elder healthcare. The aim of this study is to build a novel paradigm of modeling living habits and detecting and recognizing abnormal behaviors. We propose a classification algorithm to transform footprints into daily activities with quite mild assumptions, and then put forward a concept of activity matrix and vector for quantified living habits. By clustering activity vectors, we achieve abnormal behavior recognition.

2 Classification Algorithm Based on POI

2.1 Point of Interest

The human activity has a close relationship with the area where he or she is in. Thus, it is the basis of activity recognition by connecting human activity with the context of environment. For example, sitting at a desk is likely to show that the person is reading, and sitting on a sofa is likely to show that the person is watching TV. In general, all human activities are performed in the certain area. To simplify the problem, this paper assumes that the activities in an area are all direct with unique meanings. For example, a person sitting at a desk is most likely to read book. A person sitting at a dinner table is most likely to have a meal.

In order to achieve activity recognition, it is important to divide the indoor environment into different areas. Thus, we put forward a concept of point of interest (POI) [3], which is an area used to perform activity of daily life (ADL) [4, 5]. For example, people sitting at the desk are most likely to read book, so the desk can be identified as the POI. A house as an example is shown in Fig. 1. The house contains 10 POIs are marked in red circle. Dinner table and kitchen are regarded as the same POI because both of them are corresponding to dining.

2.2 Footprint Classification

Footprints of people are a series of coordinates who involve spatial–temporal information about daily activities, which mainly include still footprints and moving footprints. Still footprints are the footprints that people stay at certain POI and moving footprints are the footprints that people walk between two different POIs. This paper basically focuses on the former. A preprocessing method called threshold classification is proposed to obtain still footprints, which compares distance between

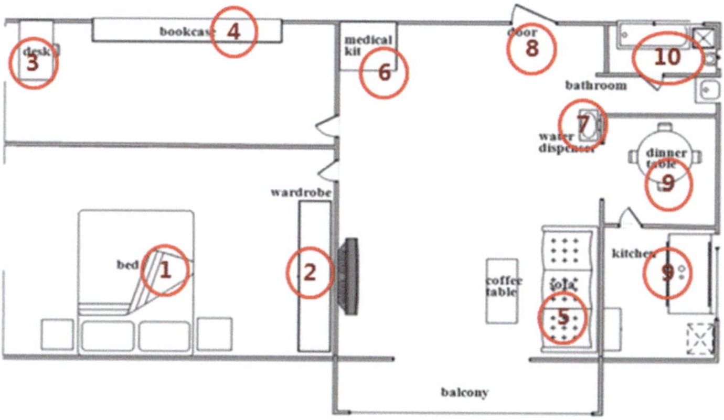

Fig. 1 The house with POIs

coordinates of adjacent footprint with the threshold computed in advance. For example, if the former is greater than the latter, footprints are recognized as moving footprints, otherwise, they are still footprints. The threshold is obtained as follows:

$$\varepsilon = \frac{1}{N}\sum_{i=2}^{N}\left[\left(x_i - x_{i-1}\right)^2 + \left(y_i - y_{i-1}\right)^2\right] \tag{1}$$

where N is the number of footprints, x_i and y_i are the abscissa and ordinate of the ith footprint respectively.

Generally speaking, still footprints that represent the same daily activity belong to the same POI, i.e., the coordinates of still footprints are close to the central coordinate of the corresponding POI. Thus, in this paper, k-means clustering algorithm [6] is used to classify still footprints. For k-means clustering algorithm, cluster number and mean of each cluster are important parameters which need to be trained. However, for the task of footprint classification, cluster number can be seen as number of POIs, and mean of each cluster is equal to central coordinate of the corresponding POI, both of them can be obtained without training.

3 Quantified Living Habits and Abnormal Behaviors Analysis

For an elder who has a regular daily life, he or she is used to perform the same behavior during similar period everyday. Thus, abnormal behaviors can be defined as sudden change of daily life, e.g., sleeping more time than usual may indicate physical or mental discomfort.

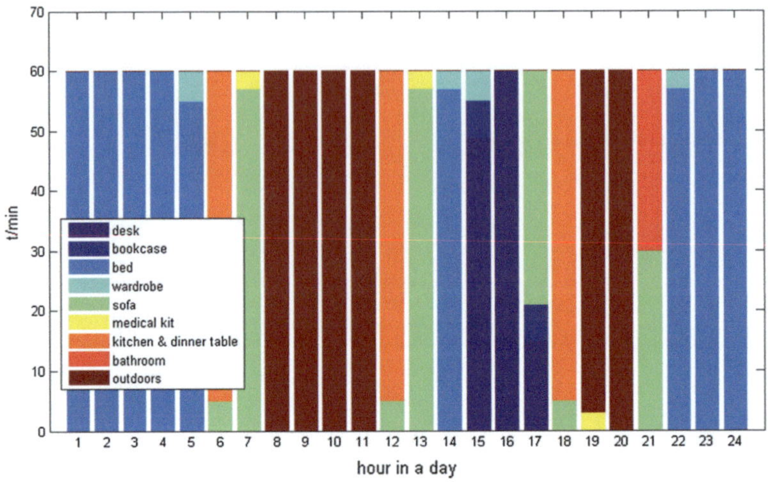

Fig. 2 The histogram of activity matrix

3.1 *Quantified Living Habits*

In this paper, activity matrix and vector is proposed to quantify living habits. Activity matrix describes the distribution of time spent when people stay at various POIs at every hour in a day. Activity vector describes accumulate time distribution of various POIs in a day.

In order to understand the concept of activity matrix and vector intuitively, this paper draw two histograms of them based on the house shown in Fig. 1. For instance, a histogram of activity matrix for an elder is shown in Fig. 2. As we have seen, the elder is in bed from about 10:00 p.m. to 6:00 a.m., which means he or she is sleeping. A histogram of activity vector can be seen in Fig. 3, which shows time spent in various POIs in a day, and the unit of the time is minutes.

3.2 *Abnormal Behaviors Analysis*

In general, for an elder who lives regularly, he or she is in normal condition much longer than abnormal condition, and foundation of abnormal behaviors recognition using activity vectors is to classify the high similarity as a class and the low similarity as the other. Thus, we can believe that the class of vectors with high similarity and large quantity can be recognized as normal while the other with small quantity is abnormal.

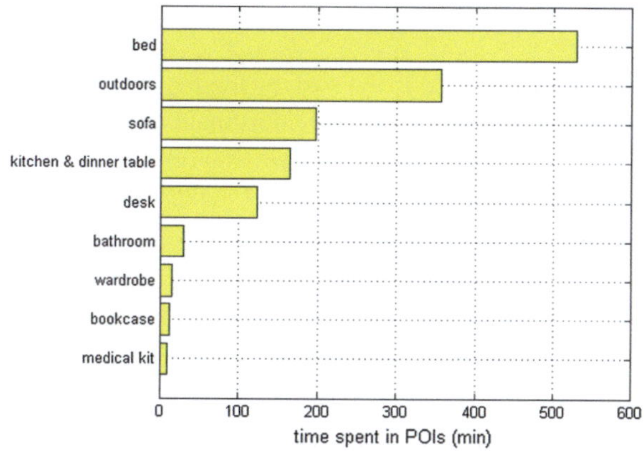

Fig. 3 The histogram of activity vector

Activity vector indicates the time distribution at various POIs, and the distance between two vectors can be used to describe the similarity of activity vectors. The closer the vectors are, the higher the similarity is. In this paper, four distance formulas are proposed as follows:

3.2.1 Euclidean Distance

The Euclidean distance is usually definition of distance, which reflects the true distance between two vectors in n-dimension space. Assuming that there are two vectors X and Y, the Euclidean distance between X and Y is defined as

$$D_{Euclidean}(X, Y) = \sqrt{\sum_{i=1}^{n} (x_i - y_i)^2} \tag{2}$$

3.2.2 Mahalanobis Distance

The Mahalanobis distance is used to indicate the covariance distance, which can calculate similarity between two unknown sample sets. The Mahalanobis distance between two vectors X and Y in n-dimension space is defined as

$$D_{Mahalanobis}(X, Y) = \sqrt{(X - Y)^T \Sigma (X - Y)} \tag{3}$$

where Σ denotes covariance matrix, it is defined as follows

$$\Sigma = Cov(X, Y) = E\left\{[X - E(X)][Y - E(Y)]^T\right\}$$

3.2.3 Block Distance

The block distance is the sum of the lengths of the projections of the line segment between the vectors onto the coordinate axes [7]. It is defined as

$$D_{Block}(X, Y) = \sum_{i=1}^{n} |x_i - y_i| \tag{4}$$

3.2.4 Minkowski Distance

The Minkowski distance between two vectors X and Y defined as

$$D_{Minkowski}(X, Y) = \sqrt[p]{\sum_{i=1}^{n} |x_i - y_i|^p} \tag{5}$$

This paper cluster the activity vector based on the above distance formulas. Clustering effect is different when using different distance formula. Thus, cophenetic coefficient is used to describe the clustering effect, it is defined as

$$c = \frac{\sum_{i<j} \left(Y_{ij} - y\right)\left(Z_{ij} - z\right)}{\sqrt{\sum_{i<j} \left(Y_{ij} - y\right)^2 \sum_{i<j} \left(Z_{ij} - z\right)^2}} \tag{6}$$

where Y denotes the distance between each vector, and Y_{ij} is the element of Y. Z denotes clustering matrix, and Z_{ij} is the element of Z, y and z are the mean value of Y and Z respectively. This paper define C_E, C_{Ma}, C_B and C_{Min} as the cophenetic coefficient to represent the clustering effect when using Euclidean distance, Mahalanobis distance, Block distance, and Minkowski distance to cluster respectively.

4 Experiment and Results Analysis

The experimental indoor environment is shown in Figs. 4 and 5, which is located in room 421 of our Electronics and Information Engineering Building. Sixteen RF nodes are located around a 4.8 m × 5 m rectangle sensing area. There exist a desk and a bed inside the sensing area, and other heavy obstructions outside. The height of the nodes is 1 m.

In the experiment, the desk and the bed are regarded as the POI in the sensing area, and we define three common daily activities which contain sleeping, reading, and walking indoors. A series of simulations of daily life based on the three activities are designed to illustrate feasibility of the methods proposed in this paper. Without loss of generality, we regard 2 min in the experiment as a day in the reality. Each sim-

Fig. 4 The experimental environment

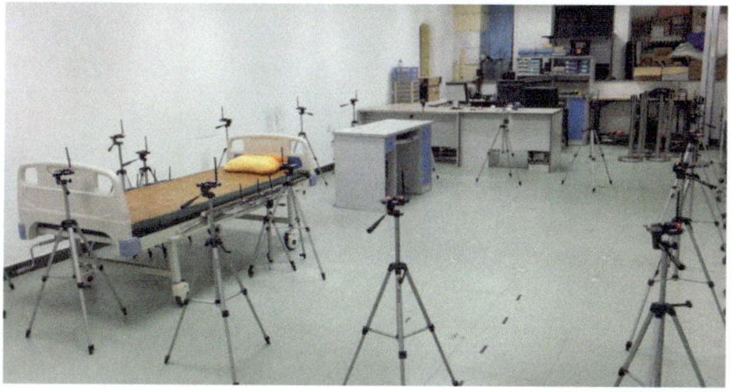

Fig. 5 The platform of experimental environment

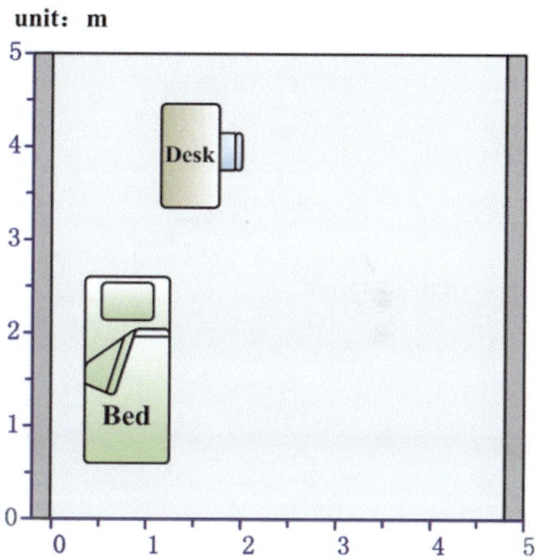

ulation lasts 2 min, and during each two-minute simulation, a person as an example would perform the three activities regularly. 15 two-minute simulations are made to represent daily life of an elder in half a month.

4.1 Classification of Footprint

In the experiment, footprint data is generated per second by running RF positioning system. Thus, each two-minute simulation has 120 footprints in all, which include still footprints and moving footprints. Footprints of a simulation are shown in Fig. 6. Still footprints and moving footprints can be obtained using threshold classification algorithm proposed, both of them are shown in Fig. 7a, b. From Fig. 7a, b, we can see that all footprints close to central coordinate of POIs are identified as still footprints, while a few footprints that move slowly are identified as still footprints incorrectly. Overall, the algorithm of classification is with high accuracy.

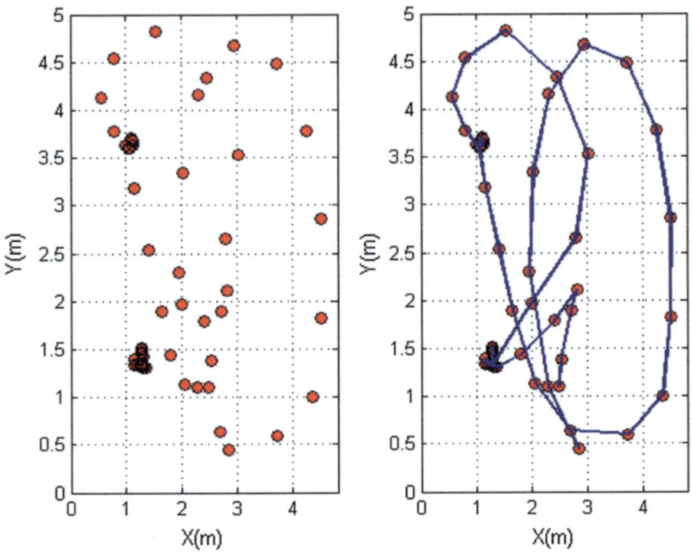

Fig. 6 Raw footprints

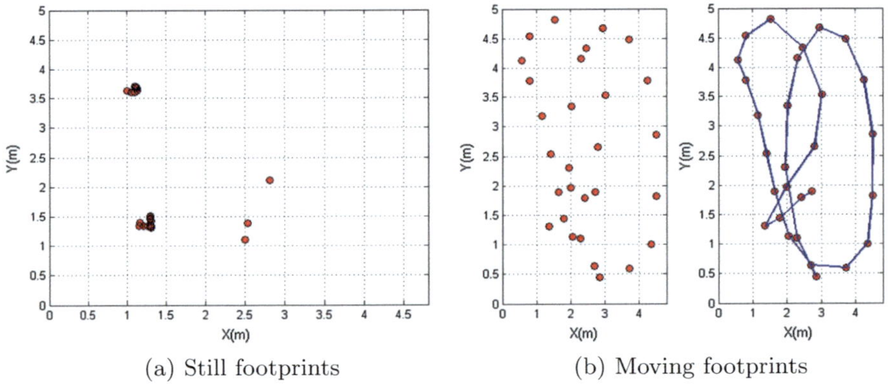

Fig. 7 Classified footprints

In the experiment, moving footprints only represent that the person is walking indoors, while still footprints contain footprints in bed and footprints beside the desk, which indicate sleeping and reading respectively. The two kinds of footprints can be separated using k-means clustering algorithm proposed. POIs central coordinate is able to be measured before the experiment, e.g., if the POI is desk, its central coordinate can be seen as barycentre coordinate of the rectangle. Then, k-means clustering algorithm is used to cluster still footprints, the result is shown in Fig. 8.

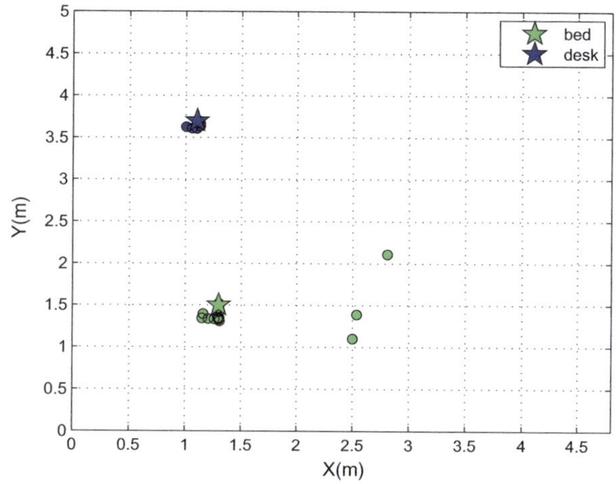

Fig. 8 Footprints after clustering

4.2 Clustering of Activity Vector

Based on the fact that footprint data is generated per second, we can count number of footprints at different POIs to obtain time spent in corresponding activities. Footprints of a two-minute simulation can be transformed into an activity vector. Then the Euclidean distance, Mahalanobis distance, Block distance, and Minkowski distance between each activity vector should be calculated to determine the similarity among the vectors. After clustering activity vectors based on similarity, the cophenetic coefficient proposed is used to evaluate the clustering effect of the four distance formulas proposed above. We can get $C_E = 0.9673$, $C_{Ma} = 0.9046$, $C_B = 0.9703$, $C_{Min} = 0.9673$. Then we find out that the clustering effect of block distance is the best, and the clustering tree based on block distance is shown in Fig. 9. The X-axis represents the number of activity vector, and the Y-axis represents the distance between the vectors. From Fig. 9, we can see that the distance between 8th and 9th activity vector is relatively small, but the distance between them and others is large. Then we can draw a conclusion that the elder is in abnormal condition during 8th and 9th simulation while he is in normal condition during others.

4.3 The Evaluation of Abnormal Behavior

After recognizing the abnormal behaviors, further study must be done to find reason why abnormal behaviors happen. Thus, an abnormal degree is put forward to describe the degree of abnormal behavior deviating from the corresponding normal behavior. The abnormal degree is defined as follows:

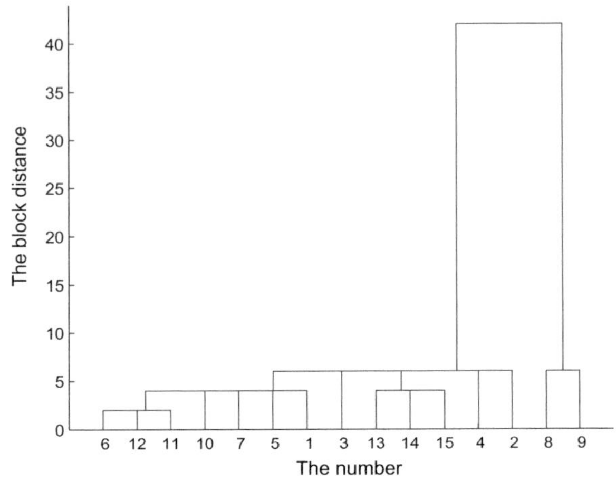

Fig. 9 Clustering tree

$$A_i = \frac{W_i - M}{M} \tag{7}$$

$$M = \frac{1}{m}\sum_{i=1}^{m} V_i$$

where V_i denotes the ith activity vector in normal condition, W_i denotes the ith activity vector in abnormal condition. M denotes the mean of activity vectors in normal condition, m denotes the number of activity vectors in normal condition, In this paper, $m = 13$. The results are as follows:

$$M = [61.69, 29.23, 29.23]$$

$$A_8 = [0.42, -0.45, -0.45]$$

$$A_9 = [0.47, -0.48, -0.52]$$

From the 8th and 9th abnormal degrees, we can find out that increase of sleeping and decrease of reading and walking indoors are the causes of abnormal behaviors. Specifically, in the 8th and 9th simulation, time spent in sleeping is increased by 42 % and 47 % respectively; time spent in reading is decreased by 45 % and 48 % respectively; and time spent in walking indoors is decreased by 45 % and 52 % respectively. We can infer reasonably that the elder may suffer from some diseases in these simulations. Thus, it is helpful for nursing staff to provide more appropriate services.

5 Conclusions

We have addressed the problem of recognizing the abnormal habit of elders, which is an important foundation to provide more appropriate services for them. Our main contribution is introducing a concept of activity matrix and vector to quantify daily activities, and proposing a method to detect abnormal behaviors properly by clustering activity vectors. An experiment using real footprint data obtained by RF localization system are made to verify effectiveness of the methods.

Acknowledgments This work was supported by the National Natural Science Foundation of P.R.China under Grant No.61375080, and the Key Program of Natural Science Foundation of Guangdong, China under Grant No.2015A030311049.

Appendix 1: Radio Tomographic Imaging (RTI): Loss Model

A wireless sensor network with objects using radio frequency (RF) nodes is shown in Fig. 10a. In RF sensor network, the received signal strength (RSS) $y_i(t)$ of link i at time t is described as [8]

$$y_i(t) = P_i - L_i - S_i(t) - F_i(t) - n_i(t) \tag{8}$$

where P_i is transmitted power, S_i is shadowing loss due to objects who attenuate the signal, $F_i(t)$ is fading loss that occurs from constructive and destructive interference of narrowband signals in multipath environment, L_i is static losses due to distance, antenna patterns, etc. $n_i(t)$ is measurement noise, and the unit is decibels.

The shadowing loss $S_i(t)$ can be approximated as a sum of attenuation that occurs in each voxel. Since the contribution of each voxel to the attenuation of a link is different for each link, a weighting is applied. it is described as

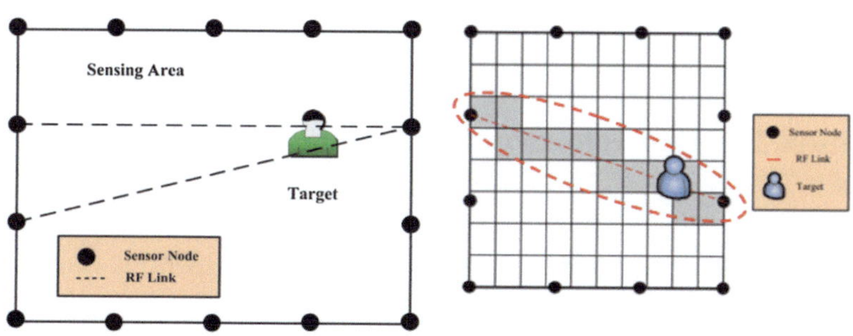

(a) RF sensor network (b) Weight model

Fig. 10 RTI model

$$S_i(t) = \sum_{j=1}^{N} w_{ij} x_j(t) \tag{9}$$

where $x_j(t)$ is the attenuation occuring in voxel j at time t, and w_{ij} is the weighting of pixel j for link i

Imaging only the changing attenuation simplifies the problem, since all static losses can be removed over time. The change in RSS Δy_i from time t_a to t_b is

$$\Delta y_i = \sum_{j=1}^{N} w_{ij} \Delta x_j + \Delta N_i \tag{10}$$

where ΔN_i is the grouping noise, it is defined as

$$\Delta N_i = F_i(t_b) - F_i(t_a) + n_i(t_b) - n_i(t_a)$$

and Δx_j is the difference in attenuation at pixel j from time t_a to t_b, it is defined as

$$\Delta x_j = x_j(t_b) - x_j(t_a)$$

Considering all links in the network, the system of RSS equations can be described in matrix form as

$$\mathbf{\Delta y} = \mathbf{W} \mathbf{\Delta x} + \mathbf{n} \tag{11}$$

where

$$\mathbf{\Delta y} = [\Delta y_1, \Delta y_2 \cdots \Delta y_M]^T$$

$$\mathbf{\Delta x} = [\Delta x_1, \Delta x_2 \cdots \Delta x_N]^T$$

$$\mathbf{n} = [n_1, n_2 \cdots n_M]^T$$

$$[\mathbf{W}]_{i,j} = w_{i,j}$$

Appendix 2: Radio Tomographic Imaging (RTI): Weight Model

An ellipsoid with foci at each node location can be used as a model to determine the weighting for each link in the network [9]. The model is shown in Fig. 10b.

If a particular voxel falls outside the ellipsoid, the weighting for that voxel is set to zero, if a particular voxel is within the ellipsoid, its weighting is set to be inversely proportional to the square root of the link distance. The model is described as [10]

$$w_{ij} = \frac{1}{\sqrt{d}} \begin{cases} 1 & \text{if} \quad d_{ij}^{t} + d_{ij}^{r} < d + \lambda \\ 0 & \text{otherwise} \end{cases} \tag{12}$$

where d is the distance between the two nodes, d_{ij}^{t} and d_{ij}^{r} are the distances from the center of voxel j to the two node locations for link i, and λ is a tunable parameter describing the width of the ellipse.

References

1. Nazerfard E, Rashidi P, Cook DJ (2011) Using association rule mining to discover temporal relations of daily activities. In: Toward useful services for elderly and people with disabilities. Springer, Berlin Heidelberg, pp 49–56
2. Van Kasteren TLM, Englebienne G, Krose BJA (2010) An activity monitoring system for elderly care using generative and discriminative models. Pers Ubiquit Comput 14(6):489–498
3. Andrienko G, Andrienko N, Wrobel S (2007) Visual analytics tools for analysis of movement data. ACM SIGKDD Explor Newsl 9(2):38–46
4. Rashidi P, Mihailidis A (2013) A survey on ambient-assisted living tools for older adults. IEEE J Biomed Health Inf 17(3):579–590
5. Soulas J, Lenca P, Thpaut A (2013) Monitoring the habits of elderly people through data mining from home automation devices data. In: Progress in artificial intelligence. Springer, Berlin, Heidelberg, pp 343–354
6. Rogers S, Girolami M (2011) A first course in machine learning. CRC Press
7. Huang B, Tian G, Wu H et al (2014) A method of abnormal habits recognition in intelligent space. Eng Appl Artif Intell 29:125–133
8. Wilson J, Patwari N (2010) Radio tomographic imaging with wireless networks. IEEE Trans Mobile Comput 9(5):621–632
9. Patwari N, Agrawal P (2008) Effects of correlated shadowing: connectivity, localization, and RF tomography. In: International conference on information processing in sensor networks, pp 82–93
10. Agrawal P, Patwari N (2009) Correlated link shadow fading in multi-hop wireless networks. IEEE Trans Wirel Commun 8(8):4024–4036

Decoupled Tracking Control for a Flexible Multi-body Satellite with Solar Panels and Manipulator

Chaoyi Shi and Yong Wang

Abstract This paper studies the tracking control of a robotic manipulator mounted on a rigid satellite with flexible solar panels. By designing a decoupled feedback controller, the manipulator can track planned paths in the presence of the disturbances from the flexural modes of the panels, and meanwhile, the attitude dynamics of the satellite are stabilized. Stability analysis is proposed based on the Floquet theory for periodic linear systems. Finally, numerical simulations are carried out to validate the controller for the nonlinear model.

Keywords Flexible multi-body systems · Feedback linearization · Stability

1 Introduction

The dynamics and control of flexible multi-body systems have been a focus of research for a long while, due to their importance in many engineering applications, with spacecraft being one of the critical areas. For example, space manipulators on a space station are critical to many important tasks and/or missions, such as loading/unloading equipments and supplies from the cargo ships, as well as assembling/desembling space modules. A flexible multi-body satellite is typically composed of robotic arms, solar panels, and the rigid main body. When a manipulator is moving a relatively massive load around, it can disturb the orientations of the satellite and excite the flexural modes of the solar panels. Moreover, the dynamics of the satellite can become unstable due to nonlinear coupling of rigid body dynamics of the satellite and the manipulator, with those low frequency flexible modes of the panels. Therefore, during the operations of the space manipulators, feedback controls are necessary not only to stabilize the orientation of the satellite, but also to suppress the vibrations of the solar panels. Accordingly, design of robust controllers

C. Shi · Y. Wang (✉)
College of Engineering, Peking University, Beijing
100871, People's Republic of China
e-mail: yongwang@pku.edu.cn

Y. Jia et al. (eds.), *Proceedings of 2016 Chinese Intelligent Systems Conference*, Lecture Notes in Electrical Engineering 405,
DOI 10.1007/978-981-10-2335-4_48

to ensure steady and reliable operation of multi-body satellite systems has become a critical technology.

The dynamic modeling of a multi-body system typically uses the finite element technique, which is based on Newton's laws, or the Hamilton variational principle, or Euler-Lagrange equations [1–3]. In general, the dynamics of a flexible spacecraft do not allow analytical solutions, but qualitative analysis can be employed to understand the dynamic characteristics. Optimal control has been widely used [4, 5], which enables performance indices to be optimized. In [6], stability and performance of closed-loop systems via PID controllers were analyzed based on the frequency design technique. Crassidis et al. [7] used sliding-mode controller design to achieve global stabilization. It should be pointed out that the aforementioned controller designs are based on simplified models, without taking into account of the influence of the dynamics of the manipulator and flexible solar panels.

In this paper, we focus on decoupled tracking control and stability analysis for a flexible multi-body satellite. Comparing with other research work, we consider the manipulator tracking control, together with stabilization of satellite attitude dynamics, all under the disturbances originating from the vibration of solar panels. Controllers are designed to stabilize the orientation of the main body, and to achieve asymptotic path tracking for the manipulator.

2 Problem Statement

In this section, the planar motion of a rigid satellite with a pair of flexible solar panels and a three-link manipulator is simplified to a $(6 + 2n)$-degree-of-freedom (DOF) system, which contains 3-DOF of the main body (2-DOF translational model and 1-DOF rotational mode), 3-DOF of the manipulator, and the first n flexural modes for each of the flexible solar panels. The dynamics of the system are constructed via the Euler-Lagrange equations. We also present the path planning of the manipulator, and the control objectives.

2.1 Model

The planar multi-body model is shown in Fig. 1. Where $[O, i, j]$ denotes the inertial reference frame and $[C, e, f]$ denotes the body frame, which C is the center of mass of the satellite's main body. In the figure, BE and DF are solar panels, with AP, PQ and QR being robotic arms. We assume the manipulator is controlled by electric motors in the joints. We also assume the torques and forces for controlling the main body are generated at point C and G. We assume D and B are located on the e axis, while A and G are located on the f axis. The manipulator end-effector can grasp a load m at point R. The vibrational dynamics of the panels are treated as external disturbances to the dynamics of the main body and the manipulator.

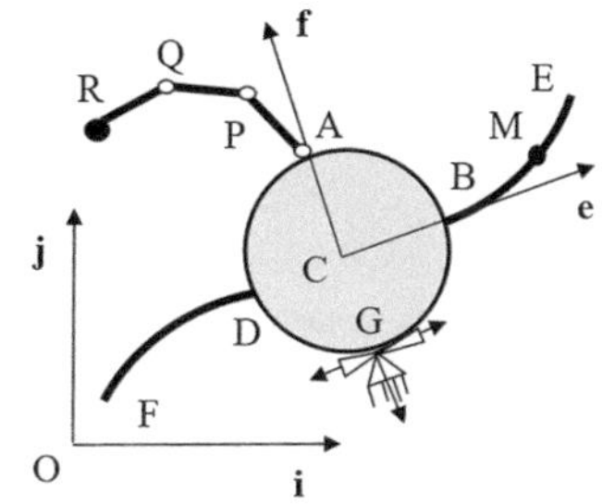

Fig. 1 The flexible multi-body model

Define the state vector

$$q = [x \quad y \quad \theta \quad \theta_1 \quad \theta_2 \quad \theta_3 \quad \boldsymbol{a}^T \quad \boldsymbol{b}^T]^T,$$

where $[x, y]$ is the main body's center of mass position, orientation coordinate θ is the angle between e axis and i axis, and θ_1, θ_2 and θ_3 are relative rotational angles of the manipulator's links. Here $\boldsymbol{a}^T$ and $\boldsymbol{b}^T$ are n-dimensional vectors, representing the amplitudes of the first n flexural modes of the solar panels BE and DF, respectively. Due to practical considerations, the following assumption is made:

Assumption 1 The relative rotational angles of manipulator's links are limited in the open interval $(-\pi, \pi)$.

Define the control vector

$$\boldsymbol{u} = [u_x \quad u_y \quad u_\theta \quad u_{\theta_1} \quad u_{\theta_2} \quad u_{\theta_3} \quad \boldsymbol{0}^T \quad \boldsymbol{0}^T]^T,$$

where u_{θ_1}, u_{θ_2} and u_{θ_3} are the control torques in manipulator's joints. We assume the floating solar panels are not directly controlled. $[u_x \quad u_y \quad u_\theta]^T$ is given by

$$\begin{bmatrix} u_x \\ u_y \\ u_\theta \end{bmatrix} = \begin{bmatrix} \cos\theta & -\sin\theta & 0 \\ \sin\theta & \cos\theta & 0 \\ R & 0 & 1 \end{bmatrix} \begin{bmatrix} T_e \\ T_f \\ T_\theta \end{bmatrix},$$

where T_e and T_f are the thrusts along e and f axis, which are exerted at point G. We assume the torque T_θ is generated by the fly-wheels in the main body.

By writing out the kinetic and potential energies of the flexible satellite, we obtain the Lagrangian L, and the Euler-Lagrangian equations of motion. The $n + 1$ and higher flexural modes of the solar panels are truncated. The generalized equations of motion can be written as

$$M(\boldsymbol{q})\ddot{\boldsymbol{q}} + f(\boldsymbol{q}, \dot{\boldsymbol{q}}) = \boldsymbol{u}, \tag{1}$$

where $M(\boldsymbol{q}) \in \mathbb{R}^{(2n+6)\times(2n+6)}$ is the symmetric inertia matrix, $f(\boldsymbol{q}, \dot{\boldsymbol{q}}) \in \mathbb{R}^{(2n+6)}$ is the nonlinear force, which consists of centrifugal, gyroscopic, and Coriolis forces. The details of the equations are given in the appendix.

2.2 Path Planning

We assume the end-effector of the manipulator grasps a load m along a path from one point to another over time t_f in the body frame. By denoting h: $\mathbb{R}^3 \rightarrow \mathbb{R}^2$ as the direct kinematics map, the path of the end-effector Q can be expressed as

$$z(t) = h(\theta_1(t), \theta_2(t), \theta_3(t)), \tag{2}$$

where $z(t) \in \mathbb{R}^2$ is tracking path in the body frame, with $z(0)$ and $z(t_f)$ being given. It is clear that there is redundancy in the rotating angles of the links, the pass planning of $\theta_k(t)\,(k = 1, 2, 3)$ may be selected by optimization.

Suppose the tracking signals of the rotational angle of the links are given by $\Theta_k(t)\,(k = 1, 2, 3)$,

$$\begin{cases} \Theta_1(t) = \dfrac{1}{2}(\Theta_{10} + \Theta_{1f}) + \dfrac{1}{2}(\Theta_{10} - \Theta_{1f}) \cos \dfrac{\pi t}{t_f}, \\ \Theta_2(t) = \dfrac{1}{2}(\Theta_{20} + \Theta_{2f}) + \dfrac{1}{2}(\Theta_{20} - \Theta_{2f}) \cos \dfrac{\pi t}{t_f}, \\ \Theta_3(t) = \dfrac{1}{2}(\Theta_{30} + \Theta_{3f}) + \dfrac{1}{2}(\Theta_{20} - \Theta_{2f}) \cos \dfrac{\pi t}{t_f}, \end{cases} \tag{3}$$

where $\Theta_{k0}(k = 1, 2, 3)$ are initial angles at $t = 0$, and $\Theta_{kf}(k = 1, 2, 3)$, are the final angles at $t = t_f$. The specific forms of the path given above is one of many possible choices that guarantee the angular velocities of the links relative to the body frame is zero at $t = t_0$ and $t = t_f$, which will not require impulsive torques from the actuators.

2.3 Control Objectives

The objective of the controller design is for the manipulator to achieve asymptotic tracking the planned path (3), and to stabilize the orientation of the satellite.

(i) In the body frame, the manipulator end-effector moves along the desired path (2), and the rotation angles of the links asymptotically tracks the planned path given by (3)
(ii) In the inertial frame, the controllers stabilize the orientation and translational motion of the main body, and meanwhile, the disturbances generated by the vibration of the solar panels can be effectively suppressed.

3 Controller Design

The proposed controller aims at stabilization, asymptotic tracking, and disturbance rejection. The design process consists of three steps. First, a feedforward controller for the manipulator to track the desired path. Second, Eq. (1) is linearized around the planned path. And finally, a linear decoupling feedback controller is designed to linearly stabilize the system.

3.1 *Feedforward Controller*

The feedforward controller guarantees the system tracks the desired path, which is given by

$$\bar{\boldsymbol{q}}(t) = [0 \quad 0 \quad 0 \quad \Theta_1(t) \quad \Theta_2(t) \quad \Theta_3(t) \quad \boldsymbol{0}^T \quad \boldsymbol{0}^T]^T.$$

By substituting the planned path $\bar{\boldsymbol{q}}(t)$ into Eq. (1), we obtain the feedforward controller $\bar{\boldsymbol{u}}(t)$, which is given by

$$\bar{\boldsymbol{u}}(t) = M(\bar{\boldsymbol{q}}(t))\ddot{\bar{\boldsymbol{q}}}(t) + f(\bar{\boldsymbol{q}}(t), \dot{\bar{\boldsymbol{q}}}(t)),$$

where

$$\bar{\boldsymbol{u}} = [\bar{u}_x \quad \bar{u}_y \quad \bar{u}_\theta \quad \bar{u}_{\theta_1} \quad \bar{u}_{\theta_2} \quad \bar{u}_{\theta_3} \quad \boldsymbol{0}^T \quad \boldsymbol{0}^T]^T.$$

3.2 *Decoupling Feedback Controller*

The objective of feedback control is to stabilize the system around the planned trajectory. First, we linearize the system around the desired path $\bar{\boldsymbol{q}}(t)$. Let $\xi = \boldsymbol{q}(t) - \bar{\boldsymbol{q}}(t)$ denote the state vector error signal, and let $\widetilde{\boldsymbol{u}}(t)$ denote feedback input. The linearization of the nonlinear dynamics Eq. (1) near the desired state $(\bar{\boldsymbol{q}}(t), \dot{\bar{\boldsymbol{q}}}(t))$ is given by

$$M(\bar{\boldsymbol{q}})\ddot{\xi} + P(t)\dot{\xi} + R(t)\xi = \widetilde{\boldsymbol{u}}(t), \tag{4}$$

where

$$P(t) = \frac{\partial}{\partial \dot{\boldsymbol{q}}}(\bar{\boldsymbol{q}}(t), \dot{\bar{\boldsymbol{q}}}(t)),$$

$$R(t) = \frac{\partial f}{\partial \boldsymbol{q}}(\bar{\boldsymbol{q}}(t), \dot{\bar{\boldsymbol{q}}}(t)) + \frac{\partial}{\partial \bar{\boldsymbol{q}}}[M(\bar{\boldsymbol{q}}(t))\ddot{\bar{\boldsymbol{q}}}(t)].$$

For the linear system (4), a decoupling controller is constructed as follows,

$$\widetilde{\boldsymbol{u}}(t) = -M(\bar{\boldsymbol{q}}(t))(K_p\xi + K_d\dot{\xi}), \tag{5}$$

where K_p and K_d are diagonal positive matrices, and the controller produces a decoupled linear second order closed-loop system. Furthermore, by substituting Eq. (5) into (4), and letting $\boldsymbol{\eta} = \dot{\xi}$, then the closed-loop system can be written as

$$\begin{bmatrix} \dot{\xi} \\ \dot{\eta} \end{bmatrix} = A(t) \begin{bmatrix} \xi \\ \eta \end{bmatrix}, \tag{6}$$

where

$$A(t) = \begin{bmatrix} 0 & I \\ -\left(K_p + M(\bar{\boldsymbol{q}}(t))P(t)\right) & -(K_d + M(\bar{\boldsymbol{q}}(t))R(t)) \end{bmatrix}$$

is periodic with period $2t_f$, i.e., $A(t) = A(t + 2t_f)$ for all t. Here I denotes the $(6 + 2n) \times (6 + 2n)$ identity matrix.

4 Stability Analysis

The stability of (6) can be determined via Floquet theory [8]. Let $\Phi(t)$ be a fundamental solution matrix of (6), then there exists a constant matrix Q such that

$$\Phi(t + T) = \Phi(t)Q, t \geq 0. \tag{7}$$

Furthermore, if the spectral radius of Q satisfies $\rho(Q) < 1$, then the linear periodic system (6) is exponentially asymptotically stable.

In general, the fundamental solution matrix $\Phi(t)$ cannot be obtained explicitly. Usually one can evaluate Q numerically. Note that by (7) we have

$$\Phi(T) = \Phi(0)Q.$$

Let $\Phi(0) = I$, then we have $Q = \Phi(T)$. This suggests a convenient yet efficient way to compute Q by well developed numerical integration methods.

5 Simulation

We use numerical simulations to validate the effectiveness of the controller design proposed in this paper. We retain the first three flexural modes for each of the solar panels and set load $m = 50$ kg. The system parameters are listed in Table 1. The initial conditions are given by

Table 1 System parameters

Parameters	Value (kg)	Parameters	Value (m)	Parameters	Value	Parameters	Value
$\widetilde{m}_0$	1000	R	1.0	m_3	10 kg	L_3	0.5 m
m_0	50	L	20	α	$1000[0.16\ 0.04\ 0.01]^T$ kg m		
m_1	50	L_1	1.5	β	$20[1\ 1\ 1]^T$ kg m		
m_2	20	L_2	1.0	K	diag{3000 6000 9000}		

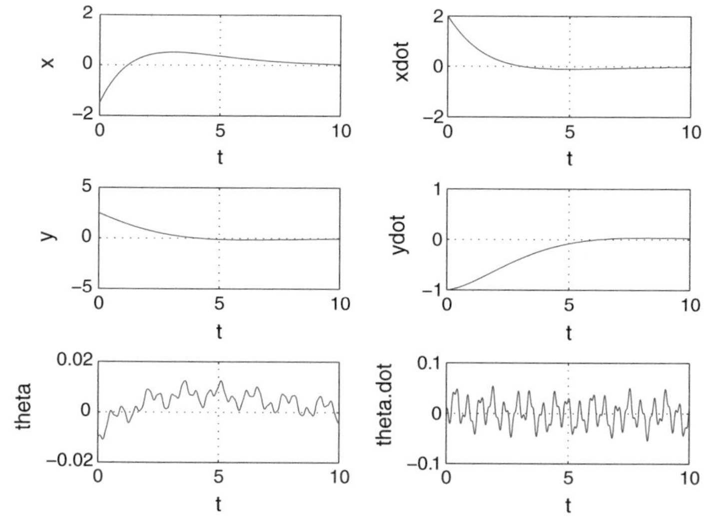

Fig. 2 The orientation and position of main body

$$\xi(0) = \begin{bmatrix}1.5 & 2.5 & -0.01 & \frac{\pi}{5} & \frac{\pi}{3} & \frac{2\pi}{5} & 0.2 & -0.3 & 0.1 & 0.2 & -0.3 & 0.1\end{bmatrix}^T,$$
$$\eta(0) = [2 \quad -1 \quad 0.01 \quad 0 \quad 0 \quad 0 \quad 0.02 \quad -0.01 \quad 0.01 \quad 0.02 \quad -0.01 \quad 0.01]^T,$$

with the initial conditions for the panels' flexural modes being randomly selected. The feedback controller is given by (6) with the feedback gains given by

$$K_p = \text{diag}[0.3 \quad 0.3 \quad 0.15 \quad 1 \quad 2 \quad 3 \quad 0 \quad 0 \quad 0],$$

$$K_d = \text{diag}[0.6 \quad 0.45 \quad 0.5 \quad 0.5 \quad 0.7 \quad 0.9 \quad 0 \quad 0 \quad 0],$$

which were selected by trial and error. The simulation results are shown in Figs. 2 and 3. The figures show the orientation and position of the main body, as well as the positions and angular velocities of manipulators links and control torques and forces. We can see that the manipulator is able to track the desired path and the orientation

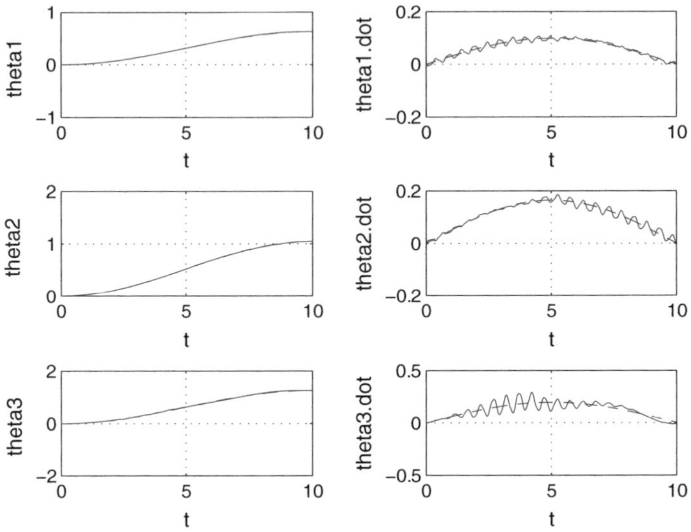

Fig. 3 The positions and angular velocities of manipulators joints

and position of the main body are stable for the controlled system. The maximum error of main body orientation is found to be less than 0.02 rad. Using the Floquet theory, the stability of (6) can be ascertained by computing the spectral radius of Q, and we get $\rho(Q) = 0.3353$. Hence the linearized system (6) is exponentially asymptotically stable.

6 Conclusions

In this paper, modeling and control design for a multi-body satellite system with flexible solar panels and manipulator is proposed. A decoupled controller is designed to achieve asymptotical stability and asymptotic tracking for the manipulator. The parameterized control law is constructed with both feedforward and feedback controllers. The stability of the closed-loop system is proved based on the Floquet theory. To verify the performance of the proposed controller, numerical simulations are performed. The results showed that the closed-loop system achieved the control design objective.

Acknowledgments This work was supported by NSFC (No. 61273111), National Basic Research Program of China (973 Program, No. 2012CB821200).

Appendix

Symbol	Item	Symbol	Item
$\widetilde{m}_0$	Mass of the main body	R	Radius of the main body
m_0	Mass of solar panel *BE*, *DF*	L	Length of *BE*, *DF*
m_1	Mass of robotic arm *AP*	L_1	Length of robotic arm *AP*
m_2	Mass of robotic arm *PQ*	L_2	Length of robotic arm *PQ*
m_3	Mass of robotic arm *QR*	L_3	Length of robotic arm *QR*
ρ	Density of the solar panels	$\boldsymbol{\gamma}$	$\gamma_k = \int_0^L \rho\xi\Lambda_k(\xi)d\xi$
K	Stiffness matrix of *BE*, *DF*	$\boldsymbol{\beta}$	$\beta_k = \int_0^L \rho\Lambda_k(\xi)d\xi$
$\boldsymbol{\Lambda}$	First *n* flexural modes of *BE*, *DF*	$\boldsymbol{\alpha}$	$\alpha_k = \int_0^L \rho\Lambda_k(\xi)^2d\xi$

$$M_0 = \widetilde{m}_0 + 2m_0 \qquad J_0 = \frac{\widetilde{m}_0 R^2}{2} + 2m_0(R^2 + RL + \frac{L^2}{3}) \quad \kappa_1^2 = \frac{1}{3}L_1^2$$

$$\kappa_2^2 = \frac{1}{3}L_2^2 \quad \kappa_3^2 = L_3^2(\frac{1}{3}m_3 + m)/(m_3 + m) \qquad \widetilde{\theta} = \theta + \theta_1 + \theta_2 + \theta_3$$

$$M(1,1) = M_0 + m_1 + m_2 + m_3 + m \qquad M(1,2) = 0$$

$$\begin{aligned} M(1,3) = &-\left(\frac{m_2}{2} + m_3 + m\right) L_2 \cos\left(\theta + \theta_1 + \theta_2\right) - \left(\frac{m_3}{2} + m\right) L_3 \cos\widetilde{\theta} \\ &- \left(m_1 + m_2 + m_3 + m\right) R\cos\theta - \left(\frac{m_1}{2} + m_2 + m_3 + m\right) L_1 \cos\left(\theta + \theta_1\right) \end{aligned}$$

$$\begin{aligned} M(1,4) = &-\left(\frac{m_1}{2} + m_2 + m_3 + m\right) L_1 \cos\left(\theta + \theta_1\right) - \left(\frac{m_2}{2} + m_3 + m\right) \\ &\left[L_2 \cos\left(\theta + \theta_1 + \theta_2\right)\right] - \left(\frac{m_3}{2} + m\right) L_3 \cos\widetilde{\theta} \end{aligned}$$

$$\begin{aligned} M(1,5) = &-\left(\frac{m_2}{2} + m_3 + m\right) L_2 \cos\left(\theta + \theta_1 + \theta_2\right) \\ &- \left(\frac{m_3}{2} + m\right) L_3 \cos\widetilde{\theta} \end{aligned}$$

$$M(1,6) = -\left(\frac{m_3}{2} + m\right) L_3 \cos\widetilde{\theta}$$

$$M(2,2) = M_0 + m_1 + m_2 + m_3 + m$$

$$\begin{aligned} M(2,3) = &-\left(\frac{m_2}{2} + m_3 + m\right) L_2 \sin\left(\theta + \theta_1 + \theta_2\right) - \left(\frac{m_3}{2} + m\right) L_3 \sin\widetilde{\theta} \\ &- \left(m_1 + m_2 + m_3 + m\right) R\sin\theta - \left(\frac{m_1}{2} + m_2 + m_3 + m\right) L_1 \sin\left(\theta + \theta_1\right) \end{aligned}$$

$$\begin{aligned} M(2,4) = &-\left(\frac{m_1}{2} + m_2 + m_3 + m\right) L_1 \sin\left(\theta + \theta_1\right) \\ &- \left(\frac{m_2}{2} + m_3 + m\right) L_2 \sin\left(\theta + \theta_1 + \theta_2\right) - \left(\frac{m_3}{2} + m\right) L_3 \sin\widetilde{\theta} \end{aligned}$$

$$M(2,5) = -\left(\frac{m_2}{2} + m_3 + m\right) L_2 \sin\left(\theta + \theta_1 + \theta_2\right) - \left(\frac{m_3}{2} + m\right) L_3 \sin\widetilde{\theta}$$

$$M(2,6) = -\left(\frac{m_3}{2} + m\right) L_3 \sin\widetilde{\theta}$$

$$\begin{aligned} M(3,3) = {} & J_0 + m_1\left(R^2 + \kappa_1^2\right) + m_2\left(R^2 + L_1^2 + \kappa_2^2\right) \\ & + 2\left(\frac{m_1}{2} + m_2 + m_3 + m\right) RL_1 \cos\theta_1 + \left(m_3 + m\right)\left(R^2 + L_1^2 + L_2^2 + \kappa_3^2\right) \\ & + 2\left(\frac{m_2}{2} + m_3 + m\right) L_2\left[R\cos\left(\theta_1 + \theta_2\right) + L_1\cos\theta_2\right] \\ & + 2\left(\frac{m_3}{2} + m\right) L_3\left[R\cos\left(\theta_1 + \theta_2 + \theta_3\right) + L_1\cos\left(\theta_2 + \theta_3\right) + L_2\cos\theta_3\right] \end{aligned}$$

$$\begin{aligned} M(3,4) = {} & m_1\kappa_1^2 + \left(\frac{m_1}{2} + m_2 + m_3 + m\right) RL_1\cos\theta_1 + \left(m_3 + m\right)\left(L_1^2 + L_2^2 + \kappa_3^2\right) \\ & + m_2\left(L_1^2 + \kappa_2^2\right) + \left(\frac{m_2}{2} + m_3 + m\right) L_2\left[R\cos\left(\theta_1 + \theta_2\right) + 2L_1\cos\theta_2\right] \\ & + \left(\frac{m_3}{2} + m\right) L_3\left[R\cos\left(\theta_1 + \theta_2 + \theta_3\right) + 2L_1\cos\left(\theta_2 + \theta_3\right) + 2L_2\cos\theta_3\right] \end{aligned}$$

$$\begin{aligned} M(3,5) = {} & m_2\kappa_2^2 + \left(m_3 + m\right)\left(L_2^2 + \kappa_3^2\right) \\ & + \left(\frac{m_2}{2} + m_3 + m\right) L_2\left[R\cos\left(\theta_1 + \theta_2\right) + L_1\cos\theta_2\right] \\ & + \left(\frac{m_3}{2} + m\right) L_3\left[R\cos\left(\theta_1 + \theta_2 + \theta_3\right) + L_1\cos\left(\theta_2 + \theta_3\right) + 2L_2\cos\theta_3\right] \end{aligned}$$

$$\begin{aligned} M(3,6) = {} & \left(m_3 + m\right)\kappa_3^2 \\ & + \left(\frac{m_3}{2} + m\right) L_3\left[R\cos\left(\theta_1 + \theta_2 + \theta_3\right) + L_1\cos\left(\theta_2 + \theta_3\right) + L_2\cos\theta_3\right] \end{aligned}$$

$$\begin{aligned} M(4,4) = {} & m_2\left(L_1^2 + \kappa_2^2\right) + \left(m_3 + m\right)\left(L_1^2 + L_2^2 + \kappa_3^2\right) + 2\left(\frac{m_2}{2} + m_3 + m\right) \\ & L_1L_2\cos\theta_2 + 2\left(\frac{m_3}{2} + m\right) L_3\left[L_1\cos\left(\theta_2 + \theta_3\right) + L_2\cos\theta_3\right] m_1\kappa_1^2 \end{aligned}$$

$$\begin{aligned} M(4,5) = {} & m_2\kappa_2^2 + \left(m_3 + m\right)\left(L_2^2 + \kappa_3^2\right) + \left(\frac{m_2}{2} + m_3 + m\right) L_1L_2\cos\theta_2 \\ & + \left(\frac{m_3}{2} + m\right) L_3\left[L_1\cos\left(\theta_2 + \theta_3\right) + 2L_2\cos\theta_3\right] \end{aligned}$$

$$M(4,6) = \left(m_3 + m\right)\kappa_3^2 + \left(\frac{m_3}{2} + m\right) L_3\left[L_1\cos\left(\theta_2 + \theta_3\right) + L_2\cos\theta_3\right]$$

$$M(5,5) = m_2\kappa_2^2 + \left(m_3 + m\right)\left(L_2^2 + \kappa_3^2\right) + 2\left(\frac{m_3}{2} + m\right) L_2L_3\cos\theta_3$$

$$M(5,6) = \left(m_3 + m\right)\kappa_3^2 + \left(\frac{m_3}{2} + m\right) L_2L_3\cos\theta_3$$

$$M(6,6) = \left(m_3 + m\right)\kappa_3^2 \qquad M(7:(6+n),(7+n):(6+2n)) = \mathbf{0}_{n\times n}$$

$$M(1,7:(6+n)) = -\boldsymbol{\beta}^T\sin\theta \qquad M(1,(7+n):(6+2n)) = \boldsymbol{\beta}^T\sin\theta$$

$$M(2,7:(6+n)) = \boldsymbol{\beta}^T\cos\theta \qquad M(2,(7+n):(6+2n)) = -\boldsymbol{\beta}^T\cos\theta$$

$$M(3,7:(6+2n)) = \boldsymbol{\sigma}^T \qquad M(4:6,7:(6+2n)) = \mathbf{0}_{3\times 2n}$$

$$M(7:(6+n),7:(6+n)) = M((7+n):(6+2n),(7+n):(6+2n)) = diag\{\boldsymbol{\alpha}\}$$

$$
\begin{aligned}
f(1) = & \left(m_1 + m_2 + m_3 + m\right) R\dot{\theta}^2 \sin\theta + \boldsymbol{\beta}^T(\dot{\boldsymbol{b}} - \dot{\boldsymbol{a}})\dot{\theta}\cos\theta \\
& + \left(\frac{m_1}{2} + m_2 + m_3 + m\right) L_1 \left(\dot{\theta} + \dot{\theta}_1\right)^2 \sin\left(\theta + \theta_1\right) \\
& + \left(\frac{m_2}{2} + m_3 + m\right) L_2 \left(\dot{\theta} + \dot{\theta}_1 + \dot{\theta}_2\right)^2 \sin\left(\theta + \theta_1 + \theta_2\right) \\
& + \left(\frac{m_3}{2} + m\right) L_3 \left(\dot{\theta} + \dot{\theta}_1 + \dot{\theta}_2 + \dot{\theta}_3\right)^2 \sin\widetilde{\theta} \\
f(2) = & -\left(m_1 + m_2 + m_3 + m\right) R\dot{\theta}^2 \cos\theta + \boldsymbol{\beta}^T(\dot{\boldsymbol{b}} - \dot{\boldsymbol{a}})\dot{\theta}\sin\theta \\
& - \left(\frac{m_1}{2} + m_2 + m_3 + m\right) L_1 \left(\dot{\theta} + \dot{\theta}_1\right)^2 \cos\left(\theta + \theta_1\right) \\
& - \left(\frac{m_2}{2} + m_3 + m\right) L_2 \left(\dot{\theta} + \dot{\theta}_1 + \dot{\theta}_2\right)^2 \cos\left(\theta + \theta_1 + \theta_2\right) \\
& - \left(\frac{m_3}{2} + m\right) L_3 \left(\dot{\theta} + \dot{\theta}_1 + \dot{\theta}_2 + \dot{\theta}_3\right)^2 \cos\widetilde{\theta} \\
f(3) = & -\left(\frac{m_1}{2} + m_2 + m_3 + m\right) RL_1\dot{\theta}_1 \left(2\dot{\theta} + \dot{\theta}_1\right) \sin\theta_1 \\
& - \left(\frac{m_2}{2} + m_3 + m\right) RL_2 \left(2\dot{\theta} + 2\dot{\theta}_1 + \dot{\theta}_2\right) \left(\dot{\theta}_1 + \dot{\theta}_2\right) \sin\left(\theta_1 + \theta_2\right) \\
& - \left(\frac{m_2}{2} + m_3 + m\right) L_1L_2 \left(2\dot{\theta} + 2\dot{\theta}_1 + \dot{\theta}_2\right) \dot{\theta}_2 \sin\theta_2 \\
& - \left(\frac{m_3}{2} + m\right) RL_3 \left(2\dot{\theta} + \dot{\theta}_1 + \dot{\theta}_2 + \dot{\theta}_3\right) \left(\dot{\theta}_1 + \dot{\theta}_2 + \dot{\theta}_3\right) \sin\left(\theta_1 + \theta_2 + \theta_3\right) \\
& - \left(\frac{m_3}{2} + m\right) L_1L_3 \left(2\dot{\theta} + 2\dot{\theta}_1 + \dot{\theta}_2 + \dot{\theta}_3\right) \left(\dot{\theta}_2 + \dot{\theta}_3\right) \sin\left(\theta_2 + \theta_3\right) \\
& - \left(\frac{m_3}{2} + m\right) L_2L_3 \left(2\dot{\theta} + 2\dot{\theta}_1 + 2\dot{\theta}_2 + \dot{\theta}_3\right) \dot{\theta}_3 \sin\theta_3 \\
& + \boldsymbol{\beta}^T(\dot{\boldsymbol{a}} - \dot{\boldsymbol{b}})(\dot{x}\cos\theta + \dot{y}\sin\theta) \\
f(4) = & \left[\left(\frac{m_1}{2} + m_2 + m_3 + m\right) L_1 \sin\theta_1 + \left(\frac{m_2}{2} + m_3 + m\right) L_2 \sin\left(\theta_1 + \theta_2\right)\right. \\
& \left. + \left(\frac{m_3}{2} + m\right) L_3 \sin\left(\theta_1 + \theta_2 + \theta_3\right)\right] R\dot{\theta}^2 \\
& - \left(\frac{m_2}{2} + m_3 + m\right) L_1L_2 \left(\dot{\theta} + 2\dot{\theta}_1 + \dot{\theta}_2\right) \dot{\theta}_2 \sin\theta_2 \\
& - \left(\frac{m_3}{2} + m\right) L_2L_3 \left(2\dot{\theta} + 2\dot{\theta}_1 + 2\dot{\theta}_2 + \dot{\theta}_3\right) \dot{\theta}_3 \sin\theta_3 \\
& - \left(\frac{m_3}{2} + m\right) L_1L_3 \left(2\dot{\theta} + 2\dot{\theta}_1 + \dot{\theta}_2 + \dot{\theta}_3\right) \left(\dot{\theta}_2 + \dot{\theta}_3\right) \sin\left(\theta_2 + \theta_3\right) \\
f(5) = & \left[\left(\frac{m_2}{2} + m_3 + m\right) L_2 \sin\left(\theta_1 + \theta_2\right) + \left(\frac{m_3}{2} + m\right) L_3 \sin\left(\theta_1 + \theta_2 + \theta_3\right)\right] R\dot{\theta}^2 \\
& + \left[\left(\frac{m_2}{2} + m_3 + m\right) L_2 \sin\theta_2 + \left(\frac{m_3}{2} + m\right) L_3 \sin\left(\theta_2 + \theta_3\right)\right] L_1 \left(\dot{\theta} + \dot{\theta}_1\right)^2 \\
& - \left(\frac{m_3}{2} + m\right) L_2L_3 \left(2\dot{\theta} + 2\dot{\theta}_1 + 2\dot{\theta}_2 + \dot{\theta}_3\right) \dot{\theta}_3 \sin\theta_3 \\
f(6) = & \left(\frac{m_3}{2} + m\right) L_3 \Big[R\dot{\theta}^2 \sin\left(\theta_1 + \theta_2 + \theta_3\right) + L_1 \left(2\dot{\theta} + \dot{\theta}_1\right) \dot{\theta}_1 \sin\left(\theta_2 + \theta_3\right)
\end{aligned}
$$

$$+ L_2 \left(\dot{\theta} + \dot{\theta}_1 + \dot{\theta}_2\right)^2 \sin\theta_3\Big]$$

$$f(7 : (6+n)) = -\boldsymbol{\beta}\,(\dot{x}\cos\theta + \dot{y}\sin\theta)\,\dot{\theta} + K\boldsymbol{a}$$

$$f((7+n) : (6+2n)) = \boldsymbol{\beta}\,(\dot{x}\cos\theta + \dot{y}\sin\theta)\,\dot{\theta} + K\boldsymbol{b}$$

References

1. Schiehlen W (2006) Computational dynamics: theory and application of multibody systems. Eur J Mech 25(4):566–594
2. Wasfy TM, Noor AK (2003) Computational strategies for flexible multibody systems. Appl Mech Rev 56(6):533–613
3. Zhang F, Duan GR (2012) Integrated translational and rotational finite-time maneuver of a rigid spacecraft with actuator misalignment. IET Control Theory Appl 6(9):1192–1204
4. Xin M, Pan H (2010) Integrated nonlinear optimal control of spacecraft in proximity operations. Int J Control 83(2):347–363
5. Aghili F (2008) Optimal control for robotic capturing and passivation of a tumbling satellite with unknown dynamics. In: AIAA guidance, navigation and control conference and exhibit, pp 1–21
6. Moradi M (2013) Self-tuning PID controller to three-axis stabilization of a satellite with unknown parameters. Int J Non-Linear Mech 49:50–66
7. Crassidis JL, Vadali SR, Markley FL (2000) Optimal variable-structure control tracking of spacecraft maneuvers. J Guidance Control Dyn 23(3):564–566
8. Khalil HK (2000) Nonlinear systems, 3rd edn. Prentice Hall

Nonlinear Servo Motion Control Based on Unknown Input Observer

Ligang Wang, Yunpeng Li, Jing Na, Guanbin Gao and Qiang Chen

Abstract This paper presents an alternative control method based on a new unknown input observer (UIO) for servo motor systems with unknown time-varying nonlinear dynamics and disturbances. By defining auxiliary filtered variables, an invariant manifold is derived and used to design the estimation of unknown dynamics. The new observer has only one scalar to be set, and thus can be easily incorporated into the control design to achieve precise output tracking. The convergence of the proposed estimator is compared with other three well-known schemes. Comparative simulation results show the satisfactory estimation and control performance.

Keywords Servo motion control · Unknown input observer · Nonlinear systems · Disturbance observer

1 Introduction

Servo motors are a kind of widely used driving motors in the industry applications [1]. To achieve high precision motion control of such mechanisms, it is essential to derive accurate model of the whole systems. However, this is not a trivial task. In practical applications, the uncertainties that degrade the motion control performance include both internal and external disturbances such as friction, load, torque, and also modeling error. To handle such uncertainties and disturbances, there are two widely used approaches: adaptive control and disturbance observer. In the adaptive control framework, e.g., [2, 3], an important assumption is that the unknown

---

L. Wang · Y. Li · J. Na (✉) · G. Gao
Faculty of Mechanical & Electrical Engineering, Kunming University of Science & Technology, Kunming 650500, China
e-mail: najing25@163.com

Q. Chen
College of Information Engineering, Zhejiang University of Technology, Hangzhou 310023, China

Y. Jia et al. (eds.), *Proceedings of 2016 Chinese Intelligent Systems Conference*, Lecture Notes in Electrical Engineering 405,
DOI 10.1007/978-981-10-2335-4_49

dynamics should be strictly reformulated as a linearly parameterized form. To relax this assumption, some functional approximators, e.g., neural network, fuzzy system, were further incorporated into the control synthesis of nonlinear servo motion mechanisms [4–6]. However, the function approximation is only valid for continuous functions in a compact set, and only semi-global stability can be proved.

In the past decades, disturbance observer (DOB) [7, 8] was also proposed, where the disturbances and modeling uncertainties are lumped as a time-varying disturbance, which is estimated using an observer. Traditional design methods of DOB are based on frequency domain techniques so that they cannot be extended to nonlinear systems [9]. In [7], a two-stage design procedure to improve disturbance attenuation ability of linear/nonlinear controllers is proposed. The DOB-based control can compensate the unparameterizable uncertainties, and has a simplified structure. In generic nonlinear DOB design, an observer has a similar structure to original system and there are several parameters to be set. In our recent work [10], we proposed a simply yet effective Unknown input observer (UIO) to address the engine torque estimation. The convergence and robustness are also rigorously analyzed.

The aim of this paper is to exploit the idea of UIO proposed in [10] for the precision motion control of nonlinear servo systems with disturbances. First, we present the design of UIO based on available system variables to design the disturbance estimators. We also compare the estimation response of the proposed UIO to other three estimators, e.g., extended state observer (ESO) [11, 12], nonlinear disturbance observer (NDO) [13], and sliding model observer [14]. The proposed UIO is incorporated into the control design to alleviate the effects of these unknown dynamics, e.g., friction and disturbance. Comparative simulations are included to show the satisfactory control performance.

2 Problem Formulation

In this paper, the following servo motion system driven by a linear DC motor as [15] will be considered as follows:

$$\begin{cases} \dot{x}_1 = x_2 \\ \dot{x}_2 = \left(a\dot{x}_1 + u - f_f - f_r - f_l\right)/b \end{cases} \tag{1}$$

where u is the control voltage, x_1, x_2 are the motor rotation position and speed; f_f is the friction force, f_r is the ripple force, and f_l is the applied load force. The parameters a, b denote the effect of mechanical and electrical dynamics, whose nominal values are available for most physical systems.

The objective of this paper is to introduce an alternative control scheme for system (1) in the presence of unknown dynamics f_f, f_r, f_l. In particular, the UIO proposed in [10] is modified to estimate and then compensate these unknown forces, which leads to a simple but efficient two-step control design procedure.

3 Disturbance Observer Design

We first consider the estimation of the unknown dynamics using the unknown input observer. Thus, we rewrite the second equation of the system (1) as

$$\dot{x}_2 = \left[ax_2 + u - F(x_1, x_2)\right] / b \tag{2}$$

where $F(x_1, x_2) = f_f + f_r + f_l$ is the lumped unknown dynamics.

This section first presents theoretical developments of a new input observer to estimate the unknown dynamics. Without loss of generality, we assume the derivative of $F(x_1, x_2)$ is bounded, i.e., $\sup_{t \geq 0} \left|\dot{F}(x_1, x_2)\right| \leq \hbar$ holds for a constant $\hbar > 0$.

A. *Unknown Input Observer Design*

We define the filtered variables x_{2f}, u_f of x_2, u as

$$\begin{cases} k\dot{x}_{2f} + x_{2f} = x_2, & x_{2f}(0) = 0 \\ k\dot{u}_f + u_f = u, & u_f(0) = 0 \end{cases} \tag{3}$$

where $k > 0$ is a filter parameter.

An ideal invariant manifold [16] will be used to inspire the design of UIO.

Lemma 1 [10] *Consider system (2) and filter operation (3), the variable*

$$\beta = (x_2 - x_{2f}) / k - (ax_{2f} + u_f - F) / b \tag{4}$$

is ultimately bounded for any finite $k > 0$, and

$$\lim_{k \to 0} \left[\lim_{t \to \infty} \{(x_2 - x_{2f}) / k - (ax_{2f} + u_f - F) / b\} \right] = 0,$$

Proof We refer to [10] for a similar proof. ◇

The above ideal invariant manifold provides a mapping from the filtered variables x_{2f}, u_f to the unknown dynamics F. Thus, it can be used to design an estimator for F without knowing any information of $\dot{x}_2$. Based on the invariant manifold, a feasible estimator of $F(x_1, x_2)$ is given by

$$\widehat{F} = ax_{2f} + u_f - b(x_2 - x_{2f}) / k \tag{5}$$

Clearly, only the filter constant $k > 0$ should be selected by the designer.

The convergence property of the proposed observer can be summarized as

Theorem 1 *For system (2) with unknown input observer (5), the estimation error $e_F = F - \widehat{F}$ is bounded by $|e_F(t)| \leq \sqrt{e_F^2(0)e^{-t/k} + k^2 \hbar^2}$ and thus $F \to \widehat{F}$ holds for $k \to 0$ or $\hbar \to 0$.*

Proof We apply a low-pass filter $(\cdot)_f = [\cdot]/(ks+1)$ on both sides of (2), so that

$$\frac{s}{ks+1}[x_2] = \frac{a}{b} \cdot \frac{1}{ks+1}[x_2] + \frac{1}{b} \cdot \frac{1}{ks+1}[u] - \frac{1}{b} \cdot \frac{1}{ks+1}[F] \tag{6}$$

We consider (6) together with the first equation of (3) and have

$$\dot{x}_{2f} = \frac{x_2 - x_{2f}}{k} = \frac{ax_{2f} + u_f - F_f}{b} \tag{7}$$

where F_f is the filtered version of F given by $k\dot{F}_f + F_f = F$. Then it follows from (5) and (7) that $\widehat{F} = F_f$, that is, the estimator gives the filtered version of the unknown dynamics. In this case, we can prove that the estimation error can be small using sufficiently small k. For this purpose, we derive the estimation error as

$$e_F = F - \widehat{F} = \left(1 - \frac{1}{ks+1}\right)F = \frac{ks}{ks+1}[F] \tag{8}$$

To facilitate the convergence proof, we further represent the estimation error (8) in the time-domain as

$$\dot{e}_F = \dot{F} - \dot{\hat{F}} = \dot{F} - \frac{1}{k}\left(F - F_f\right) = -\frac{1}{k}e_F + \dot{F} \tag{9}$$

Select a Lyapunov function as $V = \frac{1}{2}e_F^2$, then its derivative can be given as

$$\dot{V} = e_F\dot{e}_F = -\frac{1}{k}e_F^2 + e_F\dot{F} \leq -\frac{1}{k}V + \frac{k}{2}\hbar^2 \tag{10}$$

We can calculate the solution of (10) as $V(t) \leq e^{-t/k}V(0) + k^2\hbar^2/2$, so $|e_F(t)| \leq \sqrt{e_F^2(0)e^{-t/k} + k^2\hbar^2}$. In this case, one can verify that $e_F(t) \to 0$ for $k \to 0$ and/or $\hbar \to 0$. ◇

B. *Comparison to different disturbance estimation methods*

In this subsection, we will compare the proposed UIO with other three estimators for system (2) to show their convergence and implementation.

B.1: Extended state observer (ESO)

ESO was initially proposed by Han in [11, 12], and has gained many applications [1]. The basic idea of ESO is to regard the lumped disturbances as a new state variable of the system, which can be estimated via a high-gain observer. Considering F as an extended state as $x_3 = F$, then the Eq. (2) can be rearranged as

$$\begin{cases} \dot{x}_2 = (ax_2 + u - F)/b \\ \dot{x}_3 = c(t) \end{cases} \tag{11}$$

where $c(t) = \dot{F}$ is assumed to be bounded. Thus we can design an ESO as

$$\begin{cases} \dot{z}_1 = -[z_2 - \beta_1(z_1 - x_2)]/b + u/b + ax_2/b \\ \dot{z}_2 = -\beta_2(z_1 - x_2) \end{cases} \tag{12}$$

where β_1, β_2 are the feedback gains in the observer, z_1 is the estimation of x_2 and z_2 is the estimation of F. A feasible way to determine β_1, β_2 can be given as $s^2 + \beta_1 s + \beta_2 = (s+p)^2$, where $p > 0$. As analyzed in [17], if $\dot{F}$ is bounded, then $z_1 \to x_2$ and $z_2 \to F$ hold for $p \to \infty$. In this paper, to make a trade-off between the convergence and robustness, we set $p = 1000$ in the simulations. The induced high-gain of ESO leads to a potential peaking phenomena as shown in [17], which may degrade the transient control response when the estimated state z_2 is used.

B.2: Nonlinear disturbance observer (NDO)

The authors of [13] provide a nonlinear disturbance observer to estimate the unknown disturbances. From system (2), we know $F = -b\dot{x}_2 + ax_2 + u$. Then we let $L > 0$ as the observer gain, so that a *direct* DO with exponential convergence can be formulated as

$$\dot{\hat{F}} = -L\hat{F} + L(-b\dot{x}_2 + ax_2 + u) \tag{13}$$

However, the above DO requires *prior* knowledge of acceleration signal $\dot{x}_2$, which may not be available or measured in actual systems.

To address this issue, we design an auxiliary variable $z = \hat{F} + Lbx_2$, and then design the following NDO as

$$\begin{cases} \hat{F} = z - Lbx_2 \\ \dot{z} = -Lz + L(ax_2 + u + Lbx_2) \end{cases} \tag{14}$$

Then the observer error is derived from (14) as

$$\dot{e}_F = \dot{F} - L(-z - ax_2 - u - Lbx_2 + b\dot{x}_2) = -L\left(F - \hat{F}\right) + \dot{F} = -Le_F + \dot{F} \tag{15}$$

It is interesting to find that the error dynamics of NDO shown in (15) are in the same form of that of the proposed UIO. Thus, it can be proved that the observer error e_F of (15) will converge to zero for $\dot{F} = 0$ and/or $L \to \infty$. Thus, the response of NDO is the same as UIO. However, no auxiliary variable needs to be defined in the proposed UIO.

B.3: Sliding mode observer (SMO)

We assume the disturbance $F(t)$ is bounded, i.e., $|F(t)| \leq \lambda(t)$ holds for $\lambda(t) > 0$. Then, we can define the following sliding mode observer

$$\dot{\hat{x}}_2 = \frac{1}{b}[ax_2 + u - \sigma\,\text{sign}\,(x_2 - \hat{x}_2)] \tag{16}$$

with a small positive constant $\sigma > \lambda(t)$.

Then the observer output error between (2) and (16) can be obtained as $e_f = x_2 - \hat{x}_2$, so that its derivative is

$$b\dot{e}_f = -F + \sigma\,\text{sign}\,(e_f) \tag{17}$$

Based on the sliding mode theory and the equivalent control method [14], we know that e_f will reach the sliding mode surface $e_f = 0$ in finite time, and thus $F = (\lambda + \sigma)\text{sign}(e_F)$ for any bounded disturbance F. However, a well-recognized issue in the sliding model observer is the chattering due to the signum function. To reduce the chattering, a low-pass filter is adopted to give the following estimator:

$$\widehat{F} = \frac{1}{ks+1}\left[\sigma\,\text{sign}\,(e_f)\right] \tag{18}$$

In this case, we can verify the estimator error of (18) is the same as (8). Consequently, the steady-state convergence response of the sliding mode observer (16) is comparable to those of UIO and NDO. However, the estimated dynamics may not be smooth although the high-frequency switching can be reduced by introducing the low-pass filter in (18). This will be further shown in simulations. Moreover, the upper bound $\lambda(t)$ of the unknown dynamics $F(t)$ should be known in the sliding mode observer design to determine the constant σ.

4 Control Design with Disturbance Observer

In this section, we will incorporate the proposed UIO into the control design for (1) to achieve output tracking for a given command x_{1d}. System (1) with the estimator (5) can be given as

$$\begin{cases} \dot{x}_1 = x_2 \\ \dot{x}_2 = \frac{1}{b}[ax_2 + u - \widehat{F}(x_1, x_2) - e_F] \end{cases} \tag{19}$$

We define an auxiliary variable defined as

$$p = \dot{e} + k_2 e \tag{20}$$

where e is the tracking error as $e = x_1 - x_{1d}$, and k_2 is a positive constant.
Then we get the derivative of p as

$$\dot{p} = \ddot{e} + k_2\dot{e} = \left[ax_2 + u - \widehat{F}(x_1, x_2) - e_F\right] / b - \ddot{x}_{1d} + k_2\dot{e} \tag{21}$$

The controller can be designed as

$$u = -k_1 p + \widehat{F} - ax_2 - b(k_2\dot{e} - \ddot{x}_{1d}) \tag{22}$$

where $k_1 > 0$ is the feedback gain.
Then the following theorem summarizes the main results of this paper:

Theorem 3 *For the motor system (1), the controller (22) with the estimator (5) is designed. Then, for any unknown dynamics F, the estimation error e_F and the tracking error e will converge to a small compact set around zero, whose size depends on the bound* $\sup_{t \geq 0}|\dot{F}| \leq \hbar$.

Proof Substituting (22) into (21), we have the tracking control error as

$$\dot{p} = \frac{1}{b}(-k_1 p - e_F) \tag{23}$$

Select a Lyapunov function as $V = \frac{1}{2}bp^2 + \frac{1}{2}e_F^2$, so that its time derivative can be calculated along as

$$\dot{V} = bp\dot{p} + e_F\dot{e}_F = -k_1 p^2 - pe_F - \frac{1}{k}e_F^2 + e_F\dot{F} \leq -\alpha V + \frac{\eta}{2}\hbar^2 \tag{24}$$

where $\alpha = \min\{2(k_1 - \eta/2)/b,\ 2(1/k - 1/\eta)\}$ is positive for $k_1 > \eta/2 > k/2,\ k > 0$. Thus, we can obtain from (24) that $V(t) \leq e^{-\alpha t}V(0) + \eta\hbar^2/(2\alpha)$ holds and this implies that p and e_F will exponentially converge to a compact set defined by $\Omega := \left\{p, e_F \middle| |p| \leq \sqrt{\eta\hbar^2/\alpha b}, |e_F| \leq \sqrt{\eta\hbar^2/\alpha}\right\}$. ◇

5 Simulations

This section will present comparative simulation results to demonstrate the validity of the proposed method, and to compare the estimation response of the above mentioned four estimators for F. The parameters of model (1) can be found in [15], which lead to the lumped parameters $a = -123$, $b = 0.69$. Moreover, the ripple force is given by

$$f_r = A_r \sin(2\pi x_1 / P + \varphi) \tag{25}$$

where $\omega = 2\pi / P = 314$ and $\varphi = 0.05\pi$. The friction model is given as

$$f_f = \left[f_c + (f_s - f_c) e^{-(x_2 / \dot{x}_s)^2} \right] \mathrm{sign}(x_2) + Bx_2 \tag{26}$$

where $f_s = 20$, $f_c = 10$, $\dot{x}_s = 0.1$, $B = 10$ define the effects of the maximum static friction, the coulomb friction the Stribeck effect and the viscous friction. Moreover, the external load is given as $f_l = 50\sin(2\pi t)$. In the control design, the filter parameter is $k = 0.001$, and the feedback gains used in the controller are chosen as $k_1 = 2, k_2 = 500$.

Figure 1a shows the tracking responses of the motor position and speed using the presented control (22) with the proposed UIO. It is shown that fairly smooth and

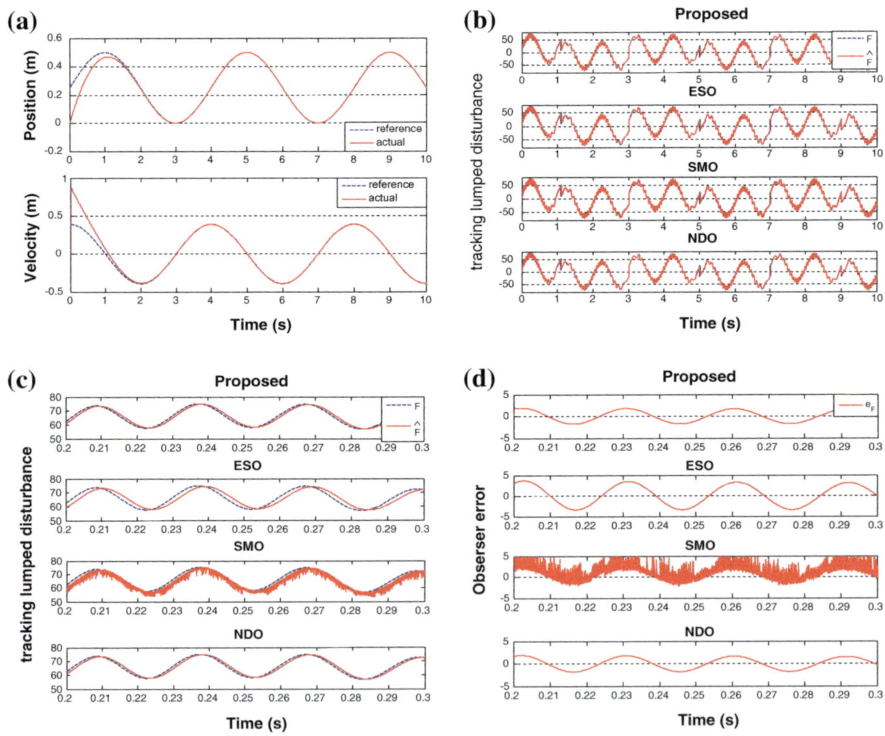

Fig. 1 Simulation results: **a** Tracking control response of the proposed control (22) with (5); **b** Estimation performance of UIO (5), ESO (12), NDO (14) and SMO (18); **c** The zoom-in plot of (**b**); **d** Estimator errors of UIO (5), ESO (12), NDO (14) and SMO (18)

satisfactory control performance can be obtained. The profiles of the estimated disturbances are given in Fig. 1b, c. The first picture of Fig. 1b shows the estimation response of F using the four estimators, and its zoom-in view of the estimation between 0.2 and 0.3 s is shown in Fig. 1c. We can see the major trends of F can be accurately captured, although there is a small phase delay (about 0.001 s). This phase delay comes from the introduced low-pass filter (3). It is noted that k should be chosen as a trade-off between the estimation performance and robustness.

Moreover, we compare their estimation error responses in Fig. 1d. It can be found that the performance of NDO is indeed very similar to that of UIO, which are all better than that of ESO and SMO. In particular, the phase delay of NDO is smaller than that of ESO. Moreover, the implementation of the proposed UIO is simpler than that of ESO. On the other hand, as we stated in Sect. 3, SMO creates oscillated estimation results. The estimation errors of all these four different estimators shown in Fig. 1d further confirm the above analysis.

6 Conclusion

In this paper, we propose a new nonlinear disturbance observer for servo mechanisms by extending the principle of a recently proposed unknown input observer. This new UIO has only one constant to be selected and a simpler structure, while its convergence response is comparable to that of generic NDO, ESO and SMO. The proposed estimator is incorporated into the feedback control design to achieve precision motion control. The closed-loop system stability including the UIO can be rigorously proved. Simulations are given to verify the theoretical analysis. The results demonstrate that the proposed UIO can achieve a superior estimation compared to ESO and SMO. Future work will focus on the robustness analysis for the proposed UIO and other estimators.

Acknowledgments This work is supported by National Natural Science Foundation of China (NSFC) under grant No. 61573174, 61433003.

References

1. Li S, Zong K, Liu H (2011) A composite speed controller based on a second-order model of permanent magnet synchronous motor system. Trans Inst Meas Control 33:522–541
2. Yao B, Xu L (2002) Adaptive robust motion control of linear motors for precision manufacturing. Mechatronics 12:595–616
3. Yang Z-J, Hara S, Kanae S, Wada K, Su C-Y (2010) An adaptive robust nonlinear motion controller combined with disturbance observer. IEEE Trans Control Syst Technol 18:454–462
4. Huang S, Tan KK, Lee TH (2002) Adaptive motion control using neural network approximations. Automatica 38:227–233

5. Na J, Chen Q, Ren X, Guo Y (2014) Adaptive prescribed performance motion control of servo mechanisms with friction compensation. IEEE Trans Ind Electron 61:486–494
6. Na J, Ren XM, Herrmann G, Qiao Z (2011) Adaptive neural dynamic surface control for servo systems with unknown dead-zone. Control Eng Pract 19:1328–1343
7. Chen W-H, Ballance DJ, Gawthrop PJ, O'Reilly J (2000) A nonlinear disturbance observer for robotic manipulators. IEEE Trans Ind Electron 47:932–938
8. Sariyildiz E, Ohnishi K (2015) Stability and robustness of disturbance-observer-based motion control systems. IEEE Trans Ind Electron 62:414–422
9. Chen W-H, Yang J, Guo L, Li S (2015) Disturbance observer-based control and related methods: an overview
10. Na J, Herrmann G, Burke R, Brace C (2015) Adaptive input and parameter estimation with application to engine torque estimation. In: 2015 54th IEEE conference on decision and control (CDC), pp 3687–3692
11. Han J (1995) The "Extended State Observer" of a class of uncertain systems. Control Decis 1
12. Han J (2009) From PID to active disturbance rejection control. IEEE Trans Ind Electron 56:900–906
13. Liu ZL, Svoboda J (2006) A new control scheme for nonlinear systems with disturbances. IEEE Trans Control Syst Technol 14:176–181
14. Edwards C, Spurgeon S (1998) Sliding mode control: theory and applications. CRC Press, Boca Raton
15. Tan K, Huang S, Lee T (2002) Robust adaptive numerical compensation for friction and force ripple in permanent-magnet linear motors. IEEE Trans Magnet 38:221–228
16. Astolfi A, Ortega R (2003) Immersion and invariance: a new tool for stabilization and adaptive control of nonlinear systems. IEEE Trans Autom Control 48:590–606
17. Guo B-Z, Zhao Z-L (2011) On the convergence of an extended state observer for nonlinear systems with uncertainty. Syst Control Lett 60:420–430

H_∞ Filtering for a Class of Discrete-Time Markovian Jump Systems with Missing Measurements

Yunyun Liu and Jinxing Lin

Abstract This paper is concerned with the H_∞ filtering problem for a class of discrete-time Markovian jump system with missing measurements. The measurement missing assumed to occur in random way and the missing probability for each sensor is governed by an individual random variable satisfying a certain probabilistic distribution over the interval [0 1]. Our attention is focused on the design of a filter such that, for the admissible random measurements, missing the error of filtering process is stochastically stable. Using the Lyapunov function combined with projection Lemma is established that the filtering error system is stochastically stable and a guaranteed H_∞ performance constraint is achieved. A numerical example is given to illustrate the feasibility and effectiveness of proposed filter.

Keywords H_∞ filtering · Markovian jump system · Missing measurement · LMI

1 Introduction

As a class of hybrid system, discrete-time Markovian jump systems (DMJs) are popular in modeling many practical systems whose structure is subject to random abrupt variation [1–5]. During the past decades, DMJs have been attracted more and more increasing research interest. To mention just a few, in [6], the filtering for a class of DMJs was investigated; in [7], the control problem of nonlinear DMJs with general transition probabilities allowed to be known, uncertain, and unknown was proposed; and in [8], the author studied the DMJs with partially known transition matrix.

Y. Liu (✉) · J. Lin (✉)
College of Automation, Nanjing University of Posts and Telecommunications,
Nanjing 210023, China
e-mail: lynn996699@126.com

J. Lin
e-mail: jxlin2004@126.com

Y. Jia et al. (eds.), *Proceedings of 2016 Chinese Intelligent Systems Conference*, Lecture Notes in Electrical Engineering 405,
DOI 10.1007/978-981-10-2335-4_50

On another research front, studies on the H_∞ filtering problem have been gaining considerable momentum over the past few years. The main aim is to design a filter such that the mapping from the external input to the filtering error is minimized or is less than a prescribed level according to the H_∞ norm. Up to now, much progress has been made in the study of the H_∞ filtering problem for Markovian jump systems [9–12]. In particular, the H_∞ filtering has been investigated in [13] and [14] for a class of discrete-time Markovian jump systems with missing measurements.

The missing measurement phenomenon usually occurs in networked control systems and has attracted considerable attention during the past few years. A model for the missing measurements phenomena is usually a Markovian jumping process. A jump linear estimator in terms of a predictor is proposed in [14], which can select a corrector gain at each time. Another Bernoulli distribution is adopted to describe the missing behavior, which is considered to be first proposed in [15], wherein an optimal recursive filter is presented for systems with missing measurements. However, most of existing literature [16–18] has implicitly assumed that the measurement signal is neither completely intact (with missing probability 0) nor completely missing (with missing probability 1), and all the sensors have different missing probabilities which according to a certain probabilistic distribution in the interval [0 1].

It should be pointed out that the problem of filtering with missing measurements for DMJs has not fully investigated, while the fault probability of each actuator is individually quantified, and is governed by an individual random variable satisfying a certain probabilistic distribution in the interval [0 1], which are still scarce up to now. The main difficulty is to deal with a problem which how to incorporate the probabilistic missing measurement into a estimation framework. Thus, the intuition motivates this paper to initiate the research on considering missing measurements in the state estimation for DMJs.

This paper is concerned with the problem of H_∞ filtering for DMJs with missing measurements. The missing measurements considered in this paper follow an individual random variable satisfying certain probabilistic distribution in the interval [0 1]. Such a probabilistic distribution could be any commonly used discrete distributions. We have an interest in designing a filter such that the overall filtering error system is stochastically stable. Using the a mode-dependent Lyapunov, the linear matrix inequality (LMI) method and sufficient conditions are derived to ensure the resulting filtering error system is stochastically stable and guaranteed H_∞ performance constraint is achieved. Illustrative a example is given to show the effectiveness of the proposed method.

Notation. The notations used are standard. The subscript "T" denotes matrix transposition. $\mathfrak{R}^n$ denotes the n-dimensional Euclidean space. The notion '*' in a matrix always denotes the symmetric block in the matrix. *diag* $\{\ldots\}$ stands for a block-diagonal matrix. $\lambda_{\max}(P)(\lambda_{\min}(P))$ denotes the largest (smallest) eigenvalue of matrix P. For a real matrix $P, P>0$ means that P is symmetric and positive definite. $E\{x\}$ denotes the mathematical statistical expectation of a stochastic variable x and $E\{x\,|\,y\}$ means the conditional expectation of x givens y.

2 System Description

Consider a probability space (Ω, F, P), where Ω, F and m represent, respectively, the sample space, the algebra of events, and the probability measure, the following discrete-time Markovian jump systems(DMJs) is considered:

$$\begin{cases} x_{k+1} = A(r_k)x_k + B(r_k)w_k \\ y_k = C(r_k)x_k + D(r_k)w_k \\ z_k = E(r_k)x_k \end{cases} \tag{1}$$

where $x(k) \in \Re^n$ is the state vector, $y(k) \in \Re^m$ is the process output, $z(k) \in \Re^p$ is the signal to be estimated, $w(k) \in \Re^q$ is the external noise which belongs to $l_2(0, \infty)$. $\{r_k,\ k \geq 0\}$ is a discrete-time Markovian chain, which takes values in finite set $S = \{1, 2, \ldots, N\}$ (N represent positive integer) with mode transition probabilities (TPS):

$$\Pr\left(r_{k+1} = j \mid r_k = i\right) = \pi_{ij}$$

where $\pi_{ij} \geq 0\ \forall i, j \in S$, and $\sum_{j=1}^{N} \pi_{ij} = 1$. The set S contains N modes of system (1) and for $r_k = i \in S$, the system matrices of ith mode are denoted by $\{A_i, B_i, C_i, D_i\}$, which are considered here to be known with appropriate dimensions.

In this paper, the measurement with sensor data missing is paid special attention, where the multiples are described as

$$\begin{aligned} \hat{y}_k &= \Xi C(r_k)x(k) \ + D(r_k)w(k) \\ &= \sum_{i=1}^{m} \xi_i \Lambda_i C(r_k)x(k) D(r_k)w(k) \end{aligned} \tag{2}$$

where $\hat{y}_k \in \Re^m$ is the measured output vector, $\Xi = dig\,\{\xi_1, \ldots, \xi_m\}$ with ξ_i being m unrelated with $w(k)$. It is assumed that the probabilistic density function has $p_i(s)$ $(i = 1, \ldots, m)$ on the interval [0 1] with mathematical expectation μ_i and variance σ_i^2. Λ_i is defined by

$$\Lambda_i = \left\{ \underbrace{0 \ldots 0}_{i-1},\ 1,\ \underbrace{0 \ldots 0}_{m-1} \right\}$$

Defining $\overline{\Xi} = E\{\Xi\}$, for a matrix $M > 0$, it follows from [16] that

$$\begin{cases} E(\Xi - \overline{\Xi}) = 0 \\ E(\Xi - \overline{\Xi})^T M (\Xi - \overline{\Xi}) = \sum_{l=1}^{m} \sigma_l^2 \Lambda_i^T M \Lambda_i \end{cases} \tag{3}$$

Remark 1 Equation (2) describes the measurement with multiple sensors, in which the diagonal matrix Ξ represents the whole missing status and the missing status and the random variable ξ_i corresponds to the ith sensors $(i = 1, \ldots, m)$. ξ_i can take values on the interval [0 1] and the probability for ξ_i to take different values which may be different. In this paper, we are interested in designing a filter of the following structure:

$$\begin{cases} \hat{x}_{k+1} = A_f(r_k)\hat{x}_k + B_f(r_k)\hat{y}_k \\ \hat{z}_k = E_f(r_k)\hat{x}_k \end{cases} \tag{4}$$

where $\hat{x}_k \in \Re^n$ is the state vector of the filter, $\hat{z}_k$ is the estimation of z_k, and it is mentioned that $A_f(r_k)$, $B_f(r_k)$ and $E_f(r_k)$ are mode-dependent H_∞ filter parameters to be determined.

Defining the filtering error as $e_k = z_k - \hat{z}_k$, the filtering error system is obtained as

$$\begin{cases} \bar{x}_{k+1} = (\overline{A}(r_k) + \widetilde{A}(r_k))\bar{x}_k + \overline{B}(r_k)w_k \\ e_k = \overline{E}(r_k)\bar{x}_k \end{cases} \tag{5}$$

where

$$\bar{x}_k = \begin{bmatrix} x_k \\ \hat{x}_k \end{bmatrix}, \quad \overline{A}(r_k) = \begin{bmatrix} A(r_k)0 & \\ B_f(r_k)\overline{\Xi}C(r_k) & A_f(r_k) \end{bmatrix}, \quad \widetilde{A}(r_k) = \begin{bmatrix} 0 & 0 \\ B_f(r_k)(\Xi - \overline{\Xi})C(r_k) & 0 \end{bmatrix}$$
$$\overline{B}(r_k) = \begin{bmatrix} B(r_k) \\ B_f(r_k)D(r_k) \end{bmatrix}, \quad \overline{E}(r_k) = \begin{bmatrix} E(r_k) & -E_f(r_K) \end{bmatrix}$$

Definition 1 [19] The filtering error system (5) is said to be stochastically stable if the case of $w_k \equiv 0$ for every initial condition $x_0 \in \Re^n$ and $r_0 \in S$, the following inequality holds:

$$E\left\{ \sum_{k=0}^{\infty} \|\bar{x}_k\|^2 \mid \bar{x}_0, r_0 \right\} < \infty \tag{6}$$

Definition 2 [19] Given a scalar $\gamma > 0$, the filtering error system (5) is said to be stochastically stable and has H_∞ noise attenuation, and for all nonzero $w_k \in l_2[0, \infty)$ the following inequality holds:

$$E\left\{ \sum_{k=0}^{\infty} \|\bar{e}_k\|^2 \right\} < \gamma^2 \sum_{k=0}^{\infty} \|w_k\|^2 \tag{7}$$

The purpose of this paper is formulated as follows: Given system (1) with missing measurement, design a filter (4) such that the resulting filtering error system (5) is stochastically stable with a prescribed H_∞ noise performance index.

Lemma 1 (Schur Complement) *Given constant matrices* S_1, S_2, S_3 *where* $S_1 = S_1^T$ *and* $0 < S_2 = S_2^T$, *then* $S_1 + S_3^T S_2^{-1} S_3 < 0$ *if and only if*

$$\begin{bmatrix} S_1 & S_3^T \\ S_3 & -S_2 \end{bmatrix} < 0 \quad \text{or} \quad \begin{bmatrix} -S_2 & S_3 \\ S_3^T & S_1 \end{bmatrix} < 0$$

3 H_{∞} Performance Analysis

In this section, we will provide H_{∞} performance analysis results for the filtering error system (5), which will be used for the filter design in the next section.

Theorem 1 *Consider the filtering error system (5) with given filter parameters. If there exist positive definite matrices* $P_i > 0$, $(i = 1 \ldots m)$ *such that the following matrix inequalities hold*:

$$\begin{bmatrix} -\overline{P}_i & 0 & G & \overline{P}_i\overline{A}_i & \overline{P}_i\overline{B}_i \\ * & -\overline{P}_i & 0 & \overline{E}_i & 0 \\ * & * & -Q_i & 0 & 0 \\ * & * & * & -P_i & 0 \\ * & * & * & * & -\gamma^2 I \end{bmatrix} < 0, \quad i = 1, 2, \ldots, s \tag{8}$$

where

$$G: = [\sigma_1 \overline{B}_1^T \overline{P}_i, \ldots, \sigma_m \overline{B}_m^T \overline{P}_i]^T, \; Q_i: = dig\{\overline{P}_i, \ldots, \overline{P}_i\}, \; \overline{P}_i = \textstyle\sum_{j=1}^{s} \lambda_{i,j} P_j$$

then the filtering error system (5) stochastically stable and H_{∞} *performance constraint is satisfied.*

Proof Define the following Lyapunov functional candidate for system (4):

$$V(\bar{x}(k), k) = \bar{x}_k^T P_i \bar{x}_k \tag{9}$$

Then

$$\begin{aligned} E[\Delta V(\bar{x}_k, k)]: &= E[V(\bar{x}_{k+1}, k+1) \,|\, \bar{x}_k, r_k - V(\bar{x}_k, k)] = \bar{x}_{k+1}^T \overline{P}_i \bar{x}_{k+1} - \bar{x}_k^T P_i \bar{x}_k \\ &= \bar{x}_k^T (\overline{A}_i^T \overline{P}_i \overline{A}_i + \widetilde{A}_i^T \overline{P}_i \widetilde{A}_i - P_i)\bar{x}_k + \bar{x}_k^T \overline{A}_i^T \overline{P}_i \overline{B}_i w_k + \bar{w}_k^T \overline{B}_i^T \overline{P}_i \overline{A}_i \bar{x}_k + \bar{w}_k^T \overline{B}_i^T \overline{P}_i \overline{B}_i w_k \\ &= \bar{x}_k^T (\overline{A}_i^T \overline{P}_i \overline{A}_i + \sum_{i=1}^{s} \sigma_i^2 B_l^T \overline{P}_i B_l - P_i)\bar{x}_k + \bar{x}_k^T \overline{A}_i^T \overline{P}_i \overline{B}_i w_k + \bar{w}_k^T \overline{B}_i^T \overline{P}_i \overline{A}_i \bar{x}_k + \bar{w}_k^T \overline{B}_i^T \overline{P}_i \overline{B}_i w_k \end{aligned}$$

where

$$\widetilde{A}_i^T \overline{P}_i \widetilde{A}_i = \sum_{i=1}^{m} \sigma_i^2 \begin{bmatrix} 0 & 0 \\ B_f c & 0 \end{bmatrix}^T \bar{P}_i \begin{bmatrix} 0 & 0 \\ B_f c & 0 \end{bmatrix}, \quad B_l = \begin{bmatrix} 0 & 0 \\ B_f c & 0 \end{bmatrix}$$

Therefore, with $w_k = 0$, by Schur complement, we can obtain from (8) that $E[\Delta V(\bar{x}_k, k)] < 0$

$$E[\Delta V(\bar{x}_k, k)]: = \bar{x}_k^T \left(\overline{A}_i^T \overline{P}_i \overline{A}_i + \sum_{i=1}^{s} \sigma_i^2 B_l^T \overline{P}_i - P_i \right) \bar{x}_k \leq -\mu \bar{x}_k^T \bar{x}_k < 0 \tag{10}$$

while $\mu = \lambda_{\min}\left\{ -(\overline{A}_i^T \overline{P}_i \overline{A}_i + \sum_{i=1}^{m} \sigma_i^2 B_l^T \overline{P}_i B_l - P_i) \right\}$ as the minimum eigenvalue of $-\left(\overline{A}_i^T \overline{P}_i \overline{A}_i + \sum_{i=1}^{m} \sigma_i^2 B_l^T \overline{P}_i B_l - P_i \right)$

Then,

$$E\left[\sum_{k=0}^{M} \Delta V(\bar{x}_k, i) \right] = E\left[V(\bar{x}_{M+1}, i) | \bar{x}_k, i \right] - E[V(\bar{x}_0, 0)] \leq -\mu E\left[\sum_{k=0}^{M} \bar{x}_k^T \bar{x}_k \right] \tag{11}$$

Thus,

$$E\left[\sum_{k=0}^{M} \bar{x}_k^T \bar{x}_k \right] \leq -\frac{1}{\mu} \{ E[V(\bar{x}_k, 0) - V(\bar{x}_{k+1}, k+1) | \bar{x}_k, k] \} \leq -\frac{1}{\mu} E[V(\bar{x}_0, k)] < \infty \tag{12}$$

The filtering error system (5) is stochastically stable from Definition 1.

Now, to establish the H_∞ performance for the system, consider the following performance index:

$$J: = E\left\{ \sum_{k=0}^{\infty} \|\bar{e}_k\|^2 - \gamma^2 \|w_k\|^2 \right\}$$

Under zero initial condition $V(\bar{x}(k), r_k) \mid_{k=0} = 0$, we have

$$J \leq E\left\{ \sum_{k=0}^{\infty} \left[\|\bar{e}_k\|^2 - \gamma^2 \|w_k\|^2 \right] + \Delta V(\bar{x}_k, k) \right\} = \sum_{k=0}^{\infty} \left[\bar{e}_k^T \bar{e}_k - \gamma^2 w_k^T w_k + \Delta V(\bar{x}_k, k) \right] = \sum_{k=0}^{\infty} \eta_k^T \Phi_i \eta_k$$

where

$$\eta_k = \begin{bmatrix} \bar{x}_k^T & w_k^T \end{bmatrix}^T,$$

$$\Phi_i = \begin{bmatrix} \overline{A}_i^T \overline{P}_i \overline{A}_i + \sum_{i=1}^{s} \sigma_i^2 B_l^T \overline{P}_i B_l - P_i + \overline{E}_i^T \overline{E}_i & \overline{A}_i^T \overline{P}_i \overline{B}_i \\ * & \overline{B}_i^T \overline{P}_i \overline{B}_i - \gamma^2 I \end{bmatrix} \tag{13}$$

By Schur complement, we can obtain from (8) that $\Phi_i < 0$, which mean $\|\bar{e}_k\|^2 < \gamma^2 \|w_k\|^2$ and $J < 0$. The proof is completed.

Remark 2 Note that it is hard to use Theorem 1 to design the desired filter due to the cross coupling of matrix product terms among different system operation modes. To overcome this difficulty, the technique using slack matrix developed in [20] can be adopted here to obtain the following improved BRL for system (5).

Theorem 2 *Consider the discrete-time markovian jump system (1). The filtering error system (5) is stochastically stable and satisfies the H_∞ performance index $\gamma > 0$, if there exist matrices $P_i > 0$, M_i, X_i $(i = 1, \ldots, m)$ satisfying the following LMIs*:

$$\begin{bmatrix} -X_i - X_i^T + \overline{P}_i & 0 & G & X_i\bar{A}_i & X_i\overline{B}_i \\ * & -X_i - X_i^T + \overline{P}_i & 0 & \overline{E}_i & 0 \\ * & * & -M_i - M_i^T + Q_i & 0 & 0 \\ * & * & * & -P_i & 0 \\ * & * & * & * & -\gamma^2 I \end{bmatrix}, \tag{14}$$
$$i = 1, 2, \ldots, s$$

where $\overline{P}_i = \sum_{j=1}^{s} \lambda_{i,j} P_j$, $M_i = \{X_i \ldots X_i\}$.

Proof Define a transformation matrix T as

$$T = \begin{bmatrix} \overline{A}_i^T & I & 0 & I & 0 \\ \overline{B}_i^T & 0 & 0 & 0 & I \\ 0 & 0 & I & 0 & 0 \end{bmatrix} \tag{15}$$

Performing the congruence transformation T to (14) leads to

$$\begin{bmatrix} \overline{A}_i^T \overline{P}_i \overline{A}_i + \sum\limits_{i=1}^{s} \sigma_i^2 B_l^T \overline{P}_i B_l - P_i & \overline{A}_i^T \overline{P}_i \overline{B}_i & \overline{E}_i \\ * & \overline{B}_i^T \overline{P}_i \overline{B}_i - \gamma^2 I & 0 \\ * & * & -I \end{bmatrix} < 0 \tag{16}$$

Appling Schur complement in (16), we can obtain (13). Then, according to Theorem 1, the filtering error system (5) is stochastically stable and satisfies the given H_∞ performance index $\gamma > 0$.

4 H_∞ Filter Design

In the above theorem, we have analyzed the stochastically stable and H_∞ performance of the filtering error system by assuming all the filtering parameters are given and known. In the following, we will propose the design method.

Theorem 3 *Consider the DMJs (1) with missing measurements. The filtering error system (5) is stochastically stable and satisfies a prescribed H_∞ performance index $\gamma > 0$, if there exist matrices P_{1i}, P_{2i}, P_{3i}, A_{Fi}, B_{Fi}, C_{Fi}, U_i, S_i, R_i $(i = 1, \ldots, m)$ and scalar $\xi_i > 0$ satisfying the following LMIs:*

$$\begin{bmatrix} \Pi_{11i} & \Pi_{12i} & 0 & 0 & 0 & G & \Pi_{17i} & \Pi_{18i} & \Pi_{19i} \\ * & \Pi_{22i} & 0 & 0 & 0 & 0 & \Pi_{27i} & \Pi_{28i} & \Pi_{29i} \\ * & * & \Pi_{33i} & \Pi_{34i} & 0 & 0 & 0 & 0 & 0 \\ * & * & * & \Pi_{44i} & 0 & 0 & E_i & -E_f & 0 \\ * & * & * & * & \Pi_{55i} & \Pi_{56i} & 0 & 0 & 0 \\ * & * & * & * & * & \Pi_{66i} & 0 & 0 & 0 \\ * & * & * & * & * & * & -P_{1i} & -P_{2i} & 0 \\ * & * & * & * & * & * & * & -P_{3i} & 0 \\ * & * & * & * & * & * & * & * & -\gamma^2 I \end{bmatrix}, \quad i = 1, 2, \ldots, s \tag{17}$$

$$\begin{bmatrix} P_{1i} & P_{2i} \\ * & P_{3i} \end{bmatrix} > 0 \tag{18}$$

where

$$\begin{aligned} &\Pi_{11i} = \Pi_{33i} = -U_i - U_i^T - \sum_{j=}^{s} \lambda_{ij} P_{ij} \quad , & &\Pi_{12i} = \Pi_{34i} = -R_i - S_i^T - \sum_{j=}^{s} \lambda_{ij} P_{2j} \\ &\Pi_{22j} = \Pi_{44j} = -R_i - R_i^T - \sum_{j=}^{s} \lambda_{ij} P_{3j} \quad , & &\Pi_{17i} = U_i A_i + R_i B_{fi} \overline{\Xi} C_i \\ &\Pi_{18i} = \Pi_{28i} = R_i A_{fi} \quad , & &\Pi_{27i} = S_i A_i + R_i B_{fi} \overline{\Xi} C_i \\ &\Pi_{29i} = s_i B_i + R_i B_{fi} D_i \quad , & &\Pi_{19i} = U_i B_i + R_i B_{fi} D_i \end{aligned} \tag{19}$$

$$\Pi_{55i} = \left\{ -U_i - U_i^T - \sum_{j=1}^{s} \lambda_{ij} P_{1j} \ldots -U_i - U_i^T - \sum_{j=1}^{s} \lambda_{ij} P_{1j} \right\}$$

$$\Pi_{56i} = \left\{ -R_i - S_i^T - \sum_{j=1}^{s} \lambda_{ij} P_{2j} \ldots -U_i - U_i^T - \sum_{j=1}^{s} \lambda_{ij} P_{2j} \right\}$$

$$\Pi_{66i} = \left\{ -R_i - R_i^T - \sum_{j=1}^{s} \lambda_{ij} P_{3j} \ldots -R_i - R_i^T - \sum_{j=1}^{s} \lambda_{ij} P_{3j} \right\}$$

Moreover, if the LMIs have a feasible solution, the parameters of an admissible filter can be given by

$$A_{fi} = R^{-1} A_{Fi}, \; B_{fi} = R^{-1} B_{Fi}, \; E_{fi} = E_{F_i} \tag{20}$$

Proof Assume the matrices P_i and X_i in Theorem 2 in the form of

$$P_i = \begin{bmatrix} P_{1i} & P_{2i} \\ * & P_{3i} \end{bmatrix}, \quad X_i = \begin{bmatrix} U_i & R_i \\ S_i & R_i \end{bmatrix} \tag{21}$$

where R_i is assumed to be nonsingular. After substituting the above matrices into (14) and replacing with $A_{Fi} = R_i A_{fi}$, $B_{Fi} = R_i B_{fi}$ and $E_{Fi} = E_{fi}$, it is easily shown that (18) and (19) are equivalent to (14). Therefore, the filtering error system is guaranteed to be stochastically stable and achieves the prescribed H_∞ performance constraint, and the proof is finally concluded.

5 Numerical Example

In this section, a numerical example is presented to demonstrate the effectiveness and the feasibility of proposed filter. Consider the DMJs with two subsystems and the following parameters:

$$\begin{aligned}
&A_1 = \begin{bmatrix} -2.0 & 0 \\ 0 & -1.9 \end{bmatrix}, \quad B_1 = \begin{bmatrix} 0.15 & 0.4 \\ 0.2 & 0.3 \end{bmatrix}, \quad C_1 = \begin{bmatrix} 0.2 & 0 \\ 0 & 0.2 \end{bmatrix}, \quad D_1 = 0.8 \\
&A_2 = \begin{bmatrix} 0.5 & 0 \\ 0 & 0.3 \end{bmatrix}, \quad B_2 = \begin{bmatrix} 0.15 & 0.4 \\ 0.2 & 0.31 \end{bmatrix}, \quad C_2 = \begin{bmatrix} 0.1 & 0 \\ 0 & 0.1 \end{bmatrix}, \quad D_2 = 0.9 \\
&E_1 = \begin{bmatrix} 0.4 & 0 \\ 0 & 0.4 \end{bmatrix}, \quad E_2 = \begin{bmatrix} 0.4 & 0 \\ 0 & 0.1 \end{bmatrix}
\end{aligned} \tag{22}$$

The transition probability matrix is

$$\Pi = \begin{bmatrix} 0.6 & 0.4 \\ 0.3 & 0.7 \end{bmatrix} \tag{23}$$

In addition, we assume the probabilistic density function of ξ_1 and ξ_2 in [0 1] described by

$$p_2(s_1) = \begin{cases} 0.8 & s_1 = 0 \\ 0.1 & s_1 = 0.5 \\ 0.1 & s_1 = 1 \end{cases}, \quad p_2(s_2) = \begin{cases} 0.7 & s_2 = 0 \\ 0.2 & s_2 = 0.5 \\ 0.1 & s_2 = 1 \end{cases} \tag{24}$$

from which the expectations and variance can be easily calculated as $\mu_1 = 0.15$, $\mu_2 = 0.2, \sigma_1^2 = 0.1025$ and $\sigma_2^2 = 0.11$.

Using Matlab LMI control toolbox to solve the LMIS in (18) and (19), we can calculate that the filter parameters are as follows:

$$A_{f1}=\begin{bmatrix}-2.3001 & 0.1251\\ -0.8294 & -2.2813\end{bmatrix},\quad B_{f1}=\begin{bmatrix}0.2963 & -0.0867\\ 0.1346 & 0.3373\end{bmatrix},\quad E_{f1}=[-0.1000 \quad -2.000]$$
$$A_{f2}=\begin{bmatrix}-0.5087 & -0.5697\\ -0.8539 & -0.7003\end{bmatrix},\quad B_{f2}=\begin{bmatrix}0.9043 & -0.4008\\ 0.7765 & -0.2489\end{bmatrix},\quad E_{f2}=[-0.0996 \quad -0.1999]$$

6 Conclusions

In this paper, the analysis and design of H_∞ filtering for discrete-time Markovian jump systems with missing measurements are investigated. The missing measurements are assumed to occur in a random way, and the missing probability for each sensor is governed by an individual random variable satisfying a certain probabilistic distribution in the interval [0 1]. We have designed a filter for the missing measurements, and the error dynamics of filtering process is stochastically stable. The numerical example implies the effectiveness and feasibility of the proposed approach.

Acknowledgments This work was supported by the National Natural Science Foundation of China (61473158) and the Natural Science Foundation of Jiangsu Province (BK20141430).

References

1. Yang R, Ng BP, Ng GW (2009) Multiple model multiple hypothesis filter with sojourn-time-dependent semi-markov switching. IEEE Signal Process Lett 16(6):529–532
2. Mazor E, Averbuch A, Bar-Shalom Y et al (1998) Interacting multiple model methods in target tracking: a survey. IEEE Trans Aerosp Electron Syst 34(1):103–123
3. Seah CE, Hwang I (2009) Stochastic linear hybrid systems: modeling, estimation, and application in air traffic control. IEEE Trans Control Syst Technol 17(3):563–575
4. Li XR, Bar-Shalom Y (1993) Design of an interacting multiple model algorithm for air traffic control tracking. IEEE Trans Control Syst Technol 1(3):186–194
5. de Farias DP, Geromel JC, do Val JBR et al (2000) Output feedback control of Markov jump linear systems in continuous-time. IEEE Trans Autom Control 45(5):944–949
6. Wu ZG, Shi P, Su H et al (2014) Asynchronous filtering for discrete-time stochastic Markov jump systems with randomly occurred sensor nonlinearities. Automatica 50(1):180–186
7. Shen M, Park JH, Ye D (2015) A separated approach to control of Markov jump nonlinear systems with general transition probabilities. IEEE Trans Cybern 1
8. Xue A, Wang H, Lu R (2016) Event-based control for discrete Markov jump systems. Neurocomputing
9. Dong H, Wang Z, Gao H (2013) Distributed filtering for a class of Markovian jump nonlinear time-delay systems over lossy sensor networks. IEEE Trans Ind Electron 60(10):4665–4672
10. Dong H, Wang Z, Ho DWC et al (2011) Robust filtering for Markovian jump systems with randomly occurring nonlinearities and sensor saturation: the finite-horizon case. IEEE Trans Signal Process 59(7):3048–3057
11. Morais CF, Braga MF, Oliveira RCLF et al (2015) State feedback control for MJLS with uncertain probabilities. Automatica: J IFAC Int Fed Autom Control 52:317–321

12. Niu Y, Dong W, Ji Y (2015) Robust filtering for discrete-time Markov jump linear system with missing measurements. Math Probl Eng 2015:1–9
13. Lu R, Li H, Zhu Y (2012) Quantized filtering for singular time-varying delay systems with unreliable communication channel. Circ Syst Signal Process 31(2):521–538
14. Nahi N (1969) Optimal recursive estimation with uncertain observation. IEEE Trans Inf Theory 15(4):457–462
15. Smith SC, Seiler P (2004) Estimation with lossy measurements: jump estimators for jump systems. IEEE Trans Autom Control 48(12):2163–2171
16. Li W, Jia Y, Zhang JDJ (2014) Distributed multiple model estimation for jump Markov linear systems with missing measurements. Int J Syst Sci 45(7):1484–1495
17. Wei G, Wang Z, Shu H (2009) Robust filtering with stochastic nonlinearities and multiple missing measurements. Automatica 45(3):836–841
18. Peng C, Fei MR, Tian E (2013) Networked control for a class of T-S fuzzy systems with stochastic sensor faults. Fuzzy Sets Syst 212(1):62–77
19. Yin Y, Shi P, Liu F et al (2014) Filtering for discrete-time nonhomogeneous Markov jump systems with uncertainties. Inf Sci 259(3):118–127
20. Zhang L, Boukas EK (2009) Mode-dependent filtering for discrete-time Markovian jump linear systems with partly unknown transition probabilities. Automatica 45(6):1462–1467

Cascade STATCOM Power Factor Automatic Compensation System

Yongdong Guo

Abstract Cascaded H bridge inverter is the key component of the static synchronous compensator (STATCOM). It is widely used in power factor compensation, for its fast response, high reliability, and safety. In this paper, we introduce a sampling system of phase-sensitive circuit into the model cascade STATCOM switch, and propose a grid power factor automatic compensation system. Model of the system operation is then presented. We also analyze the cascade STATCOM switch in normal working condition by applying the switching function of double Fourier transform technique. A simulation platform is constructed to verify the feasibility, correctness, and reliability of the system.

Keywords Phase-sensitive circuit · Cascade STATCOM · Open-circuit fault of power switch · Analytical model · Double Fourier transform

1 Introduction

As static synchronous compensator (STATCOM) is a representative of the new types of reactive power compensator, it has many advantages compared with the traditional reactive power generator [1]. It is an important part of flexible ac transmission, and the recent research also develops towards the aspect of high voltage and large capacity [2]. Due to the pressure limit of the power tube (such as IGBT), most medium-voltage high-power cascade multilevel STATCOM uses H bridge structure [3, 4]. In the actual operation, with the emergence of cascaded H bridge, high-voltage grid power factor automatic compensation derives lower cost, higher reaction speed, and provides more convenience to new users [5]. This paper applies the H bridge power grid aiming to increase the tested module-phase-sensitive circuit. According to state grid power factor, phase-sensitive circuit real-time monitoring grid, working conditions, work gives the STATCOM pulse adjustable wide PMW

Y. Guo (✉)
Henan Polytechnic University, Jiaozuo 454000, China
e-mail: guoyongdong@hpu.edu.cn

Y. Jia et al. (eds.), *Proceedings of 2016 Chinese Intelligent Systems Conference*, Lecture Notes in Electrical Engineering 405,
DOI 10.1007/978-981-10-2335-4_51

signal, in order to realize the reasonable compensation capacitance of high voltage power grid power factor.

2 Phase-Sensitive PWM Circuit

Phase-sensitive circuit is taking A phase CT signal, BC line voltage signal output.

I_A and U_{BC} phase is shown in Fig. 1.

I_A as the load properties and phase change of U_{BC}, due to the changes of I_A and the phase of U_{BC}, the expression is:

$$i_a = \frac{U_a}{(R+L+C)_a} \tag{1}$$

$$i_a = \frac{A\sin(\omega_0 t+\theta)}{(R+L+C)_a} \tag{2}$$

θ as the intersection point of I_A and t and the starting position of the I_A is half shaft, A is range.

2.1.1 the load is purely resistive, as shown in (3), pure resistive load I_A and U_{BC} 90°, at this time

$$i_a = \frac{A\sin\omega_0 t+\theta}{(R)_a} \tag{3}$$

The reactive power, the smallest $\cos\varphi = 1$, the power factor is the largest.

2.1.2 load for the inductive load, when I_a is ahead of U_{BC} voltage 0–90°

$$i_a = \frac{A\sin(\omega_0 t+\theta)}{(R+L)_a} \tag{4}$$

The perceptual load I_{A} phase current and U_{BC} voltage waveform and phase.

2.1.3 load a capacitive resistor, such as:

The load is capacitive I_A phase current waveform and phase with U_{BC}.

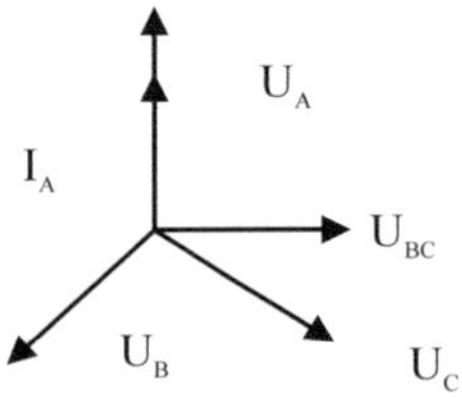

Fig. 1 I_A and U_{BC} phase is shown

When I_A is U_{BC} 90° in advance, then the phase-sensitive output is less than zero, for compensation.

$$i_a = \frac{A\sin(\omega_0 t + \theta)}{(R + c)_a} \tag{5}$$

2.2 phase-sensitive circuit output resolution, phase-sensitive circuit is composed of CT signals and PT. PT signal circuits refers to the BC line voltage signal.

Set the angular frequency of PT input signal EI to 0, omega

$$e_i = E_i e^{JW_0 t} \tag{6}$$

On the type of E_I for the input signal amplitudes, CT and PT in the circuit, a square wave has the same frequency, practical use, are represented as

$$e_i = \begin{cases} -1 & -T/2 + nT < t < nT \\ 1 & nT < t < T/2 + nT \end{cases} \tag{7}$$

Among them, $T = \frac{2\pi}{\omega_0}, n = 1, 2, 3 \ldots$.

Will be launched for the ER spectrum. It provides leaf series, the spectrum. It provides transformation for

$$F_n(n) = \frac{1}{\mathrm{T}} \int_{-\frac{T}{2}}^{\frac{T}{2}} e_i e^{-j\omega_0 t} = \frac{1}{2\pi n j} \tag{8}$$

ER for CT, A phase current signals, thus

$$\begin{aligned} e_r &= \sum_{n=-\infty}^{\infty} \mathcal{F}_n e^{jn\omega_0 t} \\ &= \sum_{n=-\infty}^{\infty} \frac{1}{2\pi n} e^{jn\omega_0 t} \end{aligned} \tag{9}$$

So the output multiplier for

$$e_s = e_i e_r = \sum_{n=-\infty}^{\infty} \frac{\mathrm{E}_r}{2\pi n j} e^{j\omega_{0(n+1)} t} \tag{10}$$

The above structure is a reverse amplifier, ER for CT signal amplitude, through the reasonable collocation R,

Op-amp output

$$F_s = V_i e^r = \begin{cases} Vi & e^r = 0 \\ -Vi & e^r = 1 \end{cases} \tag{11}$$

The signal through the filter circuit processing, due to the noisy circuit EI, so

$$e_i = e_n + e_s \tag{12}$$

Set

$$e_n = \sum_{R=-\infty}^{\infty} F_n(\kappa) e^{j\omega_0 \kappa t} \tag{13}$$

Due to

$$e_0 = e_n \cdot e_r + e_s \cdot e_r \tag{14}$$

$$e_n \cdot e_r = \sum_{n=-\infty}^{\infty} \sum_{\kappa=-\infty}^{\infty} \frac{\mathcal{F}_n(\kappa)}{2\pi nj} e^{j\omega_0 (n+\kappa)t} \tag{15}$$

Only when $n = k$, predominate is output and result as follows:

$$\sum_{\kappa=-\infty}^{\infty} \frac{\mathcal{F}_n(\kappa)}{2\pi \kappa j} \tag{16}$$

System noise is only about 5 % of the original.

3 The Normal Circumstance Cascade STATCOM Working State

Type cascade STATCOM star topology structure as shown in Fig. 1, each phase is composed of multiple series inverter bridge. The control system includes output current closed-loop control and dc voltage balance control.

Literature [6] type cascade STATCOM mathematical model is given as follows:

$$\begin{cases} \mathcal{C} \frac{\mathrm{d}u_{j(t)}}{\mathrm{d}_t} = i(t) \cdot \mathcal{F}_{\mathcal{W}}(t,j) \\ \qquad j = 1, 2, \ldots \\ \sum_{j=1}^{\mathcal{N}} \mathcal{U}_j(t) \cdot \mathcal{F}_{\mathcal{W}}(t \cdot j) + \mathcal{R} \cdot i(t) \\ \qquad + \mathcal{L} \frac{\mathrm{d}i(t)}{\mathrm{d}_t} = \mathcal{V}_s(t) \end{cases} \tag{17}$$

Type of $U_j(t)$ for the first j, an inverter dc capacity of the bridge voltage; $I(t)$ for the inverter current, rated current flowing into the direction of the STATCOM is positive $v_s(t) = V_s \sin(\omega_0 t + \delta)$ For the system phase voltage; for the inverter output voltage with the system voltage angle.

$F_W(t,j)$ for the first j a switching function of inverter bridge, when the output voltage of the power unit $u_j(t)$, when $F_W(t,j) = 1$. The output voltage of $u_j(t)$, when $F_W(t,j) = 1$.

Output zero electricity at ordinary times, $F_W(t,j) = 1$.

3.1 Normal Junior League STATCOM Working Condition

Ignore the line loss, only reactive current of STATCOM current exists, and flow direction of STATCOM is positive, the device current is as follows:

$$i(t) = \sqrt{2}I_a(t)\cos(\omega_0 t) + i_A(t) \tag{18}$$

Type of $I_Q(t)$ as the fundamental reactive current component; $i_a = i(t)$ for the current high-order harmonic component.

3.2 Capacitor Voltage

Substitute (18), (10) into (17) and available capacitance voltage function is as follows:

$$\mathcal{U}_{i(t)} = \frac{1}{C}\int i(t) \cdot e_0 d(t) \tag{19}$$

Type of $\overline{U_j}$ as the capacitor voltage dc component. $u_i(t)$ as the capacitor voltage high-frequency component. In the normal working conditions by (4), the capacitor voltage low-frequency component only contains second harmonic component.

Due to $e_n \cdot e_r$ in (14) is noise, when not to consider the noise, to be seen by (11)

$$\begin{aligned} \mathcal{U}_{i(t)} &= \frac{1}{C}\int i(t) \cdot \mathcal{V}_i \mathrm{d}(t) \\ &= \frac{1}{C}\int \mathcal{V}_i[I_\alpha(t)\sqrt{2}\cos(\omega_0 t) + i_a(t)]\mathrm{d}(t) \end{aligned} \tag{20}$$

4 Phase-Sensitive Circuit Without Power Tube Cascade STATCOM Working Condition

Because of the STATCOM output current for the reactive current, the current phase and the inverter output voltage phase difference is of 90°. STATCOM compensation inductive reactive power, for example, the electric currents generated by H bridge leading voltage is 90°. For single polarity PWM modulation, H bridge with six kinds of working condition is shown in Fig. 2. S1 to S4 conduction, current flows through the S1, S4, S1S4 conduction, current d1, d4, S2, S3 conduction, current flows through the S2, S3.

S1 and S3 conduction or S2, S4 conduction S1 and S3 conduction or S2, S4 conduction.

The working status of Fig. 3 STATCOM power unit (Fig. 4).

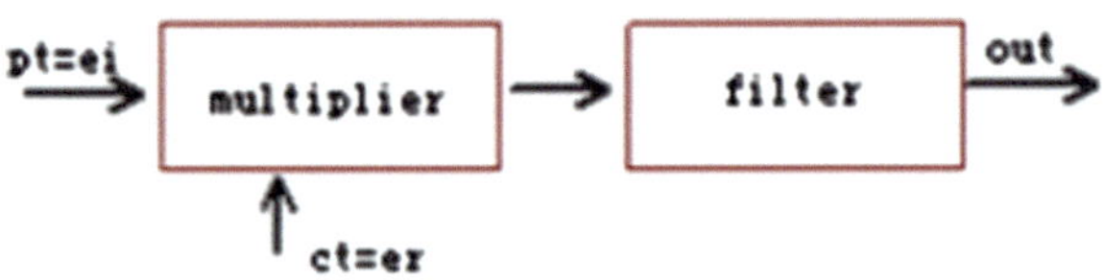

Fig. 2 Phase-sensitive circuit structural model

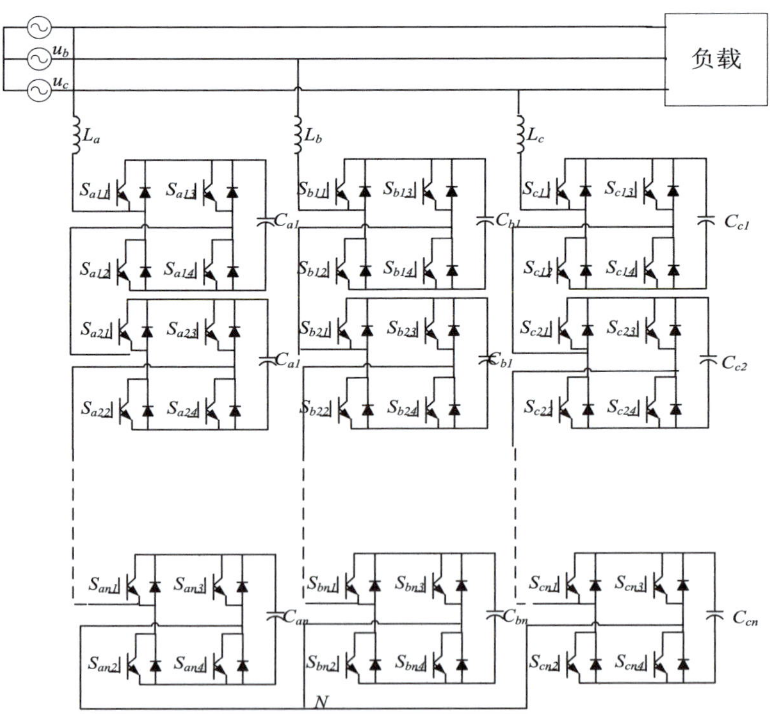

Fig. 3 Schematics of cascade STATCOM

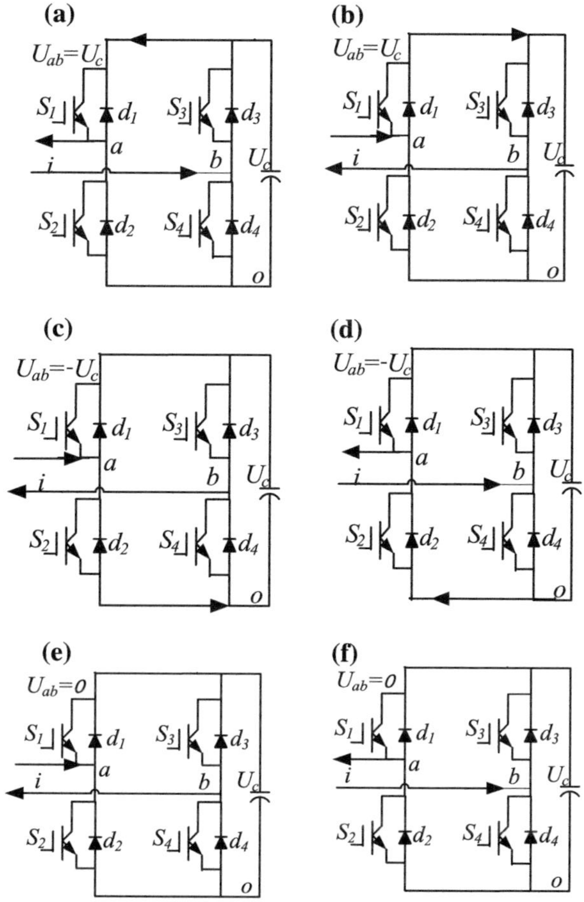

Fig. 4 Working status of H bridge. **a** $2k\pi < \omega_0 t < 2k\pi + \frac{\pi}{2}$. **b** $2k\pi + \frac{\pi}{2} < \omega_0 t < 2k\pi + \pi$. **c** $2k\pi + \pi < \omega_0 t < 2k\pi + \frac{3\pi}{2}$. **d** $2k\pi + \frac{3\pi}{2} < \omega_0 t < 2k\pi + 2\pi$. **e** $2k\pi - \frac{\pi}{2} < \omega_0 t < 2k\pi + + \frac{\pi}{2}$. **f** $2k\pi + \frac{\pi}{2} < \omega_0 t < 2k\pi + \frac{3\pi}{2}$

In Fig. 2e, f to zero level output, and a grid cycle by S1 and S3 conduction time and S2, the zero level S4 of conduction time is equal to zero level.

4.1 Switch Function Analysis

If S1 tube open-circuit fault occur, its equivalent circuit is shown in Fig. 5.

At this time for switching function $F(t,j)$ in $2k\pi < \omega_0 t < 2k\pi + \pi/2$ period, the current through S_1 tube, thus can only flow from S_2 of parallel diode, this period of time switching function $F(t,j) = 0$. Similarly, $F(t,j)$ in $2k\pi - \pi/2 < \omega_0 t < 2k\pi + \pi/2$, originally by S1 and S3 conduction of zero level will become a current flow through the d_2, S_3, $-U_{DC}$. At this time of a grid cycle switch function double Fourier decomposition.

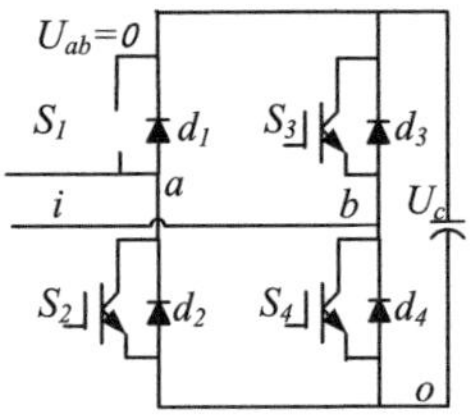

Fig. 5 Power tube open-circuit equivalent circuit

$$\begin{aligned}\mathcal{F}_{\mathcal{W}}(t,j) = & \frac{2m_a+\pi}{8\pi}+\frac{3}{4}m_a\sin(\omega_0 t)-\frac{m_a}{2\pi}\cos(\omega_0 t)\\ & +\sum_{\ell=2,4,6}^{\infty}[A_{0P}\cos(\mathcal{P}\omega_0 t)+\mathcal{B}_{0\mathcal{P}}\sin(\mathcal{P}\omega_\alpha t)]\\ & +\sum_{\ell=1}^{\infty}\sum_{\mathcal{P}=-\infty}^{\infty}[\mathcal{A}_{\ell\mathcal{P}}\cos(\omega_0 t+\mathcal{P}\omega_0 t)\\ & +\mathcal{B}_{\ell\mathcal{P}}\sin(\ell\omega_0 t+\mathcal{P}\omega_{\alpha 0}t)]\end{aligned} \tag{21}$$

Type in the

$$A_{0\mathcal{P}}+jB_{0\mathcal{P}}=-\frac{m_a}{\pi}\int_0^{\frac{\pi}{2}}\sin y e^{j\mathcal{P}y}dy$$

$$A_{\ell\mathcal{P}}+jB_{\ell\mathcal{P}}=\frac{2e^{j\frac{\pi}{2}y}}{j\ell\pi^2}\int_0^{\frac{\pi}{2}}[e^{j\frac{\pi}{2}\ell m_a\sin y+j\mathcal{P}y}-e^{j\frac{\pi}{2}\ell m_a\sin y+j\mathcal{P}y}]\mathrm{d}y$$

4.2 The Results of Simulation and Analysis

In this section, the simulation verifies the correctness of the above derivation. The simulation parameters are as follows:

Cascade number $N = 6$, the grid line voltage RMS $U_s = 10$ kV, load $R = 26\ \Omega$, $L = 30$ μh. Capacitance voltage directive value is 1800 V, size of 1500 μm, F output inductance $L = 30$ μh. Rated modulation than $m_a = 0.535$, rated output current effective value of 70.7 A.

Figures 6 and 7 respectively after the compensation system of the current waveform and spectrum can be seen in the controller system which under current THD dropped to 3.83 %, the waveform is basically consistent and system voltage waveform (Figs. 8 and 9).

Obviously, the control system will not allow such a large fluctuations, it will be by adjusting the switching function to achieve a smaller steady-state value fluctuations. 50 Hz wave simulation results to 51.33 V (note: as the delta changed little,

Fig. 6 The system current waveform beforn compensation

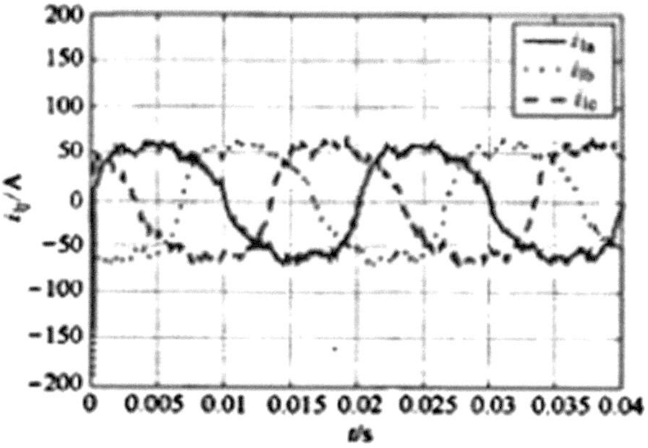

Fig. 7 The system current spectrum before compensation

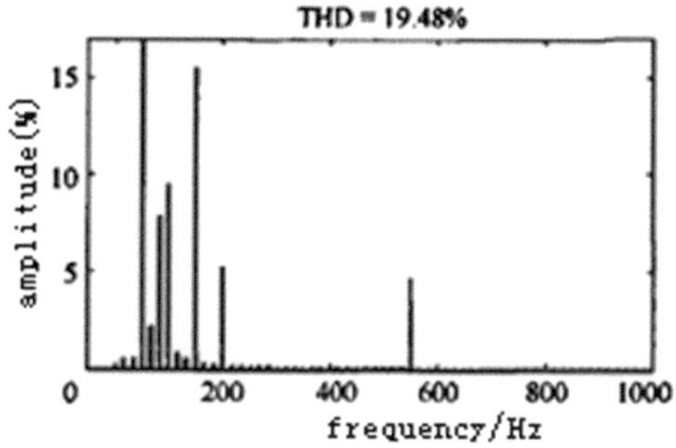

Fig. 8 The system current waveform after compensation

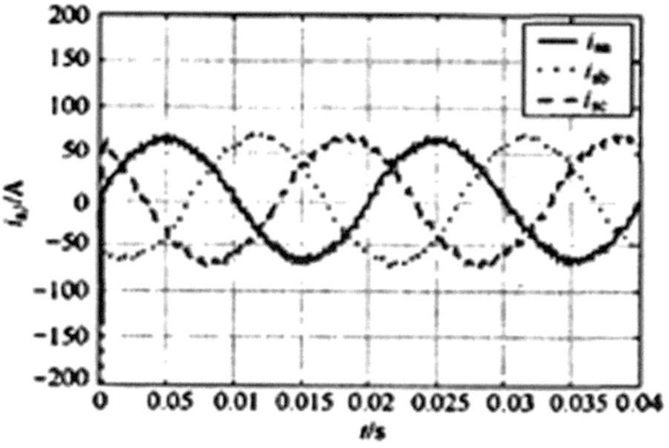

ignore the influence of the delta) here. At the same time, the fault unit about 40 V dc side voltages is rising, stable average in the vicinity of 1840 V is consistent with theoretical analysis. Can the two features position error units. At the same time, through the simulation, power unit to any power tube will be open.

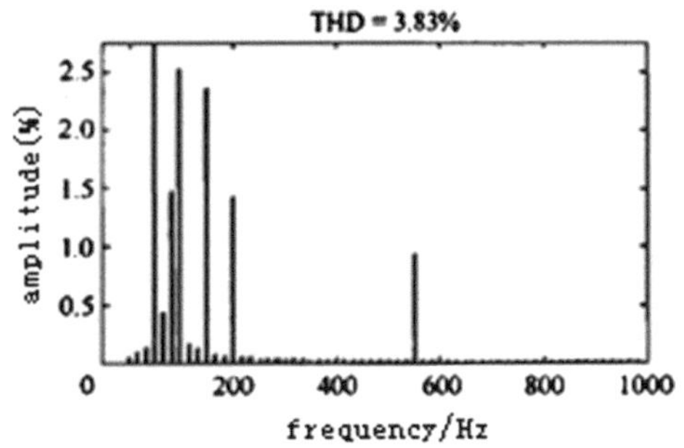

Fig. 9 The system current spectrum afford compensation

5 The Feasibility Analysis and Summarize

This paper investigates the power factor compensation system from theory on the essence of research, but in practice must be revised, through hardware bandpass filter, fully able to identify the 50 Hz voltage fluctuation and the rise of the dc voltage, thus can be used in many slopes on the type of STATCOM.

This method has the following advantages: (1) using transistors as switching element make the grid power factor adjustment control, convenient, simple, and reliable; (2) use of STATCOM device for sensor, and less devices.

References

1. Peuget R, Courtine S, Rognon JP (1998) Fault detection and isolation on a PWM inverter by knowledge-based model. IEEE Trans Ind Appl 34(6):1318–1326
2. Yang Z, Wu Z, Li H (2008) Invester fault diagnose based on detecting DC side current. Proc CSEE 27:005
3. Hu W, Xia L, Xiang D et al (2011) Rotor fault diagnosis based on DC-side current and fusion algorithm for induction motors. Electr Power Autom Equip 8:008
4. Xiao L, Li R (2006) Research on the open-circuit. In: Proceedings of the Chinese society of electrical engineering, vol 26, no 4, pp 99–101
5. Geng J, Liu W, Yu X et al (2003) Modeling of cascade STATCOM. Proc CSEE 23(6):66–70
6. Zhang X, Chan D (1997) A phase-sensitive detection circuit. Journal of Guangxi Normal University

Voltage-Balanced Control for a Cascaded 3H-Bridge Rectifier

Ziyi Fu, Boxiang Zhang and Xuejuan Xiong

Abstract This paper studies the CHBR (Cascaded H-Bridge Rectifier)'s DC side capacitor voltage balance control algorithm, introduces single-phase DQ coordinate transformation, deduces the mathematical model in this coordinate. This uses feed-forward decoupled control strategy to keep voltage and current of AC side in the same phase, puts forward an improved algorithm to balance DC capacitor voltage even if the loads of CHBR are imbalanced. Finally, establishes a simulation model of three cells' CHBR. The simulation results indicate that this improved algorithm can reduce the switching frequency effectively, and balance the DC voltage of every module much better.

Keywords Cascaded H-bridge rectifier · Bridge circuits · Power factor · Voltage control · Simulink

1 Introduction

The cascade multilevel converter is used to control the high voltage and large-capacity electric drive. It is easy to control and has modular structure, good expansibility. The H-bridge circuit is the basic power conversion unit. Several units are cascaded and become a simple cascaded circuit [1–4]. H-bridge cascaded multilevel converter DC side is connected with capacitor, and isolated from each other. So there are multigroup independent DC bus voltages. Furthermore, each H-bridge converter unit is a nonlinear, multivariable, strong coupling system, and each of the converter units has different parallel loss, switching loss, and modulation. Those problems make DC capacitor voltage of each converter unit unbalance. So how to balance the DC capacitor voltage is a research focus of cascaded H-bridge multilevel topology [5–8].

Z. Fu (✉) · B. Zhang · X. Xiong
School of Electrical and Electronic Engineering, Henan Polytechnic University, Jiaozuo 454003, China
e-mail: fuzy@hpu.edu.cn

Y. Jia et al. (eds.), *Proceedings of 2016 Chinese Intelligent Systems Conference*, Lecture Notes in Electrical Engineering 405,
DOI 10.1007/978-981-10-2335-4_52

The balanced control of DC capacitor is related to the output waveform's quality, active, and reactive power. Because of the cascaded topology, current of each power unit is equal and it means that DC output voltage must be adjusted by the same current. This feature makes it difficult to control the DC output voltage, what is more may appear as system stability problems in some cases [9, 10].

There are two main methods to solve this problem: complex modulation method and closed-loop control method. Complex modulation method contains several kinds of ways such as two-dimensional modulation method, space vector pulse width modulation method, exchange balanced modulation method, and so on [11–14]. Those methods can balance the DC capacitor quickly, accurately, and effectively. But the modulation strategies need to be designed according to the specific converter structure. So it is hard to be extended to multi-cascaded H-bridge topology. Another one is closed-loop control method. Its basic thought is to closed-loop control the capacitor voltage of each H-bridge unit separately. Common methods are based on proportional integral (PI) regulators to control. In this paper, an improved control algorithm based on exchange balanced modulation is proposed. First, control the total input voltage of the CHBR, introduce the single-phase DQ coordinate transformation to separate active and reactive power. Separately control active and reactive power, synthesis desired voltage waveforms of AC side. Sort the DC capacitor voltage of each module to determine the charge and discharge status of them. This algorithm can balance the DC capacitor voltage simply and effectively. It also can reduce switching frequency so that the energy loss of the system is less, furthermore the life of the switch becomes longer.

2 Feed-Forward Decoupling Control of CHBR

Figure 1 shows the basic topology of CHBR, it is composed of series of n power units, each power unit contains four switches with anti-parallel diodes, DC side is connected with DC bus capacitors and loads.

According to Kirchhoff's voltage law (KVL) and Kirchhoff's current law (KCL), the mathematical model of the duty cycle of CHBR is established on the basis of the topology

$$\begin{cases} U_s = \sum_{i=1}^{n} D_i U_{ci} + L\frac{di_s}{dt} + R_s i_s \\ D_i i_s = C_i \dfrac{dU_{ci}}{dt} + \dfrac{U_{ci}}{R_i} \\ U_{an} = \sum_{i=1}^{n} D_i U_{ci} \end{cases} ; \quad i = 1, 2, \ldots, n \tag{1}$$

where $U_s i_s R_s L D_i U_{ci} C_i R_i$ are grid voltage, input current, and equivalent resistance and inductance, duty ratio of each power unit, DC side capacitor voltage

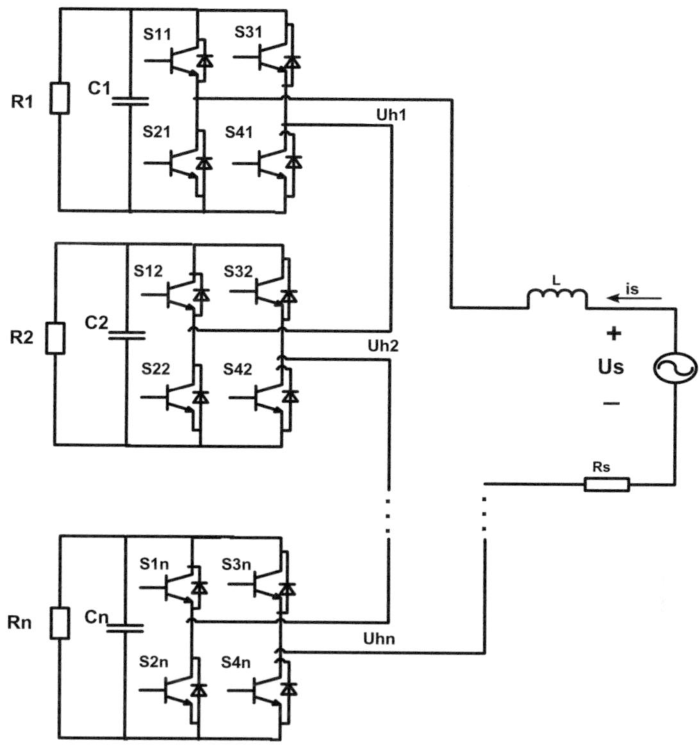

Fig. 1 Topology of Cascaded H-bridge converter contains n power units

capacitance, and load resistance respectively. Where U_{an} is the average value of the CHBR AC side voltage in a switching period.

In three-phase PWM rectifier system, through Clark and Park transformation, the sine wave signal in the three-phase stationary symmetrical coordinate is transformed into the DC signal in the two-phase synchronous rotating coordinate. But each power unit of CHBR is a single-phase rectifier, there is only one DC value, cannot do coordinate transformation directly. But construct a vertical dummy variable, which can also be the stationary coordinate of sinusoidal signals are transformed into DC signal in synchronous rotating coordinate by coordinate transformation.

Arbitrary sinusoidal value $X_{(t)} = X_m \sin(\omega t - \varphi)$ can be decomposed by trigonometric function transformation

$$X_{(t)} = X_m \sin(\omega t - \varphi) = X_m \cos\varphi \ \sin\omega t - X_m \sin\varphi \ \cos\omega t \tag{2}$$

Define $X_d = X_m \cos\varphi X_q = X_m \sin\varphi$, so (2) become

$$X_{(t)} = X_d \sin\omega t - X_q \cos\omega t \tag{3}$$

In this way X_d, X_q can be regarded as two orthogonal direct value in the DQ synchronous rotating coordinate, then control them independently. Compare with the direct control. This control method can eliminate the static error and improve the anti-interference ability. For each CHBR power unit, it is necessary to construct a virtual current i_x to carry out Park transform. Input current i_s is converted to direct current. Virtual current is shown in (4)

$$\begin{cases} i_s(t) = I_m \cos(\omega t - \varphi) \\ i_x(t) = -I_m \cos\left(\omega t - \varphi - \frac{\pi}{2}\right) \end{cases} \tag{4}$$

Then $i_s(t)$ and $i_x(t)$ can be written as the form of (2)

$$\begin{bmatrix} i_s \\ i_x \end{bmatrix} = \begin{bmatrix} \sin\omega t & -\cos\omega t \\ \cos\omega t & \sin\omega t \end{bmatrix} \begin{bmatrix} i_d \\ i_q \end{bmatrix} \tag{5}$$

Then

$$\begin{bmatrix} i_d \\ i_q \end{bmatrix} = \begin{bmatrix} \sin\omega t & \cos\omega t \\ -\cos\omega t & \sin\omega t \end{bmatrix} \begin{bmatrix} i_s \\ i_x \end{bmatrix} \tag{6}$$

The values in (1) can be expressed as following formula:

$$\begin{cases} U_s(t) = U_m \sin\omega t = u_d \sin\omega t \\ i_s(t) = i_d \sin\omega t - i_q \cos\omega t \\ U_{an}(t) = U_{and} \sin\omega t - U_{anq} \cos\omega t \end{cases} \tag{7}$$

Then take those values to (1), the mathematical model of CHBR under synchronous rotating coordinate system is obtained

$$\begin{cases} U_m = L\frac{di_d}{dt} + Ri_d + \omega L i_q + U_{and} \\ L\frac{diq}{dt} + Ri_q - \omega L i_d + U_{anq} = 0 \end{cases} \tag{8}$$

As is shown in (8), controlled variables i_d, i_q are coupled with each other, so it is hard to control them independently. This paper draws lessons from feed-forward decoupling control strategy of three-phase PWM rectifier, and applies in this model. The control equation is deduced

$$\begin{cases} U_{and} = -\left(K_p + \frac{K_i}{s}\right)\left(i_d^* - i_d\right) - \omega L i_q + U_m \\ U_{anq} = -\left(K_p + \frac{K_i}{s}\right)\left(i_q^* - i_q\right) + \omega L i_d \end{cases} \tag{9}$$

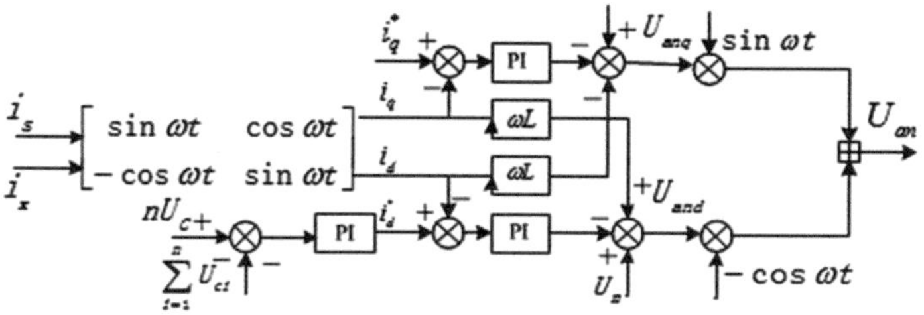

Fig. 2 Feed-forward decoupling block diagram of CHBR

where $i_d^* i_q^*$ are desired values of active current and reactive current respectively. Take (9) into (8), then get the control equation

$$\begin{cases} L\frac{di_d}{dt} = (K_p + \frac{K_i}{s})\left(i_d^* - i_d\right) - Ri_d \\ L\frac{di_q}{dt} = (K_p + \frac{K_i}{s})\left(i_q^* - i_q\right) - Ri_q \end{cases} \tag{10}$$

Through this transformation, i_d and i_q are decoupled. The value of i_d^* is given by the voltage outer loop, the value of i_q^* is set to 0 to keep the system's high power factor. The feed-forward decoupling strategy of CHBR is shown in Fig. 2.

3 CHBR DC Capacitor Voltage-Balanced Control Algorithm

As is shown in Fig. 1, the operating mode "0" corresponds to the conduction of bottom switches (S2, S4). In modes "+1" and "−1" the diagonal switches (S1, S4) and (S2, S3) are turned on respectively. Define S_i as switching function, S_i gets three cases: "1" "−1" "0", so the voltage of AC side can be expressed as

$$U_s = U_{h1} + U_{h2} + \cdots + U_{hn} \tag{12}$$

$$U_{hi} = S_i \cdot U_{Ci};\ i = 1,\ 2,\ 3\ldots \tag{13}$$

where U_{hi}, U_{Ci}, S_i are the AC side voltage, DC capacitor voltage and switch function of the *i*th H-bridge. If $U_{C1} = U_{C2} = \cdots = U_{Cn} = U_C(U_C$ is the desired value of DC capacitor voltage), so *n* H-bridge power units can synthesis *2n + 1* different voltages. According to Kirchhoff's law, get the topology model based on instantaneous value

$$\begin{cases} U_{in} = U_0 + L\frac{dI_{in}}{d_t} \\ i_{hi} = S_i \cdot i_{in} \end{cases};\ i = 1,\ 2,\ 3\ldots,\ n \tag{14}$$

where i_{hi} is the current of the ith H-bridge unit. When $i_s > 0$, if $S_i = 1$, AC side voltage source will charge for capacitor of ith H-bridge unit, then the corresponding DC capacitor voltage increases; if $S_i = -1$, AC side voltage source will discharge for capacitor of ith H-bridge, then the corresponding DC capacitor voltage decreases. When $i_s < 0$, the situation is just reversed. If $S_i = 0$, the AC side voltage source has no effect on the capacitor voltage of the DC side of the ith H-bridge unit. Therefore the CHBR should balance the DC side capacitor voltage by controlling the switching status.

3.1 Traditional Capacitor Voltage-Balanced Algorithm

As for a CHBR contains n H-bridge units, the input voltage U_s is divided into 2n equal sections with the scale of U_c (U_c is the reference of primary DC capacitor voltage). As is shown in Fig. 3.

Region K is where the magnitude of input voltage, Us, lies between (K − 1)UC and KUC. The voltage region K is defined as follows:

$$(K-1) \cdot U_c < |U_S| < K \cdot U_c \tag{15}$$

The traditional capacitor voltage-balanced control algorithm is shown in Fig. 4.

3.2 Improved Capacitor Voltage-Balanced Control Algorithm

Traditional algorithm balances the DC capacitor voltage by changing the switching status constantly. That will increase the switching loss, and generate a lot of heat. The heat is bad for the life of switches. What is more, when the loads are serious imbalance, traditional algorithms might fail. To solve this problem, this paper proposes an

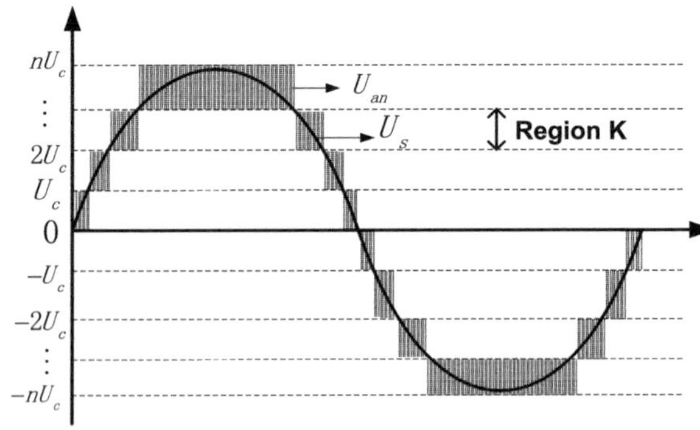

Fig. 3 Definition of voltage regions for K = 1, …, N

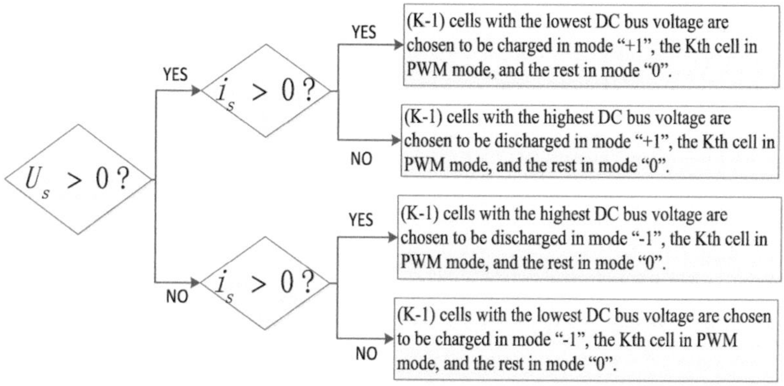

Fig. 4 Flow chart of traditional capacitor voltage-balanced control algorithm

improved algorithm, which will add a certain fault tolerance rate (Fault tolerance rate) to the traditional capacitor voltage control target, which is defined as F

$$F_i = \frac{|U_{ci}|}{U_c}; \quad i = 1, 2, 3\ldots, n \tag{16}$$

The improved algorithm process is shown in Fig. 5.

The control goal of traditional algorithm is to make each module voltage equal to the reference value. And improved algorithm determines whether the fault tolerance rate F exceeds the fault tolerance limit $F_{\max}$ first. If F does not exceed $F_{\max}$, the switch status is controlled to be the same as last sampling period. Otherwise still use the traditional balance control algorithm. Improved algorithm allows the DC capacitor voltage of each module to have a small gap, thereby reducing frequency of the H-bridge input and cut out frequency, thus reduce the switch loss, improve the system efficiency.

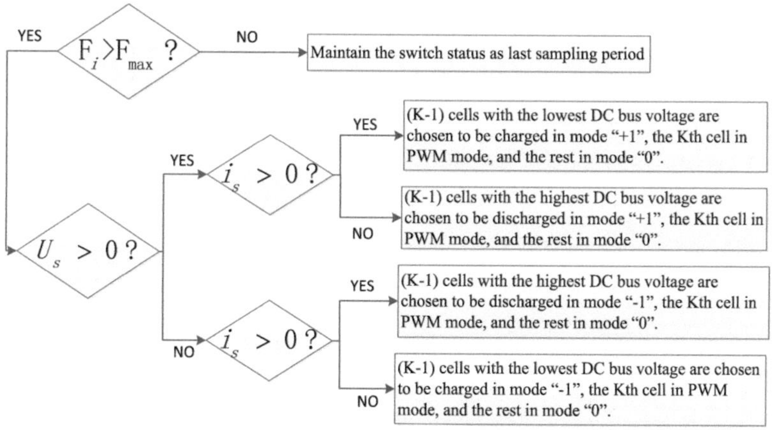

Fig. 5 Flow chart of improved capacitor voltage-balanced control algorithm

4 Simulation and Results Analysis

In order to verify the effectiveness of the improved algorithm, this paper builds a CHBR of three cells cascaded simulation model in MATLAB.

In this simulation, where AC side input voltage peak is set to 311 V, the input DC voltage reference value is set to 200 V, and the AC side inductance is set to 12 mH. All DC bus capacitor values are set to 2800 μF. The three module load R1, R2, R3 are set to 20, 25, and 30 Ω. The traditional algorithm and the improved algorithm are used in the simulation respectively, the Fault tolerance rate of improved algorithm is set to 1.003 and the results are shown in Fig. 6:

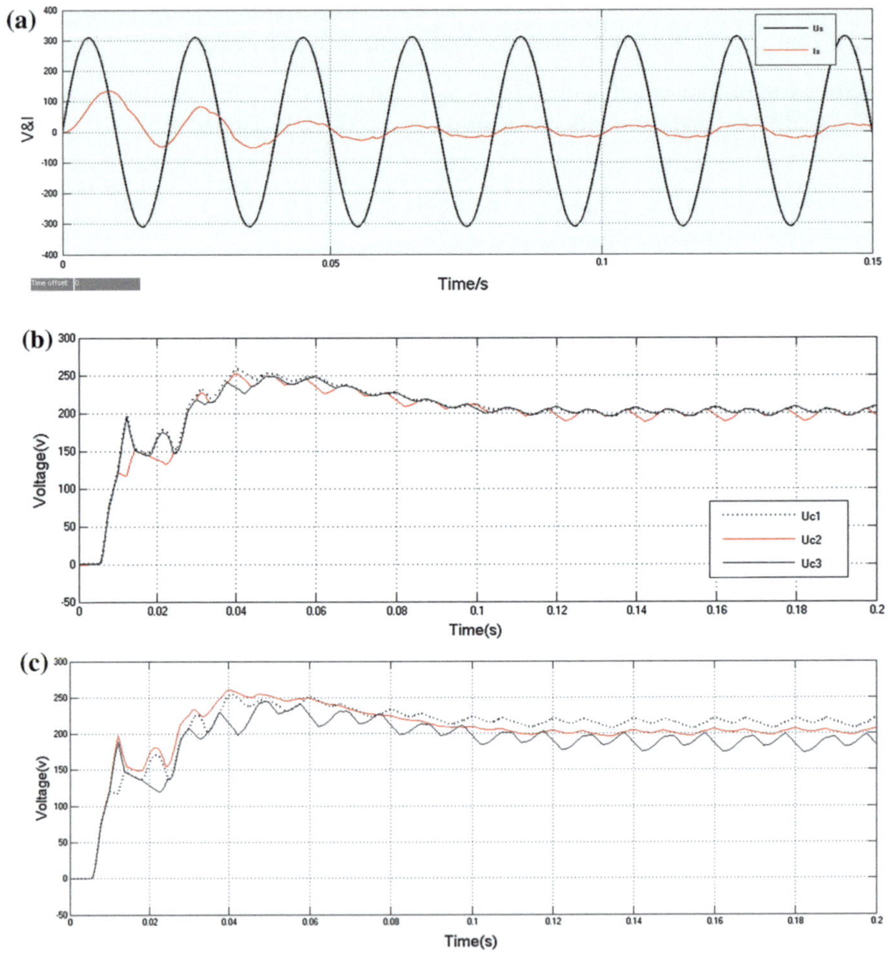

Fig. 6 Simulate results. **a** AC side voltage and current waveform. **b** DC capacitor voltages waveform of three modules (traditional algorithm). **c** DC capacitor voltages waveform of three modules (improved algorithm)

As it is shown in Fig. 6a, it is found that the phase of the AC side voltage and current can be the same phase. (b) (c) show that the improved algorithm can balance the DC capacitor voltage of three power units very well. But traditional algorithms cannot balance the DC capacitor voltages.

5 Conclusions

This paper studied CHBR and established mathematical model by analyzing the topology. Researched CHBR's modulation mode and capacitor voltage-balanced control algorithm based on capacitor voltage ranking to determine capacitor's charging and discharging state. In order to solve the problem of switch's higher frequency, put forward an improved DC capacitor voltage-balanced control algorithm, the control algorithm reducing the action times of the switches, so that the switching losses are reduced. The correctness and feasibility of the proposed control algorithm are verified by simulation.

References

1. Rodriguez J, Franquelo LG, Kouro S et al (2009) Multilevel converters: an enabling technology for high-power applications. Proc IEEE 97(11):1786–1817
2. Kai D, Zou YP, Zhan W et al (2004) A novel hybrid diode-clamp cascade multilevel converter for high power application, vol 2, pp 820–827
3. Chiasson JN, Tolbert LM, Mckenzie KJ et al (2003) Control of a multilevel converter using resultant theory. IEEE Trans Control Syst Technol 11(3):345–354
4. Kumar J, Agarwal P, Das B (2010) Implementation of cascade multilevel inverter-based STATCOM. IETE J Res 56(2):119–128
5. Hatano N, Ise T (2008) A configuration and control method of cascade H-bridge STATCOM. In: Power and energy society general meeting—conversion and delivery of electrical energy in the 21st century. IEEE 2008:1–8
6. Woodhouse ML, Donoghue MW, Osborne MM (2001) Type testing of the GTO valves for a novel STATCOM convertor. IET Digit Libr 2001:84–90
7. Ying C, Qiang G, Dianguo X (2012) Control and performance of a medium-voltage cascade H-bridge STATCOM. In: 2012 7th international power electronics and motion control conference (IPEMC), pp 2995–2999
8. Barrena JA, Marroyo L, Vidal MAR et al (2008) Individual Voltage Balancing Strategy for PWM Cascaded H-Bridge Converter-Based STATCOM. IEEE Trans Ind Electron 55 (1):21–29
9. Tao X, Li Y, Song Y et al (2012) H bridge cascaded rectifier DC voltage balance control algorithm. High Volt Eng 38 (2)
10. Leon JI, Vazquez S, Portillo R et al (2010) Two-dimensional modulation technique with DC voltage control for single-phase two-cell cascaded converters. In: IEEE international conference on industrial technology. 2010:1365–1370
11. Leon JI, Vazquez S, Watson AJ et al (2009) feed-forward space vector modulation for single-phase multilevel cascaded converters with any DC voltage ratio. IEEE Trans Industr Electron 56(2):315–325

12. Iman-Eini H, Farhangi S, Schanen JL (2008) A modular AC/DC rectifier based on cascaded H-bridge rectifier. In: Power electronics and motion control conference, 2008. EPE-PEMC 2008. 13th. IEEE 2008, pp 173–180
13. Jian WU, Shi J, Zhang Z (2011) Research on voltage and power balance control for threephase cascaded modular solid-state transformer. IEEE Trans Power Electron 26(4):1154–1166
14. Aquila AD, Liserre M, Monopoli VG et al (2008) Overview of PI-based solutions for the control of DC buses of a single-phase H-bridge multilevel active rectifier. IEEE Trans Ind Appl 44(3):857–866

Autonomous Navigation for Spacecraft Around Mars Based on Information Fusion with Cross-Correlation Noise

Jianjun Li and Dayi Wang

Abstract This paper presents an innovative autonomous navigation scheme based on information fusion for spacecraft orbiting Mars using observations of Phobos and Deimos. The optimal fusion (OF) algorithm and the covariance intersection (CI) fusion algorithm are then applied to solve the information fusion problem of multiple subsystems which are all cross-correlation. Simulation results demonstrate the superiority that the OF algorithm and the CI fusion algorithm have in handling the fusion problem with cross-correlation noise compared with Federated Filter.

Keywords Autonomous navigation · Mars · Information fusion · Cross-correlation

1 Introduction

Autonomous orbit determination of an interplanetary spacecraft is the process by which the orbit parameters are autonomously estimated using the onboard measuring devices without the supports of ground-based monitoring stations. In fact, the current process can only be operated successfully with the assistance from the ground-based facilities. Those facilities will become more complex, demanding, and expensive as the number of interplanetary missions increase and the navigation constellation will not work properly when the communication fails between spacecraft and ground-based facilities [1, 2]. Huge time delay from earth to asteroid

J. Li · D. Wang (✉)
Beijing Institute of Control Engineering, 100090 Beijing, China
e-mail: dayiwang@163.com

J. Li
e-mail: 474931055@qq.com

D. Wang
Science and Technology on Space Intelligent Control Laboratory, 100090 Beijing, China

Y. Jia et al. (eds.), *Proceedings of 2016 Chinese Intelligent Systems Conference*, Lecture Notes in Electrical Engineering 405,
DOI 10.1007/978-981-10-2335-4_53

and planets also makes it difficult to estimate real-time state of interplanetary spacecraft only depending on ground-based stations. Therefore, autonomous navigation must be studied to achieve the interplanetary exploration missions successfully in the future. Fortunately, information fusion is one of effective means to develop the technology for autonomous navigation. So far, multiple new sensors have been developed to improve performance of onboard navigation system such as optical camera and X-ray detectors [3, 4]. Those sensors are suitable to be applied to interplanetary missions. There also have been many studies about navigation schemes based on those measurements, some of which have been applied to real interplanetary missions successfully and performance show perfectly [5]. However, it is known that all these sensors have advantages and drawbacks. So combining different sensors using a reasonable fusion algorithm can increase the spacecraft's autonomous capabilities.

Fusing algorithm plays a critical role in increasing accuracy of parameters estimation for navigation system. It has two basic forms: centralized fusion and distributed fusion. Distributed fusion is more considerable challengingly because of its fault-tolerant performance and reliability [2]. Under the assumption of cross-uncorrelated sensor, a distributed Kalman filtering fusion was proposed in [6]. But often is the case that there may be cross-correlations between the sensor measurement noises in practical applications. This is because dynamic process is always observed in a common noisy environment. The Federated Filter ignores the cross-correlations among every filter and actual situation cannot satisfy such strict hypothesis conditions because of the common prior estimation and process noise. So it will not get a better information fusion result using the Federated Filter algorithm. Based on those analyses above, an innovative scheme of autonomous navigation for spacecraft around Mars based on information fusion is proposed in this paper. The paper is organized as follows: Sects. 2 and 3 describe the system models in detail, including dynamic model and measurement model. Section 4 develops the distributed fusion algorithms with cross-covariance. In the Sect. 5, simulation results and analysis are presented. And conclusion will be given in Sect. 6.

2 Autonomous Navigation System Model

The state model is built based on orbital dynamics. Here two-body model is chosen as dynamic model in J2000 Mars inertial coordinate system. In this model, only the perturbation caused by the second zonal harmonics in the Mars's gravitational field is taken into account and others are treated as noise.

$$\begin{cases} \frac{dx}{dt} = \nu_x \\ \frac{dy}{dt} = \nu_y \\ \frac{dz}{dt} = \nu_z \\ \frac{d\nu_x}{dt} = -\mu\frac{x}{r^3}\left[1 - J_2\left(\frac{R_m}{r}\right)\left(7.5\frac{z^2}{r^2} - 1.5\right)\right] + \alpha_x + w_{\nu_x} \\ \frac{d\nu_y}{dt} = -\mu\frac{y}{r^3}\left[1 - J_2\left(\frac{R_m}{r}\right)\left(7.5\frac{z^2}{r^2} - 1.5\right)\right] + \alpha_y + w_{\nu_y} \\ \frac{d\nu_z}{dt} = -\mu\frac{z}{r^3}\left[1 - J_2\left(\frac{R_m}{r}\right)\left(7.5\frac{z^2}{r^2} - 1.5\right)\right] + \alpha_z + w_{\nu_z} \end{cases} \quad (1)$$

$$r = \sqrt{x^2 + y^2 + z^2}$$

The state vector $x = [x, y, z, v_x, v_y, v_z]^T$ denotes the spacecraft's state, including positions and velocities of three axes. μ is the gravitational constant of the Mars. J_2 is the second zonal coefficient. R_m is the Mars's radius. α_x, α_y, α_z are other perturbations, such as the gravity of sun and other planets, sunlight pressure, atmospheric drag, and so on. Then the state equation of orbit dynamic model can be expressed as

$$\dot{x}(t) = f(x(t), t) + w(t) \quad (2)$$

where $w(t)$ is process noise, considered as zeros-mean Gaussian white noise.

3 Measurement Models

Phobos and Deimos, Martian moon can be easily observed by the spacecraft orbiting Mars. Navigation camera can get high-accuracy navigation information relative to targets [7]. That is to say, the spacecraft position can be determined accurately by observing the lines of sight Phobos and Deimos [8]. And angles between them can be also calculated indirectly. The relationships are shown in Fig. 1.

For navigation camera system, image processing algorithms must be created to extract navigation data from a raw image. But the focus of the paper is the image information containing target body that can be used to estimate the spacecraft position directly, image processing has been finished before.

The line of sight is direction of unit vector from spacecraft to target celestial body. It can be calculated using pixels obtained by navigation cameras. For convenience, the relationship describing the projection of three-dimensional world onto a camera detection plane is assumed a pinhole camera model. Then the line of sight of celestial body in the frame attached to camera is given by

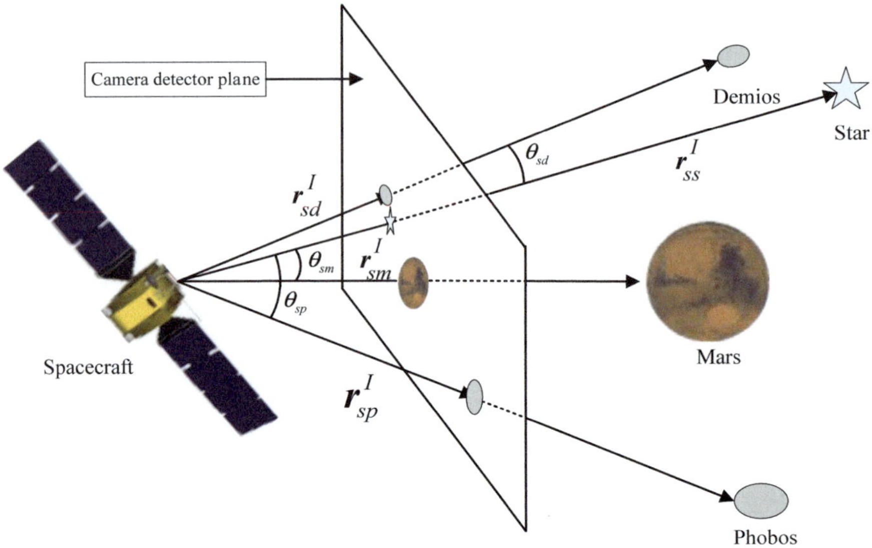

Fig. 1 The geometric position of Mars, Phobos and Deimos

$$r_{sc}^{c} = \frac{1}{\sqrt{x_d^2 + y_d^2 + f^2}} \begin{bmatrix} x_d \\ y_d \\ -f \end{bmatrix} \tag{3}$$

where f is the camera focal length, x_d and y_d are the coordinates of the observed point on the detector plane. This solution can be achieved through being rotated from the inertial frame by

$$r_{sc}^{c} = A_{cb} A_{bi} r_{sc}^{I} \tag{4}$$

where A_{cb} and A_{bi} are transfer matrixes from the spacecraft body frame to the camera frame and from the inertial frame to the spacecraft body frame. r_{sc}^{c} and r_{sc}^{I} are the lines of sight of target celestial body in the camera frame and the inertial frame, respectively. Subscript c donates the target celestial body. Let the lines of sight r_{sc}^{c} to be the observation vector, then the observation can be derived

$$y_r = h_r(x) + v_{r_{sc}} \tag{5}$$

Here, $h_r(x)$ can be expressed by the right part of Eq. (4). $v_{r_{sc}}$ denotes the measurement noise of the line of sight for the navigation camera.

4 Distributed Fusion Algorithms for State Parameter

The Federated Filter (FF) is always considered first when dealing with multiple information fusion (Fig. 2). For Federated Filter, global optimal estimation will be got by weighing all results from subsystems, where all weight coefficients are invariant instead of changing adaptively. And it is assumed that the all filters are uncorrelated using variance upper bound.

But actually, it is impossible for actual situation to satisfy this hypothesis due to the effect of process noise and the common prior estimation. Hence, it may not get a better information fusion result using the Federated Filter. Aimed at improving accuracy of navigation system, two different distributed fusion algorithms are used in the paper, including the (OF) [9, 10] and CI fusion [11]. The cross-correlations among subsystems are considered in two algorithms. Based on steady-state Kalman filter of linear time-invariant system, the comparison of accuracy relations among them was conducted in [11].

(1) The OF algorithms

The OF algorithm with matrix weight is given as [9]

$$\hat{x} = \sum_{k=1}^{L} M_k \hat{x}_k \tag{6}$$

Here, the weighting matrix M_k satisfies the constrain equation

$$\sum_{k=1}^{L} M_k = I_n \tag{7}$$

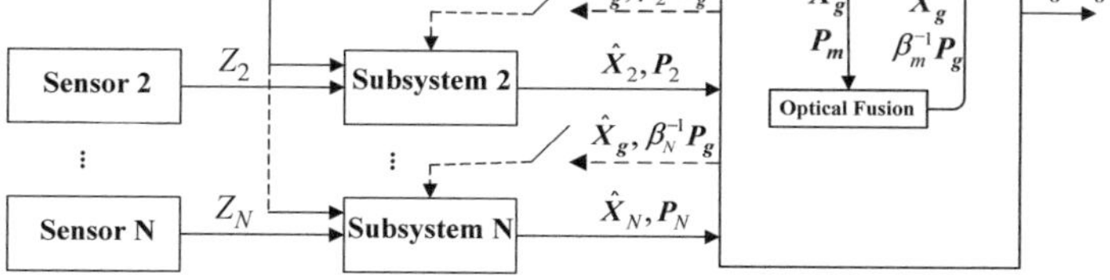

Fig. 2 The structure of Federated Filter

With the expression

$$[M_i, \ldots, M_i] = (e^T P^{-1} e)^{-1} e^T P^{-1} \tag{8}$$

where $P = (P_{ki})_{nL \times nL}$, $e = [I_n, \ldots I_n]^T$. n and L denotes the number of state and subsystems, respectively. And cross-covariance among local filters needs to be known.

(2) Covariance intersection (CI) fusion algorithm

Unlike the optimal fusion (OF) algorithms before, the covariance intersection (CI) fusion method was present for systems with unknown cross-covariance. Its basic principle is that is uses the convex combination weighting method. The CI fuser is obtained via the optimization of a nonlinear performance index function and its form is expressed as [12]

$$\hat{x} = P \sum_{i=1}^{L} \omega_i P_i^{-1} \hat{x}_i \tag{9}$$

$$P^{-1} = \sum_{i=1}^{L} \omega_i P_i^{-1} \tag{10}$$

where the $\omega_i \in [0, 1]$, and $\sum_{i=1}^{L} \omega_i = 1$.

(3) Computation cross-covariance

As mentioned before, the cross-covariance among local filters needs to be calculated in OF algorithm. The state cross-covariance P_{ij} of the local filtering errors can be defined by the Lyapunov equation (15)

$$P_{ij} = \boldsymbol{\Psi}_i P_{ij} \boldsymbol{\Psi}_j^T + \boldsymbol{\Delta}_{ij}, i, j = 1, \ldots, L \tag{15}$$

$$\boldsymbol{\Psi}_i = [I_n - K_i H_i] \boldsymbol{\Phi}_i \tag{16}$$

$$\boldsymbol{\Delta}_{ij} = [I_n - K_i H_i] \boldsymbol{\Gamma} Q \boldsymbol{\Gamma}^T [I_n - K_j H_j]^T, i, j = 1, \ldots, L \tag{17}$$

Here, we define $P_{ii} = P_i$.

5 Prototype and Test Results

In order to study the performance of different fusion algorithms to deal with correlation noise, average RMSE indicator is defined here.

$$\text{Position\_ARMSE} = \frac{1}{K}\sum_{k=1}^{K}\sqrt{\frac{1}{N}\sum_{n=1}^{N}\left(\tilde{x}_n(k)^{\mathrm{T}}\tilde{x}_n(k)\right)}$$

$$\text{Velocity\_ARMSE} = \frac{1}{K}\sum_{k=1}^{K}\sqrt{\frac{1}{N}\sum_{n=1}^{N}\left(\tilde{v}_n(k)^{\mathrm{T}}\tilde{v}_n(k)\right)}$$

The initial parameters of filters are listed in Table 1.

Under the above-mentioned condition, the simulation for the autonomous navigation schemes for spacecraft around the Mars, based on the OF and the CI fusion under cross-correlation, are conducted compared with the federated filter respectively. To study the effect of correlated noise clearly, simulations were conducted under several different conditions, including No effect from Correlated Measurements and Process Noise (NCMPN), Under effect of Correlated Process Noise (UCRN), Under effect of Correlated Measurement Noise (UCMN), and Under Cross-correlated Measurements and Process Noise (UCMPN). Results are presented in the follows figures.

(1) Comparing the difference of several simulation conditions based on the Federated Filter (FF)
(2) Under effect of Correlated Process Noise (UCPN)
(3) Under effect of Correlated Measurement Noise (UCMN)
(4) Under Cross-correlated Measurements and Process Noise(UCMPN)

Figure 3 is average RMSEs of position and velocity based on Federated Filter under several conditions. It obviously shows that correlated noise affect markedly on the accuracy of fusion algorithms. For deep space missions, it would be better to take into account the effect of correlated noise.

Figures 4, 5 and 6 are the simulation results based on the three fusion algorithms under different conditions. It can be concluded that the accuracy of OF and CI algorithm is higher than that of Federated Filter no matter under what conditions. Therefore the two fusion algorithms are very suitable for navigation system for deep space missions and can greatly enhance the reliability and accuracy without the supports from earth.

Table 1 The initial parameters of filters

Filters statistics	Value
Measurement noise R	$(1\times 10^{-5})^2 I_{3\times 3}$
State noise Q	$\begin{bmatrix} O_{3\times 3} & O_{3\times 3} \\ O_{3\times 3} & (1\times 10^{-6})^2 I_{3\times 3} \end{bmatrix}$
Initial state P_0	$\begin{bmatrix} (1\times 10^{-3})^2 I_{3\times 3} & O_{3\times 3} \\ O_{3\times 3} & I_{3\times 3} \end{bmatrix}$
Cross-covariance P_0^{ij}	$(1\times 10^{-4})^2 I_{6\times 6}$

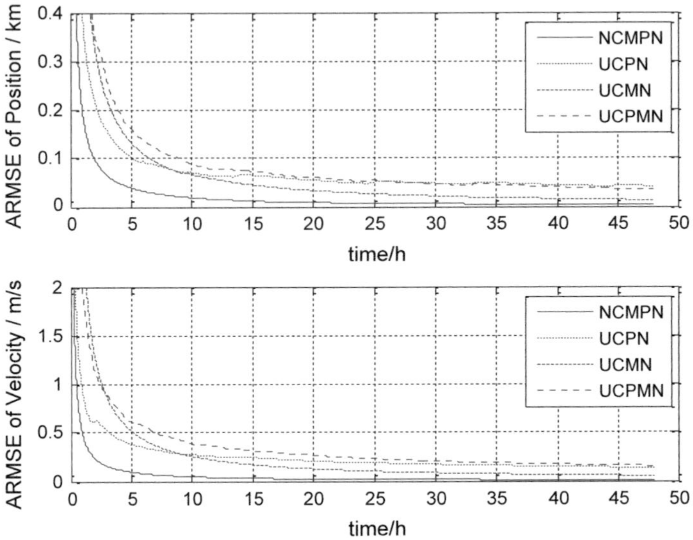

Fig. 3 The ARMSE difference of different simulation conditions

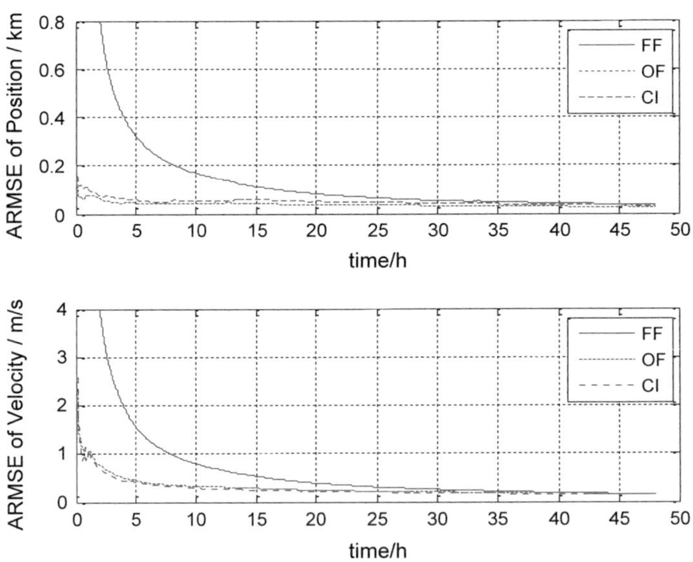

Fig. 4 Under effect of Correlated Process Noise

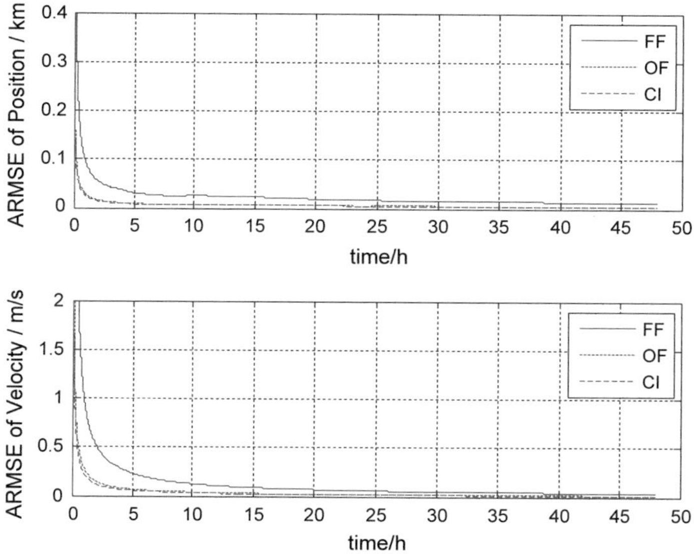

Fig. 5 Under effect of Correlated Measurement Noise

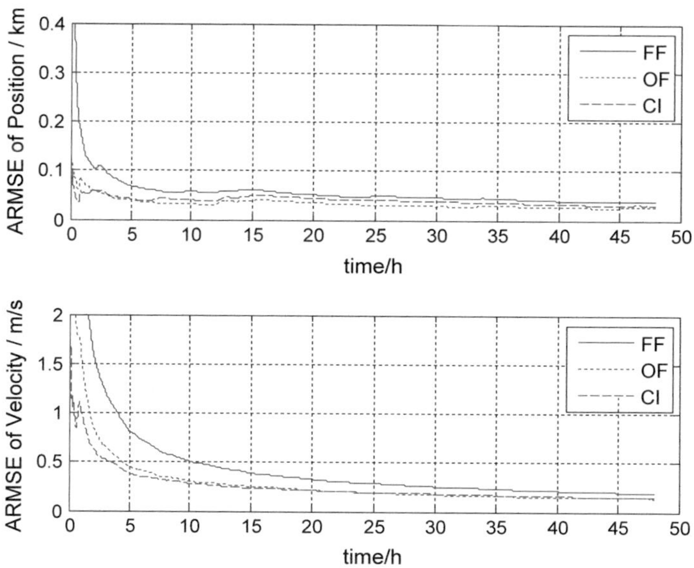

Fig. 6 Under Cross-correlated Measurements and Process Noise

6 Analysis of Experimental Results

In this paper, an autonomous navigation scheme for spacecraft orbiting Mars based on information fusion is proposed. By taking the line of sight of Martian moon, Phobos and Deimos, as observed variables, autonomous navigation scheme was built based on orbit dynamics. Meanwhile, the common process noise was also considered in fusion algorithms. Simulation results indicate that OF and CI fusion excel in dealing with those fusion problems under cross-correlation noise. Finally, simulation experiment was conducted and results show perfectly. The accuracy of the autonomous navigation is dramatically improved. Conclusion demonstrates that the scheme proposed in the paper is reasonable and can provide high-precision navigation for spacecraft around Mars.

References

1. Muñoz S, Lightsey EG (2011) A sensor driven trade study for autonomous navigation capability. AIAA guidance, navigation and control conference, Portland, Oregon
2. Christian JA, Lightsey EG (2008) A review of options for autonomous cislunar navigation. AIAA guidance, navigation and control conference, Honolulu, Hawaii
3. Sheikh S (2005) The use of variable celestial x-ray sources for spacecraft navigation. PhD thesis University of Maryland
4. Riedel JE, Bhaskaran S, Desai S et al (2000) Deep space 1 technology validation report-autonomous optical navigation. JPL Publication, Jet Propulsion Laboratory, Pasadena, CA, USA
5. Frauenholz RB, Bhat RS, Chesley SR et al (2008) Deep impact navigation system performance. J Spacecraft Rockets 45(1):39–56
6. Chong CY, Chang KC, Mori S (1987) Distributed tracking in distributed sensor networks. American control conference, Seattle, WA
7. Qian HM, Sun L, Cai JN, Huang W (2014) A starlight refraction scheme with single star sensor used in autonomous satellite navigation system. Acta Astronaustica. 96:45–52
8. Xin M, Jiancheng F, Xiaolin N (2013) An overview of the autonomous navigation for a gravity-assist interplanetary spacecraft. Prog Aerosp Sci 63:56–66
9. Shuli S (2004) Multi-sensor optimal information fusion Kalman filters with applications. Aerosp Sci Technol 8:57–62
10. Zili D, Peng Z, Wenjuan Q, Yuan G, Jinfang L (2013) The accuracy comparison of multisensory covariance intersection fuser and three weighting fusers. Inf Fusion 14:177–185
11. William JF III, Chidambar G (2009) Generalized Chernoff fusion approximation for practical distributed data fusion. In: 12th international conference on information fusion, Seattle, WA, USA
12. Carrillo-Arce LC, Nerurkar ED, Gordillo JL, Roumeliotis SI (2013) Decentralized multi-robot coop-erative localization using covariance intersection. In: 2013 IEEE/RSJ international conference on intelligent robot and systems, Tokyo, Japan

The CUDA-Based Multi-frame Images Parallel Fast Processing Method

Yiyao An, Maoyun Guo, Yi Chai and Haoxin Liang

Abstract This paper purposes a fast parallel processing method for multi-frame images based on CUDA by Nvidia employing the Sobel edge detection operator as example. To utilize the CUDA's high parallel computing capability of dense numeric calculation, the paper optimizes the data structure of multi-frame images, combines the multi-frame images into "one image" which reduces the complexity of method. And the experiment result shows that the average running time of the method based on CUDA, which is 499.7 ms, is about 15 % as much as that based on CPU when processing the 64 frames of 512 × 512 pixels images with 8-digit grayscale. The method can utilize the CUDA's computing capability greatly.

Keywords CUDA · Sobel operator · Multi-frame images · Parallel computing

1 Introduction

As a typical application of dense data, video processing hungers for high-performance computing. The traditional method is processing the frames of the video one by one from each video channel of video system. With the development of multiprocessor and cloud computing, the above processing can be parallelized. But these approaches waste lots of the logic computing resources, because there is much more numerical calculating than logic calculating in video processing.

The GPU-based general purpose heterogeneous parallel computing technology, such as Nvidia's CUDA (Compute Unified Device Architecture) affords a high-efficiency computing platform to the dense numeric calculating [1]. It is very suitable for image processing such as image segmentation and edge detection.

Now, many researchers have carried out the researches on the image processing based on CUDA [2–7]. Zuo hao rui et al. provided a parallel computing algorithm

Y. An · M. Guo (✉) · Y. Chai · H. Liang
School of Automation, Chongqing University, Chongqing 400044, China
e-mail: gmy@cqu.edu.cn

Y. Jia et al. (eds.), *Proceedings of 2016 Chinese Intelligent Systems Conference*, Lecture Notes in Electrical Engineering 405,
DOI 10.1007/978-981-10-2335-4_54

of the Sobel operator for single image based on CUDA [2]. Meng Xiaohua et al. released an algorithm of Laplace edge detection method with CUDA [3]. Luo et al. gave out a CUDA-based algorithm for Canny edge detection [5]. And all of these methods can improve the compute speed of the image edge detection.

Based on the CUDA, this paper will carry out the research on the parallel processing method for the multi-frame images from multiple channels of the video system employing the Sobel operator as example.

2 The Classic Sobel Edge Detection Operator

The classic Sobel edge detection operator is composed of two templates, $S1$ and $S2$, corresponding to the horizontal edge and the vertical edge. Shown in Fig. 1, the pixel $P(n, m)$ should employ $S1$ and $S2$, whose results are $R1$ and $R2$ correspondingly. And the result is $|R1| + |R2|$, where $|\bullet|$ represents the absolute value. The $P(n-1, m-1)$, $P(n-1, m)$, $P(n-1, m+1)$, $P(n, m-1)$, $P(n, m+1)$, $P(n+1, m-1)$, $P(n+1, m-1)$ and $P(n+1, m+1)$ are the neighbor pixels of the $P(n, m)$ [2].

So in Fig. 1, a pixel needs 11 adding operations with the Sobel operator regardless of the absolute terms. So if there are $N \times N$ pixels in one image, the total of add operations is $11\ N^2$. So if there are M frames, it will take $11\ MN^2$ add operations.

3 The CUDA-Based Multi-frame Images Edge Detection Method

Some researcher has applied CUDA-based Sobel operator parallel to image, which is mainly for the single frame image [2]. In order to make full use of CUDA on parallel processing, this paper proposes a parallel processing method for multi-frame images from multichannel videos.

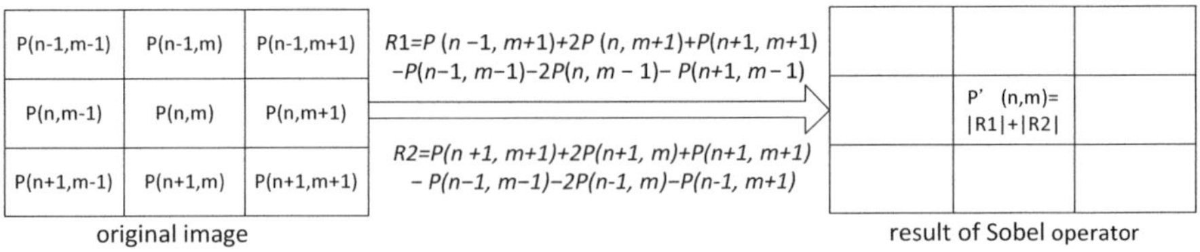

Fig. 1 The operation of Sobel edge detection operator

3.1 The Programming Model of the CUDA-Based Multi-frame Images Edge Detection Method

In CUDA programming model, CPU serves as host which controls the whole serial logic and task scheduling of programming. And GPU serves as device, which achieves data calculation with threads (usually as kernels) in highly parallel [1].

Figure 2 illustrates the execution flow of multi-frame images edge detection based on CUDA. The host (CPU) mostly completes the process of inputting the images data, launching the CUDA's threads to execute the image processing and other serial operations. And the device (GPU) generates threads grid containing thread blocks (kernels, which contain Sobel operator processing) and memory blocks (each of which contains one frame image from the channel of video) and runs the kernels (threads) in parallel [1].

3.2 The CUDA's Parallel Computing Organization

In order to achieve the parallel processing of the data, CUDA organizes the threads, which process data, with a one-dimensional or two-dimensional grid. And in the grid, the threads are organized with thread blocks which correspond to the memory blocks in CUDA. All threads in each thread block are located in the same processor core and share the limited memory resources of the core [1]. Generally, the number

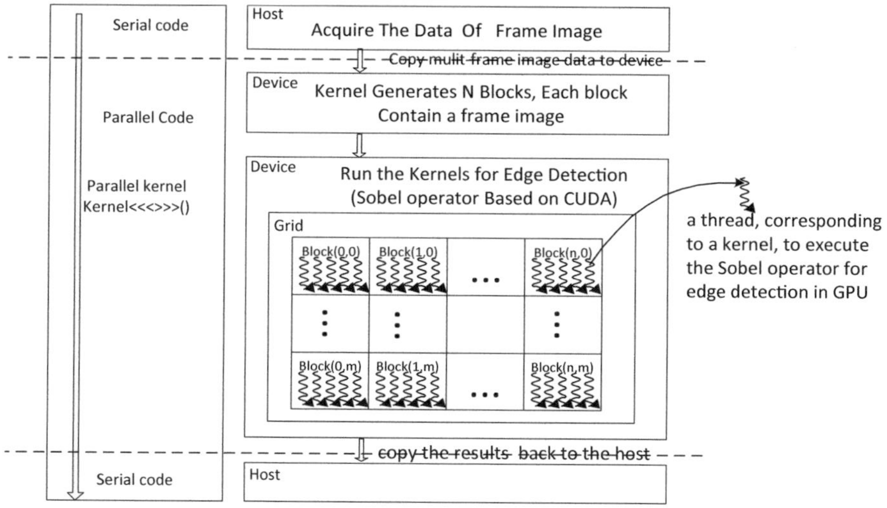

Fig. 2 The CUDA-based program execution flow

of blocks within a grid is decided by the amount of data rather than the number of processors. The former is often far more than the latter. At present, a kernel (a thread in blocks corresponding to parallel processing in GPU) can be executed by more than one same threads.

3.3 The CUDA-Based Multi-frame of Images Sobel Operator Method

As the Fig. 3 shows, the main idea of the method for processing multi-frame images from multichannel videos is to combine the multi-frame images into "one image" and employs the Sobel operator as edge detection which is executed as the parallel thread (the kernel) in GPU to utilize the GPU parallel computing resources.

In order to realize the parallel processing of multi-frame images, GPU will first allocate memory space according to the number of video channels. As shown in Fig. 3, here supposed the channels number of video system is $N = 512$, GPU can allocate N memory blocks and a thread grid containing N thread blocks. Each frame of image corresponds to one memory block. Each memory block corresponds to one thread block. And each pixel of a frame of image corresponds to one of the

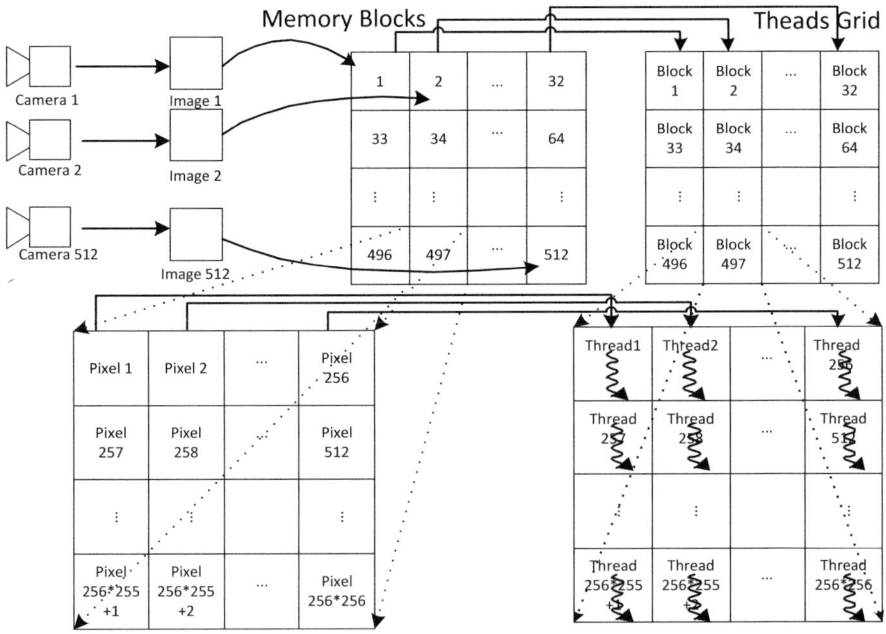

Fig. 3 The CUDA-based pixel and thread operation of multi-frame image

threads in the corresponding thread block. Because the threads of GPU are highly parallel in CUDA, the multi-frame images processing are highly parallel. So the method stated in the paper could save a lot of time.

To compare with the traditional method which based on CPU, here the image pixel's number is also set to $N \times N$ and the frames number is also set to M. Ideally, if GPU could provide enough memory and threads resources to storage and process M frames images, each with $N \times N$ pixels, all calculations of the Sobel operators of the method could be completed at one step (one computing cycle). That is, the method will only take 11 add operations to complete all operations. However, the resources of GPU are limited, so it will take more times than 11 add operations do. This will be given at the experiment part in detail.

4 Experiment and Analysis

For verifying the above method, the paper has carried on the experimental verification. The experimental environment is as follows: CPU is intel core i5-3230m 2.6 Hz. Display card is Nvidia GT635M. Internal storage is 4 GB Operating system is win8.1. CUDA version is CUDA 6.0. The size (pixels number) of the each image is 512 × 512.

4.1 Experiment Results

The experiment will run the proposed method (based on CUDA) and the traditional method (based on CPU) 10 times and gets average processing time as the running time. The experiment results are showed in the Table 1.

Table 1 The running time of the program based on CPU and CUDA

Frames of images	Time based on CUDA (ms)	Time based on CPU (ms)
1	7.19	48.15
2	14.35	95.63
4	28.703	203.62
6	43.064	301.85
8	57.398	406.84
16	114.84	786.12
24	172.20	1151.32
32	229.63	1517.94
64	449.71	2993.03

4.2 Analysis of Experimental Results

From the results, it can be seen that the running time of the paper proposing method is less than that of the traditional method based on CPU. The former is about 15 % of the latter. But it is different from the theoretical analysis in the time complexity analysis in the previous chapter.

The reason is that the theoretical time complexity analysis is a very ideal. It need not consider the actual hardware (GPU) limitations, which means that the GPU can always process all the pixels of all images with Sobel operator once (one computing cycle). In fact, the maximum number of threads GPU can run at one computing cycle is limited. When GPU cannot process the data in one computing cycle, it will execute the processing threads group by group, whose threads number depends on the GPU hardware, serially until completing the above processes. So the actual running time of the CUDA-based method is much slower than that of theoretical analysis in the paper.

5 Conclusion

With the above analysis, it can be concluded that the speed of method based on CUDA is faster than that based on CPU. The running time of the former is about 15 % of the latter. So the application of CUDA for multi-frame images processing method has a very good processing performance in speed. It is more suitable for video processing.

References

1. Nvidia. NVIDIA CUDA Programming Guide version 1.1[EB/OL]. http://www.nvidia.com/object/cuda_home.html.2007-11
2. Zuo H, Zhang Q, Yong X, Zhao R (2009) Fast Sobel edge detection algorithm based on GPU. Opto-Electr Eng 36(1):9–12
3. X Meng, Liu J, Ou Y et al (2012) Laplacian edge detection algorithm based on CUDA. Comput Eng 38:191–193
4. Xiao H (2011) Research on high efficiency heterogeneous parallel computing based on CPU + GPU in image matching. Wuhan University
5. Hou G (2013) Design and implementation of parallel algorithms image segmentation for CUDA. Dalian University of Technology
6. Luo Y, Duraiswami R (2008) Canny edge detection on NVIDIA CUDA. In: Computer vision and pattern recognition workshops, 2008, CVPRW'08. IEEE computer society-conference, pp 1–8
7. Galizia A, D'Agostino D, Clematis A (2015) An MPI-CUDA library for image processing on HPC architectures. J Comput Appl Math 273(1):414–427

Blind Source Separation Based on Mixed Integer Programming

Xiaocan Fan and Lizhen Shao

Abstract The underdetermined blind source separation problem is a common problem in our daily life, but it is difficult to solve because of its underdetermined. In the literature, sparse component analysis which exploits the sparsity of sources in a predefined sparse dictionary has been proposed to solve it. Usually, sparse component analysis uses a two-stage approach. The first stage is to estimate the mixing matrix and the second stage is to reconstruct sources. In fact, the second stage is a sparse optimization problem. In this paper, we model the problem of reconstructing sources as a bi-objective optimization problem. We take the error and sparsity as the two optimization objectives, and propose an iterative algorithm based on mixed integer programming to solve the bi-objective source reconstructing problem. Experimental results show the accuracy and effectiveness of our proposed algorithm.

Keywords Sparse component analysis · Bi-objective · Underdetermined blind source separation · Mixed integer programming

1 Introduction

Blind source separation (BSS) is a special kind of signal processing. It is a process that the sources $S \in \mathbb{R}^{n \times T}$ and the mixing matrix $A \in \mathbb{R}^{m \times n}$ are reconstructed using the mixed signals $X \in \mathbb{R}^{m \times T}$. Generally, we assume that the mixtures are linear combinations of the sources, i.e., $X = AS$. Moreover, any information about the sources and the mixing matrix are all unknown [1].

Blind source separation is a common problem in our daily life. Many independent component analysis methods have been proposed to solve it in determined case. Independent component analysis mainly exploits the independence of sources to obtain the separation matrix, i.e., the inverse of mixing matrix. However, sources can not be reconstructed successfully even if the mixing matrix has been estimated in under-

---

X. Fan · L. Shao (✉)
University of Science and Technology Beijing, 100083 Beijing, China
e-mail: lshao@ustb.edu.cn

Y. Jia et al. (eds.), *Proceedings of 2016 Chinese Intelligent Systems Conference*, Lecture Notes in Electrical Engineering 405,
DOI 10.1007/978-981-10-2335-4_55

determined case [2]. Therefore, some priori conditions of sources and the mixing matrix should be set. For instance, some signals can be sparse represented in a predefined sparse dictionary. Sparse component analysis (SCA) arises at the historic moment. It can exploit the sparsity of signal to reconstruct signal. It has gradually developed into the mainstream algorithm of blind source separation, because it overcomes the shortcoming of independent component analysis.

A two-stage approach [3] has been often used for sparse component analysis since both the mixing matrix and sources are unknown. The two-stage method, as the name suggests, is to divide the algorithm into two steps. First the mixing matrix is estimated, then sources are recovered.

In the abundant literatures with sparse component analysis, much work has been dedicated to solve blind source separation problem. In [4], Georgiev et al. formulated conditions under which a given matrix X can be uniquely represented as a multiplication of matrices A and S. In [5], Bobin et al. proposed a novel sparsity-enforcing BSS method called adaptive morphological component analysis. In [6], nine different sparse optimization problems are formulated based on l_0, l_1 or l_∞ data misfit measures. In [1], Syed et al. proposed a hierarchical 0–1 mixed integer programming to solve BBS problem. In [7], Li et al. proposed a revised version of MOEA/D based on iterative thresholding algorithm for sparse optimization. All of the above blind source separation methods are based on sparse models that have only one objective.

In this paper, we use a two-stage method to solve blind source separation problem. First of all, a linear clustering method is used to estimate the mixing matrix. Then, we propose an iterative mixed integer programming (MIP) algorithm to construct sources in the second stage. Finally, the performance of our proposed algorithm will be verified through experiments.

The rest of this paper is organized as follows. In Sect. 2, we introduce sparse optimization and bi-objective optimization. Section 3 explains our mixed integer optimization model and introduces the iterative MIP algorithm to solve bi-objective blind source separation problem. In Sect. 4, experimental results and discussions are presented. Finally, we conclude this work in Sect. 5.

2 Sparse Optimization and Bi-objective Optimization

In sparse optimization, we need to find a sparse signal $s \in \mathbb{R}^{n\times 1}$ with respect to $x = As$, where $A \in \mathbb{R}^{m\times n}$ and $x \in \mathbb{R}^{m\times 1}$ are known. Usually, there may be many possible solutions. However, if some conditions, such as restricted isometry property, have been satisfied, the equation has a unique K-sparse solution. What is more, if signals are not sparse, we can use some underlying transforms to sparse signals [8].

Generally, l_0 norm is used to measure the sparsity of a vector; the sparsity is the number of nonzero elements in a vector. So, we can seek the sparsest solution as follows:

$$\begin{aligned} &\min \ \|s\|_0 \\ &\text{s.t.} \ \ x = As. \end{aligned} \tag{1}$$

However, Eq. (1) is a NP-hard problem. Besides, we need list C_n^K kinds of linear combinations about nonzero positions of s to solve it. So, it is not easy to solve the problem. In fact, we can formulate sparse optimization problem as a bi-objective optimization problem, in which error function $\|x - As\|_2^2$ and penalty function $\|s\|_0$ are two optimization objectives. We should minimize the two objectives at the same time.

$$\min\{\|s\|_0, \|x - As\|_2^2\} \tag{2}$$

In multi-objective optimization [9], the solution of multi-objective problem is not limited to a single solution, but there may be more than one solution. The goal of multi-objective optimization is to find pareto optimal front. For two solutions x_1 and x_2, we say x_1 dominates x_2 when $\forall k \in 1, \ldots, m, f_k(x_1) \le f_k(x_2)$, and $\exists k \in 1, \ldots, m$, $f_k(x_1) < f_k(x_2)$. Therefore, if we cannot find a solution x that dominates x^*, we call x^* pareto optimal. Correspondingly, we call $f(x^*)$ nondominated. The set of all nondominated points are called pareto optimal front. Similarly, for any two solutions x_1 and x_2, we say x_1 weakly dominates x_2 when $\forall k \in 1, \ldots, m, f_k(x_1) < f_k(x_2)$. Therefore, if we cannot find a solution that weakly dominates x^*, we call x^* is weakly pareto optimal. Figure 1 shows the weakly pareto optimal front of a bi-objective optimization problem.

Suppose Fig. 1 is the pareto front for a bi-objective programming problem (2), we can see that the pareto optimal front contains a "knee" point with sparsity level K. In sparse optimization, our purpose is to find the "knee" point in pareto optimal front. In this paper, we dedicate to developing an algorithm to solve problem (2) and thus obtain the "knee" point with desired sparsity level K for problem (1).

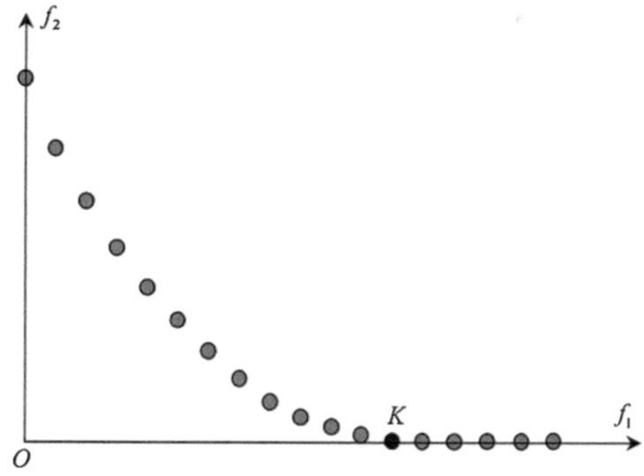

Fig. 1 Weakly pareto optimal front

3 Blind Source Separation

In this part, some assumptions should be satisfied to solve a blind source separation problem.

- any square $m \times m$ sub-matrix of the mixing matrix $A \in \mathbb{R}^{m \times n}$ should be nonsingular;
- nonzero elements of every column of the sources S should be less than $m - 1$.

3.1 First Stage: Mixing Matrix Estimation

If sources are sparse, we assume that the source $S \in \mathbb{R}^{n \times T}$ has only one nonzero component in the j-th moment. Therefore, $X_j = S_{ij} \times A_i$, namely X_j and A_i are collinear. In the same way, if the k-th moment satisfies the same condition, X_k and A_i are also collinear. So we can obtain mixing matrix A by clustering the mixed signals X, see e.g. [10].

In the case of sources are not sparse, we can sparse the mixed signals using wavelet transform. Then, the high frequency component of the transformed signals can be used as the new observation signals. Because of the linearity of wavelet transform, the mixing matrix can be obtained by clustering the observed signals in the spatial domain.

3.2 Second Stage: Source Estimation

Usually, it is not easy to solve sparse optimization problem because of the non-convexity of sparsity. Many optimization methods have been proposed to estimate sources, such as greedy algorithms and convex optimization algorithms. In order to solve bi-objective source reconstructing problem (2), we transform the problem into the following mixed integer programming problem:

$$\begin{aligned} \min \ & \|X - As\|_2^2 \\ \text{s.t.} \ & \|s\|_0 = K. \end{aligned} \tag{3}$$

where $K \in \{0, \ldots, n\}$.

Here, we introduce a binary variable z_j to express whether the j-th component is zero or not.

$$z_j = \begin{cases} 0, & \text{nonzero} \\ 1, & \text{zero.} \end{cases} \tag{4}$$

Thus, the mixed integer programming source reconstructing model is

$$\begin{aligned} \min\ & \|X_i - AS_i\|_2^2 \\ \text{s.t.}\ & |S_{ij}| \leq Qz_j, i = 1, 2, \ldots, T \\ & \textstyle\sum_j^n z_j = K, j = 1, 2, \ldots, n \\ & z_j = 0 \text{ or } 1 \end{aligned} \tag{5}$$

where Q is a large number, X_i and S_i represent the i-th column of X and S, respectively.

Now, we are going to propose an iterative MIP method to solve the bi-objective source reconstruction problem (2). Since we want to obtain the minimal sparsity value under the circumstance of the error is minimal, we do not need to find the entire pareto front, only need to find part of the pareto front which includes the "knee" point. The iterative MIP procedure can be described as Algorithm 1.

Algorithm 1 (The iterative MIP procedure)

Initialization.

(i1) Randomly generate a sequence of sparsity $Sp = \{s_1, \ldots, s_{pop}\}$
$P = \emptyset$, and $Q = \emptyset$

(i2) Input: the number of population pop; the mixed signals x;
the mixing matrix A

Iteration steps.

1: Solve MIP

for $i = 1$ to pop do
calculate the optimal solution for the problem in Eq. (5) when $K = s_i$,
and save the solution to the i-th column of P
end

2: Seek the Weakly Pareto optimal Solution

compare all solutions in P with each other, seek the weakly pareto optimal solution and put them in Q

3: Sparsity Update

3.1: order Sp, $s_{i_1} < s_{i_2} < \ldots < s_{i_{pop}}$

3.2: make a new set by interpolation method
$Spnew = \{\lfloor s_{i_j} + (s_{i_{j+1}} - s_{i_j})/2 \rfloor \mid s_{i_{j+1}} - s_{i_j} \geq 2, j = 1 : pop - 1\}$

3.3: take place a sparsity in Sp if $|Spnew| \neq 0$
$s_{i_1} \leftarrow$ randomly choose a sparsity in $Spnew$ if $|Q| > 0.5pop$
$s_{i_{pop}} \leftarrow$ randomly choose a sparsity in $Spnew$ if $|Q| \leq 0.5pop$

3.4: move sparsity in Sp if $|Spnew| = 0$
$s_{i_1} \leftarrow \min\{s_{i_{pop}}, s_{max}\}$,if $|Q| = pop$
$s_{i_{pop}} \leftarrow \max\{s_{i_{pop}}, s_{min}\}$,if $|Q| = 1$

3.5: if the "knee" point with sparsity level K is in Sp, output; otherwise go to Step 1

Output: the sparse solution P, the sparsity Sp

In Algorithm 1, we determine the set of all weakly pareto optimal solutions in P and save them to Q. The method of [11] is used to check the "knee" point of the pareto front. Therefore, we can be aware of whether Sp contain the sparsity K or not from the size of Q. As the algorithm carries out, we want to get a continuous sparse sequence. However, even if *pop* consecutive sparsity levels have been obtained, we may change it since sparsity K that we need may be excluded. Then, we can know on which side the sparsity may locate through the size of Q. For instance, if the size of Q is equal to *pop*, all members of *Sp* are on the left side of K.

4 Experimental Results

Our algorithm is carried out in the MATLAB environment, and CPLEX is used to solve the MIP problem. In order to demonstrate the performance of our proposed method, we use a synthetic instance and only consider the underdetermined case, namely $m < n$. First, all the sources are artificially generated. Figure 2 represents the original source signals. Then, data points $X \in \mathbb{R}^{3\times1000}$ have been mixed from the randomly generated mixing matrix $A \in \mathbb{R}^{3\times4}$ and the sources $S \in \mathbb{R}^{4\times1000}$.

$$A = \begin{bmatrix} 0.5456 & 0.3588 & 0.9683 & 0.4594 \\ 0.8279 & 0.7100 & 0.0231 & 0.4767 \\ 0.1297 & 0.6060 & 0.2487 & 0.7495 \end{bmatrix}$$

We use the algorithm proposed in [10] to estimate the mixing matrix, then the matrix A_r is obtained. In order to compare, A and A_r are normalized. Then, we can

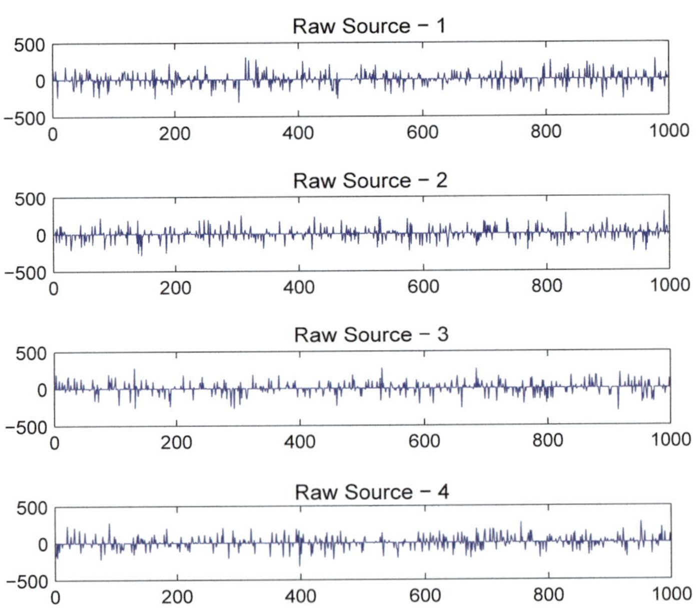

Fig. 2 Original sparse sources

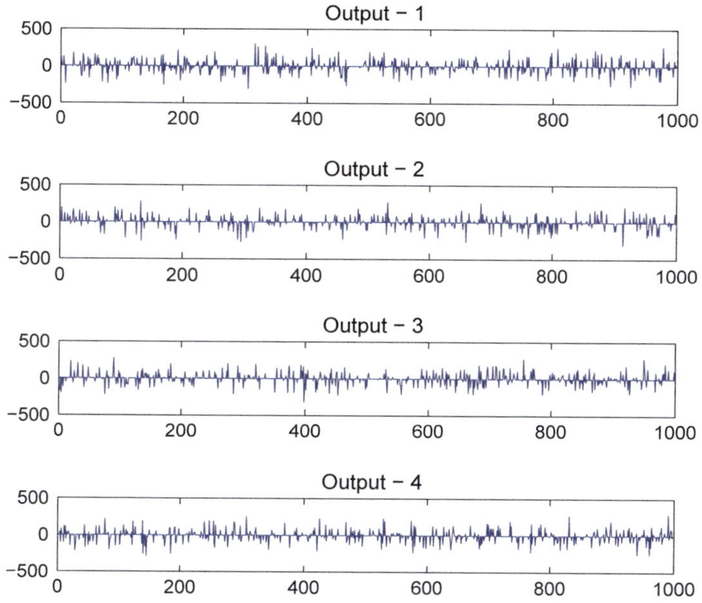

Fig. 3 Reconstructed sources

see that A_r and A differ only by permutation of the columns, which meets the characteristic that the sequence of signal reconstruction is uncertain.

$$A_r = \begin{bmatrix} 0.5456 & 0.9683 & 0.4594 & 0.3588 \\ 0.8279 & 0.0231 & 0.4767 & 0.7100 \\ 0.1297 & 0.2487 & 0.7495 & 0.6060 \end{bmatrix}$$

In addition, in order to find the source signals, we run Algorithm 1, taking every i-th moment mixture X_i and matrix A_r as input and we find the desired sparsity level K and S_i. Finally, we get the recovered sources S, see Fig. 3. It shows the practicability of our algorithm.

5 Conclusions

In this paper, we have formulated the problem of constructing source as a bi-objective optimization problem, in which error and sparsity are two objectives. Then, we propose an algorithm based on mixed integer programming to solve the problem. At each iteration, we solve a certain number of mixed integer programs to find some points in the pareto optimal front. We demonstrate the proposed algorithm by an example. This work indicates the feasibility of our algorithm.

References

1. SySed MN, Georgiev PG, Pardalos PM (2014) A hierarchical approach for sparse source blind signal separation problem. Comput Oper Res 41:386–398
2. Wang F, Li R, Wang Z et al (2015) Improved blind recovery algorithm for underdetermined mixtures by compressed sensing. Adv Swarm Comput Intell 405–412
3. He Z, Cichocki A, Li Y et al (2009) K-hyperline clustering learning for sparse component analysis. Signal Process 89(6):1011–1022
4. Georgiev PG, Theis F, Cichocki A (2005) Sparse component analysis and blind source separation of underdetermined mixtures. IEEE Trans Neural Netw 16(4):992–996
5. Bobin J, Rapin J, Larue A et al (2015) Sparsity and adaptively for the blind separation of partially correlated sources. IEEE Trans Signal Process 63(5):1199–1213
6. Bourguignon S, Ninin J, Carfantan H et al (2016) Exact sparse approximation problems via mixed-integer programming: formulations and computational performance. IEEE Trans Signal Process 64(6):1405–1419
7. Li H, Su X, Xu Z et al (2012) MOEA/D with iterative thresholding algorithm for sparse optimization problems. In: Parallel problem solving from nature-PPSN XII, pp 93–101
8. Amini F, Hedayati Y (2016) Underdetermined blind modal identification of structures by earthquake and ambient vibration measurements via sparse component analysis. J Sound Vib 366:117–132
9. Ehrgott M (2005) Multicriteria optimization. Springer Science & Business Media
10. Yu XC, Cao TT, Hu D et al (2010) Blind image separation based on wavelet transformation and sparse component analysis. J Beijing Univ Posts Telecommun 33(2):58–63
11. Branke J, Deb K, Dierolf H, et al. (2004) Finding knees in multi-objective optimization. In: International Conference on Parallel Problem Solving from Nature, pp 722–731

A Visual Feedback Model-Free Design for Robust Tracking of Nonholonomic Mobile Robots

Hui Chen, Hua Chen, Yibin Wang and Fang Yang

Abstract This paper considers the problem of designing a visual feedback control law for robust tracking of nonholonomic mobile robots. The control approach developed in this work with uncalibrated visual parameters, unknown control directions, and external disturbances. Using incomplete information of the moving objects to be tracked to propose a model-free, self-support control algorithm to ensure the tracking error can be driven into a prespecified neighborhood of zero. Global stability of the corresponding closed-loop system of tracking error is proved by the Lyapunov stability theory. Finally, the simulation results demonstrate the effectiveness of the proposed controller design method.

Keywords Visual feedback · Nonholonomic mobile robots · Model-free · Robust tracking

1 Introduction

In recent few years, Wheeled Mobile Robot (WMR) plays an important role in several fields as industry, agriculture, terrorism, explosion, space exploration, safety. They can replace human to complete son dangerous or impossible task.

H. Chen · H. Chen (✉)
Mathematics and Physics Department, Hohai University, Changzhou Campus, Changzhou, China
e-mail: chenhua112@163.com

H. Chen
e-mail: 1215352549@qq.com

H. Chen · H. Chen · Y. Wang
College of Mechanical and Engineering, Hohai University, Changzhou, China
e-mail: wangyb8000@163.com

F. Yang
School of Science, Ningbo University of Technology, Ningbo, China
e-mail: liusha_02@163.com

Y. Jia et al. (eds.), *Proceedings of 2016 Chinese Intelligent Systems Conference*, Lecture Notes in Electrical Engineering 405,
DOI 10.1007/978-981-10-2335-4_56

However, as it is known to all that it is impossible to stabilize any nonholonomic system to a point by a static state feedback law according to Brockett's conclusion [1]. Therefore, a great deal of research is done to resolve the stabilization problem of such systems. And lots of reasonable and ingenious methods have been proposed in order to overcome this design difficulty [2–5], which contains four cases: continuous time-varying feedback control laws [6–8], discontinuous feedback control laws [9–11], hybrid feedback control laws [12], and optimal control laws [13–15].

Since the stable problem has been solved by the stabilization methods, it is hard to ensure that these methods mentioned above are equally valid in the trajectory tracking problem.

In fact, research of the tracking controller design has received increasing attention from many researchers for recent decades, for example, two knowledge-based tracking controllers have been put forward to solve the problem of a computed torque nonlinear in [16]. In [17] to overcome the difficulties of using the brushed direct current motors actuates the uncertain wheeled mobile robots, a robust tracking control law is proposed. Reference [18, 19] gives some non-smooth finite-time tracking control laws for improving the convergence rate of the wheeled mobile robots. In addition, a robust control law based on the visual servoing feedback for nonholonomic mobile robots with uncalibrated camera parameters is proposed in [20].

This paper considers the robust tracking problem of nonholonomic model-free mobile robots with uncertain parameters with the unknown trajectory based on the generalized principle of self-support (GPSS) control strategy [21, 22]. The main innovations and results can be summarized as the following two respects.

(a) Based on the principle of GPSS, the control inputs can be regarded as a natural part of the robotic system itself, so the visual feedback controller can be presented.
(b) Through the research of the existing visual servoing control of nonholonomic model-free mobile robots [23–26], a new control law is proposed with uncertain directions, unknown visual parameters and external disturbances.

The structure of the paper is as follows: Sect. 2 gives a formalization of the problem considered in the paper. Section 3 states our main results. Section 4 provides an illustrative numerical example and the corresponding simulation results of the proposed methodology. Finally, a conclusion is shown in Sect. 5.

2 Problem Statement

As shown in Fig. 1, the mass center p can be measured. Assuming that the geometric center point and the mass center point of the robot are the same, the dynamic model for the nonholonomic wheeled mobile robot can be described by the following differential equations [27, 28]:

Fig. 1 Nonholonomic wheeled mobile robots

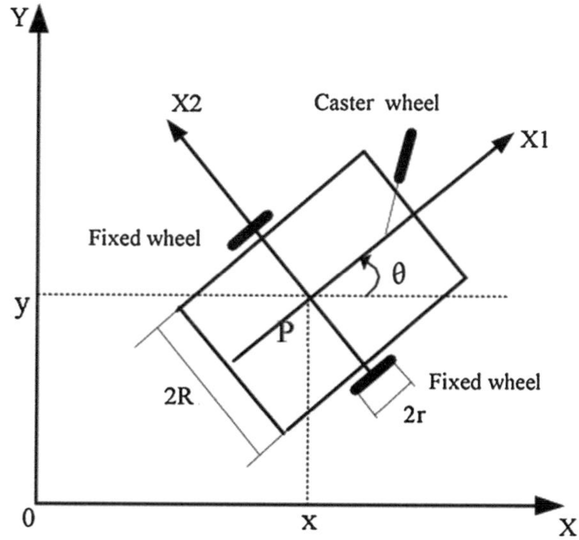

$$\begin{cases} \dot{x} = v\cos\theta \\ \dot{y} = v\sin\theta \\ \theta = \omega \\ \dot{v} = \tau_1 \\ \dot{\omega} = \tau_2 \end{cases} \tag{1}$$

where (x, y) is the position coordinates of the mass center p of the robot moving, θ denotes its heading angle from the horizontal axis, ω is steering velocity, and v is the forward velocity.

As shown in Fig. 2, using a camera fixed to measure the movement of the nonholonomic wheeled mobile robot, the corresponding kinematic equation can be described as follows from (1):

$$\begin{cases} \dot{x} = \alpha_1 v\cos(\theta - \theta_0) \\ \dot{y} = \alpha_2 v\sin(\theta - \theta_0) \\ \theta = \omega \\ \dot{v} = \beta_1(t)\tau_1 + d_1(t) \\ \dot{\omega} = \beta_2(t)\tau_2 + d_2(t) \end{cases} \tag{2}$$

where α_i is uncalibrated visual parameter, $\beta_i(t) > 0$ is the unknown control direction, θ_0 is the unknown angle of coordinate system and $d_i(t)$ is the external disturbance. τ_i is the force or torque inputs to be designed.

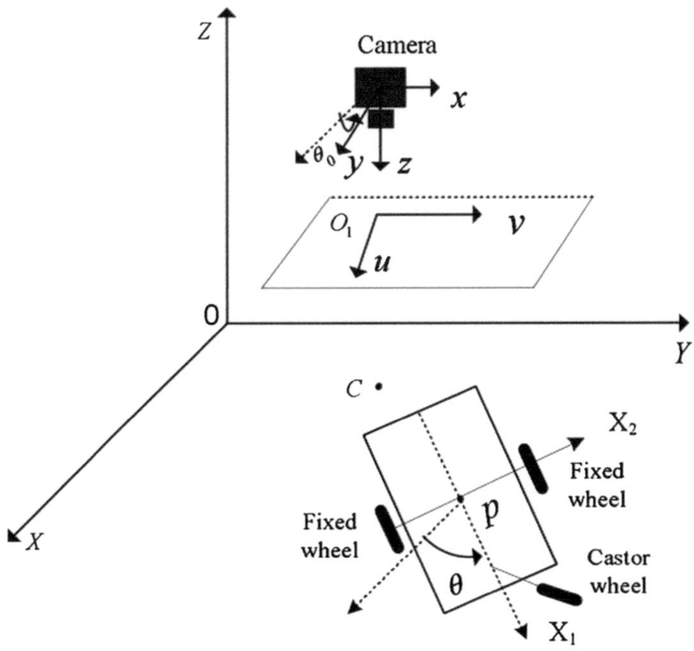

Fig. 2 Nonholonomic wheeled mobile robots with a fixed camera

Defining the output tracking error as follows:

$$\begin{cases} x_e = x - x_d \\ y_e = y - y_d \end{cases} \tag{3}$$

where x_d, y_d are the desired position point. However, we may not measure all the information of x, y and x_d, y_d because of random perturbation, complex uncertain parameter, etc., hence, we make the following assumptions:

Assumption 1 The estimation of error measurement (x_e, y_e) can be denoted as follows:

$$\tilde{x}_e = x_e - \int_0^t \omega_1(t)\mathrm{d}t \tag{4}$$

$$\tilde{y}_e = y_e - \int_0^t \omega_2(t)\mathrm{d}t \tag{5}$$

where the bounded estimated error function $\omega_i(t)$ satisfies that

$$|\omega_i(t)| \le b_i, \left| \int_0^t \omega_i(t)\mathrm{d}t \right| \le c_i \tag{6}$$

Assumption 2 Where the uncalibrated visual parameter α_i, the unknown control direction $\beta_i(t)$, the unknown angle θ_0 and the external disturbance $d_i(t)$ are assumed to be bounded, i.e.,

$$\begin{aligned} 0 < \alpha_i^- \le \alpha_i \le \alpha_i^+, 0 < \beta_i^- \le \beta_i \le \beta_i^+ \\ 0 < \theta_0^- \le \theta_0 \le \theta_0^+, 0 < d_i^- \le d_i \le d_i^+ \end{aligned} \tag{7}$$

Assumption 3 The velocities, accelerations are $\dot{x}, \dot{y}, \ddot{x}, \ddot{y}$ not directly available, but it is within a known bounded region:

$$|\dot{x}| \le m, |\ddot{x}| \le m_1, |\dot{y}| \le n, |\ddot{y}| \le n_1, |\dot{\theta}| \le r \tag{8}$$

Remark 1 What was different from other research about visual feedback of non-holonomic wheel mobile robots is that all the visual parameters are assumed to be unknown and we discussed a more complex uncertain dynamic model here.

Remark 2 We assume that there is an error estimation function in (4) and (5) because that (x, y) and (x_d, y_d) may not be got directly because of some in some complex occasions.

The control priority in this paper is to design a robust tracking visual feedback law for robot system under Assumptions 1–3.

3 Main Results

The design results will be stated as the following theorem:

Theorem 1 *Under Assumptions* 1–3, *for the dynamic system* (2), *taking The following feedback law*

$$\begin{cases} \tau_1 = -k_1 \tilde{s}_1(t) \\ \tau_2 = -k_2 \tilde{s}_2(t) \end{cases} \tag{9}$$

where $\tilde{s}_i(t)$ is the estimated error feedback signal for controller as follows:

$$\begin{cases} \tilde{s}_1(t) = \dot{\tilde{x}}_e + \lambda_1 \tilde{x}_e \\ \tilde{s}_2(t) = \dot{\tilde{y}}_e + \lambda_2 \tilde{y}_e \end{cases} \tag{10}$$

where k_i is a design parameter that satisfies

$$k_1 > \frac{2(\frac{mm_1}{(\alpha_1^-)^2} + \frac{nn_1}{(\alpha_2^-)^2}) + d_1^+}{\beta_1^- \lambda_1 \varepsilon} > 0,\ k_2 > \frac{r + d_2^+}{\beta_2^- \lambda_2 \varepsilon} > 0 \tag{11}$$

where $\lambda_i, \varepsilon > 0$ denote the scale index of convergent region of the tracking error given in advance. Then the real tracking error can be driven into the following prespecified neighborhood of zero

$$D_\varepsilon = \left\{ (x_e, y_e) \colon |x_e| \le \frac{\lambda_1 c_1 + b_1}{\lambda_1} + \varepsilon,\ |y_e| \le \frac{\lambda_2 c_2 + b_2}{\lambda_2} + \varepsilon \right\} \tag{12}$$

Proof Take a Lyapunov function candidate $V(t) = \frac{1}{2}\left(x_e^2 + y_e^2\right)$ for each x_e and y_e. Calculating the time derivative of V(t)

$$\dot{V}(t) = x_e \dot{x}_e + y_e \dot{y}_e \tag{13}$$

According to (4) and (5) in Assumption 1, we have

$$x_e = \tilde{x}_e + 0 \int_0^t \omega_1(t)\mathrm{d}t,\ \dot{x}_e = \dot{\tilde{x}}_e + \omega_1(t) \tag{14}$$

$$y_e = \tilde{y}_e + \int_0^t \omega_2(t)\mathrm{d}t,\ \dot{y}_e = \dot{\tilde{y}}_e + \omega_2(t) \tag{15}$$

Substituting (14) and (15) into (13), we have

$$\dot{\mathrm{V}}(t) = x_e(\dot{\tilde{x}}_e + \omega_1(t)) + y_e(\dot{\tilde{y}}_e + \omega_2(t)) \tag{16}$$

From (10), we have

$$\begin{cases} \dot{\tilde{x}}_e = \tilde{s}_1(t) - \lambda_1 \tilde{x}_e \\ \dot{\tilde{y}}_e = \tilde{s}_2(t) - \lambda_2 \tilde{y}_e \end{cases} \tag{17}$$

Substituting (4), (5), (9) and (17) into (16), we have

$$
\begin{aligned}
\dot{V}(t) &= x_e(\tilde{s}_1(t) - \lambda_1\tilde{x}_e + \omega_1(t)) + y_e(\tilde{s}_2(t) - \lambda_2\tilde{y}_e + \omega_2(t)) \\
&= x_e(\tilde{s}_1(t) - \lambda_1(x_e - \int_0^t \omega_1(t)\mathrm{d}t) + \omega_1(t)) + y_e(\tilde{s}_2(t) \\
&\quad - \lambda_2(y_e - \int_0^t \omega_2(t)\mathrm{d}t + \omega_2(t)) \\
&= -\lambda_1 x_e^2 - \lambda_2 y_e^2 + x_e(\tilde{s}_1(t) + \lambda_1 \int_0^t \omega_1(t)\mathrm{d}t + \omega_1(t)) \\
&\quad + y_e(\tilde{s}_2(t) + \lambda_2 \int_0^t \omega_2(t)\mathrm{d}t + \omega_2(t)) \\
&= -\lambda_1 x_e^2 - \lambda_2 y_e^2 + x_e(-\frac{\tau_1}{k_1} + \lambda_1 \int_0^t \omega_1(t)\mathrm{d}t + \omega_1(t)) \\
&\quad + y_e(-\frac{\tau_2}{k_2} + \lambda_2 \int_0^t \omega_2(t)\mathrm{d}t + \omega_2(t))
\end{aligned} \tag{18}
$$

According to system (2) the estimated feasible bound of τ_i can be presented as follows:

$$
v = (\frac{\dot{x}}{\alpha_1})^2 + (\frac{\dot{y}}{\alpha_2})^2 = \frac{\dot{x}^2}{\alpha_1^2} + \frac{\dot{y}^2}{\alpha_2^2} \quad \dot{v} = \frac{2\dot{x}\ddot{x}}{\alpha_1^2} + \frac{2\dot{y}\ddot{y}}{\alpha_2^2} \tag{19}
$$

$$
\omega = \theta \quad \dot{\omega} = \dot{\theta} \tag{20}
$$

$$
\tau_1 = \frac{\dot{v} - d_1(t)}{\beta_1(t)} = \frac{\frac{2\dot{x}\ddot{x}}{\alpha_1^2} + \frac{2\dot{y}\ddot{y}}{\alpha_2^2} - d_1(t)}{\beta_1(t)} \tag{21}
$$

$$
\tau_2 = \frac{\dot{\omega} - d_2(t)}{\beta_2(t)} = \frac{\dot{\theta} - d_2(t)}{\beta_2(t)} \tag{22}
$$

From (7) and (8) in Assumptions 1 and 3, we have

$$
\begin{cases}
|\tau_1| \le \dfrac{\left|\frac{2\dot{x}\ddot{x}}{\alpha_1^2} + \frac{2\dot{y}\ddot{y}}{\alpha_2^2} - d_1(t)\right|}{|\beta_1(t)|} \le \dfrac{2(\frac{mm_1}{(\alpha_1^-)^2} + \frac{nn_1}{(\alpha_2^-)^2}) + d_1^+}{\beta_1^-} \\
= \tau_{1\max} |\tau_2| \le \dfrac{|\dot{\theta} - d_2(t)|}{|\beta_2(t)|} \le \dfrac{r + d_2^+}{\beta_2^-} = \tau_{2\max}
\end{cases} \tag{23}
$$

According to (6), (11), and (18), we have

$$\begin{aligned}
\dot{\mathrm{V}}(t) = & -\lambda_1 x_e^2 - \lambda_2 y_e^2 + x_e(-\frac{\tau_1}{k_1} + \lambda_1 \int\limits_0^t \omega_1(t)\mathrm{d}t + \omega_1(t)) + y_e(-\frac{\tau_2}{k_2} \\
& + \lambda_2 \int_0^t \omega_2(t)dt + \omega_2(t)) \\
& \le -\lambda_1 x_e^2 - \lambda_2 y_e^2 + |x_e| \left| (\frac{\tau_1}{k_1} + \lambda_1 \int\limits_0^t \omega_1(t)\mathrm{d}t + \omega_1(t)) \right| \\
& + |y_e| \left| (\frac{\tau_2}{k_2} + \lambda_2 \int\limits_0^t \omega_2(t)\mathrm{d}t + \omega_2(t)) \right| \\
& \le -\lambda_1 x_e^2 - \lambda_2 y_e^2 + |x_e| (\frac{\tau_{1\max}}{k_1} + \lambda_1 c_1 + b_1) \\
& + |y_e| (\frac{\tau_{2\max}}{k_2} + \lambda_2 c_2 + b_2)
\end{aligned} \tag{24}$$

If $|x_e| > \frac{\lambda_1 c_1 + b_1}{\lambda_1} + \varepsilon, |y_e| > \frac{\lambda_2 c_2 + b_2}{\lambda_2} + \varepsilon$, from (24), we have

$$\begin{aligned}
\dot{\mathrm{V}}(t) \le & -\lambda_1 x_e^2 - \lambda_2 y_e^2 + |x_e| (\frac{\tau_{1\max}}{k_1} + \lambda_1 c_1 + b_1) + |y_e| (\frac{\tau_{2\max}}{k_2} + \lambda_2 c_2 + b_2) \\
= & -|x_e| (\lambda_1 (\frac{\lambda_1 c_1 + b_1}{\lambda_1} + \varepsilon) - \frac{\tau_{1\max}}{k_1} - \lambda_1 c_1 - b_1) \\
& -|y_e| (\lambda_2 (\frac{\lambda_2 c_2 + b_2}{\lambda_2} + \varepsilon) - \frac{\tau_{2\max}}{k_2} - \lambda_2 c_2 - b_2) \\
= & -|x_e| (\lambda_1 \varepsilon - \frac{\tau_{1\max}}{k_1}) - |y_e| (\lambda_2 \varepsilon - \frac{\tau_{2\max}}{k_2}) \\
= & -\frac{|x_e|}{k_1} (k_1 \lambda_1 \varepsilon - \tau_{1\max}) - \frac{|y_e|}{k_2} (k_2 \lambda_2 \varepsilon - \tau_{2\max}) \\
\le & \, 0
\end{aligned} \tag{25}$$

By Lemma 2 and (25), $x_e(t) \to 0, y_e(t) \to 0$ as $t \to \infty$, hence $x_e(t), y_e(t)$ will be driven into D_e.

On the other hand, once $|x_e| \le \frac{\lambda_1 c_1 + b_1}{\lambda_1} + \varepsilon, |y_e| \le \frac{\lambda_2 c_2 + b_2}{\lambda_2} + \varepsilon$, from (9) and (10), Eqs. (14) and (15) we have

$$\begin{aligned}
|\dot{x}_e|+|\dot{y}_e| &= |\dot{\tilde{x}}_e+\omega_1(t)|+|\dot{\tilde{y}}_e+\omega_2(t)| \\
&= |\tilde{s}_1(t)-\lambda_1\tilde{x}_e+\omega_1(t)|+|\tilde{s}_2(t)-\lambda_2\tilde{y}_e+\omega_2(t)| \\
&= \left|-\frac{\tau_1}{k_1}-\lambda_1(x_e-\int_0^t\omega_1(t)\mathrm{d}t)+\omega_1(t)\right| \\
&\quad+\left|-\frac{\tau_2}{k_2}-\lambda_2(y_e-\int_0^t\omega_2(t)\mathrm{d}t)+\omega_2(t)\right| \\
&\le \left|\frac{\tau_{1\max}}{k_1}\right|+\lambda_1(|x_e|+\left|\int_0^t\omega_1(t)\mathrm{d}t\right|)+|\omega_1(t)| \\
&\quad+\left|\frac{\tau_{2\max}}{k_2}\right|+\lambda_2(|y_e|+\left|\int_0^t\omega_2(t)\mathrm{d}t\right|)+|\omega_2(t)| \\
&\le \lambda_1\varepsilon+\lambda_1(\frac{\lambda_1c_1+b_1}{\lambda_1}+\varepsilon+c_1)+b_1+\lambda_2\varepsilon \\
&\quad+\lambda_2(\frac{\lambda_2c_2+b_2}{\lambda_2}+\varepsilon+c_2)+b_2 \\
&\le 2(\lambda_1\varepsilon+\lambda_1c_1+b_1)+2(\lambda_2\varepsilon+\lambda_2c_2+b_2)
\end{aligned} \tag{26}$$

According to (26), it is obvious to obtain

$$|\dot{x}_e|+|\dot{y}_e|\le 2(\lambda_1+\lambda_2)\varepsilon\to 0_+ \text{ as } b_1=b_2=c_1=c_2=0 \tag{27}$$

Then according to the Lyapunov stability theory, x_e, y_e will stay in D_ε.

The similar proof process for the case of $|x_e|>\frac{\lambda_1c_1+b_1}{\lambda_1}+\varepsilon,\ |y_e|\le\frac{\lambda_2c_2+b_2}{\lambda_2}+\varepsilon;\ |x_e|\le\frac{\lambda_1c_1+b_1}{\lambda_1}+\varepsilon,\ |y_e|>\frac{\lambda_2c_2+b_2}{\lambda_2}+\varepsilon$ can be easily obtained, thus omitted and this completes the proof.

4 Simulations

In this section, some numerical simulations results will be used to show how to present the model-free controllers from Theorem 1. By Assumptions 1–3 and Theorem 1, we suppose $b_1=b_2=1.0$, $c_1=c_2=1.5$, $\omega_1(t)=0.8e^{-0.02t}\cos t$, $\omega_2(t)=0.5e^{-0.03t}\sin t$, $\varepsilon_0=0.3$, $\lambda_1=\lambda_2=1.0$, $d_1=d_2=2.0, \alpha_i^-=0.6, \alpha_i^+=1.2, \beta_i^-=0.5, \beta_i^+=2.5, d_i^-=1.0, d_i^+=3.0, (i=1,2)$, $\theta_0^-=\frac{\pi}{20}$, $\theta_0^+=\frac{\pi}{4}$, $m=1.0$, $m_1=0.5$, $n=1.2$, $n_1=0.4$, $r=0.2$, $k_1=-10$, $k_2=-15$ The initial conditions of tracking error are $x_e(0)=-2.5$, $y_e(0)=1.2$.

Observing from Figs. 3 and 4, we find that the practical tracking error can be driven into the prespecified fixed neighborhood of zero with given size at about $\varepsilon_0 = 0.3$ within $t \leq 5$ s.

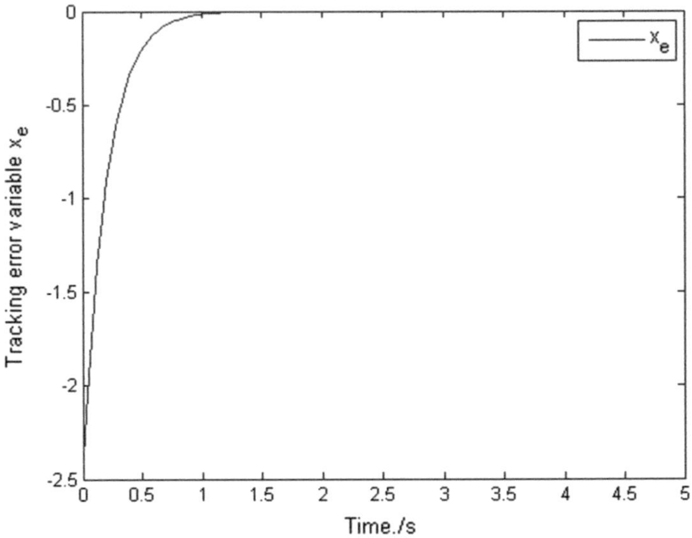

Fig. 3 The response of tracking error x_e with respect to time

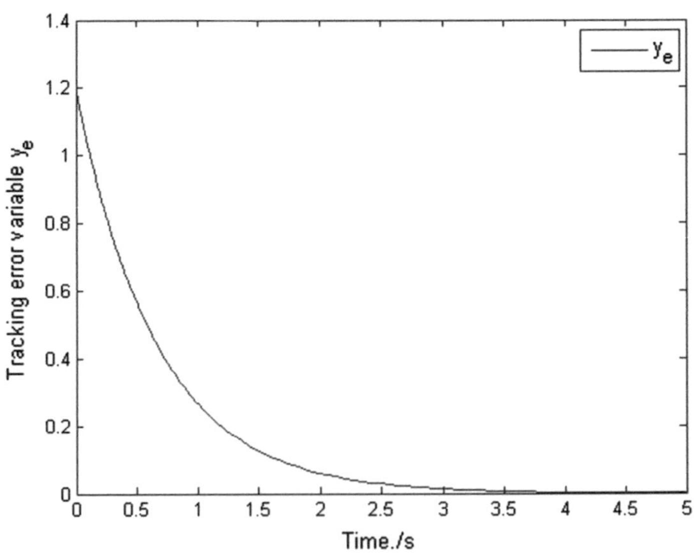

Fig. 4 The response of tracking error y_e with respect to time

5 Conclusion

In this paper, a visual feedback robust tracking of nonholonomic mobile robots has been discussed. A new idea (GPSS) is proposed to solve the problem that designs the control law with uncalibrated visual parameter, unknown control direction, unknown angle, and external disturbance. Moreover the efficiency of the control strategy is verified through the numeric simulations.

Acknowledgment This work was supported by the Natural Science Foundation of China (61304004, 61503205), the Foundation of China Scholarship Council (201406715056), the China Postdoctoral Science Foundation funded project (2013M531263), the Jiangsu Planned Projects for Postdoctoral Research Funds (1302140C), the Project Supported by the Foundation (No. CZSR2014005) of Changzhou Key Laboratory of Special Robot and Intelligent Technology, P.R. China, and the Changzhou Sci&Tech Program (CJ20160013).

References

1. Brockett RW (1983) Asymptotic stability and feedback stabilization. In: Brockett RW, Millman RS, Sussmann HJ (eds) Differential geometric control theory. Birkhauser, Boston, pp 181–208
2. Tian YP, Li S (2002) Exponential stabilization of nonholonomic dynamic systems by smooth time-varying control. Automatica 38(7):1139–1146
3. Hussein II, Bloch AM (2008) Optimal control of underactuated nonholonomic mechanical systems. IEEE Trans Automat Control 53(3):668–682
4. Ge SS, Wang Zhuping, Lee TH (2003) Adaptive stabilization of uncertain nonholonomic systems by state and output feedback. Automatica 39(8):1451–1460
5. Yuanyuan Wu, Yuqiang Wu (2010) Robust stabilization of delayed non-holonomic systems with strong nonlinear drifts. Nonlinear Anal Real World Appl 11(5):3620–3627
6. Murray RM, Sastry SS (1993) Nonholonomic motion planning: Steering using sinusoids. IEEE Trans Autom Control 38(5):700–716
7. Chen H, Wang C, Yang L, Zhang D (2012) Semiglobal stabilization for nonholonomic mobile robots based on dynamic feedback with inputs saturation. J Dyn Syst Meas Control 134 (4):041006.1–041006.8
8. Teel A, Murry R, Walsh G (1992) Nonholonomic control systems: from steering to stabilization with sinusoids. Proc IEEE Conf Decis Control 2:1603–1609
9. Astolfi A (1996) Discontinuous control of nonholonomic systems. Syst Control Lett 27:37–45
10. Bloch AM, Drakunov S (1994) Stabilization of a nonholonomic systems via sliding modes. Proc IEEE Conf Decis Control 3:2961–2963
11. de Wit CC, SZrdalen OJ (1992) Exponential stabilization of mobile robots with nonholonomic constraints. IEEE Trans Autom Control 37(11):1791–1797
12. Sordalen OJ, Egeland O (1995) Exponential stabilization of nonholonomic chained systems. IEEE Trans Autom Control 40(1):35–49
13. Soueres P, Balluchi A, Bicchi A (2001) Optimal feedback control for line tracking with a bounded-curvature vehicle. Int J Control 74(10):1009–1019
14. Hussein II, Bloch AM (2008) Optimal control of underactuated nonholonomic mechanical systems. IEEE Trans Autom Control 53(3):668–682
15. Qu Z, Wang J, Plaisted CE, Hull RA (2006) Global-stabilizing near-optimal control design for nonholonomic chained systems. IEEE Trans Autom Control 51(9):1440–1456

16. Keighobadi J, Menhaj MB (2012) From nonlinear to fuzzy approaches in trajectory tracking control of wheeled mobile robots. Asian J. Control 14(4):960–973
17. Chang Y-C, Yen H-M, Wang P-T (2012) An intelligent robust tracking control for a class of electrically driven mobile robots. Asian J. Control 14(6):1567–1579
18. Wang Z, Li S, Fei S (2009) Finite-time tracking control of a nonholonomic mobile robot. Asian J Control 11:344–357
19. Ou M, Du H, Li S (2012) Finite-time tracking control of multiple nonholonomic mobile robots. J Franklin Inst 349:2834–2860
20. Liang Z, Wang C (2011) Robust stabilization of nonholonomic chained form systems with uncertainties. Acta Automatica Sina 37(2):129–142
21. Novakovic ZR (1992) The principle of self-support in control systems. Elsevier Science Ltd
22. Chen H, Chen YQ (2015) Fractional-order generalized principle of self-support (FOG PSS) in control systems design. arXiv:1509.06043
23. Chen H, Zhang J, Chen B, Li B (2013) Global practical stabilization for nonholonomic mobile robots with uncalibrated visual parameters by using a switching controller. IMA J Math Control Inf. doi:10.1093/imamci/dns044
24. C H, Chen B, Li B, Zhang J (2013) Practical stabilization of uncertain nonholonomic mobile robots based on visual servoing model with uncalibrated camera parameters. Math Prob Eng. doi:10.1155/2013/395410
25. Chen H, Wang C, Liang Z et al (2014) Robust practical stabilization of nonholonomic mobile robots based on visual servoing feedback with inputs saturation. Asian J Control 16 (3):692–702
26. Chen H, Ding S, Chen X et al (2014) Global finite-time stabilization for nonholonomic mobile robots based on visual servoing. Int J Adv Robot Syst 11:1–13
27. Chang Y-C, Yen H-M, Wang P-T (2012) An intelligent robust tracking control for a class of electrically driven mobile robots. Asian J Control 14(6):1567–1579
28. Liang Z, Wang C (2011) Robust stabilization of nonholonomic chained form systems with uncertainties. Acta Automatica Sina 37(2):129–142

Short-Term Solar Irradiance Forecasting Using Neural Network and Genetic Algorithm

Anping Bao, Shumin Fei and Minghu Zhong

Abstract Solar irradiance is a vital factor for a solar plant because the inaccurate prediction can increase the risk and the cost of operation. To reach a high-prediction accuracy, a model for short-term direct normal irradiance prediction is proposed in this paper. First, the inputs of the model were discussed and included historical data, such as direct normal irradiance, air temperature, pressure, and the wind velocity. Then, the model was constructed and optimized by genetic algorithm. Model validation was conducted using data from the National Renewable Energy Laboratory's India Solar Resource Data. The result shows that the forecast skill of the proposed model improved 80 % over the persistence model and also better than that of some published models.

Keywords DNI forecasting · Neural network · Genetic algorithm · Short-term forecasting

1 Introduction

Due to environmental pollution and climate change caused by the use of conventional fossil fuels, renewable energy has received more and more attention [1–4]. As one of the major renewable energy, solar power penetration is growing rapidly [5]. Since most of the power grid variability is based on the load side and the fluctuations of solar irradiance are rapid and random, photovoltaic power fore-

Sponsored by Qing Lan Project.

A. Bao (✉) · S. Fei
Key Laboratory of Measurement and Control of CSE (School of Automation, Southeast University), Ministry of Education, Nanjing 210096, China
e-mail: baoanpingseu@163.com

A. Bao · M. Zhong
Nanjing College of Information and Technology, Nanjing 210096, China

Y. Jia et al. (eds.), *Proceedings of 2016 Chinese Intelligent Systems Conference*, Lecture Notes in Electrical Engineering 405,
DOI 10.1007/978-981-10-2335-4_57

casting is important for the solar power grid. The accurate prediction of the direct normal irradiance (DNI) is vital for a concentrated solar thermal plant. The inaccurate forecasting can result in the increasing of the risk and cost of operation for a solar plant [6].

Therefore, a great deal of models has been developed to predict the DNI. Some are based on physical models and spectral information [7, 8]. Some are using statistical methods, such as models of the time series [9]. The statistical models could be classified into two groups, linear models and nonlinear models. For the linear model, autoregressive (AR) and autoregressive moving average (ARMA) [10] are easily used to forecast the DNI. For the nonlinear model, artificial neural network (ANN) [11] and support vector machine (SVM) [12] are applied widely. However, the linear model cannot fully describe the information of the DNI series, because some of them are nonlinear. ANN, as a nonlinear model, may be able to predict DNI precisely with the appropriate structure. However the weights and the structure of an ANN model are chosen difficultly [13].

To overcome the disadvantages of ANN models, in this paper, an ANN-based model is developed for forecasting DNI, and its initial weight values are optimized with genetic algorithm (GA). GA is an algorithm that could solve the optimization problem of complex systems. So, the forecasting results of ANN with the weights optimized by GA would be stable and accurate.

The rest of this paper is constructed as the following: Sect. 2 introduces Artificial Neural Networks. The details of the input variables and the structure of the ANN-based model with GA are discussed in Sect. 3. The performance of the proposed model is evaluated with different experiments in Sect. 4. Conclusions are presented in Sect. 5.

2 ANN Models

Artificial Neural Network (ANN) is an artificial intelligence system developed based on modern neurobiology research, which simulates the mechanism of human brain to process information [11]. It not only has the ability of general calculation and processing data, but also can deal with knowledge of thinking, learning and memory ability like humans. Due to its great learning ability and inherent generalization ability, it has been widely used in the fields of classification, modeling, and forecasting.

The multilayer perceptron (MLP) is one of the most neural networks for forecasting in solar engineering applications. A MLP with one output and three layers can be expressed as

$$y = g(X) = \omega_2 f\left(\sum_{i=1}^{d} \omega_{1i} x_i\right) + \theta \tag{1}$$

where X is a vector of input variables; ω is the weights of each neuron; θ is the threshold value of the output layer; and $f(\cdot)$ is an activation function, such as a sigmoid function [14].

For the structure of ANN, MLP is a kind of the feed-forward neural networks. Besides it, the radial basis function neural network is also a common feed-forward neural network. In the process of training an ANN model, it will be combined with learning algorithms, such as gradient descent algorithm, Newton method, and so on.

3 ANN Model with GA

In this section, the input variables are determined first, and then an ANN-based model for forecasting solar irradiance is constructed. Finally, the GA is used to optimize the original weights of the ANN model.

3.1 Data Collection

The DNI values used in this paper were all downloaded from http://rredc.nrel.gov/solar/new_data/India. Figure 1 shows the hourly evolution of the direct normal irradiance measured on horizontal surface for 2011 at 70.05 °E and 24.05 °N. In addition, the air temperature, the pressure, and the wind velocity were also downloaded under the same conditions.

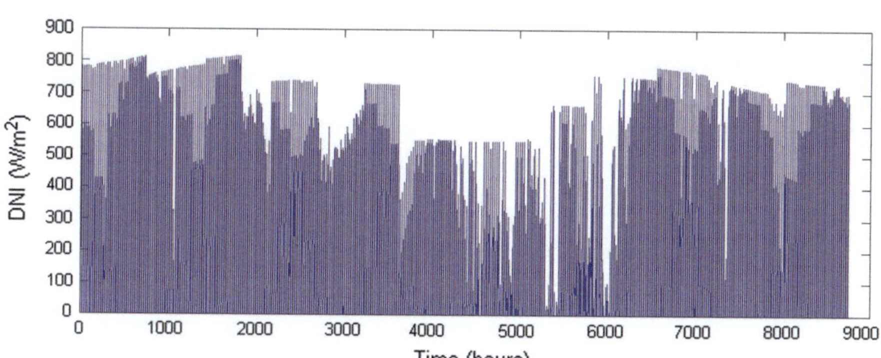

Fig. 1 Hourly direct normal irradiance for 2011

3.2 ANN-Based Forecasting Model

For predicting the solar irradiance, a feed-forward neural network with back-propagation algorithm was constructed. The key factor of the ANN model is the neural network's structure. First, the inputs should be discussed, which determine the node number of the input layer of the neural network. In our model, the partial autocorrelation coefficients of DNI data for 1 year were used to determine the input dimension referring to the AR model. Figure 2 shows the results of the partial autocorrelation function of $DNI(t)$ and $DNI(t-k)$ (k = 0, 1, …, 20) in the DNI time series. $DIN(t)$, $DNI(t-1)$, $DNI(t-2)$ and $DNI(t-24)$ were selected as the inputs of the neural network. And other meteorological data at the same moment were also as the inputs of the neural network [15].

The output is the DNI value of the next hour, so the node number of the output layer is one. Some study published shows that a three-layer neural network can fit with any continuous bounded function. Consequently, the proposed ANN model has one hidden layer. If the node number of the hidden layer is too little, the forecasting accuracy will be low; if the node number of the hidden layer is too large, the constructed model will be complex and there will be an over-fitting phenomenon. An empirical formula was used to determine the node number of the hidden layer, and it was defined as

$$N = \sqrt{I+O} + a \tag{2}$$

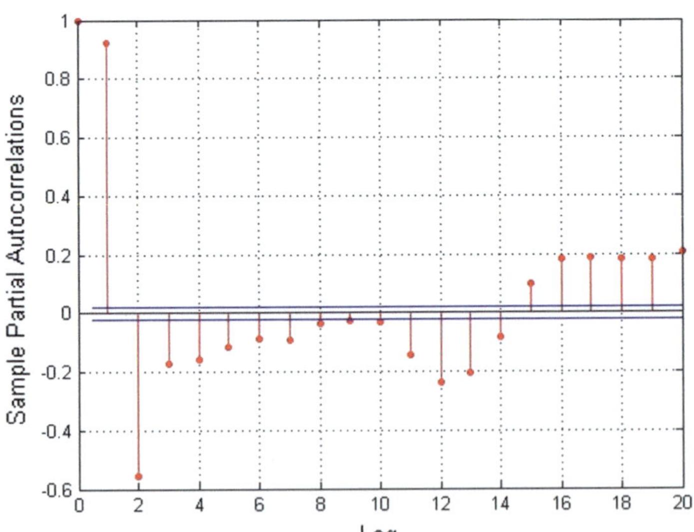

Fig. 2 Results of the partial autocorrelation function of hourly direct irradiance

where I is the dimension of the inputs and O is the dimension of the outputs, and a is a constant between one and ten. According to the cross validation, the node number of the hidden layer can be finally determined.

3.3 *Genetic Algorithm for Model Optimization*

GA was developed by John Holland et al. for solving optimization problems. It was an optimization algorithm inspired by the evolution mechanism of biological system, "survival of the fittest". The main idea of GA is that: the variables, which need to be optimized, are encoded to generate the first generation of population; then according to the evolution principle, the more appropriate population will be generated, which means the more appropriate will be achieved. In every generation, the individual will be selected from one generation based on its fitness, and then the selected individuals will be crossed and mutated to produce the next generation depending on genetic operators. Finally, the last population will be decoded to achieve the optimization solution of the original problem [16, 17].

Comparing to the general optimization algorithms, such as the gradient method, dynamic programming, branch and bound method, the genetic algorithm has the following characteristics:

(1) Adaptively. It searches the optimization solution automatic once the encoding method, the function of calculating fitness and genetic operators are determined.
(2) Parallelism. It is not limited to search a local minimal value, but search the global optimal point by searching a number of points simultaneously.
(3) Generality. It is used with the objective function and fitness function and does not need any other auxiliary information.

In this paper, the GA was used to optimize the initial weights of ANN model, and the main steps are as the following:

step 1 defining the parameters of GA such as the max iterations, population size, crossover probability, and mutation probability;
step 2 constructing the fitness function;
step 3 encoding and producing the initial population;
step 4 calculating the individual fitness;
step 5 generating the next generation of new population by selection, crossover, and mutation operation;
step 6 decoding to get the weight if the fitness of the new population meets the defined requirement, or else going to the step 5.

4 Simulation Experiments

To investigate the performance of the proposed model, some experiments were carried out with data introduced in Sect. 3. First, the four statistical metrics, normalized mean bias error, normalized mean absolute error, normalized root mean square error, and forecast skill were introduced for assessing the model's performance. Then the proposed model was used to predict DNI inter-hour ahead. Finally, the comparative experiment with other models published was carried out.

Four metrics used to assess the performance of the model are described as follows:

- normalized Mean Bias Error (nMBE): $\mathrm{nMBE} = \frac{1}{N}\sum_{i=1}^{N}(D_{ci} - D_{mi})/\overline{D} \times 100\,\%$
- normalized Mean Absolute Error (nMAE): $\mathrm{nMAE} = \frac{1}{N}\sum_{i=1}^{N}|D_{ci} - D_{mi}|/\overline{D} \times 100\,\%$
- normalized Root Mean Square Error (nRMSE): $\mathrm{nRMSE} = \sqrt{\frac{1}{N}\sum_{i=1}^{N}(D_{ci} - D_{mi})^2}/\overline{D} \times 100\,\%$
- forecast skill (s): $s = \frac{\mathrm{nRMSE}_p - \mathrm{nRMSE}}{\mathrm{nRMSE}_p} \times 100\,\%$

where N denotes the total number of data points, D_{ci} is the clear-sky DNI computed by different models, D_{mi} is the measured value by a pyrheliometer and D is the mean of the measured values $\mathrm{nRMSE_p}$ is nRMSE of the persistence model.

4.1 Short-Term Prediction

In order to evaluate the performance of the proposed model, the 90 % of the data were set as a training set and the remain were set as a testing set. Table 1 lists the results of the ANN model and the optimized model for forecasting DNI 1 h ahead. The nRMSE of the GA-ANN model is only 9.65 %, which is very low. And the forecast skill of the proposed model improved to 23 % over that of the ANN model. It is clear that the forecasting accuracy has been improved by optimized with GA. Figure 3a shows the comparison between the measured and predicted DNI by the GA-ANN model for the testing set. The predicted DNI is in agreement with the measured values in most instances.

Table 1 Results of the proposed model for forecasting DNI for the testing set

	nMBE (%)	nMAE (%)	nRMSE (%)
ANN	2.97	6.01	12.53
GA-ANN	−0.57	4.98	9.65

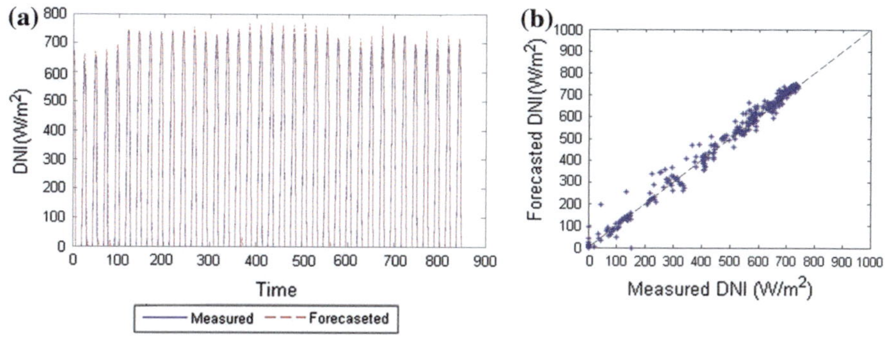

Comparison between measured and predicted DNI

Comparison between forecasted values and measurements

Fig. 3 Experiment results of GA-ANN model for the testing set

For further understanding the performance of the proposed model, the distribution of its forecasting error is shown in Fig. 3b. The distribution of the errors of all the testing set and data excepting the zero points both approximates a normal distribution. The errors are most in [−30, 30], which is under 6 % of the average of the daily solar irradiance.

4.2 Comparative Experiment

To illustrate the properties of the proposed model, the comparative experiment was carried out. Table 2 shows the comparison between the proposed model and some models mentioned in Sect. 2. The forecasting skill of the AR model to improved 58.98 % over the persistence model, although its performance is lower than that of the other models in Table 2. It means that there are linear relationship between the future solar irradiance and the historical data. The ANN model performs significantly better than the persistence model and the AR model, but it is inferior to the GA-ANN model. This shows that the way of GA optimizing the weights of the ANN model can achieve a higher accuracy of forecasting solar irradiance. The forecasting accuracy of the LS-SVM is similar to that of the proposed model. It is

Table 2 Performance statistics (nMBE %) of the proposed model and others models for six days

Model	nMBE (%)	nMAE (%)	nRMSE (%)	s (%)
Persistence	0.36	27.49	49.97	0
AR	2.04	12.86	20.50	58.98
ANN	2.97	6.01	12.53	74.92
LS-SVM	2.24	4.35	9.39	81.87
GA-ANN	1.17	4.50	9.06	81.21

hard to determine which of them is better for the term of the forecasting accuracy. But the performance of the LS-SVM is affected seriously by the model's parameters, while they are not easy to be defined and usually by experience. However, the weights of the GA-ANN model are easier by GA optimizing. So, the GA-ANN model is better to some extent than the LS-SVM model for the operation.

5 Conclusion

Several models were introduced for 1 h ahead forecasts, and an ANN-based model optimized by genetic algorithm was proposed in this paper. The genetic algorithm is used to optimize the weights of the ANN, and the order selection criteria of the time series is used to determine the input dimension. Several experiments were carried out to evaluate the performance of the proposed model with the data from a public database of India. The results show that the optimized ANN model with GA outperformed the unoptimized model. The proposed model's error is very low and its forecast skill is above 80 % level over the persistence model. Moreover, the performance of the proposed model is also better than the other two models published. In the future, the cloudiness index will be taken into consideration to improve the accuracy of forecasting solar irradiance.

References

1. Ali SMH, Zuberi M, Jibran S, Tariq, MA, Baker D, Mohiuddin A (2015) A study to incorporate renewable energy technologies into the power portfolio of Karachi. Renew Sustain Energy Rev 47(14):14–22
2. Kousksou T, Allouhi A, Belattar M, Jamil A, El Rhafiki T, Arid A, Zeraouli Y (2015) Renewable energy potential and national policy directions for sustainable development in Morocco. Renew Sustain Energy Rev 47:46–57
3. Guolian H, Tongyue S, Furong H, Pengcheng J, Jianhua Z (2014) Short-term wind speed prediction based on BP neural network with wavelet and time-series. ICIC Express Lett 8 (3):867–874
4. Ke T, Chen M, Luo H (2015) Short-term prediction method of wind farm output power based on the dynamic adjustment strategy. ICIC Express Lett Part B Appl 6(6):1601–1606
5. IEA (2015) Tracking clean energy progress. IEA input to the clean energy ministerial. www.iea.org
6. Law EW, Prasad AA, Kay M, Taylor RA (2014) Direct normal irradiance forecasting and its application to concentrated solar thermal output forecasting-a review. Solar Energy 108:287–307
7. Gueymard CA (2003) Direct solar transmittance and irradiance predictions with broadband models. Part I Detailed Theoret Perform Assess 74(5):355–379
8. Gueymard CA (2003) Direct solar transmittance and irradiance predictions with broadband models. Part II: validation with high-quality measurements. Solar Energy 74(5):381–395
9. Reikard G (2009) Predicting solar radiation at high resolutions: a comparison of time series forecasts. Solar Energy 83(3):342–349

10. Chen P, Pedersen T, Bak-Jensen B, Chen Z (2010) ARIMA-based time series model of stochastic wind power generation. IEEE Trans Power Syst 25(2):667–676
11. Alam S, Kaushik SC, Garg SN (2006) Computation of beam solar radiation at normal incidence using artificial neural network. Renew Energy 31(10):1483–1491
12. Jianwu Z, Wei Q, Eleuch A, Benghanem H, Elaoun M, Pavan C, Massi A (2013) Short-term solar power prediction using a support vector machine. Renew Energy 52:118–127
13. Tymvios FS, Michaelides SC, Skouteli CS (2008) Estimation of surface solar radiation with artificial neural networks. Modeling solar radiation at the Earth's surface. Springer, pp 221–256
14. Haikun W (2005) The theory and method of neural network structure design. National Defence Industry Press, Beijing
15. Zhang G, Peter A, Eleuch H, Benghanem M, Elaoun C, Pavan A (2003) Massi. Time series forecasting using a hybrid ARIMA and neural network model. Neurocomputing. 50:159–175
16. Golberg DEA, Eleuch H, Benghanem M, Elaoun C, Pavan AM (1989) Genetic algorithms in search, optimization, and machine learning. Addison Wesley
17. Holland JHA, Eleuch H, Benghanem M, Elaoun C, Pavan AM (1992) Adaptation in natural and artificial systems: an introductory analysis with applications to biology, control, and artificial intelligence. MIT Press

MIX
Papier aus verantwortungsvollen Quellen
Paper from responsible sources
FSC® C105338

If you have any concerns about our products,
you can contact us on
ProductSafety@springernature.com

In case Publisher is established outside the EU,
the EU authorized representative is:
Springer Nature Customer Service Center GmbH
Europaplatz 3, 69115 Heidelberg, Germany

Printed by Libri Plureos GmbH
in Hamburg, Germany